COMPUTATIONAL METHODS
IN WATER RESOURCES X

Water Science and Technology Library

VOLUME 12

COMPUTATIONAL METHODS IN WATER RESOURCES X

Volume 2

edited by

ALEXANDER PETERS

IBM Heidelberg, Germany

GABRIEL WITTUM

Universität Stuttgart, Germany

BRUNO HERRLING

Universität Karlsruhe, Germany

UDO MEISSNER

Technische Hochschule Darmstadt, Germany

CARLOS A. BREBBIA

Wessex Institute of Technology, UK

WILLIAM G. GRAY

University of Notre Dame, USA

and

GEORGE F. PINDER

University of Vermont, USA

SPRINGER-SCIENCE+BUSINESS MEDIA, B.V.

Library of Congress Cataloging-in-Publication Data

Computational methods in water resources X / editors, Alexander Peters
... [et al.].
       p.   cm. -- (Water science and technology library ; v. 12)
   Edited proceedings of the Tenth International Conference on
Computational methods in Water Resources, held at Universität
Heidelberg, Germany, Jul. 1994.

   1. Hydrology--Data processing--Congresses. 2. Hydrology-
-Mathematical models--Congresses.   I. Peter, A. (Alexander), 1956-
. II. International Conference on Computational Methods in Water
Resources (10th : 1994 : Universität Heidelberg) III. Series.
GB656.2.E42C653   1994
551.49'0285--dc20   94-17902

ISBN 978-94-010-9206-7   ISBN 978-94-010-9204-3 (eBook)
DOI 10.1007/978-94-010-9204-3

Printed on acid-free paper

EDITORS:

A. Peters
IBM Heidelberg STSS
Vangerowstr. 18
69020 Heidelberg
Germany

G. Wittum
ICA / Numerik
Universität Stuttgart
Pfaffenwaldring 27
70550 Stuttgart
Germany

B. Herrling
Institut für Hydromechanik
Universität Karlsruhe
Kaiserstr. 12
76131 Karlsruhe
Germany

U. Meissner
Technische Hochschule Darmstadt
Institut für Numerische Methoden und Informatik im Bauwesen
Petersenstr. 13
64287 Darmstadt
Germany

C.A. Brebbia
Wessex Institute of Technology
University of Portsmouth
Ashurst Lodge
Southhampton SO4 2AA
UK

W.G. Gray
University of Notre Dame
Department of Civil Engineering and Geological Sciences
Notre Dame, IN 46566-0767
USA

G.F. Pinder
College of Engineering and Mathematics
University of Vermont
101 Votey Building
Burlington, VT 05405
USA

PREFACE

These volumes constitute the edited proceedings of the Tenth International Conference on Computational Methods in Water Resources (formerly Finite Elements in Water Resources), held at Universität Heidelberg, Germany in July 1994. The biennial series began in 1976 at Princeton University, U.S.A., as a forum for researchers in the expanding field of applications of finite element methods to problems in water resources. Alternating between the U.S.A. and Europe, meetings have been held at Imperial College, U.K. (1978); the University of Mississippi (1980); Hannover University, Germany (1982); the University of Vermont (1984); the Laboratorio Nacional de Engenharia Civil, Portugal (1986); the Massachusetts Institute of Technology (1988); the Giorgio Cini Foundation, Italy (1990); and the University of Colorado at Denver (1992). The Heidelberg conference is organized jointly by Interdisziplinäres Zentrum für Wissenschaftliches Rechnen (Interdisciplinary Center for Scientific Computing) and Sonderforschungsbereich 359 of Universität Heidelberg and Institute of Supercomputing and Applied Mathematics of IBM Heidelberg.

The 1994 proceedings present the work of authors from 23 countries. Numerical methods, mathematical modeling and applications to subsurface and surface hydrology are covered by a wide variety of papers. Issues of formation description and modeling, including parameter estimation, heterogeneity, and scaling up continue to attract the attention of a large number of researchers. It is significant to mention that several papers edited in this book concern the solution of the Navier-Stokes equations.

The organizers of the Heidelberg meeting greatly appreciate the efforts of featured lecturers A.J. Baker, M.A. Celia, W. Jäger, and W. Kinzelbach. We wish to thank the invited speakers P. Ackerer, H.G. Bock, J. Carrera, G. Dagan, H. Daniels, R.E. Ewing, E.O. Frind, P. Gresho, W. Hackbusch, I. Herrera, G. Gambolati, P. Knabner, H. Kobus, M. Kawahara, S.P. Neuman, K. Pruess, R. Rannacher, J. Troesch, T. van Genuchten, P. Wesseling, J.J. Westerink, and W.G. Yeh. We are also indebted to Anja McKellar, who did most of the secretarial work.

The papers appearing in this volume have been reproduced directly from material submitted by the authors, who are wholly responsible for their content.

The Editors

CONTENTS

VOLUME 1

1. GROUNDWATER AND POROUS MEDIA

2. SUBSURFACE TRANSPORT

3. SCALING AND HETEROGENEITY

4. GEOSTATISTICS

5. REACTIVE FLOW

6. FRACTURED POROUS MEDIA

7. PARAMETER ESTIMATION

VOLUME 2

8. REMEDIATION OPTIMIZATION

9. SUBSURFACE MULTIPHASE FLOW

10. SALTWATER INTRUSION

11. SHALLOW WATER EQUATIONS

12. FLOW AND TRANSPORT IN RIVERS

13. NAVIER-STOKES EQUATIONS

14. COASTAL FLOW

15. SEDIMENT TRANSPORT

16. ALGEBRAIC METHODS

8. REMEDIATION AND OPTIMIZATION

AN ALGORITHM FOR APPROXIMATE SOLUTION OF THE GROUNDWATER CONTAMINANT REMEDIATION PROBLEM

D.P. AHLFELD and A. ZAFIRAKOU
Dept. of Civil Engineering, U-37
University of Connecticut
Storrs, CT 06269
USA

A numerical method is proposed for the design of hydraulic controls to alter the groundwater flow field as part of a plume remediation strategy. The method consists of a computationally efficient approach to combining simulation of both groundwater flow and contaminant transport modeling with optimization techniques. While numerous approaches have been proposed for optimizing well locations and pump rates, the computational challenges of incorporating contaminant transport remain significant. The novel feature of the approach proposed here is the use of an iterative scheme in which the computationally-easy hydraulic control problem is solved repeatedly. At each iteration of the scheme the constraints of the hydraulic model are modified according to feedback information provided by forward solution of a finite-difference contaminant transport model. The result provides an approximate solution to the computationally-hard contaminant remediation optimization problem. The underlying principle of the approach used is that the hydraulic control problem will be modified so that the resulting design comes as close as possible to satisfying requirements on concentration. An approach based on linear extrapolation of changes in concentration as a result of changes in the gradient imposed on the hydraulic control problem is used for the feedback mechanism. The approach is tested on a series of sample problems and shown to be effective.

I. INTRODUCTION

The objective of the remediation design problem is the minimization of the cost of operation of a remediation system. This problem can be viewed as consisting of the optimal selection of the locations and the pump rates of the remediation wells, so that the contaminated plume will be under control within a short period of time.

The solution of the problem depends upon the formulation of mathematical models which simulate both groundwater flow and contaminant transport, combined with appropriate optimization techniques. Models combining groundwater flow simulation and optimization have been proposed for management problems in aquifer dewatering [Aguado et al., 1974], in head gradient control [Molz and Bell, 1977], or for rapid removal of a groundwater contaminant plume [Lefkoff and Gorelick, 1985]. Gorelick and Voss [1984] used a finite element groundwater flow and contaminant transport simulation, combined with nonlinear optimization for the rehabilitation of aquifers that have been subjected to chemical contamination. Heidari et al. [1987] applied a groundwater management model, including a solute transport model, in the Equus Beds aquifer in Kansas, using velocity

A. Peters et al. (eds.), Computational Methods in Water Resources X, 833–840.
© 1994 *Kluwer Academic Publishers. Printed in the Netherlands.*

control as a tool for optimal plume containment. Ahlfeld et al. [1988] proposed two nonlinear optimization formulations that employ a two-dimensional Galerkin finite element simulation model of steady state groundwater flow and transient convective-dispersive transport. Finally, Bogacki and Daniels [1989] introduced another nonlinear method to optimize convective-dispersive mass-transport problems using the embedding method.

This paper involves the development of a computationally efficient algorithm that has the potential to work successfully without the high computational costs and nonlinearities present in formulations currently in the literature, and provide a practical code for use by practitioners.

II. MODEL FORMULATION

A. Groundwater Flow

A valuable tool for designing a hydraulic control system is a groundwater simulation model. The U.S. Geological Survey code MODFLOW [McDonald and Harbaugh, 1988], widely used by practitioners, is used to solve the three-dimensional partial differential equation:

$$\nabla \cdot K \nabla h + \sum_{k=1}^{n_p} q_k \, \delta(x_k, y_k, z_k) = 0 \tag{1}$$

using the average Darcy velocity vector:

$$v = -K \nabla h \tag{2}$$

B. Contaminant Transport

Contaminant transport is calculated using MT3D [C.Zheng, 1992] which can be solved on the same grid as MODFLOW. This model solves the three-dimensional partial differential governing equation:

$$\frac{\partial c}{\partial t} = \frac{\partial}{\partial x_i}\left(D_{ij}\frac{\partial c}{\partial x_i}\right) - \frac{\partial}{\partial x_i}(v_i c) + \frac{q_s}{\theta} c_s + \sum_{k=1}^{N} R_k \tag{3}$$

C. Optimization Formulation

The key to the use of the optimization approach to the hydraulic control design problem is the formulation of the design criteria as objectives and constraints in the context of an optimization formulation. The optimization problem is stated as in Ahlfeld and Heidari [1994], that is, minimizing the sum of the pump rates Σq_i as the objective function, and imposing linear constraints on head to prevent excessive drawdown and mounding:

$$h_i(q) \le h_{i,u}^* \quad \text{or} \quad h_i(q) \ge h_{i,l}^* \tag{4}$$

and linear constraints on head differences to enforce desired gradients:

$$h_{i1}(q) - h_{i2}(q) \le g_{i,u}^* \quad \text{or} \quad h_{i1}(q) - h_{i2}(q) \ge g_{i,l}^* \tag{5}$$

One additional set of constraints are placed on the concentration at the end of the simulation time (T), at specific observation points:

$$c_{i,T}(q) \le c_i^* , \, i \in I_c \tag{6}$$

All constraints, with the exception of the last one, form a linear optimization problem with respect to the decision variables (q_i's) for the case of fixed aquifer boundaries. To take advantage of the fact that linear programs can be solved relatively easy, demand shorter computational time, and have a global optimal solution, the last equation will be relaxed in our algorithm and imposed externally to the linear program.

Given a particular solution to the hydraulic control problem, a simulation of the effect of the particular pumping scheme on contaminant transport can be evaluated. If the concentration constraints in (6) are violated, then the linear hydraulic problem can be modified and resolved. The modification is accomplished by adjusting the constraints on head difference (or head gradient). This approach can be viewed as using the results of the transport simulation to calibrate the constraints on the hydraulic control model so that desirable transport behavior can be achieved. More specifically, the algorithm proceeds as follows:

Step 1: Initialization.
Define k as the index on transport iterations. Set k=1; select a head gradient g^1.
Step 2: Solution of the Hydraulic Control Model.
Determine the pump rates (q^k) as a linear function of the gradient constraints (g^k).
Step 3: Solution of the Transport Model.
Determine the concentrations (c^k) based on the hydraulic control constraints (q^k).
Step 4: Test for Convergence.
If the difference between the new pump rates (q^k) and the previous values (q^{k-1}) is small, the algorithm stops; in any other case, the iterative approach continues.
Step 5: Update of Gradient Requirements.
Determine the new head gradients (g^{k+1}) as a function of the concentrations (c^k).
Step 6: Iteration.
Set k=k+1; go to Step 2.

Step 5 is the most important part of the algorithm. It connects head gradients with concentrations at specific locations. Since g, q, and c are related by $q=L(g)$ and $c=S(q)$, where L is the linear program operator and S is the transport simulation operator, it is also true that $c=S(L(g))$. The approach is implemented by assigning a number of concentration observation points to a single gradient constraint. By Taylor series with first order truncation, a single element of the vector of observed concentrations which is assigned to the ith gradient constraint can be approximated by

$$c_l^{k+1} = c_l^k + \frac{\partial c_l}{\partial g_i}\left(g_i^{k+1}-g_i^k\right) \tag{7}$$

where l refers to the observation point at which concentration is calculated. Equation (6) requires $c_l^{k+1} \le c_l^*$. Substituting (7) into (6) and rearranging, yields

$$g_i^{k+1} \le g_i^k + \frac{\left(c_l^*-c_l^k\right)}{\left(c_l^k-c_l^{k-1}\right)}\left(g_i^k-g_i^{k-1}\right) \tag{8}$$

Considering all the concentration observation points which may be near to the gradient location, two cases can occur: the concentration can be greater, or less than that required. When the concentration is higher, the maximum gradient change suggested by (8) is used to alter the groundwater flow, whereas when the concentration is lower, the minimum gradient change suggested by (8) is used.

The value of the new head gradient that will be ultimately chosen to contain the plume will be derived from the comparison of four different scenarios based on the formula given above (8), in which not only the concentrations ($c_l^*-c_l^k$) have to be compared, but the ratio of gradients over concentrations for consecutive iteration steps ($g_i^k-g_i^{k-1}$) / ($c_l^k-c_l^{k-1}$) as well. So the gradient may receive the smallest or largest increase of decrease according to the relationship between the concentrations.

As long as the candidate wells are located close to the gradient points, and the gradient requirements implied by concentration requirements are not unrealistic, this algorithm responds correctly. If the updating step produced an infeasible set of gradient requirements, then all gradient changes are reduced by a factor α, with $0 \le \alpha \le 1$. Because of the nonlinearity of the concentration response to pumping, several iterations may be necessary to achieve feasibility.

III. VERIFICATION AND APPLICATION

The algorithm is tested on two different scenarios, that demonstrate the applicability and flexibility of the model.

The numerical model consists of 23 rows, 50 columns and 2 layers, with a regular spacing of 100 ft in all three directions. Both layers are assumed to be confined aquifers under the steady-state flow condition. The model parameters used in the simulation are listed below:

> cell width along rows and columns = 100 ft
> layer thickness = 100 ft
> hydraulic conductivity = 10 ft/day
> porosity = 0.25
> longitudinal dispersivity = 45
> horizontal transverse dispersivity = 17.5
> vertical transverse dispersivity = 17.5
> simulation time = 2-4 years

The field is surrounded only by specified-head (Dirichlet) boundaries on all sides. An initial concentration field is assumed everywhere in the model, due to a continuing source of 1000 ppb located at row 12, column 6 and layer 1. The aquifer is relatively thick, so that instantaneous vertical mixing cannot be assumed, and the transport of the solute away from the point source is considered three-dimensional.

In the first scenario, the original plume was generated after fifteen years of contaminant injection through this one source as shown in Figure 1. In order to decrease the concentration of the solute at specific locations, a single extraction well location is offered, as shown on Figure 2 with a large circle. The well is located in layer 1 and has an upper bound value of pump rate 500000 cubic feet per day. A head gradient constraint, connected with this well, directs the flow upgradient, with a starting magnitude of 0.0095. Three observation points in the same layer as the well and the gradient (numbered 1, 2, and 3 in Table 1, and shown as triangles in Figure 2), and three at the exact positions in the layer below (numbered 4, 5, and 6 in Table 1), are chosen to meet the concentration constraints so that the contamination at all observation points is no more than 20 ppb. After pumping for four years, concentration at these specific locations is less than 20 ppb, as shown in Figure 2 and Table 1, despite the continuing polluting source. Table 1, shows the convergence of the algorithm in terms of the head gradient, pump rates, and concentration. The algorithm is constructed so that, among the observation points surrounding the well, one that has the highest initial concentration becomes the binding constraint for the remediation problem; in this case it is the fifth one at the lower layer that reaches the value of 20 ppb after five iterations, giving an optimal pump rate of 29593.1 cubic feet per day.

In the second scenario the locations of the observation points are exactly the same (Figure 3), whereas the pumping well and the corresponding gradient pair are located at the same xy-coordinates but at the lower layer. For the required concentration of 30 ppb at these observation points, the remediation has to continue for only two years. The binding constraint for this scenario is the concentration at the second observation point, which is located at the upper layer (Table 2). This particular example offers the opportunity to demonstrate the convergence of the concentration at these observation points. Note that at the second observation point the concentration reaches a value of less than 30 ppb at the second and third iteration step. The algorithm corrects for this, as reflected in the updated gradient values, and returns the concentration to 30.

In each of the two cases shown, the algorithm moves close to the solution in one step. Several more steps are required to meet convergence criteria. The functional dependence suggested by (7) is strong enough to drive the algorithm. The binding constraints for each of these two scenarios are the measurements of the concentration at the observation points at the opposite layer from that where the pumping and the head difference are imposed.

IV. CONCLUSIONS

A numerical method has been introduced for the design of remediation strategies for contaminated groundwater aquifers, combining flow and transport simulation with linear optimization techniques. A computationally efficient algorithm was constructed for the solution of the minimization problem based on an iterative scheme of convergence to the optimal well locations and pump rates. The algorithm is shown to be effective in a series of simple hypothetical problems. One transport simulation is required for each iteration of the algorithm. The rapid convergence demonstrated here requires many fewer transport simulations than would be required using direct non-linear programming solution. Based on the preliminary results presented here, it is expected that the method will behave similarly in larger scale and real world problems.

Nomenclature

K = hydraulic conductivity tensor, (L^3/T)
h = hydraulic head, (L)
n_p = number of pumps
q_k = pump rate for well located at point (x_k, y_k, z_k)
$\delta(x_k, y_k, z_k)$ = Dirac delta function evaluated at point (x_k, y_k, z_k), (L/T^3)
$\mathbf{v}$ = average Darcy velocity vector, (L/T)
θ = porosity of aquifer medium, (dimensionless)
c = contaminant concentration dissolved in groundwater, (M/L^3)
x_i = distance along the respective Cartesian coordinate axis, (L)
D_{ij} = hydrodynamic dispersion coefficient, (L^2/T)
v_i = seepage or linear pore water velocity, (L/T)
q_s = volumetric flux of water per unit volume of aquifer, representing sources (positive) or sinks (negative), $(1/T)$
c_s = concentration of sources or sinks, (M/L^3)
R = retardation coefficient, (dimensionless)
$\sum_{k=1}^{N} R_k$ = chemical reaction term, (M/L^3T)
q_i = withdrawal and recharge rates at location i
$h_{i,l}{}^*, h_{i,u}{}^*$ = specified lower and upper values of head at location i
$g_{i,l}{}^*, g_{i,u}{}^*$ = specified lower and upper bounds on head differences at location i

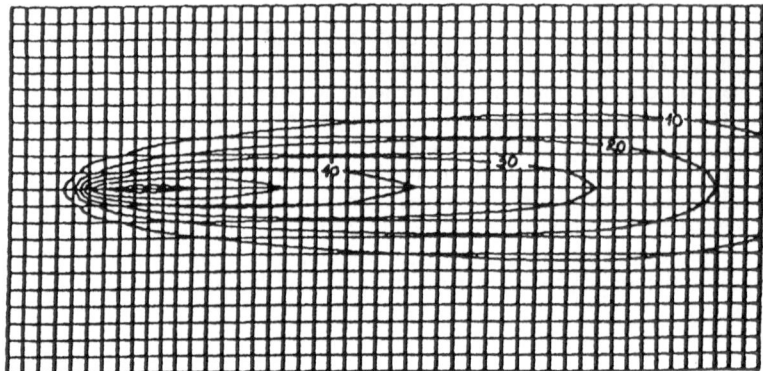

Figure 1. The initial plume after 15 years' contamination, superimposed
on finite-difference grid.

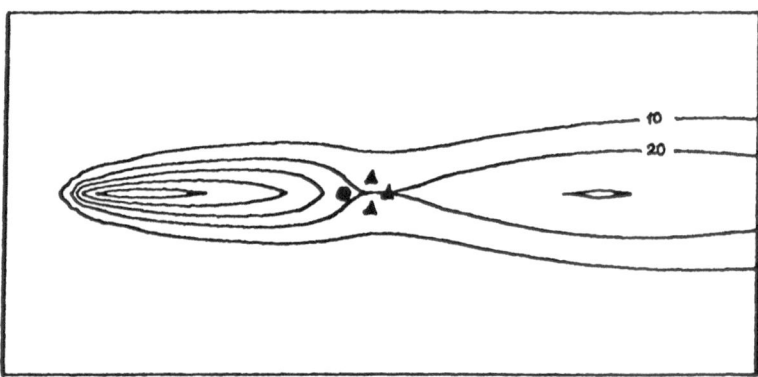

Figure 2. The plume after 4 years' remediation - First scenario. Circle
indicates well location; triangles, concentration constraint
locations.

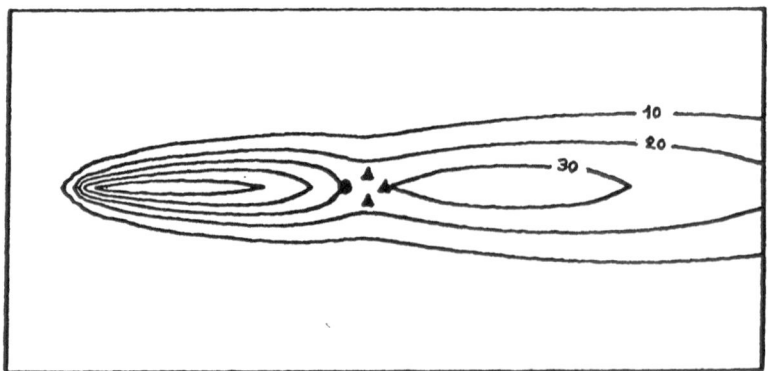

Figure 3. The plume after 2 years' remediation - Second scenario.
Circle indicates well location; triangles, concentration
constraint locations.

Iteration No.	Gradient	Pump Rate	Concentration Values					
			Obs. Pt. 1	Obs. Pt. 2	Obs. Pt. 3	Obs. Pt. 4	Obs. Pt. 5	Obs. Pt. 6
1	0.0095	11252.6	28.715	31.287	28.716	28.811	32.128	28.813
2	3.0415	27946.1	18.231	20.443	18.231	18.49	20.889	18.491
3	3.2813	29266.5	17.586	19.738	17.587	17.844	20.173	17.845
4	3.3393	29585.5	17.434	19.571	17.434	17.692	20.004	17.692
5	3.3407	29593.1	17.431	19.567	17.431	17.688	20	17.689

Table 1. Values of gradient constraint, pump rate optimal for corresponding linear program, and resulting concentrations at observation points for scenario 1.

Iteration No.	Gradient	Pump Rate	Concentration Values					
			Obs. Pt. 1	Obs. Pt. 2	Obs. Pt. 3	Obs. Pt. 4	Obs. Pt. 5	Obs. Pt. 6
1	0.0095	11252.6	31.879	36.255	31.881	30.641	34.714	30.642
2	2.3665	24229.7	24.875	29.987	24.875	23.865	29.12	23.866
3	2.3616	24202.7	24.887	29.997	24.888	23.877	29.129	23.877
4	2.3602	24194.7	24.891	30.001	24.892	23.88	29.132	23.881
5	2.3605	24196.7	24.891	30	24.891	23.88	29.132	23.88

Table 2. Values of gradient constraint, pump rate optimal for corresponding linear program, and resulting concentrations at observation points for scenario 2.

c_i^* = specified maximum concentration bound at node i
$c_{i,T}(q)$ = the concentration at node i at the last time period of the simulation, as a function of pump rates
I_c = the set of nodes at which system concentration behavior will be observed

References

Aguado, E., Remson, I., Pikul, M.F., and Thomas, W.A. (1974) "Optimal Pumping for Aquifer Dewatering", Journal of the Hydraulics Division, ASCE, 100(HY7), 869-877.

Ahlfeld, D.P., Mulvey, J.M., Pinder, G.F., and Wood, E.F. (1988) "Contaminated Groundwater Remediation Design Using Simulation, Optimization, and Sensitivity Theory 1. Model Development", Water Resources Research, 24(3), 431-441.

Ahlfeld, D.P., and Heidari, M. (1994) "Applications of Optimal Hydraulic Control to Groundwater Systems", Journal of Water Resources Planning and Management, ASCE, (in press).

Bogacki, W., and Daniels, H. (1989) "Optimal design of well location and automatic optimization of pumping rates for aquifer cleanup", Contaminant Transport in Groundwater, 363-370.

Gorelick, S.M., Voss, C.I., Gill, P.E., Murray, W., Saunders, M.A., and Wright, M.H. (1984) "Aquifer Reclamation Design: The Use of Contaminant Transport Simulation Combined with Nonlinear Programming", Water Resources Research, 20(4), 415-427.

Heidari, M., Sadeghipour, J., and Drici, O. (1987) "Velocity control as a tool for optimal plume management in the Equus Beds aquifer, Kansas", Water Resources Bulletin, 23(2), 325-336.

Lefkoff, L.J., and Gorelick, S.M. (1985) "Rapid removal of a groundwater contaminant plume", Groundwater Contamination and Reclamation, American Water Resources Association, 125-131.

McDonald, M.G., and Harbaugh, A.W. (1988) "A Modular Three-Dimensional Finite-Difference Ground-Water Flow Model", Techniques of Water-Resources Investigations of the United States Geological Survey, U.S. Geological Survey, Reston, Virginia, USA.

Moltz, F.J., and Bell, L.C. (1977) "Head Gradient Control in Aquifers Used for Fluid Storage", Water Resources Research, 13(4), 795-798.

Zheng, C. (1992) A Modular Three-Dimensional Transport Model for Simulation of Advection, Dispersion, and Chemical Reactions of Contaminants in Groundwater Systems, S.S. Papadopulos & Associates, Inc. , Bethesda, Maryland, USA.

GROUNDWATER QUALITY MANAGEMENT USING A 3-D NUMERICAL SIMULATOR AND A CUTTING PLANE OPTIMIZATION METHOD

George P. Karatzas and George F. Pinder
*College of Engineering & Mathematics, University of Vermont,
Burlington, VT 05405, USA*

ABSTRACT

In our previous work, the 'outer approximation method', a global optimization technique for minimizing a concave function over a closed convex set of constraints, was combined with a two-dimensional numerical simulator for the solution of groundwater quantity and quality management problems. In the present study, the concept of the outer approximation method, based on the work presented by Thieu et al., (1983), is expanded to consider problems with a non-convex set of constraints and it is combined with a three-dimensional numerical simulator to solve a groundwater quality management problem of a multi-layer hypothetical contaminated aquifer. Several pumping scenarios are examined to study how the total remediation cost is affected by pumping from different layers. Also, a comparison is conducted with the solution obtained using a two dimensional numerical simulator.

BACKGROUND INFORMATION

In the past few years, the problem of groundwater management has been approached by several methodologies and techniques, including the classical linear/nonlinear programming methods (Gorelic et al., 1984, Ahlfeld et al., 1988), simulated annealing (Dougherty et al., 1991), neural network and genetic algorithms (Rogers et al., 1994, Rizzo et.al., 1994), and the outer approximation method (Karatzas and Pinder, 1993). All these methods have the same main objectives, to achieve robustness and speed of performance.

The outer approximation method combined with a 2-D numerical simulator was first presented by the authors (1993), to solve groundwater management problems formulated as minimization problems involving a continuous concave function over a convex set of constraints. The objective function included treatment and installation costs which were expressed in an exponential form.

The resulting algorithm was applied to groundwater quantity and quality management problems; and the results and the performance were compared with existing solutions of the same problems using different optimization techniques. The optimization theory of the proposed methodology was based on the work presented by Thieu et al., (1983).

Since convexity does not occur in all the cases of groundwater quality management problems the aforementioned research was extended to problems with a concave feasible region. The proposed methodology, will be presented in the first part of this paper. The theory is based on the work presented by Hillestad and Jacobsen (1980) and Horst and Tuy (1990).

A. Peters et al. (eds.), Computational Methods in Water Resources X, 841–848.
© 1994 *Kluwer Academic Publishers. Printed in the Netherlands.*

In the second part, the use of a 3-D numerical simulator combined with the Outer Approximation method to solve groundwater management problems, is presented. In the past, most of the proposed optimization techniques used a 2-D numerical simulator to represent the aquifer system. The use of a 3-D numerical simulator is unquestionably a better representation of the physical system, and becomes a necessity for groundwater management problems where drains and slurry walls are incorporated in the decision making process. A hypothetical three layer aquifer was considered and the results were compared by solving the same problem using a 2-D numerical simulator. The discussion of these results will conclude the present work.

THE OUTER APPROXIMATION METHOD ALGORITHM FOR NON-CONVEX PROBLEMS

Definition of the problem:

$$\text{minimize } f(\mathbf{x})$$
$$\text{such that } \mathbf{x} \in D$$
$$\mathbf{x} \geq 0,$$

where $f : R^n \to R$ is a real-valued concave function defined throughout R^n, and D is a closed non-convex subset of R^n.
The objective function f is continuous and D is defined by a set of m constraints in the form:

$$g_i(\mathbf{x}) \leq 0, \qquad\qquad (i = 1, 2...m)$$

where $g_i, i = 1, 2, ..., m$ are continuous real-valued functions with at least one being non-convex.

Step 0
Definition of the initial enclosing polytope D_1.
Step $k = 1,2,...$
Step k starts with an enclosing polytope D_k defined by the set of vertices V_k. The vertex $\mathbf{x}^k$ that minimizes the objective function is determined such that:

$$\mathbf{x}^k = min\{f(\mathbf{x}) : \mathbf{x} \in V_k\} \tag{1}$$

If all the constraints are satisfied, i.e. $g_i(\mathbf{x}^k) \leq 0$ for all $i = 1, ..., m$, then $\mathbf{x}^k \in D$ and is an optimal solution.
Otherwise, we determine the most violated constraint $g^k(\mathbf{x})$ as:

$$g^k(\mathbf{x}) = max\{g_i(\mathbf{x}^k)\}, \qquad i = 1, ..., m. \tag{2}$$

If $g^k(\mathbf{x})$ is convex, then the methodology presented by Karatzas and Pinder (1993) is applied; otherwise we introduce the cutting hyperplane as follows:
Assume vertex $\mathbf{x}^k$ is nondegenerate, i.e. in an n-dimensional space exactly n line segments are emanating from $\mathbf{x}^k$, in n different directions. A 2-D example is shown in fig.1, where $\mathbf{x}^k = v^0$.

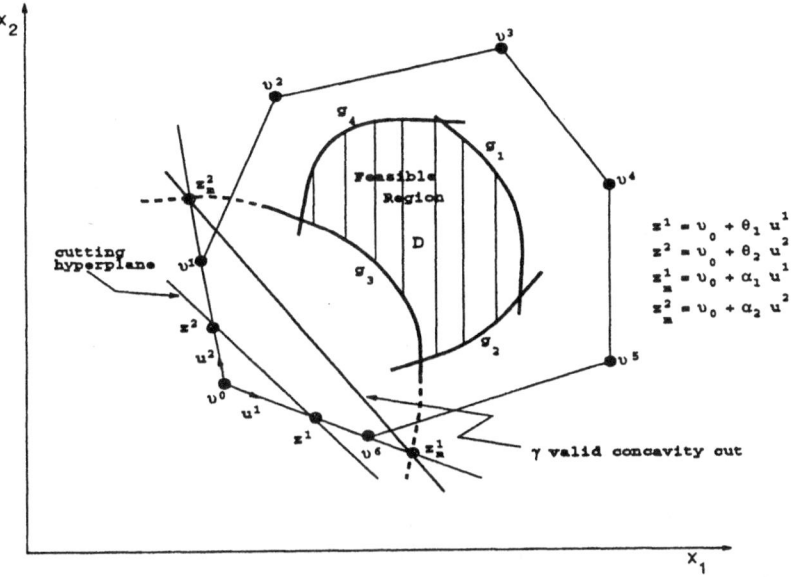

Figure 1: The Outer Approximation Method Concept for Concave Domain in a 2-D Problem

Denote those n directions $u^1, u^2, ...u^n$ defined as

$$\mathbf{u}^i = \mathbf{y}^i - \mathbf{x}^k, \tag{3}$$

where $\mathbf{y}^i$ is the i-th adjacent vertex to vertex $\mathbf{x}^k$, and $\mathbf{u}^i$ is the direction of the line segment connecting $\mathbf{x}^k$ and $\mathbf{y}^i$. The cone emanating from $\mathbf{x}^k$ in the directions $\mathbf{u}^1, \mathbf{u}^2, ..., \mathbf{u}^n$, is an n-dimensional cone with exactly n edges. Define on each edge a point $z^i \neq x^k$ such that:

$$\mathbf{z}^i = \mathbf{x}^k + \theta_i \mathbf{u}^i \tag{4}$$

and

$$g^k(\mathbf{z}^i) \geq 0 \qquad \text{for all } i = 1, 2, ..., n \tag{5}$$

where θ_i is a greater than zero constant.

All the points z_i are located in the convex subset G, between $\mathbf{x}_k$ and the reverse convex constraint $g^k(\mathbf{x})$, where G is defined as:

$$G = \{\mathbf{x} \epsilon R^n : g^k(\mathbf{x}) \geq 0\}, \tag{6}$$

The subset G is a convex subset since $g^k(\mathbf{x}) \leq 0$ is a reverse convex constraint.

Next define the $n \times n$ matrix $\mathbf{D}$ as:

$$\mathbf{D} = \left[\mathbf{z}^1 - x^k, \mathbf{z}^2 - x^k, ..., \mathbf{z}^n - \mathbf{x}^k \right],$$

where $\mathbf{z}^i - \mathbf{x}^k$ is the i-th column of $\mathbf{D}$.

$\mathbf{D}$ is a nonsingular matrix, since its columns are linearly independent vectors (n vectors emanating from $\mathbf{x}^k$ in n different directions), therefore the inverse of $\mathbf{D}$ exists.

Then the equation of the hyperplane $h(\mathbf{x})$ passing through points $\mathbf{z}^i$, $i = 1, 2, ..., n$ can be shown to be:

$$\mathbf{eD}^{-1}(\mathbf{x} - \mathbf{x}^k) = 1, \tag{7}$$

where $\mathbf{e}$ is a row vector of ones, i.e. $\mathbf{e} = (1, 1, ..., 1)$ [Hillestad and Jacobsen (1980), Horst and Tuy (1990)]. The "strongest cut" to the polytope D_k by the hyperplane $h(\mathbf{x})$ is the one that determines n points along the n directions emanating from $\mathbf{x}^k$, for which the concave constraint is satisfied by equality. This has been called γ-*valid concavity cut* [Horst and Tuy (1990)].

The hyperplane $h(\mathbf{x})$ (in this case a line) devides the space into two subspaces: 1) the subspace that includes point $\mathbf{x}^k$ and 2) the subspace that does not include point $\mathbf{x}^k$. It is easy to verify that for any point in subspace 1 the expression

$$\mathbf{eD}^{-1}(\mathbf{x} - \mathbf{x}^k)$$

is less or equal to one, and for points in subspace 2 greater or equal to one.

The new set of vertices V_{k+1} is defined as the the union of the set of the new vertices, created from the intersection of the hyperplane $h(\mathbf{x})$ and the polytope D_k, and the subset of vertices of V_k for which the equation of the hyperplane $h(\mathbf{x})$ is greater than one.

It should be noted that with the above procedure no derivatives of the constraints with respect to the decision variables is required.

After the new set V_{k+1} is defined, the new step $k + 1$ starts and the process is repeated until an optimal solution satisfying all the constraints is obtained.

AN APPLIACTION OF A GROUNDWATER QUALITY MANAGEMENT PROBLEM USING THE OUTER APPROXIMATION METHOD AND A 3-D NUMERICAL SIMULATOR

A three layer hypothetical homogeneous, isotropic, unconfined aquifer was considered with dimensions: Length $= 870m$, Width $= 870m$ and Depth $= 30m$. The 3-D numerical simulator PTC (Princeton Transport Code) Babu et al.,(1993) was employed for the representation of the aquifer. PTC is a hybrid finite element/finite difference groundwater flow and contaminant transport simulator. The physical and numerical parameters used in the simulations are presented in Table 1.

It was assumed that two contaminant sources, located at nodes 425 and 485 (the black circles in fig. 2) in the second layer, contaminated the aquifer for 15 years prior to remediation. The concentrations were normalized and considered equal to one at the sources. The boundary conditions for the flow and transport are that there is no water flux or dispersive flux across the boundaries and no applied fluxes occur anywhere within the domain. A plan view of the aquifer with the initial and boundary conditions is presented in fig. 2.

There were twenty three observation points (constraints), in each of the three layers, where it was requested that the normalized concentration becomes less or equal to .05 by the end of a remediation period of five years, and three points where the concentration had to remain less or equal to .1. All these points are indicated by a black triangle in figures 2 through 5. Also, it was assumed that the sources had been removed when the remediation started.

Four different remediation scenarios were examined: (1) Ten potential pumping wells pumping from the first layer (bottom layer) (fig. 3), (2) Ten potential pumping wells pumping

Number of layers	3
Number of nodes	900
Number of elements	841
Hydraulic conductivity, K_x	0.125 m/hour
Hydraulic conductivity, K_y	0.125 m/hour
Hydraulic conductivity, K_z	0.0125 m/hour
Longitudinal dispersivity, α_L	18 m
Transverse dispersivity, α_T	1.8 m
Vertical dispersivity, α_V	0.18 m
Diffusion coefficient, d	0.00001
Porosity, θ	0.2
Time step, Δt	4 months (2928 hours)
Number of time steps	15

Table 1: Physical and Numerical Parameters of the Hypothetical Aquifer

from the second layer (middle layer) (fig. 3), (4) Ten potential pumping wells pumping from the third layer (top layer) (fig. 5), and (4) The problem was solved using the 2-D version of the numerical simulator PTC (fig. 6). In all the above scenarios the potential well locations were considered at the same nodal locations. For all the scenarios the problem was formulated as:

$$min \quad \sum_{i=1}^{n} \alpha_i q_i + \alpha_i^0 (1 - e^{-b q_i}), \qquad i \in I \tag{8}$$

$$s.t. \quad c_j(\mathbf{q}) \le c_j^*, \qquad j \in J \tag{9}$$

$$0 \le q_i \le q_i^*, \qquad i \in I \tag{10}$$

α_i is the unit pumping cost,
α_{0_i} is the fixed cost of installing a well at node i,
q_i is the pumping rate at node i,
$c_j(\mathbf{q})$ is the concentration at node j at the end of the remediation period,
c_j^* is the upper bound of c_j,
q_i^* is the fixed upper bound of q_i,($= 20 \ m^3/hr$ for all the wells in this problem)
b is a scaling factor($=1000$ for this problem);
I is the set of points with the number of elements
  equal to the number of parameters (pumping wells),
J is the set of points with the number of elements
  equal to the number of constraints (observation nodes).

The numerical results for all the four scenarios are summarized in Table 2, and a graphical representation is shown in figures 2 through 5; active wells are indicated by a black bar with its size to be proportional to the pumping rate.

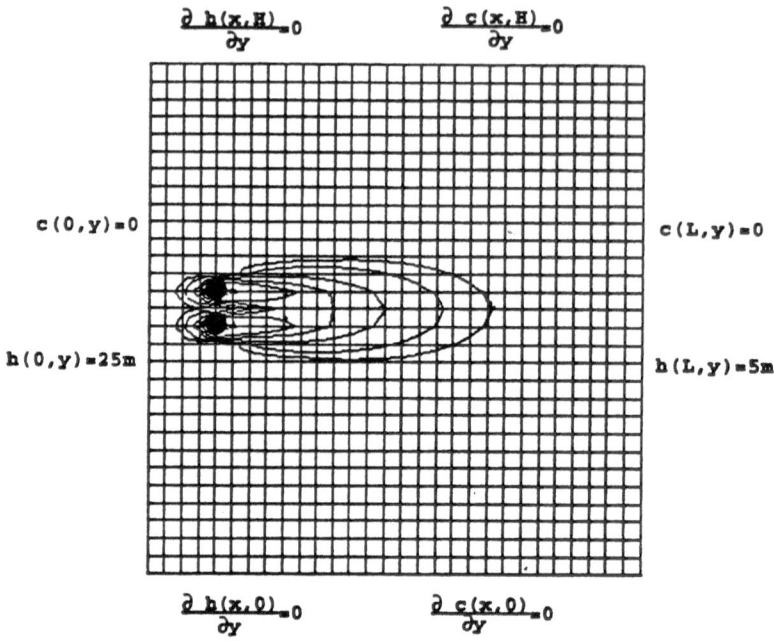

Figure 2: A Plan View of the Example Problem with the Initial and Boundary Conditions

CONCLUSIONS AND DISCUSSION

The concept of, and an application of the outer approximation method for problems with a non-convex feasible region herein is presented. This work, combined with our previous papers (convex feasible region) integrates the study of the groundwater management problem with convex and non-convex constraint functions. Since the concept of the method is based on combinatorial geometry, the computational effort increases as the number of decision variables increases. Presently the algorithm has been tested for a maximum number decision variables of 28 and several hundred constraints and the focus of future research is on expanding the methodology for the solution of large groundwater management problems with a few hundreds of decision variables and constraints.

Conclusions regarding the use of the 3-D numerical simulator can be obtained by studying Table 2. It is shown that the pumping rates and the well locations are affected by the depth

	Pumping rates (m^3/hour)										Total
Well #	1	2	3	4	5	6	7	8	9	10	Pumping
Node #	463	497	502	467	404	525	435	499	439	471	m^3/hr
Layer #3	20.0	-	-	16.1	1.8	-	-	-	-	5.0	42.8
Layer #2	20.0	-	-	15.1	-	-	2.7	-	-	5.1	42.8
Layer #1	20.0	-	-	13.8	-	4.9	-	-	-	5.2	43.9
2-D model	20.0	4.4	-	-	20.0	20.0	20.0	-	-	-	84.4

Table 2: Optimal Solutions for the Four Different Scenarios

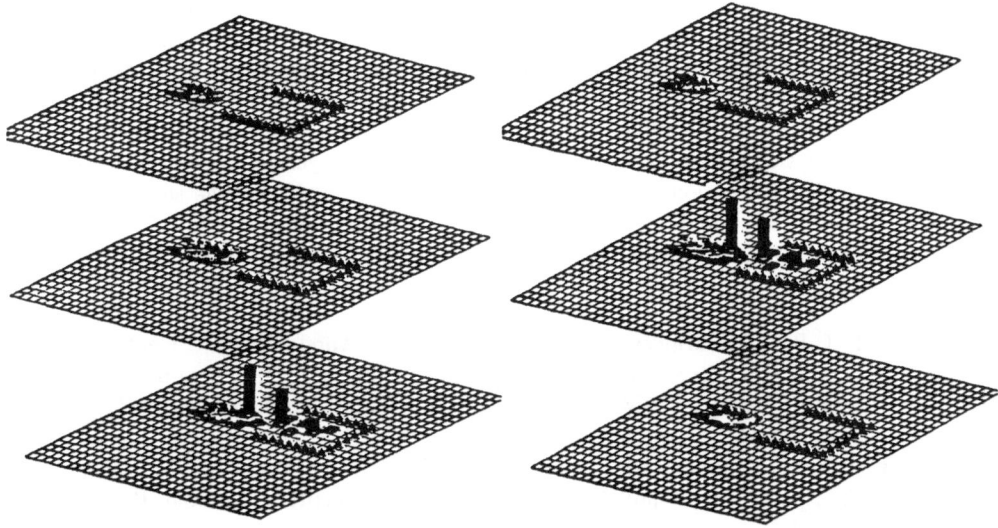

Figure 3: Scenario 1: Pumping from
the First Layer

Figure 4: Scenario 2: Pumping from
the Second Layer

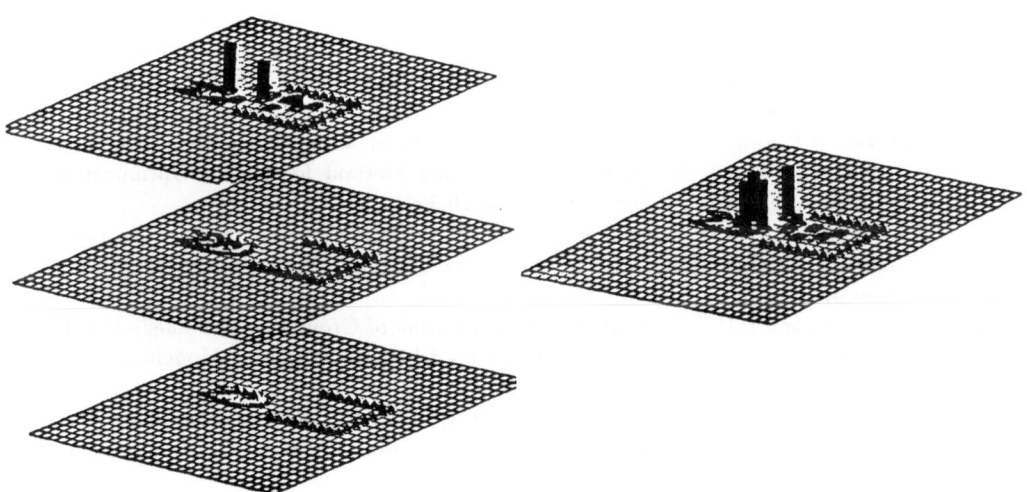

Figure 5: Scenario 3: Pumping from
the Third Layer

Figure 6: Scenario 4: Using the 2-D
Numerical Simulator

from which pumping takes place, in relation with the source location. Locating the wells in the middle layer (where the source is) results in a total pumping volume less than pumping from the bottom layer and no difference than pumping from the top layer. Also, the issue of pumping from different depth becomes important in cases where the installation cost varies between the layers due to the difference in geological formations. In the present study it was assumed that the well installation cost was the same for any layer. A significant difference in pumping and well locations was obtained using the 2-D model for the solution of the problem. Since the 3-D simulator could be considered a better representation of the aquifer systema, using a 3-D model should generate confidence in the numerical results.

REFERENCES

1. Ahlfeld, D.P., J.M. Mulvey, G.F. Pinder and E.F Wood (1988), 'Contaminated Groundwater Remediation Design Using Simulation, Optimization, and Sensitivity Theory. 1. Model Development', *Water Resources Research*, vol.24, no.3, 431-441.

2. Babu, D.K., G.F. Pinder, A. Niemi, D.P. Ahlfeld, S.A. Stothoff (1993), 'Princton Transport Code, Report 84-WR-3 (revised)', University of Vermont, Burlington, VT, USA.

3. Dougherty, D.E., and Marryott, R.A. (1991), 'Optimal Groundwater Management. 1. Simulated Annealing.', *Water Resources Research*, vol.27, no.10, 2493- 2509.

4. Hillestad, R.J. and S.E. Jacobsen (1980), 'Reverse Convex Programming', *Applied Mathematics and Optimization*, 6, 63-78.

5. Horst, R. and H. Tuy (1990), *Global Optimization*, Springer-Verlag, Berlin. Heidelberg.

6. Gorelick, S.M., C.I. Voss, P.E. Gill, W. Murray, M.A. Saunders and M.H. Wright (1984), 'Aquifer Reclamation Design: The Use of Contaminant Transport Simulation Combined with Nonlinear Programming', *Water Resources Research*, vol.20, no.4, 415-427.

7. Karatzas, G.P. and G.F Pinder (1993), 'Groundwater Management Using Numerical Simulation and the Outer Approximation Method for Global Optimization', *Water Resources Research*, vol.29, no.10,3371-3378.

8. Rizzo, D.M. and D.E. Dougherty (1994), 'Characterization of Aquifer Properties Using Artificial Neural Networks, Neural Kriging', *Water Resources Research*, in press.

9. Rogers, L.L. and F.U. Dowla (1994), 'Optimization of Groundwater Remediation Using Artificial Neural Networks with Parallel solute Transport Modeling', *Water Resources Research*, vol.30, no.2,457-481.

10. Thieu, T.V., B.T. Tam, and V.T. Ban (1983), 'An Outer Approximation Method for Globally Minimizing a Concave Function over a Compact Convex Set', *Acta Mathematica Vietnamica*, 8, 21-40.

GROUNDWATER QUALITY MANAGEMENT USING NUMERICAL SIMULATION AND A PRIMAL OPTIMIZATION TECHNIQUE

George P. Karatzas(*), Tullio Tucciarelli(**) and George F. Pinder(*)
(*)*College of Engineering & Mathematics,*
University of Vermont, Burlington, VT 05405, USA
(**)*Dipartimento di Meccanica dei Fluidi ed Ingegneria Offshore,*
Facolta' di Ingegneria, Reggio Calabria, Italy

ABSTRACT

A new methodology is presented for the solution of the groundwater quality management problem. The method uses the concepts of the cutting plane theory in combination with a line search that moves from a feasible point towards the solution of a subproblem with linear constraints and the same objective function of the original problem. The required computational effort regarding the derivatives of a single concentration with respect to the pumping rates is minimized by using sensitivity analysis and solving one linear system at each step. The method has been tested with analytical examples and shows that for problems with a linear objective cost function the computational growth rate is linearly dependent on the number of decision variables. Also, the method was used to solve a groundwater quality management problem related to the Woburn (MA) aquifer.

INTRODUCTION

In most of the developed algorithms for the solution of the groundwater quality problem, the main concern is the computational effort required to evaluate the concentrations and their derivatives at the control points. Since convexity does not occur in most of the cases this implies that, even if the objective function is convex, several points can exist that satisfy the local minimum conditions. Determination of the global minimum, when the feasible domain is not convex, is a difficult task, and a tremendous numerical effort is required when the number of decision variables (pumping rates) is more than few tens (Dougerty and Marryott (1991), Tucciarelli (1994), Karatzas and Pinder (1993)). The rate of growth in computer time as the dimensionality increases has been estimated as N^m, where N is the number of decision variables and m is an exponent larger then 2 (Culver and Shoemaker, 1992). In the present study the exponent m is drastically reduced.

The proposed algorithm is a combination of the cutting plane technique (Rao, (1984)) and the primal method (Luenberg, (1984)). The concept of the cutting plane technique is that at each step the feasible region is approximated by an enclosing polytope, that is defined by a set of linear constraints. This approximation can be improved in the next step by introducing a suitable cutting hyperplane to the polytope, that cuts off a part of it, and defines a new polytope that is a better approximation of the feasible region. Therefore, the above technique reduces the original problem to a series of subproblems with linear constraints. Each subproblem i has the same objective function and differs from subproblem i − 1 by having an additional constraint (the new cutting hyperplane).

The concept of the primal method is that at each line search it moves from one feasible point to another feasible point and always remains in the feasible region.

A. Peters et al. (eds.), Computational Methods in Water Resources X, 849–857.

DESCRIPTION OF THE ALGORITHM

Problem Formulation: In most of the cases, the groundwater quality problem is formulated as:

$$minimize \ f(\mathbf{q}) \tag{1}$$

$$subject \ to \quad c_j(\mathbf{q}) \leq c_j^{max} \qquad j = 1, k \tag{2}$$

$$0 \leq q_i \leq q_i^{max} \qquad i = 1, N \tag{3}$$

where q_i is the injection/extraction pumping rate of the i well, c_j is the concentration at the point $\mathbf{x_j}$ at the end of the remediation period, c_j^{max} is the upper concentration limit of the contaminant at $\mathbf{x_j}$, q_i^{max} is the upper pumping limit at the ith well, k is the number of observations wells, N is the number of pumping wells and $f(\mathbf{q})$ is the total cost of pumping (injection or extraction) and is a convex monotonically increasing function. The concentration c_j are related to the well discharge and recharge by means of the steady-state flow equation and the time-dependent transport equations.

The basic idea of the algorithm is to increase the rate of convergence of the primal method by using an approximating polytope, similar to the one generated by the 'cutting plane' method, in order to evaluate the direction of the line search.

STEP = 0

Find a feasible point; a possible choice for this is usually the point $\mathbf{q} = \mathbf{q^{max}}$ (Tucciarelli (1994)). Set $i = 1$ and perform a line search from point $\mathbf{q^{max}}$ towards the origins; along this line search determine a point P_i that satisfies all the constraints and make active (satisfies by equality) at least one of the constraints (fig.1).

STEP = 1

Call r the index of this constraint and define $h(\mathbf{q}) = c_r - c_r^{max}$. Compute the derivatives $\frac{\partial h}{\partial \mathbf{q}} \equiv \frac{\partial c_r}{\partial \mathbf{q}}$ of the function $h(\mathbf{q})$ evaluated at point P_i.

STEP = 2

These derivatives along with the point P_i define the tangent plane that is the constraint number i of the following subproblem *SUBPROBLEM ONE*

$$minimize \ f(\mathbf{q}) \tag{4}$$

$$subject \ to \quad g_p \leq 0 \qquad p = 1, i \tag{5}$$

$$0 \leq q_j \leq q_j^{max} \qquad j = 1, N \tag{6}$$

where

$$g_p = \frac{\partial h}{\partial \mathbf{q}} \mid_{\mathbf{q}^{P_p}} (\mathbf{q} - \mathbf{q}^{P_p}) \tag{7}$$

The point P_i is on the tangent plane number i and therefore, due to convexity, it is feasible for subproblem one. Thus, $f(\mathbf{q}^{P_i^m}) \leq f(\mathbf{q}^{P_i})$ (with equality when $P_i^m \equiv P_i$).

If P_i^m is feasible, the feasible domain of the original problem is non convex. In this case, move from point P_i^m towards the origin and define along this direction the point P_{i+1} that

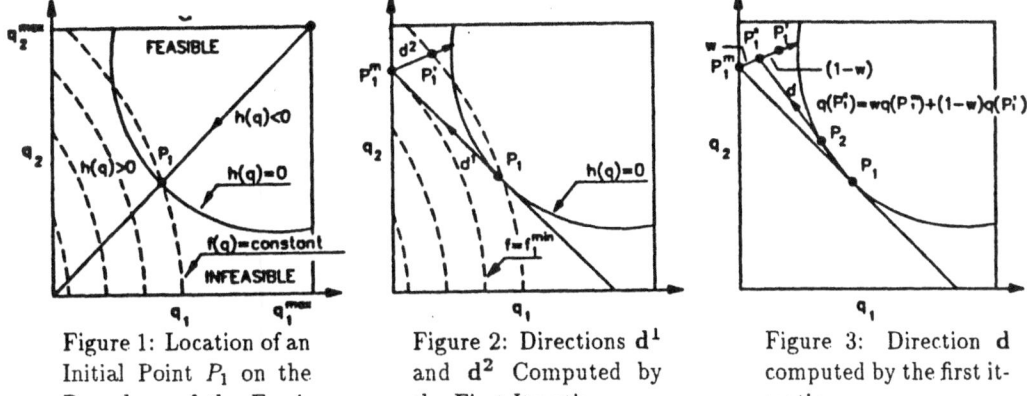

Figure 1: Location of an Initial Point P_1 on the Boundary of the Feasible Region.

Figure 2: Directions $\mathbf{d}^1$ and $\mathbf{d}^2$ Computed by the First Iteration

Figure 3: Direction $\mathbf{d}$ computed by the first iteration

is located on the outer boundary of the feasible region. This point, P_{i+1}, has an objective function less than point P_i, since the cost function increases monotonically. Discard all the previous cutting hyperplanes and start the algorithm from the point P_{i+1}.

STEP = 3

If P_i^m is feasible, define the direction $\mathbf{d}^1 = \mathbf{q}^{P_i^m} - \mathbf{q}^{P_i}$. From point P_i, moving along $\mathbf{d}^1$ it is sometimes possible to find a point P_{i+1} that has an objective function smaller than $f(\mathbf{q}^{P_i})$ and satisfies the condition $h(\mathbf{q}) = 0$ (fig.4 for $i = 3$). Employing the direction $\mathbf{d}^1$ it is possible to fail to improve the objective function; since, due to the convexity of the constraints at P_i, when P_i^m lies on the same tangent plane where the point P_i is located, the direction $\mathbf{d}^1$ is not feasible and we would find $P_{i+1} \equiv P_i$ In such a case, it is necessary to combine direction $\mathbf{d}^1$ with any feasible direction $\mathbf{d}^2$ (fig. 2). The optimal choice for direction $\mathbf{d}^2$ is defined as follows:

When the point P_i^m is on the tangent plane number i, the strategy is to choose a direction $\mathbf{d}^2$ that moves from the point P_i^m towards a point P'' that

a) is feasible for the constraint given by equation (5) number i

b) converges to the minimum of the problem one along with P_i and P_i^m. For an optimal choice for the direction $\mathbf{d}^2$ see Tucciarelli et al. (1994).

STEP = 4

Compute the point P_i' (fig. 3) along the direction $\mathbf{d}^2$ from P_i^m that has the same objective function value as P_i. Thereafter, compute the location of a new point P_i^c as:

$$\mathbf{q}^{P_i^c} = \mathbf{q}^{P_i^m}(1 - w) + w\mathbf{q}^{P_i'} \quad and \quad 0 \le w \le 1 \tag{8}$$

The reason for this is to determine a point P_i^c that improves the objective function, i.e. $f(\mathbf{q}^{P_i^c}) < f(\mathbf{q}^{P_i})$. The direction $\mathbf{d}$ that reduces the objective function and retains feasibility is determined by connecting points P_i and P_i^c. If P_i^c is infeasible, move from P_i^c to P_i until you find the point P_{i+1} on the boundary of the feasible region.

STEP = 5

Check for convergence; if convergence is not satisfied go back to *Step 1* and repeat the

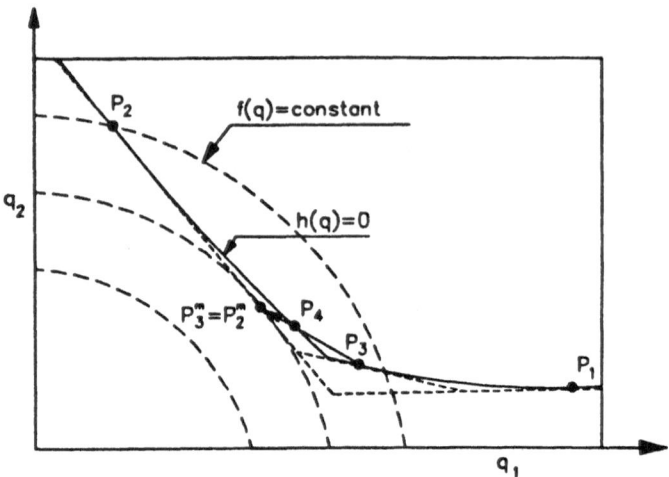

Figure 4: Location of the Point P_4 along the direction $\mathbf{d} = \mathbf{d}^1$ from P_3

process, otherwise an optimal solution is obtained and the algorithm is terminated. For the proof of convergence see Tucciarelli et al. (1994).

The Sensitivity Analysis

The most efficient minimization algorithms require the computation of the Jacobian matrix, which contains the derivatives of the constraint functions specified in the left hand side of equation (2) with respect the decision variables (pumping rates). These derivatives can be computed using one of the two approaches of sensitivity analysis as presented by Ahlfeld (1988).

The first approach computes, for each time step, the Jacobian $\mathbf{J}$ by solving the linear problem

$$\mathbf{AJ} = \mathbf{B} \tag{9}$$

where $\mathbf{A}$ contains the derivatives of the discretized equations of flow, Darcy's Law, and mass transport with respect the state variables (heads, velocities and concentrations) and $\mathbf{B}$ is a matrix with N columns.

The second approach computes, for each time step, the sensitivity matrix $\boldsymbol{\Phi}$, as the solution of the linear problem

$$\mathbf{A}^t\boldsymbol{\Phi} = \mathbf{C} \tag{10}$$

where $\mathbf{A}^t$ is the transpose of $\mathbf{A}$ and $\mathbf{C}$ is a matrix with k columns.

The solution of problems (9) or (10) is equivalent to the solution of N or k linear systems, respectively. For groundwater management problems usually leads to the first approach, since the number of pumping wells (N) in most of the problems is smaller than the number of constraints (k). Therefore, the numerical effort required for the computation of the derivatives at each step is usually proportional to the dimensionality of the problem. With the new method it is possible to use the second approach, assuming $k = 1$, since only the derivatives of the constraint active at each point P_i are required, and the computational effort becomes constant irrespectively of the dimensionality of the problem.

TESTING THE ALGORITHM

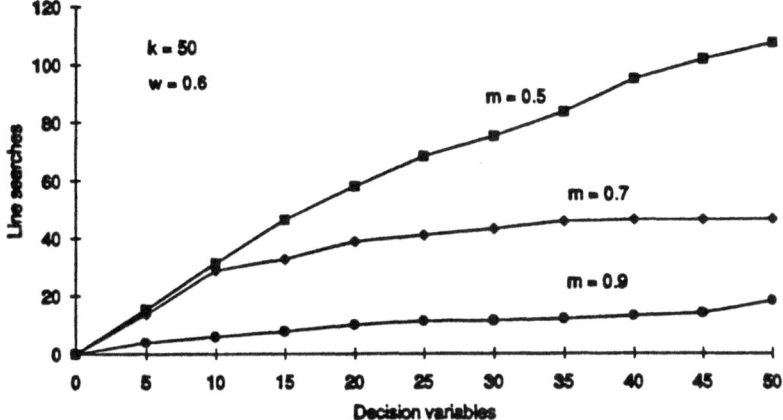

Figure 5: Number of Line Searches in the Test of 100 Cases.

Analytical examples

For the first test analytical examples were used with quasilinear constraints (Tucciarelli, 1994) and linear objective functions. The constraints were in the form of:

$$b_i - \sum a_{ij} x_j^m \leq 0 \qquad i = 1, k \quad j = 1, N \tag{11}$$

where m was allowed to vary between 0 and 1. Approximately 100 solutions were obtained using values for the coefficients a and b that were selected randomly from a range between 0 and 1 and for three selected values for the parameter m: 0.9, 0.7 and 0.5. It was assumed $N = 1, 2, \ldots, 50$, $k = 50$ and a relative error of $1x10^{-3}$.

Three measurements were performed to determine the efficiency of the algorithm: (1) the number of times that any constraint has been evaluated, (2) the number of times that the derivative of any constraint has been evaluated and (3) the number of line searches required to obtain a solution. In fig. 5 it is shown the number of line searches for the three different values of m.

The same problems were solved using MINOS 5.1 (Mutragh and Saunders, (1987)). In fig. 6 it is shown the number of times that the set of equations (9) or (10) must be solved to compute the constraints and its derivatives, for both MINOS 5.2 and the Primal Method. For MINOS 5.1 this is equal to $N+1$ times the subroutine calls and for the new method is the summation of the first two measurements. Assuming that the majority of the computational effort is takes up in these calculations, it is evident that ussing the new method this increases linearly and nonlinearly using MINOS 5.1.

A groundwater management problem. The Woburn (MA) aquifer

The Woburn aquifer was used as an application of the method to a groundwater quality management problem. The above aquifer has been studied in the past by Ahlfeld (1987), and the present problem formulation was based on his work. The aquifer was modeled using a vertically averaged two-dimensional finite elements code. The ranges of the hydraulic and

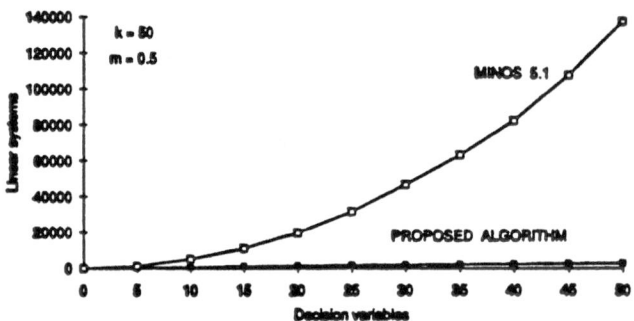

Figure 6: Number of Linear Systems Solved by MINOS and by the Primal Method in the Test of 100 Cases.

numerical parameters were (after Ahlfeld (1987)):
Number of nodes : 835, Number of elements : 767
Horizonatl Hydraulic conductivity, K : 20-1200 ft/day
Longitudinal dispersivity, α_L : 3-90 ft
Transverse dispersivity, α_T : 3-90 ft
Porosity, θ : 0.2
Saturated Thickness, b : 5-120 ft
Diffusion coefficient, d : 0.00001
Time step, Δt : 3 months (2190 hours)
Number of time steps : 12
The finite element mesh of the modeling domain isshown in fig. 7. The initial and boundaries condition at the beginning of the remediation period are shown in fig. 8. Fixed head conditions are indicated by circles, third type conditions are indicated by black triangles and specified flux conditions with value zero are indicated by diamonds. The existing river is indicated by the dashed line through the center of the figure. Also, the forty potential well are indicated by empty squares and the three observation locations by black squares. The problem was formulated as:

$min \quad \sum_{i=1}^{n} \alpha_i^+ q_i^+ + \alpha_i^- q_i^- \qquad i \in I$
$s.t. \qquad c_j(\mathbf{q}) \le c_j^*, \qquad j \in J$
$0 \le q_i^+ \le q_i^{+up}$
$0 \le q_i^- \le q_i^{-up}$
$q_i = q_i^+ - q_i^-,$

where $c_j(\mathbf{q})$ is the concentration at node j, c_j^* is the upper bound of c_j, α_i^+ is the unit injection cost, α_i^- is the unit extraction cost, q_i^+ is the injection rate at node i, q_i^- is the extraction rate at node i, q_i^{+up} is the fixed upper bound of q_i^+, q_i^{-up} is the fixed upper bound of q_i^-, and the last equation shows that at each well either injection or extraction was allowed. It was requested that the concentrations at the three observation points to remain less or equal to 50 ppb by the end of the three years remediation period.

The solution of the problem, using the Primal Method, is shown in fig. 9 and the results are summarized in Table 1. The extraction wells are indicated by a '-' sign and the injection by a '+' sign. Only six wells, from the 40 potential wells, were activated; five of them were

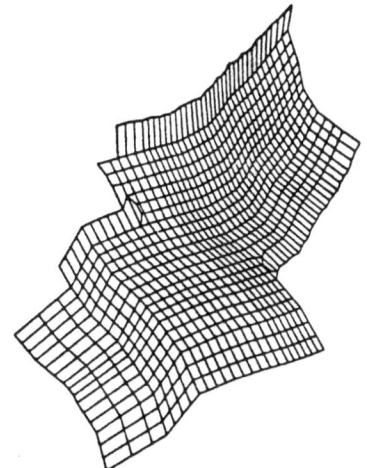

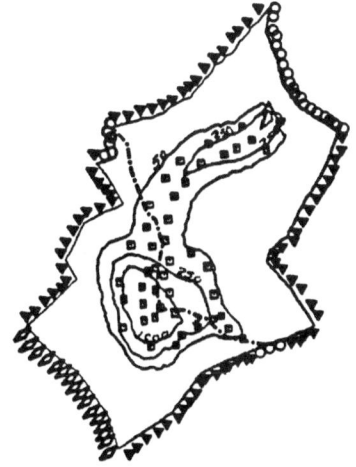

Figure 7: The Finite Element Mesh for the Woburn Aquifer

Figure 8: Initial and Boundary Conditions of the Woburn Aquifer at the Beginnig of the Remediation Period.

extracting and one was injecting. Also, in Table 1 the solution of the same problem using MINOS is included, as presented by Ahlfeld (1987). Regarding computing time, using the Primal Method the reported 'user time' was 39 minutes and using MINOS 5.1 no results were obtained even 32 hours later. All the above runs were performed on a 24 specfp92 Silicon Graphics R3000 Indigo with 80 Megabytes of memory.

PRIMAL METHOD								
	Pumping rates (f^3/sec)						Total Pumping ft^3/sec	
Well #	1	2	3	4	5	6		
Node #	90	120	123	139	188	227	Injection	Extraction
	-.114	+.05	-.157	-.194	-.156	-.404	.05	1.025
MINOS								
Well #	1	2	3	4	5	6		
Node #	142	120	123	139	188	227		
	-.342	+.05	-.571	-.512	-.75	-.75	.05	2.925

Table 1: Optimal Solution for the Woburn Aquifer.

Figure 9: Optimal Solution for the Woburn Aquifer Using the Primal Mathod

CONCLUSIONS

A new method for the solution of the groundwater quality management problem has been presented. Assuming that most of the computational effort lies on the evaluation of the constraint functions and their derivatives, the computer time required by the new method appears to grow sub-linearly with the number of decision variables. In the solution of test problems the performance of the algorithm has been shown to be better than the performances of MINOS 5.1.

REFERENCES

1. Ahlfeld, D.P., J.M. Mulvey, G.F. Pinder and E.F Wood (1988), 'Contaminated Groundwater Remediation Design Using Simulation, Optimization, and Sensitivity Theory. 1. Model Development', *Water Resources Research*, vol.24, no.3, 431-441.
2. Ahlfeld, D.P., (1987), 'Designing Contaminated Groundwater Remediation Systems Using Numerical Simulation and Nonlinear Optimization', Ph.D. Dissertation, Princeton University.
3. Culver, T.B., and Shoemaker, C.A. (1992), 'Dynamic Optimal Control Groundwater Remediation with Flexible Management Periods.', *Water Resources Research*, vol.28, no.3, 629-641.
4. Dougherty, D.E., and Marryott, R.A. (1991), 'Optimal Groundwater Management. 1. Simulated Annealing.', *Water Resources Research*, vol.27, no.10, 2493- 2509.
5. Karatzas, G.P. and G.F Pinder (1993), 'Groundwater Management Using Numerical Simulation and the Outer Approximation Method for Global Optimization', *Water Resources Research*, vol.29, no.10,3371-3378

6. Luenberg, D.G. (1984), Linear and Nonlinear Programming,Addisson Wesley.
7. Murtagh, B.A., and M.A. Saunders (1987), 'MINOS 5.1 Users Guide.' Technical Report SOL 83-20R, Department of Operations Research, Stanford Univ., CA.
8. Rao, S.S (1984), Optimization, Theory and Applications, Halsted Press, New York.
9. Tucciarelli, T. (1994), 'Solving the Groundwater Quality Management Problem: a Global Approach', Ph.D. Dissertation, Princeton University. curvature.
10. Tucciarelli, T., G. P. Pinder and G. P. Karatzas (1994), 'A Primal Method for the Solution of the Groundwater Management Problem', *Water Resources Planning and Management*, in press.

GROUNDWATER RESOURCE MANAGEMENT MODELS: A COMPARISON OF GENETIC ALGORITHMS AND NONLINEAR PROGRAMMING

Daene C. McKinney, Gregory B. Gates and Min-Der Lin
Department of Civil Engineering
University of Texas, Austin, Texas 78712
USA

The design of remedial actions for cleaning up contaminated aquifers is an important problem in the management of groundwater resources. Many remediation design problems involve the solution of mathematical programming problems which have highly nonlinear and frequently discontinuous objective functions and constraint sets. This paper examines the feasibility of combining groundwater flow and mass transport simulation models with genetic algorithms, search algorithms based on the mechanics of natural selection, to help search for optimal aquifer remediation designs. We present the methodology, model development and results of applying the genetic algorithm approach to solve an aquifer remediation problem. A minimum cost pump-and-treat system design problem has been solved by incorporating a groundwater simulation model into a genetic algorithm. The results show that genetic algorithms can effectively and efficiently be used to obtain globally optimal solutions to this groundwater management problem. The computational time required for the solution of genetic algorithm groundwater management models increases with the complexity of the problem. Results of speedup attainable by solving genetic algorithm problems on massively parallel computers are presented for problems where the simulation time required to complete each generation is high. The genetic algorithm results are compared to nonlinear programming results and show a significant improvement in computational efficiency by using the genetic algorithm method.

INTRODUCTION

The remediation of sites contaminated with hazardous wastes often includes some form of groundwater clean-up. Pump-and-treat technology is an accepted and widely used method for groundwater remediation. Currently pump-and-treat systems are designed in a somewhat piece meal fashion. Generally a well field is designed to intersect a contaminant plume so that the radius of influence of the pumping wells will overlap the boundaries of the contaminate plume and effectively isolate and remove the contaminant. Once the well field is designed and the flow rates, concentrations, and expected life are estimated, a treatment system is designed to meet regulatory restrictions over a specified time period. To take into account the interaction of these two systems and to determine an "optimal" design, computer modeling of the aquifer system, well field, and treatment system is employed.

Optimization models have been formulated which minimize the capital and operation and maintenance costs of the treatment system and well field [1]. These models typically contain several regulatory and aquifer state variable constraints. Unfortunately, nonlinear

A. Peters et al. (eds.), Computational Methods in Water Resources X, 859–866.
© 1994 *Kluwer Academic Publishers. Printed in the Netherlands.*

programming (NLP) solutions to such highly nonlinear and nonconvex problems have no guarantee of globally optimal values. Recently, nongradient-based methods which do not simply identify local optima have been developed and applied to groundwater management problems. These methods include simulated annealing [2], neural networks [3], and genetic algorithms (GAs) [4]. GAs find a solution to a random and relatively large sample of the possible solutions, it is more likely that the final GA solution approaches a true global optimum. In this paper, we examine the NLP and GA results when applied to a "real world" site with limited information about the spatial distribution of aquifer properties.

MODEL FORMULATION

The models used in this research are made up of (1) a treatment process model, (2) a groundwater flow and contaminant transport simulation model, and (3) a NLP or GA model. The treatment process model is used to design an air stripping tower for clean-up of the groundwater. This model uses the flow rates and contaminant concentrations produced by the simulation model to determine the stripping tower volume. A standard Petrov-Galerkin finite element model is used to descritize the governing equations of groundwater flow and dissolved contaminant transport [1].

The objective of the optimization model is minimize the capital and operating costs of the well field and the treatment system. The optimization model constraints are: (1) all extracted water is treated and reinjected, (2) maximum drawdown is limited to $\alpha\%$ of the original head, (3) maximum and minimum extraction and injection rates, (4) regulatory limit on concentration of contaminant in treated effluent, and (5) regulatory limit on final concentration of contaminant in the aquifer. This management model can be written as

$$Min \quad C_{treat} + \sum_{i=1}^{E} C_{extract,i} + \sum_{j=1}^{I} C_{inject,j} \tag{1}$$

subject to

$$C_{treat} = a_1 (ZA)^{b_1} + a_2 (Q)^{b_2}$$

$$C_{extract,i} = a_3 (d_i)^{b_3} + a_4 (Q_{E,i})^{b_4} (d_i - h_i)^{b_5} + a_5 Q_{E,i} (d_i - h_i) \tag{2}$$

$$C_{inject,j} = a_3 (d_j)^{b_3} + a_4 (Q_{I,j})^{b_4} (h_{oj} - h_j)^{b_5} + a_5 Q_{I,j} (h_{oj} - h_j)$$

$$\sum_{i=1}^{E} Q_{E,i} = \sum_{j=1}^{I} Q_{I,j} \tag{3}$$

$$s_i \le \alpha h_{oi} \tag{4}$$

$$Q^{min} \le Q_i \le Q^{max}, \qquad i = 1, \cdots, E; \ i = 1, \cdots, I \tag{5}$$

$$c_E = c_I \left\{ \frac{(s-1)/s}{exp[ZK_L aA / Q(s-1)/s] - 1/s} \right\} \le c^* \tag{6}$$

$$c_{i,T} \leq c^* \tag{7}$$

where Z (L) and A (L^2) are the tower height and surface area, Q (L^3/T) is the tower liquid flow rate, $K_L a$ (1/T) is the tower overall mass transfer coefficient, S is the tower stripping factor, d_i (L), h_i (L), and h_{0i} (L) are the depth of a well, the hydraulic head, and the initial hydraulic head at point i, respectively, $Q_{E,k}$ and $Q_{I,k}$ (L^3/T) are the extraction or injection rate at point k, a_i and b_i, $i=1,...,5$ are constants, s_i (L) is the drawdown and α is the percent of the original head allowed for drawdown at point i, c_I and c_E (M/L^3) are the influent and effluent concentrations of the tower, c^* (M/L^3) is the target regulatory concentration both in the tower effluent and the aquifer.

The NLP model uses a generalized reduced gradient (GRG) solution algorithm for solution [5]. Because problems of this type are nonlinear and nonconvex a global optimum can not be guaranteed. To determine a solution more likely to be a global optimum, the model was solved with a genetic algorithm (GA). The GA does not require the calculation of the derivatives of the highly nonlinear equations such as (6) and (7) which are implicitly linked to the groundwater simulation model. GAs use a random search procedure inspired by biological evolution, where trial designs are cross-breed and the fittest designs are allowed to survive and propagate to successive generations [4]. A string of randomly generated designs is generated. Each design is input into the groundwater simulation model and its overall performance is evaluated based on objective function value and constraint violations. The best or "fittest" designs are passed on to the next round of trials or generation and again evaluated in the simulator along with newly generated designs. The generation contains randomly generated designs as well as those obtained through crossover and mutation of the best designs of the previous generation. This procedure continues until an optimal design is reached. The treatment process and groundwater simulation models were combined with either a NLP or a GA model to produce the following results.

MODEL APPLICATION

The site used in the computations here is a former gasoline service station where four 3,000 gallon underground storage tanks (USTs) were removed in 1988 [6]. During the tank removal, no evidence of soil or groundwater contamination above regulatory limits was discovered. Groundwater samples collected from three soil borings (soil borings b1, b2, and b3) indicated TPH and BETX concentrations exceeding regulatory requirements of 50 µg/L benzene, 500 µg/L BETX, and 1 mg/L TPH. Subsequently, three more borings were completed as groundwater monitoring wells (monitoring wells MW-4, MW-5, and MW-6). Groundwater samples obtained from each of the monitoring wells excluding MW-5 indicated TPH and BETX concentrations exceeding regulatory limits. Concentrations of dissolved benzene exceeded regulatory limits in monitoring wells MW-5 and MW-6. The benzene concentration levels in the groundwater samples are listed in Table 1. Hydraulic head and concentration data from these six borings were used to estimate the general groundwater gradient and dissolved benzene concentrations for the site.

A 16 by 16 node (91.44 m by 91.44 m) rectangular finite element grid with nodes evenly spaced at 6.1 m intervals was used to represent the site in the simulation model. In both the NLP and GA model runs, the solutions were constrained to meet a regulatory benzene concentration of 5 µg/L in the air stripping tower effluent and in the aquifer at the end of remediation and a drawdown of less than or equal to 10 percent of the original head at selected observation nodes. Pumping and injection rates were constrained between a minimum of 0 and a maximum of 6.2 L/s. The remediation time period was two years with a 10 percent capital recovery interest rate. All costs are present value adjusted by the ENR Construction Cost Index for May of 1988 (4522). Observation nodes were picked to ensure regulatory benzene concentration compliance at the pumping well locations and to ring the contaminant plume so that a solution to the posed problem would not be to push the contaminant offsite. In each of the NLP and GA runs, 15 wells were available for pumping. In both the NLP and the GA solution 28 nodes were used as compliance nodes. The different NLP solutions reflect the different initial guesses input to the model. The aquifer properties used in the model are listed in Table 2.

Table 1. Groundwater Sampling Results.

Water Samples	Benzene (μg/L)
B-2	1900*
B-3	9600*
MW-4	20
MW-5	240
MW-6	5700*

* - exceeds regulatory limit

Table 2. Aquifer Properties

Property	Value
Porosity	0.35
Aquifer thickness	4.05 m.
Hydraulic conductivity	0.0001 m/sec
Bulk density	219.7 kg/m3
Longitudinal dispersivity	25.0 m
Transverse dispersivity	3.12 m

RESULTS

The site conditions were input to either the NLP or GA management models and several runs were made for different conditions. The results of these cases are presented in Table 3 The results of the GA cases were also used as starting conditions for the NLP model for further refinement. In this scenario the NLP algorithm begins at a point that is very close to an optimal solution. Further refinement reflects minor adjustments in tower influent flow rate and concentration, which in turn produces lower overall costs.

For case NPL-0 the initial extraction or injecting rates were guessed at 10 of the 15 possible well locations. The remaining 5 wells were begun at zero extraction and zero injection, but were available for the algorithm to use to meet the constraints. This scenario was defined to determine the effects of small pumping and injection rates over a large area of the site. It was also interesting to see how the algorithm dealt with this starting condition. Given that adding or subtracting a well from the system is a large jump in the system cost (approximately $11,000-$12,000 per well), the model does not easily turn off pumping wells if they have been input as part of the solution even if they are not necessary for the overall design. For cases NLP-1 to NLP-4, nonzero initial pumping and injection rates were guessed at two wells that seemed likely to meet the constraints. With each pair of wells, one injection well up-gradient and one extraction well down-gradient, initial guesses, including a relatively small pumping rate and a much larger pumping rate, were examined. Again, the remaining 13 wells were begun at zero extraction and zero injection, but were available for the algorithm to use to meet the constraints. For case NLP-3 no feasible solution could be found with that small of an initial pumping and injection rate. The other small pumping rate scenario, case NPL-1 likewise could not find a feasible solution with that small of initial pumping and injection rates and thus a third well was initiated by the model. Each of the larger pumping rate scenarios were able to find feasible solutions for the initial well locations. The solution conditions column of Table 3 contains information on the certainty to which a run was terminated. For the NLP algorithm, Kuhn-Tucker conditions imply the highest level of certainty with at least a local optimum and continuous first partial derivatives. This condition is followed in level of confidence by 15 steps probably cycling and then by failed to find a better point.

Table 3. Results of NLP and GA Runs

Case	Node	Initial Rate		Final Rate		Solution Conditions	Total System Cost ($)	CPU Time (min.)
		Inject (L/s)	Pump (L/s)	Inject (L/s)	Pump (L/s)			
NLP-0	23		0.708		0	15 steps probably cycling	133,526	15.3
	37	0.708		0.371				
	41		0.708		1.79			
	67	0.708		0				
	86	0.708		0.667				
	88	0.708		0.648				
	101		0.708		0			
	103		0.708		0.149			
	104		0.708		0.504			
	118	0.708		0.731				
NLP-1	23		0.708		1.07	failed to find a better point	77,767	11.4
	135	0.708		0.410				
	67	0.0		0.665				
NLP-2	23		5.66		1.52	failed to...	69,823	9.7
	135	5.66		1.52				
NLP-3	88		0.708	N.A.	N.A.	15 steps...	N.A.	0.1
	118	0.708		N.A.	N.A.			
NLP-4	88		5.66		2.51	failed to...	77,373	1.7
	118	5.66		2.51				
GA-1	37	NA	NA		1.28	300 gen.	91,701	68.46
	71	NA	NA		1.64			
	120	NA	NA	2.92				
GA-2	23	NA	NA		1.64	273 gen.	70,711	62.30
	103	NA	NA	1.64				
GA-3	37	NA	NA		0.548	412 gen.	69,810	94.02
	103	NA	NA	0.548				
GA-4	88	NA	NA	3.11		246 gen.	81,022	56.14
	135	NA	NA		3.11			

In the GA runs, each starting point is chosen at random and well extraction or injection rates are increased or decreased by the model until a better solution is found. Each successive generation of trial pumping or injection rates includes the best solutions from the previous generation so that the algorithm converges on the "optimal" solution . In the GA runs the following parameters were used: population size = 64, string length = 96, generations = 500, crossover probability = 0.9, mutation probability = 0.02, convergence criteria (generations) = 60, number of children per generation = 16.

The best NLP solution, based on total cost and ability to meet the contaminant and head constraints, is solution NLP-2 ($69,823). The best GA solution by the above criteria is GA-2 ($70,711). The differences in stripping tower costs are due solely to changes in well field design. The location of the wells determines the concentrations seen by the air stripper at various times. Different pumping rates are also required at different well locations to ensure that regulatory constraints are met. The combination of these factors must be considered when determining the well field placement as their effect on the cost of the air stripper and thus the total cost is evident.

As noted above, runs of the NLP algorithm were made using the four solutions to the GA as starting points. The results of these runs are presented in Table 4. As can be seen form these results, there is considerable improvement in the solutions. For the combined models run in series the best solution based on total cost and ability to meet constraints is GA-2.2 ($63,790). An air stripping tower is designed by the model for each run of either algorithm. The capital and operation costs of the air stripping tower as well as input data for the air stripping tower design are presented in Table 5.

Table 4. Results of Combined GA and NLP Runs

Case	Node	Initial Rate		Final Rate		Solution Conditions	Total System Cost ($)	CPU Time (min.)
		Inject (L/s)	Pump (L/s)	Inject (L/s)	Pump (L/s)			
GA-1.2	37		1.28		0.48	15 steps...	76,836	71
	71		1.64		0.43			
	120	2.92		0.91				
GA-2.2	23		1.64		0.80	failed to...	63,790	65
	103	1.64		0.80				
GA-3.2	37		0.548		0.77	failed to...	64,269	97
	103	0.548		0.77				
GA-4.2	88	3.11		2.34		failed to...	76,528	59
	135		3.11		2.34			

CONCLUSIONS

The results reported here indicate the difficulty of obtaining accurate and meaningful solutions to pump-and-treat remediation design problems using NLP methods. The use of GAs to solve these problems provides an efficient and improved alternative which results in globally optimal solutions to these difficult problems with little "fine tuning" of the solution algorithm and no consideration of alternative initial guesses.

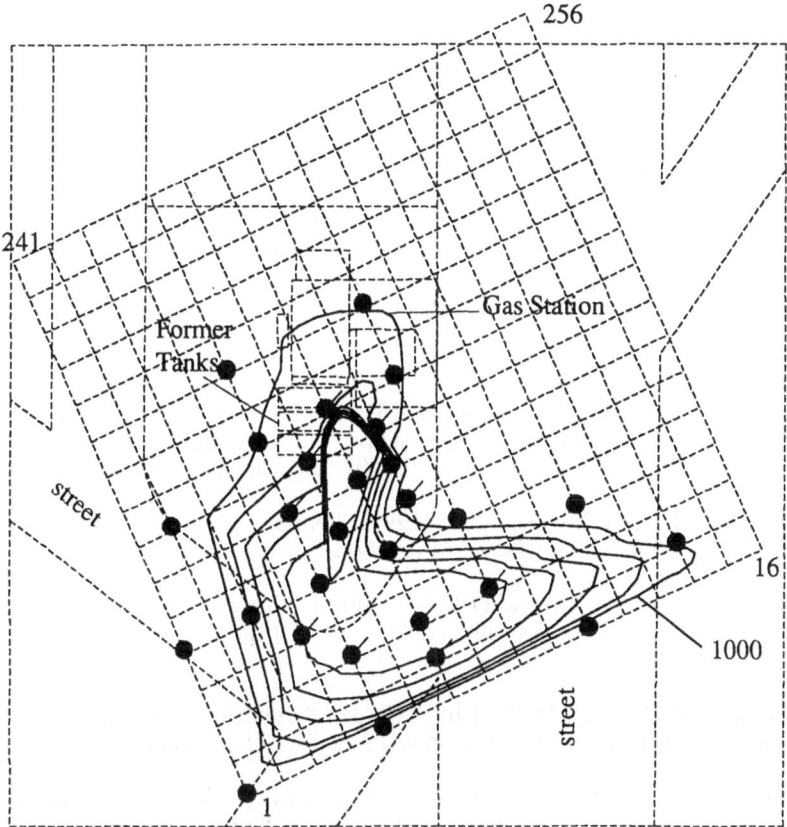

● compliance node

◖ compliance and possible pumping or injection node

Contour intervals are 1000 μg/L.

Figure 1. Site, finite element grid, and initial benzene contours.

ACKNOWLEDGMENTS

This research was partially funded by the Gulf Coast Hazardous Substances Research Center which is supported under cooperative agreement R 815197 with the United States Environmental Protection Agency, the Texas Hazardous Waste Research Center, and the Department of Energy (DE-FG05-92ER25142) HPCC Program.

Table 5. Air Stripping Tower Design and Results for Combined Runs

	GA-1.2	GA-2.2	GA-3.2	GA-4.2
Area (ft^2)	0.72	0.64	0.61	1.87
Height (ft)	28.4	26.0	28.7	26.3
Volume (ft^3)	20.4	16.7	17.5	49.1
Tower FLow Rate (L/s)	0.91	0.80	0.77	2.34
Inf. conc. ($\mu g/L$)	3290	1754	3508	1905
Eff. conc. ($\mu g/L$)	3.9	3.6	3.9	3.7
Tower Capital cost ($)	36,485	34,927	35,368	44,775
Tower O&M cost ($)	6,066	5,687	5,925	6,963

REFERENCES

[1] McKinney, D.C., and M-D. Lin, (1992). "Design Methodology for Efficient Aquifer Remediation Using Pump and Treat Systems," in Mathematical Modeling in Water Resources, T. Russell et al., (eds.) pp. 695-702, Elsevier Science Publishers, London.

[2] Dougherty, D.E., and R.A. Marryott, (1991). "Optimal groundwater management: 1. Simulated annealing," Water Resour. Res., 27(10), 2493-2508.

[3] Rogers, L.L. and F.U. Dowla, (1992). "Optimization of groundwater remediation using artificial neural networks with parallel solute transport modeling," Water Resour. Res., (accepted).

[4] McKinney, D.C., and M-D. Lin, (1993). "Genetic Algorithm Solution of Groundwater Management Models," Water Resour. Res., (accepted).

[5] Lasdon, L. S., A. D. Warren, A. Jain, and M. Ratner, (1978). "Design and testing of a generalized reduced gradient code for nonlinear programming," ACM Transactions on Mathematical Software, 4(1), 34-50.

[6] Sheridan, P.J., (1992). "A Remediation System Design for a Leaking Petroleum Storage Tank (LPST) Site in Austin, Texas," Report, Dep. of Civ. Eng., The Univ. of Texas, Austin, TX.

RECOMMENDATIONS FOR THE DEWATERING OF LETLHAKANE DIAMOND MINE (BOTSWANA) UTILIZING WATER BALANCE, MODELLING AND OPTIMIZATION TECHNIQUES

H. JANSE VAN RENSBURG[*], R.A. BUSH[*] AND G.J. VAN TONDER[**]

[*] CIVIL ENGINEERING DEPARTMENT
ANGLO AMERICAN CORPORATION
OF SOUTH AFRICA
P.O. BOX 61587
MARSHALLTOWN
2107
SOUTH AFRICA

[**] INSTITUTE FOR GROUNDWATER-STUDIES
UNIVERSITY OF THE ORANGE FREE STATE
P.O. BOX 339
BLOEMFONTEIN
9300
SOUTH AFRICA

ABSTRACT

The Letlhakane Diamond Mine of the Debswana Diamond Company (PTY) LTD. is a typical example of one of the biggest open cast mines in Southern Africa. In order to keep mining operations dry and to maintain slope stability it is essential to dewater the sandstone aquifer surrounding the impervious diamond bearing Kimberlite pipe. The aquifer in which the Mine is located forms part of a greater aquifer system which also supplies water to the nearby Orapa Mine town and several other villages. It is thus inevitable that the dewatering from Letlhakane Mine will to some extent influence hydrogeological conditions in the nearby wellfields.

In order to upgrade the dewatering infrastructure at Letlhakane Mine to keep pace with mining activities and determine the influence of increased dewatering on surrounding wellfields the need for hydrogeological modelling was recognised. Accordingly a finite element groundwater flow model was established for both the dewatering system at Letlhakane Mine and the adjacent wellfield. The calibrated flow model was coupled to a response matrix in order to achieve the required drawdown constraints while using a minimum number of new boreholes within an optimized extraction layout.

Optimization runs were performed assuming two recharge from rainfall scenarios. First, a 5mm per annum recharge from rainfall value, established by means of water balance calculations for the investigated period. The second scenario assumed no recharge from rainfall. The modelling results from these scenarios provide the Mine Management of Letlhakane Mine with optimally determined abstraction rates required to achieve dry mining operations and ensure slope stability in the open cast mine.

The effect of incremental abstraction from Letlhakane Mine on the nearby Wellfield Six has also been examined and found to have a relatively small impact.

INTRODUCTION

Letlhakane Mine, situated in Central Botswana about thirty kilometres south-east of Orapa (Figure 1), is a typical example of one of the biggest open cast diamond mines in Southern Africa. Whilst it is essential to dewater the sandstone aquifer surrounding the diamond bearing Kimberlite pipe in order to keep mining operations dry and ensure slope stability, the Mine, adjacent village of Letlhakane and mining town of Orapa are all

867

A. Peters et al. (eds.), Computational Methods in Water Resources X, 867–874.
© 1994 *Kluwer Academic Publishers. Printed in the Netherlands.*

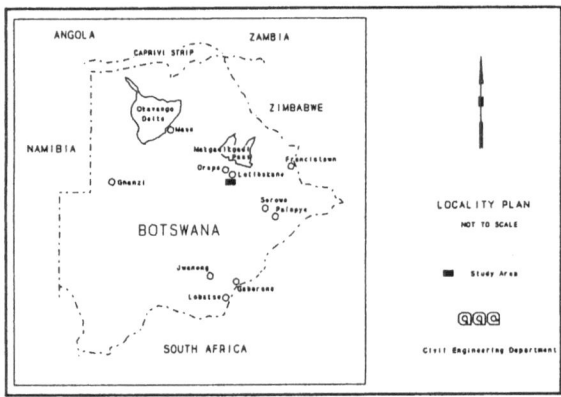

Figure 1 - Locality Map of the Study Area.

dependent on groundwater both for mining operations and domestic supply. The need to upgrade the dewatering infrastructure at Letlhakane Mine and determine the influence of any such increased abstraction on the greater groundwater system has necessitated the development of an integrated hydrogeological and decision making model for the system. Accordingly a finite element groundwater flow model coupled to a response matrix was established for both the dewatering system at the Mine and adjacent wellfield.

OBJECTIVES and MODUS OPERANDI

The objectives of the study were to:
i)   Determine the annual abstraction necessary to lower water levels as measured in all production and observation boreholes by at least 15 metres per year for the period January 1993 to December 1999, at Letlhakane Mine.
ii) Determine the feasibility of achieving the required abstraction scenarios using existing and a minimum number of additional boreholes.
iii) Determine the influence of proposed increased abstraction on adjacent wellfields.

In order to achieve the above objectives the following plan of action was implemented:
i)   Conduct a water balance study for the area under investigation (Wellfield Six and the Letlhakane mining area) to obtain:
     - estimates of the storage for the area
     - estimates of recharge from rainfall for the area.
ii) Construct a dynamic groundwater flow model using as input the storativity and recharge values obtained from the water balance study.
iii) Calibrate the groundwater flow model using historical water levels and abstraction figures.
iv) Use the calibrated groundwater flow model to determine the optimal extraction pattern from abstraction boreholes around the Letlhakane Mine Pit to effect a minimum drawdown of 105 metres (15 metres per annum) over a planning period of seven years (1993 to 1999) at all pumping and observation boreholes.
v)   Should the water balance calculations prove that groundwater recharge from rainfall is taking place, incorporate recharge into the calculations.

WATER BALANCE EQUATION

The classical hydrogeological waterbalance equation for a groundwater reservoir with an impermeable base (Kirchner and Van Tonder, 1989) for a time increment of

$$\Delta t = t_2 - t_1,$$

is given by:

$$I - O + GR - Q = \Delta W \tag{1}$$

where

$$I = \frac{(I_1 + I_2)}{2} = \textit{mean lateral inflow } (m^3/d)$$

$$O = \frac{(O_1 + O_2)}{2} = \textit{mean lateral outflow } (m^3/d)$$

GR = *effective ground water recharge into the reservoir* (m^3/d)

Q = *discharge out of (or into) reservoir in* m^3/d

$_\Delta W$ = *change in ground water volume* (m^3) = $S*_\Delta V$, *where*

S = *specific yield or effective porosity*

$\Delta\ V$= *change in saturated volume aquifer material* $(V_2 - V_1)$

Equation (1) can be rewritten as:

$$GR + I - O - \Delta V * S = Q \qquad (2)$$

This equation is the general groundwater balance equation and can be applied for a number of specific conditions (Bredenkamp *et al.*, 1989):
a)   If the inflow and outflow terms are equal, the change in groundwater storage is zero. This provides the necessary conditions to derive safe yield estimates and to predict recharge from precipitation as given in Equation (3).

$$GR = O - I + Q \qquad (3)$$

b)   By incorporating the "no recharge" recession (GR = 0), when the change in saturated volume presents a maximum over a given time period t, Equation (2) reduces to:

$$I - O - \Delta V * S = Q \qquad (4)$$

from which S can be calculated as

$$S = \frac{I - O - Q}{\Delta V} \qquad (5)$$

c)   If the aquifer is bounded by impervious dykes or by groundwater divides, I and O in (2) are zero. For this case

$$GR - \Delta V * S = Q \qquad (6)$$

from which the groundwater recharge can be calculated if S is known. If the groundwater recharge is assumed to be zero for the maximum change in saturated volume, a mean S-value for the aquifer can be calculated.

The computer program, SVF, can be used to calculate the saturated-volume-fluctuation (SVF) for an aquifer on a month to month basis (Van Tonder *et al.*, 1993). This allows:
i)   a mean S-value to be obtained which corresponds to the maximum change in saturated volume and implying zero groundwater recharge (Equation (6)).
ii)   By substitution of this S-value into the water balance equation, recharge and its relationship to rainfall can be obtained.

iii) The aquifer's saturated-volume-fluctuation (SVF) is subsequently simulated using the established rainfall-recharge relationship and varying S to obtain the best match between the simulated and the observed saturated volumes. Along the same lines the storativity can be assumed and the best rainfall-recharge relationship can be obtained.

WATER BALANCE EQUATION APPLIED TO THE LETLHAKANE AREA

Differences in groundwater flow regimes between Wellfield Six and the Mine Pit were recognised due to the cone of dewatering that exists around the Mine Pit. As interpolation techniques are used in the calculation of saturated volumes in the water balance equation it was necessary to conduct separate water balance calculations for Wellfield Six and the Mine Pit.

The monthly saturated-volume-fluctuations for Wellfield Six were determined for the period February 1989 to December 1992. This showed a decline in the saturated volume for the period February 1989 to September 1990, an increased rate of decline for the period September 1990 to June 1991 and a decrease in the rate of decline for the period June 1991 to December 1992.

If the groundwater recharge is assumed to be zero for the maximum change in saturated volume, a mean S-value for the aquifer can be calculated from Equation (6).

The three months with the largest change in saturated volume for Wellfield Six were recorded from the SVF calculations and S-values calculated assuming recharge to be zero. Table 1 shows the change in saturated volume, corresponding abstraction and the calculated S-values for the three months described above.

Table 1 - Calculated S-values for Wellfield Six.

CHANGE IN SVF (x 10^6 m³)	ABSTRACTION (m³)	CALCULATED S-VALUE
-228,75	36600	0,000160
-221,60	57240	0,000258
-163,89	108450	0,000662

Probable Mean S-value   0,000360
Average recharge using probable mean S-value: 4,3 mm/a. (1,1 % of rainfall).

The probable mean S-value as approximated above was then substituted back into the water balance equation and recharge calculated. The average recharge for the period under investigation was found to be 4,3 mm/a. Average rainfall over the same period was 391 mm/a. The ratio of recharge to rainfall is thus 1,1 %.

From the water balance equation it is clear that if inflow, outflow and recharge are zero, the S-value of a system can be obtained by plotting the abstraction against the change in saturated volume. The S-value corresponds to the slope of the line fitted through the plotted points according to the modified Hill-method proposed by Bredenkamp *et al.* (1989).

In the case of Wellfield Six, inflow and outflow were assumed to balance because no evidence exists to infer that these components of the balance equation differ significantly. The plot of the yearly abstractions versus the corresponding yearly change in saturated volume (monthly, twelve month running mean) are shown in Figure 2. This graph shows a constant relationship between change in saturated volume and abstraction until month 30 or June 1991. From this point in the record onwards this trend changes and for a constant abstraction the rate of change in the saturated volume becomes smaller.

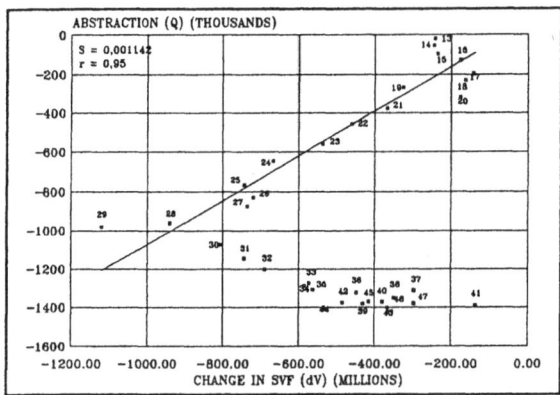

Figure 2 - Wellfield Six - Abstraction vs. dV.

Analyses for the period to month 30.

The early part of the data (up to month 30) yields an S-value of 0,001142 (correlation coefficient r = 0,95). As a straight line correlation with little scatter describes this part of the data set it is assumed that recharge was either zero or constant as varying recharge would have caused a scatter of the plotted points around the line.

Analysis for the period following month 30.

Although no direct temporal relationship between recharge and rainfall could be found with the water balance calculations, possible rainfall/recharge relationships were further investigated using the broadly accepted principle that a relationship exists between the cumulative rainfall departures (CRD) from the mean rainfall and water level behaviour within an aquifer. The cumulative rainfall departures from the mean rainfall were therefore calculated from the rainfall record for Letlhakane. The CRD together with the water levels for the observation boreholes were plotted against time. From these graphs it was observed that some of the peaks in the CRD and those in the water level records would correspond exactly if the CRD graph was shifted forward by four months (Figure 3).

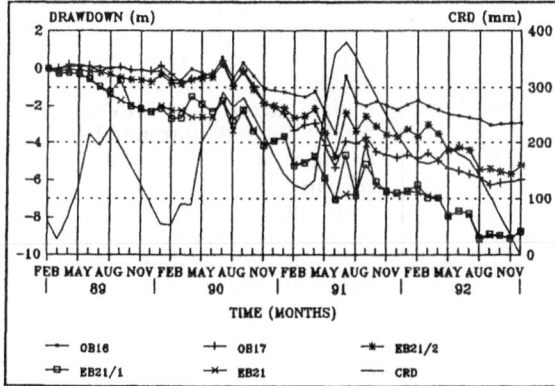

Figure 3 - Groundwater Level Fluctuations and CRD Lagged by Four Months.

The lag of four months between groundwater recharge and rainfall explains why a direct temporal correlation between recharge and rainfall was not initially evident. The CRD graph shows that greater than average rainfall accumulated to produce three distinctive anomalies, the last producing the largest cumulative departure from the mean rainfall and the first the smallest. The water level fluctuations indicate that only the second and third anomaly result in recharge to the aquifer with the third anomaly serving as the most probable reason for the change in the saturated volume recorded since June 1991.

In order to quantify recharge from the CRD, use was again made of the modified Hill-method where, by assuming no inflow, outflow or recharge, the S-value can be obtained by plotting abstraction (Q) against the change in the saturated volume (dV). If recharge however does occur, the S-value can still be obtained by plotting recharge (GR) minus abstraction (Q) against the change in saturated volume (dV). Recharge within the aquifer is then calculated by obtaining the best fit between GR-Q and dV where CRD is taken as a percentage of CRD minus a certain cut-off value. The best fit between GR-Q vs. dV are found to occur with a recharge/CRD relationship of:

GR(mm) = 0,0125*(CRD-250 mm) (Figure 4).

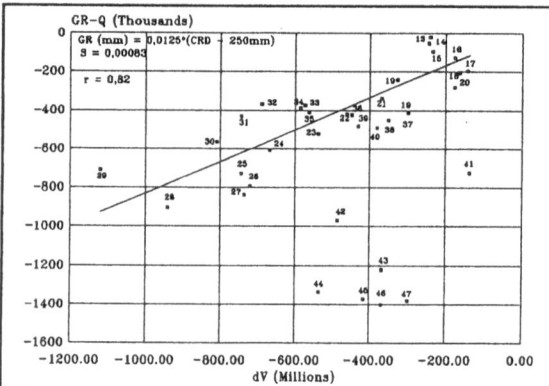

Figure 4 - Groundwater Recharge minus Abstraction vs. dV.

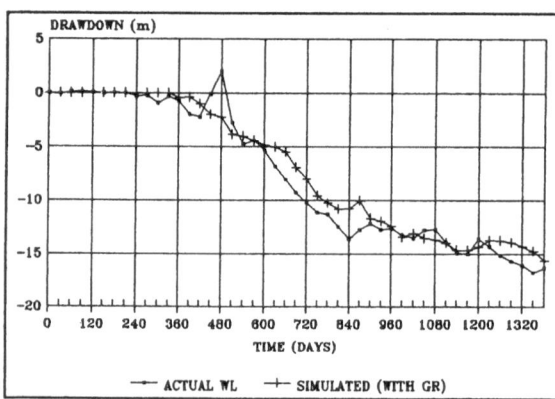

Figure 5 - Borehole EB25 - Actual vs. Simulated Water Levels.

For the investigation period the recharge amounts to 5,6 mm and the effective corresponding rainfall 501 mm. Therefore the percentage recharge above the threshold value calculated for the investigation period amounts to 1,11%.

DYNAMIC MODEL OF THE LETLHAKANE WELLFIELD SIX AND MINE PIT AREA

The AQUA-NET package (Staats, 1993) was used to construct the finite element grid for the study area whilst the AQUA-INV and AQUA programs (Van Tonder et al., 1993) were used respectively to calibrate and simulate groundwater flow in the compartment. The finite element model (AQUA) is based on the Galerkin solution of the two-dimensional partial differential equation that describes flow subject to specific boundary and initial head conditions. The results obtained through the water balance study, namely S and recharge values were used in the flow model. With these input values, acceptable correlations between the actual and simulated water levels were obtained by minor alterations to the transmissivity (T) and storage in the aquifer (Figure 5).

OPTIMIZED MANAGEMENT RECOMMENDATIONS REGARDING THE DEWATERING OF THE LETLHAKANE MINE

The key behind the response matrix method used is the description of the aquifer by a system of linear equations enabling the influence of each source or sink to be calculated separately and then superimposed to compute the complete distribution of stresses over space and time under any pumping schedule.

The response matrix is an assemblage of coefficients, each of which relates pumping at one location to drawdown at another (Gorelick, 1987). If a pumping system consists of n wells distributed in any manner over the aquifer, the hydrodynamic laws for a porous medium show that the drawdowns, s_i, occurring at every point i of the aquifer due to pumping Q_j, (j=1,---,n) are linear functions of these yields, as indicated by equation 7.

$$s_i = \sum_{j=1}^{n} a_{ij} Q_j \qquad (7)$$

The coefficient a_{ij} is the drawdown at point i, due to a unit abstraction rate at well j and is called the response matrix (Van Tonder, et al., 1993).

In the Letlhakane Mine case study the AQUA-MAN program (Van Tonder et al., 1993) was used to link the distributed parameter groundwater flow simulation model, AQUA, with mathematical optimization methods using the response matrix technique. The linear management problems formulated to achieve the goals of the investigation were solved by a simplex routine, SIMPLEX, (Van Tonder et al., 1993).

Based on planning information received from the Mine, optimization runs were performed to establish the optimum extraction rates from the production boreholes to achieve certain minimum drawdowns after specified periods of pumpage. The goal functions were therefore specified to maximize drawdown initially using existing borehole capacity subject to the following constraints:
1) a minimum drawdown in the Mine Pit area of 15 metres after each year of pumping
2) abstraction from Wellfield Six and other boreholes set equal to historical values
3) maximum abstraction from the Mine Pit boreholes set according to maximum yields.
The maximization of drawdown in the model for each year is to ensure that maximum existing pumping capacity is utilized for each year. If, for example, a drawdown of more than 15 metres was achieved for a specific year then the following year only a drawdown of that amount less 15 metres is necessary. Only if the minimum required drawdown of 15 metres is not achieved will a new borehole be drilled. This will ensure that a minimum number of boreholes will have to be utilized to achieve the required drawdown.

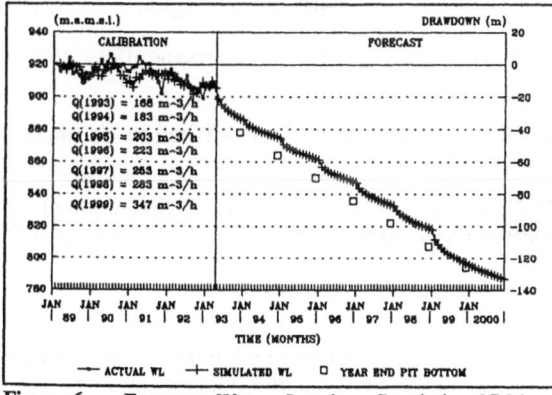

Figure 6 - Forecast Water Level - Borehole OB04 at Letlhakane Mine.

Two optimization scenarios were performed. The first without recharge from rainfall and the second with a specified recharge of 5mm per annum. Theoretical borehole positions, located outside of the maximum extraction pit were chosen to replace all present dewatering boreholes which will be destroyed by waste stripping operations during 1999.

The modelling results for these scenarios indicate that the following quantities abstracted from the following number of additional boreholes will have to be pumped to effect the required drawdown (Table 2). Figure 6 shows the forecast water level response if the recommended abstraction rates are pumped from the Mine Pit.

THE EFFECT OF LETLHAKANE MINE PIT DEWATERING ON WELLFIELD SIX

Wellfield Six forms part of the greater water supply system to Orapa town. As this wellfield is located close to Letlhakane Mine, the likely effect of increased abstraction from Letlhakane Mine on water levels in Wellfield Six needed to be considered.

The calibrated groundwater flow model predicted an incremental average drawdown of about 10 metres in Wellfield Six after an additional seven years of pumping at the recommended rates for the no-recharge scenario at Letlhakane Mine. This is considered to have only a small influence on the abstraction capabilities of the present wellfield whose available drawdown is in the order of 100 metres.

Table 2 - Summary of Borehole requirements for each Year with required Abstraction Quantities to achieve Drawdown Constraints.

YEAR	NO RECHARGE		RECHARGE=5mm/a	
	#Additional Boreholes	Required Abstraction (m³/h)	#Additional Boreholes	Required Abstraction (m³/h)
1993	0	168	0	183
1994	0	183	4	223
1995	2	203	4	263
1996	2	223	2	313
1997	3	253	1	323
1998	3	283	5	373
1999	24*	347	18**	425

* Yield per borehole required now 10,21 m³/h not 10 m³/h.
** Yield per borehole required now 12,50 m³/h not 10 m³/h.

BIBLIOGRAPHY

Bredenkamp, D.B., Van Rensburg, H.J., Van Tonder, G.J., Cogho, V.E., 1989. Quantitative estimation of aquifer storativity and recharge by means of a water balance and incorporating a finite element network. Proc. Symp., Groundwater quantity and quality, Benidorm, Spain.

Gorelick, S.M., 1987. AQ-MAN: Linear and Quadratic Programming Matrix Generator using Two-Dimensional Groundwater Flow Simulation for Aquifer Management Modelling. U.S.G.S. Report 87-4061, Washington D.C., 1987.

Kirchner, J. and Van Tonder, G.J., 1989. Exploitation potential of Karoo aquifers. Water Research Commission Contract.

Staats, S., 1993. Triangular Finite Element Mesh Generation. Unpublished M.Sc. thesis, U.O.F.S., Bloemfontein.

Van Tonder, G.J., Van Rensburg, H.J., Cogho, V.E., 1993. The Development of Groundwater Management Models. Progress report to the meeting of the Water Research Commission Steering Committee.

ACKNOWLEDGEMENTS

Permission granted by Debswana Diamond Company to use and publish information gained from Letlhakane Mine is greatfully acknowledged as is the financial support from Anglo American Corporation of South Africa and the Water Research Commission.

USE OF HIGH PERFORMANCE COMPUTING TO EXAMINE THE EFFECTIVENESS OF AQUIFER REMEDIATION

A. F. B. TOMPSON[1], S. F. ASHBY[2], R. D. FALGOUT[2], S. G. SMITH[2], T. W. FOGWELL[3], and G. A. LOOSMORE[4]

[1]Earth Sciences Division, L-206; [2]Computing and Mathematics Research Division, L-316, Lawrence Livermore National Laboratory, University of California, PO Box 808, Livermore, CA 94551, USA
[3]IT Corporation, Martinez, CA 94553, USA
[4]Department of Applied Science, University of California–Davis, Livermore, CA 94550, USA

Large-scale simulation of fluid flow and chemical migration is being used to study the effectiveness of pump-and-treat restoration of a contaminated, saturated aquifer. A three-element approach focusing on geostatistical representations of heterogeneous aquifers, high-performance computing strategies for simulating flow, migration, and reaction processes in large three-dimensional systems, and highly-resolved simulations of flow and chemical migration in porous formations will be discussed. Results from a preliminary application of this approach to examine pumping behavior at a real, heterogeneous field site will be presented. Future activities will emphasize parallel computations in larger, dynamic, and nonlinear (two-phase) flow problems as well as improved interpretive methods for defining detailed material property distributions.

INTRODUCTION

Predictive modeling of fluid flow, chemical migration, and transformation processes in the subsurface is widely applied within the oil industry and in the management of environmental contamination problems [8]. Computational models are routinely used to characterize gross rates of groundwater flow, contaminant migration and chemical transformation in the subsurface, design, analyze, or optimize the performance of remediation projects, estimate associated health risks, or demonstrate compliance with federal and state cleanup regulations.

In subsurface contamination problems, the porous soil environments are typically large (0.1-10km), complex, and nonuniform, and differ from the smaller (1m), idealized, and controlled laboratory systems where flow and transport phenomena in porous media are typically studied. This had led to increasing concern about the influence of aquifer heterogeneity on flow, transport, and remediation processes and how they are represented in field modeling applications [4,6]. Uncertainty and complexity in flow or transport behavior can be produced by small-scale (1m) variability in formation materials and their associated properties, constraints on the type and quantity of measurements available to characterize real systems, or oversimplified representations of the fundamental processes present. Failure to recognize or incorporate the effects of small-scale process interactions in field scale models could ultimately lead to unreliable conclusions about system behavior.

A. Peters et al. (eds.), Computational Methods in Water Resources X, 875–882.
© 1994 Kluwer Academic Publishers. Printed in the Netherlands.

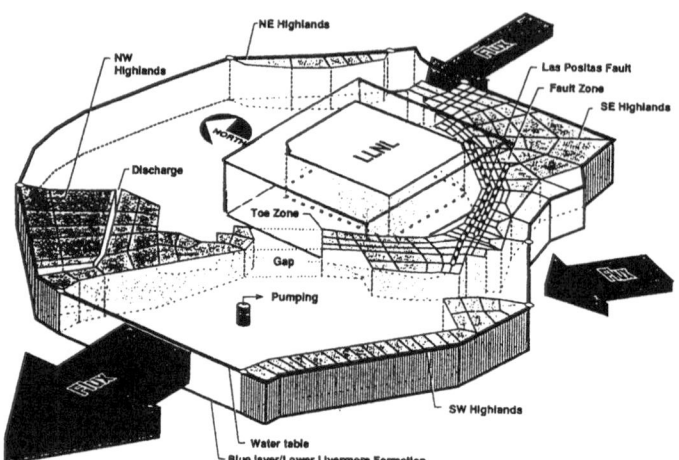

Figure 1: Perspective conceptualization of the upper aquifer beneath LLNL showing alluvial materials (white), less permeable uplifted zones (dark), a nearby fault zone, and the sample volume used for detailed simulation.

Hence, we are developing and applying highly-resolved computational models of flow, migration, and transformation phenomena to study the impacts of medium heterogeneity on aquifer-scale behavior and remediation efficiency. As a complement to theoretical and experimental analysis, detailed calculations have been used to validate the accuracy of scaling or homogenization theories, benchmark the performance of intermediate or field scale experiments, or test and simulate the efficacy of specific remediation schemes within nonuniform environments [1,7,9,10,14,15]. Modeling capabilities that address behavior in three dimensions, resolve important small-scale property variations, and incorporate specific remediation technologies or processes are typically required. Such problems may have upwards of 10^9 nodes and may require solution approaches on massively parallel computers [3,5].

EXAMPLE FIELD APPLICATION

As an example application, we consider the groundwater flow and chemical migration processes in the upper alluvial materials beneath the Lawrence Livermore National Laboratory (LLNL) in California. The groundwater in this area was contaminated with some 8 volatile organic compounds (VOC) 50 yeas ago when the lab site was occupied by a naval airfield. The existing saturated zone contamination extends from beneath LLNL to over a mile west, is slowly moving toward municipal wells in downtown Livermore, and is the focus of a large pump-and-treat remediation project [12]. In Figure 1, a schematic of the upper aquifer beneath LLNL shows the alluvial materials, several less permeable uplifted zones, a nearby fault zone, and the location of the downtown wells. The lower boundary of the aquifer overlays a relatively consistent, impermeable blue clay formation throughout the region. Flow in the alluvium goes mainly from the east and southeast toward the west.

Several 2-D modeling studies of flow and transport in the regional LLNL system have been carried out, under both natural and remedial conditions [17,18]. These have been based on simplified, or "effective" representations of flow and transport processes, a small number of physical and chemical property measurements, a reasonable amount of geologic interpretation, and some limited application of stochastic flow theory [16]. We seek here to develop a more focused 3-D modeling study in which the basic material heterogeneities are resolved and their subsequent impacts on flow and transport can be more carefully approximated. In this sense, uncertainties about the representation of effective behavior can be reduced by increasing the computational effort and property resolution.

SIMULATION APPROACH

Simulation Domain. For simplicity, our initial study volume is comprised of a 12,700 ft square prismatic block of alluvium "carved" out of the upper 320 feet of saturated aquifer surrounding the LLNL site (Figure 1). The orgin of a coordinate system with +x to the east, +y to the north, and +z up is at the lower southwest corner. The lower portion of the block intersects the bounding clay formation, while its southeast corner is crossed by the fault zone. The upper portion corresponds to the water table. Material properties such as the hydraulic conductivity, medium porosity, or mineral sorptive capacity will be specified separately for the alluvial materials, the lower clay layer, and the fault zone. As motivated by interpretive arguments below, we have chosen a rather coarse discretization involving $128 \times 128 \times 64$ (or over 10^6) nodes for this initial study, where $\Delta x = \Delta y = 100$ ft, and $\Delta z = 5$ ft.

Balance Laws. Steady, saturated flow in this system is described by [4]

$$\epsilon \mathbf{v} = -K\nabla h, \tag{1}$$

where $\mathbf{v}(\mathbf{x})$ is the average groundwater seepage velocity [L/T], $h(\mathbf{x})$ is the hydraulic head [L], K is the medium hydraulic conductivity, and ϵ is the medium porosity. In saturated and nondeforming media, h satisfies

$$\nabla \cdot (K\nabla h) = -Q, \tag{2}$$

where $Q > 0$ [1/T] represents a loss of fluid due to pumping. Physical medium heterogeneities may be reflected in spatially dependent values of K or ϵ.

The migration of one or more dissolved, sorbing, and neutrally-buoyant chemicals in the groundwater is described by [15]

$$\frac{\partial \rho_i}{\partial t} + \nabla \cdot \left(\frac{\mathbf{v}\rho_i}{\mathcal{R}_i}\right) - \nabla \cdot \left(\epsilon \mathbf{D}_i \cdot \nabla \left(\frac{\rho_i}{\epsilon \mathcal{R}_i}\right)\right) = -\left(\frac{\rho_i}{\epsilon \mathcal{R}_i}\right) Q \tag{3}$$

where $\rho_i(\mathbf{x}, t)$ is the total (liquid + sorbed) concentration [M/L^3] of chemical i at a point in the medium. The quantity $\mathbf{D}_i \approx \alpha_T |\mathbf{v}| \mathbf{I} + (\alpha_L - \alpha_T)\mathbf{v}\mathbf{v}/|\mathbf{v}|$ is the velocity-dependent hydrodynamic dispersion tensor [L^2/T], and α_L and α_T are the local longitudinal and transverse dispersivities [L]. In (3), we have assumed that sorption onto the the mineral phase occurs reversibly and in an equilibrium fashion such that the the aqueous concentration (c_i) may be inverted from $\rho_i = \epsilon c_i \mathcal{R}_i$. Here, the quantity $\mathcal{R}_i$ represents the local partitioning or retardation capacity of the soil, and may generally be concentration dependent. In the experiments below, we consider a nonreactive tracer ($i = 1$) and a linearly sorbing compound ($i = 2$) such that $\mathcal{R}_1 = 1$ and $\mathcal{R}_2 = 1 + \epsilon k_d/(1 - \epsilon)$, where k_d is a dimensionless sorption coefficient. In this case, medium heterogeneity may be reflected in spatially dependent values of α_L, α_T, or k_d.

Specification of Material Properties. Because we cannot specify detailed property distributions from measurement alone, simulations will made in one or more realizations of the system that preserve or recreate estimated statistical patterns of heterogeneity in the alluvium, while retaining the larger clay and fault zones as bulk features with constant properties. In the alluvium, heterogeneity in the hydraulic conductivity (K) will be developed from an approximate spectral (random field) model [4,13]. Assuming $\ln K(\mathbf{x}) = F + f(\mathbf{x})$, preliminary analyses of site data indicate that the geometric mean conductivity $K_G = e^F \approx 4.0$ ft/day, log-K variance $\sigma_f^2 \approx 3.2$, and exponential model correlation scales $\lambda_i \approx (200, 200, 10)$ ft, respectively [16], leading to the coarse discretization above. Although this is a common and simple model of heterogeneity, alternative representations are also being pursued. In the lower clay and fault materials, K will be assigned constant values of 0.004 and 0.4 ft/day, respectively. The soil sorptivity will be specified from a direct correlation with conductivity through $\ln \epsilon k_d \approx -0.86 - 0.32 \ln K$ [15]. The quantities ϵ, α_L, and α_T will be held constant and set to representative local values of 0.3, 1.0 ft, and 0.1 ft, respectively.

Solution of Equations. Solutions of (2) were determined in the study volume using a 7-point finite difference spatial discretization; velocity fields were then derived from (1). Inversion of the 10^6 finite difference equations was performed with a parallel preconditioned conjugate gradient (PCG) technique implemented within the PARFLOW code [2]. Current results are based on 2-step Jacobi preconditioning. In the future, we will employ polynomial and multigrid preconditioners. PARFLOW is being implemented on several distributed memory MIMD machines with message passing, including the nCUBE/2, Cray T3D, and Meiko CS-2, as well as on workstation clusters. The code achieves portability by using a communications layer called AMPS, which is ported on top of either PVM [3], the REACTIVE KERNEL/COSMIC ENVIRONMENT [11], or vendor message passing primitives. The main source code (written in C) is completely portable. The problem data is distributed onto a logical process mesh; physical mappings are handled by AMPS. Specifically, we distribute the unknown pressure heads and the underlying 7-point PDE stencil across the processes, along with auxiliary data. Realizations of the random hydraulic conductivity distribution are generated dynamically within PARFLOW using a parallel, C version of the turning bands TURN3D code [13]. This avoids input/output bottlenecks when full sets of parametric data are distributed from disk to the processors. Smaller sets of data representing fault or clay zone properties are read in from disk following the spectral generation. Once initialized, steady head solutions for our study volume were generated using 864 dedicated nodes of an nCUBE/2. The solution of a single problem with 10^6 mesh points required approximately 3500 PCG iterations and 7.5 minutes of wall-clock CPU. Since PARFLOW demonstrates good scalability (that is, solution times remain constant as larger problems are mapped onto larger processor arrays), the resolution of our study volume is limited only by the size of the parallel machine.

Solutions of (3) were determined with a serial version of the particle-grid code SLIM, modified to account for mass capture by pumping wells [15]. In this model, the distribution of total mass is distributed among a number of particles that are allowed to move through the flow field in small time increments in response to advective, diffusive, and retardation forces. When extraction wells are present, particles are removed from pre-defined capture zones as a function of the time step and retardation strength. The particle-grid algorithm is well suited for efficient solution of transport problems involving small point sources of mass in large, highly resolved flow domains.

SOME RESULTS

Realization #1 of the conductivity distribution for the study volume is shown in Figure 2a. The orientation is similar to the cut-out location indicated in Figure 1. In Figure 2, a block of the volume has been removed to show the internal structure, including the fault and lower clay zones. Two separate flow solutions were obtained for this problem. Solution "1a" corresponds to ambient conditions, with no-flux (or confined) conditions enforced at the top and bottom, and specified heads along the four sides. The head values were interpolated from regional measurements of the piezometric surface and were assumed constant over depth. Solution "1b" includes a pumping well at node (65,65,33), close to the center of the volume. This was specified by maintaining a fixed head value of 530 ft at the node. The resulting extraction rate was determined from a mass balance to be 24,000 ft^3/day, or 0.28 ft^3/sec. The head and velocity magnitude distributions for solution 1b are also shown in Figure 2, the effect of the well being marginally detectable near the corner of the cut-out block. Four other pairs of flow realizations (not shown) were developed in the same way. A sixth pair of flow fields (not shown) were developed by setting the alluvial conductivity uniformly equal to the geometric mean conductivity, $K_G = 4.0$ ft/day. Figure

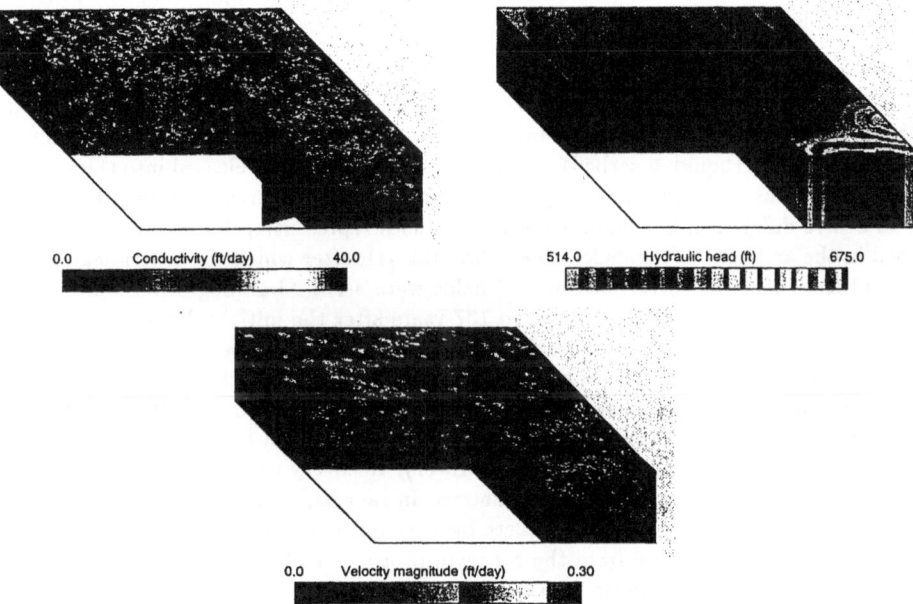

Figure 2: Distribution of hydraulic conductivity, head, and velocity magnitude for problem 1b

3a shows averages of the velocity components (and velocity magnitude) in the alluvium for

solution 1a. These were determined in a $4\Delta x$-wide y-z slab of the block, as a function of its location on the x-axis. Similar results averaged over all 5 "a" runs do not show much difference. Groundwater generally moves toward the northwest, albeit faster in the western locations where there is a converging flow effect. In Figure 3b, the variance of the velocity components (and their magnitude) as determined in the same moving slab are shown for the single run 1a and ensemble "a" runs. Here, the trend is for larger variances in the west where the mean velocities are larger as well. The high values at either end are an artifact of the fixed boundary conditions. Several transport problems were developed that focused

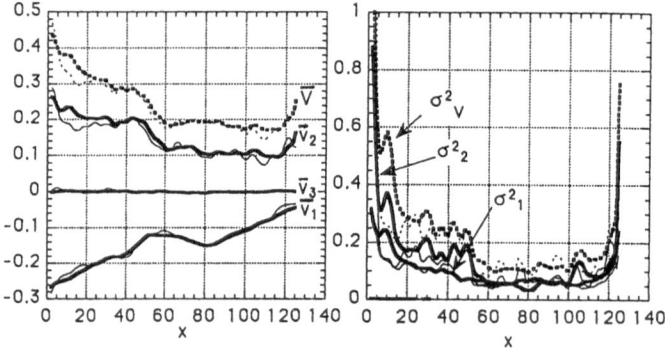

Figure 3: "y-z" averages and variances of the alluvium velocity components, as a function of the x-coordinate; thin lines = problem 1a, thick lines = ensemble over 5 replicate "a" runs.

on the migration of small pulses of the tracer and reactive compounds placed near node (95,30,45), in the southeast part of the block, to the north of the fault zone. Roughly 6.8 kg (liquid) and 10.2 kg (liquid + sorbed) of components 1 and 2 were released into the system, both corresponding to 1 ppm initial aqueous concentrations. Different flow field pairs were used to evolve different migration scenarios. Particles representing both compounds were released in the ambient "a" fields for 20000 days (55 yr), after which the central extraction well was turned and the corresponding "b" fields were used. Mass migration and capture were monitored for the next 30000 days, to 137 years after the initial release.

In Figure 4a, the average x-y mass distribution derived with flow solution 1a is shown for both components in a bird's eye perspective at t = 20000 days (closed circles represent the nonreacting tracer). Portions of both components have moved past the well by this time. Figure 1b shows the mass recovered after the well has been turned on, as a function of time (solid lines represent the nonreacting tracer). Recovery curves obtained with the 4 other pairs of flow flow realizations are also shown. In general, recovery of the reactive mass (1020 kg) takes longer, but is more complete because of the timing of turning the well on. Differences in the results arise from the separate property realizations, most importantly from their local effect on the initial mass distributions.

In Figure 4b, the same set of figures is shown for the case where flow solution 6 (based upon a uniform alluvial conductivity of K_G) is used. Here we see no macrodispersion arising from heterogeneity. We also see a slower effective plume velocity, caused because the "effective" conductivity of the heterogeneous system in problem 1 should be greater than K_G [1,6,16]. No recovery of component 2 is found.

In Figure 4c, the problem in flow field 6 is run again, except that the values of α_L and α_T in the transport model are increased to 265.0 and 26.5 ft., respectively, to approximate their macroscopic counterparts estimated from stochastic theory [16]. Here we see more significant spreading, some of which moves upstream of the source. Although the ratio α_L/α_T is too small, the longitudinal value may actually reproduce the effects in Fig. 4a if a more reasonable effective velocity is used. Mass recovery is poor because the excessive transverse spreading moves mass outside the capture zone of the well.

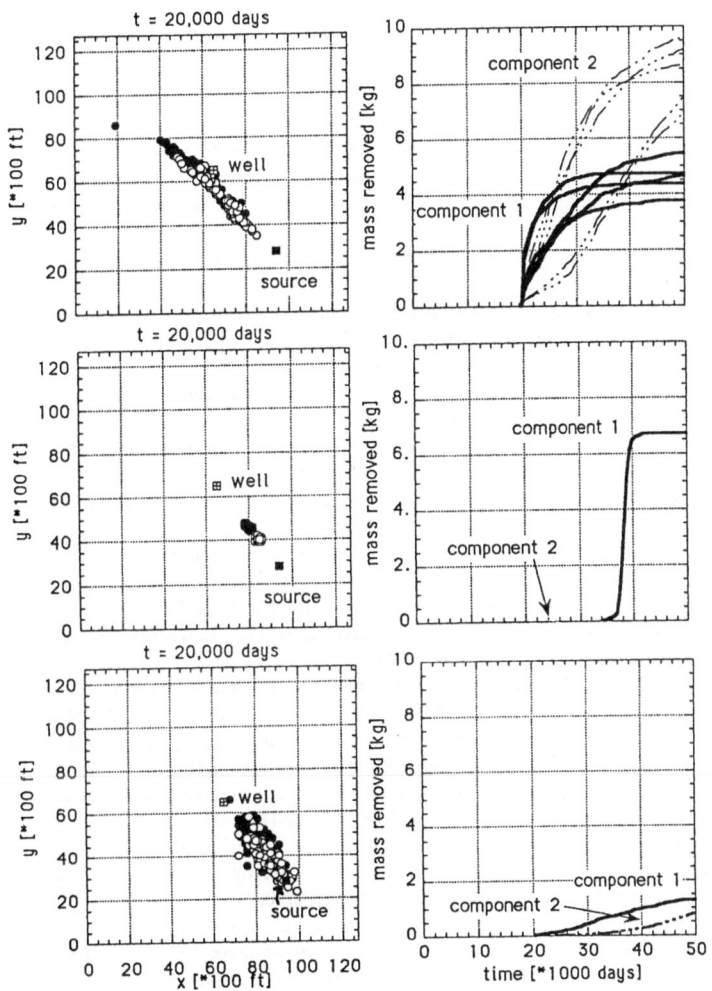

Figure 4: Particle and recovery plots (see text)

Acknowledgements. This work was conducted under the auspices of the U. S. Department of Energy by Lawrence Livermore National Laboratory under contract W-7405-Eng-48. It was supported, in part, by the Laboratory Directed Research and Development program and the Strategic Environmental Research and Development Program of the U. S. Department of Defense, through the cooperation of the USAE Waterways Experiment Station, Vicksburg, MS.

REFERENCES

1 Ababou, R., D. B. McLaughlin, L. W. Gelhar, Three-Dimensional Flow in Random Porous Media, Massachusetts Institute of Technology (R. M. Parsons Laboratory Report #318), 1988.

2 Ashby, S. F., R. D. Falgout, S. G. Smith, and A. F. B. Tompson, *Modeling Groundwater Flow on MPPs*, Proc. 1993 Scalable Parallel Libraries Conference, October 6-8, 1993, Mississippi State University; also Lawrence Livermore National Laboratory, Livermore CA (UCRL-JC 115602), 1993

3 Beguelin, A., J. Dongarra, G. A. Geist, R. Manchek, and V. Sunderam, *A users' guide to PVM: Parallel Virtual Machine*, Oak Ridge National Laboratory (ORNL/TM-11826), July, 1991.

4 Dagan, G., *Flow and Transport in Porous Formations*, Springer Verlag, 1989

5 Dougherty, D. E., *Hydrologic Applications of the Connection Machine CM-2*, Wat. Resour. Res., 27, pp 3137–3147, 1991

6 Gelhar, L. W., *Stochastic subsurface hydrology: from theory to applications*, Water Resources Res., 22(9), p 135s, 1986

7 Kinzelbach, W., *The random walk method in pollutant transport simulation*, in Groundwater Flow and Quality Modelling, E. Custodio, A. Gurgui, and J. P. Lobo Ferreira, eds., D. Riedel, Dordrecht, The Netherlands, pp 227–246, 1987

8 National Research Council, Ground water models: Scientific and Regulatory Applications, National Academy Press, Washington, 1990

9 Polmann, D., McLaughlin, D., S. Luis, L. Gelhar, and R. Ababou, *Stochastic modeling of large scale flow in heterogeneous, unsaturated soils*, Water Resources Res., 27(7), 1447–1458, 1991

10 Schäfer, W. and W. Kinzelbach, *Stochastic modeling of in-situ bioremediation in heterogeneous aquifers*, J. Contam. Hydrol. 10, pp 47–73, 1992

11 Seitz, C. L., J. Seizovic, and W.-K. Su, *The C Programmer's Abbreviated Guide to Multicomputer Programming*, California Institute of Technology, (Caltech-CS-TR-88-1), January, 1988.

12 Thorpe, R. K., W. F. Isherwood, M. D. Dresen, and C. P. Webster-Scholten, Eds., *CERCLA Remedial Investigations Report for the LLNL Livermore Site*, Lawrence Livermore National Laboratory, Livermore, CA (UCAR-10299), 1990.

13 Tompson, A. F. B., R. Ababou, and L. W. Gelhar, *Implementation of the three-dimensional turning bands random field generator*, Wat. Resour. Res., 25(10), pp 2227-2243, 1989

14 Tompson, A. F. B. and L. W. Gelhar, *Numerical simulation of solute transport in randomly heterogeneous porous media*, Wat. Resour. Res., 26, pp 2541–2562, 1990

15 Tompson, A. F. B., *Numerical simulation of chemical migration in physically and chemically heterogeneous porous media*, Wat. Resour. Res., 29, pp3709–3726, 1993.

16 Tompson, A. F. B., *Flow and Transport within the Saturated Zone Beneath Lawrence Livermore National Laboratory: Modeling Considerations for Heterogeneous Media*, Lawrence Livermore National Laboratory, Livermore CA (UCID 21828), 1990.

17 Tompson, A. F. B., E. M. Nichols, P. F. McKereghan, and M. C. Small, *Summary of Preliminary Ground Water Simulations in the Livermore Regional Modeling Study – CFEST Finite Element Code*, Lawrence Livermore National Laboratory, (UCRL-AR-107049), 1991.

18 Tompson, A. F. B., E. M. Nichols, and P. F. McKereghan, Eds., *Preliminary Simulation of Contaminant Migration at the Lawrence Livermore National Laboratory*, Lawrence Livermore National Laboratory, (UCRL-AR-report in preparation), 1994.

WEIGHTED FEEDBACK CONTROL OF GROUNDWATER REMEDIATION UNDER MULTIPLE UNCERTAINTIES

G. J. Whiffen, C. A. Shoemaker, and J. A. Denu
School of Civil and Environmental Engineering
Cornell University
Ithaca, New York 14853
U.S.A.

The "Nonlinear Weighted Feedback Method" for addressing uncertainty was presented by *Whiffen and Shoemaker* (1993). This paper will briefly summarize extensions of this method to more complex problems in groundwater remediation. The examples consider a) a much larger range of uncertainty in parameter values, b) multiple types of uncertainty, and c) a comparison of feedback to parameter updating and re-optimization. Full detail about the extensions, including figures for numerical results can be found in *Whiffen, Denu, and Shoemaker* (1994 in preparation).

A feedback law is a rule for modifying system control as a function of the difference between predicted and observed system state. In our analysis, the feedback law generated by a constrained Differential Dynamic Programming algorithm with penalty functions is used as a basis of feedback laws. We test the weighted feedback procedure for cases where there is uncertainty in both the hydraulic conductivity and initial contaminant concentrations. Optimal policies are calculated using a given or "measured" set of hydraulic conductivities and contamination distributions. The optimal policies (with and without feedback) are applied using the same finite element model with a second or "true" set of conductivities and contamination distributions. Systematic bias range of $\pm 50\%$ and a standard deviation of 50% in hydraulic conductivities are considered as well as underestimation of contamination by 25%. In our examples, feedback policies cost as much as 80% less than the cost of applying the calculated optimal policies without using a feedback law. The advantages of using feedback increase as model uncertainty increases.

INTRODUCTION

Groundwater contamination is a widespread and serious threat to drinking water sources. Groundwater contamination is a world wide problem that has been

A. Peters et al. (eds.), Computational Methods in Water Resources X, 883–890.
© 1994 *Kluwer Academic Publishers. Printed in the Netherlands.*

identified in most of the industrialized nations. The cost for cleaning up contaminated groundwater is conservatively estimated at $100 billion in the U.S. alone. The most common method for detoxifying contaminated groundwater is to locate a system of wells in the contaminated area, pump the contaminated water out of the ground, and filter or treat it chemically to detoxify it. This is referred to as "pump and treat".

Both simulation and optimization models have been used to design pump and treat groundwater remediation systems. The design parameters include where to locate pumping wells and how much to pump from each well in each time period. Optimization methods used in combination with groundwater simulation models can effectively search a much larger number of possible design alternatives than is possible with use of simulation models alone. Examples of deterministic optimization approaches to groundwater remediation include: *Gorelick et al.* (1984), *Chang et al.* (1992), *Culver and Shoemaker* (1992; 1993) and *Marryott et al.* (1993).

Our focus in this paper will be on the development of groundwater remediation system designs that remain robust and cost efficient under the considerable parameter uncertainty that exists in characterizing groundwater aquifers. Other approaches to management under uncertainty include *Wagner and Gorelick* (1987), *Ranjithan et al.* (1990), *Lee and Kitanidis* (1991) and *Tiedeman and Gorelick* (1993).

The weighted feedback method described in *Whiffen and Shoemaker* (1993) and tested in this paper incorporate uncertainty in a different way than other approaches. In particular, a feedback law is used, which modifies the optimal solution to a corresponding deterministic problem by a nonlinear function of the difference between the observed and the predicted values of hydraulic head and contamination distributions. The coefficients in the feedback function are a nonlinear function of the model parameters and weights of the penalty function. The essential task of the weighted feedback procedure is selecting weights to insure that the feedback policy is robust over a wide range of multivariate probability distributions representing parameter uncertainty and error. The advantages of this method over previous approaches is that it is does not require the assumption of zero bias in all measurements (as in *Lee and Kitanidis,* 1991), and the approach incorporates new observations made over time (unlike *Wagner and Gorelick,* 1987). In addition the method does not require that a probability distribution for the parameter error be exactly known.

OPTIMAL CONTROL PROBLEM FORMULATION

Optimal policies were calculated to solve the following constrained optimal control problem,

$$J = \min_{u_t} \sum_{t=1}^{N} g^t(x_t, u_t) \qquad Cost \qquad (1)$$

where x_t is a state vector of dimension n containing the head and concentration at each active node of the finite element mesh at time step t, u_t is a control vector of dimension m containing the pumping rates at each potential remediation well at time t, and $g^t(x_t, u_t)$ is the cost incurred at time t due to state x_t and control u_t. The total number of decision time periods is N. The cost function can be any nonlinear twice differentable function of x_t and u_t. The minimization is constrained by the finite element model's prediction of the state of the aquifer at a future time step given the control and state at the current time step,

$$x_{t+1} = T^t(x_t, u_t) \qquad t = (1, 2, ..., N-1) \qquad Transition\ Function \qquad (2)$$

The initial state of the aquifer before any remediation is

$$x_{t=1} = x_o. \qquad Initial\ Condition \qquad (3)$$

Further constraints were added to ensure only extraction is allowed at each withdrawal site (this is optional), to ensure total pumping at each time step does not exceed waste removal treatment capacity U_{max}, and to ensure a water quality standard $C_{standard}$ is met at all observation wells by the final time step, $t = N$:

$$u_t \leq 0 \qquad t = (1, 2, ..., N) \qquad Extraction\ Only \qquad (4)$$

$$- \sum_{Wells\ i} u_t(i) \leq U_{max} \qquad t = (1, 2, ..., N) \qquad Treatment\ Capacity \qquad (5)$$

$$c(x_{t=N}) \leq C_{standard} \qquad t = N \qquad Drinking\ Standard \qquad (6)$$

The function $c(x_{t=N})$ represents the concentrations at each observation well at the final time step, N. The constrained optimal control problem, (1) through (6), can be approximated by an unconstrained optimal control problem by introducing a penalty function into the objective function,

$$J' = \min_{u_t} \sum_{t=1}^{N} [g^t(x_t, u_t) + \sum_{j \in \{r_t\}} Y_j^t(x_t, u_t, w_j)] \qquad (7)$$

$$x_{t+1} = T^t(x_t, u_t) \qquad t = (1, 2, ..., N-1), \qquad (8)$$

$$x_{t=1} = x_o \qquad (9)$$

where Y_j^t is the penalty assessed for a violation of constraint j in time step t, $\{r_t\}$ is the set of constraints from (4) - (6) that are in effect at time step t, and w_j is the weighting or strength assigned to the jth constraint. The penalty functions, Y_j, in (7) do not approximate physical or economic penalties for constraint violation. In this equation, they are introduced as a numerical construct to solve the original constrained optimal control problem, (1) through (6). The solution of the constrained problem, (1) through (6) can be obtained to arbitrary precision by solving the penalty function formulation, equations (7) through (9) and letting w_j become arbitrarily large as the algorithm converges.

The form of the penalty function used in the numerical examples is the same as used by *Whiffen and Shoemaker* (1993).

$$Y_j(f_j(x_t, u_t), w_j) = \begin{cases} \xi_j & \text{if } \xi_j \leq \xi_\circ \\ a(\xi_j)^n + b(\xi_j)^m + c & \text{if } \xi_j \geq \xi_\circ \end{cases} \tag{10}$$

where

$$\xi_j = \sqrt{w_j^2 f_j^2 + \epsilon_j^2} + w_j f_j. \tag{11}$$

The parameter ϵ_j is a shape parameter of the hyperbolic function ξ_j. The function $f_j(x_t, u_t)$ is the jth constraint of the constraint vector where the constraints (4) to (6) are written in the form $f(x_t, u_t) \leq 0$. Hence $f_j(x_t, u_t)$ is the right hand side of each constraint equation, (4) - (6), minus the left hand side where j ranges over all constraints to be enforced at time t. The constants a, b, and c are chosen so that $Y_j, \frac{\partial Y_j}{\partial \xi_j}$, and $\frac{\partial^2 Y_j}{\partial \xi_j^2}$ are continuous at $\xi_j = \xi_\circ$. In particular, the values $\xi_\circ = 1$, $n = 2$, and $m = \frac{1}{2}$ were used which lead to $a = \frac{1}{3}$, $b = \frac{4}{3}$, and $c = -\frac{2}{3}$.

The cost function $g^t(x_t, u_t)$ used in the numerical examples is also the same as that used by *Chang et al.* (1992) and *Culver and Shoemaker* (1992; 1993).

$$g^t(x_t, u_t) = \sum_{i \in \Omega} a_1 u_{t,i} + \sum_{i \in \Omega} a_2 u_{t,i}(h_* - h_{t+1,i}) \tag{12}$$

for constants a_1 and a_2. In equation (12), the first term on the right hand side represents water treatment cost, and the second term reflects the pumping cost. The operation cost at each well is assumed to be the product of the extraction rates u_t and the lift $(h_* - h_{t+1,i})$. The summations in (12) are over the set of all possible pumping sites Ω.

Many other forms of the objective function and penalty function can be used. All that is required of the cost function, g^t, and the penalty functions, Y_j^t, is that they are each twice continuously differentiable in both x_t and u_t.

Differential Dynamic Feedback Laws

Once optimal policies are established for estimated parameter values feedback laws can be generated. DDP generates a feedback law during each iteration and uses this feedback law to establish an improved policy. The form of the feedback law generated by DDP is

$$u_t^{act} - \overline{u_t} = \delta u_t = \beta_t \delta x_t + \alpha_t, \qquad where \quad \delta x_t = x_t - \overline{x_t} \qquad (13)$$

The term u_t^{act} is the policy that is implemented under feedback and $\overline{u_t}$ is the optimal policy. The term $\delta x_t = x_t - \overline{x_t}$ is the vector difference between the current observed state x_t after applying (13) from time step 1 to $t - 1$, and the predicted state $\overline{x_t}$. The term β_t is an $m \times n$ matrix and α_t is an m-vector which are both functions of the first and second derivatives of $T^{t'}$, $g^{t'}$, and $Y^{t'}$ with respect to $x_{t'}$ and $u_{t'}$ where $t' \in \{t, t+1, ..., N\}$. For a description of the DDP algorithm including the calculation of β_t and α_t appearing in (13) see *Whiffen and Shoemaker* (1993). Motivation and derivation of β_t and α_t can be found in *Yakowitz and Rutherford* (1984).

Control constraints and the application of feedback

When model errors are large, we expect that the difference δx_t after application of feedback will be large. Since the feedback law (13) is based on a quadratic approximation around $\overline{x_t}$, the constraints (4), (5), and (6) may be violated in the implementation.

Violation of the control or pumping constraints, (4) and (5), within a time step can be avoided entirely at each step by truncating and/or scaling the feedback policy, u_t^{act}, after it is calculated from (13) and before it is applied to the groundwater system or simulation model. Any of a number of schemes will be possible for each set of control constraints.

For our example, the following truncation and scaling was used to ensure that the feedback pumping policy for time step t, u_t, satisfies both the extraction-only constraint, $u_t \leq 0$, and the treatment capacity constraint, $-\sum_{Wells\ i} u_t(i) \leq U_{max}$, equation (5). First calculate

$$\overline{\gamma} = \frac{-U_{max}}{\sum_{Wells\ i} \min(\beta_t \delta x_t + \alpha_t + \overline{u_t}, 0)_i}. \qquad (14)$$

Next, $\gamma = \min(\overline{\gamma}, 1)$ is the factor each extracting well rate needs to be scaled by to ensure the treatment capacity, U_{max}, is not exceeded. Thus to ensure that both the treatment and extraction-only constraints are satisfied, choose

$$u_t = \gamma \min(\beta_t \delta x_t + \alpha_t + \overline{u_t}, 0). \qquad (15)$$

Modeling uncertainty

Whiffen and Shoemaker (1993) developed feedback policies that are robust under a range of uncertainty. In particular, weights were selected based on Monte Carlo simulations of the feedback laws applied to 270 aquifer realizations, drawn from 9 different probability distributions. Errors in the mean value of the hydraulic conductivity (called a bias) were 0%, +10% and -10%. The bias errors were coupled with standard deviations (scatter) of 5%, 10%, and 25%. The combination of three means and three standard deviations lead to 9 ($= 3 \times 3$) probability distributions. Thirty realizations were generated for each probability distribution for a total of 270 realizations.

Selecting weights based on uncertainty

The constants α and β in (13) depend (in a nonlinear way) on the weights w_j, which appear in (10). Hence, for each value of w_j, there is a corresponding feedback law β_t, and α_t. *Whiffen and Shoemaker* (1993) selected the value of w_j associated with their law (# 9) based on Monte Carlo simulations using the 270 realizations of conductivity.

NUMERICAL RESULTS

In our current analysis we test feedback law (#9) in cases where there is much greater error: error in initial concentration (not considered in *Whiffen and Shoemaker,* 1993) and a much wider range of error in hydraulic conductivity. Conductivity uncertainty was modeled by generating inhomogeneous isotropic conductivity fields from a log-normal distribution using the program given by *El-Kadi* (1986). An exponential correlation distance with constant 100 meters was chosen.

Uncertainty in initial pollution conditions was modeled by generating "excess" pollution, which is randomly added to the initial, assumed contamination distribution. Excess pollution was modeled in three ways: a) normally distributed, b) concentrated clumps, and 3) small, distributed clumps.

The numerical results indicate that the feedback policy performs much better than the deterministic policy without feedback. The figures describing these results are in *Whiffen, Denu, and Shoemaker* (1994). These figures show that the performance index (which includes cost functions and constraint violations) increases rapidly (performs worse) for the no feedback policy with increases in all types of error considered (bias and standard deviation in conductivity; and the three types of error in initial contamination concentrations). By contrast the feedback policy performs well (in some cases near-optimal) over the entire range of error, with relatively small increases in the performance index as error increases. It should be recalled that the feedback law used here was selected based on a relatively small range of error considered in *Whiffen and Shoemaker* (1993) however, it performed very well on a wide range of error and uncertainty. This result demonstrates the method is not sensitive to the probability distributions chosen to represent parameter error and uncertainty.

CONCLUSIONS

There is a great deal of uncertainty and error associated with characterizing aquifers. Thus any design procedure for remediation which uses modeling must take this error and uncertainty into account. Further the method should not be sensitive to assumed error distributions. Our numerical results indicate that Nonlinear Weighted Feedback is an effective way to generate pumping strategies that will be robust in the face of substantial error and uncertainty in both hydraulic conductivity and initial concentration distribution. In all the cases we tested we found that feedback did remarkably well in minimizing the degradation of remediation performance due to error and uncertainty.

REFERENCES

Chang, L.-C., C. A. Shoemaker, and P. L.-F. Liu, Application of a constrained optimal control algorithm to groundwater remediation, *Water Resour. Res., 28*(12), 3157-3173, 1992.

Culver, T., and C. A. Shoemaker, Dynamic Optimal Control for Groundwater Remediation with Flexible Management Periods, *Water Resour. Res., 28*(3), 629-641, 1992.

Culver, T., and C. A. Shoemaker, Optimal Control for Groundwater Remediation by Differential Dynamic Programming with Quasi-Newton Approximations, *Water Resour. Res., 29*(4), 823-831, 1993.

El-Kadi, A. I., A computer program for generating two-dimensional fields of autocorrelated parameters, *Groundwater, 24*(5), 663-667, 1986.

Gorelick, S. M., C. I. Voss, P. E. Grill, W. Murray, M. A. Saunders, and M. H. Wright, Aquifer Reclamation Design: The Use of Contaminant Transport Simulation Combined with Nonlinear Programming, *Water Resour. Res., 20*(4), 415-427, 1984.

Lee, S-I and P. K. Kitanidis, Optimal Estimation and Scheduling in Aquifer Remediation with Incomplete Information, *Water Resour. Res.*, *27*(9), 2203-2217, 1991.

Marryott, R. A., D. E. Dougherty, and R. L. Stollar, Optimal Groundwater Management, 2, Application of Simulated Annealing to a Field-Scale Contamination Site, *Water Resour. Res.*, *29*(4), 847-860, 1993.

Ranjithan, S., J. W. Eheart, and J. H. Garrett Jr., Application of Neural Network in Groundwater Remediation Under Conditions of Uncertainty, *Proceedings of the International Workshop on New Uncertainty Concepts in Hydrology and Water Resources*, Madralin, Poland, September 24-26, 1990

Tiedeman, C. and S. M. Gorelick, Analysis of Uncertainty in Optimal Groundwater Contaminant Capture Design, *Water Resour. Res.*, *29*(7), 2139-2153, 1993.

Wagner, B. J., and S. M. Gorelick, Optimal Groundwater Quality Management Under Parameter Uncertainty, *Water Resour. Res.*, *23*(7), 1162-1174, 1987.

Whiffen, G. J., and C. A. Shoemaker, Nonlinear Weighted Feedback Control of Groundwater Remediation Under Uncertainty, *Water Resour. Res.*, *29*(9), 3277-3289, 1993.

Whiffen, G. J., J. A. Denu, and C. A. Shoemaker, A Test of Nonlinear Weighted Feedback Control of Groundwater Remediation Under Uncertainty, 1994 - In prparation for *Water Resour. Res.*

Yakowitz, S. and B. Rutherford, Computational Aspects of Discrete-time Optimal Control, *Appl. Math. Comput.*, *15*, 29-45, 1984.

OPTIMAL PUMPING DESIGN FOR THE REMEDIATION OF A GROUNDWATER CONTAMINATION SITE

Y. XIANG, J.F. SYKES, N.R. THOMSON
Department of Civil Engineering
University of Waterloo
Waterloo, Ontario N2L 3G1
Canada

This paper presents an optimization analysis of the remedial pumping design for a contaminated municipal aquifer located in Elmira, Canada. The remediation task is to remove two groundwater contaminant species, NDMA and chlorobenzene, to such an extent that the specified groundwater quality standards are satisfied. The contaminants, NDMA and chlorobenzene, have different initial plume concentrations, retardation characteristics, and required quality standards.

INTRODUCTION

Pump-and-treat has been employed as one of the major technologies to remediate contaminated groundwater aquifers. For cases when the remediation target is a single contaminant species, few of the proposed optimization models have been applied to tackle field scale problems: Ahlfeld et al. (1988), Marryott et al. (1993), and Tiedeman and Gorelick (1993). Generally, when the pump-and-treat method is used for groundwater plume removal, the remediation requirement may not be limited to one single contaminant species. When multiple contaminant species are to be removed to restore the groundwater to specified quality levels, a desirable pumping scheme should be optimal for the simultaneous removal of all the targeted contaminants.

This paper presents an application of an optimization model to the remedial pumping design for a groundwater contamination site located in Elmira, Canada. The remediation task is to remove two groundwater contaminant species, NDMA (N-nitrosodimethylamine) and chlorobenzene, to a sufficient extent in order to restore the groundwater quality standards for the municipal aquifer. The contaminants, NDMA and chlorobenzene, have different initial plume concentrations and retardation characteristics. The objective function to be minimized is the sum of the extraction pumping rates, and the constraints are composed of groundwater quality requirements on the maximum and the spatially averaged residual concentrations for both species.

A. Peters et al. (eds.), Computational Methods in Water Resources X, 891–898.

THE CONTAMINATION SITE AND ITS SIMULATION

In a regional groundwater flow analysis for the Elmira area (Sykes, 1991), six distinct stratigraphic geologic units were delineated. Descending from the ground surface, these are: surficial till, upper aquifer, upper till, municipal aquifer, lower till, bedrock. In the present study, a two-dimensional depth-integrated model consisting of the municipal aquifer is used, with the vertical leakages from the confining upper and lower tills being included. This conceptualization is based upon the study conducted by Sykes (1991). A finite element model is utilized to solve the flow and transport state variables governed by the following equations:

$$\frac{\partial}{\partial x_i}(bK\frac{\partial h}{\partial x_j}) = -\sum_l q_l \delta_l - w \tag{1}$$

$$\frac{\partial}{\partial x_i}(\phi b D_{ij}\frac{\partial c_\beta}{\partial x_j}) - \phi b v_i \frac{\partial c_\beta}{\partial x_i} - \phi b R_\beta \frac{\partial c_\beta}{\partial t} = -\sum_l q_l \delta_l(c_{\beta l}^q - c_\beta) - w(c_{\beta s} - c_\beta) \quad \forall \beta$$
$$\tag{2}$$

where, Einstein's convention for tensor notations applies to indices i and j ($i, j = 1, 2$ designate x, y respectively); $\beta = 1, 2$ refers to NDMA and chlorobenzene respectively; h = hydraulic head, $[L]$; c_β = contaminant concentration of the species β, $[ML^{-3}]$; D_{ij} = tensor of dispersion coefficients, $[L^2 T^{-1}]$; ϕ = porosity of the aquifer, dimensionless; R_β = retardation factor, dimensionless; q_l = the pumping rate of well l, $[L^3 T^{-1}]$; $c_{\beta l}^q$ = concentration of species β in injected or extracted water by the well l; w = the leakage flux into (positive) or out of (negative) the aquifer, $[LT^{-1}]$; $c_{\beta s}$ = contaminant source distribution, $[ML^{-3}]$.

For the selection of the modeling area, referring to Figure 1 (the numbers are in meters, the plots shown henceforth are concerned with the transport domain only, and are magnified in the spatial scale), the municipal wells and the remedial wells will impact the regional groundwater flow, and therefore a local zone to which the impact of pumping is restricted cannot be defined. Thus, the groundwater flow simulation is considered at the regional scale, whereas a subdomain is used for the transport simulation. For boundary conditions, according to Sykes (1991), the eastern and western boundaries have no-flow, and the constant heads of $372m$ and $335m$ are assigned along the northern and southern boundaries respectively. The parameters for the flow simulation were obtained by calibrating a three-dimensional groundwater flow simulation model (SWIFT III) with steady-state field data (Sykes, 1991). The boundary conditions for the transport simulation are: zero-concentration along $x = 1870m, 6280m$ and $y = 7330m$, and zero-dispersive-flux along $y = 2150m$. A constant time step of 0.5 years is used.

It is assumed that both NDMA and chlorobenzene are nonreactive, and that NDMA is conservative, whereas chlorobenzene has a retardation factor of 2.8. The initial NDMA and chlorobenzene plumes targeted for remediation are shown in Figures 2 and 3. The parameters for the groundwater-contaminant system are summarized in Table 1.

Table 1: Parameters of The Groundwater-Contaminant System

aquifer porosity	0.3
longitudinal dispersivity (m)	50
transverse dispersivity (m)	10
hydraulic conductivity (m/sec)	$0.20 \times 10^{-4} \sim 0.25 \times 10^{-2}$
aquifer thickness (m)	$0.50 \sim 30.72$
leakage parameter of the upper till (sec^{-1})	$0.691 \times 10^{-10} \sim 0.339 \times 10^{-8}$
leakage parameter of the lower till (sec^{-1})	$0.111 \times 10^{-9} \sim 0.862 \times 10^{-7}$
head in upper aquifer (m)	$337.50 \sim 376.89$
head in lower aquifer (m)	$336.41 \sim 371.92$
retardation factor for NDMA	1.0
retardation factor for chlorobenzene	2.8

ANALYSIS OF OPTIMAL REMEDIATION DESIGN

Considering extraction pumping only, a decision vector of pumping rates is defined as $\mathbf{q}^- = [q_1^- \cdots q_n^-]^T$, where $q_i^- = -q_i \geq 0$ stands for the pumping rate of the potential extraction well i, and n denotes the total number of potential remediation wells. In order to determine the pumping scheme which can be optimal for the simultaneous removal of both contaminant species, the following optimization formulation is used:

$$min \ W(\mathbf{q}^-) = \sum_{i=1}^{n} q_i^- \tag{3}$$

subject to

$$0. \leq q_i^- \leq \bar{q}_i^- \quad i = 1 : n \qquad 0. \leq \sum_{i=1}^{n} q_i^- \leq q_{max}^- \tag{4a,b}$$

$$c_{\beta a}(\mathbf{q}^-; t_s) \leq c_{\beta a}^* \qquad c_{\beta max}(\mathbf{q}^-; t_s) \leq c_{\beta max}^* \qquad \forall \beta \tag{5a,b}$$

where, $\bar{q}_i^-$ = the pumping limit on well i; q_{max}^- = the allowable total extraction rate; and $c_{\beta a}$ and $c_{\beta max}$ are the spatially averaged and peak residual concentrations of the species β at the end of the remediation period t_s; these residual concentrations are to be reduced to the specified levels $c_{\beta a}^*$ and $c_{\beta max}^*$ respectively. The constraints of types (5a,b) are specifically devised to design cleanup schemes. The optimization package NPSOL developed by Gill et al. (1986) is employed to solve the above optimization problem. A state sensitivity method is used to compute the gradients.

In the following analysis, referring to Figure 4, the municipal wells E-5, E-7 and E-10 are shutdown, whereas the wells E-5A, E-6 and E-8 are pumping at constant rates of -0.0145,-0.00926, -0.00976 m^3/sec. The municipal wells E-2 and E-9 together with

Table 2: Parameters for The Optimization Formulation

$c_{1a}^*(ppb)$	$c_{2a}^*(ppb)$	$c_{1max}^*(ppb)$	$c_{2max}^*(ppb)$	$\bar{q}_i^-(m^3/sec)$	$q_{max}^-(m^3/sec)$
0.09	0.9	0.01	80	0.04	2.

the assumed potential wells (indicated by the triangles) are considered as the possible remedial wells for optimization (initial pumping rate: $-0.01\ m^3/sec$). Assume a remediation period of $t_s = 20years$. The parameters for the optimization formulation are summarized in Table 2.

It is assumed that the source control wells PW-1,3,4 pumping at the rates of -0.00368, -0.00158, -0.00158 m^3/sec can effectively prevent the contaminants from migrating into the aquifer outside the source area. At optimality, the only binding constraint is c_{1max}, whereas other concentration constraints are significantly less than the specified remediation levels. The initial NDMA plume occupies a larger spatial extent than the chlorobenzene plume does, and the specified cleanup requirement is much more stringent than that for chlorobenzene. Therefore, an optimal pumping scheme determined by explicitly considering NMDA only may also satisfy the cleanup requirement for chlorobenzene; however, such a scheme can not be guaranteed to be optimal for the removal of both species. Figure 5 depicts the steady-state hydraulic heads under the operation of the identified optimal pumping wells (W1, W2, W3, W4), the active municipal wells (E-5A, E-6, E-8), and the source control wells (PW-1,3,4). The drawdown cone with the lowest head of approximately 339.60 meters is centered around the wells W3 and W4.

If the contaminant source control is imposed by constraining the hydraulic gradients across the source boundary inward along the control directions depicted in Figure 4 (neglecting the process of hydrodynamic dispersion for the source contaminants), for the given possible well locations, the specified pumping limits, and the rather stringent cleanup requirements for NDMA, no feasible solution could be found. For designing remedial wells for contaminated groundwater cleanup, the optimization technique generally tends to identify a set of drawdown cones centered around a few remedial wells. Obviously, contaminant source control via constraining the hydraulic gradients across the source boundary dictates a set of drawdown cones centered within or around the source area, with at least one pair of local adjacent drawdown cones across the source boundary. Unfortunately, with the present intial setup, this scenario could not be realized. Other potential patterns for both the source control wells and the plume remediation wells should be examined.

CONCLUSIONS

An analysis of the optimal remediation design for a contaminated groundwater site was conducted. This analysis indicated: (1) In general, multiple species should be

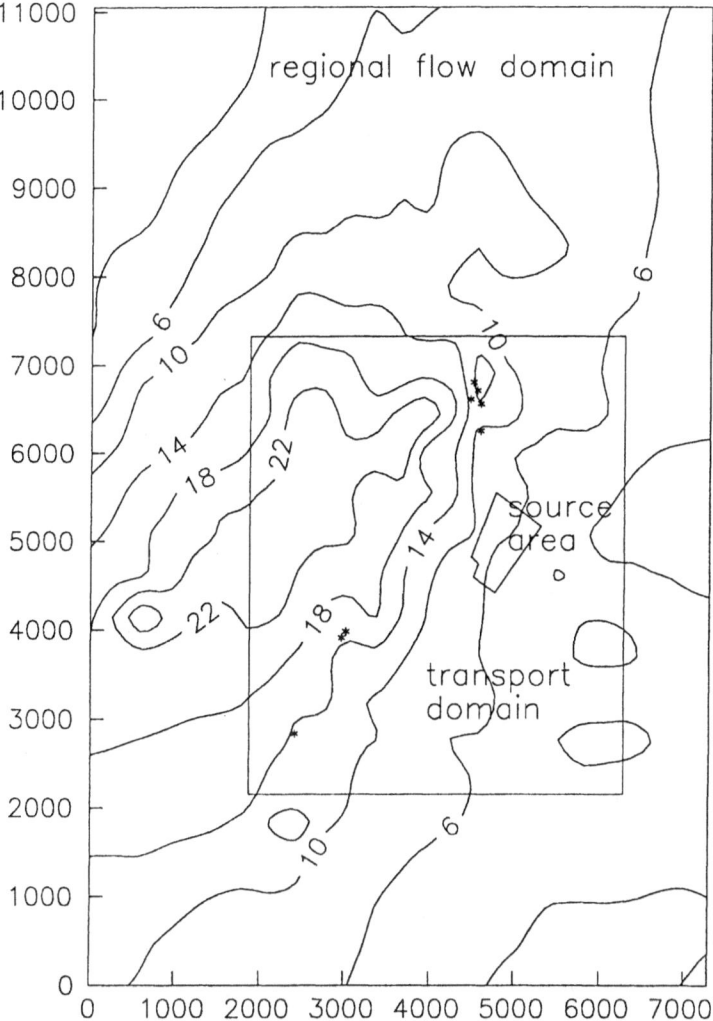

Figure 1: The regional domain for groundwater flow simulation, the subdomain for contaminant transport simulation, the contaminant source area, the municipal wells (indicated by stars), and the contour of thickness of the municipal aquifer

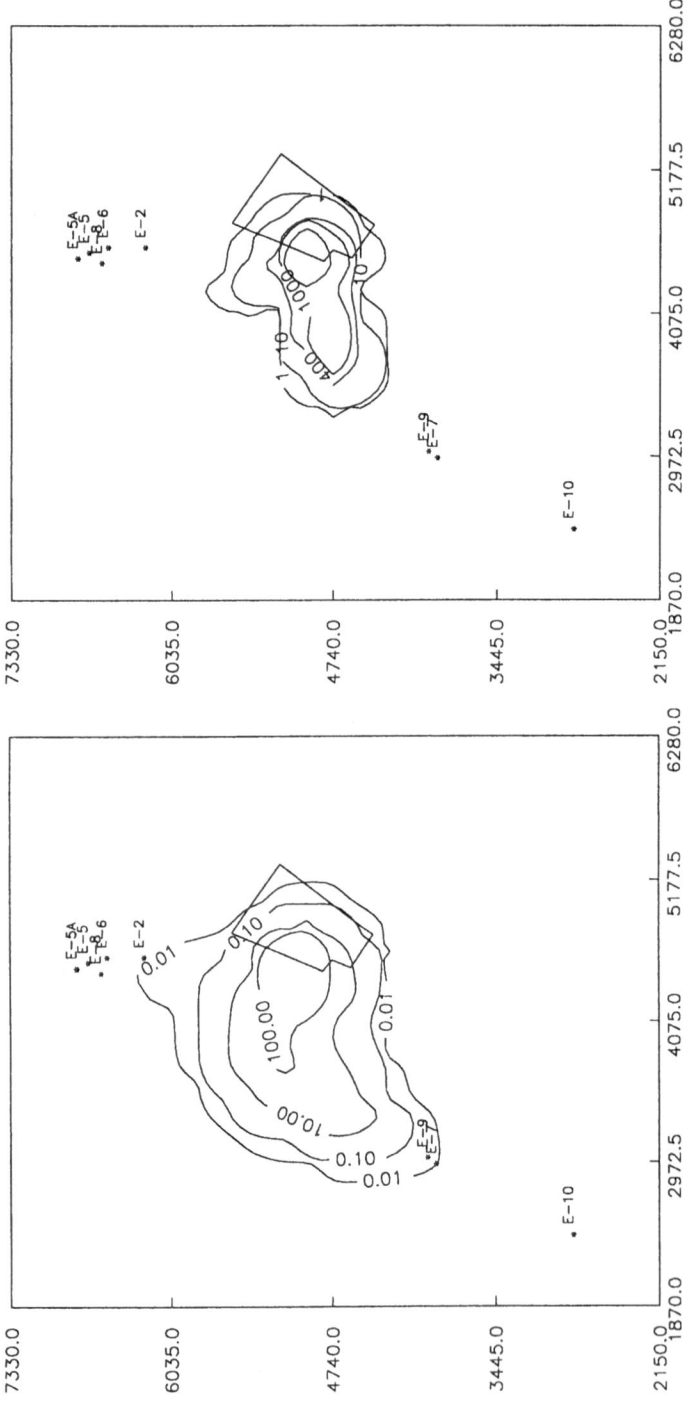

Figure 3: The initial chlorobenzene plume (in *ppb*)

Figure 2: The initial NDMA plume (in *ppb*)

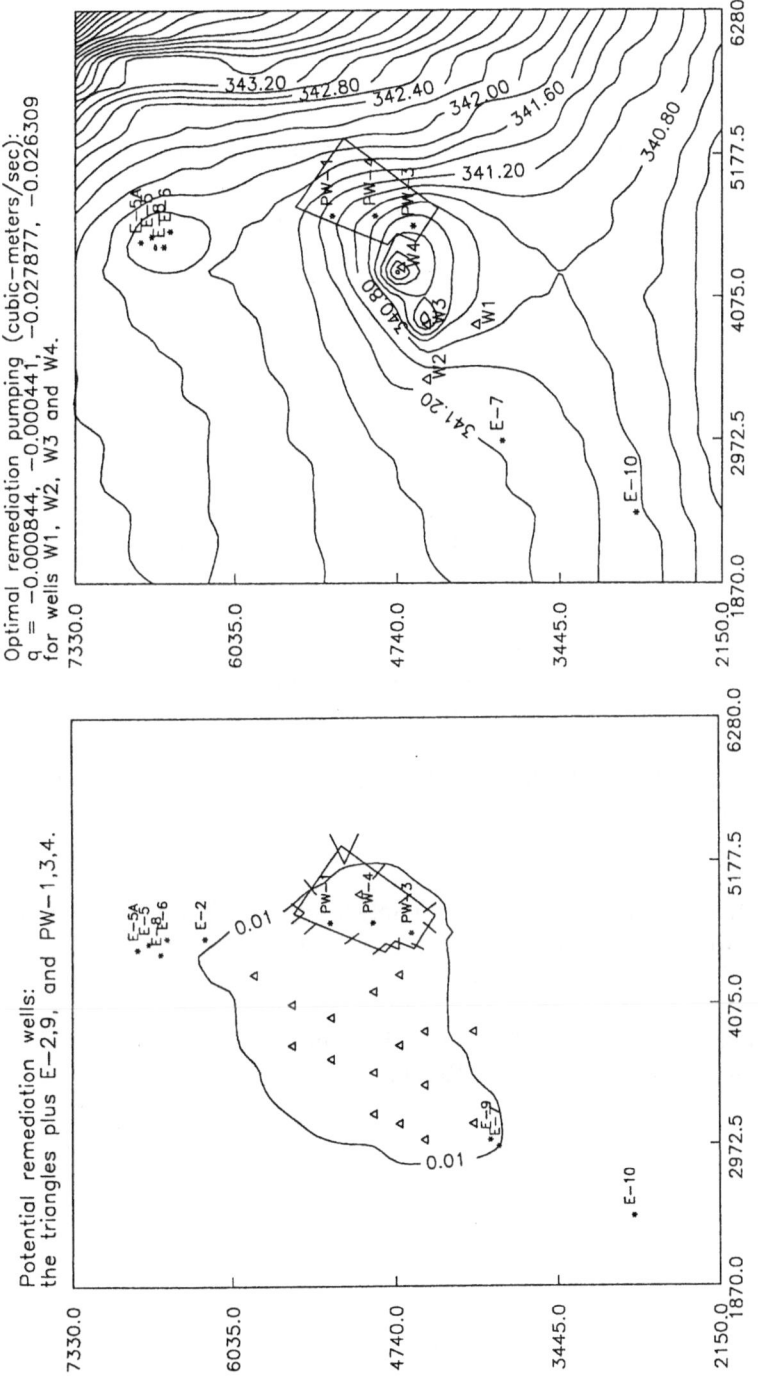

Figure 5: The steady-state hydraulic heads induced by the optimal pumping scheme

Figure 4: The potential remediation wells, the hydraulic gradient directions for source control

included in the transport simulation to find a remedial pumping scheme optimal for removing all the targeted species; but, if the initial plume of one species occupies a larger spatial extent than other species do, and the specified cleanup requirement for the spatially most extent species is more stringent than those for other species, an optimal pumping scheme determined by explicitly considering the spatially most extent species only may also satisfy the cleanup requirements for other species; however, such scheme can not be guaranteed to be optimal. (2) By taking advantage of the steady-state assumption used for flow simulation, different scales for groundwater flow and contaminant transport simulations could be employed to save computer effort. (3) With the potential wells and pumping limits used in this paper, no feasible solution could be identified if both contaminant source control via constraining the hydraulic gradients across the source boundary and stringent cleanup standards are imposed.

REFERENCES

Ahlfeld, D. P., J. M. Mulvey, and G. F. Pinder (1988), Contaminated groundwater remediation design using simulation, optimization, and sensitivity theory: 2. analysis of a field site, Water Resour. Res., 24, 443-452.

Gill, P. E., W. Murray, M. A. Saunders, and M. H. Wright (1986), User's guide for NPSOL: A FORTRAN package for nonlinear programming, Technical Report SOL 86-2, Department of Operation Research, Stanford University.

Marryott, R. A., D. E. Dougherty, R. L. Stollar (1993), Optimal groundwater management: 2. Application of simulated annealing to a field-scale contamination site, Water Resour. Res., 29, 847-860.

Sykes, J. F. (1991), Elmira/St. Jacobs Water Supply Study: Analysis of Groundwater Flow and Solute Migration, Report for CH2M Hill Engineering Ltd., Project No. ONT29307M0.01.

Tiedeman, C., and S. M. Gorelick (1993), Analysis of uncertainty in optimal groundwater contaminant capture design, Water Resour. Res., 29, 2139-2153.

AN EXTENDED IDENTIFIABILITY APPROACH FOR EXPERIMENTAL DESIGN IN GROUNDWATER MODELING

W.W-G. YEH and N.Z. SUN
Department of Civil and Environmental Engineering
UCLA
Los Angeles, California 90024
U.S.A.

Using the management equivalence identifiability (MEI) criterion, a new methodology of experimental design is developed. In this methodology, two sets of models are generated: set A consists of all admissible models that are equivalent among each other from the management perspective, while set B consists of all admissible models that are non-identifiable by existing data. Procedures for generating these two sets of models are outlined. The optimal experimental design problem is then formulated as a nonlinear programming problem with MEI as a constraint. The developed methodology has been used to study the water resources management problem in the Hemet Basin, Riverside, California. The results obtained from the study clearly demonstrate that the developed methodology is very useful for either increasing model reliability or reducing the cost of data collection.

INTRODUCTION

When a simulation model is used for management purposes, it must be accurate and reliable. The accuracy of a simulation model is determined, to a certain extent, by the accuracy of the inverse solution (Yeh, 1986) which in turn is dependent on the quantity and quality of observations. Various experimental design criteria for groundwater modeling were reviewed by Yeh (1992). The classical design criteria, such as the D-optimality and the A-optimality, are useful tools which can be used to select a design among all competing designs. The design criterion based on the management equivalence identifiability (Sun and Yeh, 1990), on the other hand, produces reliable designs for model applications while the experimental cost is minimized. The best compromise is sought to balance the management risk and the cost of data collection.

In Sun and Yeh (1990), the problem of judging whether a design is sufficient for management equivalence identifiability (MEI) is formulated into a nonlinear programming problem. This problem, however, is usually difficult to solve in practice. In the proposed research, we assume that prior information and some existing data are available. A simplified approach that avoids the solution of a complex nonlinear programming problem is presented for testing the MEI. This approach has been successfully used to study a water resources management problem of the Hemet Basin, Riverside, California, in which uncertainties of the calibrated simulation model are estimated and then the impacts of these uncertainties to management decisions are

A. Peters et al. (eds.), Computational Methods in Water Resources X, 899–906.

analyzed. We have found that prior information and existing head and concentration observations of the basin are almost sufficient for the stipulated water resources management objectives. The results obtained from the Hemet Basin study clearly show that the developed methodology is very useful for either increasing model reliability or reducing the cost of data collection.

THEORY

If the purpose of building a model is to find the optimal decisions for a given management problem, the most important problem is to determine the error of the optimized decision variables due to uncertainty of the identified parameters. Assume that we have a management objective J with decision vector $\mathbf{q}_J$ and accuracy requirements described by a weighting matrix C_J. In Sun and Yeh (1990), the management equivalence identifiability (MEI) is defined as follows:

If there exists a design D and a number $\delta > 0$, such that

$$
\begin{aligned}
&\left\| \mathbf{u}_D(\mathbf{p}) - \mathbf{u}_D(\mathbf{p}_0) \right\|_{C_D} < \delta && \text{implies} \\
&\left\| \mathbf{q}_J(\mathbf{p}) - \mathbf{q}_J(\mathbf{p}_0) \right\|_{C_J} < 1 && \text{for any } \mathbf{p} \in P_{ad}
\end{aligned}
\tag{1}
$$

then parameter $\mathbf{p}_0$ is said to be δ-management equivalence identifiable with respect to design D and weighting matrices C_J and C_D, where C_D is a given weighting matrix defining the norm of the observation space, P_{ad} the admissible set of the identified parameters.

The sufficiency of a design D for MEI can be confirmed by solving the following optimization problem:

$$
\min F(\mathbf{p}, \mathbf{p}_0) = \left\| \mathbf{u}_D(\mathbf{p}) - \mathbf{u}_D(\mathbf{p}_0) \right\|_{C_D}^2
\tag{2}
$$

subject to

$$
\left\| \mathbf{q}_J(\mathbf{p}) - \mathbf{q}_J(\mathbf{p}_0) \right\|_{C_J}^2 \geq 1, \qquad \mathbf{p} \in P_{ad}, \ \mathbf{p}_0 \in P_{ad}.
$$

If the minimum of this problem is greater than $4\eta^2$, where η is the upper bound of observation errors, then we can conclude that design D is sufficient for MEI (Sun and Yeh, 1990). Unfortunately, the optimization problem (2) is very difficult to solve in practice. Moreover, in the above definition (1) of MEI, it is assumed that the model structure error can be ignored. For practical problems, however, the model structure error can not be ignored and, in many instances, may dominate the observation error. Thus, it is necessary to develop a simplified method for testing the MEI while taking into consideration the model structure error.

In practice, we can always assume that some prior information and historical observations are available and a simulation model can be found by solving the following optimization problem:

$$
\min E(M) = \left\| \mathbf{u}(M) - \mathbf{u}^{obs} \right\|_{C_D}, \ M \in M_{ad}
\tag{3}
$$

where vector $\mathbf{u}(M_B)$ is the model output corresponding to existing observation vector $\mathbf{u}^{obs}$, M_{ad} the admissible model set determined by prior information. Let the identified model be M_B and the minimum of (3) be $E(M_B) = \delta$.

Now, we assume that M_A is the true model and we want to estimate the "distance" between M_A and M_B. Using triangle inequality, we obtain

$$\left\| \mathbf{u}_D(M_A) - \mathbf{u}_D(M_B) \right\|_{C_D} \leq \left\| \mathbf{u}_D(M_A) - \mathbf{u}_D^{obs} \right\|_{C_D} + \left\| \mathbf{u}_D(M_B) - \mathbf{u}_D^{obs} \right\|_{C_D} \tag{4}$$

Since M_A is the true model, we have

$$\left\| \mathbf{u}_D(M_A) - \mathbf{u}_D^{obs} \right\|_{C_D} \leq \eta \tag{5}$$

where η is the upper bound of observation error, and thus

$$\left\| \mathbf{u}_D(M_A) - \mathbf{u}_D(M_B) \right\| \leq \delta + \eta \tag{6}$$

Now we can conclude that any model $M_A \in M_{ad}$, if it satisfies inequality (6), can not be rejected by existing data. In other words, inequality (6) defines a model set M_{dat}, which is a subset of M_{ad}. Each member of M_{dat} should be considered as an acceptable model candidate based on existing data. Note that the structure of model M_A may be different from the structure of model M_B.

On the other hand, we can determine a model set M_{mag} from the given management problem such that all members of M_{mag} are management equivalent. If $M_{dat} \subset M_{mag}$, we conclude that existing data are sufficient for MEI; otherwise, we need to collect more data to make the set M_{dat} smaller, until condition $M_{dat} \subset M_{mag}$ is satisfied. The optimal data collection strategy should effectively decrease the size of M_{dat} for satisfying this condition.

Next we will give an example to demonstrate the application of the above procedure in the solution of a water resources management problem.

A CASE STUDY

The Hemet Basin is located in the western part of Riverside County, California, about 70 miles southeast of the city of Los Angeles. The area of the Hemet Basin is about 45 square miles. Water supplies to the basin for municipal and industrial (M&I) and agricultural (AG) uses are provided by three agencies (East Municipal Water District, Lake Hemet Municipal Water District, and City of Hemet) and private wells. Available water resources include local groundwater, imported water from the Metropolitan Water District of Southern California (MWD), imported water from the San Jacinto Basin, imported water from the San Jacinto River, and reclaimed water. Since the Hemet Basin is a developing area, the population in this area will significantly increase in the coming 20 years. As a result, the M&I use is projected to increase accordingly, from 15,300 AF/year in 1993 to 28,800 AF/year in 2010.

A proposed water resources management problem

A multi-user, multi-source management problem is presented based on discussions with decision makers of the basin. The proposed management problem can be formulated into a nonlinear programming problem. The 20-year management period is divided into seven subperiods. The number of decision variables is 546, consisting of the following variables in each subperiod:

- Extraction rates of all public wells;

- Recharge rates of specified recharge locations;

- Amounts of imported water from different sources;

- Allocations of reclaimed water produced by two plants.

The management objective is to minimize the total operational cost which consists of the cost of extraction, the cost of imported water, the cost of reclaimed water and the benefit of increasing the ending storage. The total operational cost depends not only on decision variables but also on head and TDS concentration distributions.

The total number of constraints is 698. Besides various capacity constraints, there are safe yield constraints, water quality constraints and minimum recharge constraints which are designed to prevent poor quality water from entering the basin from a neighboring basin. Some of these constraints are dependent on head and/or concentration distributions. Since head and concentration distributions in turn depend on decision variables, the proposed management problem is indeed a nonlinear programming problem.

Existing data and model calibration

Existing geologic and hydrogeologic data of the Hemet Basin include:

- 38 well logs in the Hemet Basin and surrounding area.

- Complete precipitation records.

- Land use records and the estimate of water demand & return coefficients.

- Extraction records of public wells and estimated amounts of extraction of private wells.

- Incomplete water level records of 17 observation wells from year 1968 to year 1990. Since these wells are concentrated in the central part of the basin, head observations near the inflow and outflow boundaries are lacking.

- Incomplete TDS concentration measurements of 17 observation wells from year 1968 to year 1990.

A conceptual model was built according to prior information. All unknown parameters, sink/source terms and boundary conditions were calibrated by existing data (historical observations). The criterion of calibration is to minimize the following weighted sum:

$$E = w_h E_h + w_C E_C \tag{7}$$

where

$$E_h = \left\| h_D^{\text{cal}} - h_D^{\text{obs}} \right\|_{C_D}, \quad E_C = \left\| C_D^{\text{cal}} - C_D^{\text{obs}} \right\|_{C_D} \tag{8}$$

in which w_h, w_C are weighting coefficients.

The identified model parameters are listed in Table 1, where $K_1 \sim K_{10}$ are hydraulic conductivities (ft/day) of 10 inhomogeneous zones, S the storage coefficient, θ the porosity, α_L and α_T are the longitudinal and transverse dispercivities (ft), respectively. The identified infiltration factor $\beta = 0.23$, the averaged boundary inflow $Q_{\text{in}} = 1,934$ (AF/year), and the averaged boundary outflow $Q_{\text{out}} = 6,842$ (AF/year).

Table 1. Identified Model Parameters and Their Uncertainties

Parameter	Identified Value	Lower Bound	Upper Bound	Uncertainty Cost ($)
K_1	87	36	150*	89,100
K_2	107	58	150*	43,600
K_3	30	10	100*	2,200
K_4	180	30	200*	8,700
K_5	33	8	100*	70,100
K_6	120	21	150*	51,100
K_7	0.5	0.2	1.5	14,000
K_8	5	1.0*	10*	13,500
K_9	10	2.0*	20*	58,900
K_{10}	10	2.0*	20*	16,600
S	10^{-4}	10^{-5}	2×10^{-4}	6,600
θ	0.1	0.08	0.16	103,400
α_L	600	120	800*	82,900
α_T	200	30	400*	32,900

MEI test

The minimal values of E_h and E_C obtained in the above identification problem are 7.5 (ft) and 106 (mg / l), respectively. Thus, model M_B and the associated residuals, δ, are determined. If we take $\eta = 1$ (ft) for head observations and $\eta = 50$ (mg / l) for concentration observations, then any model M_A that satisfies

$$\left\| h_D(M_A) - h_D(M_B) \right\|_{C_D} \leq 8.5 \quad \text{and} \quad \left\| C_D(M_A) - C_D(M_B) \right\|_{C_D} \leq 156 \tag{9}$$

cannot be rejected by existing data. Parameter ranges that satisfy condition (9) are listed in Table 1, where superscript * denotes that the value is a predetermined bound. For each model M_A in which the parameter values are within these ranges, we can calculated the minimal operational cost and the optimal decisions by solving the proposed management problem. The minimal operational cost for model M_B is $67,968,000. The uncertainties of the minimal operational cost caused by parameter uncertainties are also listed in Table 1. We find that the total uncertainty is less than 1%. Moreover, the optimal management decisions do not change at all with these parameter uncertainties.

On the other hand, however, the minimal operational cost is sensitive with respect to the uncertainties of β, Q_{in} and Q_{out}. All possible combinations of these three parameters should be considered in the uncertainty analysis. Table 2 shows all possible combinations and the corresponding minimal operational costs, where x means an impossible combination. Values of inflow and outflow are represented by percentages of the identified Q_{in} and Q_{out} respectively.

Table 2. Minimal Operational Costs (10^3) for Different Values of β, Q_{in} and Q_{out}

Inflow	Lower Bound of Outflow	Operational Cost	Upper Bound of Outflow	Operational Cost
		$\beta = 0.1$		
20%	x	x	52%	71,058
80%	x	x	66%	75,079
150%	x	x	83%	80,122
200%	x	x	93%	84,172
		$\beta = 0.15$		
20%	x	x	78%	73,865
80%	x	x	89%	76,705
150%	59%	63,485	103%	80,453
200%	67%	66,148	112%	84,142
		$\beta = 0.20$		
20%	567%	59,784	98%	74,404
80%	64%	60,813	109%	77,073
150%	74%	62,212	122%	80,396
200%	83%	65,136	130%	83,650
		$\beta = 0.25$		
20%	69%	58,149	116%	73,783
80%	80%	59,962	127%	76,587
150%	93%	62,175	138%	79,095
200%	102%	65.324	144%	81,536
		$\beta = 0.30$		
20%	88%	58,116	134%	73,320
80%	99%	60,097	144%	75,696
150%	113%	62,419	x	x
200%	123%	65,702	x	x

Table 3 gives error ranges of the minimal operational cost caused by various uncertainties associated with the identified model. These error ranges may be acceptable from the management viewpoint; otherwise, we need to collect more data to decrease the uncertainty associated with the management objective.

Table 3. Ranges of the Minimal Operational Costs (10^3) Caused by Various Uncertainties

Uncertainty	Lower Bound of Cost	Upper Bound of Cost	Maximal Relative Error
All parameters listed in Table 1	67,670	68,267	1%
Inflow: Q_{in}	67,176	73,428	8%
Outflow: Q_{out}	62,768	79,307	17%
Infiltration factor: β	64,222	79,706	17%
Combinations of β, Q_{in} and Q_{out}	58,116	84,172	24%

Data collection strategy

Table 3 clearly shows that the most effective data collection strategy is to decrease the uncertainties associated with boundary outflow Q_{out}, infiltration factor β, and boundary inflow Q_{in}. Unfortunately, such data can not be gathered from a short-term pumping test. In fact, if there are observation wells near the inflow and outflow boundaries and continuous observations are available, the accuracy of the identified model should be significantly improved. Through calculating sensitivity coefficients $\dfrac{\partial h}{\partial Q_{in}}, \dfrac{\partial h}{\partial Q_{out}}$ and $\dfrac{\partial h}{\partial \beta}$ for all nodes, we can find the best observation locations for increasing the accuracy of Q_{in}, Q_{out} and β. A new long-term observation design has been proposed for the Hemet Basin study that includes the determination of key observation locations and appropriate observation frequency. Four local pumping tests are designed to verify the boundary condition and to determine the recharge capacity.

CONCLUSION

An extended identifiability approach is presented for model reliability analysis and experimental design. This approach is superior to the Monte Carlo method in three aspects: 1, Assumptions on statistical distributions of the unknown parameters are unnecessary; 2, Different model structure can be considered; and 3, The data collection strategy is directly linked to model applications. The proposed methodology has been successfully applied to a real case study.

ACKNOWLEDGMENT

Research leading to this paper was supported by (1) the National Science Foundation under award MSS-9213963 and (2) the University of California, Water Resources Center, as part of Water Resources Center Project UCAL-WRC-W-767.

REFERENCE

Sun, N-Z. and Yeh, W.W-G. (1990) "Coupled inverse problems in groundwater modeling, 2. Identifiability and experimental design," *Water Resources Research*, 26(10), 2527-2540.

Yeh, W.W-G. (1986) "Review of parameter identification procedures in groundwater hydrology: The inverse problem," *Water Resources Research*, 22(1), 95-108.

Yeh, W.W-G. (1992) "Systems analysis in ground-water planning and management," *Journal of Water Resources Planning and Management,* ASCE, 118(3), 224-237.

EFFECT OF BOUNDARY CONDITIONS ON DIPOLE FLOW

V.Zlotnik[1] and G.Ledder[2]
University of Nebraska-Lincoln
[1]Department of Geology, Lincoln, NE 68588-0340, USA
[2]Department of Mathematics and Statistics, Lincoln, NE 68588-0323, USA

Dipole flow involves vertical circulation of groundwater induced by simultaneous injection and extraction through separated chambers in a single borehole. In this paper, the dipole influence radius is applied as a criterion for selection of proper boundary conditions in boundary value problems on dipole use. Solutions for infinite and confined semi-infinite aquifers (Zlotnik and Ledder, 1994) are extended to unconfined aquifers of infinite depth.

INTRODUCTION

Dipole flow is widely applied in different technologies for groundwater remediation (Herrling and Stamm, 1992). Kabala and Xiang (1992) and Kabala (1993) suggested the use of a dipole for determining aquifer parameters using Hantush's (1961) model of a confined or leaky aquifer. Implicitly they reduced their approach to another method of hydraulic conductivity measurement at large scale (on the order of the aquifer saturated thickness). However, the dipole influence rapidly vanishes with distance, and transient effects disappear in short time due to small spatial scale of the flow zone. Therefore the dipole technique can be used for mesoscale measurements of hydraulic conductivity (at the scale of 1 m) (Zlotnik and Ledder, 1994).

Instead of finite-element-based estimation of a "radius of influence" (Herrling and Stamm, 1992) a technique using analytical determination of drawdown and Stokes' stream function allows much simpler evaluation of the region of influence (Zlotnik and Ledder, 1994). Since dipole-induced flow also depends on location of aquifer boundaries near the region of influence, the concept of region of influence can significantly simplify numerical and analytical computations of the dipole. Particularly it can provide rationale for selection of a hydrodynamic aquifer model in forward or inverse problem solution (Kabala, 1993; McDonalds and Kitanidis, 1993).

A. Peters et al. (eds.), Computational Methods in Water Resources X, 907–914.

Since the region of dipole influence has limited vertical extent, the effect of only one boundary needs to be considered for meso-scale hydraulic conductivity measurements, using a model of a semi-infinite aquifer. Thus, after the cases of a vertically-infinite aquifer and a confined semi-infinite aquifer (Zlotnik and Ledder, 1994) are revisited, the new solution for water table conditions will be presented. These formulas can be used for measurements of hydraulic conductivity at meso-scale.

THE REGION OF INFLUENCE OF A DIPOLE

We consider the problem of finding the drawdown s_I in an infinite aquifer due to a dipole whose center is located at $(r, z) = (0, 0)$. The dipole consists of a single well with two screened sections, an upper (extraction) chamber and a lower (injection) chamber, each screened over a length 2Δ (Figure 1A). The chamber centers are located at $z = \pm L$. Each screen section is idealized as a uniform line source or sink with total flow rate Q. The physical model has cylindrical symmetry; thus, the problem to be solved is given below in cylindrical coordinates. To eliminate anisotropy, we replace the physical radial coordinate r with the scaled radial coordinate ρ, defined by

$$\rho = r/a, \quad a^2 = K_r/K_z, \tag{1}$$

with K_r and K_z the horizontal and vertical hydraulic conductivities. With this change, the drawdown s is governed by the equations:

$$K_z \nabla^2 s = S_s \frac{\partial s}{\partial t}, \tag{2}$$

$$\lim_{\rho \to 0} \rho \frac{\partial s}{\partial \rho} = \frac{L s_r}{2\Delta} \begin{cases} -1, & |z - L| < \Delta \\ 1, & |z + L| < \Delta \\ 0, & \text{otherwise} \end{cases}, \quad s_r = \frac{Q}{2\pi K_r L}, \tag{3}$$

where S_s is the storage coefficient, S_y is the specific yield, and s_r is a reference value for the induced drawdown. Equations (2) and (3) and the initial condition $s(\rho, z, t) = 0$ will apply to the confined and unconfined aquifers as well as the infinite aquifer; other boundary conditions are prescribed by processes at the aquifer boundaries.

For the infinite aquifer, the drawdown is also subject to the conditions of no drawdown as $\rho \to \infty$ or $z \to \pm\infty$; thus,

$$s(\rho, \pm\infty, t) = s(\infty, z, t) = s(\rho, z, 0) = 0, \tag{4}$$

The solution of the problem (2–4) is given by Zlotnik and Ledder (1994):

$$s_I(\rho, z, t) = \frac{L s_r}{8\Delta} \left[M\left(\frac{1}{T}, \frac{z_{-+}}{\rho}\right) - M\left(\frac{1}{T}, \frac{z_{--}}{\rho}\right) + M\left(\frac{1}{T}, \frac{z_{+-}}{\rho}\right) - M\left(\frac{1}{T}, \frac{z_{++}}{\rho}\right) \right], \tag{5}$$

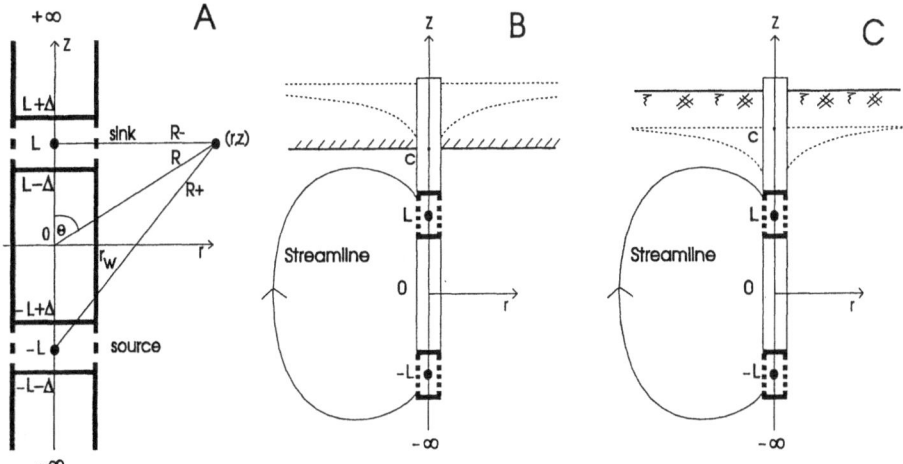

Figure 1: Dipole configuration: (A) infinite aquifer, (B) confined semi-infinite aquifer, (C) unconfined semi-infinite aquifer.

where M (Hantush, 1964), T, and $z_{\pm\pm}$ are given by

$$M(u,\beta) = \int_u^\infty x^{-1} e^{-x} \operatorname{erf}(\beta\sqrt{x})\,dx, \quad T = \frac{4K_z t}{S_s \rho^2}, \quad z_{\pm\pm} = z \pm L \pm \Delta. \qquad (6)$$

Of particular interest is the steady-state drawdown (Zlotnik and Ledder, 1994)

$$s_I(r,z,\infty) = \frac{L s_r}{4\Delta}\left[\sinh^{-1}\left(\frac{z_{-+}}{\rho}\right) - \sinh^{-1}\left(\frac{z_{--}}{\rho}\right) + \sinh^{-1}\left(\frac{z_{+-}}{\rho}\right) - \sinh^{-1}\left(\frac{z_{++}}{\rho}\right)\right]. \qquad (7)$$

Steady-state streamlines in the ρz-plane are given by level curves of the stream function ψ_I given by (Milne-Thomson, 1960)

$$\psi_I(\rho,z) = \frac{Q}{8\pi\Delta a^2}\left[\frac{1}{R_{-+}} - \frac{1}{R_{--}} + \frac{1}{R_{+-}} - \frac{1}{R_{++}}\right], \quad R_{\pm\pm}^2 = \rho^2 + z_{\pm\pm}^2. \qquad (8)$$

The region of influence can be defined as the region that contains 90% of the dipole-induced flow; the boundary of this region is given by the equation $\psi_I(\rho,z) = Q/(20\pi a^2)$ (Zlotnik and Ledder, 1994). Figure 2A illustrates the flow field with $\Delta = 0.5L$. The dashed curves in the figure are lines of constant drawdown; the solid curves are the streamlines marking out 70%, 80%, and 90% of the flow. The influence region of the dipole, in geometric coordinates, is approximately given as

$$0 < r < 10aL, \quad -4L < z < 4L. \qquad (9)$$

Note that the bounds given in (9) are for the region of the aquifer that is strongly influenced by the presence of the dipole. It is tempting to assume that any boundary located outside the region of influence of the dipole may be neglected; however, as we shall see in the following section, this is not a valid interpretation of the region of influence.

In the sequel, we will assume that the lower boundary of a finite aquifer is located at a great distance from the dipole center. We may then examine the influence of the upper aquifer boundary only.

DIPOLE FLOW IN A CONFINED SEMI-INFINITE AQUIFER

Suppose the dipole is placed so that the upper impermeable boundary of a confined semi-infinite aquifer is located at $z = c$ (Figure 1B). The dipole-induced drawdown s_C in the confined aquifer satisfies (2–3), but the boundary conditions (4) must be modified to account for the no-flow boundary at $z = c$:

$$s(\rho, -\infty, t) = s(\infty, z, t) = s(\rho, z, 0) = 0, \tag{10}$$

$$\frac{\partial s}{\partial z}(\rho, c, t) = 0. \tag{11}$$

The confined aquifer problem (2–3), (10–11) has been solved by Zlotnik and Ledder (1994) using s_I, a mirror dipole at $z = 2c$, and the principle of superposition:

$$s_C(\rho, z, t) = s_I(\rho, z, t) - s_I(\rho, z-2c, t). \tag{12}$$

Figure 2B illustrates the steady-state streamlines for a semi-infinite aquifer with confining layer at $z = 10L$, with the streamlines (from Figure 2A) for an infinite aquifer included for comparison. The boundary exerts only a slight influence on the dipole region of influence. Thus, we may safely neglect any horizontal boundary outside the zone $-10L < z < 10L$ in most calculations.

Figure 2C illustrates the steady-state flow field with the confining layer at $z = 4L$. Note that the 90% streamline is affected by the boundary, but the 70% streamline is essentially unaffected. An aquifer boundary at the edge of the region of influence of the dipole has a noticeable effect on the far field of the dipole-induced flow. Horizontal boundaries that are near the influence region of a dipole should not always be neglected even if they are somewhat outside the region of influence.

DIPOLE FLOW IN AN UNCONFINED SEMI-INFINITE AQUIFER

In this case the equations (2–3) and (10) hold, but the no-flow condition at (11) must be replaced by the linearized free-surface boundary condition at the water

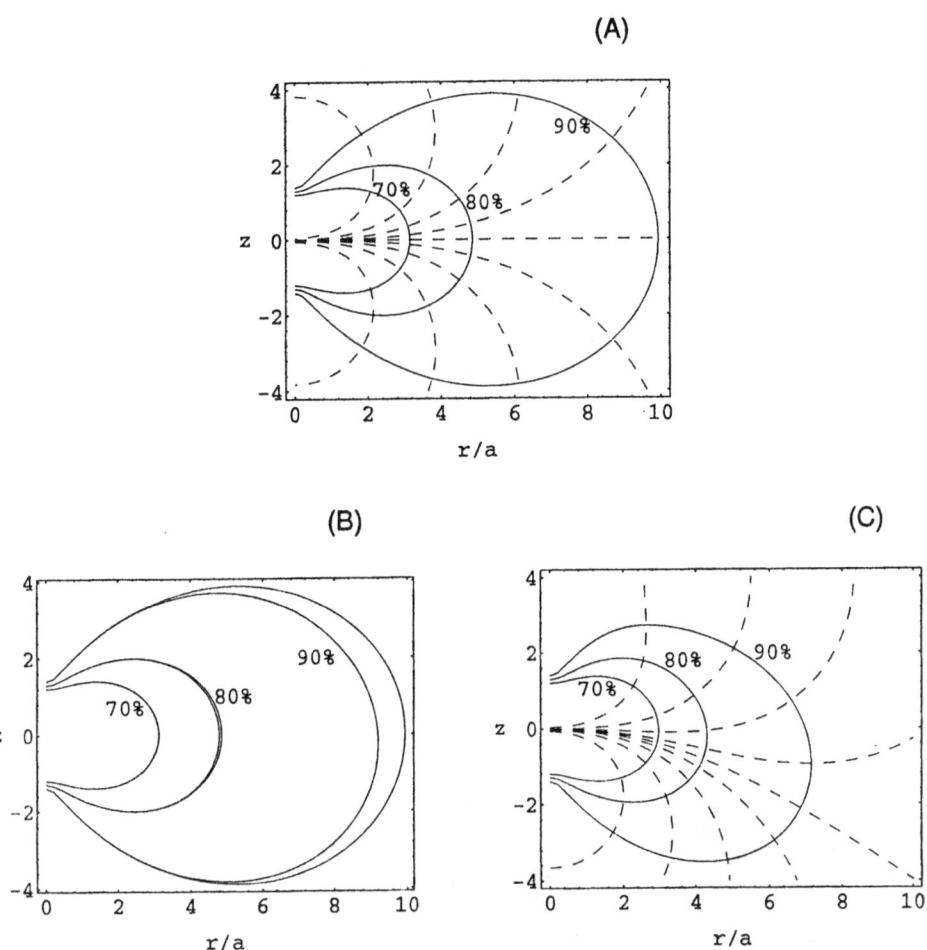

Figure 2: Steady-state streamlines (solid) and equipotentials (dashed) around a dipole: (A) infinite aquifer, (B) semi-infinite aquifer with $c = 10L$ and infinite aquifer, (C) semi-infinite aquifer with $c = 4L$.

table (Neuman, 1974):

$$K_z \frac{\partial s}{\partial z}(\rho, c, t) + S_y \frac{\partial s}{\partial t}(\rho, c, t) = 0. \tag{13}$$

The similarity of this problem to that for the confined aquifer may be exploited to aid in the solution of (2–3), (10), (13). Let

$$s_2 = s_U - s_C \tag{14}$$

be the correction to s_C due to the water table. s_2 satisfies (2), (10), and the additional conditions

$$\lim_{\rho \to 0} \rho \frac{\partial s_2}{\partial \rho} = 0, \tag{15}$$

$$K_z \frac{\partial s_2}{\partial z}(\rho, c, t) + S_y \frac{\partial s_2}{\partial t}(\rho, c, t) = -S_y \frac{\partial s_C}{\partial t}(\rho, c, t) = -2 S_y \frac{\partial s_I}{\partial t}(\rho, c, t) \tag{16}$$

The right-hand side of (16) can be evaluated from (5). For simplicity, the present treatment is limited to the special case $\Delta \to 0$. A more general treatment will appear in a subsequent paper. Differentiating (5), setting $z = c$, taking limits as $\Delta \to 0$, and substituting the result into (16) gives the water table boundary condition as

$$K_z \frac{\partial s_2}{\partial z}(\rho, c, t) + S_y \frac{\partial s_2}{\partial t}(\rho, c, t) = -\frac{L\sqrt{S_s} s_r}{2\sqrt{\pi K_z t^3}} e^{-\frac{S_s \rho^2}{4 K_z t}} \left[e^{-\frac{S_s(c-L)^2}{4 K_z t}} e^{-\frac{S_s(c+L)^2}{4 K_z t}} \right]. \tag{17}$$

After the introduction of dimensionless variables using the definitions

$$\bar{t} = \frac{t}{t_r}, \quad t_r = \frac{L^2 S_s}{K_z}, \quad \sigma = \frac{L S_s}{S_y}, \quad \bar{\rho} = \frac{\rho}{L}, \quad \bar{z} = \frac{z}{L}, \quad \bar{c} = \frac{c}{L}, \quad \bar{s} = \frac{s_2}{s_r}, \tag{18}$$

the problem given by (2), (10), (15), (17) becomes

$$\nabla^2 \bar{s} = \frac{\partial \bar{s}}{\partial \bar{t}}, \tag{19}$$

$$\bar{s}(\bar{\rho}, -\infty, \bar{t}) = \bar{s}(\infty, \bar{z}, \bar{t}) = \bar{s}(\bar{\rho}, \bar{z}, 0) = 0, \tag{20}$$

$$\lim_{\bar{\rho} \to 0} \bar{\rho} \frac{\partial \bar{s}}{\partial \bar{\rho}} = 0, \tag{21}$$

$$\frac{\partial \bar{s}}{\partial \bar{z}}(\bar{\rho}, \bar{c}, \bar{t}) + \frac{1}{\sigma} \frac{\partial \bar{s}}{\partial \bar{t}}(\bar{\rho}, \bar{c}, \bar{t}) = -\frac{1}{2\sigma\sqrt{\pi \bar{t}^3}} e^{-\frac{\bar{\rho}^2}{4\bar{t}}} \left[e^{-\frac{(\bar{c}-1)^2}{4\bar{t}}} - e^{-\frac{(\bar{c}+1)^2}{4\bar{t}}} \right]. \tag{22}$$

Solution of (19)–(22) is obtained using Hankel and Laplace transforms:

$$s_2 = \int_0^\infty \frac{y J_0(\bar{\rho} y)}{2A} [E(-A_-, 1) - E(-A_-, -1) + E(A_+, 1) - E(A_+, -1)] \, dy, \tag{23}$$

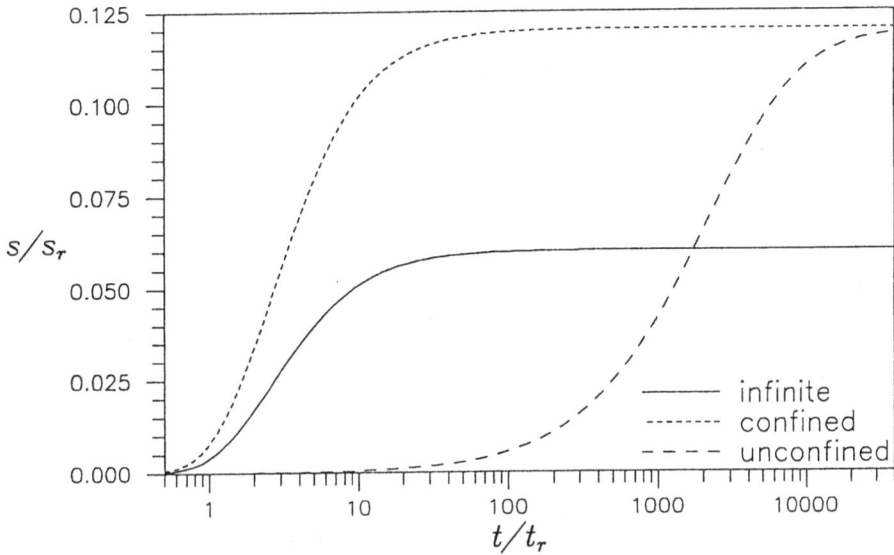

Figure 3: Transient drawdown for infinite and semi-infinite aquifers with upper boundary at $c = 4L$.

with

$$E(x, k) = e^{x(2\bar{c} - \bar{z} + k + \sigma\bar{t})}\text{erfc}\left(\frac{2x\bar{t} + 2\bar{c} - \bar{z} + k}{2\sqrt{\bar{t}}}\right), \quad A = \sqrt{y^2 + \sigma^2/4}, \quad A_\pm = A \pm \frac{\sigma}{2}.$$

Figure 3 illustrates the time history of the drawdown at a point located at $\rho = L$, $z = 4L$ for each of the three aquifer types considered here. For the semi-infinite cases, the aquifer boundary is at $c = 4L$. Of particular interest is the observation that the confined and unconfined aquifers behave very differently for short times, but behave the same for sufficiently long times. Typically, $t_r < 10^{-6}$ days, so the right edge of Figure 3 is reached within one hour. Hence, the effect of the water table boundary is very short-lived, and transient effects for a confined aquifer are even shorter.

CONCLUSIONS

Choice of boundary conditions in dipole problems depends on the relative locations of the observation point, the aquifer boundaries, and the dipole. If the nearest horizontal boundary is located at a distance more than $10L$ from the dipole center, the model of an infinite aquifer can be safely used for all observation points. If one horizontal boundary is located at a shorter distance $(4L-10L)$, confined or water table models for a semi-infinite aquifer are recommended; however, in the immediate vicinity of the dipole the potential field and streamlines are relatively unaffected by the presence of the boundary.

The steady regime in infinite and confined semi-infinite aquifers and the quasi-steady regime in unconfined aquifers are rapidly achieved. Steady conditions in an unconfined aquifer are achieved later in accordance with the theory of delayed yield.

ACKNOWLEDGEMENTS

This work was supported in part by the Water Center of the University of Nebraska-Lincoln and IBM. Our thanks to A. Zlotnik and C. Rudnick for help with the diagrams.

REFERENCES

Carslaw, H.S. and Jaeger, J.C. (1959) Conduction of Heat in Solids. Clarendon Press, Oxford.

Hantush, M.S. (1964) "Hydraulics of Wells", in V.T. Chow (ed.), Advances in Hydroscience, vol.1, Academic Press, New York, pp. 281-432.

Herrling, B., and Stamm, J. (1992) "Numerical results of calculated 3D vertical circulation flows around wells with two screen sections for in situ on-site aquifer remediation", in T.F.Russel, R.E.Ewing, C.A.Brebbia, W.G.Gray, and G.F.Pinder (eds.), Computational Methods in Water Resources IX, vol.I, Numerical Methods in Water Resources, Elsevier Applied Science, New York, pp. 483-492.

Kabala, Z.J. (1993) "Dipole flow test: a new single-borehole test for aquifer characterization", Water Resources Res. 29, 99-107.

Kabala, Z.J., and J.Xiang. (1992) "Skin effect and its elimination for single borehole aquifer tests", in T.F.Russel, R.E.Ewing, C.A.Brebbia, W.G.Gray, and G.F.Pinder (eds.), Computational Methods in Water Resources IX, vol.I, Numerical Methods in Water Resources, Elsevier Applied Science, New York, pp. 467-474.

McDonald, T.P. and Kitanidis, P.K. (1993) "Modeling of the free surface of an unconfined aquifer near recirculation well", Ground Water 31, 774-780.

Milne-Thomson, L.M. (1960) Theoretical Hydrodynamics, McMillan, New York.

Neuman, S. (1974) "Effect of partial penetration on flow in unconfined aquifers considering delayed gravity response", Water Resour. Res. 10, 303-312.

Zlotnik, V. and Ledder, G. (1994) "Mezo-scale theory of dipole flow in uniform anisotropic aquifers", submitted to Water Resour. Res.

9. SUBSURFACE MULTIPHASE FLOW

EVAPORATION IN A THREE DIMENSIONAL LIQUID-GAS MODEL

C. APPERT([1]),V. POT([2]) and S. ZALESKI([3])

([1]) Laboratoire de Physique Statistique, CNRS,
E.N.S., 24 rue Lhomond, 75231 Paris Cedex 05, France.
([2]) Laboratoire de Géologie appliquée
and
([3]) Laboratoire de Modélisation en Mécanique, CNRS URA229,
Université Pierre et Marie Curie, 4 Place Jussieu,
75252 Paris Cedex 05, France.

Numerical simulations of evaporation in porous media with the liquid-gas model developed by Appert and Zaleski (1990) were performed. The observed results are qualitatively similar to those of laboratory experiments. We observed transport through the liquid phase present in wetting films. This transport is at least as important as transport through the gas phase. The liquid-gas model is a numerical method that allows to represent details occuring at the pore scale. Unlike the FHP model (Frisch et al, 1986), the liquid-gas model spontaneously separates into a light phase (the gas phase) and a dense phase (the liquid phase). We have extended this model to three dimensions (3D). We have used the algorithm of Somers and Rem (1992) and added interactions at a distance which are responsible for phase separation. The 3D model shows good agreement with thermodynamics for equilibrium pressures of a curved interface. Surface tension has been measured and partially predicted (Appert et al, 1993). In order to simulate a more realistic evaporation, we are constructing a multi-species 3D model. The model simulates a liquid phase mostly made of species 1 and evaporating in an atmosphere of species 2. A third species plays the role of a tracer.

INTRODUCTION

The process of evaporation in humid soils takes place at the soil surface. In arid soils it takes place deeper, about 1 or 2 meters beneath the surface (Fontes et al, 1986). When a soil evaporates, water flows up to the surface (where evaporation occurs) under pressure gradients created by capillarity effects. Water moves in the liquid phase, in the vapor phase or both (Rose, 1963).

Vapor flux starts in the unsaturated zone, where pores are partially filled with water. Isotopic studies provide arguments to determine the importance of vapor flux relative to liquid flux in the unsaturated zone. We call liquid flux the flux in

A. Peters et al. (eds.), Computational Methods in Water Resources X, 917–924.
© 1994 Kluwer Academic Publishers. Printed in the Netherlands.

condensed pores and the surface flux. Heavy isotopes of water, namely O^{18} and D, are used as tracers of evaporation. O^{18} and D isotopic profiles in soil water show a maximum isotopic enrichment near surface. This maximum is called the evaporating front: above it the water flux is predominantly a vapor flux, beneath it isotopes diffuse either in the liquid phase or in both the liquid and vapor phases. Fontes et al. (1986) showed that evaporating flux estimations double if diffusion in the vapor phase is considered. These estimations were made following the isotopic model of Barnes and Allison (1983) and Craig and Gordon (1965).

To improve the estimates of evaporation flux, a better understanding of the evaporation process at the pore level is necessary. In a partially saturated medium this process may be very complex. Philip and de Vries (1957) noticed that water in saturated throats may form a barrier to vapor diffusion. However it is not a barrier to water transport but rather a short–circuit, because of the higher density of water. If this process is dominant most of the liquid flux in a partially saturated medium is in these throats. Surface flux is then negligible and only reduces the effective size of the pore (Quénard and Sallee, 1992).

Our aim in this paper is to understand better these numerous phenomena occuring during soil drying using numerical simulations. We use the lattice gas method, a kind of cellular autamaton where position, velocity and time are all discretized (Frisch, 1986). Numerical simulations of evaporation in porous media were performed with the 2D liquid gas model developped by Appert and Zaleski (1990) (Pot et al, 1993). These simulations can be compared qualitatively with laboratory experiments. A regular front is observed when the geometry of the porous medium is regularly distributed. Irregular patterns appear on microscopic length (several pores), controlled by capillary effects : large pores empty first because of a weak capillary pressure. Conversely small pores remain saturated longer.

A possible analog model for a complex porous medium is an ensemble of capillary channels. This motivated us to use the the 2D liquid gas model to simulate evaporation in a simple capillary channel. We observed an important surface flux relative to the vapor flux. Obviously the relative importance of vapor and surface flux would depend on pore geometry and the precise physical nature of the wetting films. However the real pore geometry is very different from both the simple 2D channel and the the more complex 2D models used in the litterature (Pot et al, 1993).

For several reasons, space dimensionality plays an important role. For instance, in 2D is it impossible to have a continous wall surface except in the simplest case of a channel wall. This motivated us in the work presented here to add a third dimension to the model.

In what follows, we describe the 3D liquid gas model and its hydrodynamic and thermodynamic behavior. We present numerical simulations of evaporation in a model porous medium.

3D LIQUID-GAS MODEL

The 3D liquid gas model is an extension of the 2D liquid-gas model. The underlying lattice is the projection of a pseudo-four dimensional lattice : the face-centered-hypercubic (FCHC) lattice (Frisch, 1987). Each node or site is connected to 24 nearest neighbors. The simulated fluid is made of discrete particles with unit velocities travelling from site to site on the lattice. At most 24 particles occupy a site, and the exclusion principle is obeyed: there is no more than one particle with its fixed velocity vector at a site. Particles collide when they meet at a node. We use the reduced collision table of Somers and Rem (1992). Then we add a table of interactions at a distance. This step accounts for the attractive long-range forces effective over several molecular diameters. The table is the generalization in 3D of the interactions at a distance of the 2D minimal liquid gas model (Appert et al, 1993). Particles with opposite velocities are sent back towards each other. The particles are separated by a distance r called the range. A spontaneous phase separation between a light phase (the gas phase) and a dense phase (the liquid phase) starts when $r \leq 6$ sites. The resulting equation of state presents a Van der Waals form, with a plateau being the region of coexistence of liquid and gas phases, and a metastable region. We studied the model with range $r = 8$ sites.

The solid phase is easily implemented with lattice gases. A site becomes a solid site when its particles are removed and when it turns to be an obstacle for the other particles travelling on the lattice. Various geometries constrained by the underlying lattice are drawn putting together solid sites. Then we may construct walls with a no-slip condition : particles are sent the opposite direction they collide the wall.

Solid sites also exert long-range attractive forces over particles. The strength of this force varies: from 0 to 100% of the particles located at a distance r are attracted by the walls. We control the probability g_w for a particle near the wall to be attracted by the solid. Non-wetting fluid, partial wetting fluid and complete wetting fluid should be simulated by varying g_w. The size of the solid phase is constrained by the distance of interaction. Actually, all the particles in a pore are attracted by a wall if the pore width is smaller than r_w. To avoid this effect, we use a rather small value $r_w = 3$.

VALIDATION OF THE 3D LIQUID GAS MODEL

To understand the numerous phenomena occuring in evaporation processes, it is important to know how our 3D model obeys simple hydrostatic and thermodynamic laws. We measure in an unsaturated pore liquid and gas pressure on both sides of a meniscus. The pore is limited by two rectangular walls located at at a distance $2R$ from each other. Let κ be the meniscus curvature and θ the wetting angle. Then $1/R = \kappa/\cos\theta$

The liquid phase wets the grains and therefore is on the outside of the curved

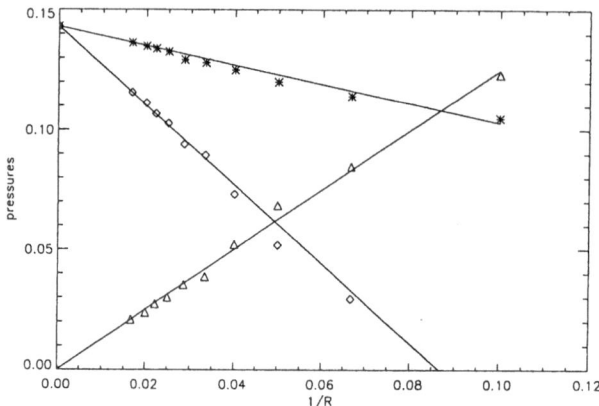

Figure 1: Equilibrium pressures against curvature of int erface. Stars represent gas pressure, diamonds liquid pressure and triangles capillary pressure. Straight lines are least square fits.

meniscus. The Gibbs-Thomson relations are:

$$P_G = P_{eq} + (1 - \frac{\rho_L}{\rho_L - \rho_G})\sigma\kappa \qquad (1)$$

$$P_L = P_{eq} - \frac{\rho_L}{\rho_L - \rho_G}\sigma\kappa \qquad (2)$$

P_G, P_L and P_{eq} are respectively the gas pressure,the liquid pressure and the equilibrium pressure for a flat interface. ρ_G and ρ_L are the gas density and the liquid density. These relations come from the equality of chemical potentials of the gas phase on the inside of a curved interface to the chemical potential of the liquid phase on outside (Rocard, 1952). A linear slope relates our measured gas and liquid pressures to $1/R$ (Fig 1.). New Gibbs-Thomson equations for the model write :

$$P_G = P_{eq} + (1 - \alpha\frac{\rho_L}{\rho_L - \rho_G})\sigma\kappa \qquad (3)$$

$$P_L = P_{eq} - \alpha\frac{\rho_L}{\rho_L - \rho_G}\sigma\kappa \qquad (4)$$

We introduced a factor, α, to take into account the lack of internal energy of lattice gases. It is close to unity (namely 1.35). This number is similar to what we found with the 2D liquid gas model (Pot et al., 1993). Actually our model does not take into account internal energy that leads to the Gibbs-Thomson relations. The incorrect slope reveals this infavourable feature of the liquid gas model. However, it does not prevent it from a correct qualitative behavior.

Capillary pressure against curvature is deduced from our measured liquid and gas pressures (Fig. 1). The relation is linear according to hydrostatic Laplace's law :

$$P_C = \Delta P = P_G - P_L = \sigma\kappa \qquad (5)$$

with P_C the capillary pressure. It leads to $\sigma cos\theta = 1.15$ and to a surface tension: $\sigma = 1.4 \pm 0.7$. The important error bar comes from the graphical estimation of the contact angle θ between the meniscus and the grains ($\theta = 35 \pm 10$). Measurements were done with the wetting range equal to the distance of interaction between particles. Like the 2D model, our 3D model has a satisfactory thermodynamic and hydrodynamic behavior.

NUMERICAL SIMULATIONS OF EVAPORATION

We performed numerical simulations of evaporation in a 3D porous medium. We chose spheres as building blocks. We randomly distributed the position of the centers of the spheres. The sphere radii were also randomly chosen between two fixed limits. The spheres may intersect or entirely overlap. The obtained porous medium then presents areas of macroporosity. A two dimensional section of the model shows the spatial distribution of the spheres (Fig. 2). The solid fraction is 73% and is well below the percolation threshold. This ensures that the void space is connected. Figure 3 shows this 3D evaporating porous medium. The lattice is $80 \times 80 \times 120$ sites and contains 1200 spheres of radius ρ. We drew the sphere radius in the interval $1 < \rho < 10$ measured in sites. Boundary conditions are periodic in the three dimensions. It means that a particle that leaves a side of the box re-enters the opposite side of the box. Therefore, wall at the bottom of the box separates the saturated zone from the gas phase. Isothermal evaporation is performed at the top of the model as described : on a horizontal layer 3 sites wide we imposed the density of the particles to be zero. Of course, this "evaporation" is somewhat similar to evaporation in a vacuum. The size of the porous medium is not large enough to see the pattern formed by the evaporating front. However our 3D views show that the front seems to regularly progress to the bottom of the model.

The model is implemented on several workstations. It reaches a cpu speed of 30000 site updates per second on a Solbourne station and 180000 sites per second on a IBM Risc 6000 model 370 workstation.

CONCLUSION

We displayed in this paper preliminary simulations of evaporation in a 3D porous medium. These computations should be followed by more quantitative measurements and the simulation of a greater variety of porous media. For instance one should quantitatively measure water and gas flux, as well as compare them to simple theoretical models. Model porous media could be extended to include double or multiple porosity. In fact one of the great strengths of our model is the ease with which solid boundary conditions may be changed.

On the methodological side, we feel that the model still has room for improvement of its computational efficiency. Such improvements would allow one to simulate larger model media, including more grains. The numerical efficiency of the present

t=5000

Figure 2: Two dimensionnal section of an evaporating 3D porous medium. The lattice is 80×120 sites. Solid phase is grey, liquid phase is black and gas phase is white. Evaporation is performed at the top of the box.

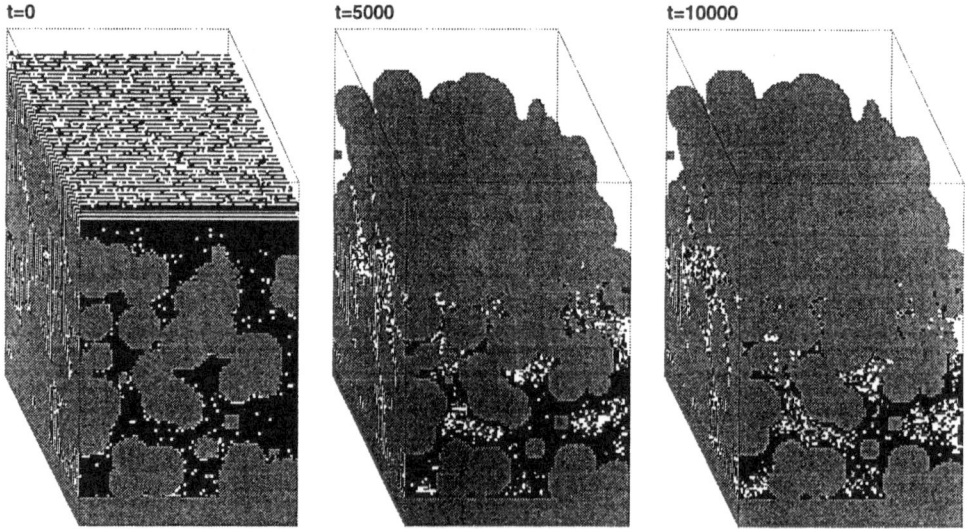

Figure 3: Evaporation on a $80 \times 80 \times 120$ lattice. Total porosity is 27%. $g_w = 0.5$, $r_w = 3$ sites and $r = 8$ sites.

model is limited by the relatively large distance of interaction. Here we suggest a simple way to improve the model. Take a given site O. The attractive force exerted by O reaches the 24 sites located at r sites from O, corresponding to the 24 directions of the lattice. One could improve the model by letting O interact with all the sites M such that $R_{min} < OM < R_{max}$ where R_{min} and R_{max} are the minumum and maximum range of interaction in our new model. In our new mode the interaction is both stronger and more isotropic.

Another ambitious program is to work out a multi-species 3D lattice gas model in order to study the changes of state. It is an interesting but complex model as it introduces diffusion process.

REFERENCES

Appert C. and Zaleski S. (1990) "A lattice gas with a liquid-gas transition", Phys. Rev. Lett. 64, 1-4.

Appert C., d'Humieres D. and Zaleski S. (1993) "Lattice gas with minimal attractive intractions", C.R. Acad. Sci. Paris 316, Série II, 569-574.

Appert C., Pot V. and Zaleski S. (1993) "Liquid-gas models on 2D and 3D lattices", preprint, to appear in the proceedings of NATO conference on Pattern Formation in Lattice-Gas Automata, June 1993.

Barnes C.J. and Allison G.B. (1983) "The distribution of D and O^{18} in dry soils 1-Theory", J. of Hydrology 60, 141-156.

Craig H. and Gordon L.I. (1965) "D and O^{18} variations in the ocean and the marine atmosphere", in Conference on Stable Isotopes in Oceanographic Studies and Paleotemperatures, edited by V. Lishi e Giglu, Laboratorio di Geologia Nucleare, Pisa, Italy, 9-130.

Fontes J.Ch., Yousfi M. and Allison G.B. (1986) "Estimation of long-term, diffuse groundwater discharge in the northern Sahara using stable isotope profiles in soil water", J. of Hydrology 86, 315-327.

Frisch U., Hasslacher B. and Pomeau Y. (1986) "Lattice-gas automata for the Navier-Stokes equation", Phys. Rev. Lett. 56, 1505-1508.

Frisch U., d'Humieres D., Hasslacher B., Lallemand P., Pomeau Y. and Rivet J.P. (1987) "Lattice-gas hydrodynamics in two and three dimensions", Complex Systems 1, 649-707.

Philip J.R. and De Vries D.A. (1957) "Moisture movement in porous materials under temperature gradient", Trans. Amer. Geophys. Un. 38, 222-232.

Pot V., Appert C., Melayah A., Rothman D.H. and Zaleski S. (1993) "Modelling water flow in unsaturated porous media by interacting lattice gas–cellular automata:

Evaporation", preprint.

Quénard D. and Sallee H. (1992) "Water-vapour adsorption and transfer in cement-based materials : a network simulation", Materials and Structures 25, 515-522.

Rocard Y. (1952) Thermodynamique, Masson et Cie, Paris.

Rose D.A.(1963) "Water movement in porous materials" "Part1 - Isothermal vapour transfer", Brit. J. Appl. Phys. 14, 256-262. "Part2 - The separation of the components of water movement", Brit. J. Appl. Phys. 14, 491-496.

Somers J.A. and Rem P.C. (1992) "Obtaining numerical results from the 3D FCHC-lattice gas", in Springer Proceedings on Physics, Workshop on Numerical Methods for the Simulation of Multi-Phase and Complex Flow, Springer-Verlag, 59-78.

MULTIPHASE FLOW SIMULATION WITH VARIOUS BOUNDARY CONDITIONS

Z. CHEN, R. E. EWING, and M. ESPEDAL
Department of Mathematics and
the Institute for Scientific Computation
Texas A&M University
College Station, TX 77843
USA

Multiphase flow simulation with various nonhomogeneous boundary conditions in groundwater hydrology and petroleum engineering is considered. The phase flow equations are given in a fractional flow formulation, i.e., in terms of a saturation and a global pressure. It is shown that most commonly used boundary conditions for groundwater hydrology and petroleum engineering problems can be incorporated into the pressure-saturation formulation.

INTRODUCTION

In petroleum reservoir simulation the governing equations that describe fluid flow are usually written in a fractional flow formulation, i.e., in terms of a saturation and a global pressure [5], [9]. The main reason for this fractional flow approach is that efficient numerical methods can be devised to take advantage of many physical properties inherent in the flow equations. However, this pressure-saturation formulation has not yet achieved wide application in groundwater hydrology. In petroleum reservoirs total flux type boundary conditions are conveniently imposed and often used, but in groundwater reservoirs boundary conditions are very complicated.

In this paper multiphase flow simulation with various nonhomogeneous boundary conditions in groundwater hydrology and petroleum engineering is considered. We show that most commonly encountered boundary conditions in groundwater reservoirs can be incorporated in the fractional flow formulation. A numerical method based on use of a mixed finite element method for the global pressure and a standard Galerkin method for the saturation is presented.

DIFFERENTIAL EQUATIONS

Let $\Omega \subset \mathbb{R}^d$, $d \leq 3$, be a porous medium. The usual equations describing two-phase flow

A. Peters et al. (eds.), Computational Methods in Water Resources X, 925–932.

in Ω are given by the mass balance equation and Darcy's law for each of the fluid phases

$$(1.1) \qquad \frac{\partial(\phi\rho_\alpha s_\alpha)}{\partial t} + \nabla \cdot (\rho_\alpha v_\alpha) = f_\alpha, \quad x \in \Omega, t > 0,$$

$$(1.2) \qquad v_\alpha = -\frac{kk_{r\alpha}}{\mu_\alpha}(\nabla p_\alpha - \rho_\alpha g), \qquad x \in \Omega, t > 0,$$

where $\alpha = w$ denotes the wetting phase (e.g. water), $\alpha = a$ indicates the nonwetting phase (e.g. air or oil), ϕ and k are the porosity and absolute permeability of the porous system, ρ_α, s_α, p_α, v_α, and μ_α are the density, saturation, pressure, volumetric velocity, and viscosity of the α-phase, f_α is the source/sink term, $k_{r\alpha}$ is the relative permeability of the α-phase, and g is the gravitational, downward-pointing, constant vector. Impose the customary property that the fluid fills the volume:

$$(1.3) \qquad s_a + s_w = 1.$$

Also, define the capillary pressure function p_c by

$$(1.4) \qquad p_c(s_w) = p_a - p_w.$$

For notational convenience, introduce the phase mobility function

$$(1.5) \qquad \lambda_\alpha = \frac{k_{r\alpha}}{\mu_\alpha}, \quad \alpha = w, a.$$

Now, substitute (1.3)–(1.5) into (1.1) and (1.2) to obtain the usual two-pressure equation formulation. The most commonly encountered boundary conditions for the two-pressure equations are of first-type, second-type, third-type, and "well" type [1], [5]. Let $\partial\Omega$ be a set of three disjoint regions Γ_i, $i = 1, 2, 3$, and let $\Gamma_3 = \cup_j \Gamma_{3,j}$ where each $\Gamma_{3,j}$ is connected. Then we consider for $\alpha = w, a$ and $s = s_w$,

$$(1.6) \qquad p_\alpha = p_{\alpha D}(x, t), \qquad\qquad\qquad x \in \Gamma_1, t > 0,$$

$$(1.7) \qquad v_\alpha \cdot \nu + b_\alpha(x, t, s)p_\alpha = g_\alpha(x, t, s), \quad x \in \Gamma_2, t > 0,$$

$$(1.8a) \qquad \int_{\Gamma_{3,j}} (v_w + v_\alpha) \cdot \nu = g_j(t), \qquad x \in \Gamma_{3,j}, t > 0,$$

$$(1.8b) \qquad p_\alpha = p_{\alpha D}(x, t) + d_j(t), \qquad x \in \Gamma_{3,j}, t > 0,$$

where $p_{\alpha D}$, b_α, g_α, and g_j are given functions, d_j is an arbitrary scaling constant, and ν is the outer unit normal to $\partial\Omega$. The initial condition is given as

$$(1.9) \qquad s_w(\cdot, 0) = s_w^0, \quad x \in \Omega.$$

The model is now completed.

Pressure-saturation formulation

To devise our numerical method, as mentioned in the introduction we rewrite (1.1) and (1.2) in a pressure-saturation formulation. For this, define the global pressure [5]

$$
(1.10) \qquad
\begin{aligned}
p &= \frac{1}{2}(p_w + p_a) + \frac{1}{2} \int_{s_c}^{s} \frac{\lambda_a - \lambda_w}{\lambda} \frac{dp_c}{d\xi} d\xi \\
&= p_w + \int_0^{p_c(s)} \left(\frac{\lambda_a}{\lambda}\right) \left(p_c^{-1}(\xi)\right) d\xi,
\end{aligned}
$$

where $\lambda = \lambda_w + \lambda_a$ and $p_c(s_c) = 0$, and the total velocity

$$
(1.11) \qquad v = -k\lambda \left(\nabla p - G_\lambda\right),
$$

where

$$
G_\lambda = \frac{\lambda_w \rho_w + \lambda_a \rho_a}{\lambda} g.
$$

Then it can be easily seen that

$$
(1.12\text{a}) \qquad v_w = q_w v + k\lambda_a q_w \nabla p_c - k\lambda_a q_w \delta\rho g,
$$
$$
(1.12\text{b}) \qquad v_a = q_a v - k\lambda_w q_a \nabla p_c + k\lambda_w q_a \delta\rho g,
$$

where $q_\alpha = \lambda_\alpha/\lambda$, $\alpha = w, a$, and $\delta\rho = \rho_a - \rho_w$. Consequently,

$$
(1.13) \qquad v = v_w + v_a.
$$

Add (1.1) with $\alpha = w$ and $\alpha = a$ to give the pressure equation

$$
(1.14) \qquad \nabla \cdot v = -\frac{\partial \phi}{\partial t} - \sum_{\alpha=w}^{a} \frac{1}{\rho_\alpha}\left(\phi s_\alpha \frac{\partial \rho_\alpha}{\partial t} + v_\alpha \cdot \nabla \rho_\alpha - f_\alpha\right).
$$

Substitute (1.12a) into (1.1) with $\alpha = w$ to obtain the saturation equation

$$
(1.15) \qquad
\begin{aligned}
\phi \frac{\partial s_w}{\partial t} &+ \nabla \cdot \left(q_w v + k\lambda_a q_w(\nabla p_c - \delta\rho g)\right) \\
&= -s_w \frac{\partial \phi}{\partial t} - \frac{1}{\rho_w}\left(\phi s_w \frac{\partial \rho_w}{\partial t} + v_w \cdot \nabla \rho_w - f_w\right).
\end{aligned}
$$

The capillary diffusion term $D(s)$ in this saturation equation is clearly defined by

$$
D(s) = -k\lambda_a q_w \frac{dp_c}{ds}.
$$

We now have the pressure-saturation equations (1.11), (1.14), and (1.15). Let $\Gamma_{p,i} = \Gamma_i$, $i = 1, 2, 3$, $\Gamma_{s,2} = \Gamma_2$, and $\Gamma_{s,1} = \Gamma_1 \cup \Gamma_3$. Then the boundary conditions for the pressure-saturation equations become

$$(1.16) \qquad p = p_D(x,t), \qquad\qquad\qquad\qquad\qquad x \in \Gamma_{p,1},\, t > 0,$$

$$(1.17) \qquad v \cdot \nu + b(x,t,s)p = G(x,t,s), \qquad\qquad x \in \Gamma_{p,2},\, t > 0,$$

$$(1.18a) \qquad \int_{\Gamma_{p,3,j}} v \cdot \nu = g_j(t), \qquad\qquad\qquad\qquad x \in \Gamma_{p,3,j},\, t > 0,$$

$$(1.18b) \qquad p = p_D(x,t) + d_j(t), \qquad\qquad\qquad x \in \Gamma_{p,3,j},\, t > 0,$$

$$(1.19) \qquad s = s_D(x,t), \qquad\qquad\qquad\qquad\qquad x \in \Gamma_{s,1},\, t > 0,$$

$$(1.20) \qquad (q_w v + k\lambda_a q_w (\nabla p_c - \delta \rho g)) \cdot \nu$$
$$\qquad\qquad\qquad + b_w(x,t,s)p = G_w(x,t,s), \quad x \in \Gamma_{s,2},\, t > 0,$$

where p_D and s_D are the transforms of p_{wD} and p_{aD} by (1.10) and (1.4), and

$$b = b_w + b_a,$$

$$G = g_w + g_a - b_a p_c + b \int_0^{p_c(s)} q_a\left(p_c^{-1}(\xi)\right) d\xi,$$

$$G_w = g_w + b_w \int_0^{p_c(s)} q_a\left(p_c^{-1}(\xi)\right) d\xi.$$

The initial condition is the same as in (1.9).

Petroleum reservoirs

The flow of two incompressible fluids (e.g. water and oil) in a porous medium Ω has been extensively studied by petroleum engineers for many decades. In this case the pressure-saturation equations (1.11), (1.14), and (1.15) reduce to the following simplified equations:

$$(1.21) \qquad \nabla \cdot v = -\frac{\partial \phi}{\partial t} + f, \qquad\qquad\qquad\qquad x \in \Omega,\, t > 0,$$

$$(1.22) \qquad v = -k\lambda\left(\nabla p - G_\lambda\right), \qquad\qquad\qquad x \in \Omega,\, t > 0,$$

$$(1.23) \qquad \phi \frac{\partial s_w}{\partial t} + \nabla \cdot (q_w v + k\lambda_o q_w(\nabla p_c - \delta \rho g)) = -s_w \frac{\partial \phi}{\partial t} + \tilde{f}_w, \quad x \in \Omega,\, t > 0,$$

where $f = f_w/\rho_w + f_o/\rho_o$ and $\tilde{f}_w = f_w/\rho_w$. We remark that $\frac{\partial \phi}{\partial t}$ is quite small, and is usually neglected. Typical examples of the relative permeability functions $k_{r\alpha}$, $\alpha = w, o$, the capillary pressure function p_c, the fractional flow function q_w, and the capillary diffusion function D for an oil-water system are plotted in Figure 1.

Groundwater hydrology

We now consider an air-water system where the water is assumed to be incompressible, but the air is supposed to be compressible. Furthermore, the air density is assumed to be a function of air pressure. Then we see from (1.10) that

$$\frac{1}{\rho_a}\frac{\partial \rho_a}{\partial t} = c_a\left(\frac{\partial p}{\partial t} + q_w\frac{\partial p_c}{\partial t}\right),$$

$$\frac{1}{\rho_a}\nabla \rho_a = c_a\left(\nabla p + q_w\nabla p_c\right),$$

where compressibility c_a is defined by

$$(1.24) \qquad c_a = \frac{1}{\rho_a}\frac{d\rho_a}{dp_a}.$$

Apply these equations in (1.11), (1.14), and (1.15) to obtain

$$(1.25) \quad s_a c_a\frac{dp}{dt} + \nabla\cdot v = -\frac{\partial \phi}{\partial t} - s_a c_a q_w\frac{dp_c}{dt} + f, \qquad\qquad x\in\Omega,\, t > 0,$$

$$(1.26) \quad v = -k\lambda\left(\nabla p - G_\lambda\right), \qquad\qquad x\in\Omega,\, t > 0,$$

$$(1.27) \quad \phi\frac{\partial s_w}{\partial t} + \nabla\cdot\left(q_w v + k\lambda_a q_w(\nabla p_c - \delta\rho g)\right) = -s_w\frac{\partial \phi}{\partial t} + \tilde{f}_w, \quad x\in\Omega,\, t > 0,$$

where

$$\frac{d}{dt} = \phi\frac{\partial}{\partial t} + \frac{v_a}{s_a}\cdot\nabla.$$

Note that the pressure equation is a parabolic equation in the present situation. Thus we need the initial condition

$$p(\cdot, 0) = p^0, \quad x\in\Omega.$$

An example of the air density function is given by the relation [13]

$$\rho_a = \rho_{0a}\left(1 + \frac{p_a}{p_{0a}}\right),$$

where ρ_{0a} is the density of the air phase at the pressure p_{0a}. Typical examples of the relative permeability functions $k_{r\alpha}$, $\alpha = w, a$, the capillary pressure function p_c, the fractional flow function q_w, and the capillary diffusion function D for an air-water system are plotted in Figure 2. The initial and boundary conditions in (1.9) and (1.16)–(1.20) remain the same in the two situations above.

FINITE ELEMENT METHOD

We now develop a finite element approximation procedure for numerically solving (1.21)–(1.23) and (1.25)–(1.27). We only consider the latter case, i.e., an air-water system; the

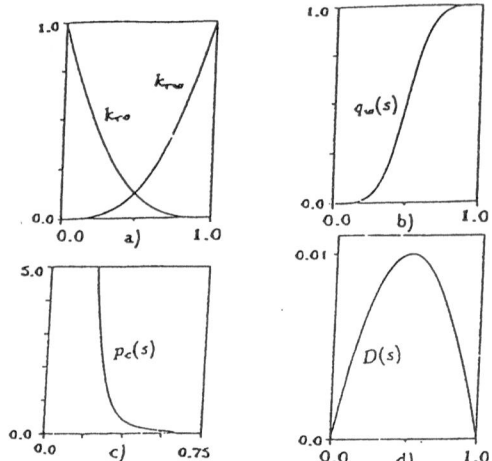

Fig. 1. a) Relative permeability
b) fractional flow function c) capillary pressure
d) capillary diffusion function for an oil-water system.

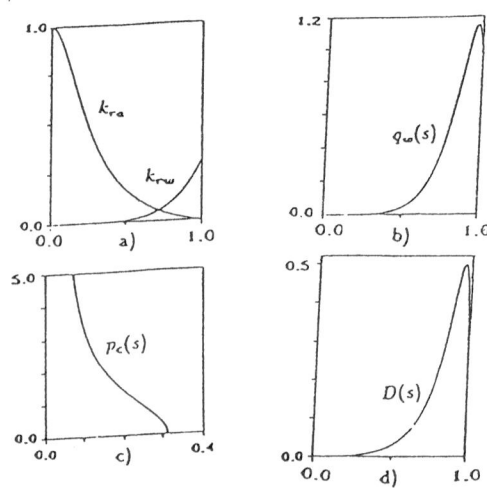

Fig. 2. a) Relative permeability
b) fractional flow function c) capillary pressure
d) capillary diffusion function for an air-water system.

former case is simpler. Let $(\cdot, \cdot)_R$ denote the $L^2(R)$-inner product (we omit R if $R = \Omega$), and set $S = \{\psi \in H^1(\Omega) : \psi|_{\Gamma_{s,1}} = 0\}$. For $0 < h_p < 1$ and $0 < h < 1$, let T_{h_p} and T_h be quasiregular partitions of Ω. For each $R \in T_{h_p}$, let $V_h(R) \times W_h(R)$ denote some standard mixed finite element space for second order elliptic problems (see, e.g., [2], [3], [4], [6], [10], and [12]). Then we define

$$V_h = \{\xi \in (L^2(\Omega))^d : \xi|_R \in V_h(R) \text{ for each } R \in T_{h_p}\},$$
$$W_h = \{\psi \in L^2(\Omega) : \psi|_R \in W_h(R) \text{ for each } R \in T_{h_p}\},$$
$$L_h = \left\{r \in L^2\left(\bigcup_{e \in \partial T_{h_p}} e\right) : r|_e \in V_h \cdot \nu|_e \text{ for each } e \in \partial T_{h_p}\right\}.$$

Finally, let $S_h \subset S$ be a standard C^0 finite element space [7] associated with T_h. Our finite element method is formulated as follows. The mixed finite element solution of the pressure equation is $\{v_h, p_h, l_h\} : (0, \infty) \to V_h \times W_h \times L_h$ satisfying

$$\left((1 - s_h)c_a(s_h, p_h)\frac{dp_h}{dt}, \psi\right) + \sum_{R \in T_{h_p}} (\nabla \cdot v_h, \psi)_R$$
$$= -\left(\frac{\partial \phi}{\partial t} + (1 - s_h)c_a(s_h, p_h)\frac{dp_c(s_h)}{dt} - f(s_h, p_h), \psi\right), \quad \forall \psi \in W_h,$$

$$((k\lambda(s_h))^{-1}v_h, \xi) - \sum_{R \in T_{h_p}} \left((p_h, \nabla \cdot \xi)_R - (l_h, \xi \cdot \nu_R)_{\partial R \setminus \{\Gamma_{p,1} \cup \Gamma_{p,3}\}}\right)$$
$$= (G_\lambda(s_h, p_h), \xi) - (p_D, \xi \cdot \nu)_{\Gamma_{p,1}} - \sum_j (p_D + d_j, \xi \cdot \nu)_{\Gamma_{p,3,j}}, \quad \forall \xi \in V_h,$$

$$\sum_{R \in T_{h_p}} (v_h \cdot \nu_R, r)_{\partial R \setminus \Gamma_{p,1}} = (G(s_h) - b(s_h)p_h, r)_{\Gamma_{p,2}} + \sum_j (g, r)_{\Gamma_{p,3,j}}, \quad \forall r \in L_h,$$

$$p_h(\cdot, 0) = p_h^0, \quad x \in \Omega,$$

and the finite element solution of the saturation equation is $s_h : (0, \infty) \to S_h + s_D$ satisfying

$$\left(\phi\frac{\partial s_h}{\partial t}, \psi\right) - (q_w(s_h)v_h + k(\lambda_a q_w)(s_h)(\nabla p_c(s_h) - \delta\rho(s_h, p_h)g), \nabla\psi)$$
$$= -(G_w(s_h) - b_w(s_h)p_h, \psi)_{\Gamma_{s,2}} + \left(\tilde{f}_w - s_h\frac{\partial \phi}{\partial t}, \psi\right), \quad \forall \psi \in S_h,$$

$$s_h(\cdot, 0) = s_h^0, \quad x \in \Omega,$$

where p_h^0 and s_h^0 are some approximations in W_h and S_h of p^0 and s^0, respectively. We conclude with three remarks. First, while, for completeness, the standard finite element method is considered for the saturation equation here, due to its convection-dominatedness feature it can be solved using more efficient numerical approaches such as characteristic

Petrov-Galerkin methods based on operator splitting [8], transport diffusion methods [11], and other characteristic based methods. Second, the Lagrange multipliers over edges or faces are here used. The reasons for this are that the linear system arising from this unconstrained mixed formulation leads to a symmetric, positive definite system for the Lagrange multipliers, which can be easily solved, and that the boundary conditions (1.16)–(1.20) can be easily incorporated in this formulation. Finally, note that we have a coupled nonlinear system for the velocity v_h, the pressure p_h, and the saturation s_h. In general, the boundary data depend on the saturation, which makes the whole system even more difficult to solve. Our future work will be concentrated on development of computer programs based on the pressure-saturation formulation for physically reasonable data and on extension of the present techniques to three-phase flow.

ACKNOWLEDGEMENT

This work is partially supported by the Department of Energy under contract DE-ACOS-840R21400. Also, the authors thank Professor Michael Celia for his valuable suggestions.

REFERENCES

1. T. Arbogast, *Two-phase incompressible flow in a porous medium with various nonhomogeneous boundary conditions*, IMA Preprint Series #606 (1990).
2. F. Brezzi, J. Douglas, Jr., R. Durán, and M. Fortin, *Mixed finite elements for second order elliptic problems in three variables*, Numer. Math. **51** (1987), 237–250.
3. F. Brezzi, J. Douglas, Jr., M. Fortin, and L. D. Marini, *Efficient rectangular mixed finite elements in two and three space variables*, RAIRO **21** (1987), 581–604.
4. F. Brezzi, J. Douglas, Jr., and L. D. Marini, *Two families of mixed finite elements for second order elliptic problems*, Numer. Math. **47** (1985), 217–235.
5. G. Chavent and J. Jaffré, *Mathematical Models and Finite Elements for Reservoir Simulation*, North-Holland, Amsterdam, 1978.
6. Z. Chen and J. Douglas, Jr., *Prismatic mixed finite elements for second order elliptic problems*, Calcolo **26** (1989), 135–148.
7. P. Ciarlet, *The Finite Element Method for Elliptic Problems*, North-Holland, Amsterdam, 1978.
8. N. S. Espedal and R. E. Ewing, *Characteristic Petrov-Galerkin subdomain methods for two phase immiscible flow*, Comput. Methods Appl. Mech. Eng. **64** (1987), 113–135.
9. R. E. Ewing and M. A. Celia, *Multiphase flow simulation in groundwater hydrology and petroleum engineering,*, Computational Methods in Subsurface Hydrology, G. Gambolati, ed., Computational Mechanics Publications, Boston (1990).
10. J. C. Nedelec, *Mixed finite elements in R^3*, Numer. Math. **35** (1980), 315–341.
11. O. Pironneau, *On the transport-diffusion algorithm and its application to the Navier-Stokes equations*, Numer. Math. **38** (1982), 309–332.
12. P. A. Raviart and J. M. Thomas, *A mixed finite element method for 2nd order elliptic problems* (1977), Lecture Notes in Math. 606, Springer-Verlag, Berlin, 292–315.
13. J. Touma and M. Vauclin, *Experimental and numerical analysis of two-phase infiltration in a partially saturated soil*, Transport in Porous Media **1** (1986), 27–55.

MULTI-PHASE FLOW MODELING OF AIR INJECTION IN GROUND WATER

M.I.J. VAN DIJKE and S.E.A.T.M. VAN DER ZEE
Department of Soil Science and Plant Nutrition
Wageningen Agricultural University
P.O. Box 8005
6700 EC Wageningen
The Netherlands

We model air injection in ground water in a homogeneous three-dimensional axially symmetric domain, with emphasis to the developing steady state. Numerically, the mixed form of Richards equation for both incompressible thought phases is solved. Analytically, a similarity solution for the steady state is found assuming that water is immobile and air saturations are small. The resulting maximal air / water interface position in terms of the physical parameters compares favourably with the numerical approximation.

1 Introduction

Organic liquids (solvents, gasoline) present in ground water at residual saturation can not be removed by water flushing. Air injection may enhance microbial degradation and volatilization losses.

Air injection in the saturated zone is a flow process of two immiscible phases. In a homogeneous three-dimensional axially symmetric domain, Richards equations for air and water are solved using a finite element model with modified Picard iterations, where both air and water are considered as incompressible.

In this paper we assess whether it is reasonable to expect a steady state. Supported by an analytical solution, we furthermore characterize the steady state solution in terms of the width of the air cone and the air volume stored in this cone.

A. Peters et al. (eds.), Computational Methods in Water Resources X, 933–940.

2   Problem statement and numerical method

To model air injection we use Darcy's law for both air (a) and water (w), which in combination with the mass balance equations results in the mixed form of Richards equations [2, 3]:

$$\phi \frac{\partial S_f}{\partial t} + \nabla \cdot (\frac{K_{abs}\, k_{rf}}{\mu_f} \nabla (p_f + \rho_f\, g\, z)) = 0, \quad f = a, w, \tag{1}$$

where K_{abs} is soil absolute permeability, ϕ soil porosity, S_f effective fluid saturation, k_{rf} fluid relative permeability, μ_f fluid viscosity, p_f fluid pressure, ρ_f fluid density and g gravity. Both fluids are considered as incompressible.

The set of equations (1) is completed by the (known) constitutive relationships $S_a + S_w = 1$, $p_c = p_a - p_w$, $S_f = S_f(p_c)$ and $k_{rf} = k_{rf}(S_f)$ and is solved for the unknowns S_w, S_a, p_w and p_a.

For the relationships p_c, k_{rw} and k_{ra} we use

$$p_c(S_w) = \frac{\rho_w\, g}{\alpha} (S_w^{-\frac{1}{m}} - 1)^{1-m} \tag{2}$$

$$k_{rw}(S_w) = S_w^{\frac{1}{2}} (1 - (1 - S_w^{\frac{1}{m}})^m)^2 \tag{3}$$

$$k_{ra}(S_a) = S_a^{\frac{1}{2}} (1 - (1 - S_a)^{\frac{1}{m}})^{2m}, \tag{4}$$

where $m \in (0,1)$ and $\alpha > 0$ are scaling parameters [3, 4]. With a numerical model for two-phase flow we solve Equations (1).

Space is discretized by triangular finite elements and the time derivative is treated with a backward Euler method. The resulting algebraic equations are solved by the modified Picard method [2].

An axially symmetric domain with radial coordinate r and vertical coordinate z directed upward is considered. The grid size is most refined close to the source in the r-direction (13 nodes) and is equidistant in z (22 nodes). Injecting air through a filter of 1.0 m length and diameter 0.1 m, about 4.5 m below the water table requires a computational domain as shown in Figure 1: the boundary at $r = 0.05$ m is impermeable, except for the filter where the air velocity $q_{ar} = q_{in}$ is constant. The boundary at the base is impermeable and the remaining boundaries have a designated constant water based hydraulic head $H_f = \dfrac{p_f}{\rho_w\, g} + \dfrac{\rho_f}{\rho_w} z$, $f = a, w,$, such that air pressure is zero at 1.5 m above the water table. The hydraulic heads are also prescribed as initial values for the entire domain.

The flow problem is characterized by the domination of convection terms, due to large density and viscosity ratios of air and water and requires a careful numerical treatment. In addition the numerical solution takes a large CPU time.

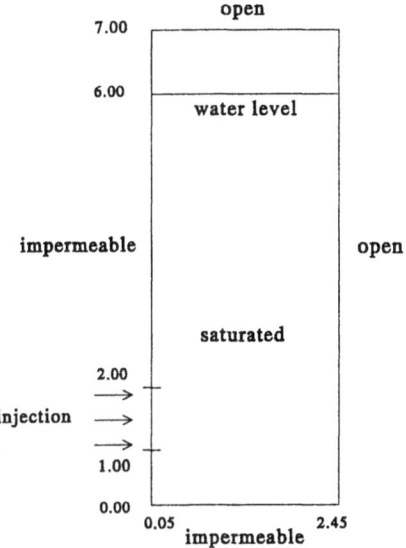

Figure 1: Domain for numerical treatment of Richards equations

3 An analytical steady state solution

Numerical computations (Section 4) indicate that water velocity is negligible in the steady state. Thus, Problem (1) reduces to a single-phase problem (air) that still involves capillary forces between air and water

$$p_c = p_a - p_w = p_a + \rho_w \, g \, z, \tag{5}$$

where the reference pressure for water is chosen zero. We define the injection rate $Q = q_{in} \, A$, where A is the filter surface. For an analytical assessment variables are made dimensionless by the redefinitions

$$r := \frac{r}{h} \qquad z := \frac{z}{h}, \qquad A := \frac{A}{h^2}$$
$$\underline{q}_a := \frac{q_a}{q_{in}}, \qquad p_c := \frac{p_c}{\rho_w \, g / \alpha},$$

where h is the distance between the water table and the centre of the injection filter. Furtermore, dimensionless gravity and capillary numbers are defined as

$$N_g := \frac{K_{abs} \, \Delta \rho \, g \, h^2}{\mu_a \, q_{in} \, A}, \quad N_c := \frac{K_{abs} \, \rho_w \, g \, h}{\mu_a \, q_{in} \, A \, \alpha}.$$

Hence, renaming $S := S_a$ and $k := k_{ra}$, we must solve

$$
\begin{cases}
\nabla \cdot k \, \nabla \left(N_c \, p_c - N_g \, z \right) = 0 & \text{for } r > \varepsilon \; z < 0 \\[2mm]
q_{ar} = 1 & \text{for } r = \varepsilon, \; -h_2 < z < -h_1 \\[2mm]
\dfrac{\partial S}{\partial r} = 0 & \text{for } r = \varepsilon, \; z < -h_2 \text{ and } -h_1 < z < 0 \\[2mm]
S = S^b & \text{for } r > \varepsilon, \; z = 0,
\end{cases}
\tag{6}
$$

where $h_2 - h_1$ and ε denote the dimensionless filter size and S^b is a prescribed saturation.

Next, the filter diameter ε is taken zero, such that the injection rate Q remains unchanged. We assume that convection dominates flow in the z-direction at a certain distance above the source

$$
|N_g \frac{\partial k}{\partial z}| \gg |N_c \frac{\partial}{\partial z} (k \frac{\partial p_c}{\partial z})|.
$$

Neglecting the upper boundary a semi-infinite domain $\{(r,z) \, | \, r > 0, \; z > z_0\}$, $0 < z_0 \le 1$ is defined. Numerical results indicate that air saturation is significantly smaller than 1. Using this, we approximate S as a function of k by

$$
S(k) \sim m^{\frac{4m}{1+4m}} \, k^{\frac{2}{1+4m}};
\tag{7}
$$

and using (7) the diffusion coefficient $D(k) = k \dfrac{d \, p_c}{d \, k}$ is approximated by

$$
D(k) \sim \frac{2 \, (1 - m)}{4m + 1} \, m^{\frac{m-1}{1+4m}} \, k^{\frac{2(1-m)}{1+4m}}.
\tag{8}
$$

Additionally, we define the constants

$$
\begin{aligned}
p &= \frac{2 \, (1 - m)}{1 + 4m}, \quad 0 < p < 2, \\[2mm]
C &= \frac{4m + 1}{2 \, (1 - m)} \, m^{\frac{1-m}{1+4m}} \frac{N_g}{N_c}, \quad C > 0.
\end{aligned}
$$

We are interested both in the relative permeability $k(r, z)$ and in the position of the free boundary $f(z)$, which separates regions where $k > 0$ and where $k \equiv 0$. $f(z)$ is finite because $p > 0$ (see [1]). Although a boundary condition at $z = z_0$ is not known, we have an additional integral condition following from mass balance considerations:

$$
2 \pi \int_0^{f(z)} r \, q_{az} \, dr = 2 \pi \, N_g \, A \int_0^{f(z)} r \, k \, dr = A \quad \text{for } z \ge z_0.
\tag{9}
$$

Hence, we look for a positive solution of the parabolic problem

$$
\begin{cases}
C \dfrac{\partial k}{\partial z} = \dfrac{1}{r} \dfrac{\partial}{\partial r}(r\,k^{\,p}\,\dfrac{\partial k}{\partial r}) & \text{for } 0 < r < f(z),\ z > z_0 \\[2mm]
\dfrac{\partial k}{\partial r}(0, z) = 0 & \text{for } z > z_0 \\[2mm]
k(f(z), z) = 0 & \text{for } z > z_0 \\[2mm]
\displaystyle\int_0^{f(z)} r\,k\,dr = \dfrac{1}{2\pi N_g} & \text{for } z > z_0.
\end{cases}
\tag{10}
$$

A similarity solution of (10) is obtained following [1], where the level z_0 and the corresponding $f_0 = f(z_0)$ are in practice unknown. However, if we assume a point source at $(r, z) = (0, 0)$ that satisfies the mass balance condition (9), the similarity solution is valid for all $z > 0$, irrespective of the position of z_0. We use this as an approximate solution for the full steady state problem (6). Omitting details of the derivation and choosing $z_0 = 1$ we get

$$
k(r, z) = \frac{p+1}{p\,\pi\,N_g} \frac{1}{f^2(z)} \left(1 - \frac{r^2}{f^2(z)}\right)^{\frac{1}{p}}.
\tag{11}
$$

For the position of the free boundary we derive

$$
f(z) = f_0\, z^{\frac{1}{2(p+1)}}, \quad z \geq 0
\tag{12}
$$

and $f_0 = f(1)$, the approximate dimensionless air cone width at the water table is given by

$$
f_0 = \left(\frac{4}{C\,(\pi N_g)^p}\right)^{\frac{1}{2(p+1)}} \left(\frac{p+1}{p}\right)^{\frac{1}{2}}.
\tag{13}
$$

4 Results

Equations (1) are solved for p_f and S_f, $f = a, w$, simultaneously. For all computations we took

$$
\begin{array}{lll}
K_{abs} = 5.30 \cdot 10^{-11}\ \mathrm{m^2} & \rho_a = 1.24 & \mathrm{kg\,m^{-3}} \\
\mu_a = 1.77 \cdot 10^{-5}\ \mathrm{Pa\,s} & \rho_w = 1.00 \cdot 10^3 & \mathrm{kg\,m^{-3}} \\
\mu_w = 1.30 \cdot 10^{-3}\ \mathrm{Pa\,s} & g = 9.8 & \mathrm{m\,s^{-2}}.
\end{array}
$$

For α, m and q_{in} we took the values shown in Table 1, where also the dimensionless parameter f_0 (Equation (13)) is calculated ($h = 4.5\,\mathrm{m}$, $A = 0.314\,\mathrm{m^2}$).
Typical contour plots for air saturations are given in Figure 2 for case 5.
In Figure 3 total stored air volume is set out against time for cases 3, 5 and 6.

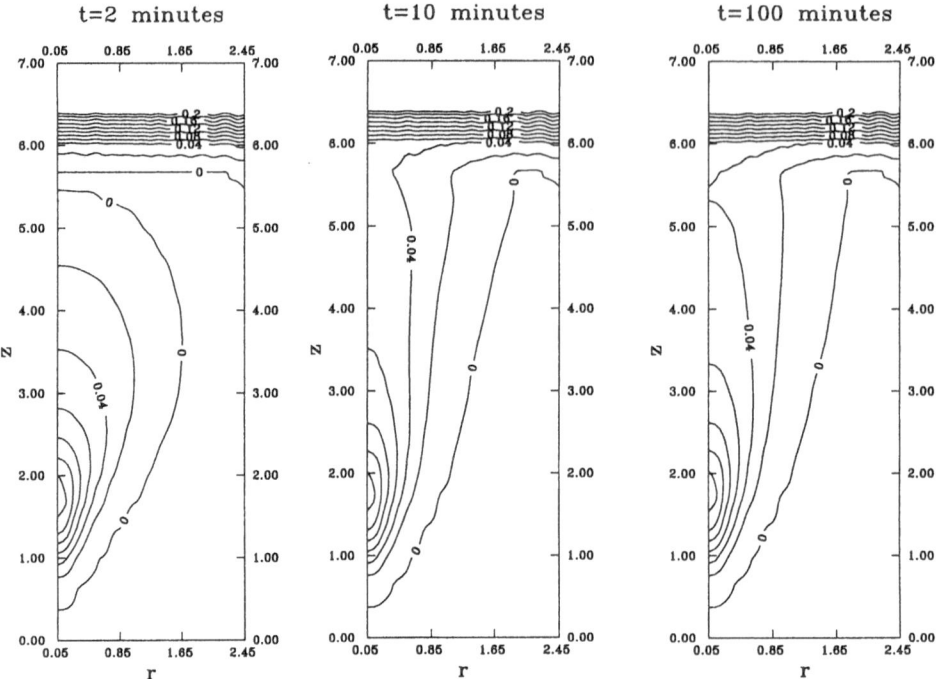

Figure 2: Air saturation contourplots

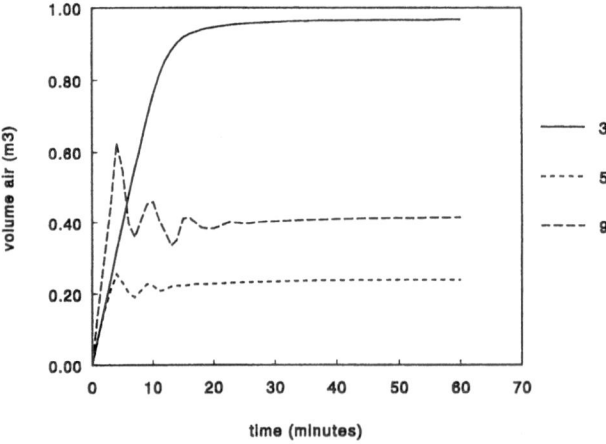

Figure 3: Stored air volume

case	α (m^{-1})	m	q_{in} $(\mathrm{m\,s}^{-1})$	f_0
1	0.500	0.750	$4.42 \cdot 10^{-3}$	1.04
2	1.00	0.500	$4.42 \cdot 10^{-3}$	0.538
3	1.00	0.750	$4.42 \cdot 10^{-3}$	0.763
4	2.00	0.750	$4.42 \cdot 10^{-3}$	0.561
5	2.00	0.500	$4.42 \cdot 10^{-3}$	0.415
6	2.00	0.667	$4.42 \cdot 10^{-3}$	0.513
7	2.00	0.800	$4.42 \cdot 10^{-3}$	0.589
8	2.00	0.500	$2.21 \cdot 10^{-3}$	0.380
9	2.00	0.500	$8.84 \cdot 10^{-3}$	0.452
10	2.00	0.500	$1.32 \cdot 10^{-2}$	0.476

Table 1: Input parameters and f_0 for different cases

For some cases oscillations occur, which may be explained by physical instabilities: when injected air reaches the water table an amount of stored air much larger than the injection volume leaves the saturated zone and this process is repeated one or more times. Numerical problems occur for large q_{in} and α: the storage rate becomes larger than the injection rate, which is physically unacceptable. Besides, computation time is very large: for example on a 486 DX 50 MHz p.c. CPU time is 1/2 to 3 days for realistic cases.

From Figure 3 it is seen that for all cases a steady state is reached within about 1 hour. The numerically obtained dimensionless steady air cone width (maximal value of r on the contourline for $S=0.001$) is shown versus f_0 in Figure 4. Along the α-curve only α is varied (cases 1, 3, 4), along the m-curve m (cases 5, 4, 6, 7) and along the q_{in}-curve q_{in} (cases 8, 5, 9, 10); the separate point denotes case 2. Cases 1 and 3 are obtained by extrapolation, because contourlines went out of the domain. For almost all cases f_0 gives a good measure for the numerically obtained air cone width. Note that f_0 is calculated on the semi-infinite domain; for the real domain the air cone width must be smaller in view of the decrease of the volume rate when air reaches the upper boundary.

Acknowledgement

Part of this project was sponsored by the National Computing Facilities Foundation (NCF) for the use of super computing facilities, with financial support from the Netherlands Organization for Scientific Research (NWO project NLS 61-251) and by the Netherlands Integrated Soil Research Programme (PCTB project 5029).

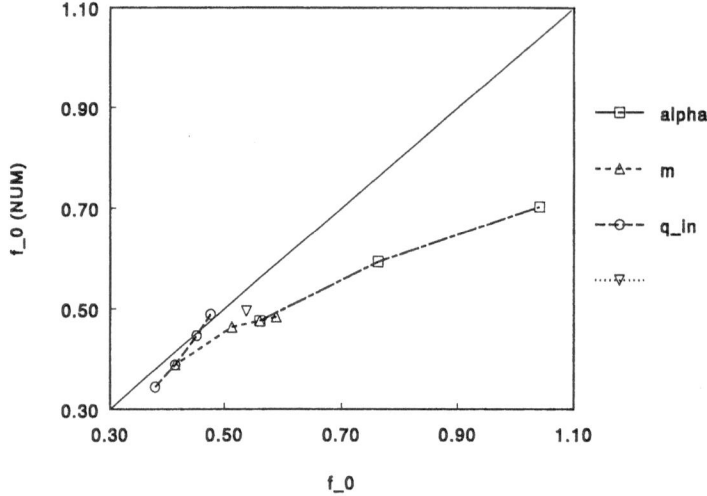

Figure 4: Air cone width against f_0

References

[1] Aronson, D.G. (1985) "The porous medium equation" in A. Dold & B. Eckmann (eds.), Lecture Notes in Mathematics, Vol. 1224 : Nonlinear Diffusion Processes, Springer Verlag, Berlin, pp. 1-46.

[2] Celia, M.A., Bouloutas, E.T. & Zarba, R.L. (1986) "A general mass-conservative numerical solution for the unsaturated flow equation", Water Resources Research 26, 1483-1496.

[3] Kaluarachchi, J.J. & Parker, J.C. (1989) "An efficient finite element method for modeling multiphase flow", Water Resoures Research, 25, 43-54.

[4] Parker, J.C., Lenhard, R.J. & Kuppusamy, T. (1987) "A parametric model for constitutive properties governing multiphase flow in porous media", Water Resources Research 23, 618-624.

SIMULATION AND INTERPRETATION OF MULTIPHASE PROCESSES IN POROUS AND FRACTURED-POROUS MEDIA

R. HELMIG; H. KOBUS and C. BRAUN
Institut für Wasserbau
University of Stuttgart
Pfaffenwaldring 61, 70550 Stuttgart
Germany

A numerical model concept for simulating multiphase flow processes is presented. The conceptual model involves the approximation of heterogeneous porous and fractured porous media by application of one–, two– and threedimensional elements in space that can be combined arbitrarily in a single model. The mathematical formulation of multiphase flow is valid for both saturated and unsaturated regions, so that dynamic front propagation behaviour as well as different geology-related retention characteristics can be accounted for in the model. For every Finite Element model problems arise when capillary pressure effects can be neglected and the governing equations are of hyperbolic type. To overcome these difficulties, the differential equations were formulated in terms of pressure and saturation as primary variables. To provide an accurate representation of the resulting saturation-front velocities and to avoid numerical oscillations a modified Petrov-Galerkin method was employed. To verify the code, the BUCKLEY–LEVERETT problem was solved, and together with an adaptive Finite Element algorithm excellent results were obtained.

1   INTRODUCTION

Contamination of ground water resources by hazardous substances has become an issue of increasing interest. Petroleum products and halogenated hydrocarbon solvents, ubiquitous in our industrialized society, are among the most serious threats to ground water systems. These organic compounds (non-aqueous phase liquids (NAPL's)) are characterized by their immiscibility with water and air, and low, but toxicologically significant air– and water–phase solubilities. In situations where water, air and NAPL phases occupy the pore space in the subsurface, complex muliphase flow systems are generated.

Modeling of multiphase flow systems involves a number of distinct steps, proceeding from conceptual to mathematical to numerical models. The first task is to develop a conceptual model of the flow system. Based on empirical observations and accepted scientific principles, the conceptual model sets forth general notions about the physical and geometric makeup of the system and its constituents, the important flow and transport processes, and the nature of expected perturbations and constraints. Subsequently a mathematical model can be developed.

A. Peters et al. (eds.), Computational Methods in Water Resources X, 941–949.

Perhaps the most critical step in a successful modelling effort is the development of the conceptual model. This must strike a proper balance between the complexity of a subsurface system, that will always be only incompletely known, and the level of detail required to address specific engineering issues. The analyst also needs to come to grips with the broad range of space and time scales that may play a role in system definition and performance.

A major issue in numerical modelling is the credibility of computer programs (codes) and simulation results. This is of special importance in technical areas that involve public acceptance. To develop credibility for model predictions, "code verification" exercises are performed. There one seeks to establish, by comparison with independently (often analytically) derived results, that a numerical simulation code does indeed provide accurate solutions for the problem that it was designed to solve. The much more ambitious and difficult task of process interpretation (validation) aims at demonstrating that a numerical model provides a valid representation of physical reality. Verification and process interpretation can only be performed for specific limited conditions; one very obvious limitation being the space and time scales over which the behaviour of a subsurface flow system can be evaluated.

The design and the accurate modelling of the multiphase flow processes e.g. of a subsurface remediation scheme require accurate data of site specific parameters like properties of the porous medium and the present fluids. Also the interactions of the NAPLs with the matrix and with other phases (e.g. water and air) are represented in the relative permeability- and capillary pressure-saturation functions and are important basic inputs although difficult to establish.

The most promising aspects to overcome these problems are offered by well controlled laboratory experiments at different sizes from column to technical scales – such as e.g. the **VEGAS**program (Research Facility for Subsurface Remediation) (Fig. 1) KOBUS ET. AL. (1993) [8].

2   GOVERNING EQUATIONS

Multiphase flow in porous or fractured–porous media is described by a set of nonlinear partial differential equations of two basic categories: the material balance or continuity equation (conservation law) and the equation of motion (Darcy's law). For multiphase fluid flow, these equations must hold for each phase. The conservation law for each phase α is given by a hyperbolic equation which relates porosity, velocity, saturation, density and mass flux. Darcy's law for each phase α is given by an elliptic equation which relates the physical properties of the fluids and pore space with the driving forces of pressure differences and changes in elevation. The multiphase *pressure* differential equation, including the continuity equation combined with Darcy's law, is widely applied for formulating hydrological problems [3].

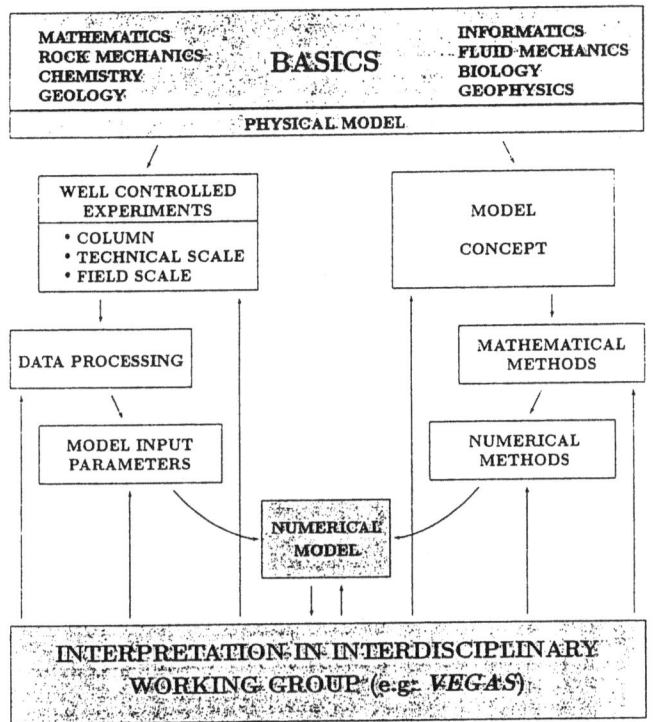

Figure 1: Interdisciplinary cooperation of physical and numerical modelers

$$L_1(p_\alpha) :=$$
$$n\varrho_\alpha \frac{\partial S_\alpha}{\partial t} + S_\alpha n \frac{d\varrho_\alpha}{dp_\alpha} \frac{\partial p_\alpha}{\partial t} - \mathrm{div}\left\{ k_{r\alpha} \frac{\varrho_\alpha}{\mu_\alpha} K \cdot (\mathbf{grad}\, p_\alpha - \varrho_\alpha g) \right\} - \varrho_\alpha q_\alpha = 0 \quad (1)$$

where n denotes the porosity, K is the permeability tensor, ϱ_α, μ_α, p_α, S_α and q_α denote density, viscosity, pressure, saturation and volumetric production rate of phase α respectively. When several fluid phases are present simultaneously in the void space of a porous medium or in a fracture, the presence of any one of these phases will interfere with the flow of all the others. Quantitatively, this mobility reduction is expressed in terms of the relative permeability factor $k_{r\alpha}$. To solve for the unknowns p_α and S_α, more equations are needed. One is given by the fact that the sum of the volume fractions at any point must add to unity:

$$\sum_{\alpha=1}^{n_{phas}} S_\alpha = 1$$

and the other is given as a relationship for capillary pressures as a function of saturation:

$$p_{ca\psi}(S_\alpha) = p_\psi - p_\alpha \qquad \psi = 1, 2, \ldots, n_{phas}; \ \psi \neq \alpha.$$

The system is nonlinear because $k_{r\alpha}$ and p_c are functions of the saturation S_α. The shapes of these two – often nonlinear – functions are important in determining the nature of the system of equations. This system has a mixed nature and may even change type (elliptic - parabolic - hyperbolic) in certain regions.

From the empirically determined relationship between capillary pressure and saturation, it is observed that the capillary pressure decreases in a strictly monotonic manner with decreasing saturation. An inverse function of the form $S_\alpha = S_\alpha(p_{ca\psi})$ therefore exists. A precondition for the solution of the system of equations is that the capillary pressure $p_{ca\psi} \neq 0$. In the case of fractures (fault zones), the capillary pressure $p_{ca\psi}$ is very small or even equal to zero. The above formulation is therefore not suitable as a mathematical basis for investigating highly heterogeneous media.

A *pressure/saturation* formulation involving the unknowns $p_1, S_2, S_3, \ldots, S_{n_{phas}}$, which is valid for an arbitrary capillary pressure, is presented as follows:

$$\mathbf{grad}\, p_\alpha \ = \ \mathbf{grad}\,(p_1 + \sum_{\Psi=2}^{n_{phas}} p_{c1\alpha}) \ = \ \mathbf{grad}\, p_1 + \sum_{\Psi=2}^{n_{phas}} \frac{dp_{c1\alpha}}{dS_\Psi}\, \mathbf{grad}\, S_\Psi \qquad (2)$$

and

$$\frac{\partial S_1}{\partial t} \ = \ \frac{\partial}{\partial t}\left(1 - \sum_{\Psi=2}^{n_{phas}} S_\Psi\right) \ = \ -\sum_{\Psi=2}^{n_{phas}} \frac{\partial S_\Psi}{\partial t} \qquad (3)$$

where the phase $\alpha = 1$ represents the fluid phase with the highest affinity to the matrix.

This yields the following expression for phase $\alpha = 1$:

$$L_1(p_1, S_\Psi) \ := \ -\mathrm{div}\left\{k_{r1}\frac{\varrho_1}{\mu_1}K \cdot \mathbf{grad}\, p_1\right\} + \mathrm{div}\left\{\varrho_1^2\frac{k_{r1}}{\mu_1}K \cdot g\right\}$$
$$+ n\left(1 - \sum_{\Psi=2}^{n_{phas}} S_\Psi\right)\frac{d\varrho_1}{dp_1}\frac{\partial p_1}{\partial t} - n\varrho_1\sum_{\Psi=2}^{n_{phas}} \frac{\partial S_\Psi}{\partial t} - \varrho_1 q_1 = 0 \qquad (4)$$

and for the remaining phases $\alpha \geq 2$:

$$L_\alpha(p_1, S_\Psi) \ := \ -\mathrm{div}\left\{k_{r\alpha}\frac{\varrho_\alpha}{\mu_\alpha}K \cdot \mathbf{grad}\, p_1\right\} + \mathrm{div}\left\{\varrho_\alpha^2\frac{k_{r\alpha}}{\mu_\alpha}K \cdot g\right\}$$

$$+nS_\alpha \frac{d\varrho_\alpha}{dp_\alpha} \frac{\partial p_1}{\partial t} + nS_\alpha \frac{d\varrho_\alpha}{dp_\alpha} \sum_{\Psi=2}^{n_{phas}} \frac{dp_{c1\alpha}}{dS_\Psi} \frac{\partial S_\Psi}{\partial t} + n\varrho_\alpha \frac{\partial S_\alpha}{\partial t}$$

$$- \text{div}\left\{ k_{r\alpha} \frac{\varrho_\alpha}{\mu_\alpha} K \cdot \sum_{\Psi=2}^{n_{phas}} \frac{dp_{c1\alpha}}{dS_\Psi} \, \text{grad}\, S_\Psi \right\} - \varrho_\alpha q_\alpha = 0. \tag{5}$$

Equations 4 and 5 represent a coupled dynamic system of differential equations which describes the simultaneous transport of two or more immiscible fluids in an unsaturated or saturated porous medium. Due to the nonlinear dependence of the capillary pressure and the relative permeability on the degree of saturation, the system of equations behaves in a highly nonlinear manner. This is further intensified by the fact that the physical parameters (such as relative permeability and degree of saturation) as well as the application of the law of motion in the matrix exhibit large differences between various parts such as fracture fault zones or systems with e.g. sand lenses.

3 NUMERICAL REALIZATION

The system of equations presented is solved using the Finite Element Method [2]. In discretizing the transient problem, it is appropriate to formulate the time discretization on the basis of an iterative concept. An implicit two-point algorithm developed for this purpose is incorporated into the Newton-type iterative concept for the consistently linearized multiphase problem. The time integration algorithm presented here employs a central difference scheme with a weighting factor of $\theta = 0,5$, in which the unknowns p_1 and S_Ψ are solved for a time step of Δt (CRANK-NICHOLSON method). With $\theta = 1$, the implicit EULER method (backward difference method) is obtained. In order to describe the multiphase problem in fractured porous media, shape functions are employed within the framework of a *standard isoparametric concept* for approximating the coordinate field (X), the pressure field of the first phase (p_1), and the saturation field (S_Ψ) of phase Ψ. The degrees of freedom of the shape functions are identified by means of discrete nodal coordinates, nodal pressures and nodal saturations.

In order to describe various complicated geological structures (e.g. flow channels, fractures and rock matrix), it is necessary to employ arbitrary combinations of finite elements of different dimensions. Moreover, the finite elements must allow a correct description of the problem defined by the multiphase flow processes.

The modelling system consists of one–, two– and three–dimensional elements with (multi-) linear shapefunctions applied to line elements, plane isoparametric quadrilateral elements and isoparametric hexahedral elements.

As the solutions are computed for each element in turn, the geological parameters may vary from one element to the next. The afore-mentioned combinations of different element types may be applied in a single model. The spatial orientation of the elements is arbitrary in order to simulate complex geological structures [7].

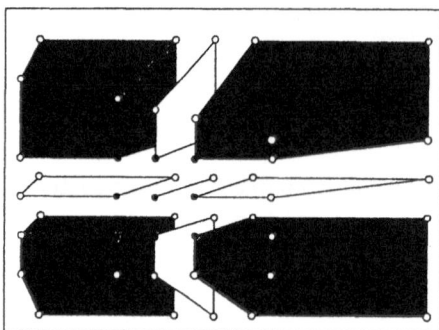

Figure 2: Arbitrary coupling of 1-D, 2-D and 3-D elements

Formally speaking, the coupled, highly nonlinear multiphase equations (Eqns. (4) and (5)) are of the parabolic type when the capillary pressure gradient is significant. If however this gradient becomes very small in relation to the size of the system to be solved, the equations become hyperbolic [4]. The numerical solution obtained from the finite difference and the STANDARD GALERKIN finite element formulation (BUBNOV–GALERKIN method) leads to invalid solutions, particularly when saturation distributions are computed for small capillary pressure effects. These solutions are characterized by physically unjustifiable oscillations of phase shifts as well as an incorrect simulation of front velocities.

In order to ensure convergence of the numerical solution, a modified PETROV–GALERKIN method [2] was developed in which the test functions are up to two polynomial degrees higher than the base function (shape function). In the following, the upwinding factors for the quadratic and cubic weighting functions are denoted by α and β, respectively. As a supplement to this, a *lumped* finite element formulation was also prepared. This results in a mathematical-numerical formulation which is specially tailored so that the transient multiphase algorithm can be improved by modifying the test function in time and space, thereby guaranteeing convergence.

This type of approach was first proposed by Westerink and Shea (1989) [6] for the predominantly advective transport problem.

4   VERIFICATION

The standard method for verifying multiphase processes in the absence of capillary pressure effects represents the Buckley and Leverett (1942) [1] problem, which describes the unsteady displacement of oil by water in a one dimensional, horizontal system. The important point is that the absence of a capillary pressure gradient leads to a formulation of shock in the saturation profile. The following investigations were carried out using 1-D elements. The saturation of the wetting phase (water) and the pressure of the nonwetting phase (oil) were prescribed as boundary conditions on the left-hand boundary, with assigned values of $S_w = 0,8$ and $p_n = 2 \cdot 10^5 \, [Pa]$, respectively. A mass flow rate of $m_n = 1,505 \cdot 10^{-3} \left[\frac{kg}{s}\right]$ leaving the system was chosen as the boundary condition on the right-hand boundary.

In order to assess the influence of the *relative permeability/saturation relationship* on the solution behaviour of the method presented, a **Brooks** and **Corey** function was applied. The investigations confirm that the presented multiphase algorithm satisfactorily solves the **Buckley-Leverett** problem. In order to ensure a good approximation of the front propagation behaviour, it is necessary to employ the quadratic *upwinding formulation*, whereby a significant reduction in the oscillation of the solution may be achieved with the aid of the lumped finite element or the cubic upwinding formulation. Under these conditions, the conservation of mass is also guaranteed.

In order to reproduce a sharp dynamic front for a nearly 100% displacement (100% saturation for both phases), it was furthermore necessary to develop for 1D elements an **adaptive Petrov-Galerkin** formulation based on the shown formulation.

A comparison of the numerical and analytical solutions (see **Welge (1952)** [5]) is given in Fig. 3.

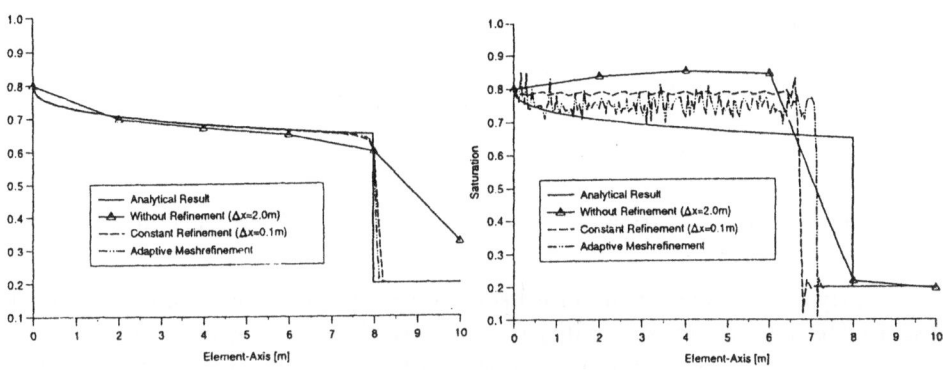

Figure 3: a) Modified Petrov Galerkin Method; b) Standard Galerkin Method

5 CONCLUSIONS

A numerical modelling system for multi-phase flow processes in porous and fractured porous media is discribed. To ensure the reproduction of the correct shape of the dynamic saturation front propagation of the phases, a modified Petrov–Galerkin method has been developed.

The numerical model must provide a quantitative description of the relevant processes and mechanisms, based on scientific principles. In cases where sufficient understanding of the important processes has not yet been achieved, as e.g. for multiphase flow in a heterogeneous nature of the subsurface, numerical modelling must be guided by investigations of those processes in laboratory or field experiments.

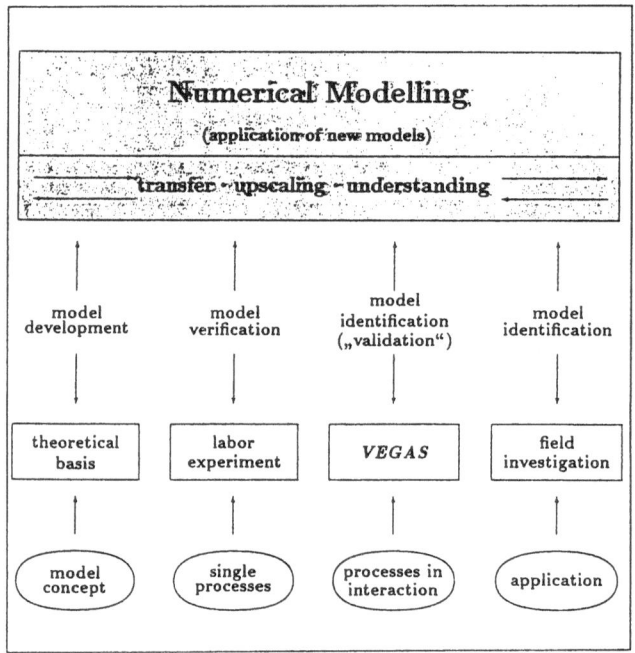

Figure 4: Interpretation of Subsurface Processes by Numerical Models

Presently no method is known or is likely to emerge in the near future, which could demonstrate the accuracy or validity of a numerical model in general "final" terms. This limitation calls for a cautious approach in model applications. Numerical modelling should be open to iterative refinement; models need to be complemented with engineering judgement, and provisions for monitoring and confirmation of system performance must be made. The test facility **VEGAS** will provide new possibilities for extending the experimental data base for checking and validating multiphase flow and transport models.

ACKNOWLEDGEMENTS

The first author gratefully acknowledges the financial support of the Deutsche Forschungsgemeinschaft (DFG). A large part of the work reported here was carried out at the Institute of Fluid Mechanics, University of Hannover, FRG and at the LBL, University of California at Berkeley, USA.

References

[1] Buckley S.E. and M.C. Leverett: Mechanism of Fluid Displacements in Sands. *Trans. AIME*, 146:107–116, 1942.

[2] Helmig R. *Theorie und Numerik der Mehrphasenströmungen in geklüftet-porösen Medien.* Bericht Nr.34, Institut für Strömungsmechanik und Elektron.Re chnen im Bauwesen, Universität Hannover, 1993.

[3] Huyakorn P.S. and G.F. Pinder: *Computational Methods in Subsurface Flow.* Academic Press, London, 1983.

[4] Peaceman D.W.: Fundamentals of Numerical Reservoir Simulation. Elsevier, Amsterdam, 1977.

[5] Welge H.J.: A Simplified Method for Computing Oil Recovery by Gas of Water Drive. *Trans. AIME 159*, 91–98, 1952.

[6] Westerink J.J. and D. Shea: Consistent Higher Degree Petrov–Galerkin Methods for the Solution of the Transient Convection–Diffusion Equation. *International Journal of Num. Methods in Engineering*, 28:1077–1101, 1989.

[7] Zielke W. und R. Helmig: Grundwasserströmung und Schadstofftransport — FE–Methoden für klüftiges Gestein. In Univ. Fredericiana Karlsruhe (TH), Editor, *Wissenschaftliche Tagung "Finite Elemente — Anwendung in der Baupraxis".* Verlag Ernst u. Sohn, Berlin, 1991.

[8] Kobus H., Cirpka, O., Barczewski, B., und Koschitzky, H.-P.: Versuchseinrichtung zur Grundwasser- und Altlastensanierung VEGAS - Konzept und Programmrahmen, Mitteilungsheft Nr.82, Institut für Wasserbau, Germany.

APPLICATION OF DOMAIN DECOMPOSITION TECHNIQUES FOR MULTIPHASE GROUNDWATER PROBLEMS

A.S. MAYER
Department of Geological Engineering, Geology and Geophysics
Michigan Technological University
Houghton, MI, 49931-1295
USA

A two-dimensional two-phase flow model is developed using finite element approximations in space. Domain decomposition schemes based on the Schwarz alternating procedure are applied to a multiphase groundwater flow problem by splitting the problem domain into a subdomain containing two phases and subdomain where only a single phase exists. Schemes are tested for accelerating the convergence of the domain iteration steps. The dimensions of the subdomains are re-defined for each time step. A refined grid relative to the remainder of the domain is used in the two-phase subdomain. The domain decomposition schemes are shown to significantly reduce computational effort when compared to full domain solutions.

INTRODUCTION

Multiphase groundwater problems can result from the introduction and subsequent migration of nonaqueous phase liquids (NAPLs). Accurate simulation of NAPL migration problems frequently requires the solution of a large number of two-phase flow equations. However, the NAPL plume often is contained within a small area relative to the full simulation domain. This type of problem is ideally suited to the application of domain decomposition techniques. The advantages of domain decomposition for this problem are (Ewing, 1990): 1) coupled, nonlinear flow equations are solved only in the subdomain containing the NAPL; 2) refined grids required to simulate NAPL flow are applied only in the subdomain containing the NAPL; and 3) the resulting algorithm is more amenable to parallelization. A wide range of domain decomposition techniques have been applied to the solution of differential equations including variations on the Schwarz alternating procedure, iterative sub-structuring methods, and methods based on optimization theory. Surveys of these techniques may be found in Glowinski *et al.* (1987) and Chan *et al.* (1988, 1989). In the following work, the Schwarz alternating procedure is optimized for the solution of the NAPL migration problem.

951

A. Peters et al. (eds.), Computational Methods in Water Resources X, 951–958.
© 1994 *Kluwer Academic Publishers. Printed in the Netherlands.*

GOVERNING EQUATIONS

Descriptions of NAPL infiltration into initially water-saturated, water-wet, porous media require solving for fluid pressures and saturations from NAPL and aqueous phase flow equations. The bulk flow equation for phase α in the absence of reactions and internal sources and sinks is

$$\frac{\partial}{\partial t}\left(\theta_\alpha \rho_\alpha\right) + \nabla \cdot \left(\theta_\alpha \rho_\alpha \mathbf{v}_\alpha\right) = 0 \tag{1}$$

where α is a phase index, θ_α is the volumetric fraction, ρ_α is the density, and $\mathbf{v}_\alpha$ is the mean pore velocity vector. Velocities may be expressed in terms of pressure head using a modified form of Darcy's law for multiphase flow:

$$\mathbf{v}_\alpha = -\frac{K_\alpha}{\theta_\alpha}\left(\nabla \psi_\alpha + \mathbf{z}\right) \tag{2}$$

where $K_\alpha = \rho_\alpha g k k_{r\alpha}/\mu_\alpha$, g is the gravitational acceleration constant, k is the intrinsic permeability, $k_{r\alpha}$ is the relative permeability, μ_α is the fluid dynamic viscosity, ψ_α is the fluid pressure head, and $\mathbf{z}$ is the unit vector in the upward, vertical direction. Two mobile phases are described with flow equations—an aqueous phase $(\alpha = a)$ and a NAPL $(\alpha = n)$.

A number of constitutive relationships are needed for closure of equation (1). The capillary pressure head is defined as $\psi_{na} = \psi_n - \psi_a$ for a two-phase system with a wetting $(\alpha = a)$ and a nonwetting $(\alpha = n)$ phase. The van Genuchten functions are used to define volumetric fraction as a function of capillary head and to define relative permeabilities as functions of volumetric saturation for a two-phase system. More detail concerning the development of the governing equations and constitutive relationships can be found in Mayer and Miller (1994).

ALGORITHMS

Overview of Solution Methods

A standard Galerkin finite-element method was used to approximate the spatial derivatives in the two-dimensional $(x\text{-}z)$ form of (1). A first-order, fully implicit finite-difference method was used to approximate the temporal derivatives. The coupled, nonlinear, two-phase flow equations are solved by Picard iteration for ψ_a and ψ_n. The aqueous phase volumetric fraction at the new time step $(l + 1)$ and iteration level $(k + 1)$ is approximated by expansion in a Taylor series about ψ_{na}, neglecting second order terms and above. The systems of equations were solved with an iterative, Orthomin solver coupled with Jacobi preconditioning. Additional details of the solution methods may be found in Mayer and Miller (1994).

The solution domain Ω with boundary $\delta\Omega$ is decomposed into subdomains Ω_i with boundaries $\delta\Omega_i$. Subdomain Ω_1 comprises the portion of Ω where only single-phase flow occurs, while Ω_2 contains the area where two-phase flow is predominant. In all of the algorithms described in this work, single-phase flow equations are solved on Ω_1, two-phase flow equations are solved on Ω_1, and the subdomain solutions are coupled through inter-subdomain boundaries, Γ_i ($\Gamma_i = \delta\Omega_i \backslash \Gamma'_i$, where $\Gamma'_i = \delta\Omega \cap \delta\Omega_i$). Figure 1 illustrates the subdomain partitioning scheme.

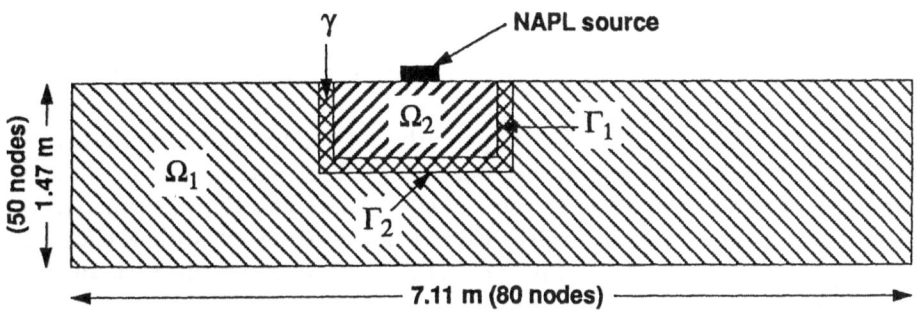

Figure 1. Subdomain Partitioning Scheme and Schematic of Example Problem.

Domain Decomposition Schemes

The Schwarz alternating procedure (SAP) was adopted as the underlying domain decomposition algorithm. The SAP involves decomposing Ω into overlapping subdomains, Ω_1 and Ω_2. The overlapping region is defined as $\gamma = \Omega_1 \cap \Omega_2$ (see Figure 1). The algorithm for the SAP is

(3a) solve equation (1) for $\psi_{a1}^{j+1,k+1}$ in Ω_1, with $\psi_{a1}^{j+1,k+1} = \psi_{a2}^{j,k+1}$ on Γ_1 and Ω boundary conditions on Γ'_1

(3b) solve equation (1) for $\psi_{a2}^{j+1,k+1}$ in Ω_2, with $\psi_{a2}^{j+1,k+1} = \psi_{a1}^{j+1,k+1}$ on Γ_2 and Ω boundary conditions on Γ'_2

where ψ_{ai} is the aqueous phase pressure head in subdomain i and the superscript j is the domain iteration index. This two-step procedure is repeated until the values of ψ_a on Γ_1 and Γ_2 converge. The domain iteration loop is stepped inside the Picard iteration loop. The SAP can be modified to accelerate the convergence of the domain iteration by relaxing the boundary updating. This algorithm can be written as (Evans and Kang, 1988)

(4a) solve equation (1) for $\psi_{a1}^{j+1,k+1}$ in Ω_1, with $\psi_{a1}^{j+1,k+1} = \psi_{a2}^{j-1,k+1}$

$\quad + \omega \left(\psi_{a2}^{j,k+1} - \psi_{a2}^{j-1,k+1} \right)$ on Γ_1 and Ω boundary conditions on Γ_1'

(4b) solve equation (1) for $\psi_{a2}^{j+1,k+1}$ in Ω_2, with $\psi_{a2}^{j+1,k+1} = \psi_{a1}^{j,k+1}$

$\quad + \omega \left(\psi_{a1}^{j+1,k+1} - \psi_{a1}^{j,k+1} \right)$ on Γ_2 and Ω boundary conditions on Γ_2'

Optimal values for the relaxation factor, ω^*, have been determined for various homogeneous, ordinary differential equations and associated boundary conditions (*e.g.* Evans and Kang, 1988), but apparently not for transient, multiphase flow equations. Two empirical approaches were adopted for determining ω^*. The first approach was to determine a fixed value of ω^* in space and time through trial and error. The second approach involves calculating a vector $\vec{\omega}^*$ with a chord-slope type of approximation. The components of $\vec{\omega}^*$ vary dynamically with space and iteration level and are determined by

$$ \vec{\omega}^* = \left(\vec{\psi}_a^{n_k,k} - \vec{\psi}_a^{0,k} \right)^T \left(\vec{\psi}_a^{1,k} - \vec{\psi}_a^{0,k} \right)^{-1} \tag{5} $$

where n_k is the number of domain iterations from the last Picard iteration level, k. The values of $\vec{\omega}^*$ were weighted by the ratio of successive time step sizes (Δt) such that $\vec{\omega}^{*\ l+1} = \vec{\omega}^{*\ l+1} \left(\Delta t^{l+1}/\Delta t^l \right)$ (Nacul and Lett, 1993).

A scheme was developed which allows for adaptation of the sizes of the sub-domains to the problem at each time level. This approach takes advantage of the fact that the region containing two phases changes with time and so the size of the subdomain where the two-phase flow problems are solved (Ω_2) can be minimized through time. The dimensions (number of rows × number of columns) comprising Ω_2 are fixed at the beginning of each time level by performing one Picard iteration over a large Ω_2 and finding the elements where two phases exist. The subdomain dimensions are fixed by adding a "buffer" of elements around the two-phase elements, since the first Picard iteration usually underestimates the extent of two-phase flow.

The grid discretization required for stable solutions of the two-phase flow problem frequently is smaller than the discretization required for the single-phase problem. Another scheme was developed where the grid size for Ω_1 was allowed to increase as integer multiples of the grid size for Ω_1 without exceeding a Courant number criteria of $Cr = v_a \Delta t/\Delta x = 1$. The values of $\psi_{a1}^{j+1,k+1}$ assigned to the slave nodes of Γ_2 in algorithm step a) were estimated with a linear interpolation scheme.

The sequential nature of the SAP prevents straightforward implementation on parallel computing architectures. There are several domain decomposition algorithms based on the SAP (*e.g.* Anderson, 1988), which avoid overlapping subdomains and are parallelizable. A variation on the SAP was developed in this work to investigate whether simple modifications could produce faster computations with a parallel machine. Steps (4a) and (4b) were decoupled by modifying (4b):

(4b') solve equation (1) for $\psi_{a2}^{j+1,k+1}$ in Ω_2, with $\psi_{a2}^{j+1,k+1} = \psi_{a1}^{j-1,k+1}$

$$+ \omega \left(\psi_{a1}^{j,k+1} - \psi_{a1}^{j-1,k+1} \right) \text{ on } \Gamma_2 \text{ and } \Omega \text{ boundary conditions on } \Gamma_2'$$

where $\psi_{a2}^{j+1,k+1}$ now depends only on ψ_{a1} from the j and $j-1$ domain iteration levels. The decoupling allows for parallel solutions, but presumably at the cost of slower convergence.

EXAMPLE PROBLEM

The problem simulated in this work involves the infiltration of NAPL from a small source area located at approximately the top center of a cross-sectional domain (see Figure 1). The full domain, Ω, was discretized into 3,871 quadrilateral elements, each of which was 0.09 m in the x-direction and 0.03 m in the z-direction. The boundary and initial conditions are described in Table 1. A 0.18-m wide segment at the top boundary, located at $2.70 \leq x \leq 2.88$ m, served as the entry region for the NAPL, which was accommodated by a constant NAPL volumetric fraction of 0.166. The simulation occurred over a period of 400 s, divided into five time steps which increased geometrically by a factor of 1.6 each time step. The properties associated with the porous medium and the immiscible fluids are listed in Table 2.

The efficiency of the domain decomposition schemes was compared by applying each scheme to the example problem (see Table 3). The first entry in Table 3, designated as "Full," represents a simulation with no domain decomposition and is provided as a basis of comparison. The next three entries, designated as "SAP," "SAPA," and "SAPAA" are results for the SAP without acceleration, SAP with fixed ω^*, and SAP with dynamic $\vec{\omega}^*$, respectively. Application of these three schemes resulted in CPU time reductions of at least two-fold over the "Full" scheme. The incorporation of relaxation within the "SAPA" scheme produced significant reductions in n_j and CPU time when compared to the results for the "SAP," where no relaxation was allowed. The value of ω^* for the "SAPA" scheme was found by trial and error to be $\omega^* = 1.28$, within the range of ω^* found by Evans and Kang (1988) for a homogeneous Poisson problem. Application of the dynamic "SAPAA" scheme resulted in a greater reduction in n_j, and a corresponding reduction in CPU time. A fixed size of Ω_2 was used in the "SAP," "SAPA," and "SAPAA" schemes.

Table 1. Boundary and Initial Conditions.

Boundary Conditions

$\psi_a =$ constant for $x = 0$, $x = 7.11$ m and $0 \leq z \leq 1.47$ m [a]

$\dfrac{\partial \psi_n}{\partial x} = 0$ for $x = 0$, $x = 7.11$ m and $0 \leq z \leq 1.47$ m

$\dfrac{\partial \psi_a}{\partial z} = 0$ and $\dfrac{\partial \psi_n}{\partial z} = 0$ for $0 \leq x \leq 7.11$ m and $z = 0$, $z = 1.47$ m [b]

Initial Conditions [b]

$v_{a,x} = 0.1$ m/day

$\theta_a = \phi$ (= porosity)

---

[a] left- and right-hand side ψ_a boundaries were subjected to constant, fully-saturated hydrostatic conditions for aqueous phase flow, resulting in the indicated initial aqueous phase velocity.

[b] except for NAPL entry segment.

Table 2. Fluid and Porous Media Properties.

Parameter	Value	Units
ϕ	4.0×10^{-1}	
k	1.0×10^{-8}	cm^2
ρ_a	9.99×10^{-1}	$g\ cm^{-3}$
ρ_n	1.46×10^{0}	$g\ cm^{-3}$
μ_a	1.0×10^{-2}	$g\ cm^{-1}\ s^{-1}$
μ_n	5.7×10^{-3}	$g\ cm^{-1}\ s^{-1}$
Parameters for ψ_{na}-θ_α-$k_{r\alpha}$ relationships:		
n	1.7×10^{0}	
α_{na}	1.92×10^{-1}	cm^{-1}
θ_{nr}	8.0×10^{-2}	
θ_{air}	1.0×10^{-2}	

The next entry in Table 3 shows results for the "SAPA/D" scheme, where the size of the subdomains was allowed to vary at each time level. This scheme produced significant reductions in CPU time when compared to the "SAP" scheme. The reduction in CPU time was due to the smaller number of equations required to solve the example problem. The size of Ω_2 ranged from 180 nodes in the first time step to 552 nodes in the last time step, which was the size of the fixed Ω_2 in the "SAP," "SAPA," and "SAPAA" schemes.

Results are also given in Table 3 for the "SAPA/G" scheme, which involved variable grid sizing for Ω_1. These results are for the optimal grid size for Ω_1 with respect to CPU time— twice the grid size for Ω_2. Larger grid sizes for Ω_1 produced large numbers of domain iterations required for convergence, outweighing the reduction in operations in solving the flow problem on Ω_1. Comparison of the results for the "SAPA" and "SAPA/G" schemes indicate that substantial reductions in CPU time can be obtained by increasing the grid size in Ω_1. Improved convergence properties in the "SAPA/G" scheme for larger grid sizes in Ω_1 could be achieved with a higher-order interpolation scheme for computing values at the slave nodes.

Application of the parallel scheme, or "SAPA/PP," resulted in longer CPU times when compared to the "SAPA" scheme. A speed-up of 1.3 (measured as the ratio of CPU time to wall clock time) was achieved, but the number of domain iterations increased significantly. However, it is likely that more efficient parallel schemes can be developed, from the standpoint of superior convergence speed-up properties.

Table 3. Simulation Results for Domain Decomposition Schemes[a]

Scheme	Total Picard Iterations (n_k)	Total Domain Iterations (n_j)	CPU Time (s)
Full	82	—	2472
SAP	76	894	1170
SAPA	76	692	887
SAPAA	76	543	681
SAPA/D	91	751	695
SAPA/G	76	732	796
SAPA/PP	76	1201	948[b]

[a] Simulations were conducted on a four-processor Cray Y-MP

[b] Wall clock time from a dedicated run over two processors

CONCLUSIONS

Domain decomposition techniques based on the Schwarz alternating procedure were tested on a multiphase groundwater problem. Domain iteration convergence was improved with a static acceleration scheme; even better improvement was observed with a dynamic scheme. Adaptive sizing of the subdomains and reduction of the grid discretization on the single phase subdomain brought about substantial reductions in computational effort. A parallel scheme did not improve upon the sequential scheme. Improvements in parallelization, convergence acceleration, and grid discretization schemes could result in more efficient schemes.

ACKNOWLEDGEMENT

The author is grateful for the computer time donated by the Cray Research Corporation.

REFERENCES

Anderson, C. R. (1988) "Domain Decomposition Techniques and the Solution of Poisson's Equation in Infinite Domains," in Chan, T. F. *et al.*, Domain Decomposition Methods, Proceedings of the Second International Symposium on Domain Decomposition Methods, SIAM, Philadelphia, pp. 129–139.

Chan, T. F., Glowinski, R., Periaux, J., and Widlund, O. B. (eds.) (1988) Domain Decomposition Methods, Proceedings of the Second International Symposium on Domain Decomposition Methods, SIAM, Philadelphia.

Chan, T. F., Glowinski, R., Periaux, J., and Widlund, O. B. (eds.), (1989) in Third International Symposium on Domain Decomposition Methods for Partial Differential Equations, SIAM, Philadelphia.

Evans, D. J., and Kang, L. (1988) "New Domain Decomposition Strategies for Elliptic Partial Differential Equations," in Chan, T. F. *et al.*, Domain Decomposition Methods, Proceedings of the Second International Symposium on Domain Decomposition Methods, SIAM, Philadelphia, pp. 173–191.

Ewing, R. E. (1990) "A Survey of Domain Decomposition Techniques and Their Implementation," Advances in Water Resources, 13, 117–125.

Glowinski, R., Golub, G., Meurant, G. A., and Periaux, J. (1988) First International Symposium on Domain Decomposition Methods, SIAM, Philadelphia.

Mayer, A. S., and Miller, C. T. (1994) "Dissolution of Nonaqueous Phase Liquids: Effects of Mass Exchange Characteristics and Porous Media Heterogeneity," Water Resources Research, in review.

Nacul, E. C., and Lett, G. S. (1993) "Under and Over Relaxation Methods for Accelerating Nonlinear Domain Decomposition Methods," 12th Symposium on Reservoir Simulation, Society of Petroleum Engineers, Richardson, Texas, pp. 105–112.

"HYSTERESIS EFFECTS IN WATER/GAS FLOW IN POROUS MEDIA"

Louis MINSSIEUX
INSTITUT FRANÇAIS DU PETROLE
Division Gisements
1 & 4, avenue de Bois Préau - B.P. 311
92506 RUEIL-MALMAISON - FRANCE

Water/gas flow properties in porous media and related hysteresis effects are illustrated by experiments carried out in 1 meter long water-wet outcrop sample.

It is shown that during water injection (imbibition step) following gas injection (drainage), hysteresis comes from reduced gas relative permeability and partial gas entrapment, that means limited increase in water saturation and relative permeability.

This phenomenon does also exist in the presence of a non aqueous residual liquid phase such as hydrocarbons, immiscible with water.

Another consequence of hysteresis associated with polyphasic flow, lies in the possibility of mobilization of a fraction of trapped hydrocarbons as shown in the case of the water alternating gas injection runs presented.

In a first approach in numerical modelling, simulations of the water/gas flow experiments are shown to match the observed cyclic differential pressure drops and fluid saturation variations accompanying the hysteresis phenomena described.

INTRODUCTION:

In case of pollution by NAPL (non aqueous phase liquid such as hydrocarbons), a preliminary task consists in determining the extension and underground movement of contaminants especially above an aquifer. Then water injection from well to well, was proposed as a first method of soil remediation.

The aquifer itself associated with its capillary air/water fringe, can also alternatively rise and fall as a function of local hydraulic environment conditions. Under all these conditions, diphasic and even triphasic fluid flow occuring in the subsurface, has to be taken under consideration.

This paper deals with the typical properties of air/water flow in porous medium, in presence or not of a contaminant hydrocarbon phase. Experiments with dynamic hysteresis effects related to fluid saturation history of the medium (1,2,3,4) have been especially described and analysed.

959

A. Peters et al. (eds.), Computational Methods in Water Resources X, 959–966.
© 1994 Kluwer Academic Publishers. Printed in the Netherlands.

Let's consider first, a porous medium such a soil, containing a liquid phase, water, and a gas phase, air.In this case, it is clear that water is the wetting phase and gas the non-wetting one.Dynamic conditions for diphasic fluid flow in porous medium can lead to one or the other following situation:

- either the wetting phase saturation (% pores volume, PV)is increasing, such a configuration will be called imbibition by water,or simply "imbibition",

- either it is decreasing(and co-existing gas saturation increasing),then this state will be called "drainage" by the non-wetting phase.

Both situations exist, for example, in the case of rising or falling movements of water table.

The object of this paper lies mainly in the comparison of gas/water flow characteristics corresponding respectively to drainage or imbibition conditions.It focuses on the important differences which distinguish them.

Before using blind natural porous media, it is useful to recall some demonstrative observations carried out in glass models, illustrating specific fluid behavior in the two situations defined above.

FLUID INTERFACE BEHAVIOR AT PORE SCALE

Under drainage conditions of a pore network, gas progresses in form of continuous channels, looking for the lowest pore threshold pressure due to capillarity ($P_c = 2 \gamma \cos\theta/r$), i.e moves via larger pore throats.Contact angle is practically nul when pore walls are covered with water films (5).

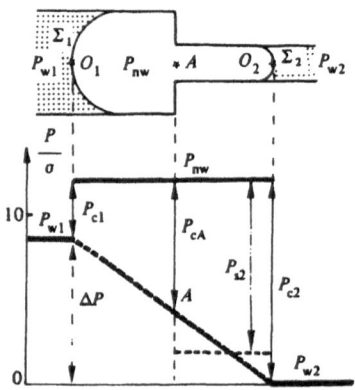

Fig.1 : Gas trapping due to pore geometry

Under imbibition conditions, water is the continuous invading fluid.Gas becomes the displaced phase and observations in capillaries or micromodels (5) (6) (7) show a clear trend to phase rupture in form of individual bubbles. Figure 1 illustrates how gas bubbles are trapped by flowing water, as a result of pore geometry (7) and high value of water/air interfacial tension.Pressure gradient necessary to overcome bubble resistance, becomes higher and higher as new bubbles are created and trapped in pore network. Angle contact hysteresis effect (5) is negligible here as advancing and receding angles at gas bubble interfaces are practically the same.

This distinct phase behavior observed at pore scale under drainage or imbibition conditions, already suggests we can expect noticeable differences between these both configurations of flow in porous media. Such observations are the basic local aspects of hysteresis macroscopic effects found in the following diphasic flow experiments performed in natural sandstone.

WATER/GAS FLOW EXPERIMENTS IN NATURAL POROUS MEDIA

Porous medium

Fluid flow experiments were carried out in a one meter long physical model made of an homogeneous outcrop sample: a Vosges sandstone with the following characteristics:

- section : 4.5 x 4.5 cm^2
- length : 1 m
- porosity : 24.7%
- water permeability : 55 mD

This sample was sealed in the core holder shown on figure 2, by means of a fusible alloy (melting point =138°C).

Fluids used

Gas was nitrogen (viscosity = 0.02 cP at 70°).

Water was a 40g/l KCl brine (viscosity = 0.41cP) in order to avoid any interaction with clays present in Vosges sandstone.

Hydrocarbon phase was a commercial close-cut hydrocarbon fraction called Soltrol 130 (i.e.a C10-C12 alcane fraction, with a viscosity of 0.70 cP at 70°C), supplied by Phillips Petroleum Co.It is considered in this work, completely water unsoluble.

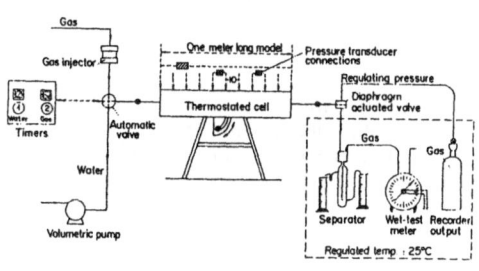

Fig.2 :   Diagram of equipment used

Operating conditions

All experiments were conducted at 70°C, in the equipment described in figure 2.Outlet pressure was set at 70 bar by means of a back pressure valve.

Water/gas flow runs were performed in the model initially saturated with water (S_w = 100% PV), by simultaneous injection of gas and water until a steady-state flow regime was reached.

Fluids were injected at constant volumetric flow rate.Water flow rate was adjusted with a pump. Gas flow rate was delivered by a special device in which a constant pressure drop was applied on a micrometer valve.

Volumes of effluents produced were measured at atmospheric pressure, after fluid separation at regulated temperature (25°C).

Pressure transducers enabled recordings of overall pressure drops or measurements only on sections of the model (See figure 2).

Table 1:Diphasic steady-state runs

Fluid flow rate (cm³/mn)		At end of step	
Water	gas	Average gas saturation (% PV)	Pressure drop (bar)
- Drainage conditions -			
1.12	2.06	11.5	1.45
0.78	2.38	14.3	1.70
0.56	2.60	14.5	1.52
0.20	2.96	18.7	1.20
0.04	3.12	20.5	0.97
0	1.90	29.3	0.48
- Imbibition conditions -			
0.07	3.26	27.6	1.6
1.40	3.53	23.9	9.2
0.50	2.66	21.2	2.9
0.86	2.30	18.9	4.1

Analysis of water/gas flow experiments

  The steady-state experiments presented (also called as the Pennstate method) involving co-current injection of gas and water, were interpreted with conventional Darcy's law for diphasic flow. The fluid flow rate ratios chosen so as to provide drainage, then imbibition conditions, are given in table 1, as well as the average gas saturation and pressure drop reached at each permanent regime.

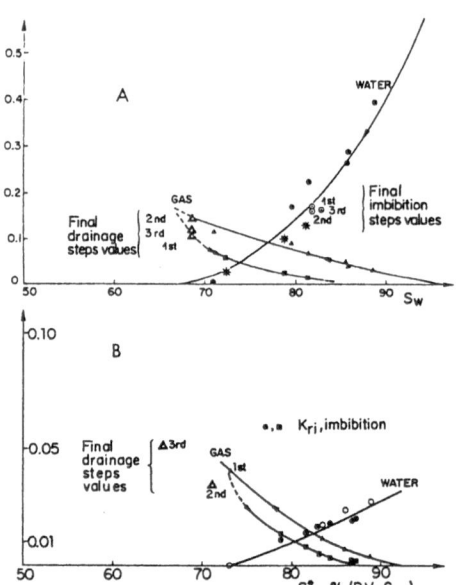

Fig.3 : Relative permeability determinations

Results of two-phase experiments (S_h = 0%):

On figure 3A, are given the water/gas relative permeability (Krw/Krg) curves, obtained respectively under first drainage then subsequent imbibition conditions.

Examination of Kr curves clearly shows differences existing between drainage and imbibition flow conditions : gas flow is not reversible and Krg clearly decreases as water saturation increases during imbibition step.

This hysteresis effect in gas flow can be explained by a relocation of gas phase within pores as water saturation begins to increase under imbibition conditions.This phase local new arrangement is especially important at water front when a gas fraction becomes trapped.Then pressure drop reaches its highest value, and remains stable along the whole imbibition step.

In the Vosges sandstone used, about 20% PV of gas could be trapped during water injection (See figure 3), even after several drainage/imbibition cycles.

Such a value in trapped or residual gas saturation is consistent with results of Chierici (2) collected in more than 200 core samples of various morphology, including unconsolidated sands, sandstones or limestones.

As to Krw, it appears to be only a function of water or wetting phase saturation, whatever the way it has been reached. This result is commonly observed (6) in water-wet porous media where hydraulic continuity of flowing water is fulfilled.

Triphasic experiments ($S_h \neq 0$):

The same tests were repeated with an average residual hydrocarbon (Soltrol) saturation of 25.2 to 28.5% in the model (See operating conditions in table 2). It has to be noted here that Vosges sandstone remains water-wet under these conditions and that Soltrol can spread on the brine used as water phase.

Figure 3B gives the correspondind Krg/Krw curves obtained.

It can be seen that water/gas flow still exhibits the same hysteresis behavior under drainage and imbibition conditions as above; here the presence of residual hydrocarbons within pores, does not change the typical features of water/gas flow properties.

Table 2: Steady-state runs with residual oil

Fluid flow rate (cm^3/mn)		At end of step	
Water	gas	Average gas saturation (% PV)	Pressure drop (bar)
- Drainage conditions -			
0.35	1.20	7.2	7.8
0.20	1.20	10.2	5.2
0.083	1.17	12.1	2.9
0	1.13	19.7	0.8
- Imbibition conditions -			
0.05	1.15	15.5	2.6
0.10	1.21	13.8	4.4
0.15	0.98	12.8	5.5
0.25	1.03		8.4
0.40	1.0	9.8	12.6

AN ILLUSTRATION OF GAS-LIQUID FLOW HYSTERESIS : WATER ALTERNATING GAS (WAG)INJECTION

WAG experiments consisted of alternate injections of gas and water slugs in the same water saturated model (Sw=100%). Operating conditions of this series of runs are gathered in table 3. Such a procedure is used in petroleum industry to improve the gas sweep of oil fields submitted to gas injection.

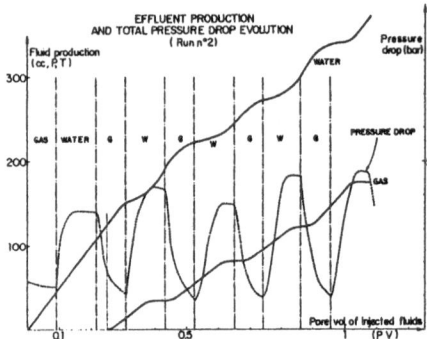

Fig.4 : Effluent production in WAG run n°2

The main characteristics of these WAG flow experiments (See Fig.4) are as follows:

- cyclic high pressure drop in imbibition steps, stabilized after water breakthrough at model outlet,
- stop-and go feature of gas and water production,
- cyclic fluid saturation variations(Fig.5) related to reproducible trapping during each imbibition step, of part of the gas present at end of drainage steps(function of slug size).

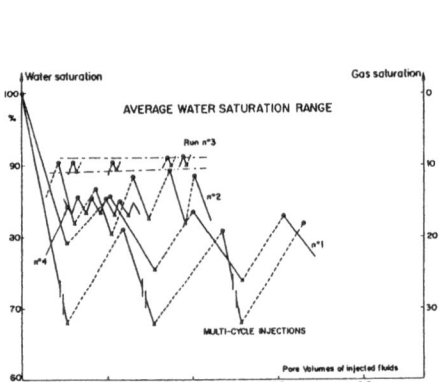

Fig.5 : Saturation variation in WAG experiments

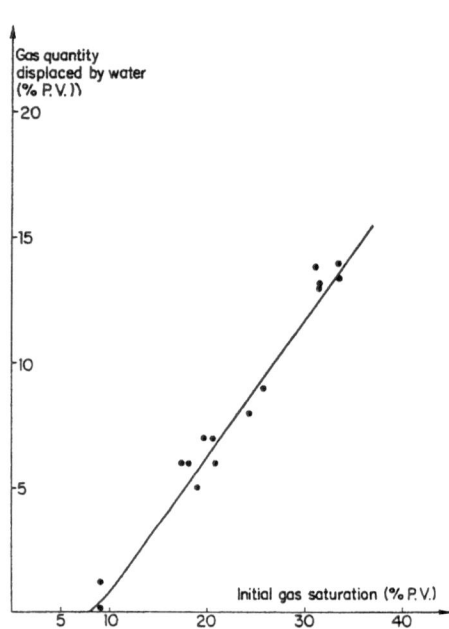

Fig.6 : Gas trapping in imbibition steps

From the series of runs carried out with different slug sizes (water/gas ratio # 1 in tests of table 3), it has been drawn the correlation of figure 6, between mobile gas displaced by water and initial gas in place, which characterizes the Vosges sandstone used.

Table 3: WAG experiments

Run n°	Slug size (% PV)		1st Gas B.T. (% PV inj.)	Gas sat. variation/ end each step (% PV)	Pressure drop (bar) at end of	
	Water	Gas			Drainage	Imb.
1	25.0	25.0	16.6	9.0	2.2	6.9
2	12.5	9.0	25.0	7.0	1.0	4.7
3*	2.4	2.2	23.0	1.3	2.0	3.9
4	3.0	5.0	24.5	2.0	2.1	4.9
1st Slug (drainage)	/	3.53 PV	15.8	31.9	0.83	/
1st imb. (second slug)	33.5	/		13.1	/	9.6
- Simulations -						
1st drainage	/	3.53 PV	19.2	32.0	0.76	/
1st imb. without Hysteresis	33.5	/	/	27.5	/	2.5
- With Hysteresis	"	/	/	13.4	/	9.9
- WAG 3* with Hyst.	2.4	2.2	19.5	1.1	1.7	4.4

Then several other WAG runs were performed with a residual hydrocarbon saturation of 33%, left after waterflood of the sample.The above WAG features still exist under these new conditions, but in addition, an oil production occurs, about 20% of oil in place in the case of run on Fig.7, carried out with the same slug size as used in run 1 (table 3). This mobilization of oil is believed to be another result of respective relocation of the three phases, combined with gas entrapment within pores, during each successive imbibition step.

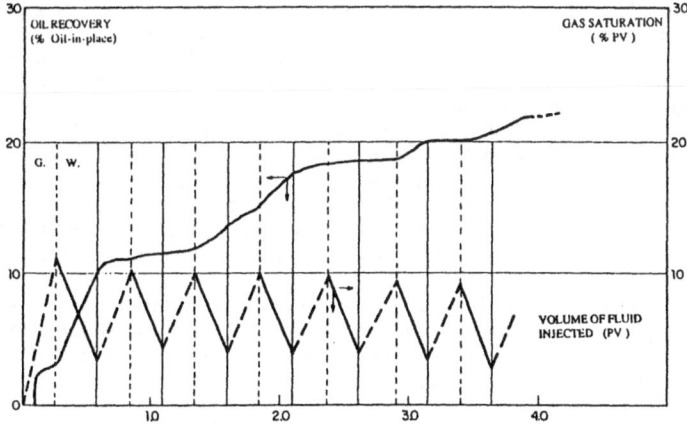

Fig.7 : WAG run with a residual hydrocarbon saturation

SIMULATION OF WAG EXPERIMENTS :

For simulation of above WAG runs without residual hydrocarbons, an in-house simulator was used, belonging to the so-called black-oil model type, common in Reservoir Engineering.

Calculations were conducted first with only one set of Krg/Krw curves.Results given in table 3, show that experimental pressure drops or average fluid saturations during successive injection steps, cannot be matched.

Then the double set of Krg/Krw curves (Fig.3A) was input in the simulator for respectively drainage and imbibition sequences.Also input was the correlation of figure 6, giving the gas trapped fraction resulting from all capillary effects at pore constrictions during water injection, as a function of gas in place at end of drainage steps.

With such an approach of hysteresis phenomena, main features of WAG experiments were reproduced (See table 3), that means acceptable match of overall pressure drops and average gas saturation variations. A similar way of simulating hysteresis in diphasic flow was proposed by Land (3) and Killough (8).Carlson (4) also published an improved calculation method based on limited experimental data needed for porous medium characterization.

A comparison of different numerical treatments of hysteresis in polyphasic flow in porous media is beyond the scope of this study. Our purpose was to emphasise the necessity for simulators to take into account hysteresis in water/gas flow in order to correctly simulate real mechanisms and experimental data when they are available.

CONCLUSIONS:

1/   Results obtained in a sandstone physical model illustrate the importance of dynamic hysteresis effects in water/gas flow, function of the history of water saturations in porous medium.

2/   Such hysteresis phenomena are to be taken into account in mathematical models used for simulating or predicting water/air transport, for example in the case of aquifer movements or implementation of remediation technics in unsaturated zones of the subsurface.

3/   Necessary parameters to determine in the laboratory as simulator input data, include water/gas drainage/imbibition relative permeability curves and gas trapping properties, representative of the porous material considered.

REFERENCES:

(1)   Colonna, M., Brissaut, J, and Millet, J.L. (1970) "Some aspects of 2 phase flow in underground storage of natural gas", SPE paper 2941, Houston.
(2)   Chierici, G.L., Ciucci, G.M. and Long, G. (1963) "Experimental research on gas saturation behind the front in gas reservoirs subjected to water drive", Proc. Sixth World Petrol. Cong., 483-498.
(3)   Land, C.S. (1968) "Calculation of imbibition relative permeability for two and three-phase flow from rock properties", Soc. Pet. Eng. J, June, 149-156.
(4)   Carlson, F.M. (1981) "Simulation of relative permeability hysteresis to the nonwetting phase", SPE paper 10157, San Antonio.
(5)   Tang Kong, V.W. and Wardlaw, N.C. (1991) "Effects of wettability and pore geometry on mobilization of oil and gas in physical models", Can. J. of Chem. Eng.69, 259-265.
(6)   Tixier, M. (1993) "Influence du coefficient d'étalement sur les écoulements triphasiques en milieu poreux", thesis on March 11th in Strasbourg Un.
(7)   Lenormand, R., Zarcone, C. and Sarr, A. (1983) "Mechanisms of the displacement of one fluid by another in a network of capillary ducts", J.Fluid Mech. 135, 337-353.
(8)   Killough, J.E. (1976) "Reservoir simulation with history dependent saturation functions", Soc. Pet. Eng.J. Feb, 37-48.

FINITE ELEMENT SIMULATION OF OIL SPILL CLEANUP USING AIR SPARGING

Rabi H. Mohtar[1], Roger B. Wallace[2] and Larry J. Segerlind[3]
Dept. Agri. & Bio. Engg[1]. Penn State University, State College, Pennsylvania 16802
Depts. Civil[2] & Agri.[3] Engg , Michigan State University, E. Lansing Michigan 48824

ABSTRACT

Air sparging, an extension of soil vapor extraction, is a technique used as a treatment of subsurface oil-spills below the water table. Unlike the latter, is effective in the saturated and the unsaturated zone where water, oil, and gas share partial saturation. It usually follows primary treatment by oil pumping. Air sparging is a relatively new technique and has not been fully explored. This paper presents a numerical procedure to solve the air sparging problem. A computer model, SPARG, is developed based on the finite element method to describe the flow of air in the presence of a static wetting fluid, water. The model solves for air pressure, saturation, and flux, amount of water drained due to air displacement, and capillary pressure. The latter variable was used to locate the air water interface. The model was verified against a linear analytical problem, and an experimental data. The model was also proven to conserve mass. The model can be used by to increase the effectiveness of the technique.

INTRODUCTION

Fuel leakage and spills are considered a major environmental threat in residential areas near service stations and storage tanks especially where many of these tanks are getting old. Primary treatment of these spill sites is to pump the Light Non-Aqueous Phase Liquid (LNAPL) mass. After such treatments large portions of the released hydrocarbon remains bounded as a residual immiscible fluid. The residual liquid serves as a continuous source of contamination for ground water supply for a long time. Classical methods to further clean contaminated sites from the residual oil include soil removal, forced percolation, encapsulation, or trenching are expensive and sometimes prohibited in populated areas.

Soil Vacuum Extraction (SVE) is an alternative technique to remediate soil contamination. It proved to be inexpensive and effective to clean Volatile Organic Contaminants (VOC) constituents from Non-Aqueous Phase Liquid (NAPL) that is located near the water table. The method is an in-site process non-destructive and less expensive than its alternatives. SVE is successful in removing VOC contaminated sites in the unsaturated zone above the top of the capillary fringe. It shows to be not effective in the saturated zone where the NAPL are located near the water table. In such regions the efficiency of SVE is controlled by how fast these liquids will diffuse from the saturated region to the zone above the capillary fringe where SVE is most effective (Cho, 1991).

As an extension to the SVE, air sparging emerge to clean VOC contaminated soil under

967

A. Peters et al. (eds.), Computational Methods in Water Resources X, 967–974.
© 1994 Kluwer Academic Publishers. Printed in the Netherlands.

saturated conditions. The method will work effectively where the LNAPLs contaminants are floating at the water surface of an aquifer. Figure 1 is a cross section of a contaminated site showing the oil spill occupying the vadose zone and the vicinity of the water table. An air sparging well is located below the water table surface forcing air into the porous media. The non-wetting fluid, air, pushes the water away from the sparging well and creates a cone of desaturation. Air flow through the desaturated region creates a rapid means for transporting the VOCs to the ground surface to be managed by either ground surface or sink pumps. The sparging well, the cone of desaturation, and air flow patterns are also shown in the figure. The sink pump may and may not be present. The cone of desaturation makes a continuum with the top of the capillary fringe line. Air sparging treatment essentially reduced to forced air ventilation when the sparging point is located in the unsaturated zone above the capillary action zone.

The theory of immiscible fluids where one or more non-wetting fluids (oil and/or air) shares saturation with a wetting fluid (water) is well developed. In such a system any of the three fluids or all may be in motion within the porous medium they occupy (Corey, 1986). Despite the fact that all the tools needed to develop a comprehensive understanding are available, there are limited number of published theoretical research dealing with the air sparging problem. Few recent studies appear on the latter technique in the consultant companies technical reports (Brown et. al. 1991, Kresge and Dacey 1991, and Marley et. al. 1992). Therefore, the implementation have been mainly dependent on experience and estimations by practioners. While such studies are valuable for the research community, lack of theory limits the understanding to the many processes and their interactions whose effects are observed in the field.

Wallace et al. (1994) developed a theoretical framework for air sparging. They presented a mathematical model that describes the phenomenon. They implemented their model using the finite element method as a numerical tool to determine flow distribution after a steady air injection. They presented an experimental procedure and reported some results comparing numerical and experimental models. While they have presented the theoretical concept of air sparging this paper will focus on the numerical formulation of the process.

The main objective of this study is to develop a numerical solution for air sparging using the finite element method. This solution will describe the flow of non-wetting fluid, air, in the presence of static wetting fluid, water. Some of the specific objectives are :

1) Develop a two dimensional finite element computer model to numerically evaluate the air flow distribution during steady air sparging. The flow parameters include: air and water pressure, air and water saturation, air flux, and the amount of water drained due to air displacement.

2) Investigate a hypothetical air sparging problem using the developed computer model. This includes identifying the location of air/water interface.

THEORETICAL DEVELOPMENT
Problem description

During sparging and at steady state, the water is static and air flow is under steady state equilibrium conditions. The oil is considered immobile and moves only in its vapor state through the air stream. Figure 2 shows a full description of the parameters that are involved in the sparging problem. The depressed capillary fringe as hypothetically drown represents the interface between a fully water saturated and non-saturated region. The underlying assumption is that the air pathways are continuous in the entire domain, shaded and non-shaded, and that the air permeability in the water saturated zone is small non-zero value. The interface between saturated and unsaturated zone (the cone of depression) will be determined from the air and water pressure values at a later stage in the computations. Due to symmetry only half of the domain is shown and considered. The boundary conditions as shown in this figure are as follows: the bottom, left hand side, and right hand side up to the water table are considered no flow boundaries. The top, and right hand side above water table are at atmospheric pressure. The domain includes the shaded and non-shaded regions.

Governing equations

The general two dimensional steady state anisotropic gas flow equation in porous media can be written as:

$$\frac{\delta}{\delta x}\left[k_x\rho\frac{\delta h_a}{\delta x}\right]+\frac{\delta}{\delta y}\left[k_y\rho\frac{\delta h_a}{\delta y}\right]=0 \tag{1}$$

where k_x and k_y are the intrinsic permeabilities of air in the x and y directions respectively (L^2), h_a is the air pressure head (L), ϱ is the density of air (M/L^2T).

According to Boyle's law the air density is associated with air pressure head according to the following relation:

$$\rho=\frac{\rho_0 p}{p_0} \tag{2}$$

where ϱ_0 and p_0 are density and pressure of air at Standard Temperature and Pressure (STP), p is the air pressure evaluated as $h_a \varrho g$.

Equations 1 and 2 are valid for a single fluid phase system. For air as the non-wetting phase in a porous media containing water, a two phase fluid system prevails. In the later system any of the two fluids could be in motion. In the present investigation water will be considered static and the pressure head of water is dependent only on the height of the water column above the point of interest. The air will be considered steady fluid flow. This scenario where air is moving and the water is static is not common in nature and is generally a man made phenomenon.

If air is flowing at a constant rate, while water has reached an equilibrium static stage, water saturation will be constant with time, although it could vary spatially. This steady state water saturation is a function of the water and air pressure head the later of which is unknown in the system. On the other hand the non-wetting fluid permeability is a function of the water saturation. That relationship generates a highly nonlinear system that can still be described by Equation 1, however, the coefficients k_x and k_y are functions of the air pressure, the variable being solved.

Similarly, the air density in Equation 1 is a function of the air pressure. Although to a lesser extent, this creates another source of non-linearity that was also been addressed.

The capillary pressure head that exist across an interface between any two fluids is defined as:

$$h_c = h_a - h_w \tag{3}$$

where h_a is the pressure head of air as in Equation 1, will be used to define the location of the interface, h_w is the water pressure head which is hydrostatic.

The main source of non-linearity in the two phase system is the non-wetting fluid permeability that is dependent on non-wetting fluid pressure head. The relationship between air permeability and capillary pressure head has been investigated (Corey 1987). This relation is described as:

$$k_{rw}=k_m\left[1-\left[\frac{h_d}{h_c}\right]^{\lambda}\right]^2\left[1-\left[\frac{h_d}{h_c}\right]^{2+\lambda}\right] \tag{4}$$

where h_d is the displacement pressure head at which saturated soil starts to drain, k_m is the maximum air permeability that occurs at minimum water saturation, λ is a soil parameter defined as the slope of pore size distribution curve.

Equation 4 is valid under the condition that $h_c >$ air entry pressure head, the pressure at which air replaces water in a soil matrix undergoing desaturation. The air entry pressure head is a slightly larger value than displacement pressure head. In-depth study of this relation shows that the curve is steep at low values of h_c, consequently low h_a, where small variation in h_c give rise to large changes in k_{aw}. This characteristic controls stability of the numerical solution at low air pressures.

FINITE ELEMENT FORMULATION

The finite element method was applied to the differential Equation 1. A brief description will be presented in this section after Segerlind (1984). Applying the Galerkin residual operator to Equation 1 through multiplying by the shape function vector and integrating over the area we get:

$$\{r^{(e)}\} = -\int_{A}\int [N]^{T}\left[\frac{\delta}{\delta x}\left(k_x \rho \frac{\delta h_a}{\delta x}\right) + \frac{\delta}{\delta y}\left(k_y \rho \frac{\delta h_a}{\delta y}\right)\right] dA \tag{5}$$

where $[N]^{T}$ is the transpose of the shape function vector. Shape functions are interpolation functions with order up to the user choice i.e. linear, quadratic, or cubic, $\{r^{(e)}\}$ is the element residual vector. The goal is to minimize the global residual or error vector towards zero. A is the area of the element over which the investigation holds. It is assumed that the permeabilities k_x and k_y in Equation 5 do not vary over the element. For that reason they can be taken outside the integration. The condition imposed on the anisotropy is that the principal directions of anisotropy coincide with the x y coordinate axes where y is the vertical direction.

Carrying out the integration above for a specified element type and writing the results in a matrix form result in:

$$[k^{(e)}]\{h_a\} - \{f^{(e)}\} = 0 \tag{6}$$

where $[k^{(e)}]$ is the element stiffness matrix that comes from the partial derivative terms, $\{h_a\}$ is the vector of the unknowns, $\{f^{(e)}\}$ is the force vector and is zero, before boundary conditions are imposed, for the case where no source terms in the general differential equation are present as is the case for our present problem defined in Equation 1.

Using the standard stiffness procedure, the element matrices are used to build a global stiffness matrix and global force vector. The resultant global system of equations has the same form as the standard finite element system of equations, similar to Equation 6 where the element matrices are replaced by global matrices. The system is then solved using a direct Gauss linear system of equations solver to determine $\{h_a\}$. In brief, the finite element formulation has transformed the system of pde (Equation 1) into a system of algebraic equations (Equation 6). The solution for the later system is readily obtained.

SOLUTION PROCEDURE

A computer model (SPARG), based on the finite element method, was developed to simulate air sparging in porous media. The model determines the pressure head and flow distribution of air in a two phase flow system namely air and water. SPARG is an extension of the linear two-dimensional finite element method based computer program TDFEILD, Segerlind (1984). SPARG is written in QUICK BASIC and handles the non-linear features of the air sparging problem discussed above. A grid generation capability was developed to generate data for the problem in question. For a rectangular domain and a specified number of divisions along the sides of the domain, SPARG will generate nodal coordinates and element connectivity that are needed for the finite element module. The finite element computations were based on TDFEILD. The later code was modified to handle nonlinear, heterogeneous, anisotropic air sparging problems. The finite element program reads data in, evaluates element matrices, builds a global matrix and modifies that global matrix for boundary conditions, and solve for h_a using the direct Gauss method.

Due to the non-linearity of the problem, air permeabilities and air density had to be initialized for all elements to a same arbitrary value. Had these initial values been correct, the calculated air pressure head would have been final. Since that is not the case, corrections are needed to achieve the final solution (h_a). After the first iteration the calculated h_a values will be used to compute new air permeability values based on Equation 4. Thus the process is iterative and the solution is said to converge when the ratio of the sum of the absolute value of the difference in h_a between the previous and current solutions to the sum of current air pressure head falls bellow a user defined error tolerance value. The error is defined as:

$$e = \frac{\sum_{i=1}^{nodes} |h_{new} - h_{old}|}{\sum_{i=1}^{nodes} h_{new}} \tag{7}$$

where new and old refer to current and previous iterations

The air permeability as it appear in Equation 1 is an element property and is a function of the air pressure head at the nodes of the element. Permeability was evaluated at each node and the harmonic average over the element is computed. Similarly, the air density over the element is a harmonic average of its nodal densities. The permeability function has limits on the capillary head values (h_c) used in Equation 4. Capillary head values less than air entry head values (h_d) are not acceptable. Water saturated region will have negative h_c values i.e. $h_c < h_d$, in that case a minimum non-zero permeability were used. Decomposing the force vector requires dividing by the element stiffness matrix that will be forced to zero if the permeability is zero. For that reason zero permeabilities are not acceptable.

After convergence SPARG will generate output information on the nodal air pressure heads, element permeabilities, nodal capillary heads, element fluxes in the x and y directions, total air flux going through the matrix, and element saturations. Nodal air pressure head and nodal capillary head values are used to generate contour plots using SURFER, A Golden Software program. On a capillary head contour map, the location of the air/water interface is defined as the contour line corresponding to the value of the displacement head, where the depressed capillary fringe lie.

RESULTS AND DISCUSSION

SPARG was used to simulate the sparging problem described in Figures 1 and 2. Since no exact analytical solution is available, some understanding of the physical system is needed to see how reasonable were the results in describing the physics of the problem. The following parameters were used for the simulation that will follow: Length of the x-axis domain: 0.5 m, length of the y-axis domain: 0.8 m, height of water table above x-axis: 0.3 m, displacement pressure head: 0.2 m, well height above x-axis: 0.4 m, soil pore size distribution index (λ): 2, maximum permeability (K_{max}): 1E^{-11} m^2 minimum permeability (K_{min}): 1E^{-17} m^2, error tolerance : 0.001 %. Sparging point pressure 10.3m. The grid was right angle triangles with vertical and horizontal side dimensions of 0.1m. Results are reported as capillary head contour plots in Figure 3. This plot was used to locate the air water interface. As air is pumped into the water saturated system and a steady state is achieved air would have displaced the water up until the capillary pressure head between the two fluids is equal to the displacement pressure head, 0.2 m in this case. The 0.2 m contour that represents the

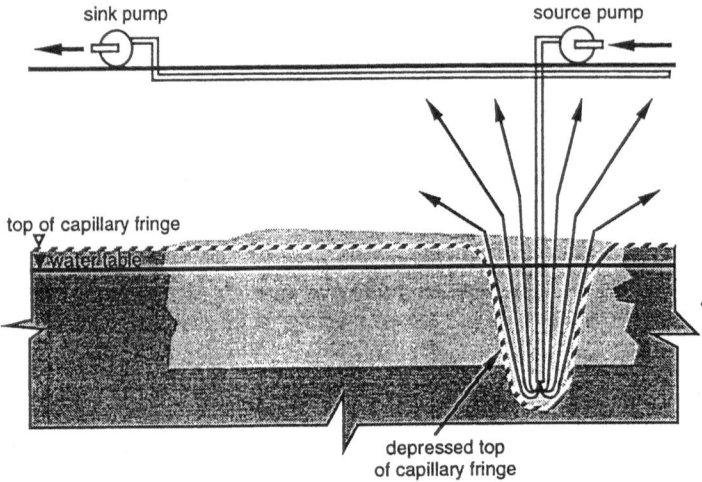

Figure 1. An NAPL contaminated site during air sparging. The contaminants are in the
vadose zone and near the water table region (light gray).

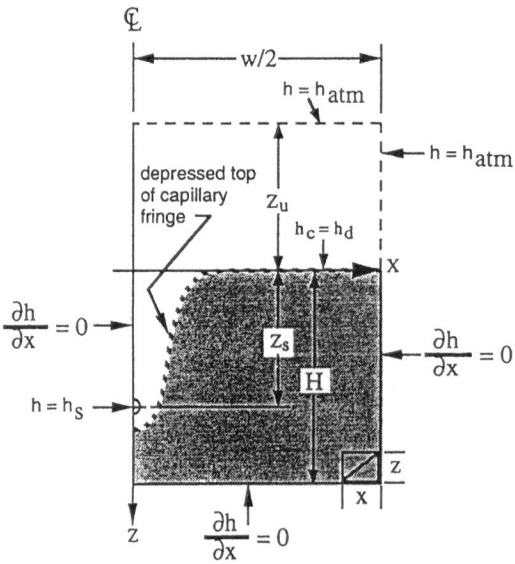

Figure 2. Air sparging problem defenition showing: problem parameters, cone of depression,
domain, and boundary conditions

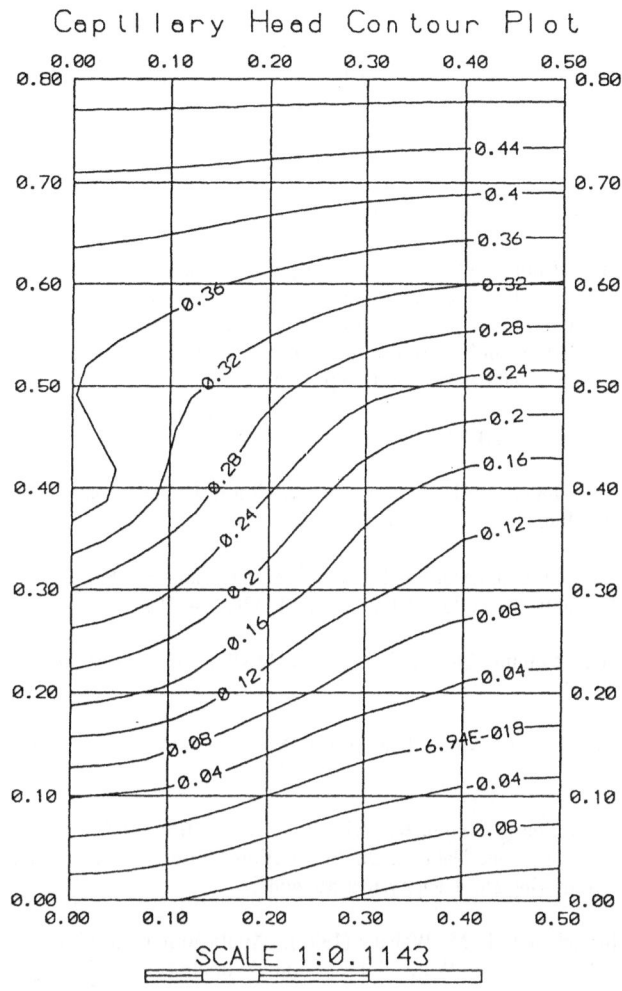

Figure 3. Capillary head contour plot (Pd = 0.2; grid .1x.1 m; well pressure 10.3m)

interface looks exactly the way we anticipated it in Figures 1 and 2. The interface approaches the undisturbed water table on the right side boundary tangentially.

SPARG was validated and tested on a variety of laboratory scenarios and performed well. Future tests against field scale experiments will be conducted.

SUMMARY AND CONCLUSIONS

A comprehensive finite element based model (SPARG) for the solution of steady air sparging technique is presented. SPARG was developed based on that technique handles a 2-dimensional heterogeneous, anisotropic, steady state air sparging problems. The model was verified for the linear problem. Results for the non-linear case agrees with the physical expectation of the problem. The technique presents a method to locate the water air interface in a two phase flow system. SPARG can be further utilized to increase our understanding of the physical air sparging problem as well as the solution procedure.

REFERENCES

Brown, R., C. Herman, and E. Henry. 1991. The Use of Aeration in Environmental Clean-ups. Groundwater Technology Inc.

Cho, Jong Soo. 1991. Forced Air Ventilation for Remediation of Unsaturated Soils Contaminated by VOC. EPA publication 60012-91/016. ADA, OK 74820.

Corey A. T. 1986. Mechanics of Immiscible Fluids in Porous Media. Water Resources Publications. Littleton Co. 80161.

Kresge, M. W. and M. F. Dacey 1991. An Evaluation of In Situ Groundwater Aeration. Groundwater Technology Inc.

Marley, M.C. (1991). Air Sparging in Conjunction with Vapor Extraction for Source Removal at VOC Spill Sites. Proceedings National Outdoor Action Conference on Aquifer Restoration, Ground Water Monitoring and Geophysical Methods.

Marley, M. C., D. J. Hazebrouck and T. M. Walsh. (1992). Application of In Situ Air Sparging as an Innovative Soils and Ground Water Remediation Technology. GWMR Spring.

Mohtar, R. H. (1992). Finite Element Analysis of the Air Sparging Problem. MS Report, Michigan State University.

Segerlind, L. J. (1984). Applied Finite Element Analysis. John Wiley and Sons, Inc.

Wallace, R. B., H. Annable, and R. H. Mohtar. 1994. Mathematical Models for Steady Air Sparging in Porous Media. Water Resources Research (In review).

Simulation of Two- and Three-Dimensional Dense Solute Plume Behavior with the METROPOL-3 Code

M. Oostrom (*), A. Leijnse (**), K.R. Roberson (*)

(*) *Pacific Northwest Laboratory[a], P.O. Box 999, MS K6-77, Richland, WA, 99352, U.S.A.*

(**) *RIVM, National Institute of Public Health and Environmental Protection, P.O. Box 1, 3720 BA, Bilthoven, The Netherlands.*

ABSTRACT

Contaminant plumes emanating from waste disposal facilities are often denser than the ambient groundwater. These so-called dense plumes sink deeper into phreatic aquifers and may, under certain conditions, become unstable. The behavior of variable density, aqueous-phase contaminant plumes in saturated, homogeneous 2-D and 3-D intermediate-scale aquifer models was investigated with the finite element code METROPOL-3. The numerical results compare, in a quantitative sense, to previously reported laboratory-scale transport experiments. The simulations show that dense plumes are more likely to penetrate deeper into aquifers and eventually become unstable with increasing density differences between the leachate solution and the ambient groundwater, and other important parameters as the saturated hydraulic conductivity of the porous medium, leakage rate of the contaminant solution, and source width. The significance of unstable behavior decreases with increasing dispersivity values. It was observed that 3-D flow patterns have a stabilizing effect on dense contaminant plume behavior.

INTRODUCTION

Many contaminant plumes originating from waste disposal facilities are denser than the ambient groundwater in natural aquifers (Freeze and Cherry, 1979). When density differences are significant, solute transport is not only a result of forced (hydraulically driven) advection and dispersion/diffusion but also of free convection. Studies reviewed by Gebhart et al. (1988) showed that, when a dense liquid overlies a less dense liquid, density gradients can introduce gravitational instabilities, giving rise to free convective transport. Such instabilities lead to enhanced mixing and dilution of contaminants. As a result of instabilities, dense contaminant plumes tend to contaminate larger areas of aquifers. Behavior of dense contaminant plumes has not been studied in great detail, and knowledge of variable density plume behavior remains incomplete. Published results of field studies are scarce (Kimmel and Braids, 1980), while only a few related laboratory studies have been reported (List, 1965; Paschke and Hoopes, 1984; Schincariol and Schwartz, 1990; Oostrom et al., 1992a,b; Hayworth, 1993). Oostrom et al. (1992a,b) studied dense plume behavior in relatively narrow flow containers by means of flow visualization and detailed salt

---

[a] Pacific Northwest Laboratory is operated for the U.S. Department of Energy by Battelle Memorial Institute under Contract DE-AC06-76RLO 1830.

A. Peters et al. (eds.), Computational Methods in Water Resources X, 975–982.
© *1994 Kluwer Academic Publishers. Printed in the Netherlands.*

concentration measurements with a dual-energy gamma radiation technique. They found that, for a given porous medium, dense plumes were either stable or unstable depending on magnitude of the horizontal flow velocity, the contaminant leakage rate, and the difference between the contaminant solution and the ambient groundwater. Hayworth (1993) performed similar experiments in wider flow containers and observed the source configuration also has a distinct effect on plume behavior. He noted that the likelihood of the occurrence of instabilities increased when the contaminant solution entered the flow container from a line source instead of a point source.

The objective of this paper is to present initial attempts to simulate dense plume behavior in both 2-D and 3-D flow domains similar to the experimental intermediate-scale laboratory flow containers used by Oostrom et al. (1992a,b) and Hayworth (1993). The METROPOL-3 results and experimental data can only be compared in a qualitative sense since the experiments were conducted in partly saturated porous media and METROPOL-3 allows saturated conditions only. Due to the strong coupling of flow and transport equations, simulations of dense contaminant plumes are not trivial. So far, only Frind (1982) and Koch and Zhang (1992) have published simulations of 2-D variable density plumes. The density effects in the plumes generated by Frind's finite element model (1982) were masked by the relatively high values of the dispersivity values used. Koch and Zhang (1992), using the code MOCDENSE, showed that, on a regional scale, 2-D dense plumes tend to sink deeper into aquifers. Instabilities were observed when the dense leachate concentrations were larger than 9000 ppm in combination with relatively small longitudinal dispersivities.

NUMERICAL METHOD

The simulations reported in this paper were conducted with the finite element code METROPOL-3 (Sauter et al., 1993). The coupled nonlinear partial differential equations for either the 2-D or 3-D flow and transport equations are solved simultaneously for liquid pressure and salt mass fraction using an approximate finite element solution. The mass balance equations for the liquid and the dissolved solute are, respectively,

$$\frac{\partial(n\rho)}{\partial t} + \nabla \cdot (\rho q) = 0 \tag{1}$$

$$\frac{\partial(n\rho\omega)}{\partial t} + \nabla \cdot (\rho q \omega) + \nabla \cdot J = 0 \tag{2}$$

where n is the porosity, ρ is the liquid density (ML^{-3}), q is the Darcy velocity (LT^{-1}), ω is the salt mass fraction, and J is the hydrodynamic dispersive flux (MT^{-1}L^{-2}). The Darcy velocity is written as

$$q = -\frac{k}{\mu} \cdot (\nabla p - \rho g) \tag{3}$$

where k is the intrinsic permeability tensor (L^2), μ is the fluid viscosity (ML^{-1}T^{-1}), p is the fluid pressure (MT^{-1}L^{-2}), and g is the gravitational acceleration (LT^{-2}). The dispersive flux is written as

$$J = -n\rho D \cdot \nabla \omega \tag{4}$$

where D is the hydrodynamic dispersion tensor. Equations (1) and (2) are nonlinear because of the dependency of the liquid properties μ and ρ on ω. The equations are coupled through q and ρ. The equations of state for the fluid properties ρ and μ are, respectively,

$$\rho = \rho_0 \exp^{\gamma\omega + \beta(p - p_r)} \tag{5}$$

$$\mu = \mu_0 (1 + 1.85\omega - 4.1\omega^2 + 44.5\omega^3) \tag{6}$$

where ρ_0 is the reference density, p_r the reference pressure, γ is a constant (0.69), and μ_0 the reference viscosity. The system of ordinary differential equations and the resulting algebraic equations are obtained after application of the Galerkin weighted residual method. The time derivatives are approximated by first-order finite differences and a fully implicit or Euler backward timestepping scheme is used. This method has the advantage that the time step size is not restricted by stability requirements. However, since the algebraic equations are non-linear, convergence requirements limit the time step size that can be used. In the solution procedure, the time derivatives are not reduced, but rather a lumped formulation has been adopted. The sets of algebraic mass balance equations can be conveniently written in matrix form as follows

$$A_p p = b_p \tag{7}$$

$$A_\omega \omega = b_\omega \tag{8}$$

where p and ω are the vector of unknowns. The matrix A_p and vector b_p are only mildly dependent on the pressure p. When the darcy velocity q and the boundary mass fluxes do not change significantly, the matrix A_ω and vector b_ω are also only slightly affected by changes in the pressure p. However, both matrices and both vectors in Eqs. (7) and (8) are strongly dependent on the salt mass fraction. Based on these considerations, a two-step iteration scheme was developed (Leijnse, 1992). First, the salt mass fraction equation is solved with a Picard iteration scheme using estimates for the pressure and the Darcy velocity from the previous time step. After convergence, the pressure equation is solved only once. New values for the Darcy velocities are calculated from the pressure distribution. The salt mass fraction equations have to be solved again if a converged solution has not been obtained.

The matrix associated with the linearized pressure equations is symmetric and is solved by a conjugate gradient method using diagonal scaling as preconditioning (Pini and Gambolati, 1990). The matrix associated with the linearized salt mass fraction equation is non-symmetric and is solved by the Bi-CGSTAB method (Van der Vorst, 1990).

The flow domains used in the simulations are all 2 m long and 1 m high. The width values used in the 3-D simulations were 0.1, 0.5 and 1.0 m. The contaminant solutions were introduced from a source on top of the saturated porous media with a constant leachate velocity. A horizontal pressure gradient was imposed on the domain, forcing the ambient groundwater to move from the left to the right. Each simulation consisted of two parts. In the first part, a tracer solution with an initial concentration of 10^{-2} g/l was introduced in the flow domain. This concentration doesn't significantly affect the density of contaminant plumes. After a steady-state neutral tracer plume has developed, the leachate concentration was increased to the desired value while everything else was kept constant. Throughout the simulations, the grid Peclet numbers in the near source area were always smaller than 2. The maximum time step was chosen a priori and was based on the expected maximum velocity for each simulation and the grid size. The Courant numbers during the simulations were always smaller than 1.

RESULTS AND DISCUSSION

2-D plumes

The simulations show that when the density difference between the contaminant solution and the ambient groundwater is increased, the liquid density gradients in the near source area increase and the plumes sink deeper into the flow domains, and are more likely to become unstable. The nearly neutral plume is transported in the upper part of the flow domain while dense unstable plumes may eventually reach the bottom of the aquifer. Examples of this behavior are shown in Fig. 1. In Fig. 1a., a nearly neutral plume (c_o=1 g/l) is shown, while in Fig. 1b. a dense plume is shown with initial concentration, c_o=50 g/l, corresponding to a relative density difference of 3.5%.

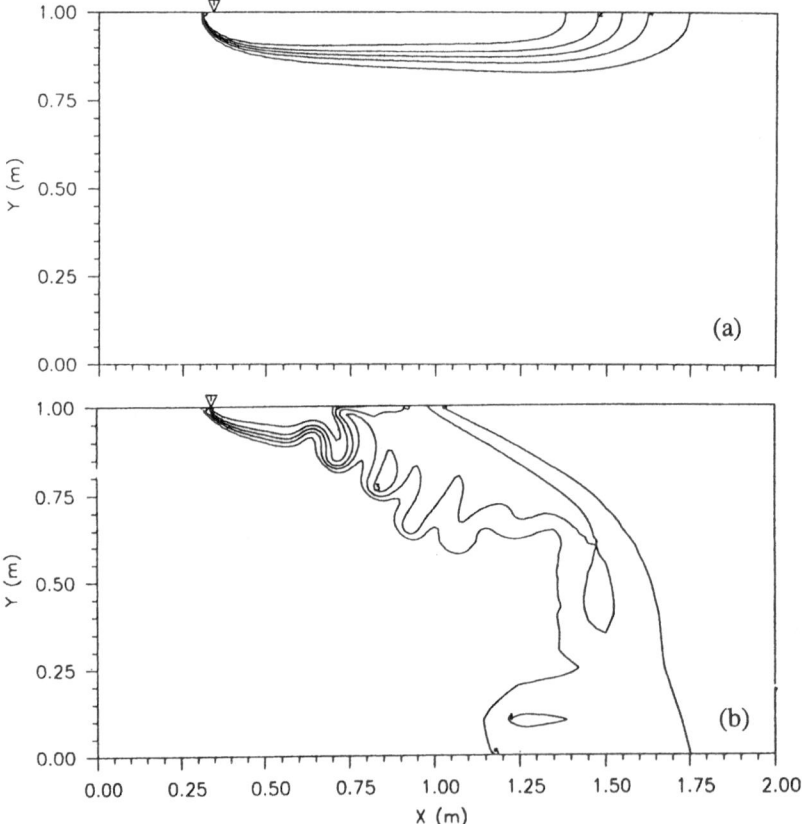

Fig. 1. Concentration contours of a) a neutral plume (c_o=1g/l), and b) a dense plume (c_o=50 g/l) after 12 hours with a horizontal Darcy velocity q_x=0.78 m/day. Shown are the 10, 30, 50, 70, and 90% contours of the c_o.

Other simulations indicated that larger Darcy velocities, as a result of higher horizontal pressure gradients, tend to stabilize dense plumes. A twofold increase in the flow rate almost completely stabilized the 50-g/l dense plume depicted in Fig. 1b, except for a small dip of the front edge of the plume (Fig. 2).

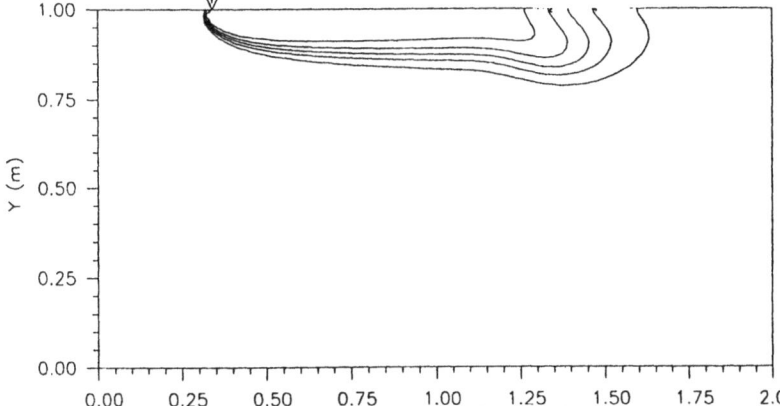

Fig. 2. Concentration contours of a 50 g/l plume with a horizontal Darcy velocity $q_x=1.56$ m/day. Shown are the 10, 30, 50, 70, and 90% contour of c_0.

Increases in the saturated hydraulic conductivity, while keeping the Darcy velocity constant, resulted in deeper-penetrating and generally more unstable plumes. These effects are related to near source behavior of the leaking contaminant solution. Since the leakage rates are the same in these simulations, dense plumes in media with larger conductivity values initially move faster in the vertical direction.

More mixing, realized in the simulations by using larger longitudinal or transverse dispersivity values, decreases the density gradients. Plumes with comparable density differences showed decreasing signs of unstable behavior with increasing dispersivity values.

Deeper-sinking plumes and an increased likelihood of unstable behavior when the density difference between the plume and the ambient groundwater increases or when the flow velocity in the flow domain decreases, have also been observed in the flow experiments described by Oostrom et al. (1992a,b) and Hayworth (1993).

3-D plumes

The dense plumes in the simulations emanated from either a centered point source or from line sources with various lengths. Plumes that developed from the point source did not show instabilities under conditions similar to the simulations in the 2-D flow domains. Contaminant concentrations exceeding 200 g/l were necessary to induce mild unstable patterns. These dense plumes did show, however, that plume penetration depth increases with increasing leaching rate and density difference. Relative-concentration contour lines for a plume resulting from a 100-g/l contaminant solution emanating from a point source into a 50-cm-wide flow domain are shown in Fig. 3a. This plume sinks rather deeply into the aquifer but remained stable at all times. A log-velocity plot of a y-z cross section (Fig. 3b) just downstream of the source shows water currents flowing downward below the plume and subsequently sideways and up. This circulatory flow pattern keeps the plume stable.

In contrast, a 100-g/l plume emanating from a 25-cm-wide line source showed instabilities in the longitudinal direction and also a totally different velocity profile downstream (Fig. 4a and b.) In this case, the ambient groundwater is unable to wrap around the plume to keep it from developing unstable patterns. These simulations clearly show the impact the source configuration has on dense plume behavior.

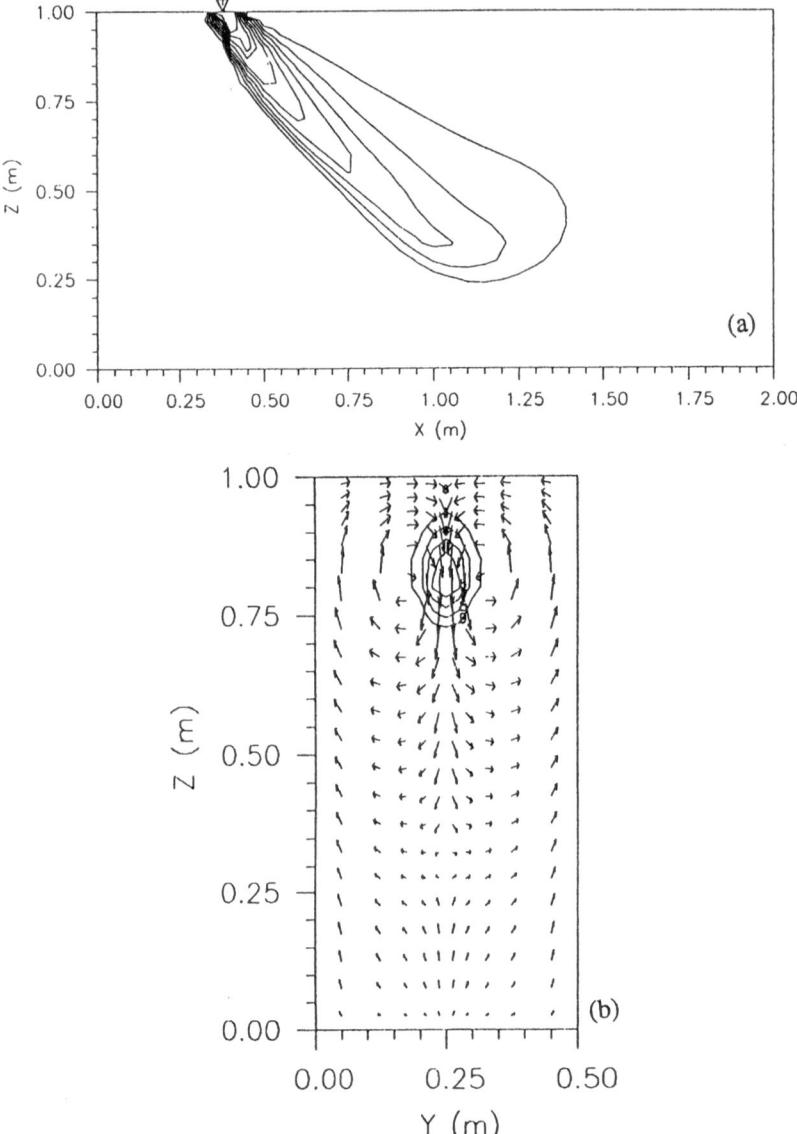

Fig. 3. a) Concentration contours, and b) log-velocity profiles at x=0.5 m of a 100-g/l plume emanating from a point source. The contours shown are the 10, 20, 30, 40, 50, 60, 70, 80, and 90% contours of c_o.

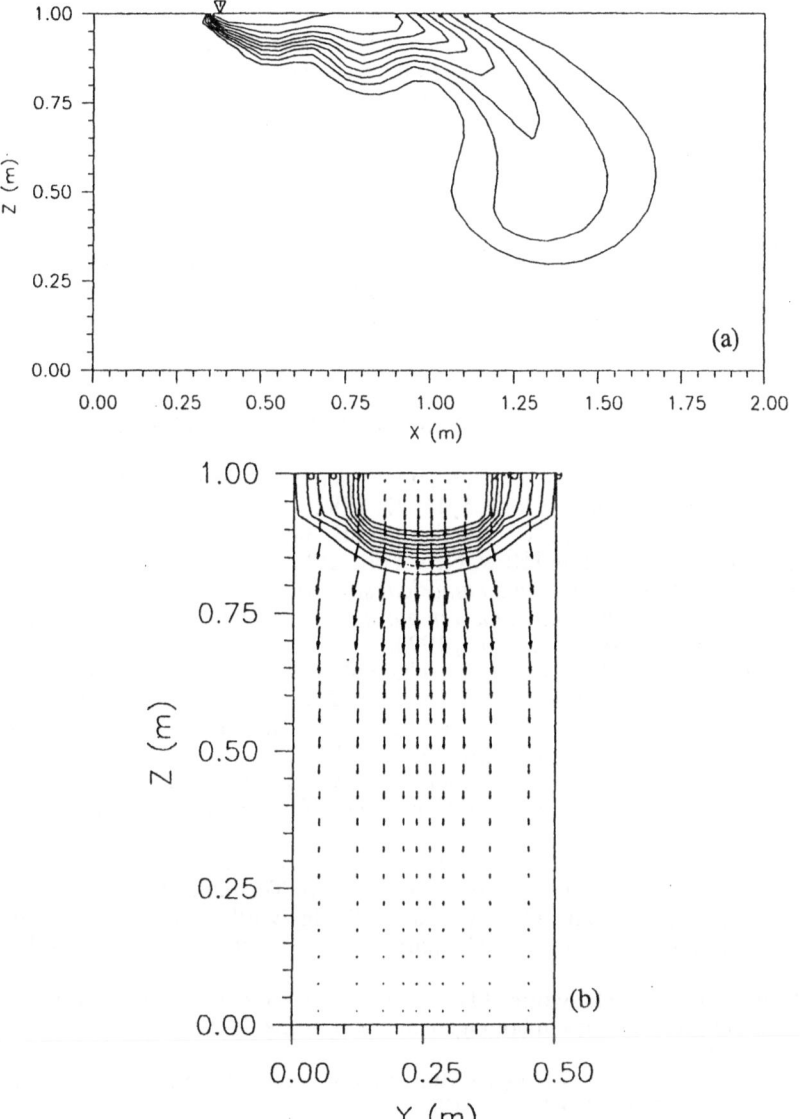

Fig. 4. a) Concentration contours, and b) velocity profiles at x=0.5 m of a 100-g/l plume
    emanating from a 25-cm-wide line source. The contours shown are the 10, 20, 30,
    40, 50, 60, 70, 80, and 90% contours of c_o.

CONCLUSIONS

The simulations in the 2-D and 3-D flow domains support in a qualitative sense the experimental results reported by Oostrom et al. (1992a,b) and Hayworth (1993). The simulations show that dense plumes are more likely to penetrate deeper into homogeneous aquifers and eventually become unstable with increasing density difference, saturated hydraulic conductivity, leakage rate of the contaminant solution, and source width. The significance of density effects decreases with increasing longitudinal and vertical dispersivity values. Large differences are observed between plumes in 2-D and 3-D flow domains under similar conditions. In general, 3-D groundwater flow patterns have a stabilizing effect on dense contaminant plume behavior.

Dense plume experiments in fully saturated homogeneous and heterogeneous porous media are currently being conducted at Auburn University, Alabama. The results will be used to further test the METROPOL -3 code. Additional numerical simulations are planned to investigate the phenomena that initiate and develop instabilities.

REFERENCES

Freeze, R.A., and J.A. Cherry, *Groundwater,* Prentice-Hall, Englewood Cliffs, New Jersey, 1979.

Frind, O.E., Simulation of long term transient density-dependent transport in groundwater, *Advances in Water Res.*, 5, 89-97, 1982.

Gebhart, B., Y. Jaluria, R.L. Mahajan, and B. Sammakia, *Buoyancy-Induced Flows and Transport*, Harper and Row, New York, 1988.

Hayworth, J.S, *A physical and numerical study of three-dimensional behavior of dense aqueous phase contaminant plumes in porous media*. Ph.D. Dissertation, Dept. of Civil Eng., Auburn University, Alabama, 1993.

Kimmel, G.E., and O.C. Braids, Leachate plumes in groundwater from Babylon and Islip landfills, Long Island, N.Y., *U.S. Geol. Surv. Prof. Paper, 1085*, 1980.

Koch, M., and G. Zhang, Numerical simulation of the effects of variable density in a contaminant plume, *Groundwater,* 30(5), 731-742, 1992.

Leijnse, A., *Three-dimensional modeling of coupled flow and transport in porous media*, Ph.D. Dissertation, Dept. of Civil Eng. and Geol. Sciences, Notre Dame, Indiana, 1992.

List, E.J., The stability and mixing of a density-stratified horizontal flow in a saturated porous medium, *Rep. KH-R-11*. Calif. Inst. of Technol., Pasadena, 1965.

Oostrom, M., J.H. Dane, O. Guven, and J.S. Hayworth, Experimental investigation of dense solute plumes in an unconfined aquifer model, *Water Resour. Res.*, 28(9), 2315-2326, 1992a.

Oostrom, M., J.S. Hayworth, J.H. Dane, and O. Guven, Behavior of dense aqueous phase leachate plumes in homogeneous porous media, *Water Resour. Res.*, 28(8), 2123-2134, 1992b.

Paschke, N.W., and J.A.Hoopes, Buoyant contaminant plumes in groundwater, *Water Resour. Res.*, 20(9), 1183-1192, 1984.

Pini, G., and G. Gambolati, Is a simple diagonal scaling the best preconditioner for conjugate gradients on supercomputers?, *Advances in Water Res.*,13, 147, 1990.

Sauter, F.J., A. Leijnse, and A.H.W. Beusen, Metropol's user's guide. *RIVM Report 725205003*, Bilthoven, the Netherlands, 1993.

Schinariol R.A. and F.W. Schwartz, An experimental investigation of variable density flow and mixing in homogeneous and heterogenous media, *Water Resour. Res.*, 26(10), 2317-2329, 1990.

Van der Vorst, H.A., Bi-CGSTAB: a fast and smoothly converging variant of Bi-CG for the solution of non-symmetric linear systems, *Preprint 63*, Dept. of Math., Univ. of Utrecht, The Netherlands, 1990.

IMPACT OF GRID RESOLUTION AND PARAMETRIC REPRESENTATION ON MULTIPHASE FLOW SIMULATIONS

KLAUS RATHFELDER and LINDA M. ABRIOLA
Department of Civil and Environmental Engineering
University of Michigan
Ann Arbor, MI 48109-2125
USA

The impact of grid resolution and parametric representation on the accuracy of numerical multiphase flow simulations was examined by sensitivity investigations. Grid resolution significantly influenced model predictions of organic liquid distribution, entrapment, and spreading distance. Model predictions were also significantly affected by the functional representation of capillary pressure - saturation relations. Results indicate that accurate prediction of long-term redistribution processes requires fine grid resolution and representation of capillary pressure gradients over the full range of capillary pressure.

INTRODUCTION

The accidental release of hazardous non-aqueous phase liquids (NAPLs) into groundwater aquifers poses health and safety problems, as well as, liability and economic concerns. Numerical models of multiphase flow in the subsurface have been developed to aid in prediction of NAPL migration pathways and fate. Relatively little attention, however, has been focused on factors influencing the numerical accuracy of multiphase flow models. Sensitivity investigations in the simulation of field scale two-phase immiscible flow were undertaken to examine the effect of grid resolution and parametric representation on model predictions.

MATHEMATICAL FORMULATION

The phase mass balance equations describing two phase flow in a porous medium may be expressed as (Abriola, 1989),

$$\phi \frac{\partial S_\alpha}{\partial t} = \boldsymbol{\nabla} \cdot \left[\frac{k k_{r\alpha}}{\mu_\alpha} \left(\boldsymbol{\nabla} P_\alpha - \gamma_\alpha \boldsymbol{\nabla} z \right) \right] \tag{1}$$

where $\alpha = w, o$ denotes the liquid phase (w=water, o=NAPL), ϕ = porosity, S_α = fluid saturation, k = intrinsic permeability tensor, $k_{r\alpha}$ = relative permeability, μ_α =

A. Peters et al. (eds.), Computational Methods in Water Resources X, 983–990.
© 1994 Kluwer Academic Publishers. Printed in the Netherlands.

dynamic viscosity, P_α = pressure, γ_α = specific gravity, and z = vertical coordinate. For simplicity, the NAPL is assumed to be completely immiscible with water, and the soil matrix and both liquids are assumed to be incompressible.

Constitutive relations for capillary pressure and relative permeability provide supplementary expressions needed to solve the system of equations above. Capillary pressure and relative permeability are empirically related to saturation of the wetting phase. There is, however, no single relation for these parameters which is used exclusively in numerical models. One common functional form for capillary pressure was developed by Brooks and Corey (1964),

$$P_c = P_d \overline{S}_w^{-1/\lambda} \tag{2}$$

where P_d = entry pressure, λ = pore size index, $\overline{S}_w = (S_w - S_{rw})/(1 - S_{rw})$, and S_{rw} = residual saturation. The Brooks and Corey function is frequently employed in the petroleum industry, and has also been used in application to contaminant hydrology problems (Faust et al., 1989; Kueper and Frind, 1991). A limitation of the Brooks and Corey function is that it is valid only for $P_c > P_d$. This constraint is generally not a concern in petroleum reservoir simulations where NAPL saturations normally remain non-zero. An absence of NAPL, however, is a common initial condition in the simulation of contaminant hydrology problems.

Various modifications have been proposed to extend applicability of (2) over the full range of $P_c(S_w)$. Su and Brooks (1975) developed a three parameter extension of (2), and Morel-Seytoux and Billica (1985) coupled (2) with an inverse tangent function. Neither of these approaches, however, is widely used in contaminant hydrology applications. Rather the model of van Genuchten (1980) has gained widespread usage. This model fits capillary pressure - saturation data with the expression,

$$P_c = \frac{1}{\alpha} \left(\overline{S}_w^{-1/m} - 1 \right)^{1/n} \tag{3}$$

where α, n, and $m = 1 - 1/n$ are fitting parameters.

The NAPL capillary pressure and relative permeability relations are hysteretic, and the degree of NAPL entrapment (S_{ot}) has been found to depend on the maximum NAPL saturation ($1-S_{w_{min}}$) reached during NAPL infiltration (primary drainage). Studies indicate hysteresis and entrapment must be included in numerical models to obtain accurate representation of NAPL migration (Essaid et al., 1993; Kueper et al., 1993). To capture these effects the model of Kaluarachchi and Parker (1992) is used in this work. The constitutive relations employed are summarized in Table 1.

NUMERICAL SIMULATIONS

To explore model sensitivity to grid and parametric representation, the infiltration and redistribution of tetrachloroethylene (ρ_o=1630 kg/m^3; μ_o=0.009 Pa-s) in the saturated zone was simulated under various conditions. Comparisons were made for

Table 1: Relative permeability and capillary pressure functions used in simulations.

Capillary Pressure $\qquad$ where:

van Genuchten $\; - \; P_c = \frac{1}{\alpha}\left(\overline{\overline{S}}_w^{-1/m} - 1\right)$ $\qquad \overline{\overline{S}}_w = \overline{S}_w + \overline{S}_{ot}$

Brooks and Corey $\; - \; P_c = P_d\overline{\overline{S}}_w^{-1/\lambda}$

$\overline{S}_{ot} = \overline{S}_{ot_{\max}}\left(\dfrac{\overline{S}_w - \overline{S}_{w_{\min}}}{1 - \overline{S}_{ot_{\max}} - \overline{S}_{w_{\min}}}\right)$

Relative permeability

$k_{rw} = \overline{S}_w^{1/2}\left[1 - \left(1 - \overline{S}_w^{1/m}\right)^m\right]^2$

$\overline{S}_{ot_{\max}} = \dfrac{1 - \overline{S}_{w_{\min}}}{1 + R\left(1 - \overline{S}_{w_{\min}}\right)}$

$k_{ro} = \left(1 - \overline{\overline{S}}_w\right)^{1/2}\left(1 - \overline{\overline{S}}_w^{1/m}\right)^{2m}$

$R = \dfrac{1 - S_{rw}}{S_{ro}} - 1$

two soil types: a uniform medium sand and a graded silt loam. Table 2 lists all soil properties and Figure 1 shows the primary drainage branch of the capillary pressure curves. The silt loam is slightly less permeable than the sand, has a greater entry pressure, and has larger capillary pressure gradients over the full range of $P_c(S_w)$. Details of the numerical model used to solve the flow equations (1) are presented in Abriola $et\ al.$ (1992).

Table 2: Soil properties used in simulations.

	Touchet Silt Loam GE 3*	#70 Silica**
ϕ	0.469	0.41
$k\ (m^2)$	3.577(-12)	8.19(-12)
S_{rw}	0.405	0.189
S_{ro}	0.3[†]	0.25[†]
$\alpha\ (cm^{-1})$	0.0062[‡]	0.0244
n	7.09	7.78
$P_d\ (cm)$	137.1	33.1
λ	4.41	3.3

* van Genuchten (1980) $\qquad$ ** Kueper and Frind (1991)
† estimated $\qquad$ ‡ scaled from air-water data

One-Dimensional Simulations

Figure 2 shows simulation results from a configuration in which NAPL is injected into a horizontal soil column at a constant and continuous specific discharge. Plot 2A shows NAPL saturation profiles in silt for a small and large specific discharge after equal volumes of NAPL have been injected. Plots 2B and 2C show the travel

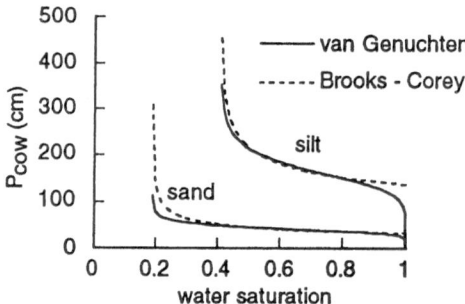

Figure 1: Capillary pressure functions used in simulations

distance and percentage of NAPL entrapped as a function of spatial discretization, respectively.

The results presented in Figure 2 can be related to the relative magnitude of advective and capillary fluxes. The magnitude of advective flow depends on the injection velocity, and in the case of non-horizontal flow, on the difference in density between the two fluids. Capillary flow is a dispersive flux which depends on the capillary pressure gradient and is related to the gradient of the retention curve and the spatial saturation gradient($\frac{dP_c}{dx} = \frac{dP_c}{dS_w}\frac{dS_w}{dx}$).

Advective fluxes are dominant at the larger specific discharge ($q = 1\ m/d$), resulting in a relatively sharp saturation front in Figure 2A. For this case, differences in the results using the van Genuchten and Brooks and Corey functions appear to be minor. At the smaller specific discharge ($q = 0.05\ m/d$) capillary fluxes increase in relative magnitude producing a more disperse saturation profile in Figure 2A. The functional form of $P_c(S_w)$, in this case, has substantial effect on lateral spreading at the leading edge of the front due to the difference in representation of dP_c/dS_w near $S_w=1$. Figure 1 indicates dP_c/dS_w near $S_w=1$ is smaller in the Brooks and Corey form, and thus, use of this capillary model results in smaller simulated capillary fluxes in this region.

Figures 2B shows the extent of NAPL penetration for equivalent injection volumes increases with increasing relative magnitude of capillary flux. This is evidenced by the greater penetration distances with the van Genuchten form at both flows, and by the greater penetration for the smaller injection rate. A consequence of greater penetration is that a larger proportion of the total NAPL volume will be entrapped as shown in Figure 2C. This is significant for accuracy of multi-dimensional simulations where the quantity of entrapment resulting from lateral capillary drive influences the volume of NAPL available for vertical drainage and pooling. Figures 2B and 2C also demonstrate that prediction of the penetration distance and NAPL entrapment are influenced by grid refinement, with larger values predicted with increasing grid size.

Figure 3 presents results from the simulation of vertical upward NAPL flow due to capillary pressure gradients. The vertical capillary flux was initiated by artificially

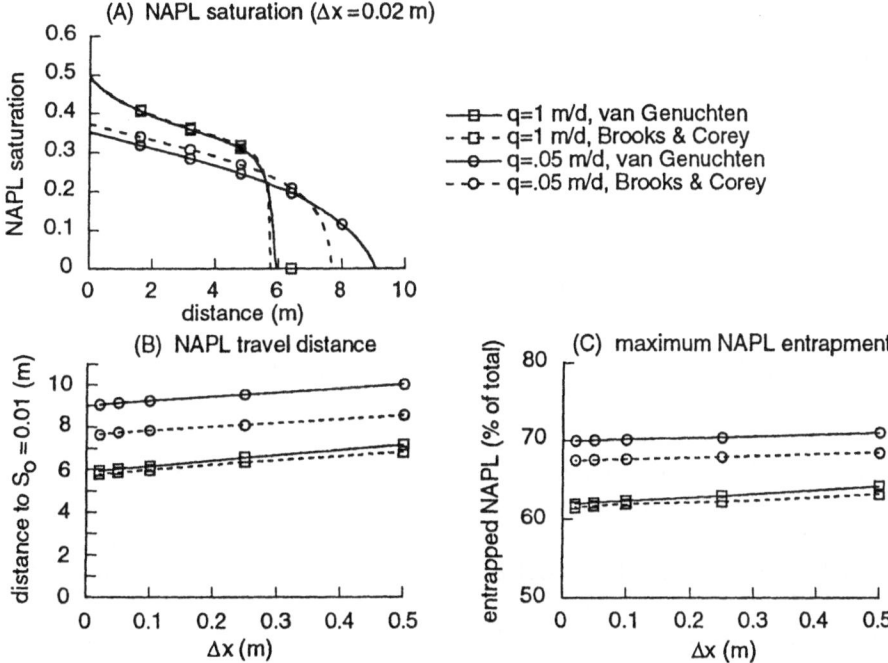

Figure 2: Results from the simulation of horizontal NAPL flow in the silt loam.

applying a constant NAPL saturation $(S_o=0.5)$ at the lower boundary; water was initially in hydrostatic equilibrium. Figure 3A shows that simulations using the van Genuchten function have a greater extent of vertical penetration. This, again, is a consequence of the representation of dP_c/dS_w near $S_w=1$. The extent of vertical migration was also found to depend strongly on grid size. Less penetration occurred with increasing grid size, and no penetration occurred at a grid size exceeding the entry pressure. Figure 3B shows the impact of predicted vertical migration on the volume of NAPL entrapment. Greater vertical penetration increases the volume of entrapped NAPL, which in the multi-dimensional scenario, translates to a reduction in the volume of mobile NAPL available for lateral spreading.

Two-Dimensional Simulations

Infiltration and redistribution of PCE was simulated in a homogeneous, saturated soil domain, bounded by a horizontal impermeable boundary at a depth of 3 m. The NAPL was released uniformly over a circular spill area with $r=0.5$ m at a rate of 50 l/d, for a total of 30 days. Subsequent redistribution and spreading was simulated for a total for 360 days. Numerical simulations were obtained on a quasi three-dimensional cylindrical $(r$-$z)$ coordinate system.

Figure 4 shows predicted NAPL saturation distribution in both soils at $t = 360$ days. Functional representation of $P_c(S_w)$ data with the van Genuchten and Brooks

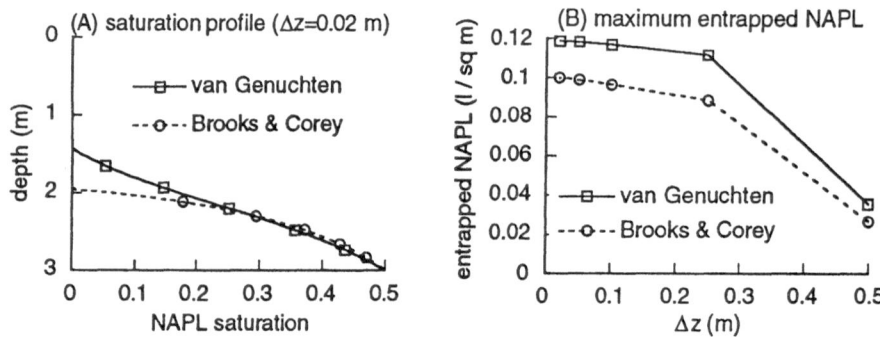

Figure 3: Results from simulation of vertical upward NAPL flow in the silt loam.

and Corey forms clearly produce different solution behavior. Solutions with the van Genuchten function exhibit greater capillary spreading, both in the lateral direction above the NAPL pool, and in the vertical direction within the NAPL pool. Solutions obtained with the Brooks and Corey function exhibit a lack of capillary spreading manifested by comparably sharp NAPL saturation fronts at the lateral and vertical boundaries of the contaminated region. These results are consistent with those previously demonstrated in Figures 2 and 3, and are attributed to the representation of capillary pressure gradients near $S_w=1$.

Grid resolution was also found to substantially affect the volume of NAPL entrapment and lateral spreading distance along the impermeable boundary. Grid refinement in the horizontal direction reduced the volume of NAPL entrapment above the NAPL pool, due to a reduction in NAPL spreading as shown in Figure 2. As a consequence, the NAPL volume in the pool increased with refinement, resulting in greater radial spreading along the impermeable stratum. This behavior occurred in solutions with either the van Genuchten or Brooks and Corey functions. Grid refinement in the vertical direction increased the volume of NAPL entrapment resulting from vertical capillary fluxes as shown in Figure 3. Because the Brooks and Corey function predicts less vertical rise, and consequently less entrapment, significantly greater radial spreading was predicted with this function.

SUMMARY AND DISCUSSION

Grid refinement in the simulation of two-phase flow problems significantly altered predicted NAPL distributions and maximum spreading distances. Sensitivity to grid resolution stems from the dependence of k_r and P_c on saturation. Smearing of saturation gradients on relatively coarse grids results in the calculation of diminished NAPL mobility; grid refinement reduces smearing of the NAPL gradients, providing a more accurate representation of NAPL mobility. When capillary forces play a significant role in NAPL migration, saturation gradients can occur over relatively small spatial scales. Thus these results imply that accurate simulation of NAPL migration requires

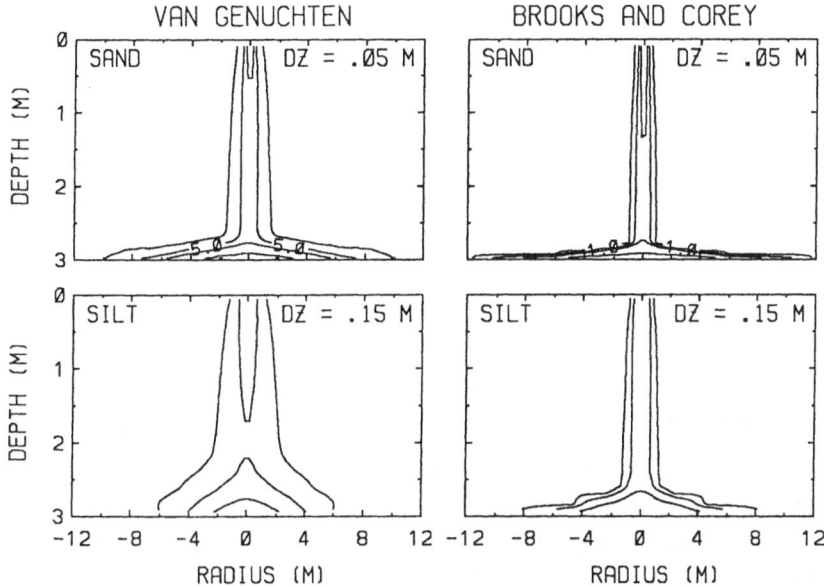

Figure 4: Predicted NAPL saturation distributions; contour intervals = 1, 5, 10, 20, and 30 % saturation.

a fine grid structure to capture small scale capillary drive which occurs during long term NAPL redistribution. Note that these conclusions are based upon numerical investigations in homogeneous media. Small scale heterogeneities would clearly add increased grid resolution requirements as was shown by Kueper *et al.* (1993).

Functional representation of $P_c(S_w)$ data with the Brooks and Corey and van Genuchten forms produced different solution behavior in the simulation of two-phase flow problems. The Brooks and Corey model produced solutions exhibiting sharper NAPL saturation fronts because capillary pressure gradients near $S_w=1$ are significantly smaller than those from the van Genuchten form. Conversely, greater capillary spreading was observed in solutions with the van Genuchten form. This was particularly evident in the prediction of vertical spreading as the NAPL migrated radially over the lower impermeable stratum. The degree of capillary spreading affects the volume of NAPL entrapment; greater spreading results in more entrapment. Thus solutions with the van Genuchten form exhibited greater NAPL entrapment, resulting in shorter radial spreading distances. Proper representation of dP_c/dS_w over the full range of P_c appears to be necessary for the accurate prediction of long-term NAPL redistribution processes.

The choice of grid resolution and representation of $P_c(S_w)$ has implications for field scale modeling capabilities. Computational demands will increase dramatically for fine grid resolution over large time scales of redistribution. Moreover the van Genuchten function is more nonlinear than the Brooks and Corey form, which will

further limit simulation time step size and increase overall computational requirements. These computational burdens, dictated by solution accuracy considerations,
point to the need for improved field scale simulators. Self-adaptive approaches to
computational resolution in space and time would appear to offer the most promise
for addressing these concerns.

ACKNOWLEDGMENTS

This work was supported by a Faculty Award for Women Scientists and Engineers
Grant from the National Science Foundation (EID 9023090).

REFERENCES

Abriola, L.M., K. Rathfelder, S. Yadav and M. Maiza (1992) *VALOR Code Version
1.0: A PC Code for Simulating Subsurface Immiscible Contaminant Transport*,
Final Report, EPRI TR-101018, Electric Power Research Institute, Palo Alto.

Abriola, L.M. (1989) Modeling migration of organic chemicals in groundwater systems
- A review and assessment, *Envir. Health Perspectives*, 83, 117-143.

Brooks, R.H. and A.T. Corey (1964) Hydraulic properties of porous media, *Hydrol.
Paper No. 3*, Colorado State University, Fort Collins.

Essaid, H.I., W.N. Herkelrath and K.M. Hess (1993) Simulation of fluid distributions
observed at a crude oil spill site incorporating hysteresis, oil entrapment, and
spatial variability of hydraulic properties, *Water Resour. Res.*, 29(6), 1753-70.

Faust, C.R., J.H. Guswa and J.W. Mercer (1989) Simulation of three-dimensional
flow of immiscible fluids within and below the unsaturated zone, *Water Resour.
Res.*, 25(12), 2449-64.

Kaluarachchi, J.J. and J.C. Parker (1992) Multiphase flow with a simplified model
for oil entrapment, *Trans. in Porous Media*, 7, 1-14.

Kueper, B.H., D. Redman, R.C. Starr, S. Reitsma and M. Mah (1993) A field experiment to study the behavior of tetrachloroethylene below the water table: spatial
distribution of residual and pooled DNAPL, *Ground Water*, 31(5), 756-66.

Kueper, B.H. and E.O. Frind (1991) Two-phase flow in heterogeneous porous media,
1. Model development, *Water Resour. Res.*, 27(6), 1049-1057.

Morel-Seytoux, H.J. and J.A. Billica (1985) A two-phase numerical model for prediction of infiltration: Application to a semi-infinite soil column, *Water Resour.
Res.*, 21(4), 607-15.

Su, C. and R.H. Brooks (1975) Soil hydraulic properties from infiltration tests, *Watershed Management Proceedings, Irrigation and Drainage Division, ASCE*, Logan,
Utah, Aug 11-13, 516-42.

van Genuchten M.T. (1980) A closed-form equation for predicting the hydraulic conductivity of unsaturated soils, *Sci. Soc. Am J.*, 44, 892-898.

APPLICATION OF THE LATTICE BOLTZMANN / LATTICE GAS TECHNIQUE TO MULTI-FLUID FLOW IN POROUS MEDIA

W.E. Soll, S.Y. Chen, K.G. Eggert, D.W. Grunau, and D.R. Janecky
Earth and Environmental Sciences Division
Mail Stop F665
Los Alamos National Laboratory
Los Alamos, NM 87545
USA

The lattice Boltzmann approach to modeling fluid flow provides an efficient and reliable method for solving the Navier-Stokes equations and studying multi-fluid flow problems. In this paper, we report state of the art capabilities of our lattice Boltzmann simulator for single- and two-fluid flows in two- and three-dimensional problems. We review the development of the code and present some of the latest results. Some of the flexibility available in the model includes arbitrary pore space descriptions, wettability effects, surface tension relations, and chemical reactivity. Simulations of two-fluid flow through a digitized micromodel geometry, and through a high resolution, digitized sample of Berea sandstone are presented. Relative permeability as a function of wettability[*] and capillary number[^] is discussed. Integration of the lattice Boltzmann approach into larger scale models to build a more powerful tool for analyzing constitutive behavior is considered.

INTRODUCTION

Over the past twenty years the study of subsurface fluid flow has grown dramatically, in almost direct relation to the growth of computing power. Problems that were intractable twenty years ago are routinely solved today. We have reached a point where computing power is no longer a major restricting factor in tackling the majority of subsurface flow and transport problems. Yet, we remain perplexed as to how to model many systems - not computationally, but conceptually. Why is this? Our computing power has far surpassed our ability to gain understanding about the systems of interest. We can compute flow in a system that has a different permeability in every grid-block, but we are unable to physically determine what permeabilities to use at such a high resolution. We can write code to compute chemical reactivity, but we do not understand the physical processes that are occurring within the reactions, and so on. Data are being collected with ever increasing success, due to development of better technologies for sampling and

---

[*] Wettability is a measure of which of the fluids present in the pore space is preferably in contact the solid.

[^] Capillary number is a non-dimensional number that characterizes the relative strengths of viscous forces to interfacial tension forces in a given systems.

A. Peters et al. (eds.), Computational Methods in Water Resources X, 991–999.
© 1994 Kluwer Academic Publishers. Printed in the Netherlands.

measurement, but it is neither feasible nor economic to continuously rely on physical measurement.

One means of potentially lessening our need for experimental data is to use a modeling approach that returns to first principles of physics. The growth of computing power has made it feasible to address questions regarding the influence of individual physical properties on fluid flow and transport behavior. Over the past eight years a computational approach that is particularly suited for studying micro-scale phenomena and relating them to continuum scale relationships has been under continuous development and extension. This approach is the "lattice gas" (LG) or "lattice Boltzmann" (LB) method.

What is a lattice gas / lattice Boltzmann? The LG method is a technique for simulating fluid flow which grew out of cellular automata theory. It is a temporally and spatially discrete method of reproducing fluid behavior. The LG method models a fluid as a collection of discrete particles moving through a space that is discretized by a regular lattice. These particles, which have unit mass and momentum, move through the lattice in a two-stage process. At each time step the particles advect to their next location ("streaming") then interact with other particles which occupy the lattice site ("collision"). Collisions are defined such that mass and momentum are conserved, and the averaged response recovers the partial differential equation of interest. For example, in the case of subsurface flow we are primarily interested in the Navier-Stokes equations. This approach is able to provide information on the microscopic scale, by looking at individual particle motion, and on the macroscopic scale, using ensemble averages of particle distribution and motion. Having the information at both scales simultaneously is important for improving our understanding of the constitutive parameters that are commonly used in large-scale simulations of flow and transport in porous media.

The LB approach is an extension of LG, in which the fluid is modeled using average particle populations, rather than discrete particles. The LB model still fully recovers the Navier-Stokes equations, however it eliminates the statistical noise inherent in the Boolean approach of the LG. Therefor, although the LB approach uses floating point operations instead of bit operations, the computational demands are balanced by the elimination of Monte Carlo-type runs to obtain a smooth, average solution.

Frisch, Hasslacher, and Pomeau, in 1986, introduced the first LG model that recovered the correct hydrodynamics. The "FHP" model is a two-dimensional, single-fluid flow model using a triangular lattice and up to six particles per lattice site (corresponding to up to one particle per link). This model was extended to three dimensions in d'Humieres et al. (1986) and Frisch, et al. (1987), using a face-centered hypercubic (FCHC) lattice with up to twenty-four particles per lattice site. The FHP model continues to be the foundation for LG codes.

Chen et al. (1992) introduced a LB model with two particles speeds that was able to utilize a true 3-D lattice which made it possible to have only fourteen nearest neighbors, as opposed to twenty-four. This significantly reduces the required computation time.

Rothman and Keller (1988) extended the single-fluid FHP model to simulate multi-fluid flows by introducing particles of two "colors". Each particle continues to follow the same collision rules as in the single fluid model except at interfaces. At interfaces the collision rules change to redistribute colors such that particles of the same color tend to remain together. Chen et al. (1991) and Somers and Rem (1991) modified the two-fluid approach, using both colored particles and colored "holes", where a "hole" denotes where a particle previously resided. This modification results in a more computationally efficient collision process at fluid-fluid interfaces.

Lattice gas automata have proven to be especially successful for describing flows through porous media because of the ease with which complex geometries and corresponding boundary conditions are handled. LG/LB methods have also been successfully applied to other subjects, including chemical reactivity (Wells et al., 1991; Gabetta and Monaco, 1991; Dawson et al.; 1993), magnetohydrodynamics (Chen et al., 1991), miscible displacement (Holme and Rothman, 1992), non-Newtonian fluids (Aharanov and Rothman, 1993), and thermohydrodynamics (Alexander et al., 1993).

THE LATTICE BOLTZMANN FLOW SIMULATOR AT LOS ALAMOS NATIONAL LABORATORY (LANL)

We believe that the LANL simulator the most complete and efficient LB model currently in existence. LANL is also fortunate enough to have massive computing power in its 1024-node massively parallel Connection Machine 5 (CM-5, by Thinking Machines) not equaled anywhere else. The LG and LB schemes are intrinsically parallel, making implementation on a massively parallel computer easy and efficient. The efficiency of implementation and storage allows very high resolution of many complex systems. This combination of code capability and computing power provides a formidable tool for studying the fundamental processes involved in multi-fluid flow and transport.

The LANL model is based on the FHP work in 2-D. The model uses the expanded 3-D capabilities of the 2-speed, 14-neighbor lattice of Chen et al. (1992). Multiple fluids are implemented based on Gunstensen et al. (1991) with modifications described by Grunau et al. (1993) to more efficiently model fluid interfaces. Further extensions have been made and reported (Grunau, 1993) that allow arbitrary viscosity and density ratios for the fluids, and arbitrary interfacial tensions.

The model is capable of simulating flow of one to three (or more) fluids in two or three spatial dimensions. Domain sizes as large as 256^3 lattice sites have been run. The 256^3 lattice corresponds to an actual high resolution digitization of a 2 cm^3 sample of Berea sandstone obtained through micro-computed tomography.

The LB model allows for arbitrary porous medium characteristics. This includes an arbitrarily complex pore geometry, including surface roughness. It is further possible, with the current code, to define arbitrary surface wettability and/or mineralogy, discretized as finely as one lattice site. Tests of the code have shown it can reproduce contact angles ranging from 20° to 180°. This range is sufficient to allow simulation in the entire range of fluid-pair wettabilities. Specification of mineralogy is of interest when modeling chemical reaction processes.

Options for boundary conditions for the inlet and outlet faces include: fixed pressure drop, fixed inlet and outlet density, and fixed input velocity (Dirichlet condition). Top and bottom boundaries are assigned to be either periodic or no-flow.

The LANL LB model is also able to implement almost arbitrary fluid characteristics. As all values in the LB methods are dimensionless, fluid characteristics are all defined in terms of ratios. Viscosity ratio, density ratio, and interfacial tension are all specifiable by the user. The only limitation on the choice of these values is in how that choice affects simulation run time. An increase in density ratio of one order of magnitude corresponds to approximately doubling the necessary run time.

The LB simulator provides a variety of output information. Both local and averaged velocities, pressures, and momenta are given. Fluid distributions and bulk saturations are provided, and relative permeabilities are computed. All averaged information is calculated as a straight summation over the probability distributions for each link of each lattice site. Relative permeabilities are computed based on the formulation given in de Marsily (1986).

Another area of effort that we have been focusing on with the LB/LG models is of coupled flow and chemical reactions. We continue to develop our LG and LB hydrodynamic models which integrate chemical transport and reaction processes for application to study at the pore scale. These models are partially distinct from the general multi-fluid models. The chemical reaction models incorporate most of the flexibility of the multi-fluid models but are currently simplified to two dimensions to allow for the increased computationally requirements. Diffusion, sorption and desorption, and solutions and dissolution are incorporated into the models. Sorption and desorption reactions are the focus of work with fixed boundary conditions between solution and solid (Wells et al., 1991; Janecky et al., 1992). Initial detailed models have been developed for sorption/desorption chemical reactions at solid surfaces (Janecky et al., 1993), including multicomponent sorption/site competition as a function of space and time in heterogeneous pore network structures, where hydrodynamic transport, solute diffusion and mineral surface processes are all treated explicitly. The present LB-chemistry model includes potential for up to 200 solute components and binary solute reactions (Dawson et al., 1993).

RESULTS FROM THE LANL LATTICE BOLTZMANN SIMULATOR

Our recent work has been primarily directed toward the application of the LB method to simulation of constitutive relationships in porous media. We began by simulating single fluid flow through a variety of porous media and computing permeabilities for the media. Both 2-D (micromodels) and 3-D (beadpacks, actual porous samples) systems were simulated. Simulations were run in digitizations of actual porous media, which made it possible to closely compare simulated and measured permeabilities. Table 1 contains values for measured and calculated permeabilities in a number of different porous media.

The agreement between simulated and measured permeabilities is excellent. There is one notable difference for the sandstone simulated in the short direction of the core. The discrepancy between the two values is most likely due to the subsample from which the digitization was made not being homogeneous at the scale of the simulation, while the larger experimental sample was homogeneous. This raises the question of how large a simulated domain must be, and to what resolution must we discretize to obtain a representative pore space. Current considerations of these questions is considered in the last section of this paper.

Porous Medium	k (exper.) [darcy]	k (simul.) [darcy]	Imaging Resolution	Comments
Micromodel-1	23	21.5	~80 μm	pore depth=50 μm
Micromodel-2	37	32	~80 μm	pore depth=74 μm
Beadpack	1100	1000	50 μm	
Berea	1	1.2	10 μm	
Sandstone	7.4	8.6	20 μm	flow in long dir
Sandstone	7.4	21.5	20 μm	flow in short dir

Table 1: Comparison of measured and calculated permeabilities for various porous media

Simulations have also been run for two-fluid flow, from which the corresponding relative permeabilities have been computed. Both 2-D and 3-D systems have been used, but figures are selected from 2-D systems for ease of graphical presentation. Examples of some of the model results are shown in Figures 1 and 2.

We are not only interested in being able to reproduce the appropriate functional response, but also in identifying what influence various physical features and flow processes have on the constitutive behavior. With the two-fluid systems we have also addressed the question of how wettability and capillary number (N_{ca}) affect displacement efficiency and permeability. Figure 1 shows fluid distributions in a digital micromodel pore space for flows at three capillary numbers. In Figure 1a the flow is in a viscous dominated regime, reflected by more frontal displacement and high displacement efficiency than the other figures. In Figure 1c the flow is in an interfacial tension dominated regime, which is reflected in the highly fingered fluid distribution, and the high residual saturations. Figure 2 shows the corresponding calculated relative permeability curves for the systems run at the larger two of capillary numbers.

The LB simulator is capable of capturing a range of rock wettability between 20° and 180°. This range is sufficient to be able to capture the entire range of wettability, in conjunction with the model ability to change the surface wettability. Figure 3 plots the displacement efficiency versus the wettability. The most efficient displacement is obtained at neutral wettability, and it is most difficult to extract oil from oil-wet rock.

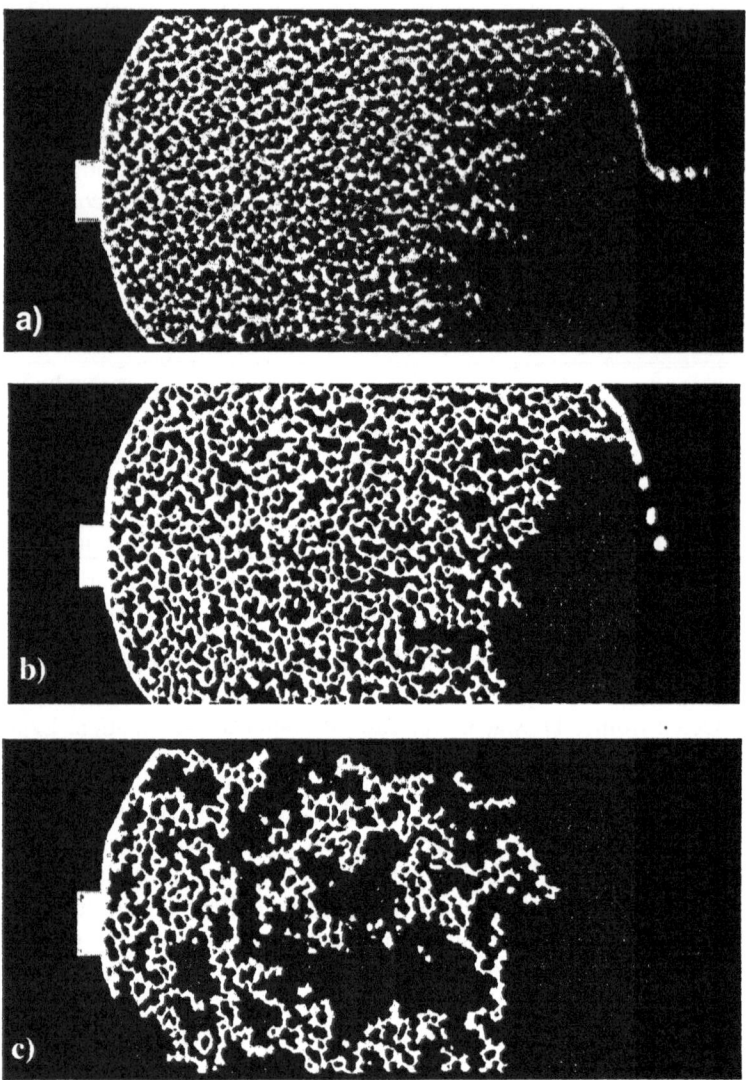

Figure 1. Comparison of immiscible displacement under different capillary number (velocity) drives. White is non-wetting. Gray is wetting. a) Nca = 2.2 x 10^{-2}; b) Nca = 5.7 x 10^{-3}; c) Nca = 1.5 x 10^{-4}.

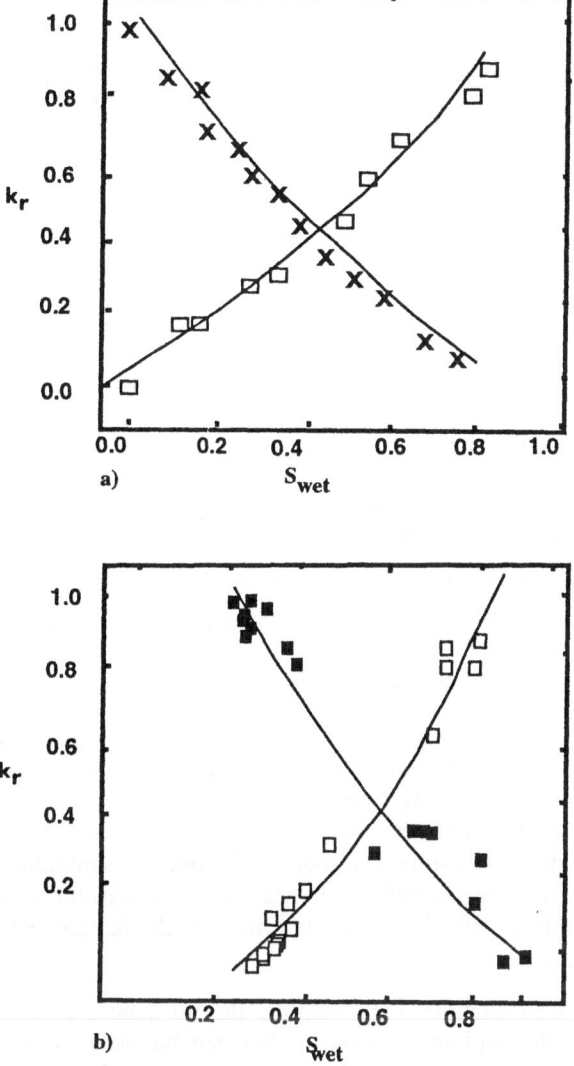

Figure 2. Relative permeabilities for micromodel simulations.
a) $N_{ca} = 10^{-2}$, b) $N_{ca} = 10^{-3}$

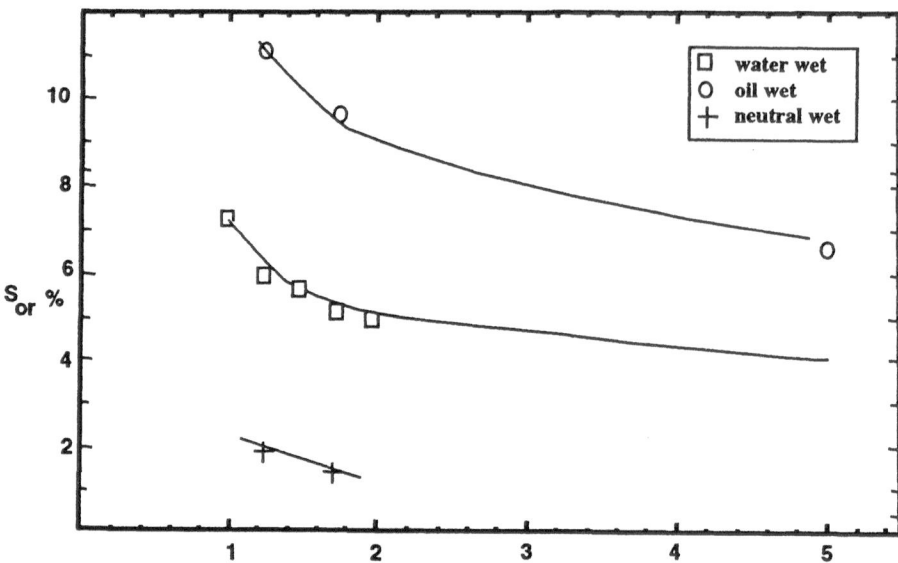

Figure 3. Effect of wettability on residual oil saturation from a simulated
waterflood in a micromodel (water displacing oil).

MOVING BETWEEN MODELING SCALES

We are beginning to address the question of moving between scales of fluid flow using
LB techniques. Although the LB method uses averaging to provide any macroscale
information, i.e. pressure and velocity fields, the resulting averaged quantities must be
representative of the simulation domain to be of value. It is not clear, at this time, whether
these averaged quantities values can be extrapolated beyond the simulation domain. We
must first determine appropriate averaging volumes and geometries for use in paralleling
classical averaging theories. It should be kept in mind that the particular REV size may
vary between different extensive variables.

We are currently trying to quantitatively describe the influences of different system
properties to arrive at descriptive parameters that can be used to synthesize new,
physically based functional forms for constitutive relationships. We are examining how
pore structure, material texture, and fluid-solid interactions affect the definition of
characteristic medium constants or responses (such as the porosity, permeability, relative
permeability, wettability, dispersivity, etc.). This will help to identify a true REV. We
anticipate paralleling common theoretical procedures by integrating, pore-scale quantities
obtained from simulations (such as specific volume, specific interfacial area, fluid mass
density, fluid momentum density, etc.) over predetermined REVs, thus computing typical
macroscopic quantities (saturation, flux, interfacial area per unit volume) that are usually
measured experimentally, and applied in field scale equations.

ACKNOWLEDGMENTS

This work has been supported by the Los Alamos National Laboratory which is operated by the University of California for the U.S. Department of Energy. The author wishes to acknowledge the Advanced Computing Laboratory of Los Alamos National Laboratory, Los Alamos, NM 87545. This work was performed on computing resources located at this facility.

REFERENCES

Aharonov, E., and DH. Rothman (1993). Non-Newtonian Flow (Through Porous-Media) - A Lattice-Boltzmann Method. Geophysical Research Letters, 20(8). pp. 679-682.

Alexander, F.J., S.Y. Chen, and J.D. Sterling (1993). Lattice Boltzmann Thermohydrodynamics. Phys. Rev. E, 47(4). pp. R2249-R2252.

Chen, S.Y., H.D. Chen, D. Martinez, and W. Matthaeus (1991). "Lattice Boltzmann Model for Simulation of Magnetohydrodynamics", Phys. Rev. Lett., 67. pp. 3776

Chen, S.Y., G.D. Doolen, K. Eggert, D. Grunau, and E.Y. Loh (1991). Local Lattice Gas Model for Immiscible Fluids. Phys. Rev. A, 43(12). pp. 7053-7056.

Chen, S.Y., Z. Wang, X.W. Shan, and G.D. Doolen (1992). Lattice Boltzmann Computational Fluid-Dynamics in Three Dimensions. J. Stat. Phys., 68(3/4). pp. 379-400.

Dawson, S.P., S. Chen, and G.D. Doolen (1993). "Lattice Boltzmann Computations for Reaction-Diffusion Equations". J. Chem. Phys., 98(2). pp. 1514-1523.

de Marsily, G. (1986). Quantitative Hydrogeology. Academic Press, Inc, New York. pp. 41-45.

Frisch, U., B. Hasslacher, and Y. Pomeau (1986). "Lattice-Gas Automata for the Navier-Stokes Equation". Physical Review Letters, 56. pp. 1505-1508.

Frisch, U., D. d'Humieres, B. Hasslacher, P. Lallemand, Y. Pomeau, and J.P. Rivet (1987). "Lattice Gas Hydrodynamics in Two and Three Dimensions". Complex Syst, 1. pp. 649-707.

Gabetta, E., and R. Monaco (1991). The Discrete Boltzmann Equation for Gases with Bi-Molecular Chemical Reactions. From Discrete Models of Fluid Dynamics, AS. Alves, ed, World Scientific. pp.22-34.

Grunau, D.W. (1993). "Lattice Methods for Modeling Hydrodynamics". PhD Dissertation, Colorado State University, Dept. of Math.

Grunau, DW., SY. Chen, and K. Eggert (1993). "A Lattice Boltzmann Model for Multi-phase Fluid Flows". Phys. Fluids A.,5(10). pp. 2557-2562.

Gunstensen, A.K., and D.H. Rothman (1991). "A Lattice Gas Model for 3 Immiscible Fluids". Physica D, 47(1/2). pp. 85-96.

Gunstensen, A.K., D.H. Rothman, S. Zaleski, and G. Zanetti (1991). "Lattice Boltzmann Model of Immiscible Fluids". Phys. Rev. A., 43(8). pp.4320-4327.

Holme, R., and D.H. Rothman (1992). "Lattice Gas and Lattice Boltzmann Models of Miscible Fluids". J. Stat. Phys., 68(3/4). pp. 409-429.

Janecky, D.R., S.Y. Chen, S. Dawson, K.G. Eggert, and B.J. Travis (1992). "Lattice Gas Automata for flow and transport in geochemical systems". In Y.K. Kharaka and A.S. Maest (Eds.), Proceedings of the 7th International Symposium on Water-Rock Interaction, A.A. Balkema, Rotterdam, Netherlands, pp. 1043-1046.

Janecky, D.R., S.Y. Chen, and K.G. Eggert (1993). "Detailed Heterogeneous Pore Scale Models for Transport Coupled to Chemical Reactions and Multi-Phase Flow". EOS, Am. Geophys. Union.

Rothman, D.H., and J.M. Keller (1988). "Immiscible Cellular-Automaton Fluids". J. Stat Phys, 52(3/4). pp.1119-1127.

Somers, J.A., and P.C. Rem (1991). "Analysis of Surface-Tension in 2-Phase Lattice Gases". Physica D, 47(1/2). pp. 39-46.

Wells, J.T., D.R. Janecky, and B.J. Travis (1991). "A Lattice Gas Automata Model for Heterogeneous Chemical Reactions at Mineral Surfaces and in Pore Networks". Physica D, 47, pp. 115-123.

MULTIPHASE FLOW AND TRANSPORT IN ROUGH-WALLED FRACTURES

N.R. THOMSON and S. J. ESPOSITO
Department of Civil Engineering
University of Waterloo
Waterloo, Ontario, N2L 3G1
CANADA

The rough-walled fracture is the basic component of a dual-continuum system; however very little is known about the behaviour of non-aqueous phase liquids in such a fracture. To improve our understanding, a finite volume formulation for multiphase flow and aqueous phase transport in a single rough-walled fracture has been developed. A hypothetical example is presented to demonstrate the utility of the developed model.

BACKGROUND

Subsurface contamination at sites which are situated on fractured clay or fractured sedimentary bedrock has resulted in part due to accidental releases of non-aqueous phase liquids (NAPLs). Once these liquids enter a fractured subsurface regime their ultimate fate will be controlled by the nature of this dual-continuum system. It is believed that the geometry of the fractures control the bulk flow with the porous matrix playing a minor role. Some of the NAPL that flows through an individual fracture will become trapped in smaller aperture regions or perhaps isolated, and hence may be considered non-mobile. These regions of essentially non-mobile NAPLs will slowly partition into the surrounding aqueous phase and therefore have the potential to create a large aqueous phase plume within the dual-continuum system.

The subsurface environments that contain these fracture systems have typically been represented by porous medium models. This porous medium representation of a fractured domain is only valid when there is a high fracture density and connectivity, and the problem domain is large enough to encompass vast numbers of fractures. For smaller domains with fewer fractures, a porous medium model cannot accurately represent the fracture flow system. In these cases, a model that represents discrete fractures must be employed. Before such a model can be utilized with confidence, the fundamental components of such a model, multiphase flow and transport within a single rough-walled fracture, must be investigated to determine the controlling mechanisms.

This paper discusses the numerical formulation that has been employed to investigate multiphase flow and transport within a single rough-walled fracture. A hypothetical example to illustrate the utility of the developed model is also presented.

A. Peters et al. (eds.), Computational Methods in Water Resources X, 1001–1008.
© 1994 *Kluwer Academic Publishers. Printed in the Netherlands.*

FRACTURE CONCEPTUALIZATION

The distribution of fracture apertures is very important; however, it is extremely difficult to accurately measure the aperture distribution within a fracture, especially if the fracture sample is not to be destroyed. Due to the difficulty in measuring actual aperture values, past studies of flow in rough-walled fractures have employed aperture distributions generated by geostatistical methods based on a lognormal aperture distribution and a specified covariance structure (e.g., *Pruess and Tsang*, 1990). The random field generator developed by *Robin* (1991) using the Fast Fourier Transform spectral techniques is adopted in this work. A fracture plane is defined and discretized laterally by the x- and y-coordinate directions. Each grid cell is assigned a unique aperture value and an elevation at the centroid of the cell.

MODEL OVERVIEW

The developed model consists of a flow and transport component which are numerically represented by the finite volume method (Patankar, 1980). The model has been constructed in such a way as to allow for the flow and transport components to be solved sequentially, or for the flow field and NAPL distribution to be established and then the transport component to be determined. In the latter situation, the flow field and NAPL distribution can be updated after a specified amount of NAPL mass has partitioned to the aqueous phase.

Time is treated explicitly. This allows the problem to be divided into three stages at each time step: pressure field determination, fluid migration, and aqueous phase transport. The pressure field is based on fundamental flow relationships, the known location of the fluid-fluid interfaces, and the imposed boundary conditions. Given the pressure field for a time t, the flows at the cell faces are calculated for the time period t to $t+\Delta t$. Knowing these flows, the saturation information is altered. Using this new saturation information along with the pressure distribution at time t, aqueous phase transport calculations are performed. Since this transport calculation allows for partitioning from the NAPL to the aqueous phase, the saturation information is updated a second time during the time step.

Each fluid-fluid interface within the solution domain is represented as a discrete pressure discontinuity and requires an exact representation of this interface location. This requirement is met by defining not only the degree of saturation of the non-wetting phase within a cell, but also the distribution of this phase within the cell. Each cell is defined in terms of a saturation shape and a drainage phase, as well as a saturation level. The saturation level S_P, of cell P represents the fraction of the cell volume occupied by the non-wetting phase (assumed to be the NAPL). The shape provides a simplified representation of phase distribution within the cell, and provide a means of determining saturation, capillary pressure and interface location. Thirty shapes have been developed and are considered to be sufficient to represent the entire range of possible non-wetting phase distributions within a cell (Murphy and Thomson, 1993). In order to make the flow problem tractable, only a single phase is allowed to flow across each face of cell P. Clearly this is an approximation; however, with a suitably fine fracture discretization, this has not been shown to significantly affect the results obtained from this model.

The adopted definition of the non-wetting phase saturations and shapes along with the wetting phase pressure distribution, are used in the wetting phase transport calculations to determine the required non-wetting/wetting phase contact area, and the wetting phase velocities. This non-wetting/wetting phase contact area is employed in the mass partitioning relationships.

The system of equations for both the pressure field distribution and for the aqueous phase concentration field at each time step is solved iteratively using a preconditioned second-degree incomplete factorization scheme with Orthomin acceleration.

Flow Formulation

The flow component of this mathematical model is based on a finite volume expression of mass continuity, the cubic law, the capillary pressure equation and a set of prescribed pressure and phase boundary conditions. It is assumed that the flow system satisfies the following requirements: isothermal, incompressible fluid, constant aperture within a cell, laminar flow, smooth and impermeable fracture walls within a cell, and no external forces. Cells are assigned an initial shape corresponding to the initial conditions of the simulation event. Pressure within a cell is represented by a value calculated for the cell's centroid. The cell pressure corresponds to the pressure within the phase that occupies the location of the cell centroid. Any capillary pressure influence within the cell is determined from the shape, phase and saturation, as well as the defined aperture value for that cell. Ignoring mass partitioning for the moment, the change in the saturation level of each fluid β over a time interval may be expressed for an individual cell P as

$$\frac{V_P(S_{P,\beta}^k - S_{P,\beta}^{k-1})}{\Delta t^k} = Q_{w,\beta}^{k-1} + Q_{s,\beta}^{k-1} - Q_{n,\beta}^{k-1} - Q_{e,\beta}^{k-1} \tag{1}$$

where the saturation of fluid β ($\beta = w$ (wetting phase), nw (non-wetting phase)) at the beginning and end of the time increment Δt^k is represented by $S_{P,\beta}^{k-1}$ and $S_{P,\beta}^k$, respectively; the volume of the fracture cell is represented by $V_p = a(\Delta x)(\Delta y)$, a is the cell aperture, and the flow rate of fluid β across cell face i ($i = w, s, n$ and e) is represented by $Q_{i,\beta}^{k-1}$. The explicit definition of flow rates in (1) allows time to be treated explicitly. Flow rates are determined based on the known pressure and phase distributions at time t^{k-1}. Since only one phase is allowed to flow across a given cell face during one time step, the flow rates are independent of the phase crossing the cell face and hence the following expression of mass conservation can be employed

$$Q_w^{k-1} + Q_s^{k-1} - Q_n^{k-1} - Q_e^{k-1} = 0 \tag{2}$$

Equation (2) forms the basis for solving the pressure field associated with a specific phase distribution at a given time t^{k-1}. The flow across cell face i at time t^{k-1} is given by

$$Q_i^{k-1} = -\frac{1}{12}\left(\frac{cl_{i,1}\mu_w + cl_{i,2}\mu_{nw}}{a_1^3 fw_1} + \frac{cl_{i,3}\mu_w + cl_{i,4}\mu_{nw}}{a_2^3 fw_2}\right)^{-1}(\Delta p_i + z_i^*) \qquad (3)$$

where the net pressure difference between the cells is represented by Δp_i, the elevation potential is represented by z_i^*, μ_β is the fluid viscosity, $cl_{i,j}$ is the dimensionless characteristic length j for cell face i, fw_i is a dimensionless flow width for cell face i, and the lower and upper coordinate cells are denoted by the subscripts 1 and 2, respectively (see Murphy and Thomson (1993) for details). The driving force between two cell centroids due to gravity incorporates the density effects of the fluids that lie between the cell centroids as expressed by

$$z_i^* = g\left(\rho_w(cl_{i,1} + cl_{i,3}) + \rho_{nw}(cl_{i,2} + cl_{i,4})\right)\Delta z_i \qquad (4)$$

which reduces to $z_i^* = \rho\, g\,\Delta z_i$ for the case of a single fluid. The net pressure difference between the cell centroids across cell face i is given by

$$\Delta p_i = p_2 - p_1 + p_{c_i} \qquad (5)$$

where the net capillary pressure difference between cell centroids 1 and 2 is represented by p_{c_i}. Individual components of p_{c_i} are calculated using $(2\sigma\cos\theta/a)$, where σ is the interfacial tension, θ is the contact angle, and a is the aperture. If no fluid-fluid interfaces lie between the cell centroids, p_{c_i} is equal to zero.

Transport Formulation

The phase saturations determined in (1) are further altered at the end of each time step since non-wetting phase mass is permitted to partition into the wetting phase. To account for this mass partitioning, the following expression can be written for each cell containing the non-wetting phase

$$\frac{V_P \rho_{nw}[(S_{P,nw}^k)' - (S_{P,nw}^k)]}{\Delta t^k} = -\sum_{nb}\lambda_{nb}(C_{eq} - C_{nb})A_{nb} \qquad (6)$$

where $(S_{P,nw}^k)'$ represents the altered saturation level in cell P after mass partitioning has taken place, nb represents the neighbouring cells involved in wetting phase transport, λ_{nb} is the mass transfer rate coefficient, C_{eq} is the solubility of the non-wetting phase in the wetting phase, C_{nb} is the wetting phase concentration in the neighbouring cell nb, and A_{nb} is the non-wetting/wetting phase contact area between cell P and cell nb. To keep this mass accounting manageable, mass partitioning is only allowed to take place into neighbouring cells, and therefore it is assumed that the wetting phase within a cell that contains the non-wetting phase does not play a role in the transport calculations. Assuming that the transport mechanisms of advection, dispersion, diffusion, dissolution, and matrix diffusion are the dominant mechanisms, then the change in the concentration level over a time interval for an individual cell P containing non-wetting phase is

$$\frac{V_P(C_P^k - C_P^{k-1})}{\Delta t^k} = J_w - J_e + J_s - J_n + J_b - J_t + \sum_{nb} \lambda_{nb}(C_{eq} - C_{nb})A_{nb} \quad (7)$$

where the wetting phase concentration at the beginning and end of the time increment Δt^k is represented by C_P^{k-1} and C_P^k respectively, and J_i $(i = w, e, n, s, t, b)$ represents the total wetting phase mass flux crossing each face of cell P. The terms J_t and J_b represent mass flux leaving/entering cell P from the surrounding porous media. The remaining four mass flux terms only exist if that neighbouring cell contains all wetting phase. If a neighbouring cell contains the non-wetting phase, then the wetting phase mass flux term between that cell and cell P does not exist. Instead, the component of the last term in (7) corresponding to that neighbour allows for mass transfer from the non-wetting phase to the wetting phase to take place across this cell face. Each of the wetting phase mass flux terms in (7) are constructed from an exponential profile of the concentration between the cell centroids which is derived by considering a steady state mass transport condition (Patankar, 1980). For example, the mass flux across the east face of cell P is given by

$$J_e = v_e \Delta y_P a_e \left[C_P + \frac{C_P - C_E}{\exp(P_e) - 1} \right] \quad (8)$$

with

$$P_e = \frac{v_e (\Delta x_E + \Delta x_P)}{2 D_e} \quad (9)$$

where v_e is the wetting phase velocity defined at the east face; Δy_P is the width of cell P in the y-direction; a_e is the harmonic mean aperture between cell E and cell P; C_P and C_E are the wetting phase concentrations in cell P and cell E respectively; P_e is the Peclet number defined at the east interface; Δx_P and Δx_E are the width in the x-direction of cell P and cell E respectively, $D_e = \alpha v_e + D^*$ is the dispersion coefficient defined at the east face, α is dispersivity, and D^* is the free water diffusion coefficient. The use of this exponential profile provides an adaptive upstream weighting control based on the Peclet number. For example, when the Peclet number is large the exponential profile simulates full upstream weighting.

EXAMPLE

To demonstrate the applicability of the developed model, a realistic scenario was constructed in which pure phase trichloroethylene (TCE) was allowed to enter a 2.56 m by 1.28 m vertical fracture initially water filled. For illustration purposes, the pure phase TCE was allowed to penetrate and come to rest in the fracture before any water phase transport was allowed to occur.

The fracture domain was discretized into a 128 node by 64 node control volume grid with a block spacing of 20 mm in both the x- and y-coordinate directions. An aperture

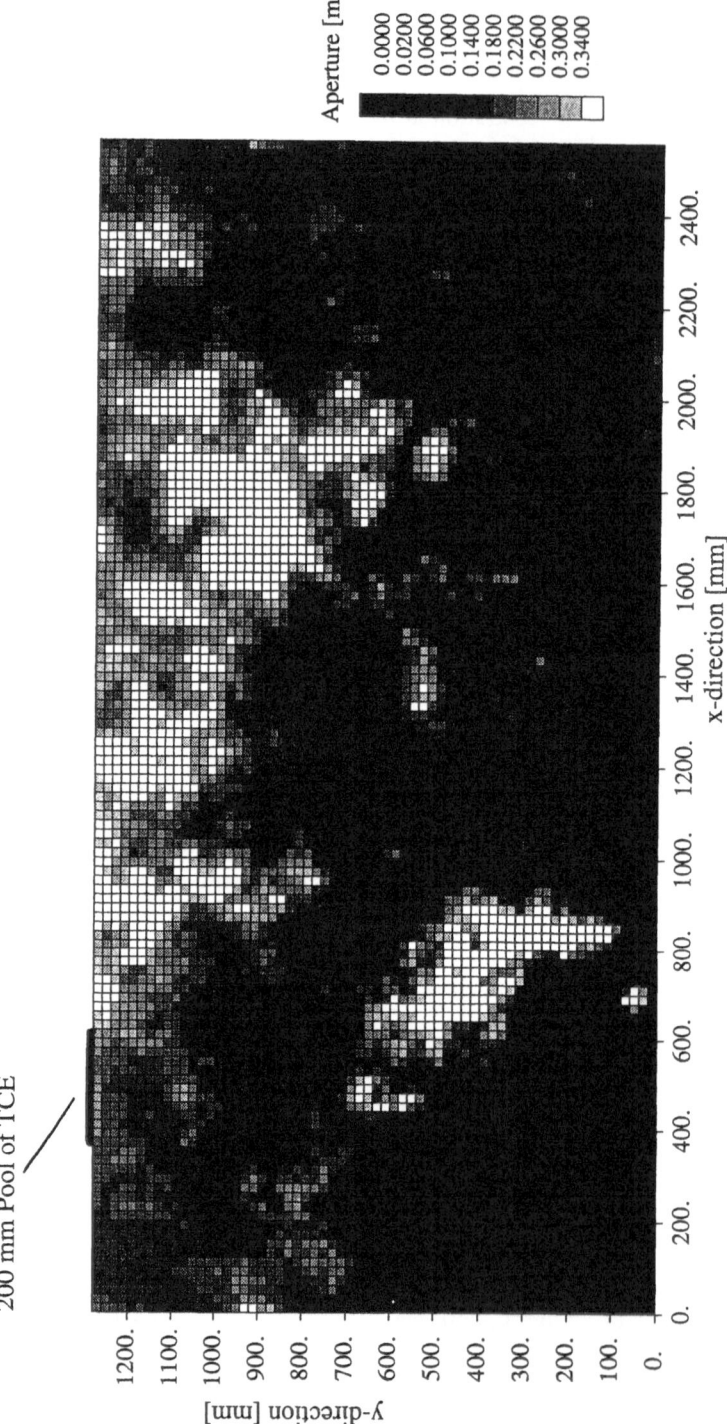

Figure 1. Fracture aperture distribution.

field was generated with a ln-mean aperture of -2.38 ln(mm), a ln-variance of 0.7 ln(mm^2), and a principle correlation length of 200 mm in both the x- and y-coordinate directions. This generated aperture field was altered to represent a fracture with a decreasing mean aperture with depth by widening the top boundary by half the mean aperture and closing the bottom boundary by half the mean aperture. Regions within the fracture domain in which apertures become negative as a result of this linear transformation where assigned zero aperture values. The resulting aperture field is shown in Figure 1.

A no-flow boundary for both fluid phases was prescribed along the bottom and top of the fracture. A 1530 mm and a 1430 mm water head was prescribed along the left and right boundaries respectively to create a water phase pressure gradient from left to right. The physical fluid properties and model parameters used in this example are: TCE (density 1.456 mg/mm^3, viscosity 0.556 mg/mm/s, surface tension 45000 mg/s^2, contact angle 30°, solubility 1.42x10^{-3} mg/mm^3); water (density 0.9982 mg/mm^3, viscosity 1.002 mg/mm/s), and dispersivity 10 mm.

A 200 mm pool of TCE was allowed to sit along a 260 mm length of the top boundary as shown in Figure 1 until approximately 20 g of TCE entered the fracture. At this point the TCE pool along the top boundary was removed and redistribution of the TCE within the fracture was permitted for 275 secs. An enlarged view of the TCE distribution directly beneath the source area is shown in Figure 2. The TCE distribution at 275 secs was employed as the initial condition for the water phase transport. In this example it is assumed that the TCE distribution is essentially non-mobile; therefore, the water phase flow field was updated whenever all the TCE within a control volume was depleted. The effects of matrix diffusion were not considered in this example. A maximum time step length of 4000 secs was employed for all transport simulations.

Figure 3 presents the temporal profile of pure phase TCE mass remaining in the fracture plane for various values of the mass transfer rate coefficient λ. Since in these simulations no mass was allowed to move into the surrounding porous media, the dissolved TCE in the aqueous phase was forced to exit the fracture plane along the right boundary. Depending on the rate coefficient, the time to completely dissolve all the pure phase TCE can vary from approximately 80 days to 300 days.

SUMMARY

A numerical model has been developed to investigate multiphase flow and aqueous phase transport in a single rough-walled fracture. This model is based on the finite volume method and incorporates an exponential profile for the aqueous phase concentration between cells for upstream weighting control. A realistic example was presented to demonstrate the potential utility of the developed model.

REFERENCES

Murphy, J. R., and Thomson, N. R., (1993) "Two-phase flow in a variable aperture fracture", Water Resour. Res. 29(10), 3453-3476.

Patankar, S. V., (1980) Numerical Heat Transfer and Fluid Flow, Hemisphere Publishing Corporation, Washington, D.C.

Pruess, K., and Tsang, Y, W., (1990) "On two-phase relative permeability and capillary pressure of rough-walled fractures", Water Resour. Res., 26(9), 1915-1926.

Robin, M. J. L., (1991), "Migration of reactive solutes in three-dimensional heterogeneous porous media", Ph.D. thesis, Earth Sciences Department, Univ. of Waterloo, ON.

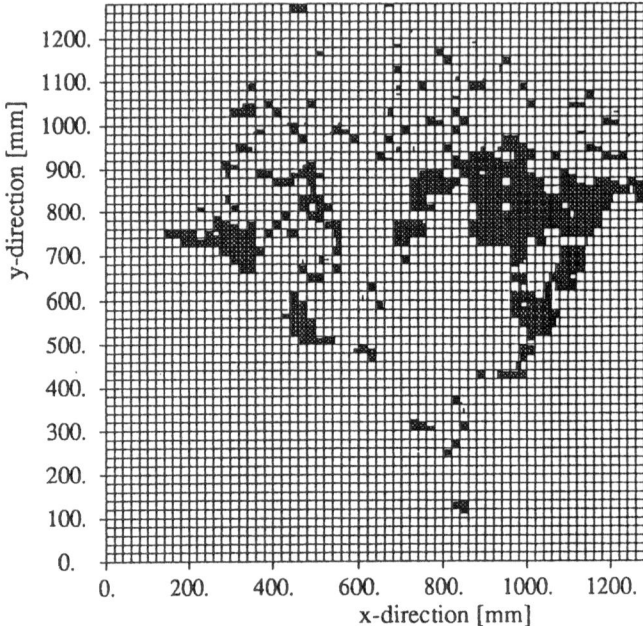

Figure 2. Enlarged view of the pure phase TCE distribution beneath the source area.

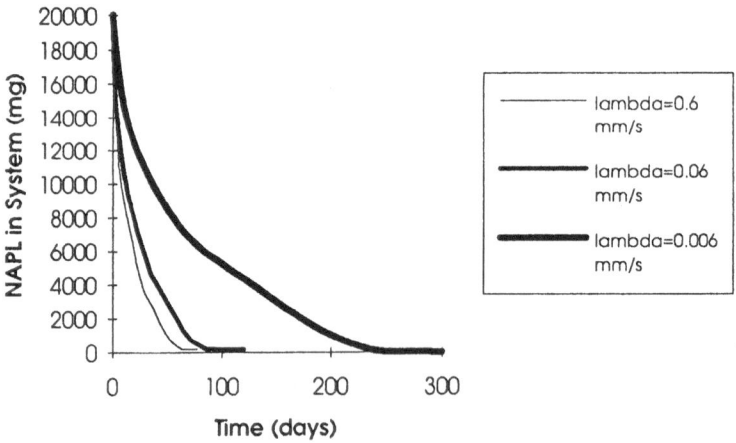

Figure 3. Temporal profile of pure phase TCE remaining in fracture plane for various mass transfer coefficients.

10. SALTWATER INTRUSION

NUMERICAL SIMULATIONS OF SEAWATER INTRUSION INTO THE NILE DELTA AQUIFER

E. HOLZBECHER and R. BAUMANN
Institut für Wasserbau und Wasserwirtschaft
Technische Universität Berlin, 10623 Berlin, Straße des 17. Juni 142-144
Germany

Increased salt concentrations in the Nile delta aquifer (Egypt) have often been reported. It could possibly be attributed to saltwater intrusion from the Mediterranean sea into the aquifer. Numerical simulations by various modelers in the 1980s predicted salt fronts intruding more than 100 of kilometers. Modeling with the FAST code in contrast show considerably smaller intrusion – being more in agreement with recent measurements. Variations of grid spacing and of the 1.order derivatives discretisation method demonstrate the severe influence of numerical dispersion on the computational results.

INTRODUCTION

The intrusion of seawater into the adjacent aquifer which often supplies coastal inhabitants with freshwater is a common problem in many seaside regions. Saltwater intrusion contaminates an aquifer that it can no longer be used to its fulliest extent, even after an active intrusion has been brought to a standstill.

To understand the processes involved with seawater intrusion, groundwater withdrawal and recharge must be considered. An accurate water balance calculation can help to determine whether we are handling a steady state or transient seawater intrusion.

The Nile delta's Pleistocene aquifer has been believed to be suffering from active saltwater intrusion. Numerous studies have been done on this phenomenon and its present reach. The transition zone from freshwater to saltwater varies according to author and source. It has been calculated to lie between anywhere from 80 to 130 km inland from the Mediterranean shoreline (AMER et al. 1981, KASHEF 1983, SHERIF et al. 1988).

A. Peters et al. (eds.), Computational Methods in Water Resources X, 1011–1018.
© 1994 *Kluwer Academic Publishers. Printed in the Netherlands.*

Within the Special Research Project 69 of the German Research Society (DFG) extensive field trips in 1991 and 1992 were carried out to get an idea about the actual extent of seawater intrusion in the Nile delta. The field measurements down to depth of 330 m indicated seawater concentrations (35000 ppm) only along the coastline. The transition zone reaches about 40 km inland. In the middle delta only freshwater concentration (less than 1000 ppm) were found through electrical conductivity measurements (BAUMANN u. MOSER 1992).

This data base has been used for calibrating a numerical model. After the calibration procedure variations from the reference model were performed in order to test the numerical influence on the flow and concentration distribution.

BASIC PRINCIPLES AND DIFFERENTIAL EQUATIONS

Conservation of fluid mass is given by the continuity equation:

$$\nabla \cdot (\rho \varphi \mathbf{u}) = 0 \qquad (I)$$

(with fluid density ρ, porosity φ and average mean velocity $\mathbf{u}$,). Darcy's law is assumed to be valid in the following form:

$$\mathbf{u} = -\frac{k\rho g}{\varphi \mu}(\nabla p + \mathbf{e_z}) \qquad (II)$$

(k permeability, g constant of gravity, μ dynamic viscosity, p pressure head, $\mathbf{e_z}$ unit vector in direction of gravity). Anisotropies and inhomogeneities within the region in question are not considered. Viscosity of the fluid is assumed to be constant and density is assumed to change linearly with respect to salt concentration (c):

$$\rho = \rho_0 + \Theta \cdot \Delta\rho \qquad (III)$$

ρ_0 is the density of fresh cold water. Θ is the normalized concentration within the model, i.e. $\Theta = (c-C_{min})/(C_{max}-C_{min})$. $\Delta\rho$ denotes the density difference from high saline water ($c=C_{max}$) to fresh water ($c=C_{min}$). In the saline case treated here $\Delta\rho$ is positive. The Boussinesq-assumption supposed to be valid. That means that density differences can be neglected in all equations with the exception of the buoyancy term in equation (II).

A general form of the equation for transport (in a flow field $\mathbf{u}$) and sorption can be stated as (see for example: BEAR/VERRUIJT 1987)

$$R \cdot \frac{\partial c}{\partial t} = -\mathbf{u} \cdot \nabla c + \nabla \cdot \mathbf{D} \cdot \nabla c + q \qquad (IV)$$

The first term on the right hand side describes advection. The second, including dispersion tensor **D**, describes dispersion and diffusion. q adds all dditional sources and sinks for the substance. R is the retardation factor. For salt transport, where Cl$^-$ ions are the main component, holds: R$\approx$1.

Because it is the intention of this paper to analyse the coupling of flow and transport, the simplifying assumption of a scalar function D is made for the saline case, i.e. dispersion or macro-dispersion is assumed to obey Fick's law for flow-vector **j**

$$\mathbf{j} = -D \cdot \nabla c \tag{V}$$

with diffusivity D $[m^2/s]$.

With the mentioned assumptions the set of governing equations 2-dimensional vertical model can be modified by introducing streamfunction Ψ and Θ as unknown variables. Using the variable transformations:

$$t \longrightarrow t \cdot D/H^2 \qquad x \longrightarrow x/H \qquad z \longrightarrow z/H \tag{VI}$$

(H is characteristic length, mostly height of the model) the variables become dimensionless and the governing equations change to:

$$\omega = Ra \cdot \frac{\partial \Theta}{\partial x} \qquad\qquad \nabla^2 \Psi = \omega \tag{a,b}$$

$$\tag{VII}$$

$$F = \nabla^2 \Theta + \frac{\partial \Theta}{\partial x} \cdot \frac{\partial \Psi}{\partial z} - \frac{\partial \Theta}{\partial z} \cdot \frac{\partial \Psi}{\partial x} \qquad R \cdot \frac{\partial \Theta}{\partial t} = F \tag{c,d}$$

ω is the vorticity and F proportional to the rate of mass of the substance which is entering at any place of the model. Elder (1967a and 1967b) derived this set of equations for the thermal case - without sorption the retardation factor R in (VIId) is not found in Elder's publications. The opposite sign in the last two terms of equation (VIIc) - compared with Elder's equations - stems from the different sign convention in $\Delta \rho$ and in streamfunction. The constant Ra, the Rayleigh-no. for porous media flow, is a dimensionless combination of parameters:

$$Ra = g \cdot k \cdot \Delta\rho \cdot H / \mu \cdot D \tag{VIII}$$

(h height of model). Variables F and ω can be removed from equations (VII) easily. There remains a set of two partial differential equations for the two variables Ψ and Θ:

$$\nabla^2 \Psi = Ra \cdot \frac{\partial \Theta}{\partial x} \tag{a}$$

$$\tag{IX}$$

$$\nabla^2 \Theta - \frac{\partial \Psi}{\partial x} \cdot \frac{\partial \Theta}{\partial z} + \frac{\partial \Psi}{\partial z} \cdot \frac{\partial \Theta}{\partial x} = R \cdot \frac{\partial \Theta}{\partial t} \tag{b}$$

A model like this is appropriate to study the general behaviour of solutions for a problem of coupled flow, transport and sorption. Only a small set of parameters remains: the dimensionless parameters Ra and R. The first includes all dynamical parameters for the fluid system, the second is related to the interaction bet-ween fluid and solid phase. Besides R and Ra remain the geometrical parameters.

A saltwater intrusion problem using a similar set of equations as (IX) was trea-ted by Henry (1964) - not using the Rayleigh-no. terminology. Basic transforma-tions of Henry's equations lead to the system (IX) (HOLZBECHER 1991).

NUMERICAL METHOD

The numerical simulations of seawater intrusion in the Nile Delta were carried out using the numerical flow and transport model FAST—C(2D) from the FAST software package which was developed at the Technical University of Berlin (HOLZBECHER 1991). FAST—C(2D) is a special modul for modeling saline or ther-mal, steady or transient convection in a porous media in two-dimensional regi-ons. Using FAST—C(2D) code the transient movement of a saltwater front was modeled for a saline disaster in Japan (HOLZBECHER and KITAOKA 1993). For this paper steady-state saline convection has been calculated in all cases.

The FAST model is based on the finite difference/finite volume method on rec-tangular grids. Standard 2.order stencils are used for spatial discretisations of 2.order terms in the differential equations. For the 1.order terms, as in (IXb), the-re is the alternative between central (CIS 'central in space') and upwind (BIS w'backard in space') scheme. Application of the 2.order CIS is restricted to ca-wses, here the grid-Peclet-No. criteria are fulfilled:

$$Pe_x := u_x \cdot \Delta x / D \leq 2 \qquad\qquad Pe_z := u_z \cdot \Delta z / D \leq 2 \qquad\qquad (X)$$

Otherwise the algorithm becomes unstable. 1.order BIS is unconditional stable, but introduces numerical dispersion, which in 1.order is given by:

$$D_x^{num} = u_x \cdot \Delta x / 2 \qquad\qquad D_z^{num} = u_z \cdot \Delta z / 2 \qquad\qquad (XI)$$

To reduce the effect of numerical dispersion, in FAST codes the option is im-plemented to change D value by D^{num} - also called truncation error correction (LANTZ 1971). This physically motivated alteration of the numerical method was proposed by Bear and Verruijt (1987). Holzbecher (1988) showed that this 'modi-fied upwind scheme' (AXELSSON/GUSTAFFSON 1979) works quite well even for the general dispersion tensor **D**. Nevertheless the correction can be motivated in that way if the grid-Peclet criteria are fulfilled (otherwise D becomes negative).

Formulas for numerical dispersion, as shown above, are derived for the Finite
Difference method. Nevertheless the problems are basically the same in the Finite Volume and in the Finite Element approach (CHRISTIE e.a. 1976).

Both equations (IX) are solved alternately (inner iteration), until changes in one
iteration step are below a predefined tolerance which was specified as 10^{-5} in
the calculations, which are presented in this paper. The same tolerance was taken for the iterative preconditioned conjugate gradient algorithm for the solution
of the linear systems. FAST–C(2D) was run on a SUN workstation.

The iterative technique for the solution of the nonlinear coupling of eq.s (IX)
may be the source for an additional numerical dispersion: D^{num}-caused errors in
solving eq. (IXb) will be followed errors in Ψ in eq. (IXa) and so on. It is the aim
of this paper to give a qualitative estimation of the additional numerical dispersion.

DISCRETISATION AND BOUNDARY CONDITIONS

The 2-dimensional-vertical numerical calculations where carried out for a section of 100 km length and a depth of 1 km reaching from the Mediterranean in
the north to the town of Tanta in the south (see fig. 1). The considered flow
domain was wide enough since first simulation runs pointed out a smaller seawater intrusion and consequently no influence on the landward boundary was
observed.

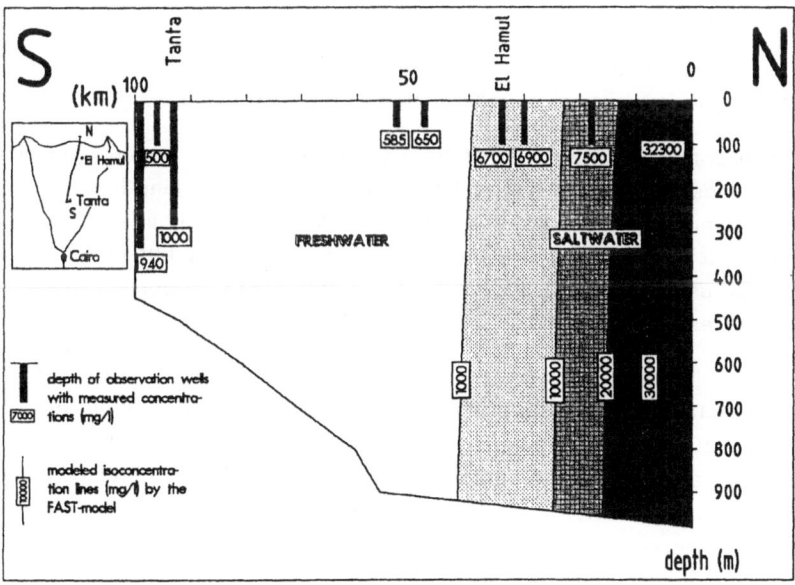

Fig.1: Cross-section location, observations and modeling results

The spatial discretisation of the reference case was set to 200 blocks in x-direction and 10 blocks in z-direction. Under these conditions the grid-Peclet-No. criteria were fulfilled at any section in the modelled area.

Fig. 2 shows the boundary conditions of the modeled area. At the upper and lower boundary the values of the streamfunction are given. No vertical components for the fluid velocity should exist on the left and right boundary. For concentration flow the upper and lower boundary is closed. The left side is set to freshwater concentration (C=0) and on the right seawater concentration is given (C=1).

Fig. 2:
Boundary
conditions

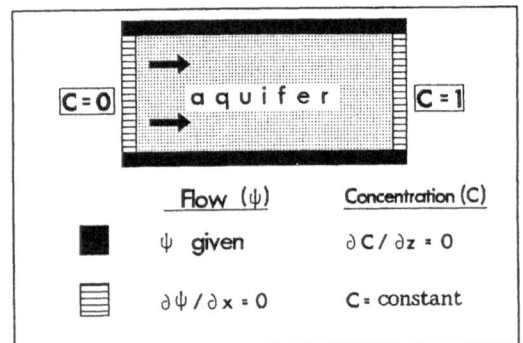

MODELING RESULTS

Calibration of the model with the database lead to a Ra-no. of 20. The inflow rate on the southern vertical boundary was derived from the water balance.

Under these conditions the result of fig. 1 is obtained. Salt concentration isolines are drawn almost vertical (it has to be considered that the length/height aspect ratio is 100/1 and that the pictures are superelevated). The seawater is not - as previously assumed - intruding more than 100 km inland, but instead reaches a maximum intrusion of 40 to 50 km south of the coast. Recorded in-situ measurements correspond with those of the model.

Fig. 3,4 and 5 illustrate the effect of grid-refinement. Here all parameters have been kept constant except x-direction grid-spacing, which has been altered in a way that 50, 200 and 800 equal blocks were specified. The values for Pe_x in the simulations are 0.88, 0.24 and 0.065. The effect of numerical dispersion is demonstrated quite clearly. In agreement with the theoretical derivations numerical dispersion increases with the grid-length. Computational output for 400 and 800 blocks show identical figures, so that fig. 5 can be considered as grid-independant. Note, that Pe_x is much smaller than prescribed by criterion (X).

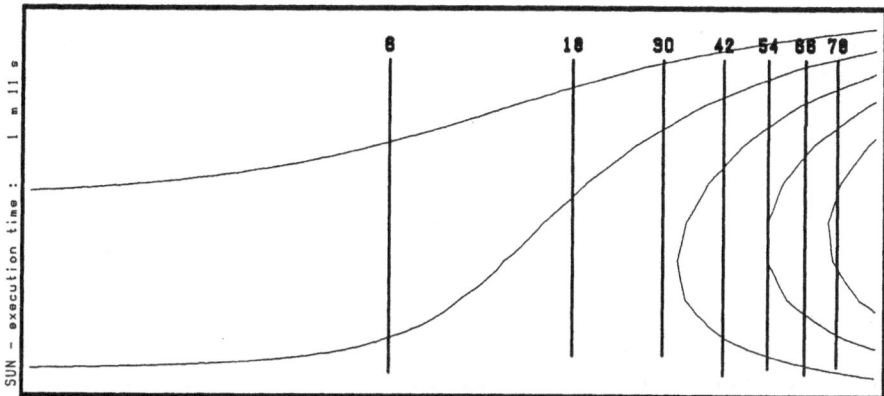

Fig.3: Seawater intrusion in a 50 X 10 grid model

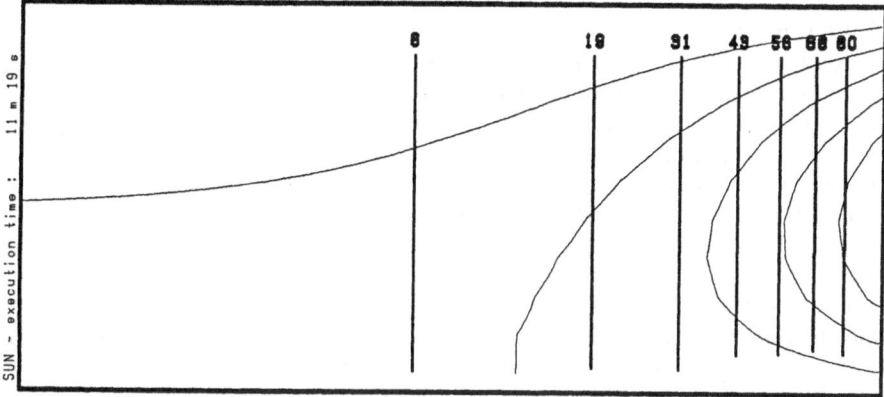

Fig.4: Seawater intrusion in a 200 X 10 grid model

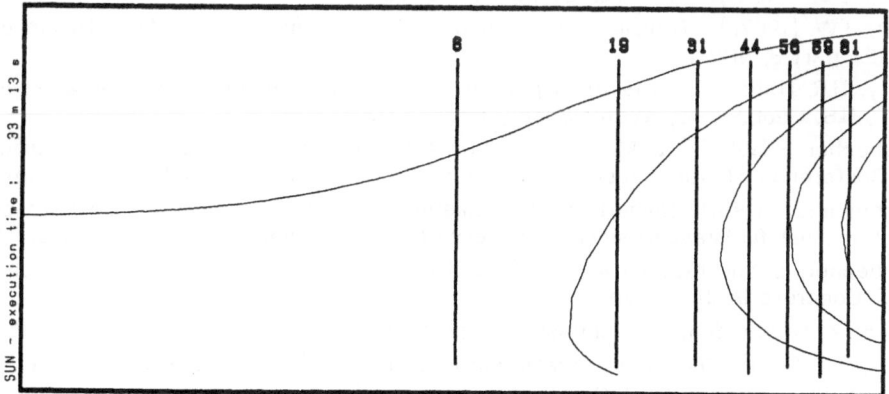

Fig.5: Seawater intrusion in a 800 X 10 grid model
(Thick lines: isohalines with % of seawater salinity, thin lines: equidistant streamlines)

In order to test the effect of numerical dispersion some variations of the discretisation method were chosen. As mentioned above the grid-Peclet-criterion is fulfilled, so that CIS and the truncation error correction for the BIS can be applied alternatively. - CIS shows almost the same results as the BIS method. BIS with truncation error correction delivers a dispersion, which is slightly smaller and nearer to the finest discretisation result. Thus the corrected BIS discretisation is the most convenient method of the three.

CONCLUSION

Saltwater intrusion in the Nile delta has been overestimated in previous models. The main reason for that is numerical dispersion resulting from coarse gridspacings.

R e f e r e n c e s

Amer, A., Moustafa, M. and Farid, M. (1981) "Calibration of a seawater intrusion model for the Nile Delta aquifer", Proc. Int. Conf. Water Res. Man., 311-325.

Axelsson, O. and Gustafsson, I. (1979) "A modified upwind scheme for convective transport equations and the use of a conjugate gradient method for the solution of non-symmetric systems of equations", J. Inst. Maths Applics 23, 321-337

Baumann, R. and Moser, H. (1992) "Modellierung der Meerwasserinvasion im Delta arider und semiarider Gebiete am Beispiel des Nildeltas", Z. dt. geol. Ges. 143, 316-324.

Bear, J. and Verruijt, A. (1987) "Modeling Groundwater Flow and Pollution"

Christie, I., Griffiths, D.F., Mitchell, A.R. and Zienkiewics, O.C. (1976), "Finite element methods for second order equations with significant first derivatives", Int. J. Num. Meth. Engng. 10, 1389-1396

Elder, J.W. (1967a) "Steady free convection in a porous medium heated from below", J. Fluid Dynamics vol. 27 part 1, 29-48

Elder, J.W. (1967b) "Transient convection in a porous medium", J. Fluid Dynamics vol. 27 part 3, 609-623

Henry, H.R. (1964) "Effects of dispersion on salt encroachment in coastal aquifers", U.S. Geol. Survey Water Supply Paper 1613-C

Holzbecher, E. (1988) "Zur Modellierung von Ausbreitungsvorgängen im Grundwasser", VII. Conference 'Mathematische Simulation der Grundwasserentnahmen', Krakov

Holzbecher, E. (1991) "Numerische Modellierung von Dichteströmungen im porösen Medium", Inst. für Wasserbau und Wasserwirtschaft, Techn. Univ. Berlin, Mitteilung 117

Holzbecher, E. and Kitaoka K. (1993) "Saline disasters and modeling approach", XXV. Congress IAHR, Tokyo

Kashef, A. (1983) "Saltwater intrusion in the Nile Delta", Groundwater 21(2), 160-167.

Lantz, R.B. (1971) "Qualitative evaluation of numerical diffusion (truncation error)", Soc. of Petr. Eng. J., 315-320

Sherif, M., Singh, V. and Amer, A. (1988) "A two-dimensional finite element model for dispersion (2D-FED) in coastal aquifers", J. Hydrology 103, 11-36.

THREE-DIMENSIONAL FINITE ELEMENT MODEL FOR SALTWATER INTRUSION INTO AQUIFERS

A. LARABI and F. DE SMEDT
Laboratory of Hydrology, Free University Brussels
Pleinlaan 2, B1050 Brussels, Belgium.

Saltwater intrusion into aquifers involves the determination of the interface separating freshwater from saltwater. Additionally, difficulties arise when dealing with sea water intrusion into unconfined aquifers, because this requires the location of the phreatic surface. A numerical procedure is developed for solving these problems in steady state with the finite element technique. The problem is solved with a fixed finite element mesh by iteratively adjusting the moving boundaries and neglecting flow in the unsaturated and saltwater zones. This is achieved without explicitly changing the coefficient matrix of the finite element equations, which makes the method computationally fast. Test problems, including saltwater intrusion into confined and unconfined aquifers, demonstrate that the technique is accurate when compared with analytical solutions and laboratory measurements.

INTRODUCTION

Sea water intrusion must be addressed in managing groundwater resources in islands and coastal areas, where it is necessary to predict the behaviour of the groundwater under various conditions, such as to determine the extent to which ingress of sea water occurs. This problem can be analyzed with two methods. The first method considers both fluids miscible and takes into account the existence of a transition zone between them (Voss and Souza, 1987), and the other one is based on the abrupt interface approximation (Bear and Verruijt, 1987). If the transition zone is thin relative to the thickness of the freshwater lens and it is immobile, then it is appropriate to assume that the freshwater and saltwater do not mix (immiscible), and the transition zone is considered to be a sharp interface. Custodio (1992) reported that both approaches produce good results if applied to the right circumstance. In this study, it is assumed that an abrupt interface exists between freshwater and saltwater at rest. A numerical procedure is developed using a fixed FE mesh technique. This approach presents the advantage that it involves solution of only the flow equation for freshwater. However, the problem remains basically nonlinear because the interface position is iteratively adjusted until a stable position is obtained. Test problems demonstrate that the technique is accurate when compared with analytical solutions available for confined and unconfined flow. Verification of the numerical model is also made with respect to observations from a 3-D laboratory box model for a phreatic flow (Sugio, 1992).

A. Peters et al. (eds.), Computational Methods in Water Resources X, 1019–1026.
© 1994 Kluwer Academic Publishers. Printed in the Netherlands.

THEORY

For equilibrium conditions, the governing equation of groundwater flow can be written as

$$\nabla \cdot (K \nabla h) = 0 \tag{1}$$

where h is the hydraulic potential, K is the hydraulic conductivity of the aquifer, which depends upon the position and the hydraulic pressure, and $\nabla = (\partial/\partial x, \partial/\partial y, \partial/\partial z)$. The hydraulic potential is defined as a freshwater head

$$h = z + \Psi = z + p/\rho g \tag{2}$$

where z is the elevation, Ψ is the freshwater pressure head, p is the pressure, g is the acceleration due to gravity and ρ is the freshwater density. In the unsaturated zone the pressure potential becomes smaller than zero, such that the saturated zone can be distinguished from the unsaturated zone by the condition $\Psi \geq 0$. To distinguish the freshwater zone from the saltwater, we can make use of the fact that the pressure in the freshwater should be lower than the pressure in saltwater that is in hydrostatic equilibrium with the sea level, or

$$\rho_s(z_s - z)/\rho - \Psi \geq 0 \tag{3}$$

where ρ_s is the saltwater density and z_s is the sea water level.

In the present approach, a fixed finite element mesh technique is adopted to discretize the entire domain, including the unsaturated and saltwater regions. A system of nonlinear equations results

$$G(h)\ h = Q \tag{4}$$

where **h** is the vector of unknown nodal heads, **G** is the conductance matrix and **Q** a vector containing the boundary conditions. The coefficients of the conductance matrix are given by

$$G_{ij} = \int_V K \nabla b_i \nabla b_j dxdydz \tag{5}$$

where b_i and b_j are basis functions respectively related to nodes i and j. The following properties are satisfied (Larabi and De Smedt, 1993a) : **G** is symmetric positive definite and all row sums are zero, or

$$G_{ii} = -\sum_{j \neq i} G_{ij} \tag{6}$$

The groundwater flow is assumed to occur only in the freshwater region. There will be no flow in the unsaturated and saltwater regions such that the water table and the interface are effectively impervious boundaries. But, their locations are unknown initially and must be obtained as part of the solution to the problem. Two contributions can be recognized in the

G_{ij} coefficients : the gradients of the basis functions refer to the geometry of the finite elements, while the K-factor refers to the hydraulic properties of the medium. It follows that the G_{ij} coefficients depend upon the position of the water table and the interface. In case the elements remain fixed, the conductivity will depend upon the pressure, as some elements will fall in the unsaturated zone or in the saltwater zone. Hence, the algebraic system is nonlinear and can only be solved iteratively. We can simplify the solution procedure by making use of Equation (6); the finite element equation can be written as

$$\sum_{i \neq j} G_{ij}(h_j - h_i) = Q_i \qquad (7)$$

which can be considered as a numerical equivalent of Darcy's law. Now, for $i \neq j$, G_{ij} can be approximated as (see Larabi and De Smedt, 1993b)

$$G_{ij} \cong k_{ij} \int_V K_s \nabla b_i \nabla b_j \, dxdydz = k_{ij} G_{ij}^* \qquad (8)$$

where K_s is the saturated hydraulic conductivity and k_{ij} a relative conductivity ($0 < k_{ij} < 1$), which depends on the status of the water in the region between nodes i and j. Notice that the saturated conductance coefficients G_{ij}^* remain fixed during the solution procedure, and only the relative conductivities have to be adapted in each iteration. Nodes in the unsaturated zone are identified by negative pressure potentials. For the saltwater zone, the pressure potentials are compared by equation (3). If equation (3) for nodes i and j is positive, the zone between nodes i and j is located in the freshwater zone and ($k_{ij} = 1$). In case equation (3) is negative for i or j or both, k_{ij} has to be updated. The same procedure holds for the unsaturated zone. This updating must be done in a smooth way when passing from the saltwater zone or from the unsaturated to the saturated flow domain. The following method (Larabi and De Smedt, 1993b) was chosen to achieve this :

$$k_{ij} = [\varphi_i k(\varphi_i) + \varphi_j k(\varphi_j)]/(\varphi_i + \varphi_j) \qquad (9)$$

where φ_i is a measure of the distance of a node to either the water table or the saltwater interface, it is given by $\varphi_i = \Psi_i$ in case of the unsaturated zone, and $\varphi_i = \Psi_i - \rho_s(z_s - z_i)/\rho$ in case of the saltwater zone. $k(\varphi)$ is a relative conductivity function, which is defined as

$$
\begin{aligned}
k(\varphi) &= 1 && \text{if } \varphi > 0 \\
k(\varphi) &= (1+\varepsilon)/2 && \text{if } \varphi = 0 \\
k(\varphi) &= \varepsilon && \text{if } \varphi < 0
\end{aligned} \qquad (10)
$$

where ε is theoretically zero, but chosen here as a small number in order to allow for the finite element equations corresponding to nodes in the unsaturated or saltwater zones to remain in the algebraic equation system, without obstructing the numerical solution procedure. The iteration process is repeated until there is no more significant change in the potential values over the freshwater domain.

For the boundary conditions, only the seepage and the outflow face at the sea pose a problem. The position of these is also initially unknown and constitutes another non-linearity of the problem. The objective is to satisfy the condition of zero pressure at the nodes on the seepage face in contact with the atmosphere; and a hydrostatic pressure condition corresponding to sea level for the nodes on the outflow face in contact with the sea. Therefore, all the nodes expected to be part of the seepage or the outflow face are treated initially as prescribed potential boundaries, with the potential equal to the elevation for the seepage face, and the potential equal to the sea water level for the outflow face corrected for density difference. After every iteration step, the flux values on the seepage face are checked. If an inflowing flux is encountered, this is set equal to zero and in the next iteration step this node is treated as an impervious boundary. On the other hand, if a positive value of pressure is encountered at a boundary node in the unsaturated zone, or a hydrostatic pressure less than the freshwater pressure at a boundary node in the saltwater zone, this node is treated in the next iteration step respectively as a seepage or an outflow face node boundary.

APPLICATION AND RESULTS

Validation with 2-D analytical solutions

The simplest way to model saltwater intrusion is to assume that the freshwater interface, and underlying saltwater are in hydrostatic equilibrium. This assumption is known as the Ghyben-Herzberg relationship. A simple stationary interface solution of the problem illustrated in Figure 1a was given by Glover (1959). Other analytical solutions for the interface problem based on the Dupuit assumption can be found in the works of Bear and Verruijt (1987) and Strack (1989), among others. Since an interface between freshwater and saltwater at rest is a streamline, also the hodograph method can be used to solve this free boundary problem. Detournay and Strack (1988) derived an analytical solution which involves the determination of a phreatic surface, an interface separating flowing freshwater from saltwater at rest, and seepage and outflow faces where freshwater discharges into the sea (Figure 1b).

The present FE approach is applied to simulate an example of confined flow which involves the interface problem (Figure 1a), in which the recharge flow rate $Q = 3.9$ cm^2/s for case 1 and $Q = 18.8$ cm^2/s for case 2. Other characteristics of the problem are $\rho = 1.0$ g/cm^3, $\rho_s = 1.029$ g/cm^3, $K = 69$ cm/s and the aquifer thickness is $D = 27$ cm. The Glover analytical solution for both cases is used for comparison with the model results. The flow domain is discretized with hexahedral elements of size 4x4x3 cm^3. This yields grids with 105, 1 and 9 elements respectively in the x-, y- and z-directions. Hence, the flow domain is discretized into 2120 nodes and 945 elements, and Figure 2a illustrates the finite element mesh. For numerical analysis, an arbitrary length of the flow domain is taken as 420 cm. The results are shown in Figure 2b for both values of Q and show a good agreement with the Glover solution. Comparison between these two figures also shows that the location, shape and extent of the interface depend, among other factors, upon the rate of discharge of groundwater to the sea.

The model is also applied to simulate saltwater intrusion interface in an unconfined aquifer (Figure 1b). The domain is truncated at the depth of 8 m (below sea level), with the assumption of an impervious aquifer bottom. Therefore, the dimensions chosen for the flow domain along the x-, y- and z-directions are respectively 6, 0.1 and 9.649 m. This flow domain is discretized into 9516 nodes and 4620 elements distributed on 60, 1 and 77 elements respectively along the x-, y- and z-directions. Other aquifer characteristics are K = 40 m/day, total freshwater outflow per unit width normal to the plane of flow Q = 40 m^2/day, $\rho = 1000$ kg/m^3, $\rho_s = 1400$ kg/m^3 (Read sea), $\alpha = 30°$ and $\beta = 15°$. Figure 3a shows the finite element grid of this problem and Figure 3b shows the phreatic surface, the interface, the seepage and outflow faces obtained by the analytical solution (Detournay and Strack, 1988) and the present FE model. This clearly shows that there is a good agreement between both solutions.

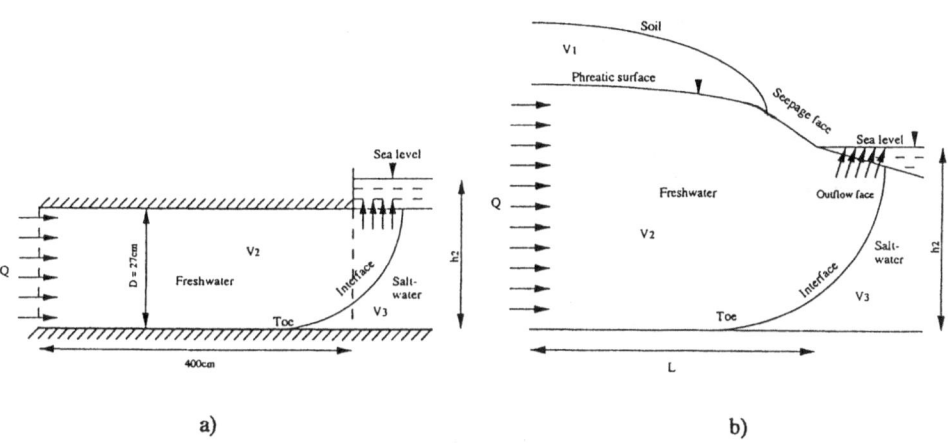

Figure 1. Problem definition, a) Confined aquifer and b) Unconfined aquifer.

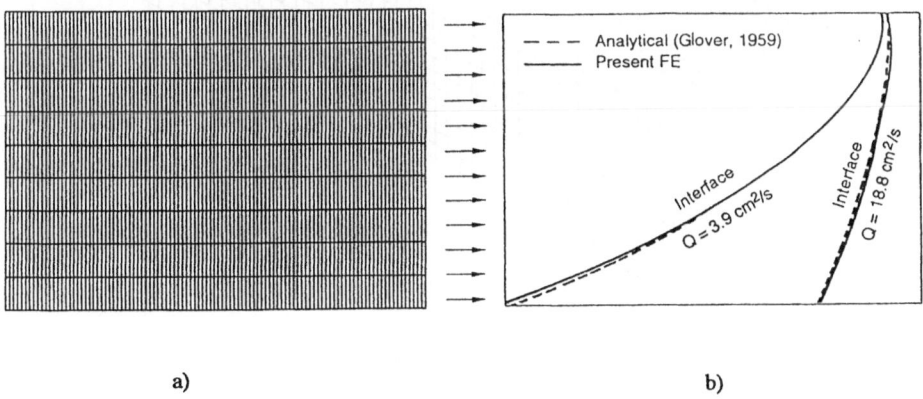

Figure 2. Confined flow, a) Finite element mesh and b) Comparison of numerical and analytical results.

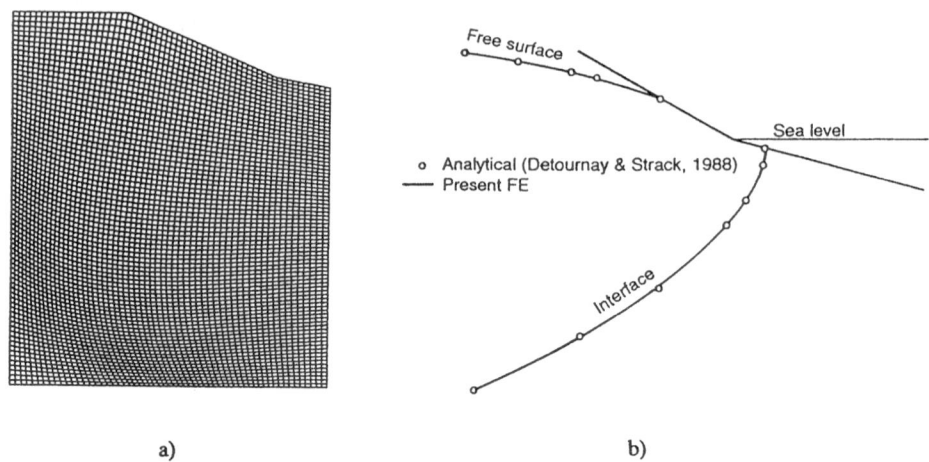

a)   b)

Figure 3. Unconfined flow, a) Finite element mesh and b) Position of the interface and free surface.

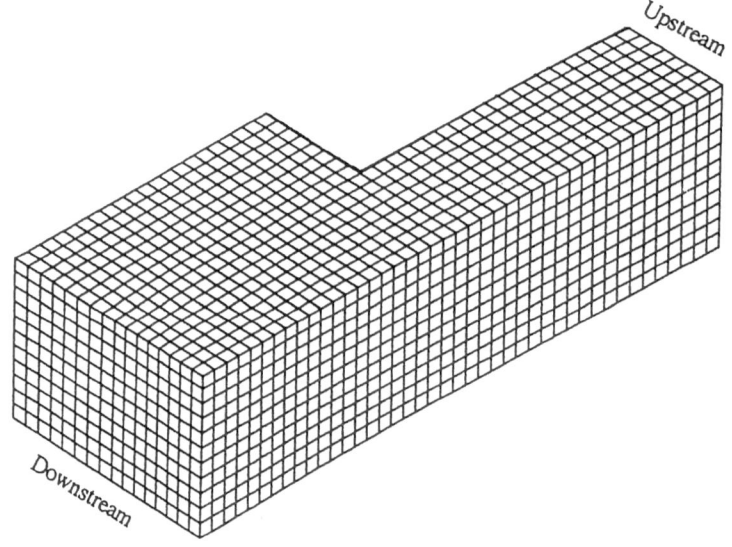

Figure 4. Three-dimensional finite mesh of the laboratory sand model.

Validation with 3-D laboratory model

To demonstrate that the numerical model developed here is capable to accurately simulate 3-D groundwater flow problems with a freshwater-saltwater interface, comparison of the results is made with respect to observations obtained from a 3-D sand model (Sugio, 1992). This laboratory model consists of a 3-D sand box model. The modelling part of the aquifer is 163.8 cm long and 47.5 cm high. The width of the model has two values; 63.2 cm from the downstream end until the length 82.3 cm, and 30 cm in the rest part. Salt was added to the freshwater and thoroughly mixed up to the density of 1.030 g/cm³ to model saltwater, and then was colored by dye in order to observe the position of the interface separating freshwater and saltwater. Sand with 0.76 mm mean diameter is used for which the hydraulic conductivity, obtained from measurements, is 1.293 cm/s. The upstream and downstream water levels are respectively $h_1 = 44.15$ cm and $z_s = 40.67$ cm, and freshwater flows in the sand model in about three hours to achieve steady state conditions. The behaviour of the freshwater-saltwater interface was measured for the front, side and bottom sections of the box model. The present numerical model is applied to this 3-D sand aquifer model using the same data. The flow domain is discretized into 8772 nodes and 7392 elements. Figure 4 shows a 3-D configuration of the FE mesh. The experimental results with regard to the position of the freshwater-saltwater interface are plotted for each section of the model together with the numerical results in Figures 5a, b and c respectively for the front, side and bottom sections. The results show that the 3-D predicted results compare very well with the observed results. The position of the free surface is also shown for both front and side sections.

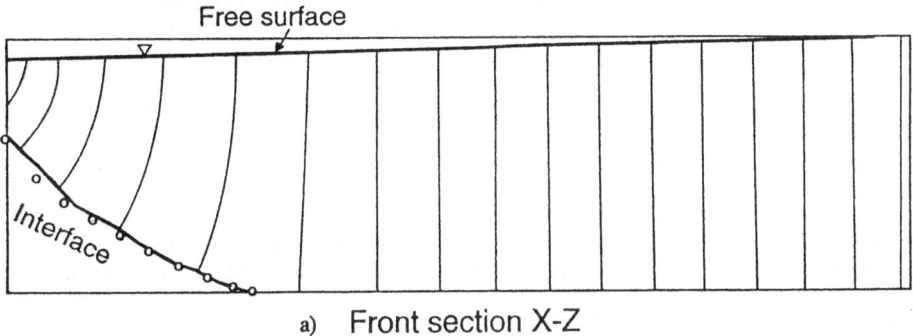

a)   Front section X-Z

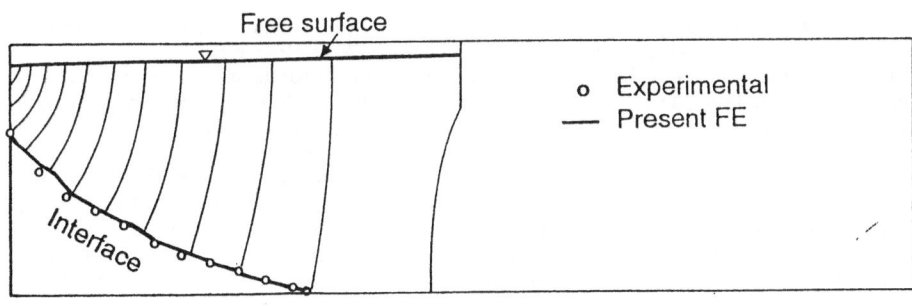

b)   Side section X-Z

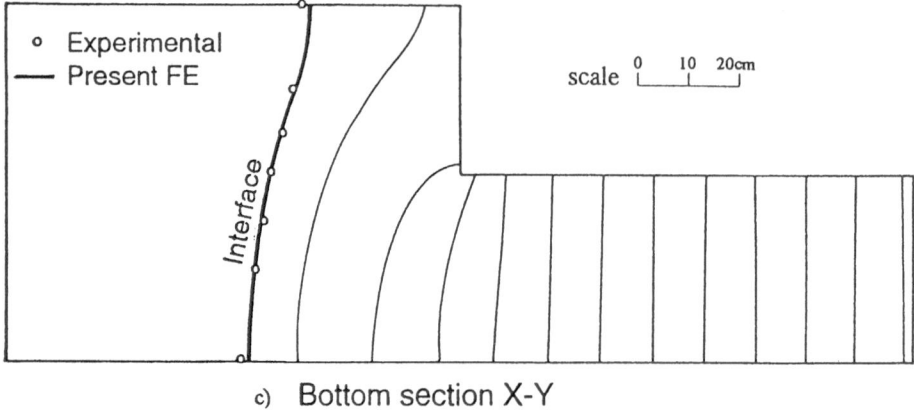

c) Bottom section X-Y

Figure 5. Comparison of numerical and experimental results for the interface position, a) Front section, b) Side section and c) bottom section (0.2cm of interval for the potential distribution).

CONCLUSIONS

A numerical procedure solution is developed using a fixed finite element mesh technique to simulate confined and unconfined groundwater flow coupled with the interface problem. The accuracy of the model was tested with available analytical solutions. Validation of the model was made with respect to observations from 3-D laboratory box model for a phreatic flow. In all these applications the numerical results show a good agreement with the analytical and experimental results.

REFERENCES

Bear, J. and Verruijt, A. (1987) "Modeling groundwater flow and pollution", Reidel, Dordrecht, 414 pp.
Custodio, E. (1992) "Study and modelling of saltwater intrusion into aquifers", Proc. 12th Salt Water Intrusion Meeting, Barcelone, pp. 3-10.
Detournay, C. and Strack, O.D.L (1988) "A new Approximate technique for the hodograph method in groundwater flow and its application to coastal aquifers", Water Resour. Res., Vol. 24, pp. 1471-1481.
Glover, R.E. (1959) "The pattern of freshwater flow in a coastal aquifer", J. Geophys. Res., Vol. 64(4), pp. 439-475.
Larabi, A. and De Smedt, F. (1993a) "Solving 3-D hexahedral finite elements groundwater models by preconditioned conjugate gradient methods", in press, accepted in Water Resour. Res.
Larabi, A. and De Smedt, F. (1993b) "Numerical finite element model for three-dimensional phreatic flow in porous media", in Proc. Moving Boundaries'93, Int. Conf. Free and Moving Boundary problems, L.C Wrobel and C.A Brebbia, eds., Computational Mechanics Publications, pp. 49-56.
Strack, O.D.L. (1989) "Groundwater mechanics", Prentice Hall, New Jersey, 732 pp.
Sugio, S. (1992) "Abrupt interface model of seawater intrusion in coastal aquifer", in HYDROCOMP'92, Int. Conf. on Interaction of Computational methods and Measurements in Hydraulics and Hydrology, J. Gayer, O. Starosolszky and C. Maksimovic, (eds), Water Resource Centre, Budapest, pp. 81-88.
Voss, C.I. and Souza, W.R. (1987) "Variable density flow and solute transport simulation of regional aquifers containing a narrow freshwater-saltwater transition zone", Water Resour. Res., pp. 1851-1866.

EVALUATION OF VERTICAL LEAKAGE SCHEMES FOR MULTILAYER SHARP-INTERFACE SALTWATER-INTRUSION MODEL

N. PARK* and Y.-S. WU**
Department of Civil Engineering*   HydroGeoLogic, Inc.**
Dong-A University   1165 Herndon Parkway, Suite 900
840 Hadan-Dong Saha-Gu   Herndon, VA 22070
Pusan, 604-714   USA
Korea

We evaluate vertical leakage schemes implemented in two general-purpose sharp interface models for simulating saltwater intrusion phenomena in multilayer aquifer systems. The two models have been developed by Essaid (1990a,b) and Huyakorn et al. (1994), respectively. Although both models were built on the same assumption of sharp interface, they differ conceptually and numerically in a number of aspects. The major conceptual difference lies in the treatment of vertical leakage. In this study we evaluate two leakage schemes and analyze limitations. A simple example problem was used to demonstrate limitations. It was found that restrictions in the vertical leakage schemes limit the applicability of Essaid's model.

INTRODUCTION

Many coastal areas rely on ground water as a source of freshwater. However, too much demand for ground water may break the delicate balance between the freshwater and saltwater flow causing the saltwater to intrude the freshwater zone. Hence, numerical modeling performed in supporting management of coastal water resources should be able to simulate the dynamics of saltwater and freshwater accurately.

In analyzing freshwater-saltwater dynamics, the most rigorous approach is the density-dependent, coupled flow and transport analysis. However, standard numerical techniques (e.g., central finite difference or Galerkin finite element method) for the solute transport equation are subject to spurious oscillations when the dispersion is relatively smaller than the advection. Oscillations can be avoided if the Peclet number criterion is satisfied by using fine discretization. But satisfying the Peclet number criterion for a practical simulation with large areas may be prohibitive on all but supercomputers. To reduce oscillations higher order numerical methods (for example, Sudicky, 1989; Park and Liggett, 1990) have been developed. However, the computational burden of these approaches may still be formidable for practical problems.

As an alternative, the sharp interface approach is often used. Major advantages of this approach are: 1) no need to satisfy the Peclet number criterion 2) no need for dispersivities which are difficult to estimate; and 3) reduction of computational dimensions by one through the vertical integration.

Extensive work has been conducted to analyze sharp interface problems in a single layer (e.g. see Essaid 1990b for a comprehensive review). However, extension of the sharp interface approach to multilayer aquifers were rather limited, mostly vertical cross sections with two aquifer layers. As extensions of the single layer approach to multilayer aquifers, only a handful of models with limited capabilities were proposed (for example Collins and Gelhar, 1971; Mualem and Bear, 1974; Rumer and

A. Peters et al. (eds.), Computational Methods in Water Resources X, 1027–1034.

Shiau, 1968; Bear and Kapuler, 1981 among others). To our knowledge Essaid (1990a,b) was the first to propose a general-purpose multilayer sharp interface numerical model, called SHARP. Later, Huyakorn et al. (1994) presented another general-purpose multilayer model, called SIMLAS, with a number of enhancements.

Although both models were built on the same principle, major difference in the two codes lies in the treatment of vertical leakage across aquitards. The treatment of vertical leakage is straightforward for single phase models, e.g., freshwater flow only models. However, for sharp interface models vertical leakage poses unique difficulty because of the instantaneous vertical equilibrium assumption and possibly a discontinuous interface profile across aquifer layers. The difficulty becomes apparent in the previous work: Collins and Gelhar (1971) prohibited downward leakage of saltwater from the upper layer. Bear and Kaupler (1981) used an impervious aquitard. Rumer and Shiau (1968) were able to avoid vertical leakage by using a continuous interface across aquifer layers. Mualem and Bear (1974) incorporated freshwater leakage into freshwater zone.

MATHEMATICAL DEVELOPMENT

Governing Equations

We consider a general situation involving a layered coastal aquifer-aquitard system containing freshwater and/or saltwater within each aquifer (Figure 1). The freshwater forms a pillow (or lens) that is variable in thickness and overlies the slightly denser saltwater. Vertical flow components in each liquid zone are assumed negligible so that the Dupuit assumption can be used. Under these assumptions, the governing equations for areal flow of freshwater and saltwater can be derived via vertical integrations of their respective three-dimensional mass balance equations (Huyakorn and Pinder, 1983, pp. 101-109). For aquifer unit m, which is overlain and underlain by aquifer units m+1 and m-1, respectively (Figure 1), the required governing equations may be written as follows:

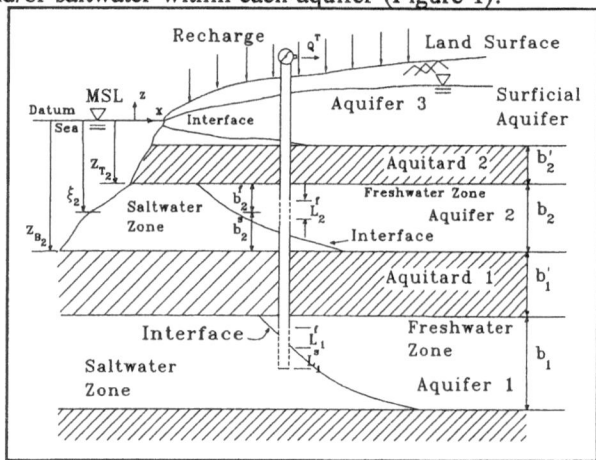

Figure 1. Saltwater intrusion in a multilayer aquifer system.

$$\frac{\partial}{\partial x_i}\left[K_{ijm}^f \, b_m^f \, \frac{\partial h_m^f}{\partial x_j}\right] + \Gamma_m^f + Q_m^f = b_m^f \, S_{sm}^f \, \frac{\partial h_m^f}{\partial t} - \theta \, \frac{\partial \xi_m}{\partial t}, \quad (i,j=1,2) \quad (1)$$

$$\frac{\partial}{\partial x_i}\left[K_{ijm}^s \, b_m^s \, \frac{\partial h_m^s}{\partial x_j}\right] + \Gamma_m^s + Q_m^s = b_m^s \, S_{sm}^s \, \frac{\partial h_m^s}{\partial t} + \theta \, \frac{\partial \xi_m}{\partial t}, \quad (i,j=1,2) \quad (2)$$

where superscripts f and s refer to freshwater and saltwater, respectively, the primes refer to aquitards, h is the vertically integrated hydraulic head, b_m^f and b_m^s are thicknesses of the freshwater and saltwater zones in aquifer unit m, K^f_{ijm} and K^s_{ijm} are hydraulic conductivities with respect to freshwater and saltwater, S^f_{sm} and S^s_{sm} are aquifer specific storage coefficients in the freshwater and saltwater zones, θ is the effective porosity, ξ_m is the elevation of the saltwater-freshwater interface, Γ^f_m and Γ^s_m are net freshwater and saltwater vertical leakage flux, respectively, to the aquifer layer m through overlying and underlying semiconfining layers, and Q^f_m and Q^s_m are volumetric fluxes of freshwater and saltwater due to pumping (or recharge).

For each aquifer layer equations (1) and (2) contain three unknowns, h^f, h^s, and ξ_m. Therefore, for closure, we need one more equation. The remaining equation is obtained from the compatibility condition of continuous pressure at the interface. Then, the condition can be written as (Bear, 1979):

$$\xi_m = \frac{1}{\epsilon^f}\left[\frac{\rho^s}{\rho^f}h^s_m - h^f_m\right] \tag{3}$$

where ρ_f and ρ_s are the freshwater and saltwater densities, respectively, and $\epsilon^f = \Delta\rho/\rho^f$ is the density difference ratio in which $\Delta\rho = \rho^s - \rho^f$. The governing equations can be solved when appropriate initial and boundary conditions are specified.

For multilayer aquifer systems the system of equations, (1), (2), and (3), are written for each aquifer layer. Connection between layers are made via vertical leakage, which is the subject of the next section.

TREATMENT OF VERTICAL LEAKAGE

In this section vertical leakage schemes implemented in the SHARP code (Essaid, 1990a,b) and the SIMLAS code (Huyakorn et al., 1994) are described. For convenience the leakage scheme used in the SHARP code is called the pressure formulation and in the SIMLAS code the hydraulic head formulation for reasons which will become clear later.

Vertical leakage calculation involves two distinct steps: First, one needs to calculate the leaky flux in the confining layer. Then, because there could be two different types of liquids in the receiving aquifer, one needs to allocate the leakage appropriately.

Pressure formulation

Essaid computed (1990b) the rate of leakage through a leaky aquitard by applying one-dimensional Darcy's law between the liquids in direct contact with the aquitard. When two liquids are of the same type, usual Darcy's law in terms of hydraulic heads can be used. However, when two liquids are of different types, Essaid used the density-dependent Darcy's law to compute the rate of vertical leakage.

For the two-aquifer system in Figure 2, one-dimensional Darcy's law for nonuniform density can be written as:

$$q = -\frac{k'}{\mu}\left(\frac{p_u - p_d}{b'} + \rho g\right) \tag{4}$$

where q is the rate of vertical leakage; k' and b' are the vertical permeability and the thickness of the confining layer, respectively; p_u and p_d are liquid pressures above and below the confining layer, respectively; and ρ is the density of the liquid in the aquitard. Essaid's scheme is called the pressure formulation because of the use of pressures.

Once the rate of the vertical leakage through a confining layer is computed, the leakage becomes a distributed source in the receiving aquifer. In distributing the vertical leakage, Essaid (1990b) proposed two different methods: the complete mixing and the restricted mixing methods. As the names imply, the two methods differ in how the leakage is distributed.

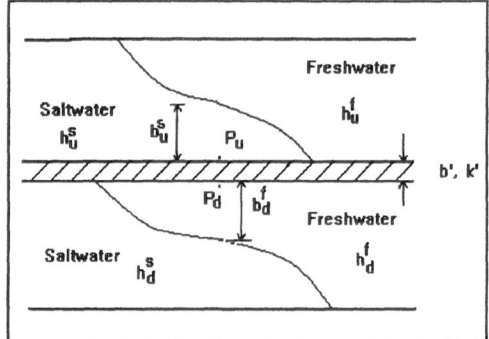

Figure 2 Two aquifer-layer system

The complete mixing scheme assumes that the amount of leakage is small relative to the amount of the water in the receiving aquifer and that the vertical leakage becomes a part of the ambient ground water regardless of the original type. This scheme may be reasonable when vertical leakage is small. However, as she correctly pointed out, the scheme cannot simulate the displacement of one type of water by the other type through vertical leakage.

The restricted mixing scheme forbids all vertical leakages between different types of liquids except for upward leakage of freshwater into the overlying saltwater zone. The fate of the freshwater then depends on the situation in the receiving aquifer.

Hydraulic head formulation

In contrast to Essaid's model, Huyakorn et al. computed the rate of vertical leakage between the same type of liquids whether or not they are in direct contact with the aquitard. Then, they allocated the leakage based solely on the original type without mixing. Their approach is consistent with the basic sharp interface assumption of immiscibility. For the situation depicted in Figure 2 the leaky flux for the lower aquifer becomes:

$$q_d^l = \alpha^l{}_T \, \lambda^l \, (h_u^l - h_d^l), \qquad l = f, s \tag{5}$$

where $\lambda^p = K'/b'$ is the leakance of the semiconfining layer for liquid type p and α's are dimensionless parameters that determine the type of leakage occurring through the aquitard.

The dimensionless leakage parameters, α_T^f and α_T^s in (5) take on values between 0 and 1. Some of the parameters may be set to zero (i.e., no leakage) for the following cases: 1. If there is no liquid to discharge, appropriate parameters are set to zero. 2. Freshwater cannot leak into a lower aquifer when it is underlain by saltwater. Likewise, saltwater cannot leak into an upper aquifer when it is overlain by freshwater. The basis of the first restriction is obvious. The second restrictions make certain that the vertical equilibrium is preserved within each aquifer by preventing the lighter liquid to sink through the heavier liquid to reach the leaky aquitard and the heavier liquid to float through the lighter liquid. The second restrictions can be expressed as follow:

$$\alpha_T^s = 0 \quad when \quad h_m^s > h_{m+1}^s \quad and \quad \xi_m < Z_{Tm} \tag{6a}$$

$$\alpha_T^f = 0 \quad when \quad h_m^f < h_{m+1}^f \quad and \quad \xi_{m+1} > Z_{Bm+1} \tag{6b}$$

and when both $\alpha^s{}_T$ and $\alpha^f{}_T$ are equal to 1, both are reset to 1/2.

For a general case involving more than two layers vertical leakage is possible through both upper and lower semiconfining layers. In that case α's for bottom leakages are determined in a similar manner.

ANALYSIS OF LEAKAGE COMPUTATION SCHEMES

In this section we analyze the leaky flux computation schemes. Since the two schemes become identical when liquids above and below a confining aquitard are of the same type, we will consider a case when they are of different types. Without the loss of generality we will only consider two-layer case as depicted in Figure 2.

Pressure formulation

Calculation of leaky flux using the pressure formulation (4) is straightforward except for the evaluation of the density term. The density, which represents the liquid in the aquitard, depends on the type of liquid in the aquitard. Essaid used an arithmetic average.

Hydraulic head formulation

Because the hydraulic head formulation treats leakages of saltwater and freshwater separately, we shall consider one type at a time. First, the freshwater leakage equation, (5) with $l=f$, becomes for the two-layer system (Figure 2) under consideration:

$$q^f = -K'^{lf} \alpha^f \left(\frac{h_u^f - h_d^f}{b'} \right) \tag{7}$$

To compare (7) with the pressure formulation (4), h_u^f is first replaced with $-\epsilon^f \xi_u$ + $(\rho^s/\rho^f)h_u^f$ obtained from (3). Then, using the hydrostatic condition, h_u^s and h_d^f are written in terms of pressures at the bottom and the top of the aquifers, respectively. Upon reorganization the following equation is obtained:

$$q^f = -K'^{lf} \left(\frac{1}{\rho^f g} \frac{p_u - p_d}{b'} + 1 - \epsilon^f \frac{b_u^s}{b'} \right) \tag{8}$$

The first two terms in the parenthesis of (8) are identical to the pressure formulation (4) if the freshwater density is used in (4). The difference between (8) and (4) is in the last term of (8). However, an order of magnitude analysis indicates that the last term is two orders smaller than the preceding terms provided that b_u^s is not significantly larger than b'.

Likewise, saltwater leakage can be expressed as:

$$q^s = -K'^{ls} \left(\frac{1}{\rho^s g} \frac{p_u - p_d}{b'} + 1 + \epsilon^s \frac{b_d^f}{b'} \right) \tag{9}$$

where $\epsilon^s = \Delta\rho/\rho^s$. Again, (9) is identical to (4) except for the small deviation term.

EXAMPLE

In this section a simple example is selected to test both numerical models, in particular, the leakage schemes. The system under consideration is a two aquifer-layer system of uniform thicknesses separated by a leaky aquitard (Figure 3). Initially, the upper aquifer is filled with saltwater and the lower aquifer with freshwater. Although the system appears quite simple, this type of situation is not uncommon in practice. For example, if an aquifer filled with freshwater extends beyond a coastline, then seawater can lie directly above freshwater as it is depicted in Figure 3.

Relevant flow parameters are as specified in Figure 3. For simplicity the system is considered completely enclosed. Since the system is in an unstable state, it would reorganize itself to a stable situation, i.e., freshwater on top of saltwater.

Analytical solution

Since there is no spatial variation, the governing equations (1), (2), and (3) reduce to a set of nonlinear algebraic and ordinary

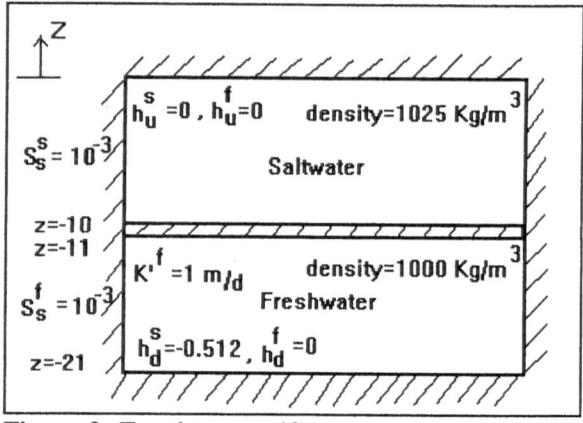

Figure 3. Two layer-aquifer system for the example

differential equations. Since (3) is an algebraic equation, only the two partial differential equations, (1) and (2), need to be solved first for any two unknowns from h^f, h^s, and ξ. Huyakorn et al. selected h^f and ξ as the primary unknowns (see Huyakorn et al., 1994 for arguments). To remove h^s from (2) equation (3) is used. Then, the freshwater equation (1) is added to the updated saltwater equation for further simplification. After some algebraic manipulation. Finally the following system of equations are obtained for the lower aquifer:

$$(Z_{dT} - \xi_d) S_{sd}^f \frac{dh_d^f}{dt} - \theta \frac{d\xi_d}{dt} = \alpha_T^f \lambda^f (h_u^f - h_d^f) \tag{9}$$

$$[b_d - (\epsilon^f - \epsilon^s + \epsilon^f \epsilon^s)(\xi_d - Z_{dB})] S_{sd}^f \frac{dh_d^f}{dt} + (\xi_d - Z_{dB}) S_{sd}^f \epsilon^f \frac{d\xi_d}{dt}$$
$$= (\alpha_T^f \lambda^f + \alpha_T^s \lambda^s \frac{\rho^f}{\rho^s}) h_u^f - (\alpha_T^f \lambda^f + \alpha_T^s \lambda^s \frac{\rho^f}{\rho^s}) h_d^f + \alpha_T^s \lambda^s \epsilon^f (\xi_u - \xi_d) \tag{10}$$

Equations for the upper aquifer are obtained similarly:

$$(Z_{uT} - \xi_u) S_{su}^f \frac{dh_u^f}{dt} - \theta \frac{d\xi_u}{dt} = -\alpha_B^f \lambda^f (h_u^f - h_d^f) \tag{11}$$

$$[b_u - (\epsilon^f - \epsilon^s + \epsilon^f \epsilon^s)(\xi_u - Z_{uB})] S_{su}^f \frac{dh_u^f}{dt} + (\xi_u - Z_{uB}) S_{su}^f \epsilon^f \frac{d\xi_u}{dt}$$
$$= -(\alpha_T^f \lambda^f + \alpha_T^s \lambda^s \frac{\rho^f}{\rho^s}) h_u^f + (\alpha_T^f \lambda^f + \alpha_T^s \lambda^s \frac{\rho^f}{\rho^s}) h_d^f - \alpha_T^s \lambda^s \epsilon^f (\xi_u - \xi_d) \tag{12}$$

The nonlinear ordinary differential equations (9)-(12) describe the transient behavior of the system. Although the equations can be solved numerically, we elect to linearize them noting that the coefficients of the nonlinear terms are orders of magnitude smaller than those of linear terms, especially for the parameters used in the present example. The linearized equations were solved easily using the Laplace transform technique. Because the solutions are bit lengthy, and because they are not crucial in the analysis, only the plot of the time varying interface positions is presented in Figure 4.

Also depicted in the same figure is the fourth order Runge-Kutta solution of the nonlinear differential equations. Comparison of the linear and nonlinear solutions indicates that the linearization error is indeed small.

Numerical solutions

The same example problem was simulated using SHARP and SIMLAS codes. For SHARP both the restricted mixing and the complete mixing methods were used. As was anticipated, neither method of SHARP was able to predict the repositioning of saltwater and freshwater. It is easy to see why the restricted mixing method would fail because it prohibits downward leakage of saltwater into the underlying freshwater zone. On the other hand, the complete mixing scheme allows leakage of all

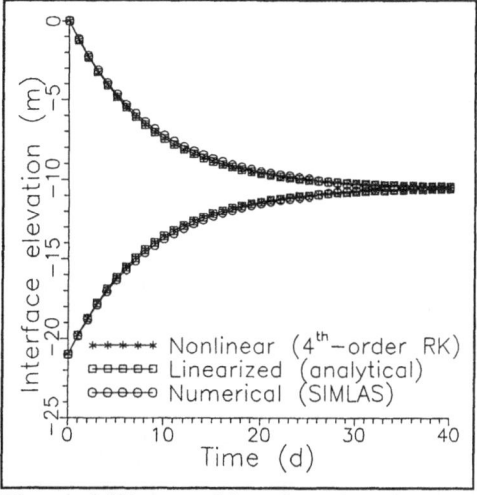

Figure 4 History of interface elevations

directions: Saltwater is allowed to leak downward, and freshwater to leak upward. However, saltwater leaked into the freshwater zone becomes freshwater, and freshwater leaked into the saltwater zone becomes saltwater due to the complete mixing assumption. Therefore, although the simulation started with an unstable state, SHARP maintains the state due to the phase transition between saltwater and freshwater.

The SIMLAS code model simulated the problem successfully. The results are presented in Figure 4 along with the approximate analytical solution. Good agreement is observed.

SUMMARY AND CONCLUSIONS

In this study, we evaluated two vertical leakage schemes implemented in two recent general-purpose sharp interface models developed by Essaid (1990a,b) and Huyakorn et al. (1994). Two aspects of vertical leakage were examined separately: The first aspect consists of the calculation of the leaky flux through a semiconfining layer. The second aspect comprises the allocation of the flux to saltwater and/or freshwater in the receiving aquifer. Essaid's model and Huyakorn et al.'s model differ in both aspects.

In calculating leaky flux in a semiconfining layer, Essaid used the density-dependent Darcy's law between two liquids in direct contact with the leaky layer from below and above. On the other hand, Huyakorn et al. calculated the leaky flux between the same type of liquids whether or not they are in direct contact with the leaky layer. Thus the two methods differ when the two liquids are of different types. The difference was shown to be small when the freshwater or saltwater zone thickness is not too large compared with the thickness of the leaky layer.

In allocating leaky flux Essaid proposed two different methods, so-called the complete mixing and the restricted mixing. The former method disregards the original type of the leaky flux and switches the type to that of the ambient ground water in the receiving aquifer. The latter method restricts all vertical leakages between different types of liquids except for the upward leakage of freshwater into an overlying saltwater zone. Then, the fate of the freshwater depends on the amount of freshwater in the computational block. In contrast, Huyakorn et al. did not allow mixing between different types of liquids. Their approach is consistent with the basic assumption of sharp interface approximation. Furthermore, leakage of any direction is allowed without any restriction.

A simple, but not uncommon in practice, example was used to assess the capabilities of both models. An analytical solution was obtained for the linearized governing equations. Huyakorn et al.'s model simulated the example successfully, and the result was in good agreement with the analytical solution. On the other hand, Essaid's model was not able to simulate the problem due to the arbitrary restrictions and mixing.

It is shown that Huyakorn et al.'s vertical leakage scheme is more flexible in handling a wide variety of field situations. The applicability of Essaid's model is restricted because of the limitations in the vertical leakage scheme. For example her model can not be used to simulate situations described in the example or situations involving saltwater upconing via vertical leakage from a deeper aquifer layer.

REFERENCES

Bear, J. (1979) Hydraulics of Groundwater, McGraw-Hill, New York.

Bear, J., and Kaupler I. (1981) "A numerical solution for the movement of an interface in a layered coastal aquifer", Journal of Hydrology 50, 273-298.

Collins, M. A. and Gelhar, L. W. (1971) "Seawater intrusion in layered aquifers", Water Resources Research 7(4), 971-979.

Essaid, H. I. (1990a) "A multilayered sharp interface model of coupled freshwater and saltwater flow in coastal systems: Model development and application, Water Resources Research 26(7), 1431-1454.

Essaid, H. I. (1990b) "The computer model SHARP, a quasi-three-dimensional finite-difference model to simulate freshwater and saltwater flow in layered coastal aquifer systems", US Geol. Survey, Water-Resources Investigations Report 90-4130, p. 181.

Huyakorn, P. S., Park, N. and Wu, Y.-S. (1994) "A multiphase approach to the numerical solution of sharp-interface saltwater intrusion problem: I. layered model formulation", submitted for publication.

Huyakorn, P. S. and Pinder, G. F. (1983) Computational Methods in Subsurface Flow, Academic Press, San Diego, California.

Mualem, Y. and Bear, J. (1974) "The shape of the interface in steady flow in a stratified aquifer", Water Resources Research 10(6), 1207-1215.

Park, N. and Liggett, J. A. (1990) "Taylor-least squares finite element for two-dimensional advection-dominated unsteady advection-diffusion problems", Int. J. for Numer. Meth. for Fluids 11, 21-38.

Rumer, R., and Shiau, J. (1968) "Salt water interface in a layered coastal aquifer", Water Resources Research 4(6), 1235-1247.

Sudicky, E. A. (1989) "The Laplace transform Galerkin technique: A time-continuous finite element theory and application to mass transport in groundwater", Water Resources Research 25(8), 1833-1846.

TWO-DIMENSIONAL MODELING OF SALTWATER INTRUSION

R.O. Strobl and G.T. Yeh
Department of Civil and Environmental Engineering
The Pennsylvania State University
University Park, PA 16802
U.S.A.

A two-dimensional finite element model for density-dependent flow and transport through saturated-unsaturated porous media has been developed from the combination and modification of two previously published, separate flow and transport models. Hence, the newly combined model can handle a wide range of real-world problems, including the simulations of saturated-unsaturated flow alone, contaminant transport alone, or combined flow and transport. In addition, the model developed was sought to give the user a vast number of numerical techniques, methods, and options at his/her disposal to make the model as flexible as possible in handling groundwater flow and transport problems. One of the many numerical options and techniques included in the model is a hybrid Lagrangian-Eulerian finite element method incorporated in the transport module, To study the reliability and versatility of the developed model, three seawater intrusion problems were investigated.

INTRODUCTION

Often when a contaminant is introduced into the groundwater system, clear changes in the groundwater density occur that may be sufficiently large to alter the flow dynamics of the system. The best-known case of such an occurrence is saltwater intrusion. The hydraulic gradients that are produced from excessive pumping may induce a flow of saline water toward the pumping well. Hence, there is a need to predict the location and movement of the saltwater interface in order to be able to protect freshwater aquifers from the possible danger of contamination. For this purpose, a 2-Dimensional Finite Element Model for density-dependent Flow And Transport through saturated-unsaturated porous media (2DFEMFAT) has been developed. This model stems from the combination and modification of two previous codes, a groundwater flow model (FEMWATER, Yeh, 1987) and a subsurface contaminant transport model (LEWASTE, Yeh, 1992). In the newly combined model, density-dependent effects are accounted for, since it is necessary to consider the seawater intrusion problem as a density-dependent flow and transport problem in order to account for the dispersed nature of the saltwater-freshwater interface and the associated saltwater circulation zones. A complete description of the model can be found in Strobl (1993).

MATHEMATICAL THEORY

The governing equation for flow is basically the modified Richards' equation and can be

1035

A. Peters et al. (eds.), Computational Methods in Water Resources X, 1035–1042.
© 1994 Kluwer Academic Publishers. Printed in the Netherlands.

stated as:

$$\frac{\rho}{\rho_0}(\alpha\rho_0 g\frac{\theta}{n_e} + \beta\rho_0 g\theta + \frac{\rho_0}{\rho}\frac{d\theta}{dH})\frac{\partial H}{\partial t} = \nabla \cdot [K \cdot \{\nabla H + (\frac{\rho}{\rho_0} - 1)\nabla z\}]$$

$$+ \frac{\rho^*}{\rho_0}q \quad (or -\frac{\rho}{\rho_0}q) \tag{1}$$

where H is the referenced hydraulic head (L), t is the time (T), K is the hydraulic conductivity tensor (L/T), z is the potential head (L), q is the internal source and/or sink (L^3/T/L), ρ is the density (M/L^3) at chemical concentration C (M/L^3), ρ_0 is the referenced density, ρ^* is the density of the injected fluid (M/L^3), g is the acceleration of gravity (L/T^2), θ is the moisture content (dimensionless), n_e is the effective porosity (L^3/L^3), α is the coefficient of consolidation of the media (LT^2/M), and β is the compressibility of the fluid (LT^2/M).

The governing equation for transport can be stated as:

$$\theta\frac{\partial C}{\partial t} + \rho_b\frac{\partial S}{\partial t} + V \cdot \nabla C = \nabla \cdot (\theta D \cdot \nabla C) - \lambda(\theta C + \rho_b S) + qC_{in} -$$

$$\left[\frac{\rho^*}{\rho}q - \frac{\rho_0}{\rho}V \cdot \nabla\left(\frac{\rho}{\rho_0}\right)\right]C - \alpha\rho_0 g\frac{\rho}{\rho_0}(\theta C + \rho_b S)\frac{\partial H}{\partial t} - \tag{2}$$

where
$$\theta K_w C - \rho_b K_s S + \left[(\alpha\rho_0 g\frac{\theta}{n_e} + \beta\rho_0 g\theta + \frac{\rho_0}{\rho}\frac{d\theta}{dH})\frac{\partial H}{\partial t} - \frac{\partial\theta}{\partial t}\right]C$$

$$S = K_d C \quad \text{for linear isotherm} \tag{3a}$$

$$S = \frac{s_{max}KC}{1 + KC} \quad \text{for Langmuir isotherm} \tag{3b}$$

$$S = KC^n \quad \text{for Freundlich isotherm} \tag{3c}$$

where C is the material concentration in the aqueous phase (M/L^3), t is the time (T), ρ_b is the bulk density of the medium (M/L^3), S is the material concentration in the adsorbed phase (M/M), V is the discharge (L/T), ∇ is the del operator, D is the dispersion coefficient tensor, λ is the decay constant (1/T), C_{in} is the material concentration of the injected fluid in the case of sources or the material concentration at the point of withdrawal in the case of sinks (M/L^3), K_w is the biodegradation rate constant through aqueous phase (1/T), K_s is the biodegradation rate constant through adsorbed phase (1/L), K_d is the distribution coefficient (L^3/M), s_{max} is the maximum concentration permitted in the medium in the Langmuir nonlinear isotherm, K is the coefficient in the Langmuir or Freundlich nonlinear isotherm, and n is the power index in the Freundlich nonlinear isotherm.

In the specific case of seawater intrusion, the constitutive relation between fluid density and concentration takes the form:

$$\rho = \rho_0[1 + (\frac{\rho_{max}}{\rho_0} - 1)c] \tag{4}$$

where c is the dimensionless chemical concentration (actually divided by maximum) and ρ_{max} is the maximum density of the fluid (M/L^3).

DEMONSTRATIVE EXAMPLES

In order to test the computer code developed in this research, the authors verified the code with three seawater intrusion examples reported in the literature [Huyakorn et al., 1987]. To simulate these examples, the computer code 2DFEMFAT was executed using the VS FORTRAN compiler licensed at The Pennsylvania State University.

Example 1: Seawater intrusion in a confined aquifer

This example is widely known as Henry's seawater intrusion problem (Henry, 1959). The confined aquifer under consideration is a uniform isotropic aquifer that is bounded below and above by impermeable strata. In addition, the aquifer is exposed on the right side by a stationary seawater body and is recharged on the left side by a constant freshwater influx. The boundary condition of the coastal side of Henry's problem, however, was modified to permit convective mass transport out of the aquifer system over the top portion of the coastal boundary. In this top portion of the seaward boundary, the normal concentration gradient was set equal to zero. This modification to Henry's problem was necessary because the original boundary condition on the coastal side dictates that the fluid must change pretty suddenly from freshwater to seawater. Such a circumstance is not only rather unrealistic but also hard to satisfy numerically.

To test and compare the model under research, a transient simulation was performed. The convergence tolerances for head and relative concentration were prescribed as 0.01 and 0.01, respectively. The longitudinal and lateral dispersivities, α_L and α_T, respectively, were set equal to zero, while the molecular diffusion coefficient D_0 was set to 6.6 x 10^{-2} m^2/day. All the parameters selected to simulate these cases were similar to those chosen by other researchers. It should be noted that the boundary condition ($\partial c/\partial n = 0 \quad 80 \le z \le 100 m$) that allows convective mass transport out of the aquifer system over the top portion is consistent with the boundary condition used in the simulations by Huyakorn et al. (1987), Frind (1982b),and Segol et al. (1975), but is a revision to the boundary condition used by others, such as Lee and Cheng (1974) and Henry (1964). In order to perform the simulation, the region of interest was discretized into 176 nodes and 150 rectangular elements. The initial concentration and reference hydraulic head were set equal to zero. The initial time step size was selected as 12 days with a time step multiplier of 1.17169. The maximum allowable time step was selected to be 600 days.

Figure 1 discloses the results of the advancement of the 0.5 isochlor for the transient-state, constant dispersion case at the end of t = 6000 days. The present analysis is compared to those of Segol et al. (1975), Frind (1982b), and Huyakorn et al. (1987). Unquestionably, all four solutions are in satisfactory agreement with each other. Segol et al.'s (1975) solution is apparently slightly shifted on the seaward side, probably due to the expense of using a more primitive groundwater flow equation than the other analyses.

In Figure 1, a peculiar bend in the 0.5 isochlor line can be noticed at the coastal boundary for the present analysis. This may be attributed to the fact that the nodal point at z = 70m on the coastal boundary was specified as a Dirichlet boundary condition (as c = 1.0), and would seem to cause the bend. The bend can be avoided by refining the grid. However, to keep consistent with the meshes of the other researchers, this was not done. The reason why the simulations of the other researchers produced a smooth 0.5 isochlor in that region was because the other researchers apparently used smoothing factors in their plotting packages.

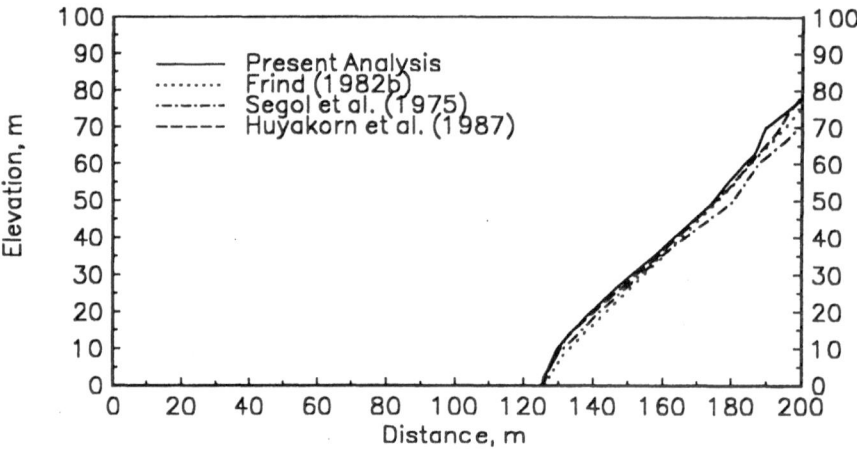

Figure 1.   0.5-isochlor distribution at t=6000days

Example 2: Seawater intrusion in a phreatic aquifer

This example involves an anisotropic unconfined aquifer that is being recharged from the top as well as from the freshwater side and invaded by seawater on the coastal side. The saturated thickness is assumed to be 50 m. In addition, the top boundary of the phreatic aquifer, which is a free surface, is assumed to be fixed at an elevation of 50 m above the base of the aquifer. Even though this approximate assumption of the free surface conditions may seem unreasonable at first glance, it can be considered a conceivable assumption for this particular example problem since the maximum rise in the water table due to recharge is not expected to exceed but a few percent of the initial saturated thickness of 50 m (Huyakorn et al., 1987). A steady-state simulation of this example problem was performed to compare the results with those of Huyakorn et al. (1987) and Galeati et al. (1992). The region of interest was discretized into 250 (10 x 25) rectangular elements and 276 nodes. The initial values of reference hydraulic head and concentration were set equal to zero. The maximum allowable iteration tolerances for head and relative concentration were assigned values of 0.01 and 0.01, respectively.

Figure 2 presents the position of the 0.5 isochlor at steady-state for comparison of the present analysis with Huyakorn et al. (1987) and Galeati et al. (1992). Clearly all three simulations compare fairly well. However, the present analysis compares somewhat better with Galeati et al.'s (1992) simulation rather than with Huyakorn et al.'s (1987) simulation, most likely because the present analysis as well as Galeati et al.'s (1992) model used the same method for computing the nodal velocities as suggested by Yeh (1981), while Huyakorn et al.'s (1987) model employed a much simpler approach for calculating the nodal velocities, which in problems involving more complex flow patterns can lead to incorrect nodal velocities. Huyakorn et al.'s (1987) main reason for developing simple formulae that compute the centroidal values of the Darcy velocity components was to reduce the computational effort to a minimum (Huyakorn et al., 1986). Nevertheless, it should be noted that the vertical components of velocity may be appreciable for phreatic aquifers receiving vertical recharge, while the vertical components of velocity are small where vertical recharge is absent, such as assumed in the treatment of confined aquifers (Henry, 1964). This reason could also explain why the 0.5 isochlor results at the coastal boundary are very similar for the present analysis and Huyakorn et al.'s (1987) analysis in Example 1, but not in Example 2. Also shown in Figure 2 is the Ghyben-Herzberg (sharp) interface as derived from the reference hydraulic head profile using the Ghyben-Herzberg relation. The Ghyben-Herzberg model apparently exhibits a much further advancement of the 0.5 isochlor in contrast to all other models.

For a clearer understanding of the dynamics of the flow in the aquifer, Figure 3 depicts the velocity vector field for the present analysis. As can be essentially predicted from the 0.5 isochlor plot (Figure 2), seawater enters through the lower left boundary, mixes with the freshwater in the aquifer, and then exits at the outlet on the top left boundary. It should be noted that seawater enters at a slower rate than the fluid that leaves the aquifer in the upper left-hand portion of the boundary. Hence, the seawater invasion is a gradual, slow process.

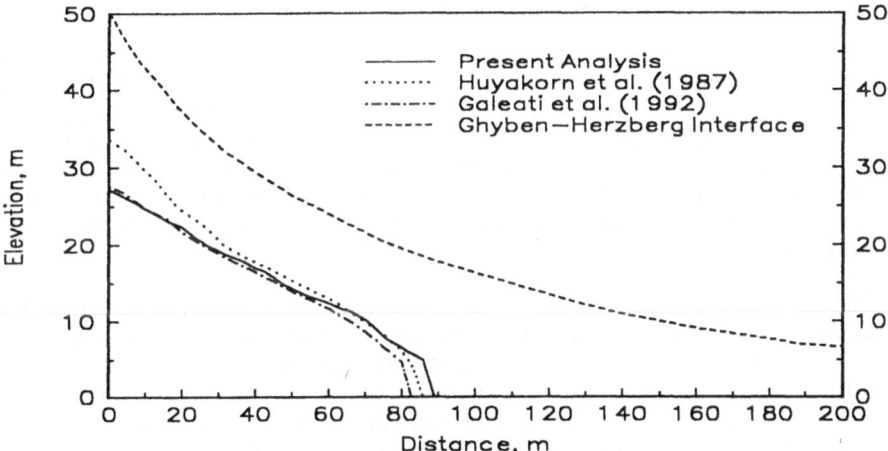

Figure 2. 0.5-isochlor distribution and Ghyben- Herzberg interface

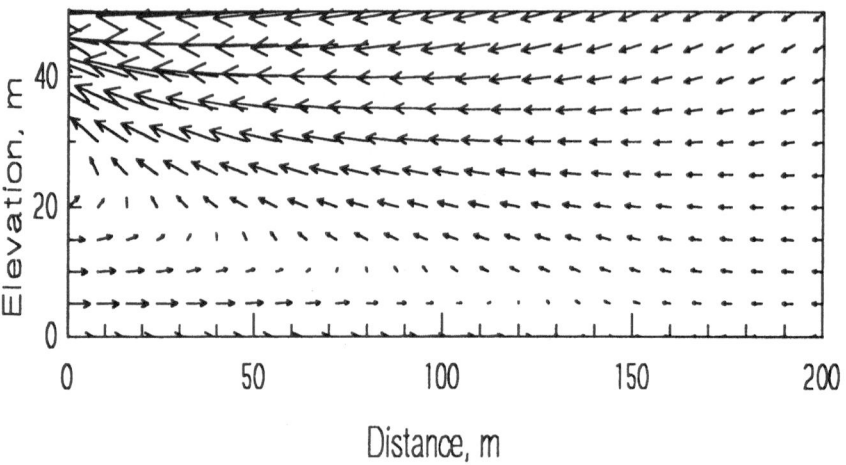

Figure 3. Velocity vector field plot of present analysis

Example 3: Seawater intrusion in an aquifer-aquitard system

In this example, an aquifer-aquitard system subject to seawater intrusion is investigated. The physical setting of this problem involves an anisotropic confined aquifer that is bounded above by an aquitard and below by an impermeable layer. The aquitard, on the other hand, is bounded above by a phreatic aquifer. Freshwater recharges the confined aquifer from land side, while seawater invades on the coastal side. At the base of the unconfined aquifer, the head was prescribed as H_T. Also, the mean sea level was used as the head datum. For comparison of the results with Huyakorn et al. (1987), a steady-state analysis of this example problem was simulated. In order to simulate this problem, the confined aquifer and aquitard were discretized into a rectangular grid of 650 elements (300 for the aquifer and 350 for the aquitard) with 714 nodes. To start the steady-state simulation, initial values for the reference hydraulic head and concentration were set equal to zero. The maximum allowable iteration tolerances for head and relative concentration were assigned values of 0.01 and 0.01, respectively.

Note that the prescribed function H_T is given in Figure 4. Also plotted in Figure 4 are the profiles of the reference head at the top of the confined aquifer for the present analysis as well as for Huyakorn et al.'s (1987) simulation. As may be seen, the comparison is good. A slight ascent of the profile of the reference head at the top of the aquifer (H_a) is seen. This observation reaffirms what one might already have suspected, namely that the fluid is flowing seawards along the top of the aquifer. This profile intersects the prescribed function H_T at roughly 9200 ft. Figure 5, on the other hand, shows selected isochlor lines for comparison between the present analysis and Huyakorn et al.'s (1987) simulation. The vertical variation of the isochlor lines for Huyakorn et al.'s (1987) analysis appears to be insignificant near the interface between the aquitard and the confined aquifer, while for the present analysis there is a "concentration jump" near the interface. This "jump" can be explained from the fact that the present model has a better resolution in computing the velocities because the Galerkin finite element was used, while Huyakorn et al.'s model used simple formulae to compute centroidal velocities. It should be noted that in order for Huyakorn et al.'s model to have a concentration "jump", the grid in their simulation needs to be refined.

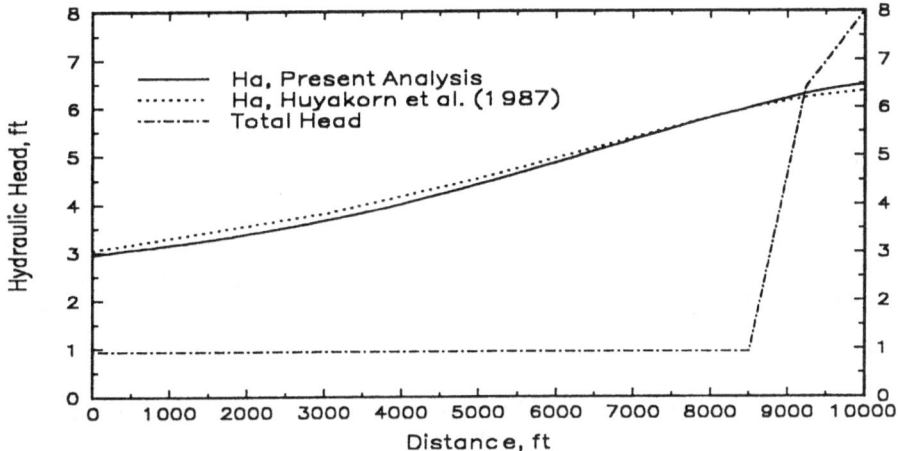

Figure 4. Prescribed total head and comparison of reference hydraulic head at the top of the confined aquifer

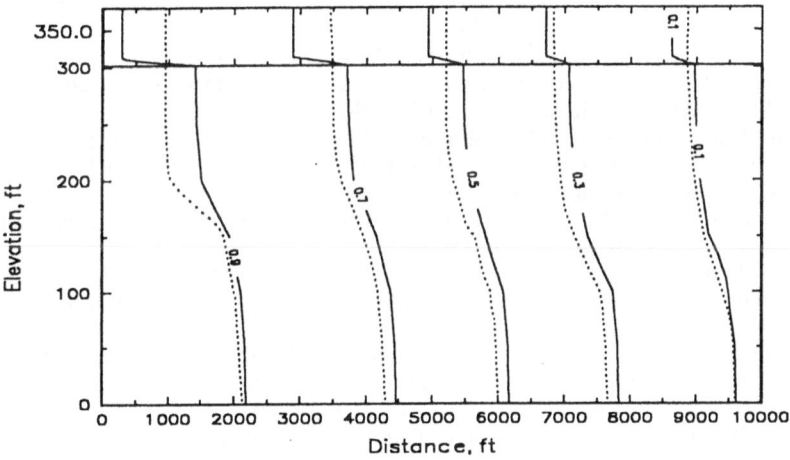

Figure 5. Comparison of selected isochlor lines for present analysis and Huyakorn et al. (1987)

CONCLUSIONS

The basic aim of this research was to develop a two-dimensional finite element model for density-dependent flow and transport through saturated-unsaturated porous media by successfully combining and modifying two separately and previously existing groundwater flow and solute transport modules. Several important points can be drawn as conclusions for the present research effort. Firstly, even though the merit of the developed model was contingent on good agreement with similar simulations of other researchers, the validity of the present model is believed to be highly probable, not only due to the production of similarly comparable results of the other models, but also due to the believed numerical soundness of the developed model. Hence, due to the model's numerous numerical strategies, such as methods for dealing with difficulty of obtaining numerical convergence, efficient finite element techniques, etc., a wider variety of problems can be dealt with in ease. The use of a dispersive model instead of a sharp interface model is widely accepted as necessary for analysis of seawater encroachment problems and proved to be very reasonable for use in the developed model by comparing the results with previously published results.

REFERENCES

Frind, E.O. (1982b) "Simulation of long-term transient density-dependent transport in groundwater", Advances in Water Resources, 5:2, 73-88.

Galeati, G., Gambolati, G. and Neuman, S.P. (1992) "Coupled and partially coupled Eulerian-Lagrangian model of freshwater-seawater mixing", Water Resources Research, 28:1, 149-165.

Henry, H.R. (1959) "Salt intrusion into fresh water aquifers", Journal of Geophysical Resources, 64:11, 1911-1919.

Henry, H.R. (1964) "Effects of dispersion on salt encroachment in coastal aquifers", Seawater in Coastal Aquifers (ed. H. H. Cooper et al.), U. S. Geological Survey Water Supply Paper 1613-C, Washington, D.C., 35-69.

Huyakorn, P.S., Springer, E.P., Guvanasen, V. and Wadsworth, T.D. (1986) "A three-dimensional finite-element model for simulating water flow in variably saturated porous media", Water Resources Research, 22:13, 1790-1808 .

Huyakorn, P.S., Andersen, P.F., Mercer, J.W. and White, H.O. (1987) "Saltwater intrusion in aquifers: Development and testing of a three-dimensional finite element model", Water Resources Research, 23:2, 293-312.

Lee, C.H. and Cheng, R.T. (1974) "On seawater encroachment in coastal aquifers, Water Resources Research, 10:5, 1039-1043.

Segol, G., Pinder, G.F. and Gray, W.G. (1975) "A Galerkin finite element technique for calculating the transient position of the salt water front", Water Resources Research, 11:2, 343-347.

Strobl, R.O. (1993), 2DFEMFAT: A two-dimensional finite element model for density-dependent flow and transport through saturated-unsaturated porous media, Master's thesis, The Pennsylvania State University, University Park, Pa..

Yeh, G.T. (1981) "On the computation of Darcian velocity and mass balance in the finite element modeling of groundwater flow", Water Resources Research, 17:5, 1529-1534.

Yeh, G.T. (1987) FEMWATER: A finite element model of water flow through saturated-unsaturated porous media, First Revision, Rep. ORNL-5567/R1, Oak Ridge Nat. Lab., Oak Ridge, Tenn., 37831, 258 pp..

Yeh, G.T., (1992) "Class notes: CE597C: Computational Subsurface Hydrology Part II", Spring Semester, The Pennsylvania State University, University Park, Pa..

11. SHALLOW WATER EQUATIONS

MODELING NEAR-BOTTOM ADVECTIVE ACCELERATION IN SURFACE WATER MODELS

ANDRÉ B. FORTUNATO
ANTÓNIO M. BAPTISTA
Center for Coastal and Land-Margin Research and Department of Environmental Science and Engineering, Oregon Graduate Institute of Science & Technology, P.O. Box 91000, Portland OR 97291-1000, USA.

ABSTRACT

This paper describes problems related to the evaluation of near-bottom horizontal velocity gradients in cartesian coordinates. An analytical model and simple scaling arguments show that computing advective accelerations in cartesian coordinates may require extremely fine horizontal grids. Numerical tests are used to confirm this conclusion and to exemplify the unrealistic velocity profiles that can be obtained due to large errors in the evaluation of the advective acceleration. It is therefore recommended that horizontal gradients of velocity be evaluated in sigma coordinates. This conclusion is in stark contrast with similar studies for the baroclinic pressure gradient, and is due to the rapid variation of the vertical gradient of horizontal velocities in the bottom boundary layer.

INTRODUCTION

The use of sigma coordinates, wherein the height of the water column is mapped into a fixed interval, has become widespread in three-dimensional hydrodynamic models developed over the last decade (see review by Cheng and Smith 1990). The adoption of the sigma, or stretched, coordinates is explained by three important advantages relative to the older z-coordinates: (a) the resolution over depth is more uniform; (b) a smooth bottom topography can be represented; and (c) the treatment of the free surface boundary condition is straightforward.

However, there is currently some concern regarding the ability of the sigma coordinates to deal with steep topographic features. Haney 1991 showed that, in the presence of strong stratification and steep slopes, the use of sigma coordinates leads to very large errors in the evaluation of baroclinic pressure gradients. These errors are particularly troubling because they can generate moderate currents in a system that should otherwise be at rest (Walters and Foreman 1992).

To avoid this problem, several modelers have proposed computing horizontal gradients of density and/or and velocity directly in cartesian coordinates, by interpolating the values needed in planes of constant z (Sheng et al. 1990, Laible 1992, Beckmann and Haidvogel 1993). Indeed, truncation error analysis suggests that this approach may

A. Peters et al. (eds.), Computational Methods in Water Resources X, 1045–1052.
© 1994 Kluwer Academic Publishers. Printed in the Netherlands.

decrease the errors in the evaluation of horizontal gradients by several orders of magnitude (Fortunato and Baptista 1994). However, even though numerical experiments support this conclusion for the baroclinic pressure gradient, it will be shown in this paper that, under certain circumstances, it may be best to evaluate horizontal derivatives in sigma coordinates.

In this paper we examine the evaluation of horizontal gradients of velocities in hydrodynamic models that explicitly solve for the vertical structure of the flow. An analytical solution and scaling arguments suggest that this evaluation will be more accurate in sigma than in cartesian coordinates. Numerical results are then used to exemplify the type of errors obtained when advective accelerations are computed in cartesian coordinates. Several formulations for the treatment of near bottom velocity gradients are presented, then compared by examining velocity profiles for a wave passing over a step. The results for all formulations present a similar unrealistic behavior, in sharp contrast with those obtained from a standard sigma coordinate formulation, which are hardly distinguishable from a reference simulation.

ANALYTICAL SOLUTION

To isolate the errors that arise solely from evaluating horizontal gradients of velocity in cartesian coordinates, we consider a case with analytical solution. To derive the analytical solution, we write the linear momentum equation for an uniform flow in a channel:

$$\frac{\partial}{\partial \sigma}(A_v \frac{\partial u}{\partial \sigma}) = gH^2\theta \tag{1}$$

where g is the gravitational acceleration, H is the total water depth, θ is the bottom slope. Following Luettich and Westerink 1991, the vertical eddy viscosity A_v is assumed to vary linearly as $A_v = A_{v0}(\sigma + 1 + \sigma_0)$. A_{v0} is scaled as Hu_* ($u_* = \sqrt{\tau_b/\rho}$ is the friction velocity), or, for a uniform flow, $A_{v0} \sim H\sqrt{Hg|\theta|}$; $\sigma_0 \equiv z_0/H$ is the dimensionless roughness height.

Integrating (1) from a generic position σ, to the free surface ($\sigma=0$) where a no-stress boundary condition is used, we get:

$$\frac{\partial u}{\partial \sigma} = \frac{gH^2\theta}{A_{v0}} \frac{\sigma}{\sigma + 1 + \sigma_0} \tag{2}$$

Equation (2) is integrated from the bottom ($\sigma=-1$) to a generic position to yield:

$$u(\sigma) = \frac{gH^2\theta}{A_{v0}}\left(\sigma + 1 + (1+\sigma_0)\, ln\left(\frac{\sigma_0}{1+\sigma_0+\sigma}\right)\right) + u_0 \tag{3}$$

where u_0 is the bottom velocity. It can be verified a posteriori that the sum of the horizontal and vertical advective accelerations is zero, so Equation (3) is also the solution of the momentum equation including advective acceleration, coupled with the continuity equation.

The horizontal derivative can be computed as:

$$\frac{\partial u}{\partial x} = -\frac{gH\theta^2}{A_{v0}}\frac{\sigma}{\sigma + 1 + \sigma_0}$$

(4)

Figure 1 - a) Parameters for the analytical solution; b) velocity profile.

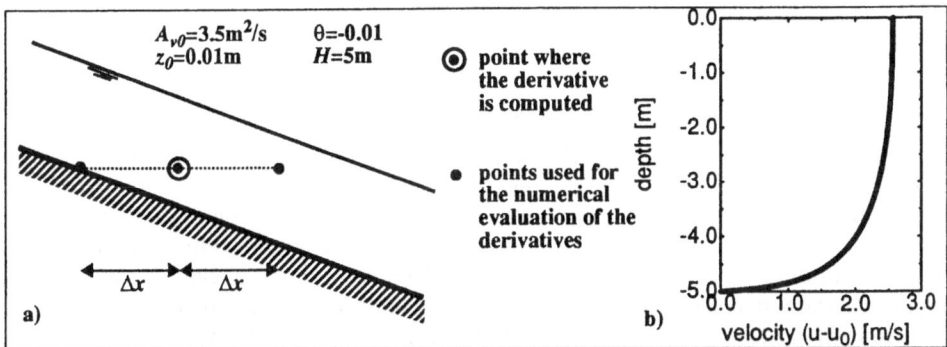

Numerical approximations were computed using centered differences and Equation (3), and compared against the exact solution, Equation (4). The physical parameters and the velocity profile are shown in Figure 1.

The exact and numerical gradients of velocity are shown in Figure 2. Also shown in Figure 2 is the error in the advective term scaled by the gravitic forcing. This error was computed assuming a no-slip condition at the bottom ($u_0=0$), and would further increase if a more common slip condition was applied.

Figure 2 - Error in the advective term scaled by the gravity forcing, and velocity gradients.

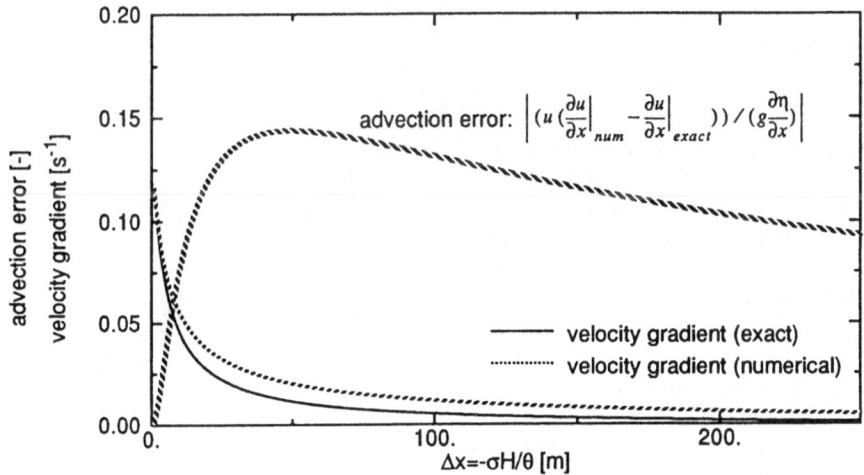

This figure shows that, unless the horizontal discretization is unreasonably fine (of the order of 10m), the horizontal gradient will be over-predicted by a very large amount.

This error arises from the rapid variation of the second vertical derivative of the horizontal velocity near the bottom, which in turn is triggered by the variation of the eddy viscosity. If a constant eddy viscosity was selected in this simplified solution, the velocity gradient in Figure 2 would be linear and the numerical and analytical solutions would coincide. However, when a more realistic eddy viscosity is chosen, the second derivative of velocity decreases sharply away from the bottom, leading to large numerical errors.

Another way to look at this problem is to recognize that we have, at the bottom:

$$\left.\frac{\partial u}{\partial x}\right|_{z \equiv constant} = \left.\frac{\partial u}{\partial x}\right|_{\sigma \equiv constant} - \frac{\sigma\theta}{H}\frac{\partial u}{\partial \sigma} \gg \left.\frac{\partial u}{\partial x}\right|_{\sigma \equiv constant} \tag{5}$$

In the analytical solution presented, this relation holds necessarily, since the derivative along a sigma plane is zero. In a more general case, it can be shown that this derivative is still zero at the bottom if the variations in the other horizontal direction can be neglected. In general, Equation (5) is expected to hold if the bottom slope is not too small, and the bottom boundary layer is represented. When relation (5) is valid, simple scaling arguments show that the discretization needed to represent the gradient on the LHS of Equation (5) is much finer than that needed to represent the RHS: Equation (5) can be scaled as:

$$\frac{U}{\Delta x} \gg \frac{U}{\Delta r} \Rightarrow \Delta r \gg \Delta x \tag{6}$$

On the other hand, the term involving a vertical derivative in Equation (5) is evaluated in the vertical grid, typically much finer than the horizontal, and therefore involves relatively small errors. Therefore, evaluating near-bottom horizontal gradients of velocity in sigma coordinates should be more accurate than in cartesian coordinates.

NUMERICAL TESTS

Alternative formulations

Figure 3 - Notation

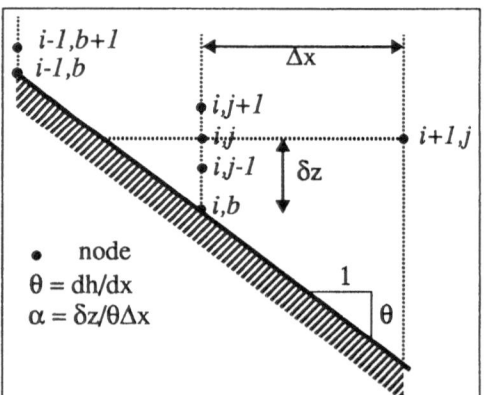

To evaluate horizontal gradients in cartesian coordinates while writing the equations in sigma coordinates, the necessary values at neighboring verticals are interpolated (e.g., see Fortunato and Baptista 1994). However, this is not possible near the bottom, and a special formulation is needed. Four alternatives are presented, along with the second order truncation errors E_2. The notation used is shown in Figure 3. The velocity at $(i+1,j)$ is assumed to be exact, even though there are errors introduced by the interpolation. These errors were studied elsewhere (Fortunato and Baptista 1994), and therefore are ignored here for simplicity. In the numerical tests, a very fine vertical grid is used near the bottom to minimize these errors.

In the first formulation (A), the velocity is interpolated along the bottom (Fortunato and Baptista 1994). The horizontal gradient is computed as:

$$\frac{\partial u}{\partial x} \approx \frac{u_{i+1,j} - (1-\alpha) u_{i,b} - \alpha u_{i-1,b}}{\Delta x (1+\alpha)} \tag{7}$$

$$E_{2A} = \frac{\Delta x}{2} \frac{1-\alpha}{1+\alpha} \frac{\partial^2 u}{\partial x^2} + \alpha \theta \Delta x \frac{1-\alpha}{1+\alpha} \frac{\partial^2 u}{\partial x \partial z} - \frac{\alpha \theta^2 \Delta x}{2} \frac{1-\alpha}{1+\alpha} \frac{\partial^2 u}{\partial z^2} \tag{8}$$

where α is defined in Figure 3. The main disadvantage of this formulation is that the derivatives are not centered. In particular, it collapses into an upwind/downwind method at the bottom node.

The second formulation (B) is similar to that proposed by Beckmann and Haidvogel 1993. A fictitious value below the bottom is obtained by extrapolation of the vertical profile. The finite difference analog and the truncation error are:

$$\frac{\partial u}{\partial x} \approx \frac{u_{i+1,j} - \dfrac{\Delta z + \Delta x (1-\alpha) \theta}{\Delta z} u_{i-1,b} + \dfrac{\Delta x (1-\alpha) \theta}{\Delta z} u_{i-1,b-1}}{2\Delta x} \tag{9}$$

$$E_{2B} = \frac{\theta \Delta x (1-\alpha)}{4} (\Delta z + \theta \Delta x (1-\alpha)) \frac{\partial^2 u}{\partial z^2} \tag{10}$$

Formulation B does not introduce horizontal numerical diffusion. However, because of the extrapolation, it is very sensitive to small errors in the nodal values when the near-bottom vertical nodal spacing is small.

Formulation C avoids both extrapolation and horizontal numerical diffusion:

$$\frac{\partial u}{\partial x} \approx \frac{u_{i+1,j} - u_{i-1,b}}{2\Delta x} - \frac{(1-\alpha) \theta}{4\Delta z} (u_{i,j+1} - u_{i,j-1}) \tag{11}$$

$$E_{2C} = -\frac{\theta^2 \Delta x (1-\alpha)^2}{4} \frac{\partial^2 u}{\partial z^2} + (\theta \Delta x (1-\alpha)) \frac{\partial^2 u}{\partial x \partial z} \tag{12}$$

Finally, formulation D is similar to formulation C, but further eliminates the vertical numerical diffusion. This may be very important when modeling bottom boundary layers where the vertical eddy viscosity is very small.

$$\frac{\partial u}{\partial x} \approx \frac{u_{i+1,j} - u_{i-1,b}}{2\Delta x} - \frac{(1-\alpha) \theta}{4\Delta z} (u_{i,j+1} - u_{i,j-1}) + $$
$$\frac{\theta^2 \Delta x (1-\alpha)^2}{4} \frac{u_{i,j+1} - 2u_{i,j} + u_{i,j+1}}{(\Delta z)^2} \tag{13}$$

$$E_{2D} = (\theta\Delta x (1 - \alpha)) \frac{\partial^2 u}{\partial x \partial z} \qquad (14)$$

Numerical results

A 2D vertical hydrodynamic model, $RITA_{2v}$ (Fortunato and Baptista 1993) was used to compare the four alternative formulations. $RITA_{2v}$ (**RI**ver and Tidal Analysis 2D vertical) is a two dimensional, laterally averaged, baroclinic, hydrodynamic model. The external mode is solved with the Generalized Continuity Wave Equation on linear finite elements, and the internal mode can accommodate either the traditional sigma coordinate system or the more flexible localized sigma coordinates. A channel with a maximum slope of 2% (Figure 4) was forced with a S2 tide. The vertical eddy viscosity is parametrized as $\kappa u_*(z+z_0)$ in the lower 20% of the water column, and as constant in the upper layer (κ is the von Kármán constant (0.4), u_* the friction velocity, and z_0 a roughness height taken as 0.005m). A quadratic friction law is used, the friction coefficient being defined as $[ln((z_0+\Delta z_b)/z_0)/\kappa]^{-2}$. The domain was discretized with 61 evenly spaced nodes in the horizontal, and the time step was set to 5s. The vertical mesh is increasingly fine near the bottom in order to represent the bottom boundary layer:

$$\sigma_i = -\frac{(i-n)^p}{(1-n)} \qquad p = 0.5, \quad i = 1, ..., n, \quad n = 40 \qquad (15)$$

Figure 4 - Bathymetry of the numerical test.

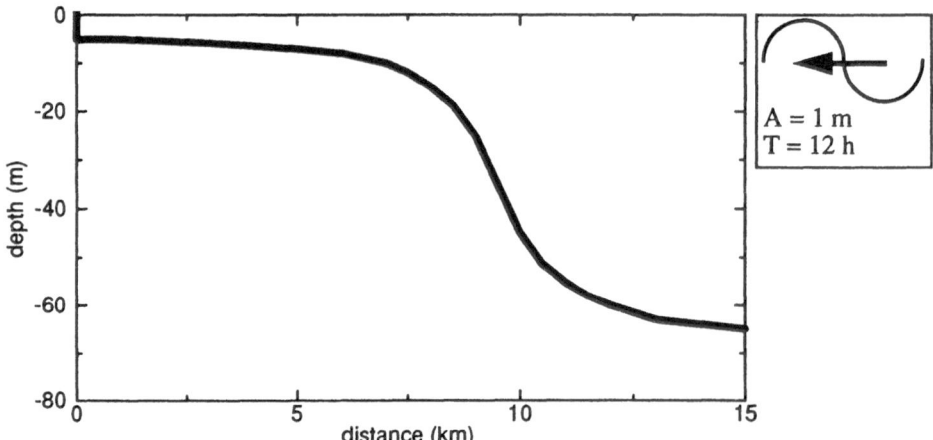

The model was run for ten tidal cycles with only the external mode, then for one more tidal cycle in 2Dv mode. The final velocity profiles at x=9.5km are shown in Figure 5a. Profiles obtained with the traditional sigma coordinates for both the same grid, and a finer grid (301 horizontal nodes, 80 vertical) are also shown for comparison (Figure 5b). All four formulations in Figure 5a exhibit the same unrealistic behavior near the bottom, while the sigma coordinate results are hardly distinguishable from the reference results. Since the four approximations are very different, these results suggest that their common characteristic, the evaluation of horizontal gradients in cartesian coordinates, is responsible for the large errors. The test was repeated computing the horizontal gradients in sigma coordinates near the bottom ($\alpha<1$), and in cartesian coordinates in the rest of the water column (formulation E). In this case, the same type of behavior is only weakly present (Figure 5b), supporting the idea that it is only near the bottom, where the vertical profile exhibits large vertical gradients, that evaluating horizontal derivatives in cartesian coordinates leads to large errors.

Figure 5 - Horizontal velocity profiles at x=9.5km.

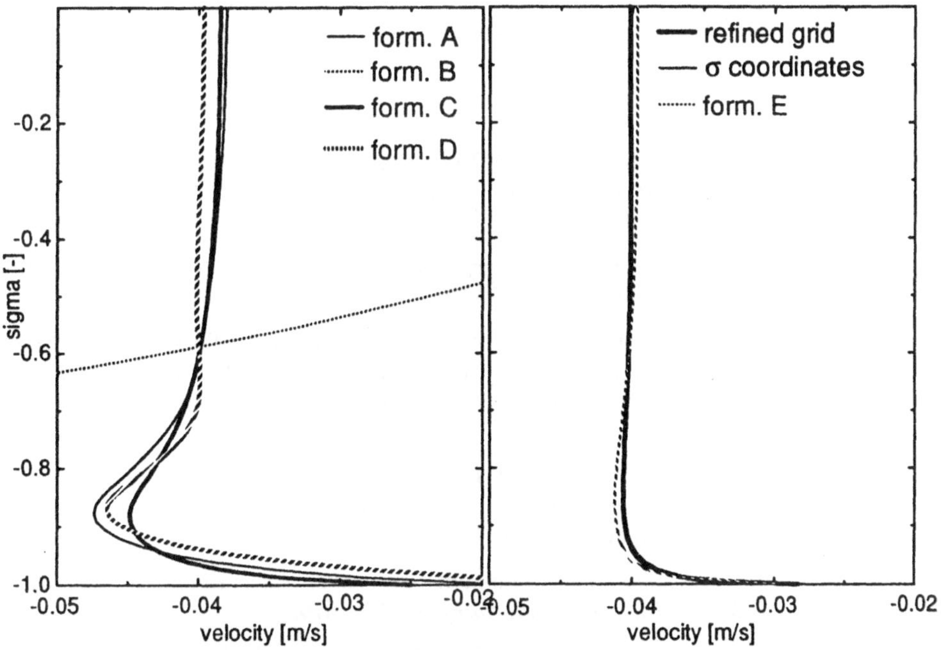

CONCLUSIONS

This paper described problems related to the evaluation of near bottom horizontal gradients of velocity directly in cartesian coordinates. Both analytical and numerical results support the conclusion that this approach will lead to unrealistic results unless extremely fine horizontal grids are used. It is therefore recommended that the horizontal velocity gradients be computed in sigma coordinates.

This conclusion is in stark contrast with recent work on the evaluation of the baroclinic pressure term (Fortunato and Baptista 1994), for which the opposite recommendation was made. A tentative interpretation is that the gradients should be computed as much as possible in the direction along which they are smaller. This direction will clearly depend on the physical process and on the specific conditions of the problem. Typically, the direction of the near bottom flow is determined by the bathymetry, whereas the density field is basically influenced by gravity. Therefore, the gradients should be computed preferentially in sigma planes for velocity, and in horizontal planes for density. However, in some particular cases the density may exhibit a behavior similar to that discussed for velocity (e.g., density gradients determined by near bottom suspended sediments), and the flow may not follow the bottom (e.g., at the tip of a salt wedge).

Acknowledgments

We thank Dr. Richard Luettich, of the Institute of Marine Sciences, and Ms Anabela Oliveira, of OGI, for reviewing early versions of this manuscript. The first author was sponsored by Junta Nacional de Investigação Científica e Tecnológica (Portugal), under grant BD-1786/91-IG.

References

Beckmann, A. and D.B. Haidvogel (1993). Numerical Simulation of Flow Around a Tall Isolated Seamount. Part I: Problem Formulation and Model Accuracy, *J. Phys. Oceanography*, 23, 1736-1753.

Cheng, R.T. and P.E. Smith (1990). A Survey of Three-Dimensional Numerical Estuarine Models, *Proc. Estuarine and Coastal Modeling*, Spaulding [ed.], ASCE, 1-15.

Fortunato, A.B. and A.M. Baptista (1993). RITA$_{2v}$ User's Manual. 2D Vertical Hydrodynamic Model for River and Tidal Analysis. Part I - Flow Model, *OGI-CCALMR Software Documentation Series SDS7, 93-3*, Oregon Graduate Institute, Portland, Oregon.

Fortunato, A.B. and A.M. Baptista (1994). Localized Sigma Coordinates for the Vertical Structure of Hydrodynamic Models, *Proc. 3rd International Conference Estuarine and Coastal Modeling*, Spaulding et al. [ed.], ASCE, submitted.

Haney, R.L. (1991). On the Pressure Gradient Force over Steep topography in Sigma Coordinate Models, *J. Phys. Oceanography*, 21, 610-619.

Laible, J.P. (1992). On the Solution of the Three-Dimensional Shallow Waters Equations Using the Wave Equation Formulation, *Proc. Computational Methods in Water Resources IX, Vol. 2*, Russel et. al. [eds.], 545-552.

Luettich, R.A. and J.J. Westerink (1991). A Solution for the Vertical Variation of Stress, Rather than Velocity, in a Three-Dimensional Circulation Model, *Int. J. for Num. Met. in Fluids*, 12, 911-928.

Phillips, N.A. (1957). A Coordinate System Having Some Special Advantages for Numerical Forecasting, *J. Meteor.*, 14, 184-185.

Sheng, Y.P, H.-K. Lee and K.H. Wang (1990). On Numerical Strategies of Estuarine and Coastal Modeling, *Proc. Estuarine and Coastal Modeling*, Spaulding [ed.], ASCE, 291-301.

Walters, R.A. and M.G.G. Foreman (1992). A 3D, Finite Element Model for Baroclinic Circulation on the Vancouver Island Continental Shelf, *J. of Marine Systems*, 3.

SHALLOW LAKE MODELLING USING QUADTREE-BASED GRIDS

C.GÁSPÁR[1], J.JÓZSA[1] and J.SARKKULA[2]
[1] VITUKI Consult Rt (Environment and Water Management Consultants)
Kvassay Jenő út 1, H-1095 Budapest, Hungary
[2] National Board of Waters and the Environment, Hydrological Office
PB 250, SF-00101 Helsinki, Finland

A relatively new and efficient way of complex shallow lake modelling is presented. The method is based on the generation of a *quadtree cell system*, which results in a non-uniform but Cartesian grid with flexibly controllable local refinements. The interpolation of initial data as well as the discretisation of the governing equation are carried out on the same quadtree grid. Both steady and unsteady problems are investigated concerning wind-driven circulations. In case of steady flow problems, the computation can be speeded up by using multigrid tools in the quadtree context. The essential steps are presented through the example of the Lake Pyhäjärvi, Finland.

INTRODUCTION

When modelling a physical phenomenon by applying some numerical technique, one has always to face requirements and restrictions which could be satisfied to the detriment of each others only. For instance, if a refined model should be applied for some reason, the computational time will necessarily grow up, which is often inadmissible, when large models or long-term simulations have to be carried out. On the other hand, if the model model has to be as fast as possible, the spatial "resolution" of the model will remain relatively low.

As it is well known, a possible solution to the above dilemma is the use of some *local refinement technique* or *curvilinear coordinates*. However, they generally make the original mathematical problem a bit more complicated. Instead, we present a relatively new method which seems to be able to unify the simplicity of the traditional finite difference methods and the flexibility of the local refinement techniques (and curvilinear coordinates). The method has been succesfully applied to various problems so far: in this paper we apply this approach to the simulation of wind-driven circulations in shallow lakes. The heart of the method is the so-called *quadtree grid generation* technique, which is a systematic subdivision of an initial rectangle resulting in a non-uniform, non-equidistant (but Cartesian) cell system (computational grid). The structure of the grid can be easily controlled in various simple ways e.g. by the geometry of the boundary (which can make the grid boundary-fitted in a certain sense) or by abrupt *changes* of the initial data (or even of some earlier approximate solution). This makes it possible to refine the

A. Peters et al. (eds.), Computational Methods in Water Resources X, 1053–1060.

grid only in the subregions of special interest, consequently, the number of the introduced unknowns can be "as small as possible". This grid generation technique is known as "unstructured grid generation" which have been used most frequently in finite element context. In contrast to the finite element approach, here we use finite difference schemes developed for non-uniform grids, which makes the discretisation much simpler.

The paper attempts to present all the important stages which are to be handled in modelling a shallow lake. First, based on the lake geometry, the computational grid is to be generated by using the quadtree algorithm. Next, the bottom topography should be given. This intermediate task generally requires a scattered data interpolation technique, which is solved separately, using the same quadtree grid. As a third step, appropriate finite difference schemes are to be defined in such a way that they are consistent with the shallow water equations. It is possible to handle both steady and unsteady problems. In the first case, a special *multigrid technique* has been also developed, which can further reduce the necessary computational work by a considerable amount. The computational advantages of the methods make it possible to simulate longer time periods (months or even years) with an acceptable, modest computational effort. We will show these essential steps via the example of a Finnish lake. We are focusing, however, on presenting the *computational methods*, without making any comparison between the computed and measured results.

GOVERNING EQUATIONS

The flow field is described by the well-known *shallow water equations (Abbott*, 1979). In a lot of practical cases in shallow lake modelling, these equations can be simplified further by neglecting both the convective and the diffusive terms, provided that these terms are small compared with the gravitational and the frictional (wind and bottom friction) terms. Dropping out the neglected terms, we obtain:

$$\frac{\partial \eta}{\partial t} + \frac{\partial p}{\partial x} + \frac{\partial q}{\partial y} = 0 \tag{1a}$$

$$\frac{\partial p}{\partial t} + gh\frac{\partial \eta}{\partial x} + Ap = B \cdot |W| \cdot W_x \tag{1b}$$

$$\frac{\partial q}{\partial t} + gh\frac{\partial \eta}{\partial y} + Aq = B \cdot |W| \cdot W_y \tag{1c}$$

(Notations: η: water surface elevation (m); h: water depth (m); p,q: specific discharges (m^2/sec); g: acceleration due to gravity; A: bottom friction term; B: wind friction coefficient; $W=(W_x,W_y)$: wind vector (m/sec)).
If, as usual, *quadratic bottom friciton* is assumed, the term A has the form:

$$A = \frac{g \cdot \sqrt{p^2 + q^2}}{k^2 h^{7/3}} \tag{2}$$

(where k stands for the Manning coefficient). A simplified assumption for the bottom friction term is the following:

$$A = \frac{r}{h} \tag{3}$$

(linear bottom friction), where r is a constant.

The *velocity components* u, v can be calculated easily from p, q:

$$u = \frac{p}{h}, \qquad v = \frac{q}{h} \tag{4}$$

It is often important to find the *steady-state* solutions of Eqs.(1). In this case, if the flow domain is not multiply connected, Eq.(1a) allows to introduce the stream function ψ by the following definition:

$$p = \frac{\partial \psi}{\partial y}, \qquad q = -\frac{\partial \psi}{\partial x} \tag{5}$$

Applying the rigid lid approximation, from (1b), (1c) and (5) we easily obtain:

$$\text{div} \frac{A}{h} \text{grad} \psi = B \cdot \left[\frac{\partial}{\partial y} \left(\frac{1}{h} \cdot |W| \cdot W_x \right) - \frac{\partial}{\partial x} \left(\frac{1}{h} \cdot |W| \cdot W_y \right) \right] \tag{6}$$

If linear bottom friction is assumed, Eq.(6) is a linear, elliptic type differential equation. As it is well known, Eqs.(1) and (6) have to be supplied with proper *boundary conditions*. Eq.(1) requires also an *initial condition*.

GRID GENERATION BY USING QUADTREES

To solve Eq.(1) and Eq.(6) respectively, our approach is to use *finite difference schemes*, but the computational grid is not supposed to be either uniform or equidistant. It is important that the local grid density (the spatial resolution of the discretisation) could be flexibly controlled in space, in order to minimize the total number of the discrete unknowns as well as the necessary computational work. Therefore we used the so-called *quadtree algorithm* for generating the computational grid. Here we briefly recall the main ideas: for further details, see *Cheng et al* (1988); *Gáspár et al* (1991).

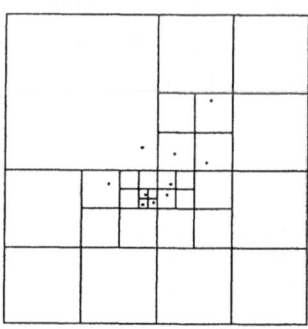

Figure 1. A QT cell system generated by 10 points

The quadtree algorithm is a recursively defined subdivision procedure, starting from an initial square and controlled by a finite set of *controlling points*. Each square is to be subdivided further into four congruent subsquares, if it contains some minimal number of controlling points. Fig.1. shows a simple example: here the algorithm is controlled by 10 points. The subsquares which are not subdivided further will form the computational cell system (grid). It is clear that the local grid density follows the local density of the controlling points, that is, multi-level local refinements can be defined in a very comfortable way. The subdivision procedure can be represented by a directed tree-like graph in a straightforward way: the graph elements represent the subsquares, while the branches correspond to the subdivision of a selected subsquare. In this representation, the computational grid is formed by the *leaves* of the tree.

A quadtree (QT-) grid is said to be *regular*, if there are no abrupt changes in cell sizes, i.e. the ratio of the sizes of two neighbouring cells is at most 2. The grid can be always regularized by performing additional subdivisions, but the regularity criterion can be built also in the recursive definition of the algorithm as well. The above property will play an essential role in the definition of difference schemes, since it keeps the number of different cell configurations under an acceptable limit.

In shallow lake modelling, it is natural to refine the grid at the vicinity of the shore of the lake, which makes the grid "boundary-fitted" in a sense. Therefore we controlled the grid generation algorithm mostly by the shoreline. See Fig.2. for a model problem (Lake Pyhäjärvi, Finland, length: cca 25 km, width: cca 10 km, average depth: 5 m).

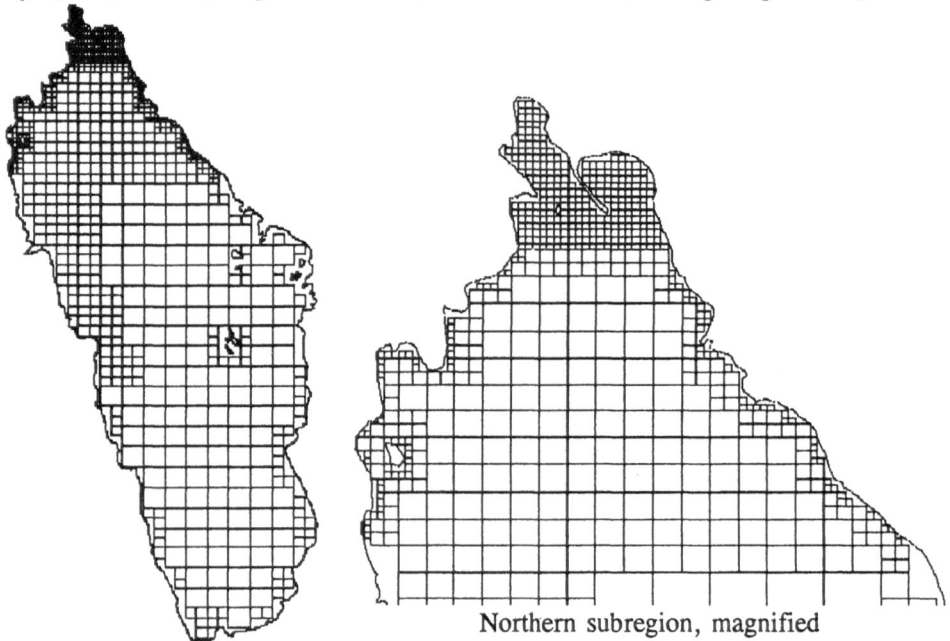

Northern subregion, magnified

Figure 2. Lake Pyhäjärvi
        QT-grid

The set of the controlling points was completed by some extra points in order to create local refinements on the subregions of special interest. The total number of the cells was below 1000: this number would have been about 20000 if a uniformly fine grid (with the finest cell size) had been applied.

DATA INTERPOLATION

The next task, which is not always mentioned explicitely, is the determination of initial data (wind field, bottom topography) on the computational grid. Since the wind field is often assumed to be uniform, it can be approximated everywhere without difficulty; the only problem is to interpolate the bottom levels (which are given initially on a set of more or less scattered points) in the cell centers or cell vertices. This is a typical example for the *scattered data interpolation* problem.

Various methods have been proposed (see e.g. *Franke*, 1982), which, however, do not exploit the special structure of the interpolation points an are often remarkably expensive from computational point of view. Here we used the algorithm proposed by *Gáspár and Simbierowicz* (1992a). The main idea is to seek the interpolation function as a solution of the *biharmonic* equation

$$\Delta\Delta f = 0 \tag{7}$$

supplied with the prescribed interpolation equations

$$f(x_j) = f_j, \qquad j=1,\dots,N \tag{8}$$

where (x_j, f_j) is the position and the value of the jth depth data. Eqs.(8) are interpreted as "inner boundary conditions". At the true boundary of the flow domain i.e. along the shoreline, the usual Neumann-type conditions

$$\frac{\partial f}{\partial n} = 0, \qquad \Delta f = 0 \tag{9}$$

are imposed. Note that (7) is always equivalent to a *pair* of simple Poisson equations:

$$\Delta f = g, \qquad \Delta g = 0 \tag{10}$$

which can be solved very efficiently by a multigrid technique, using the previously defined QT-grid. For details, see *Gáspár and Simbierowicz* (1992a).

DISCRETISATION AND SOLUTION

First, consider the steady problem (6) supplied with a Dirichlet-type boundary condition (which means the prescription of the *normal velocity components* along the boundary). Using the linear bottom friction (3), let the approximate values of ψ be attached to the cell centers. Now the task is to define cell-centered finite difference schemes which are consistent to the elliptic differential operator on the left-hand side of (6). Such schemes can be defined easily by integrating the differential equation over each cell. Assuming the function h to be constant on each cell (i.e. approximating h with such a piecewise function) and using Green's formula, we obtain that only the *fluxes*

$$\int_\Gamma \frac{\partial \psi}{\partial n} \, d\Gamma$$

through each cell side are to be approximated by finite differences. Thanks to the regularity of the QT-grid, there are only few essentially different cell configurations. For example, the flux across the eastern side Γ_E of the cell C can be approximated by the following simple formulas (see Fig.3):

$$\int_{\Gamma_E} \frac{\partial \psi}{\partial n} d\Gamma \sim \begin{cases} \psi_E - \psi_C & \text{(case } (a)) \\ \frac{4}{3} \cdot (\frac{\psi_{NE} + \psi_{SE}}{2} - \psi_C) & \text{(case } (b)) \\ \frac{2}{3} \cdot (\psi_E - \frac{\psi_N + \psi_C}{2}) & \text{(case } (c)) \end{cases} \qquad (11)$$

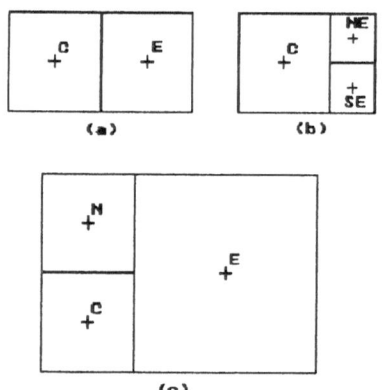

Figure 3. Cell-centered schemes: cell configurations

It can be proved that the above technique results in a stable and convergent discretisation. The natural mathematical tool of these investigations is the *Bramle-Hilbert-lemma* instead of Taylor's expansion. For further details, see *Ewing et al* (1991); *Gáspár and Simbierowicz* (1992b).

To solve the discrete equations, a special multigrid method was developed which fits the quadtree context. Recall that a multigrid method requires a set of nested grids with proper inter-grid transfer operators (restrictions and prolongations) and, of course, in each discretisation level, the original problem should be approximated in a consistent way (see e.g. *Brandt*, 1984). Since the QT-algorithm gives *automatically* a nested cell system, the multigrid idea can be applied in a very natural way, see *Gáspár and Józsa* (1992) and also *Gáspár et al* (1991); *Józsa and Gáspár* (1992).

Finally, consider the unsteady problem (1a)-(1c), in which either linear or quadratic bottom friction is allowed. Now we have to extend the well-known concept of *staggered grid* to the QT-grids. In this approach, we attach the approximate p-values (resp. q-values) to the vertical (resp. horizontal) *cell sides*. The η-values are attached to the cell centers (see Fig.4). The regularity of the QT-grid guarantees again that there is no other essentially different case.

Based on the above approximation, the following (essentially explicit) schemes can be derived:

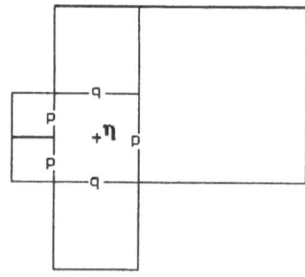

Figure 4. Staggered QT-grid

$$\eta^{n+1} := \eta^n - \Delta t \cdot (\frac{p_E^n - p_W^n}{\Delta x} + \frac{q_N^n - q_S^n}{\Delta y})$$

$$p^{n+1} := \frac{1}{1+A^n \Delta t} \cdot \left[p^n + \Delta t \cdot \left(-gh\frac{\partial \eta^n}{\partial x} + B|W|W_x \right) \right]$$

$$q^{n+1} := \frac{1}{1+A^n \Delta t} \cdot \left[q^n + \Delta t \cdot \left(-gh\frac{\partial \eta^n}{\partial y} + B|W|W_y \right) \right]$$

where the lower indices E,W,N,S refer to the spatial directions; the derivatives $\partial \eta / \partial x$, $\partial \eta / \partial y$ are approximated in a very similar way as in (11).

Fig.5 shows a computed flow field for the model example using the value $\Delta t = 12$ sec, and assuming a consant southern wind field with wind speed of 5 m/sec. The simulated time period was one day, which was long enough to reach the steady state (linear bottom friction was assumed). There was no significant difference between the above calculated steady state and the computed solution of the corresponding steady equation (6): the latter method, however, requires much less computational work.

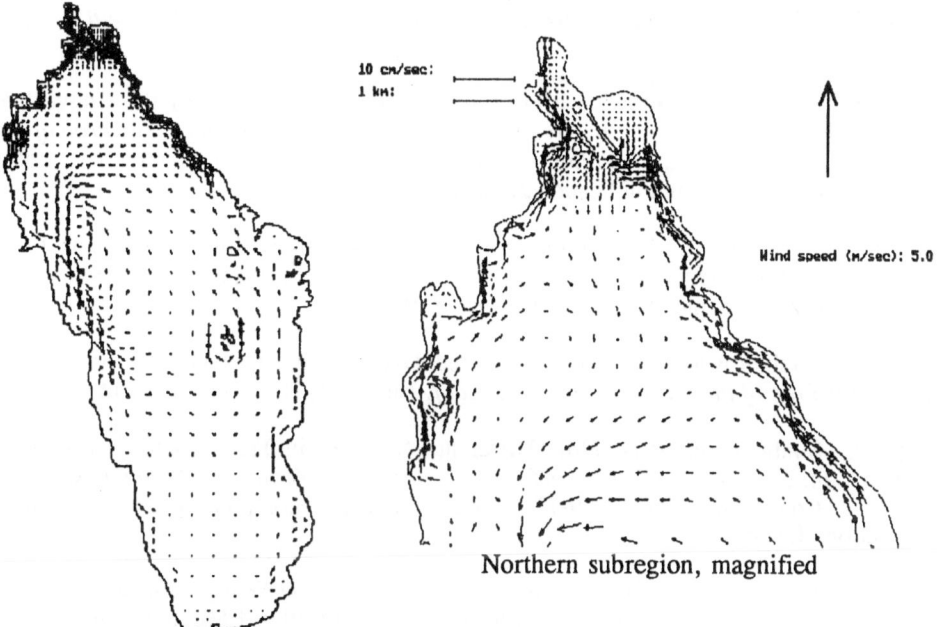

Northern subregion, magnified

Figure 5. Computed flow field

Due to expliciteness, only conditioned stability can be expected. Indeed, in our model example, the scheme became unstable when Δt exceeded the value 12 sec. However, thanks to the small number of the unknowns, it was possible to simulate longer time periods in acceptable computation time. Using a PC with 80486 processor, the ratio between the simulated and the necessary CPU time was much more than 100.

REFERENCES

Abbott M B (1979) "Computational Hydraulics. Elements of the Theory of Free Surface Flows". Pitman, London.

Brandt A (1984) "Multigrid Techniques. 1984 Guide with Applications to Fluid Dynamics". *GMD-Studien, Nr. 85*. Bonn.

Cheng J H, Finnigan P M, Hathaway A F, Kela A, Schroeder W J (1988) "Quadtree/octree meshing with adaptive analysis". In: *Numerical Grid Generation in Computational Fluid Mechanics '88* (eds S Sengupta, J Häuser, P R Eiseman, J F Thompson). Pineridge Press, Swansea, UK.

Ewing R E, Lazarov R D, Vassilevski P S (1991) "Local Refinement Techniques for Elliptic Problems on Cell-Centered Grids. I. Error analysis". *Math Comput*, Vol 56, pp 437-462.

Franke R (1982) "Scattered Data Interpolation: Test of Some Methods". *Math Comput* Vol 38 No 157, pp 181-200.

Gáspár C, Józsa J, Simbierowicz P (1991) "Lagrangian Modelling of the Convective Diffusion Problem Using Unstructured Grids and Multigrid Technique". In: *Proc of the First Int Conf on Water Pollution*, Southampton, UK, 3-5 September 1991. Computational Mechanics Publications/Elsevier.

Gáspár C, Józsa J (1992) "Two-dimensional Lagrangian Flow Simulation Using Fast, Quadtree-based Adaptive Multigrid Solver". In *Proc of 9th GAMM Conf*, Lausanne, Switzerland, 25-27 September 1991. Vieweg Verlag.

Gáspár C, Simbierowicz P (1991a) "Scattered Data Interpolation Using Unstructured Grids". In: *Proc of the HYDROCOMP '92 Conf*, Budapest, Hungary, 25-29 May, 1992.

Gáspár C, Simbierowicz P (1992b) "Difference Schemes in Tree-structured Multigrid Context". In: *Proc of the IX. Int Conf on Computational Methods in Water Resources*, Denver, Colorado, USA, 9-12 June, 1992. Computational Mechanics Publications/Elsevier.

Józsa J, Gáspár C (1992) "Fast, Adaptive Approximation on Wind-induced Horizontal Flow Patterns in Shallow Lakes Using Quadtree-based Multigrid Method". In: *Proc of the IX. Int Conf on Computational Methods in Water Resources*, Denver, Colorado, USA, 9-12 June, 1992. Computational Mechanics Publications/Elsevier.

An Optimal Control Analysis of Water Pollution Problem Using SAKAWA - SHINDO Method

Hirokazu Hirano *
Mutsuto Kawahara **
Ken-ichi Yoshida **
* Faculty of Policy Studies, Chuo University, Tokyo, Japan
** Department of Civil Engineering, Chuo University, Tokyo, Japan

1. INTRODUCTION

In general, water quality control is one of the most important social problems especially in a closed water region. If we want control this water pollution problem, we have to consider the inequality constrains. Until now, we have been studying about the dynamic optimization without constrain problem for first step. This type problem can be solved by Cojugate Gradient (CG) Method, Dynamic Programing (DP) Method, etc.. However these methods cannot be solved under inequality constrain. Therefore this paper proposed a method based the SAKAWA−SHINDO method to solve the water pollution controal problem with inequality constraints on functions of control.

In this study, the optimal control analysis has been applied to water pollution problem. The main purpose of control of water pollution problem is to reduce the contaminated water, i.e. the determination of control values to keep the water quality which satisfy a criterion of concentration at arbitary point in the flow field.

In the present optimal control analysis, the linear shallow water equation and the convection diffusion equation are used as the governing equations for flow field. The explicit Euler method is apllied to discretize the shallow water equation and the convection diffusion equation. The simulations of simple model and applications to the water pollution problems have been carried out. The results of the method are quite resonable. The application for practical problem will be carried out.

2. BASIC EQUATIONS

A optimal control analysis of water pollution problem can be expressed by the shallow water equation and the convection diffusion equation. The time is represented by t, the linear shallow water equation can be expressed as follows:

$$\frac{\partial u}{\partial t} + g\frac{\partial \zeta}{\partial x} = 0 \tag{1}$$

$$\frac{\partial v}{\partial t} + g\frac{\partial \zeta}{\partial y} = 0 \tag{2}$$

$$\frac{\partial \zeta}{\partial t} + h(\frac{\partial u}{\partial x} + \frac{\partial v}{\partial y}) = 0 \tag{3}$$

1061

A. Peters et al. (eds.), Computational Methods in Water Resources X, 1061–1066.

where u, v are mean velocity of water flow, ζ is mean water elevation, g and h are gravitational acceleration and water depth respectively. Denoting pollutant concentration as c, the equation of conservation can be expressed as follows.

$$\frac{\partial c}{\partial t} + u\frac{\partial c}{\partial x} + v\frac{\partial c}{\partial y} - \kappa(\frac{\partial^2 c}{\partial x^2} + \frac{\partial^2 c}{\partial y^2}) = \Delta\Omega \tag{4}$$

where κ is the diffusion coefficient and $\Delta\Omega$ is source term of substance.

3. OPTIMAL CONTROL THEORY

The optimal control thory employed in this paper is based on the quadratic control theory. Assume that the pollutant concentration is given all over the time duration, the control concentration can be determined by minimizing the performance function. This is called as the tracking control. The basic equations can be transformed into the dicretized form using the finite element method. On the basis of Galerkin formulation, equations (1)–(3) and (4) can be transformed into the weighted residual equations. Using the liner interpolation functino based on the triangular elements, the finite element equations can be expressed as follows:

$$M_{\alpha\beta}\dot{u}_\beta + gH^x_{\alpha\beta}\zeta_\beta = 0 \tag{5}$$

$$M_{\alpha\beta}\dot{v}_\beta + gH^y_{\alpha\beta}\zeta_\beta = 0 \tag{6}$$

$$M_{\alpha\beta}\dot{\zeta}_\beta + h(H^x_{\alpha\beta}u_\beta + H^y_{\alpha\beta}v_\beta) = 0 \tag{7}$$

$$M_{\alpha\beta}\dot{c}_\beta + K^x_{\alpha\beta\gamma}u_\beta c_\gamma + K^y_{\alpha\beta\gamma}v_\beta c_\gamma + S_{\alpha\beta}c_\alpha = \Delta\Omega \tag{8}$$

where $u_\beta, v_\beta, \zeta_\beta, c_\beta$ represent the nodal values of the concentration at the node of triangular element. The coefficient matrices in equations (5)–(8) can be expressed as:

$$M_{\alpha\beta} = \int_V (\Phi_\alpha\Phi_\beta)dV, \qquad S_{\alpha\beta} = \kappa\int_V (\Phi_{\alpha,x}\Phi_{\beta,x})dV + \kappa\int_V (\Phi_{\alpha,y}\Phi_{\beta,y})dV$$

$$K^x_{\alpha\beta\gamma} = \int_V (\Phi_\alpha\Phi_\beta\Phi_{\gamma,x})dV, \qquad K^y_{\alpha\beta\gamma} = \int_V (\Phi_\alpha\Phi_\beta\Phi_{\gamma,y})dV$$

$$H^x_{\alpha\beta} = \int_V (\Phi_\alpha\Phi_{\beta,x})dV, \qquad H^y_{\alpha\beta} = \int_V (\Phi_\alpha\Phi_{\beta,y})dV$$

where Φ_α denotes the interpolation function. The conventional finite element superposition procedure leads to the global form as follows.

For the discretization in time, the explicit Euler scheme is applied to (5)–(8). From equations (5)–(8), the state equations can be expressed as follows:

$$\dot{u}_\beta = -gM^{-1}_{\alpha\beta}H^x_{\alpha\beta}\zeta_\beta \tag{9}$$

$$\dot{v}_\beta = -gM^{-1}_{\alpha\beta}H^y_{\alpha\beta}\zeta_\beta \tag{10}$$

$$\dot{\zeta}_\beta = -hM^{-1}_{\alpha\beta}([H^x_{\alpha\beta}]u_\beta + [H^y_{\alpha\beta}]v_\beta + [H^x_{\alpha\beta}]_{c.p}w^x_\beta + [H^y_{\alpha\beta}]_{c.p}w^y_\beta) \tag{11}$$

$$\dot{c}_\beta = -M^{-1}_{\alpha\beta}(K^x_{\alpha\beta\gamma}u_\beta c_\gamma + K^y_{\alpha\beta\gamma}v_\beta c_\gamma + S_{\alpha\beta}c_\beta + \Delta\Omega) \tag{12}$$

where w_β^x and w_β^y are mean velocity of the control points, $(c.p)$ means matrix of the controal points, $([\,])$ matrix means global form matrix.

The performance of control can be evaluated in the following manner. Let the control time duration be $[t_0, t_f]$ where t_0, t_f mean starting and final times of control respectively, which are assumed fixed constants. The preformance function used in this paper is expressed in the form as follows.

$$J = \frac{1}{2} \int_{t_0}^{t_f} \left((\mathbf{c} - \mathbf{c_d})_\alpha^T \mathbf{Q}_{\alpha\beta}(\mathbf{c} - \mathbf{c_d})_\beta + w_\alpha^x \mathbf{R}_{\alpha\beta} w_\beta^x + w_\alpha^y \mathbf{R}_{\alpha\beta} w_\beta^y \right) dt \tag{13}$$

where c_d means reference values of concentration and $\mathbf{Q}, \mathbf{R}$ matrices are weighting functions. Essentially the pollutant concentration should be less than the critical value. The control points velocities w_β^x and w_β^y are inequality constraint as follows:

$$w_{min}^x \leq w^x \leq w_{max}^x \tag{14}$$

$$w_{min}^y \leq w^y \leq w_{max}^y \tag{15}$$

Then the water pollution control problem can be expressed to find the control w^x, w^y to minimize the performance function J under the state equations $(9)-(13)$ and the inequality constraint $(14)-(15)$ on the control value.

4. SAKAWA–SHINDO METHOD

To apply the optiomal control theory, it is necessary to the Hamiltonian function as:

$$H = \frac{1}{2} \left((\mathbf{c} - \mathbf{c_d})_\alpha \mathbf{Q}(\mathbf{c} - \mathbf{c_d})_\beta + w_\alpha^x \mathbf{R}_{\alpha\beta} w_\beta^x + w_\alpha^y \mathbf{R}_{\alpha\beta} w_\beta^y \right) + p_{u\alpha}\dot{u}_\alpha + p_{v\alpha}\dot{v}_\alpha + p_{\zeta\alpha}\dot{\zeta}_\alpha + p_{c\alpha}\dot{c}_\alpha \tag{16}$$

where p denotes the Lagrange Parameter. The Euler–Lagrange equation and the transversality condition can be described as follows:

$$\dot{p}_{u\beta} = -\frac{\partial H}{\partial u} = h M_{\alpha\beta}^{-1}[H_{\alpha\beta}^x]p_{\zeta\alpha} + M_{\alpha\beta}^{-1} K_{\alpha\beta\gamma}^x c_\gamma p_{c\alpha} \tag{17}$$

$$\dot{p}_{v\beta} = -\frac{\partial H}{\partial v} = h M_{\alpha\beta}^{-1}[H_{\alpha\beta}^y]p_{\zeta\alpha} + M_{\alpha\beta}^{-1} K_{\alpha\beta\gamma}^x c_\gamma p_{c\alpha} \tag{18}$$

$$\dot{p}_{\zeta\beta} = -\frac{\partial H}{\partial c} = g M_{\alpha\beta}^{-1} H_{\alpha\beta}^x p_{u\alpha} + g M_{\alpha\beta}^{-1} H_{\alpha\beta}^y p_{v\alpha} \tag{19}$$

$$\dot{p}_{c\beta} = -\frac{\partial H}{\partial c} = -Q_{\alpha\beta}(c_\beta - c_d) + p_{c\beta} M_{\alpha\beta}^{-1}(K_{\alpha\beta\gamma}^x u_\beta + K_{\alpha\beta\gamma}^y v_\beta + S_{\alpha\beta}) \tag{20}$$

$$p(t_f) = 0 \tag{21}$$

To secure the stability of the computation, the Hamiltonian function is modified in the following form.

$$K^{(i)} = H^{(i)} + (w^{x(i)} - w^{x(i-1)})[W^{(i)}](w^{x(i)} - w^{x(i-1)}) + (w^{y(i)} - w^{y(i-1)})[W^{(i)}](w^{y(i)} - w^{y(i-1)}) \tag{22}$$

where superscripted (i) means ith interation cycle and $W^{(i)}$ is a constant weighting matrix. Because $K^{(i)}$ is not restrained with respect to $w^{(i)}$, the optimality condition can be written:

$$\frac{\partial K^{(i)}}{\partial w^{(i)}} = 0$$

which leads to the optimal control as:

$$w_\beta^{x(i)} = sat((R_{\alpha\beta} + 2W^{(i)})^{-1}(2w_x^{(i-1)}W^{(i)} + hM_{\alpha\beta}^{-1}[H_{\alpha\beta}^x]_{c.p}^T p_{\zeta\beta})) \tag{23}$$

$$w_\beta^{y(i)} = sat((R_{\alpha\beta} + 2W^{(i)})^{-1}(2w_y^{(i-1)}W^{(i)} + hM_{\alpha\beta}^{-1}[H_{\alpha\beta}^y]_{c.p}^T p_{\zeta\beta})) \tag{24}$$

where the function sat(w) is defiend as follows:

$$sat(w) = \begin{cases} w_{\min} & (w < w_{\min}) \\ w & (w_{\min} \leq w \leq w_{\max}) \\ w_{\max} & (w > w_{\max}) \end{cases}$$

The optimal control of SAKAWA–SHINDO method can be summarized as follows. Assuming the appropriate stability constants $W^{(i)}$, the minmum value of J equation (13) can be obtained by equation (23),(24) on p in equations (17) − (20) with equation (21). Then the minimum value of J can be found by equation (13) solving equation system with respect to p.

5. NUMERICAL TEST

As a numerical example, a simple one dimensional channel has been computed to test the adaptability of present method. Fig.1 shows the finite element idealization and boundary conditions. On the inlet boundary S_f, this points are velocity control points, and the normal velocity component is given to be zero at horizontal walls S_c. The initial conditions, velocities $u_0 = 0.0m/s$, $v_0 = 0.0m/s$, concentration $c_0 = 6.0ppm$, are assumed over the flow field. The diffusion coefficient κ and reference value c_d are $\kappa = 0.5m^2/s$ and $c_d = 1.0ppm$ respectively. The computational time $[t_0, t_f]$ is divied in 5000steps increments with the time increment $\Delta t = 0.003sec$. The weighting matrices are selected as $Q = 1.0$ and $R = 0.01$, on the observation points. The pollution points are given uniform concentration $\Delta\Omega = 0.8(ppm\ m^3/s)$. The control function w with inequality constraints are selected as $w_{\min} = 0.0m/s$ and $w_{\max} = 10.0m/s$.

Fig.2−4 show the computed results obtained by SAKAWA–SHINDO method. Fig.2 represents the pollutant concentration and computed control concentration at the observation point ($Point − 2$), where the solid line is control concentration, the dotted line is pollutant concentration. Fig.3 is the control velocities at $Point − 1$. Fig.4 is the cost function. From these figures, relatively good results can be obtained.

6. CONCLUSION

In this paper, an optimal control analysis of water pollution problem using SAKAWA– SHINDO method has been presented. This method is able to control the state value i.e.., velocity employing the control value predicted using the information of inflow velocity in the immediate future. This method has been applied to simulation of one dimensional water pollution problem. The results of the method are quite reasonable. The application for practical problem will be carried out in future.

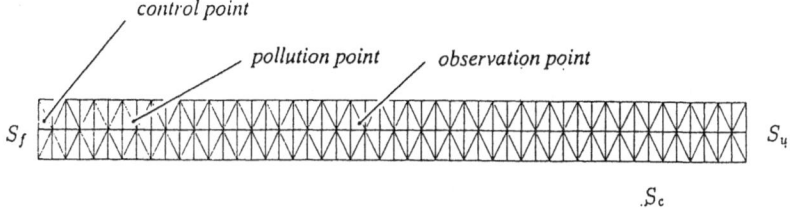

Fig.1 Finite element mesh

NX 153	Length 10.0m	$w_{max} = 10.0m/sec$
MX 200	Width 0.8m	$w_{min} = 0.0m/sec$
	Depth 10.0m	

S_f	Inlet boundary
S_u	Outlet boundary
S_c	Horizontal walls

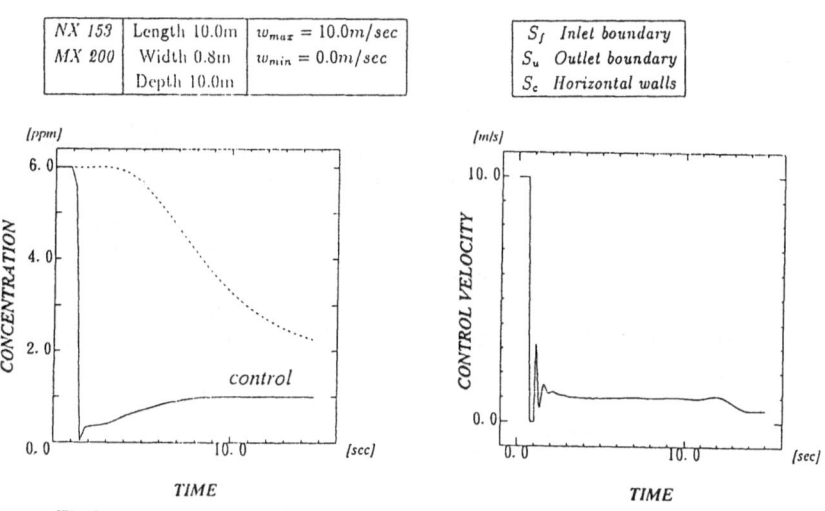

Fig.2 concentration at point–2

Fig.3 control value at point–1

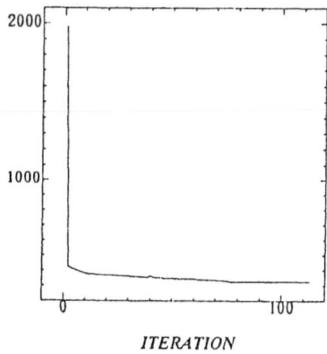

Fig.4 cost function

References

[1] Kanou, H. , "Theory and Computational Methods in Optimaization", CORONA PUB-LISHING Co., Ltd., Tokyo Japan., (1987)

[2] Sakawa, Y. and Y.Shindo, "On Global Convergence of an Algorithm for Optimal Control", IEEE., AC−25, 6, pp.1149−1153, (1980)

[3] Jarmark, B. "On Convergence Control in Differential Dynamic Programming Applied to Realistic Aircraft and Differntial Game Problem", Proc.1977 IEEE Conf. ecision Contr., pp471−479, (1977)

[4] Jarmark, B. "A New Convergencs Control Technique in Differential Dynamic Programming", Royal Inst. Technol., Stocholm, Sweden, Rep. TRITA−REG−7502, (1975)

[5] Pontryagin, L.S. etc., "The Mathematical Theory of Optimal Processes", New York:Interscience, (1962)

[6] Tanaka, Y. and M. Kawahara, "Control fo Flood Using Finite Element Method and Optimal Control Theory", Finite Element Analysis in Fluids, Eds. T.J. Chung and G.R. Karr, University of Alabama in Huntsville Press, pp.612−622, (1989)

[7] Kawasaki, T. and M. Kawahara, "A Flood Control of Dam Reservoir by Conjugat, Finite Element Analysis in Fluids, Eds. T.J. Chung and G.R. Karr, University of Alabama in Huntsville Press, pp.629−634, (1989)

[8] Umetsu, T., Y. Tanaka and M. Kawahara, "Optimal Control of Flood Using Finite Element Method", Proc. J.S.C.E., No.446/I−19, Vol.9, No.1, pp.11−23, (1992)

[9] Hirano, H. and K.Yoshida, "An Optional Control Analyaia of Water Pollution Problem", 2nd Japan−US Symposium on Finite Element Methods in Largr−Scale Computational Fluid Dynamics, (1994)

A FINITE ELEMENT STUDY OF SOLITARY WAVE BY BOUSSINESQ EQUATION

Shouichiro KATO, Akira ANJU and Mutsuto KAWAHARA
Department of Civil Engineering
Chuo University
1-13-27 Kasuga Bunkyou-ku Tokyo 112
Japan

Abstract

This paper discusses the solitary wave that is one of the finite amplitude wave in the view of the numerical analysis. In this paper, the collision of the two solitary waves is treated on the two dimensional incompressible inviscid flow. For the governing equations, the Boussinesq equation is used for the analysis. The finite element method based on the linear triangular element and the conventional Galerkin method is used to solve this equation. The Euler method is used to descretize the time, and the Leap-Frog method is also used for the comparision with the previous method in respect of the stability of the calculation.

1. INTRODUCTION

The natural disaster like the tsunami directly damages the many habitants, and the damage of the commercial harbor area is also serious problem to the nation. Therefore, the study of the nearshore wave becomes important for the prevention from the damages. However, the wave transformation is a very complicated phenomenon due to the simultaneous occurrence of several physical effects such as refraction, diffraction, reflection and so on.

In this century, the finite amplitude wave theory has progressed rapidly for the purpose of predicting the nearshore wave in the area of civil engineering. When it is tried to predict the natural phenomena, it is demanded that the phenomena is expressed by the mathematical models. Now, thanks to the progress of its theory, it became possible to describe mathematically not only the linear wave but also the non-linear one. For instance, it is well-known that the Boussinesq equation can describe the solitary wave, one of the non-linear waves. The advantage of this equation is that the wave with the developed wave front can be expressed, and the usefulness to the nearshore wave. Therefore, the Boussinesq equation is widely used in the harbor engineering. The main purpose of this study is to investigate the natures of the solitary wave by this equation, and the collision occurrence of the two solitary waves in the same direction or in the opposite direction. The reasons that the finite element method is used in this paper are because it offers its great abilities to treat the boundary, and variable-detail spatial discretizations for the tsunami analysis.

A. Peters et al. (eds.), Computational Methods in Water Resources X, 1067–1072.
© 1994 Kluwer Academic Publishers. Printed in the Netherlands.

2. GOVERNING EQUATIONS

The Boussinesq equation is employed as the governing equation to solve the problem of the shallow water flow. The equations of motion and continuity are expressed as follows:

$$\frac{\partial u}{\partial t} + u\frac{\partial u}{\partial x} + v\frac{\partial u}{\partial y} + g\frac{\partial \eta}{\partial x} = \frac{h^2}{3}\left(\frac{\partial^3 u}{\partial x^2 \partial t} + \frac{\partial^3 u}{\partial x \partial y \partial t}\right) \tag{1}$$

$$\frac{\partial v}{\partial t} + u\frac{\partial v}{\partial x} + v\frac{\partial v}{\partial y} + g\frac{\partial \eta}{\partial y} = \frac{h^2}{3}\left(\frac{\partial^3 v}{\partial y^2 \partial t} + \frac{\partial^3 v}{\partial x \partial y \partial t}\right) \tag{2}$$

$$\frac{\partial \eta}{\partial t} + \frac{\partial}{\partial x}\{(h+\eta)u\} + \frac{\partial}{\partial y}\{(h+\eta)v\} = 0 \tag{3}$$

On the boundary S_1, the velocity is assumed as:

$$u = \hat{u} \quad , \quad v = \hat{v} \qquad on \qquad S_1 \tag{4}$$

The overhat indicates the prescribed value on the boundary S_1. On the boundary S_2,

$$\eta = \hat{\eta} \qquad on \qquad S_2 \tag{5}$$

The overhat indicates the prescribed value on the boundary S_2. And the boundary condition on S_3 is as follows:

$$r = \frac{\partial}{\partial t}\left(\frac{\partial u}{\partial x} + \frac{\partial u}{\partial y}\right)l = \hat{r} \qquad on \qquad S_3 \tag{6}$$

$$s = \frac{\partial}{\partial t}\left(\frac{\partial v}{\partial x} + \frac{\partial v}{\partial y}\right)m = \hat{s} \qquad on \qquad S_3 \tag{7}$$

The overhat indicates the prescribed constant value and l,m represent the directional cosine.

3. INITIAL CONDITION

Initial condition is given as:

$$u = \hat{u}_0 \qquad at \qquad t = 0 \tag{8}$$

$$v = \hat{v}_0 \qquad at \qquad t = 0 \tag{9}$$

$$\eta = \hat{\eta}_0 \qquad at \qquad t = 0 \tag{10}$$

And the initial velocity, the initial water level are given as:

$$u = \sqrt{gh}\frac{\eta_0}{h}sech^2\sqrt{\frac{3\eta_0}{4h^3}}(x - ct) \tag{11}$$

$$\eta = \eta_0 sech^2\sqrt{\frac{3\eta_0}{4h^3}}(x - ct) \tag{12}$$

$$c = \sqrt{gh(1 + \frac{\eta_0}{h})} \tag{13}$$

where u, η and h represent the velocity, the water level and the water depth and c,g represent the wave velocity, the gravity acceleration.

4. FINITE ELEMENT FORMULATIONS

Equations (1)-(3) are transformed into the weighted residual equations employing the conventional Galerkin method and the finite element equations can be described in the following form:

$$M_{\alpha\beta}\dot{u}_\beta + \frac{h^2}{3}(K_{\alpha x\beta x}\dot{u}_\beta + G_{\alpha x\beta y}\dot{u}_\beta) + L_{\alpha\beta\gamma x}u_\beta u_\gamma + I_{\alpha\beta\gamma y}v_\beta u_\gamma + gA_{\alpha\beta x}\eta_\beta = \sum_\alpha^x \quad (14)$$

$$M_{\alpha\beta}\dot{v}_\beta + \frac{h^2}{3}(P_{\alpha y\beta y}\dot{v}_\beta + Q_{\alpha y\beta x}\dot{v}_\beta) + L_{\alpha\beta\gamma x}u_\beta v_\gamma + I_{\alpha\beta\gamma y}v_\beta v_\gamma + gZ_{\alpha\beta y}\eta_\beta = \sum_\alpha^y \quad (15)$$

$$M_{\alpha\beta}\dot{\eta}_\beta + N_{\alpha\beta x\gamma}H_\beta u_\gamma + L_{\alpha\beta\gamma x}H_\beta u_\gamma + I_{\alpha\beta\gamma y}H_\beta v_\gamma + J_{\alpha\beta y\gamma}H_\beta v_\gamma = 0 \quad (16)$$

where 'over dot' means differentiation with respect to time and the coefficient matrices can be expressed as follows:

$$M_{\alpha\beta} = \int_\Omega \Phi_\alpha\Phi_\beta d\Omega \quad , \quad K_{\alpha x\beta x} = \int_\Omega \Phi_{\alpha,x}\Phi_{\beta,x}d\Omega \quad , \quad G_{\alpha x\beta y} = \int_\Omega \Phi_{\alpha,x}\Phi_{\beta,y}d\Omega \quad (17)$$

$$L_{\alpha\beta\gamma x} = \int_\Omega \Phi_\alpha\Phi_\beta\Phi_{\gamma,x}d\Omega \quad , \quad I_{\alpha\beta\gamma y} = \int_\Omega \Phi_\alpha\Phi_\beta\Phi_{\gamma,y}d\Omega \quad , \quad A_{\alpha\beta x} = \int_\Omega \Phi_\alpha\Phi_{\beta,x}d\Omega \quad (18)$$

$$P_{\alpha y\beta y} = \int_\Omega \Phi_{\alpha,y}\Phi_{\beta,y}d\Omega \quad , \quad Q_{\alpha y\beta x} = \int_\Omega \Phi_{\alpha,y}\Phi_{\beta,x}d\Omega \quad , \quad Z_{\alpha\beta y} = \int_\Omega \Phi_\alpha\Phi_{\beta,y}d\Omega \quad (19)$$

$$N_{\alpha\beta x\gamma} = \int_\Omega \Phi_\alpha\Phi_{\beta,x}\Phi_\gamma d\Omega \quad , \quad J_{\alpha\beta y\gamma} = \int_\Omega \Phi_\alpha\Phi_{\beta,y}\Phi_\gamma d\Omega \quad (20)$$

$$\sum_\alpha^x = \int_s (\Phi_\alpha\hat{r})ds \quad , \quad \sum_\alpha^y = \int_s (\Phi_\alpha\hat{s})ds \quad (21)$$

The combination of explicit and quasi-explicit scheme is used to discretize the time, which can be derived as follows:

$$[M_{\alpha\beta} + \frac{h^2}{3}(K_{\alpha x\beta x} + G_{\alpha x\beta y})]u^{n+1} = [M_{\alpha\beta} + \frac{h^2}{3}(K_{\alpha x\beta x} + G_{\alpha x\beta y})]u^n - \Delta t\{L_{\alpha\beta\gamma x}u^n u^n$$
$$+ I_{\alpha\beta\gamma y}v^n u^n + gA_{\alpha\beta x}\eta^n\} \quad (22)$$

$$[M_{\alpha\beta} + \frac{h^2}{3}(P_{\alpha y\beta y} + Q_{\alpha y\beta x})]v^{n+1} = [M_{\alpha\beta} + \frac{h^2}{3}(P_{\alpha y\beta y} + Q_{\alpha y\beta x})]v^n - \Delta t\{L_{\alpha\beta\gamma x}u^n v^n$$
$$+ I_{\alpha\beta\gamma y}v^n v^n + gZ_{\alpha\beta y}\eta^n\} \quad (23)$$

$$\bar{M}_{\alpha\beta}\eta^{n+1} = \tilde{M}_{\alpha\beta}\eta^n - \Delta t\{N_{\alpha\beta x\gamma}H^n u^n + L_{\alpha\beta\gamma x}H^n u^n + I_{\alpha\beta\gamma y}H^n v^n + J_{\alpha\beta y\gamma}H^n v^n\} \quad (24)$$

The aforementioned method to descretize the time is the Euler method, and this time the Leap-Frog method is also used for the purpose of the computational stability as follows:

$$\dot{u} \approx \frac{1}{2\Delta t}(u^{n+1} - u^{n-1}) \quad (25)$$

$$u \approx u^n \quad (26)$$

In equation (24), $\bar{M}$ denotes the lumped coefficient of M and $\tilde{M}$ is the mixed coefficint, which

$$\tilde{M} = e\bar{M} + (1-e)M \quad (27)$$

where e is referred to as the lumping parameter.

5. NUMERICAL EXAMPLES

The following numerical examples are analized by the Euler method for the time discretizations. Fig.1 shows the finite element mesh of the channel. The water depth is set as 1.0[m] and the channel width is 2.0[m]. There are two cases for the channel length. In case of Fig.2 and Fig.3 the length is about 400[m], and in Fig.4 and Fig.5 about 700[m]. Fig.2 and Fig.3 are the cases that the two solitary waves go in the opposite direction, Fig.4 and Fig.5 in the same direction. Fig.2 is different from Fig.3 with respect to the wave height. In Fig.4 and Fig.5 the same behaviors are also shown. A dotted line indicates the case when a part of two solitary waves was particularly calculated. Table-1 shows the data used in each case, and in Table-2 the values of the mixed wave height in each case are listed.

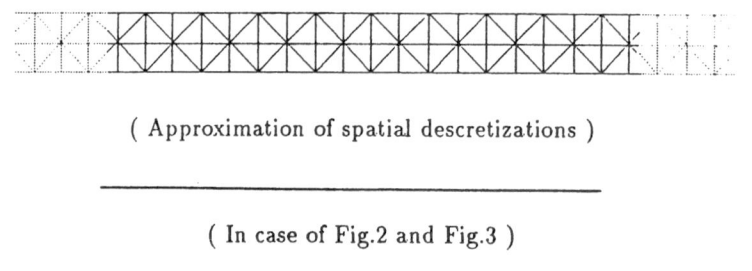

(Approximation of spatial descretizations)

(In case of Fig.2 and Fig.3)

(In case of Fig.4 and Fig.5)

Fig.1 Finite element mesh

Table-1 Data used in each case

	Fig.2	Fig.3	Fig.4	Fig.5
number of nodes	1509	1509	3009	3009
number of elements	2008	2008	4008	4008
lumping parameter	0.9600	0.9600	0.9480	0.9480
time increment[sec]	0.0250	0.0250	0.0250	0.0250
spatial increment[m]	0.8000	0.8000	0.7000	0.7000

Table-2 Mixed wave height

	Fig.2	Fig.3	Fig.4	Fig.5
mixed wave height[m]	0.02000320	0.03007947	0.26155710	0.33927959

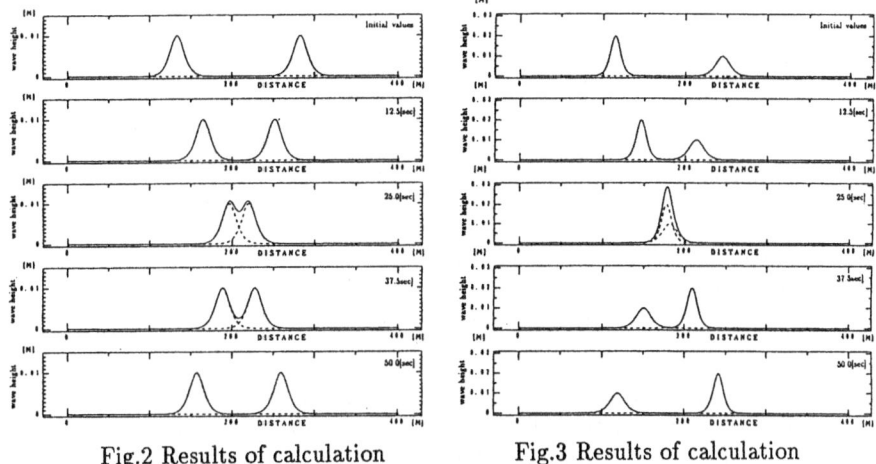

Fig.2 Results of calculation   Fig.3 Results of calculation

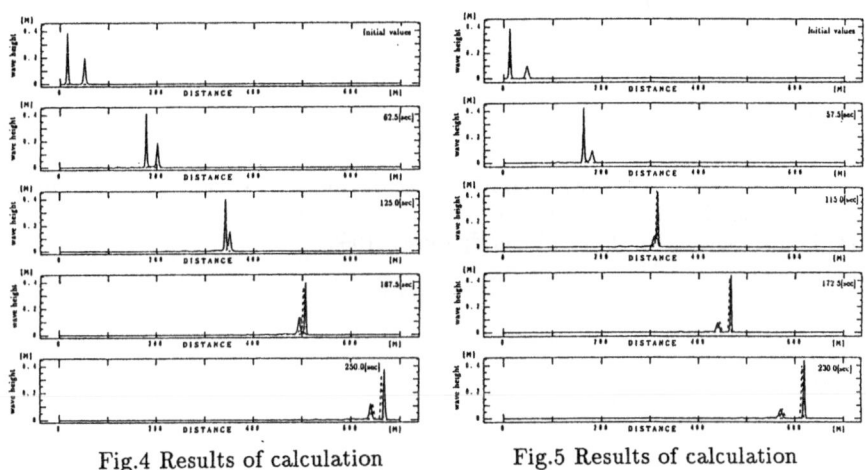

Fig.4 Results of calculation   Fig.5 Results of calculation

6. CONCLUSION

In this paper, the variation on the collision occurrence has been discussed. However, there is an important problem before its discussion. It is whether a solitary wave can preserve its shape in the propagation. The numerical results show that a solitary wave approximately preserves while the computational time is short and if the relative wave height is small enough. The propagation of the wave becomes unstable in this reverse. In case of the same direction, the relative wave height needs to be enlarged for the guarantee of the computational capacity and time, and the computational time is also longer. Therefore, not only the Euler method but also the Leap-Flog method were used for the stability of the calculation. However, the computational stability of the Euler method was superior to that of the Leap-Flog method in this paper. The numerical results indicate some remarkable things as follows.

In case of the opposite direction, the shape of the each wave is approximately preserved in the propagation. The mixed wave height nearly equals the sum of the two waves height before the collision, and after the collision the position shift isn't recognized.

In the same direction, it can not be said that the shape of the each wave is preserved. However, there are some tendencies. The mixed wave height become smaller than the previous higher one. After the collision, the higher wave is accelerated and the smaller is decelerated.

In general, it is predicted that the solitary wave perfectly passes through the other. Finally, the aforementioned problem of the stability and the application in the real phenomena will be the future subjects.

References

[1] J.Y.Cheng and M.Kawahara, "Study on Finite Element Analysis of Boussinesq equation",1993

[2] C.J.M.Fortes and J.M.A.Covas, "Harbour Resonance and Wave Refraction-Difraction Computations in Harbours Using The Finite Element Method", Vol.2,pp.1067-1080,Finite Elements in Fluids,Pineridge Press,1993.

PARAMETER IDENTIFICATION OF BOTTOM FRICTION AND EDDY VISCOSITY OF TIDAL FLOW IN TOKYO BAY

Mutsuto KAWAHARA, Akira ANJU and Kenichiro MATSUMOTO
Department of Civil Engineering
Chuo University
1-13-27 Kasuga Bunkyo-ku Tokyo 112
Japan

Abstract

This paper presents the parameter identification of the tidal flow. For the governing equation, non-linear shallow water equation including the bottom friction and the eddy viscosity terms is used. In order to minimize the performance function, the conjugate gradient method (Fletcher-Reeves method) is employed. This algorithm is simple and suitable for a large scale calculation.

1. INTRODUCTION

The shallow water equation is used as the basic equation of the flow ploblem in the coastal hydrodynamics. In general, the bottom friction and the eddy viscosity terms are included in the equation of motion as an energy decrease. These parameters affect not only on computational results but also on numerical stability. In order to know the behavior of the flow exactly in the analysis domain, it is necessary to identify these parameters exactly. It is possible to identify the parameters at the simple model, but in the pratical ploblem the computation often becomes unstable and the parameter cannot be well identified. The reason is that the error is always included in the obsevrved data, the finite element mesh is approximate and so on. Considering these, it is impossible to identify these parameters at the same time. In this study, the following identification method is used, i.e. , the bottom friction is identified with the eddy viscosity being fixed, and then the eddy viscosity is identified with the bottom friction which is identified in the previous computation is fixed.

2. BASIC EQUATION

The two dimensional unsteady and non-linear shallow water equation can be written as follows;

$$\frac{\partial u_i}{\partial t} + u_j u_{i,j} + g\eta_{,i} - \nu(u_{i,j} + u_{j,i})_{,j} + \frac{\tau}{H}(u_k u_k)^{\frac{1}{2}} u_i = 0 \tag{1}$$

$$\frac{\partial \eta}{\partial t} + \{(h + \eta)u_i\}_{,i} = 0 \tag{2}$$

where u_i, g, ν, τ, H, h, η are the velocity components of x_i(i=1,2) direction, the gravity acceleration, the eddy viscosity, the bottom friction, the average water depth in each element, the water depth and the water elevation.

A. Peters et al. (eds.), Computational Methods in Water Resources X, 1073–1080.

Boundary S consists of two types of boundaries. One is boundary S_1 on which the velocity and the water elevation are given and the other is boundary S_2 on which the surface force is specified.

$$u_i = \hat{u}_i, \qquad \eta = \hat{\eta} \qquad on \qquad S_1 \qquad (3)$$

$$t_i = \nu(u_{i,j} + u_{j,i})n_j = \hat{t}_i \qquad on \qquad S_2 \qquad (4)$$

Initial condition is that $\hat{u}_i^0$ and $\hat{\eta}_i^0$ are the functions expressing the initial velocity and water elevation at the initial state t_0.

$$u_i = \hat{u}_i^0, \qquad \eta = \hat{\eta}^0 \qquad at \qquad t = t_0 \qquad (5)$$

3. FINITE ELEMENT EQUATION

The weighted residual equation is obtained employing the convensional Galerkin method applied to the governing equation (1)–(2). The finite element equation can be derived using the linear interpolation function based on the three nodes triangular finite element as follows;

$$M_{\alpha\beta}\dot{u}_{\beta i} + K_{\alpha\beta\gamma j}u_{\beta j}u_{\gamma i} + gH_{\alpha\beta i}\eta_\beta + S_{\alpha i\beta j}u_{\beta j} + I_{\alpha\beta}\gamma_{\beta i}^b = \hat{\sum}_{\alpha i} \qquad (6)$$

$$M_{\alpha\beta}\dot{\eta}_\beta + B_{\alpha\beta\gamma i}(h_\beta + \eta_\beta)u_{\gamma i} + C_{\alpha\beta i\gamma}(h_\beta + \eta_\beta)u_{\gamma i} = 0 \qquad (7)$$

where

$$M_{\alpha\beta} = \int_V \Phi_\alpha \Phi_\beta dV, \quad K_{\alpha\beta\gamma j} = \int_V \Phi_\alpha \Phi_\beta \Phi_{\gamma,j} dV, \quad H_{\alpha\beta i} = \int_V \Phi_\alpha \Phi_{\beta,i} dV,$$

$$S_{\alpha i\beta j} = \nu(\int_V \Phi_{\alpha,j}\Phi_{\beta,i} dV + \int_V \Phi_{\alpha,k}\Phi_{\beta,k}\delta_{ij} dV), \quad I_{\alpha\beta} = \frac{\tau}{H}\int_V \Phi_\alpha \Phi_\beta dV$$

$$B_{\alpha\beta\gamma i} = \int_V \Phi_\alpha \Phi_\beta \Phi_{\gamma,i} dV, \quad C_{\alpha\beta i\gamma} = \int_V \Phi_\alpha \Phi_{\beta,i}\Phi_\gamma dV$$

$$\hat{\sum}_{\alpha i} = \int_S \Phi_\alpha \hat{t}_i dS, \quad \gamma_{\beta i}^b = (u_{\beta k}u_{\beta k})^{\frac{1}{2}}u_{\beta i}$$

The two-step explicit method is applied to discretize the finite element equation.

4. SENSITIVE EQUATION

In order to identify the parameters, it is necessary to compute the sensitive equation to use the Fletcher–Reeves method which is one of the conjugate gradient method. Generally, the bottom friction and the eddy viscosity is not the same in the whole field. The flow field is divided into several domains and these parameters are assumed to be constant in each domain. The bottom friction and the eddy viscosity are defined as follows;

$$\tau^T = \{\tau_1, \tau_2, \cdots, \tau_k\}$$
$$\nu^T = \{\nu_1, \nu_2, \cdots, \nu_k\}$$

The sensitive matrix can be obtained by differentiating with respect to each parameter as follows;

1. Differentiated with respect to the bottom friction:

$$M_{\alpha\beta}\frac{\partial u'_{\beta i}}{\partial\tau} + K_{\alpha\beta\gamma j}\left(\frac{\partial u_{\beta j}}{\partial\tau}u_{\gamma i} + u_{\beta j}\frac{\partial u_{\gamma i}}{\partial\tau}\right) + gH_{\alpha\beta i}\frac{\partial\eta_\beta}{\partial\tau} + S_{\alpha i\beta j}\frac{\partial u_{\beta j}}{\partial\tau}$$
$$+ I_{\alpha\beta}\frac{\partial\gamma^b_{\beta i}}{\partial\tau} + \frac{\partial I_{\alpha\beta}}{\partial\tau}\gamma^b_{\beta i} = \frac{\partial\hat{\Sigma}_{\alpha i}}{\partial\tau} \tag{8}$$

$$M_{\alpha\beta}\frac{\partial\dot{\eta}_\beta}{\partial\tau} + B_{\alpha\beta\gamma i}\left\{\frac{\partial\eta_\beta}{\partial\tau}u_{\gamma i} + (h_\beta + \eta_\beta)\frac{\partial u_{\gamma i}}{\partial\tau}\right\} + C_{\alpha\beta i\gamma}\left\{\frac{\partial\eta_\beta}{\partial\tau}u_{\gamma i} + (h_\beta + \eta_\beta)\frac{\partial u_{\gamma i}}{\partial\tau}\right\} = 0 \tag{9}$$

2. Differentiated with respect to the eddy viscosity:

$$M_{\alpha\beta}\frac{\partial u'_{\beta i}}{\partial\nu} + K_{\alpha\beta\gamma j}\left(\frac{\partial u_{\beta j}}{\partial\nu}u_{\gamma i} + u_{\beta j}\frac{\partial u_{\gamma i}}{\partial\nu}\right) + gH_{\alpha\beta i}\frac{\partial\eta_\beta}{\partial\nu}$$
$$+ S_{\alpha i\beta j}\frac{\partial u_{\beta j}}{\partial\nu} + \frac{\partial S_{\alpha i\beta j}}{\partial\nu}u_{\beta j} + I_{\alpha\beta}\frac{\partial\gamma^b_{\beta i}}{\partial\nu} = \frac{\partial\hat{\Sigma}_{\alpha i}}{\partial\nu} \tag{10}$$

$$M_{\alpha\beta}\frac{\partial\dot{\eta}_\beta}{\partial\nu} + B_{\alpha\beta\gamma i}\left\{\frac{\partial\eta_\beta}{\partial\nu}u_{\gamma i} + (h_\beta + \eta_\beta)\frac{\partial u_{\gamma i}}{\partial\nu}\right\} + C_{\alpha\beta i\gamma}\left\{\frac{\partial\eta_\beta}{\partial\nu}u_{\gamma i} + (h_\beta + \eta_\beta)\frac{\partial u_{\gamma i}}{\partial\nu}\right\} = 0 \tag{11}$$

The two-step explicit method is applied to discretize the sensitive equation. Initial condition of the sensitive equation is defined as follows;

$$\frac{\partial u_i}{\partial\tau} = \frac{\partial\eta}{\partial\tau} = 0, \qquad \frac{\partial u_i}{\partial\nu} = \frac{\partial\eta}{\partial\nu} = 0 \qquad at \qquad t = t_0 \tag{12}$$

5. CONJUGATE GRADIENT METHOD

The performance function is defined as follows;

$$J = \frac{1}{2}\int_{t_0}^{t_f}\sum_{\mu=1}^{N}\{\tilde{\eta}_\mu(t) - \eta_\mu(t)\}^T[S]\{\tilde{\eta}_\mu(t) - \eta_\mu(t)\}dt \tag{13}$$

where N, $\tilde{\eta}_\mu$, η_μ, [S] are the total number of the observation point, the water elevation observed at the observation point μ, the water elevation computed at the observation point μ and the weighted matrix respectively. In order to apply the conjugate gradient method, it is necessary to compute the gradient of the performance function. The sensitivity of the water elevation with respect to the bottom friction and the eddy viscosity can be formulated using differentiation of equation (8)–(11).

$$\frac{\partial J}{\partial\tau} = -\int_{t_0}^{t_f}\sum_{\mu=1}^{N}\{\frac{\partial\eta_\mu}{\partial\tau}\}^T[S]\{\tilde{\eta}_\mu(t) - \eta_\mu(\tau, t)\}dt \tag{14}$$

$$\frac{\partial J}{\partial\nu} = -\int_{t_0}^{t_f}\sum_{\mu=1}^{N}\{\frac{\partial\eta_\mu}{\partial\nu}\}^T[S]\{\tilde{\eta}_\mu(t) - \eta_\mu(\nu, t)\}dt \tag{15}$$

When the conjugate direction d^i with respect to the bottom friction and the eddy viscosity can be written as $d^T = \{d_\tau, d_\nu\}$, the step width is α which minimizes $J(\tau + \alpha d_\tau)$ and $J(\nu + \alpha d_\nu)$. Thus, J is expanded by the Taylor expansion and differentiating with respect to α, and α can be obtained as follows;

1. For the bottom friction identification:

$$\alpha^i = \frac{\int_{t_0}^{t_f} \sum_{\mu=1}^{N} \{\frac{\partial \eta_\mu}{\partial \tau} d_\tau\}^T [S] \{\tilde{\eta}_\mu - \eta_\mu(\tau, t)\} dt}{\int_{t_0}^{t_f} \sum_{\mu=1}^{N} \{\frac{\partial \eta_\mu}{\partial \tau} d_\tau\}^T [S] \{\frac{\partial \eta_\mu}{\partial \tau} d_\tau\} dt} \qquad (16)$$

2. For the eddy viscosity identification:

$$\alpha^i = \frac{\int_{t_0}^{t_f} \sum_{\mu=1}^{N} \{\frac{\partial \eta_\mu}{\partial \nu} d_\nu\}^T [S] \{\tilde{\eta}_\mu - \eta_\mu(\nu, t)\} dt}{\int_{t_0}^{t_f} \sum_{\mu=1}^{N} \{\frac{\partial \eta_\mu}{\partial \nu} d_\nu\}^T [S] \{\frac{\partial \eta_\mu}{\partial \nu} d_\nu\} dt} \qquad (17)$$

The gradient with respect to the conjugate direction can be written as follows;

1. For the bottom friction identification:

$$\beta^i = \frac{\partial J(\tau^{i+1})/\partial \tau \cdot \partial J(\tau^{i+1})/\partial \tau}{\partial J(\tau^i)/\partial \tau \cdot \partial J(\tau^i)/\partial \tau} \qquad (18)$$

2. For the eddy viscosity identification:

$$\beta^i = \frac{\partial J(\nu^{i+1})/\partial \nu \cdot \partial J(\nu^{i+1})/\partial \nu}{\partial J(\nu^i)/\partial \nu \cdot \partial J(\nu^i)/\partial \nu} \qquad (19)$$

6. COMPUTATIONAL ALGORITHM

The computational algorithm to identify the bottom friction is shown as follows;

1.   Assume initial values of τ^0, set allowance constant ϵ and i=0

2.   Compute $\eta_\mu^0(\tau^0, t)$, $\partial \eta_\mu^0 / \partial \tau$ and $J(\tau^0)$

3.   Compute $\partial J(\tau^0)/\partial \tau$ and set $d_\tau^0 = -\partial J(\tau^0)/\partial \tau$

4.   Determine α^i so as to minimize $J(\tau^i + \alpha^i d_\tau^i)$

5.   Compute $\tau^{i+1} = \tau^i + \alpha^i d_\tau^i$

6.   Compute $\eta_\mu^{i+1}(\tau^{i+1})$ and $J(\tau^{i+1})$

7.   If $|J(\tau^{i+1}) - J(\tau^i)| < \epsilon$ then stop

8.   Compute $\partial J(\tau^{i+1})/\partial \tau$

9.   Compute β^{i+1}

10.   Compute $d_\tau^{i+1} = -\partial J(\tau^{i+1})/\partial \tau + \beta^{i+1} d_\tau^i$

11. Set i=i+1 and go to 4

The algorithm of the eddy viscosity is in the same way.

7. NUMERICAL ANALYSIS OF TOKYO BAY

The parameter identification of Tokyo Bay is carried out as a pratical example. The finite element idealization is shown in Fig.1. Total numbers of nodal points and finite elements are 685 and 1216 respectively. Fig.2 shows the water depth. The flow field is divided into two domains according to the water depth, the water depth in domain1 is deeper than 15.0m, in domain2 is shallower than 15.0m. This is shown in Fig.3. The sinusoidal wave with the amplitude of 0.42m and with the period of 12 hours 25 minutes is imposed on the open boundary. The water elevation and the velocity are computed using the time increment $\triangle t$=15.0sec and lumping parameter e=0.9. On the coastline, outward normal velocities are given to be zero as the boundary condition. The measurement points and their measured values are shown in Table.1.

location	amplitude(cm)	phase delay(degree)
Yokosuka	41.24	152.77
Futtsu	45.00	148.97
Kimitsu	44.09	150.69
Yokohama	47.25	153.57
Anegasaki	56.00	156.00
Samugawa	52.20	153.10
Funabashi	48.42	154.45

Table.1

The bottom friction is identified with the eddy viscosity being fixed. The eddy viscosity is given ($\nu_1 = 100.0 m^2/s$, $\nu_2 = 500.0 m^2/s$) respectively and initial values of the bottom friction is given ($\tau_1 = \tau_2 = 0.0$). Fig.4 and Fig.5 illustrate the convergence process of the values of the bottom friction and the values of the performance function. Allowance constant ϵ is 1.0×10^{-4}. Then, the eddy viscosity is identified with the bottom friction which is identified in the previous computation is fixed. The bottom friction is given ($\tau_1 = 0.0027, \tau_2 = 0.0161$) respectively and initial values of the eddy viscosity is given ($\nu_1 = \nu_2 = 0.0$). Fig.6 and Fig.7 illustrate the convergence process of the eddy viscosity and values of the performance function. Fig.8 shows the comparison between the water elevations observed at the observation points and the water elevations computed using the bottom friction and the eddy viscosity identified.

8. CONCLUSION

The parameter identification of the pratical example has been presented in this paper. Whether the computation is stable or not depends on the finite element mesh and the division of the analysis domain. Especially the latter is the most important problem. In this study, the analysis domain is divided into two domains. Considering the computation time, this division is appropriate. In this divided domains, the water depth 15.0m is the borderline of the domain whether the computation is stable or not.

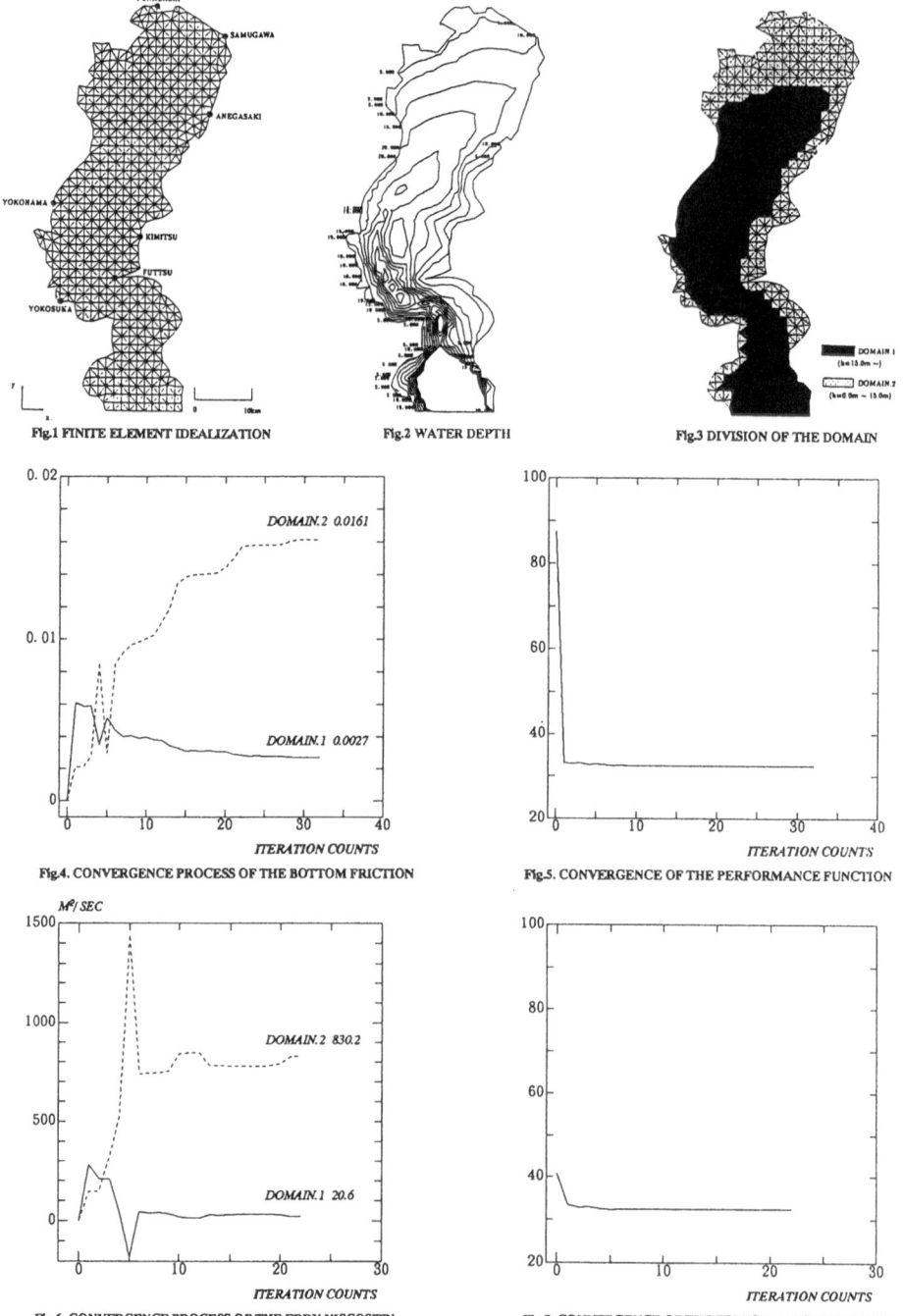

Fig.1 FINITE ELEMENT IDEALIZATION

Fig.2 WATER DEPTH

Fig.3 DIVISION OF THE DOMAIN

Fig.4. CONVERGENCE PROCESS OF THE BOTTOM FRICTION

Fig.5. CONVERGENCE OF THE PERFORMANCE FUNCTION

Fig.6. CONVERGENCE PROCESS OF THE EDDY VISCOSITY

Fig.7. CONVERGENCE OF THE PERFORMANCE FUNCTION

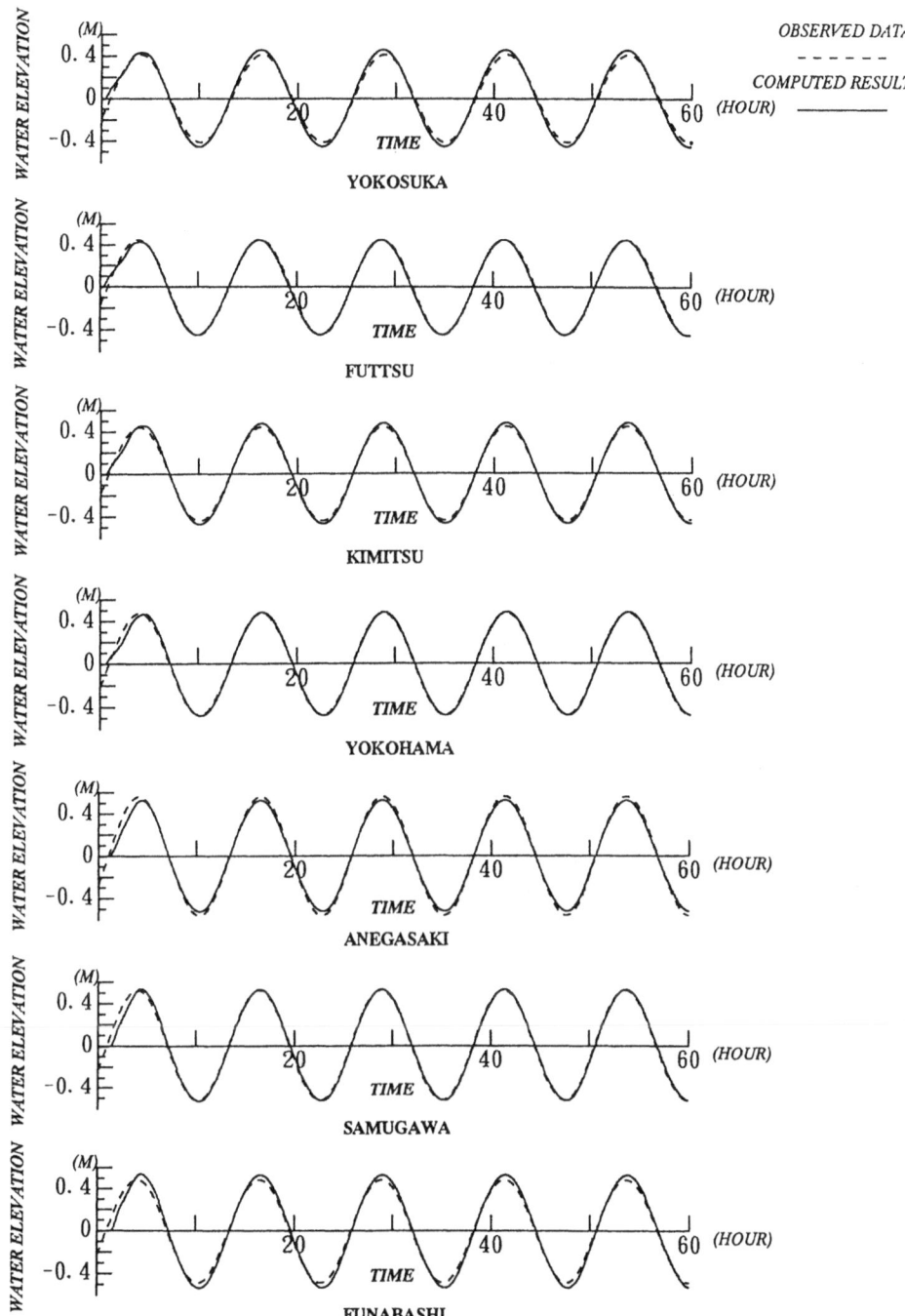

Fig.8. COMPARISON BETWEEN THE COMPUTED AND THE OBSERVED WATER ELEVATIONS

References

[1] R Goda, M Kawahara: A Fundamental Study for Parameter Ientification of Tial Flow, 1992

[2] T kodama: Study on Finite Element Analysis of Shallow Water Equation and Its Applications, Sato kogyo engineering research institute, reports of engineering research institute No.2, p162-177, 1992

[3] M Kawahara: Flow Analysis of Finite Element Method, Japan Union of Science and Engineering Publishing, Tokyo, 1985

[4] H Kano: Optlmal Theory and Optimalize of System, p53- 109, Corona Publishing, Tokyo, 1987

[5] M Hattori: Harbor engineering, p13-122, Corona Publishing, tokyo, 1987

NORMAL FLOW BOUNDARY CONDITIONS IN SHALLOW WATER MODELS - INFLUENCE ON MASS CONSERVATION AND ACCURACY

R. L. KOLAR[1], W. G. GRAY[2], and J. J. WESTERINK[2]

[1] Department of Civil and Environmental Engineering
University of New Haven, West Haven, CT 06516, USA
[2] Department of Civil Engineering and Geological Sciences
University of Notre Dame, Notre Dame, IN 46556, USA

ABSTRACT

Finite element solution of the shallow water wave equations has found increasing use by researchers and practitioners in the modeling of oceans and coastal areas. Wave equation models, most of which use equal-order, C^0 interpolants for both the velocity and the surface elevation, successfully eliminate spurious oscillation modes without resorting to artificial or numerical damping. An important question for both primitive equation and wave equation models is the interpretation of boundary conditions. Analysis of the characteristics of the governing equations shows that a single condition at each boundary is sufficient. Yet there is not a consensus in the literature as to what that boundary condition must be or how it should be implemented in a finite element code. Traditionally (partly because of limited data) surface elevation is specified at open ocean boundaries while the normal flux is specified as zero at land boundaries. In most finite element wave equation models, both of these boundary conditions are implemented as essential conditions. Our recent work focuses on alternate ways to numerically implement normal flow boundary conditions with an eye toward improving the mass-conserving properties of wave equation models. In particular, we have found that treating normal fluxes as natural conditions with the flux interpreted as external to the computational domain results in a mass conservative scheme for all parameter values. Use of generalized functions in the finite element formulation shows this is a natural interpretation. A series of two-dimensional experiments demonstrates that this interpretation also improves the accuracy of primitive equation models by eliminating some of the spurious oscillation modes.

BACKGROUND

Shallow water equations are obtained by vertically averaging the microscopic mass and momentum balances over the depth of the water column. Early finite element solutions of the shallow water equations were often plagued by spurious oscillations. Various methods were introduced to eliminate the oscillations but all included some type of artificial damping. Lynch and Gray [1] and Gray [2] present the wave continuity equation as a means to successfully suppress spurious oscillations without resorting to numerical or artificial damping of the solution. Since the inception of the wave continuity formulation in 1979, the original algorithm has been modified in a number of substantial ways: a numerical parameter was introduced to provide a more general means of weighting the primitive continuity equations [3]; viscous dissipation terms were incorporated [4, 5]; and three-dimensional simulations were realized by resolving the velocity profile in the vertical [6, 7]. The resulting algorithm has been extensively tested using analytical solutions and field data and is currently being used to model the hydrodynamic behavior of coastal and oceanic areas [8-10].

A. Peters et al. (eds.), Computational Methods in Water Resources X, 1081–1088.

In the course of some of the applications, it was discovered that when nonlinear components of the solution are significant, the wave continuity equation in its original form does not conserve mass. Two methods of mitigating the errors are presented in Kolar et al. [11]. In the first, it is shown that if G, the numerical parameter in the generalized wave continuity equation, is increased so that its value is one or two orders of magnitude larger than the bottom friction, then mass conservation is greatly improved. However, an upper bound on G exists above which the solution becomes too primitive and spurious oscillations appear in the solution. Dispersion analysis can be used as a tool to *a priori* predict the maximum value of G. The second mitigation technique reformulates the convective term in the generalized wave continuity equation so that a consistency exists between the momentum and continuity equations (e.g., both equations cast the advective terms in non-conservative form). If both mitigating measures are used in conjunction, then global mass balance errors are eliminated while errors in local regions (even individual elements) are virtually nonexistent except for regions near the open boundaries. One and two-dimensional applications demonstrate the effectiveness of the procedure. However, the persistence of errors near the open boundaries and absence of errors near land boundaries, has led to this study of the effect of boundary conditions on mass conservation and solution accuracy.

CONSERVATION EQUATIONS

Primitive forms of the balance laws are obtained by vertical averaging of the microscopic balance laws. Using operator notation, we present the primitive form of conservation of mass (continuity equation) as

$$L \equiv \frac{\partial \zeta}{\partial t} + \nabla \cdot (H\mathbf{v}) = 0 \tag{1}$$

The conservative form of conservation of momentum is given by

$$\mathbf{M}^c \equiv \frac{\partial (H\mathbf{v})}{\partial t} + \nabla \cdot (H\mathbf{v}\mathbf{v}) + \tau H\mathbf{v} + H\mathbf{f} \times \mathbf{v} + gH\nabla \zeta - \mathbf{A} - \frac{1}{\rho}\nabla \cdot (H\mathbf{T}) = 0 \tag{2}$$

and the non-conservative form of conservation of momentum is given by

$$\mathbf{M} \equiv \frac{1}{H}\left(\mathbf{M}^c - \mathbf{v}L\right) = 0 \tag{3a}$$

Substituting (1) and (2) into (3a) gives

$$\mathbf{M} \equiv \frac{\partial \mathbf{v}}{\partial t} + \mathbf{v} \cdot \nabla \mathbf{v} + \tau \mathbf{v} + \mathbf{f} \times \mathbf{v} + g\nabla \zeta - \frac{\mathbf{A}}{H} - \frac{1}{\rho H}\nabla \cdot (H\mathbf{T}) = 0 \tag{3b}$$

In operator form, the generalized wave continuity (GWC) equation is

$$W^G \equiv \frac{\partial L}{\partial t} + GL - \nabla \cdot \mathbf{M}^c = 0 \tag{4a}$$

Substituting (1) and (2) into (4a) gives

$$W^G \equiv \frac{\partial^2 \zeta}{\partial t^2} + G\frac{\partial \zeta}{\partial t} + (G - \tau)\, \nabla \cdot (H\mathbf{v}) - \nabla \cdot \left[\nabla \cdot (H\mathbf{v}\mathbf{v}) + H\mathbf{f} \times \mathbf{v} + gH\nabla\zeta \right.$$

$$\left. -\, \mathbf{A} - \frac{1}{\rho}\nabla \cdot (H\mathbf{T}) \right] - H\mathbf{v}\cdot\nabla\tau = 0 \tag{4b}$$

The wave continuity equation, as it originally appeared in Lynch and Gray [1], is obtained by setting $G = \tau$. Note that the primitive continuity equation can be viewed as a limiting form of the generalized wave continuity equation by letting $G \to \infty$.

DISCRETIZATION

Equations (1), (2), (3b), and (4b) are discretized in space using a standard Galerkin finite element approximation with linear elements. Implicit time discretization of L and W^G uses a three time level approximation centered at k. Time discretization for $\mathbf{M}$ and $\mathbf{M}^c$ uses a lumped two time level approximation centered at $k + 1/2$; the discrete equations are linearized by formulating the advective terms explicitly. Exact quadrature rules are employed. The resulting discretized equations can be found in [6]. A sequential solution procedure is adopted where the continuity equation ((1) or (4b)) is used to solve for elevations and the momentum equation ((2) or (3b)) is used to solve for the velocity field.

BOUNDARY CONDITIONS

The governing conservation equations represent a coupled hyperbolic system of partial differential equations that describe the propagation of long water waves in shallow water. As such, characteristic theory is an appropriate tool to study proper specification of boundary conditions. In particular, for the primitive conservation equations, it has been shown that one condition on each physical boundary is required (in addition to initial conditions for the "time boundary"). Drolet and Gray [12] extend the analysis of characteristic planes to the wave continuity equation and determined that a single boundary condition is still sufficient.

Mathematically, the conditions are specified as one of three types: Dirichlet (Type I) in which the value of the dependent variable (elevation or flux) is specified, Neumann (Type II) in which the value of the flux is specified, and Robin (Type III) which is a linear combination of the first two. In shallow water modeling, these types describe the physical situations of known-elevation, known-flux, or stage-discharge relations, respectively (the latter are often referred to as radiation boundary conditions). This article focuses on the first two types of conditions. In finite element vernacular, a Dirichlet condition means that the value of the dependent variable is known on the boundary; it is referred to as an essential condition. A specified-flux condition enters the right hand side vector in the set of discrete equations and is referred to as a natural boundary condition.

For a single partial differential equation, such as Laplace's equation or the diffusion equation, specified boundary conditions fall neatly into one of the above categories and implementation is unambiguous. Unfortunately, such is not the case for the coupled hyperbolic system at hand, for what may serve as an essential condition for the momentum equation may equally be interpreted as a natural condition for the continuity equation. This ambiguity has led to an inconsistent treatment of boundary conditions in the literature; to date, no consensus on the "best" way to implement the conditions exists. Complicating the matter is the fact that data often dictates what information is available at the boundary. For example, elevation data, either from global tidal models or from field measurements, is more reliable and more prevalent than is velocity data. The end result is that the researcher is often faced with the task of choosing one interpretation over the other, which, frequently, is tantamount to choosing which boundary equation to discard.

Lynch [13] seems to be the first to study the effect of boundary conditions on mass conservation in the context of the wave continuity algorithm. Using the well-known properties of linear basis functions that the sum of the functions over all elements is equal to one and the sum of the gradient of the basis functions over all elements is zero, he demonstrates that all terms of the continuity equation must be retained in order to maintain global conservation of mass, regardless of the nature of the boundary data. He refers to this interpretation of the boundary conditions as mass conservative boundary conditions. However, several open issues remain. For example, it is not clear how momentum conservation is affected by this interpretation - an equally-important consideration.

Accordingly, we undertook an extensive series of one-dimensional experiments to study various means of implementing mass conservative boundary conditions. A shallow one-dimensional channel was used for the model problem so that significant nonlinear components are generated. Conditions for the problem are:

channel coordinates	$0 \leq x \leq 50$ km
channel depth	5 m
eddy viscosity ϵ	0.093 m²/sec
Δt	25 sec
Δx	2 km
boundary conditions	$\zeta(0, t) = 1.0 \sin [2\pi t/12.42$ hrs] m
	$u(50, t) = 0.0$ m/sec
initial conditions	cold start: $\zeta(x, 0) = u(x, 0) = 0.0$
bottom friction τ	0.0001 sec⁻¹ (constant)

The x-axis is defined positive to the right. The boundary conditions describe a channel with a land boundary at $x = 50$ km being forced by an M_2 tide with 1 meter amplitude at $x = 0$ km.

Mass conservation was checked globally and locally by comparing mass accumulation with cumulative net flux for the region of interest. Details of the algorithm are presented in [11]. For model comparison, a fine grid solution using primitive balance laws, which is mass conservative, was taken as the true solution. Note that for one-dimensional simulations with constant bathymetry, spurious oscillations do not plague the solution so that a primitive solution is satisfactory.

Conventional formulation of the boundary conditions implements all boundary information through the use of essential boundary conditions. That is, specified elevation results in a reduced matrix for the continuity equation and specified velocity results in a reduced matrix for the momentum solution with the corresponding finite element equation discarded. This interpretation leads to gross mass balance errors as demonstrated in Figure 1. For perfect mass balance, the two curves should overlay one another. (The mitigating procedures discussed earlier are not implemented here so as to isolate the effect of boundary conditions.)

Results of the numerical experiments show that the key to using mass conservative boundary conditions with C^0 basis functions is interpretation. Specifically, state variables that appear in boundary integrals must be interpreted as external to the domain. To fix ideas, consider the finite element formulation of the primitive equations((1) and (3b)) for the one-dimensional model problem where Green's Theorem has been applied to the flux term in (1). The weak form is given by

$$\langle \frac{\partial \zeta}{\partial t}, \varphi_i \rangle - \langle Hu, \frac{\partial \varphi_i}{\partial x} \rangle + Q\Big|_{x_{0-}}^{x_{50+}} = 0 \tag{5}$$

Proper interpretation requires that the boundary flux term, $Q = Hu$, be viewed as external to the domain, i.e., an unknown quantity. Hence the $\pm$ designation on the limits of integration. In this way, the number of equations plus boundary conditions is equal to the number of unknowns so that all information is used; no equations are discarded. Thus, at the left boundary where ζ is specified, elevation is known so that (5) is solved for external flux. The momentum equation is then used to solve for velocity at the left boundary, u_{0+}. At the right boundary, the flux is known so it enters the finite element formulation given in (5) naturally, and the equation is solved for the unknown elevation. The momentum equation is then used to solve for the unknown velocity at the right boundary, u_{50-}. Note that such an interpretation allows for a discontinuity in velocity at the boundary.

When this external flux interpretation was applied to the GWC algorithm, simulation of the model problem results in no global mass balance error as shown in Figure 2. A large number of additional one-dimensional experiments, some looking at mixed formulations, were conducted to test the hypothesis. In one particularly interesting experiment, Green's Theorem was applied to the finite amplitude term, $g\nabla\zeta$, in the momentum equation. Consistent with the external flux interpretation, this boundary term had to be interpreted as external to the domain in order to realize stable, accurate, mass conservative results. Alternative interpretations (for example, treating specified flux as both essential in the momentum balance and natural in the continuity equation) that were tested led to unstable algorithms.

Additional evidence to support this interpretation of boundary terms comes from two sources. First, Westerink et al. [14] conducted a number of two-dimensional numerical experiments

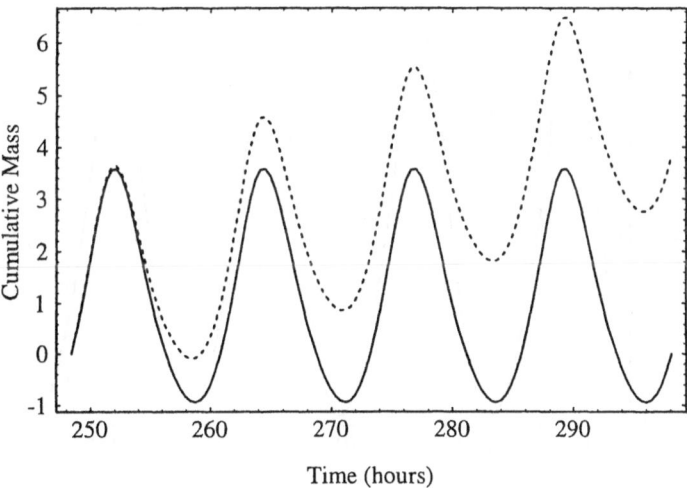

Time (hours)

Figure 1 ⋅ Global mass balance check of the GWC formulation ($G = \tau$) for the model problem, conventional interpretation of boundary conditions.

on a fictitious harbor with a known analytical solution. Both primitive and GWC algorithms were tested. One of the conclusions from the paper is that when known boundary fluxes are treated as external to the domain and implemented as natural conditions in the continuity equation, then the accuracy of the solution improves. In particular, $2\Delta x$ oscillations are damped. Experimental results were supported with dispersion analyses of both interior and boundary equations. Second, Gray and Celia [15] use generalized functions to facilitate finite element formulation of Laplace's equation. With this unique approach, it becomes clear that the boundary term resulting from application of Green's Theorem is indeed external to the computational domain. The method can also be applied to the coupled hyperbolic system of balance laws considered in this article; an outline of the essential steps in the analysis is given below.

To clarify the presentation, focus on the linearized, steady-state, primitive continuity equation with constant bathymetry. Under these assumptions, equation (1) simplifies to

$$\nabla \cdot \mathbf{v} = 0 \qquad\qquad (6)$$

Discretize all space, Ω_∞, into triangular elements (element shape is arbitrary). Approximate $\mathbf{v}$ with $\mathbf{v}_e \gamma_e$ (summation implied) where $\mathbf{v}_e$ is a piecewise polynomial approximation of $\mathbf{v}$ inside the domain of interest (Ω), γ_e is a generalized Heaviside step function that has a value of one in element e and zero outside, and the summation is over all elements in Ω_∞. Outside of Ω, the choice of the approximating function for $\mathbf{v}$ is arbitrary; a reasonable choice is to select $\mathbf{v}_e$ so that it satisfies the governing differential equation exactly.

Obtain the finite element approximation by weighting (6) with a set of polynomial weighting functions, φ_i, which are defined to be equal to 1 at node i and zero at all other nodes, and integrating the result over the domain

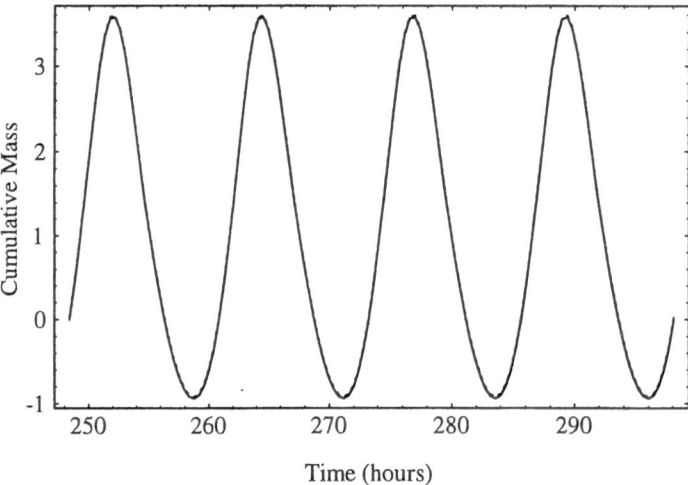

Figure 2   Global mass balance check of the GWC formulation ($G = \tau$) for the model problem, external flux interpretation of boundary conditions.

$$\sum_e \int_{\Omega_\infty} \nabla \cdot (\mathbf{v}_e \gamma_e) \, \varphi_i d\Omega = 0 \quad \text{for} \quad i = 1, \dots N \tag{7}$$

Next expand the spatial derivative in (7) and use properties of the step function to simplify the resulting terms. These properties are that the sum of γ_e over all elements is one and that the gradient of γ_e is a generalized Dirac function whose integral relations are similar to the familiar one-dimensional Dirac [16]. Thus, integrals involving $\nabla \gamma_e$ are converted to integrals over the boundary of the element where the discontinuity in γ_e occurs. Now, if E is defined to be all elements that have node i in common, then (7) becomes

$$\sum_E \left(\int_{\Omega_E} \nabla \cdot \mathbf{v}_E \varphi_i d\Omega - \int_{\partial \Omega_E} \mathbf{n}_E \cdot \mathbf{v}_E \varphi_i d\Gamma \right) = 0 \quad \text{for} \quad i = 1, \dots N \tag{8}$$

where $\mathbf{n}_E$ is a unit normal pointing out from element E.

Consider the case where i is on the boundary of the domain of interest, Ω. (For the case of i on the interior of the domain, (8) reduces to a familiar finite element approximation.) Define elements A to be those elements of E that are inside Ω and elements B those outside of Ω. Then, it can be shown that (8) simplifies to

$$-\sum_A \int_{\Omega_A} \mathbf{v}_A \cdot \nabla \varphi_i d\Omega + \int_{\partial \Omega} \mathbf{n}_A \cdot \mathbf{v}_B \varphi_i d\Gamma = 0 \tag{9}$$

In the second term, $\mathbf{v}_B$ is external to the domain so its value can be chosen to produce the best solution in Ω. An obvious choice is to select $\mathbf{v}_B$ such that $\mathbf{n}_A \cdot \mathbf{v}_B = 0$ on land boundaries and $\mathbf{n}_A \cdot \mathbf{v}_B$ is equal to the normal flux on open boundaries. Clearly in this formulation, $\mathbf{v}_B$ is interpreted as external to the computational domain.

SUMMARY

Shallow water models based on a finite element solution of the wave continuity equation have evolved as powerful tools for simulating the hydrodynamic behavior of coastal and oceanic waters. Experiments and analyses show that proper interpretation of boundary conditions is needed to improve the accuracy and the mass-conserving properties of the algorithm. In particular, boundary terms that result from the application of Green's Theorem to the weighted residual form of the governing equations should be interpreted as external to the computational domain. This interpretation can be viewed as the finite element counterpart of the use of imaginary nodes (nodes outside the boundary) in finite difference algorithms.

REFERENCES

1   Lynch, D.R. and Gray, W.G. (1979) "A Wave Equation Model for Finite Element Tidal Computations", *Computers and Fluids*, 7, 3:207-228.

2   Gray, W.G. (1982) "Some Inadequacies of Finite Element Models as Simulators of Two-Dimensional Circulation", *Advances in Water Resources*, 5, 171-177.

3   Kinnmark, I.P.E. (1986) *The Shallow Water Wave Equations: Formulation, Analysis, and Application*, Lecture Notes in Engineering, No. 15, Springer-Verlag.

4   Kolar, R.L. and Gray, W.G. (1990) "Shallow Water Modeling in Small Water Bodies", *Computational Methods in Surface Hydrology*, G. Gambolati et al. (eds.), Computational Mechanics Publications/Springer-Verlag, 149-155.

5   Lynch, D.R., Werner, F.E., Cantos-Figuerola, A., and Parrilla, G. (1988) "Finite Element Modeling of Reduced-Gravity Flow in the Alboran Sea: Sensitivity Studies", *Proceedings: Workshop on the Gibraltar Experiment*, Madrid, Spain.

6   Luettich, R.A. Jr., Westerink, J.J., and Scheffner, N.W. (1991) *ADCIRC: An Advanced Three-Dimensional Circulation Model for Shelves, Coasts, and Estuaries*, Dept. of the Army, U.S. Army Corps of Engineers, Washington, D.C.

7   Naimie, C.E. and Lynch, D.R. (1992) "Applications of Nonlinear Three-Dimensional Shallow Water Equations to a Coastal Ocean", *Computational Methods in Water Resources IX, Vol. 2: Mathematical Modeling in Water Resources*, Russell et al. (eds.) Computational Mechanics/ Elsevier, Southampton/London, 589-608.

8   Foreman, M.G.G. (1988) "A Comparison of Tidal Models for the Southwest Coast of Vancouver Island", *Developments in Water Science #35, Computational Methods in Water Resources, Vol. 1: Modeling of Surface and Subsurface Flows*, Celia et al. (eds.), Computational Mechanics/Elsevier, Southampton/Amsterdam, 231-236.

9   Westerink, J.J., Luettich, R.A., Baptista, A.M., Scheffner, N.W., and Farrar, P. (1992) "Tide and Storm Surge Predictions Using a Finite Element Model", *Journal of Hydraulic Engineering*, 118, 1373-1390.

10   Gray, W.G. (1989) "A Finite Element Study of Tidal Flow Data for the North Sea and English Channel", *Advances in Water Resources*, 12, 3:143-154.

11   Kolar, R.L., Westerink, J.J., Cantekin, M.E., Blain, C.A. (1994) "Aspects of Nonlinear Simulations Using Shallow Water Models Based on the Wave Continuity Equation", *Computers and Fluids*, in press.

12   Drolet, J., and Gray, W.G. (1988) "On the Well Posedness of Some Wave Formulations of the Shallow Water Equations", *Advances in Water Resources,* 11, 84-91.

13   Lynch, D.R. (1985) "Mass Balance in Shallow Water Simulations", *Communications in Applied Numerical Methods*, 1, 153-159.

14   Westerink, J.J., Luettich, Jr., R.A., Wu, J.K., and Kolar, R.L. (1994) "The Influence of Normal Flow Boundary Conditions on Spurious Modes in Finite Element Solutions to the Shallow Water Equation", *International Journal for Numerical Methods in Fluids*, in press.

15   Gray, W.G., and Celia, M.A. (1990) "On the Use of Generalized Functions in Engineering Analysis", *International J. of Applied Engineering Education*, 6(1), 89-96.

16   Gray, W.G., Leijnse, A., Kolar, R.L., and Blain, C.A. (1993) *Mathematical Tools for Changing Spatial Scales in the Analysis of Physical Systems*, CRC Press, Boca Raton, FL.

SOLUTION OF THE TWO-DIMENSIONAL NONLINEAR SHALLOW WATER EQUATIONS THROUGH APPLICATION OF THE FINITE ELEMENT FLUX-CORRECTED TRANSPORT SCHEME

RICHARD C. MARTINEAU and RAY A. BERRY
Idaho National Engineering Laboratory
Idaho Falls, Idaho, 83415-2110, U.S.A.

ABSTRACT

Overland flooding continues to be one of the most destructive of natural phenomena. In general, overland floods can be highly transient as in the case of sudden dam collapses and flash floods in canyons with the associated flow fields being bounded by irregularly bedded surfaces covered with a variety of vegetation, rocks, and manmade structures. The transient nature of floods in combination with irregular bounding surfaces allow floods to exhibit complex physical phenomena such as hydraulic jumps and flood fronts moving over dry surfaces. Simulation of these phenomena is a difficult computational fluid dynamics problem requiring nonlinear mathematical representation that has limited the success of previous simulation attempts. The complexity is introduced by the need for the floodplain code to be high-order accurate both temporally and spatially. Historically, high-order spatially accurate schemes exhibit nonphysical oscillations in regions of rapidly changing flow gradients for advective equations. Undershoots result in simulation instabilities when the magnitude of the undershoot is great enough to cause the simulated flow depth to be negative. It is essential not to have undershoots when simulating the propagation of flood waves over a dry surface where the flow depth in front of the wave must be zero. Avoiding the nonphysical oscillations dictates that the solution scheme be monotonic.

In this study, a high resolution finite element scheme, originally developed for the simulation of high-speed compressible flows, has been adapted to solve the two-dimensional nonlinear shallow water equations. Successful implementation of this scheme has yielded a temporally and spatially high-order accurate two-dimensional floodplain model. The finite element model employs flux-corrected transport as a nonlinear filter allowing simulation of propagating flood waves over *dry* flow fields. The two-dimensional model is capable of simulating floodplain and drainage system events caused by precipitation, channel discharge, dike failure, and tsunamis. Future plans include adding capabilities for snow melt, wind-driven flows, and hydrodynamic transport of contaminants.

INTRODUCTION

Overland flow can be approximated by the shallow water equations. With these equations, the flow normal to the bedded surface is assumed to be zero. The horizontal components of velocity are then integrated between the bed and the free surface of the

A. Peters et al. (eds.), Computational Methods in Water Resources X, 1089–1096.
© 1994 *Kluwer Academic Publishers. Printed in the Netherlands.*

flow to yield depth average velocities. The nonlinear form of the shallow water equations can approximate flow discontinuities similar to those found in the high-speed compressible gas equations. It would seem straightforward to apply any one of the modern solution schemes to the problem of overland flow. However, numerically simulating overland flow has several peculiarities which make it a difficult computational fluid dynamics problem. Floods can exhibit moving flood fronts over dry surfaces. Also, floods generally occur over a long duration compared to other transient fluid dynamics problems. Flow fields associated with flooding can be highly complex. Many flow fields contain very irregularly bedded surfaces covered with a variety of vegetation, rocks, etc.

An ideal floodplain code should be high-order accurate (second-order or higher) both temporally and spatially. However, high-order spatially accurate schemes generally exhibit nonphysical oscillations (over and undershoots) in regions of rapidly changing flow gradients for advective equations. Overshoots can result in inaccurate wave speeds due to incorrect wave heights. In addition to being inaccurate, undershoots can cause simulation instabilities when the undershoot causes the simulated flow depth to be negative. This is particularly true for simulating the propagation of flood fronts over a dry surface where the flow depth must be positive or zero. Minimizing the over and under shoots dictates that the solution scheme be nearly monotonic.

Application of the two-step Taylor-Galerkin finite element method for the solution of high-speed compressible flow problems has been extensively researched by Löhner[1-5] et al. Zalesak[6] developed a multidimensional generalization of the one-dimensional flux-corrected transport (FCT) solution scheme developed by Boris and Book[7-9] which is directly extendible to produce high resolution schemes on unstructured grids. This method combines a high-order scheme together with a low-order scheme in such a way that the high-order flux is used in regions where the solution is smooth. In regions where the solution contains high gradients or discontinuities, the high and low-order schemes are combined and limited so that the solution is monotonic and conservative. Löhner[4,5] implemented this generalized concept of FCT into the Taylor-Galerkin FEM for systems of hyperbolic equations. The solution method is known as Finite Element Flux-Corrected Transport (FEM-FCT).

The purpose of this report is to describe the ability of the FEM-FCT solution scheme to accurately solve the nonlinear shallow water equations. The governing equations for the model will be presented and the two-Step Taylor-Galerkin FEM-FCT will be discussed. Numerical results will be shown for several sudden dam collapse type problems and idealized floodplains.

THE GOVERNING EQUATIONS

In the context of this study, the conservative form of the shallow water equations is considered. The two-dimensional conservation of mass equation is

$$\frac{\partial h}{\partial t} + \frac{\partial (hu)}{\partial x} + \frac{\partial (hv)}{\partial y} = 0 \quad , \tag{1}$$

where h is the flow depth, and u and v are the depth averaged components of velocity in the x and y-directions, respectively. Assuming hydrostatic pressure, the x-component momentum equation is

$$\frac{\partial(hu)}{\partial t} + \frac{\partial}{\partial x}\left(hu^2 + \frac{gh^2}{2}\right) + \frac{\partial(huv)}{\partial y} = 0 \quad, \tag{2}$$

where g is the gravitational constant. The y-component momentum equation is

$$\frac{\partial(hv)}{\partial t} + \frac{\partial(huv)}{\partial x} + \frac{\partial}{\partial y}\left(hv^2 + \frac{gh^2}{2}\right) = 0 \quad. \tag{3}$$

The shallow water equations (1)-(3) can be written in the vector form

$$\frac{\partial U}{\partial t} + \nabla \cdot F = 0 \quad, \tag{4}$$

where U is the vector of conserved variables and F is the vector of advective fluxes.

THE TWO-STEP TAYLOR-GALERKIN FINITE ELEMENT METHOD WITH FLUX-CORRECTED TRANSPORT

The two-step Taylor-Galerkin FEM-FCT solution scheme used here for solving the nonlinear shallow water equations is an application of the solution scheme advanced by Löhner[4,5] et al. for the solution of high speed compressible flows. Temporal differencing is accomplished by successive forward-time Taylor series expansions, yielding an explicit second-order two-step Lax-Wendroff type time discretization. Spatial discretization is accomplished using the consistent mass Galerkin finite element method. FCT is a high resolution scheme designed to give enhanced resolution in areas of flow discontinuities and large changes in gradient (curvature) while limiting the nonphysical oscillations that are usually present when using a high-order accurate scheme. FCT employs a high-order scheme combined with a low-order scheme in such a way that the high-order solution dominates in smooth regions of the flow, whereas the low-ordered solution dominates near the discontinuities and high solution curvature. The low-order scheme is required to produce monotonic results in order to maintain positivity of the desired conserved variable.

The High-Order Scheme

The high-order scheme incorporated here employs a two-step form of the Taylor-Galerkin scheme as described in detail by Löhner[2] et al. Only a brief description is given here. The first step of the two-step Taylor-Galerkin scheme is to advance the solution to the half-step. This is accomplished by writing a Taylor's series for U at time $t^n + \Delta t/2$ about time t^n, yielding

$$U^{n+1/2} = U^n + \frac{\Delta t}{2} \frac{\partial U}{\partial t}\bigg|^n + O(\Delta t^2) \quad . \tag{5}$$

Substituting equation (4) into equation (5), yields

$$U^{n+1/2} = U^n - \frac{\Delta t}{2} \nabla \cdot F^n \quad . \tag{6}$$

The second step is developed by writing two Taylor's series for U^n about time $t^n + \Delta t/2$ and U^{n+1} about time $t^{n+1} - \Delta t/2$. Subtracting these two series gives

$$U^{n+1} = U^n + \Delta t \frac{\partial U}{\partial t}\bigg|^{n+1/2} + O(\Delta t^3) \quad . \tag{7}$$

Substituting equation (4) into equation (7), yields the second step of the Taylor-Galerkin scheme

$$U^{n+1} - U^n = -\Delta t \nabla \cdot F^{n+1/2} \quad . \tag{8}$$

Equation (8) can then be represented in the more compact form

$$M_c \delta U = R \tag{9}$$

where δU is the vector of conserved variable increments after a time step, M_c is the consistent mass matrix, and R is the vector of element contributions. Because M_c is diagonally dominant, equation (9) can be solved more efficiently with an iterative procedure. Iteratively, equation (9) becomes

$$M_l \delta U_{i+1} = R_i + (M_l - M_c) \delta U_i , \quad i = 1, 2, ..., \text{Niter} \tag{10}$$

where M_l denotes the lumped mass matrix, i is the iteration count, and Niter is the number of iterations. Three iterations was found to be sufficient for convergence when solving the shallow water equations.

The Low-Order Scheme: Lumped-Mass Taylor-Galerkin Plus Diffusion

As discussed before, the low-order scheme must not produce artificial overshoots and undershoots. Mass-diffusion added to the lumped-mass Taylor-Galerkin scheme will yield a monotonic scheme. Following Löhner[4], the diffusion is obtained by subtracting the lumped-mass matrix from the consistent-mass matrix for linear (the only type considered here) elements. The lumped-mass Taylor-Galerkin plus diffusion scheme is defined as

$$M_l \delta U^l = R + DIFF \tag{11}$$

where δU^l is the low-order solution and $DIFF$ is the added diffusion.

NUMERICAL RESULTS

Three numerical examples are included to demonstrate the numerical performance of the two-step Taylor-Galerkin FEM-FCT scheme in solving the nonlinear shallow water equations.

1. One-Dimensional Dam Break

The problem of a one-dimensional sudden dam break is included because there exists an analytical solution (Stoker[10]) to compare with the numerical solution. The computational mesh is shown in Figure 1. The mesh contains 2120 nodes and 4000 elements.

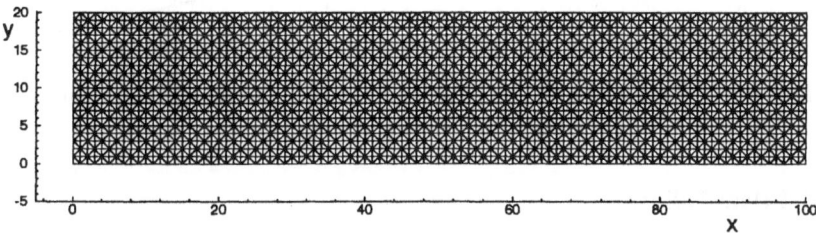

Figure 1   Computational mesh for one-dimensional dam break simulations

Figure 2 compares the analytical and numerical solutions for a time of $t = 2.5s$. The initial conditions are $h = 20$ for $0 \leq x \leq 50$ and $h = 2$ for $51 \leq x \leq 100$.

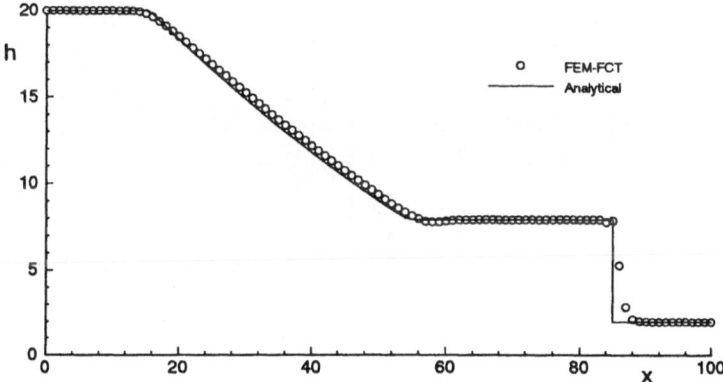

Figure 2   Analytical and numerical solutions for one-dimesional dam break simulation with a downstream depth of $h = 2$

Figure 3 compares the analytical and numerical solutions for a time of $t = 1.5s$. The initial conditions are $h = 20$ for $0 \leq x \leq 50$ and $h = 0$ for $51 \leq x \leq 100$. For this and the following dry flow field problems, when $h \leq 1 \times 10^{-8}$, u and v are not recalculated.

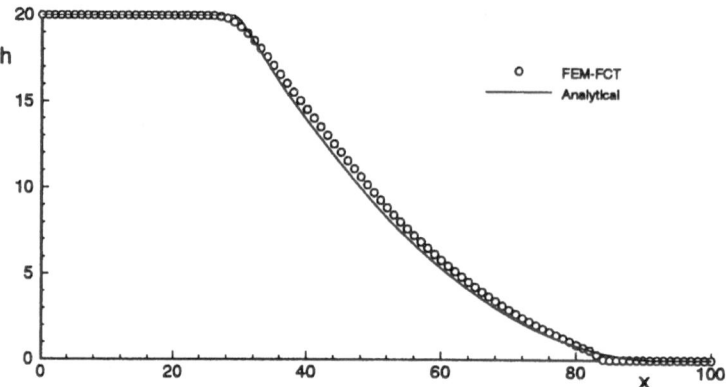

Figure 3 Analytical and numerical solutions for one-dimesional dam break
simulation with a downstream depth of $h = 0$

2. Partial Dam Failure

The computational mesh for a partial dam failure is shown in Figure 4. The reservoir
extends from $25 \leq x \leq 100$ and $25 \leq y \leq 100$. The breach extends from $75 \leq y \leq 100$.
The mesh contains 2570 nodes and 4950 elements.

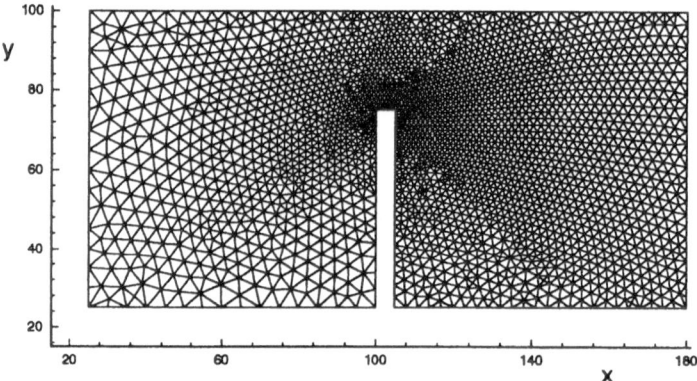

Figure 4 Comutational mesh for partial dam failure simulation

The initial conditions for this simulation are $h = 10$ in the reservoir and $h = 0$
downstream from the breach. Solid wall boundary conditions are imposed everywhere
except at $x = 180$ where non-reflective boundary conditions were imposed. Figure 5
illustrates the depth contours for a time of $t = 5s$ after the dam breaches. Note the
smooth flood front advancing onto the dry flow field.

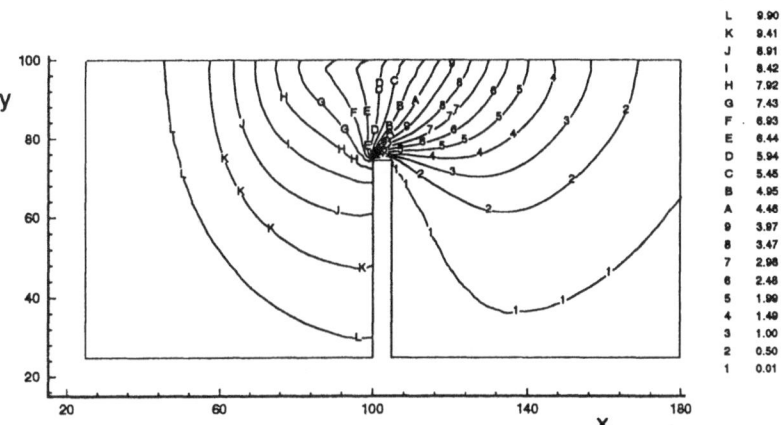

Figure 5 Flow depth contours for partial dam failure simultation

3. Supercritical Flow in a Constricted Channel

For the final example, a steady-state solution was obtained for supercritical flow in a constricted channel. Figure 6 shows the computational mesh containing 2673 nodes and 5120 elements for the channel whose cross-sectional area is constricted one-third by a triangular object extending from the side.

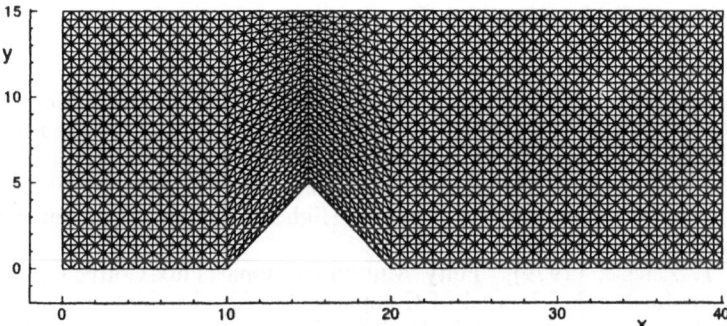

Figure 6 Comptational mesh for the constricted channel simulation

The initial conditions for this simulation are $h = 2$, $u = 8.86$, and $v = 0$ at $x = 0$. The rest of the domain was started as a dry flow field. Solid wall boundary conditions were imposed on the sides of the channel and non-reflecting boundary conditions were imposed at the exit, $x = 40$. Figure 7 shows the steady-state depth contours. The hydraulic jump in the form of a bow shock is sharply defined upstream from the constriction. Note that the depth directly behind the constriction remains nearly zero.

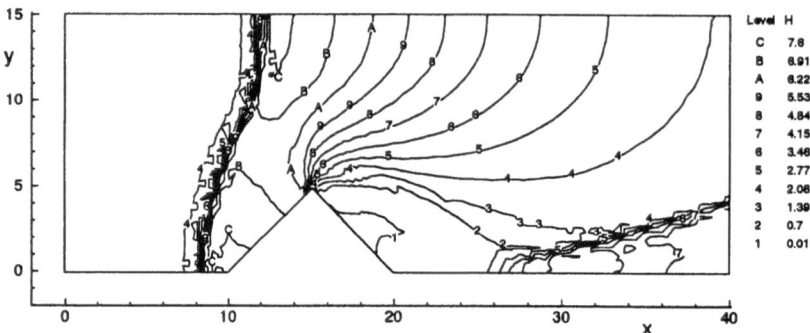

Figure 7 Flow depth contours for the constricted channel simulation

ACKNOWLEDGEMENTS

The authors would to thank Professor Rainald Löhner for his time and valuable suggestions during the course of this study.

REFERENCES

1. R. Löhner, K. Morgan, O.C. Zienkiewicz (1984) "The Solution of Nonlinear Hyperbolic Equation Systems by the Finite Element Method", International Journal for Numerical Methods in Fluids, Vol. 4.

2. R. Löhner, K. Morgan, and O.C. Zienkiewicz (1985) "An Adaptive Finite Element Procedure for Compressible High Speed Flows", Computer Methods in Applied Mechanics and Engineering, Vol. 51.

3. R. Löhner, K. Morgan, and O.C. Zienkiewicz (1985) "Finite Element Methods for High Speed Flows', AIAA, 85-1531.

4. R. Löhner, K. Morgan, J. Peraire, and M. Vahdati (1987) "Finite Element Flux-Corrected Transport (FEM-FCT) for the Euler and Navier-Stokes Equations", International Journal for Numerical Methods in Fluids, Vol. 7.

5. R. Löhner, K. Morgan, M. Vahdati, J. P. Boris, and D. L. Book (1988), "FEM-FCT: Combining Unstructured Grids with High Resolution", Communications in Applied Numerical Methods , Vol. 4.

6. Steven T. Zalesak (1979), "Fully Multidimensional Flux-Corrected Transport Algorithms for Fluids", Journal of Computational Physics, Vol. 31.

7. J. P. Boris and D. L. Book (1973), "Flux-Corrected Transport. I. SHASTA, a Transport Algorithm that Works", Journal of Computational Physics, Vol. 11.

8. D. L. Book, J. P. Boris, and K. Hain (1975), "Flux-Corrected Transport. II. Generalizations of the Method", Journal of Computational Physics, Vol. 18.

9. J. P. Boris and D. L. Book (1976), "Flux-Corrected Transport. III. Minimal-Error FCT Algorithms", Journal of Computational Physics, Vol. 20.

10. J. J. Stoker (1948) "The Formation of Breakers and Bores, The Theory of Nonlinear Wave Propagation in Shallow Water and Open Channels", Communications on Applied Mathematics, Vol. 1.

GENERALIZED FINITE DIFFERENCE METHOD AND NUMERICAL FORMULAS FOR DERIVATIVES

JUNJI MATSUMOTO
Texas Water Development Board
1700 North Congress Avenue, P.O.Box 13231, Austin, Texas 78711
U.S.A.

Mathematical derivation for a new numerical procedure called the Generalized Finite Difference (GFD) method is presented. It is based on the finite element (FE) method with linear triangular elements. A few new concepts such as equal weighting, double element and selfish approach are introduced for the derivation. The method is easy to understand and simple to apply, yet maintains the flexibility of following arbitrary geometry in creating a computational grid. Another interesting feature is that this procedure does not require a coordinate transformation. Formulas for the first order derivatives and the second order derivatives for the GFD method are presented.

INTRODUCTION

The purpose of this paper is to describe a new numerical procedure, the Generalized Finite Difference method. The concept and its mathematical details were first presented at the previous CMWR meeting (Matsumoto, 1992), in which the numerical differentiation formulas were derived by applying the FE method to linear triangular elements. Because no restriction was imposed on the configuration of linear triangular elements in the initial development, the computational mesh for the FE method and the GFD method is identical and has the flexibility as the FE method. The new GFD method presented here still uses the linear triangular elements for mathematical derivation but restricts the element configuration to a specific pattern and the resulting computational mesh is more like a finite difference (FD) grid. By restricting the element pattern, it loses some flexibility but gains a computational efficiency by enabling the application of the alternating direction implicit (ADI) scheme. The formulas presented here are more specific than the one in Matsumoto (1992) and they resemble to the standard FD formulas, at least conceptually. The main difference is that the GFD consists of two components each representing the approximation of the derivative along the orientation of the modeling axis. The initial development is briefly repeated here and then the new development is followed.

GFD METHOD

The equation used here for explanation is the two dimentional depth-averaged (x-direction) momentum equation:

A. Peters et al. (eds.), Computational Methods in Water Resources X, 1097–1104.

$$\frac{\partial U}{\partial t} + U\frac{\partial U}{\partial x} + V\frac{\partial U}{\partial y} + g\frac{\partial \zeta}{\partial x} + \tau U = 0 \qquad (1)$$

where U and V are the depth-averaged velocities in the x and y directions, ζ the water surface elevation, g the gravity, τ the bottom friction stress, t the time, x and y are the coordinates. Other terms such as Coriolis term are not included in (1) for simplicity of presentation.

For two-dimensional modeling, a linear triangular basis function ϕ_i for node i is given by

$$\phi_i = a_i + b_i x + c_i y \qquad (2)$$

where the coefficients $a_i, b_i,$ and c_i are computed by

$$a_i = (x_j y_k - x_k y_j)/2A_e \ , \ b_i = (y_j - y_k)/2A_e \ , \text{ and } c_i = (x_k - x_j)/2A_e \qquad (3)$$

where A_e is the area of triangular element e, while x_j, x_k, y_j, y_k are the coordinates of node j and node k (Pinder and Gray 1977). To emphasize the nodal location at which a FE equation is written, it is called the base node and usually subscript i is used; subscripts j and k are used to describe two other nodes of element e as shown in Figure 1. In the GFD method the coefficient b_i or c_i plays a crucial role. It represents a numerical differenciation in the x direction or y direction taking into account the orientation and the location of the neighboring nodes.

By applying the Galerkin method a FE equation for Eq.(1) can be written as

$$\sum_{e \in EL_i} \frac{\partial U_i}{\partial t}\frac{A_e}{3} + \sum_{e \in EL_i} \left(U_i \cdot DUDX_e + V_i \cdot DUDY_e + g \cdot DZDX_e + \tau_i \cdot U_i \right)\frac{A_e}{3} = 0 \qquad (4)$$

where EL_i represents a set of elements (directly) surrounding base node i, and $DUDX_e$ is the elementwise approximation of $\partial U/\partial x$ defined by

$$DUDX_e = \sum_{j \in ND_e} U_j \cdot \frac{\partial \phi_j}{\partial x} = \sum_{j \in ND_e} U_j \cdot b_j = U_i \cdot b_i + U_j \cdot b_j + U_k \cdot b_k \qquad (5)$$

where ND_e represents a set consists of three nodes comprising of element e. Other terms such as $DUDY_e$ and $DZDX_e$ are similarly defined.

By introducing the area weight $w_e = A_e / SUMA_i$ where $SUMA_i$ represents the total area of surrounding elements, and dividing both sides by $SUMA_i/3$, Eq.(4) becomes

$$\frac{\partial U_i}{\partial t} + \sum_{e \in EL_i} w_e \cdot \left(U_i \cdot DUDX_e + V_i \cdot DUDY_e + g \cdot DZDX_e + \tau_i \cdot U_i \right) = 0 \qquad (6)$$

Now we define the nodal approximation of the spatial derivative term by

$$\frac{\partial U}{\partial x}_{at\ i} \approx DUDX_i = \sum_{e \in EL_i} w_e \cdot DUDX_e \qquad (7)$$

and then we obtain the final form of the equation:

$$\frac{\partial U_i}{\partial t} + U_i \cdot DUDX_i + V_i \cdot DUDY_i + g \cdot DZDX_i + \tau_i \cdot U_i = 0 \qquad (8)$$

where $DUDY_i$ and $DZDX_i$ are the nodal approximations defined similarly to $DUDX_i$. Equation (8) can be called a GFD equation because the formulas (5) and (7) can be considered a generalization of the standard finite differences.

So far the derivation is a review of original development. As can be seen, its development is very general because triangular elements can be placed in any configuration. Before presenting a more restricted arrangement, a few new concepts are introduced. First, a double element, and then a selfish approach.

DOUBLE ELEMENT

It is illegal in the FE method to overlay an element with another element. For example, you can not place element-3 and element-4 in Figure 2b on top of element-1 and element-2 in Figure 2a as shown in Figure 2c. I call this overlapping element as a double element. In the GFD method double element is allowed. Actually, at least for the linear finite element, double element should be allowed. Mathematically there is no discrepancy as long as the weighting is properly treated.

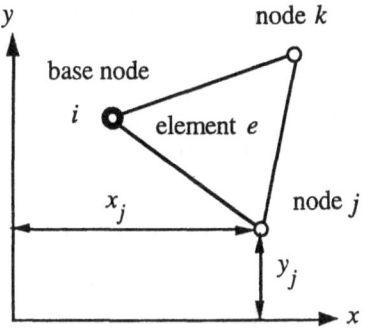

Figure 1. Node and element

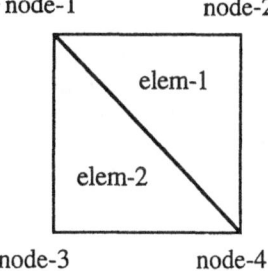

Figure 2a. Example mesh-1

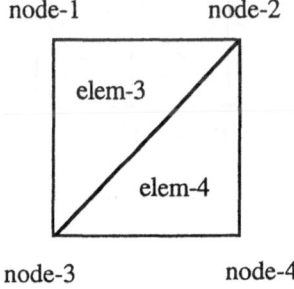

Figure 2b. Example mesh-2

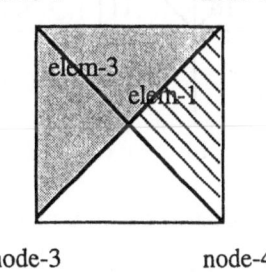

Figure 2c. Example mesh-3
Double Element: elem-3
is placed over elem-1

The idea of double element came about when I had a difficulty in making a computer model to work. Upon close examinations I noticed the model tended to crash at a corner boundary node like node-2 in Figure 2a. I thought it could be a problem of information flow. A corner node like node-2 is not communicating well with the neighboring nodes and the numerical error (or numerical noise) may be piling up at such a node.

The idea of information flow was tied to another idea that the integration process of the FE method can be thought of an averaging process. Eq.(5) for $DUDX_e$ is an elementwise approximation of $\partial U/\partial x$, while the nodal approximation, $DUDX_i$, is taken as the average of these elementwise approximations. The averaging process is indicated by Eq.(7) and it is the average by area weighting.

Based on this averaging concept, we may have as many double elements as we desire, but they need to be weighted properly. For instance, applying the derivative formula (7) to node-2 in Figure 2a, $DUDX_2 = DUDX_{e1}$, while for node-2 in Figure 2b, $DUDX_2 = (DUDX_{e3} + DUDX_{e4})/2$, assuming these elements have the same area. For node-2 in Figure 2c for double element, $DUDX_2 = (DUDX_{e1} + DUDX_{e3} + DUDX_{e4})/3$. By this we gained a better information flow. Corner node-2 in Figure 2a communicates with node-1 and node-4, but in Figure 2c it also communicates with node-3.

Another advantage of the double element is that it eliminates or at least minimizes the mesh orientation problem. A frequently used mesh pattern is shown in Figure 3a. An internal node like node-6 or node-7 has a different connectivity with neighboring nodes. Node-6 has four connections while node-7 has eight connections. The information flow is not uniform and this may cause some undesirable effect (Platzman 1981). Figure 3b is a double element mesh, every node is connected equally with the neighboring nodes and thus the flow of information is uniform.

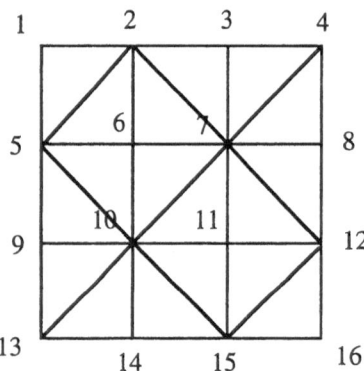

Figure 3a. Typical FE mesh

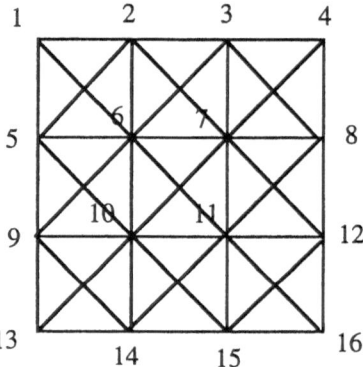

Figure 3b. Double element mesh

SELFISH APPROACH

This is another interesting idea stemmed from the double element. In Figure 3b all internal nodes are connected to its eight surrounding nodes, or all 12 elements including double elements are utilized for the gradient calculation. But, we do not have to use all these elements surrounding the base node. We can select the elements so as to achieve the uniform flow of information as well as the computational efficiency.

For example, at node-6 in Figure 3b, we may take only four elements which directly surround base node-6 as illustrated by Figure 3c. I call this selective scheme a selfish approach becasue the base node pays attention only to those elements most closely connected to it.

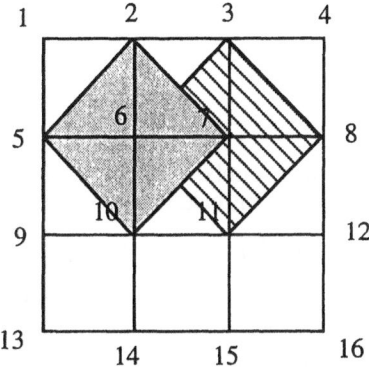

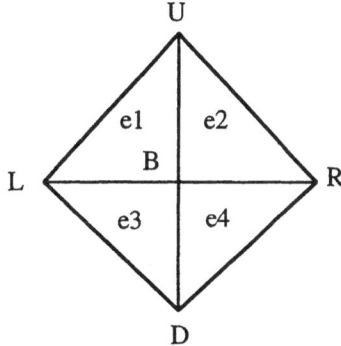

Figure 3c. Double element mesh with Selfish Approach

Figure 4. Four element configuration

Aside from the question of accuracy this selfish approach should increase the computational efficiency because the number of elements used for gradient calculation is 1/3 compared with the full double element mesh as Figure 3b or 2/3 compared with a regular FE mesh as Figure 3a. Although the detail is not presented here, the same concept (of double element and selfish approach) can be applied to three dimensional mesh of tetrahedral elements and it should help increase the computational efficiency greatly.

FIRST ORDER DERIVATIVE

Let us fix the mesh pattern to the four element configuration shown in Figure 3c or Figure 4, which looks a lot like a FD grid. A major difference between this grid and the conventional FD grid is that the GFD grid has more freedom in locating grid points as may be seen in Figure 5.

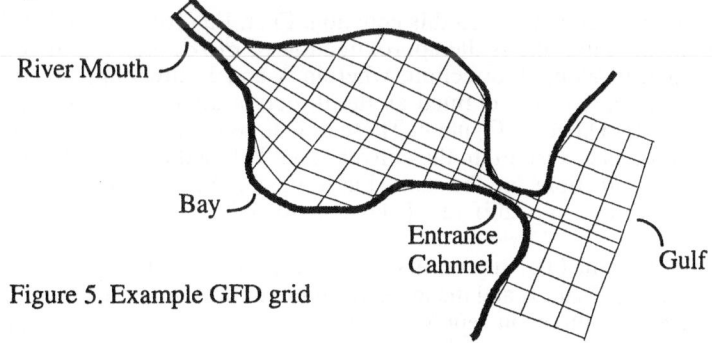

Figure 5. Example GFD grid

Let us also assume the weighting factors in Eq.(7) to be equal for all elements mainly because of brevity in presentation. Then for the four element congiguration as in Figure 3c, the nodal approximation of the velocity gradient can be calculated by

$$DUDX_i = (DUDX_{e1} + DUDX_{e2} + DUDX_{e3} + DUDX_{e4})/4 \tag{9}$$

A slightly different expression for $DUDX_e$ is needed. Since $b_i + b_j + b_k = 0$, or

$b_i = -(b_j + b_k)$, Eq.(5) for $DUDX_e$ can be rearranged as

$$DUDX_e = (U_j - U_i)b_j + (U_k - U_i)b_k \tag{10}$$

To derive a formula for the first order derivative, a new notation shown in Figure 4 is employed where B represents the base node, R the right node, L the left node, U the upper node, D the down node; e1,e2,e3, and e4 are the element numbers as before. Applying Eq.(9) and (10), a nodal approximation of the velocity gradient is calculated by

$$DUDX_B = [\{(U_R - U_B)b_{R1} + (U_U - U_B)b_{U1}\} + \{(U_U - U_B)b_{U2} + (U_L - U_B)b_{L2}\}$$
$$+ \{(U_L - U_B)b_{L3} + (U_U - U_B)b_{D3}\} + \{(U_D - U_B)b_{D4} + (U_R - U_B)b_{R4}\}]/4 \tag{11}$$

or

$$DUDX_B = \{\Delta U_R \cdot b_{R1} + \Delta U_U \cdot b_{U1} + \Delta U_U \cdot b_{U2} + \Delta U_L \cdot (-b_{L2})$$
$$+ \Delta U_L \cdot (-b_{L3}) + \Delta U_D \cdot (-b_{D3}) + \Delta U_D \cdot (-b_{D4}) + \Delta U_R \cdot b_{R4}\}/4 \tag{12}$$

where b's are the coefficient of x in the basis function (2) and the subscripts indicate the associated node and element; ΔU's are the differences of U's defined by

$$\Delta U_R = (U_R - U_B), \ \Delta U_L = (U_B - U_L), \ \Delta U_U = (U_U - U_B), \ \Delta U_D = (U_B - U_D) \tag{13}$$

Eq.(12) can be rearranged as

$$DUDX_B = \{\Delta U_R \cdot (b_{R1} + b_{R4}) + \Delta U_L \cdot (-b_{L2} - b_{L3})$$
$$+ \Delta U_U \cdot (b_{U1} + b_{U2}) + \Delta U_D \cdot (-b_{D3} - b_{D4})\}/4 \tag{14}$$

We now introduce the average coefficients of b's defined by

$$\bar{b}_R = (b_{R1} + b_{R4})/2, \bar{b}_L = -(b_{L2} + b_{L3})/2, \bar{b}_U = (b_{U1} + b_{U2})/2, \bar{b}_D = -(b_{D3} + b_{D4})/2 \tag{15}$$

then Eq.(14) becomes

$$DUDX_B = 0.5 \cdot (\Delta U_R \cdot \bar{b}_R + \Delta U_L \cdot \bar{b}_L) + 0.5 \cdot (\Delta U_U \cdot \bar{b}_U + \Delta U_D \cdot \bar{b}_D) \tag{16}$$

We can make a few observations on this equation. First, Eq.(16) has two bracketed terms or two components. First one is the approximation of $\partial U/\partial x$ along X direction and the second one is that of along Y direction, wherein X and Y directions are the modeling directions. The X direction is defined as the direction along nodes L,B,R, and the Y direction along nodes D,B,U in Figure 4. These directions differ from x and y directions of the Cartesean coordinate system. In the conventional FD method the modeling directions and the coordinate directions are the same, but in the GFD method they are different and even X and Y directions can be different from node to node.

The first component in (16) represents the average of the right side approximation associated with nodes B and R and the left side approximation associated with nodes L and B, while the second component represents the average of the upper side approximation associated with nodes U and B and the down side approximation associated with nodes B and D. The first component is the same as the formula for conventional FD method but the second component is new. When the modeling directions match the coordinate directions, the second component is zero. If the axis are rotated 90 degree conterclockwise, the first component becomes zero but the second component supplies the approximation. If the rotation is between zero and 90 degree, both components provide the approximations. The weighting among the approximations are automatically handled by the coefficients $\bar{b}'s$.

Notice how simple the equation is. There is no coordinate transformation involved. The coefficients $\bar{b}'s$ are constants and need to be computed only once at the beginning of iteration. A similar expression for $\partial U/\partial y$ is listed below.

$$DUDY_B = 0.5 \cdot (\Delta U_R \cdot \bar{c}_R + \Delta U_L \cdot \bar{c}_L) + 0.5 \cdot (\Delta U_U \cdot \bar{c}_U + \Delta U_D \cdot \bar{c}_D) \tag{17}$$

where $\bar{c}'s$ are the average coefficients of c's similar to $\bar{b}'s$ defined by

$$\bar{c}_R = (c_{R1} + c_{R4})/2, \ \bar{c}_L = (c_{L2} + c_{L3})/2, \ \bar{c}_U = (c_{U1} + c_{U2})/2, \ \bar{c}_D = (c_{D3} + c_{D4})/2 \tag{18}$$

SECOND ORDER DERIVATIVE

Although the second order derivatives such as the diffusion terms are not included in Eq.(1), they are important terms. This section presents the derivation of the numerical formula for the second order terms.

Assuming the equal weighting among elements, the second order term can be approximated by (Matsumoto 1991)

$$\frac{\partial^2 U}{\partial x^2}\bigg|_{at\ i} \approx D^2 UDX_i^2 = -0.5 \sum_{e \in EL_i} DUDX_e \cdot b_i \tag{19}$$

Eq.(19) is applicable to an arbitrary configuration of linear triangular elements. But to derive a specific formula for the four element configuration, we apply Eq.(19) for the base node B in Figure 4. Then

$$D^2 UDX_B^2 = -0.5 \ [\ (\Delta U_R \cdot b_{R1} + \Delta U_U \cdot b_{U1}) \cdot b_{B1} + \{ \Delta U_L \cdot (-b_{L2}) + \Delta U_U \cdot b_{U2} \} \cdot b_{B2}$$
$$+ \{ \Delta U_L \cdot (-b_{L3}) + \Delta U_D \cdot (-b_{D3}) \} \cdot b_{B3} + \{ \Delta U_R \cdot b_{R4} + \Delta U_D \cdot (-b_{D4}) \} \cdot b_{B4}] \tag{20}$$

To simplify the expression we substitute the b's by their avearges:

$$\bar{b}_{B14} = 0.5 \cdot (b_{B1} + b_{B4}), \ \bar{b}_{B23} = 0.5 \cdot (b_{B2} + b_{B3}),$$
$$\bar{b}_{B12} = 0.5 \cdot (b_{B1} + b_{B2}), \text{ and } \bar{b}_{B24} = 0.5 \cdot (b_{B2} + b_{B4}) \tag{21}$$

Then Eq.(20) becomes

$$D^2 UDX_B^2 = -0.5 \ [\ \{ \Delta U_R \cdot (b_{R1} + b_{R4}) \cdot \bar{b}_{B14} + \Delta U_L \cdot (-b_{L2} - b_{L3}) \cdot \bar{b}_{B23} \}$$
$$+ \{ \Delta U_U \cdot (b_{U1} + b_{U2}) \cdot \bar{b}_{B12} + \Delta U_D \cdot (-b_{D3} - b_{D4}) \cdot \bar{b}_{B34} \}] \tag{22}$$

or

$$D^2 UDX_B^2 = - \ [\ \{ \Delta U_R \cdot \bar{b}_R \cdot \bar{b}_{B14} + \Delta U_L \cdot \bar{b}_L \cdot \bar{b}_{B23} \}$$
$$+ \{ \Delta U_U \cdot \bar{b}_U \cdot \bar{b}_{B12} + \Delta U_D \cdot \bar{b}_D \cdot \bar{b}_{B34} \}] \tag{23}$$

We need one more averaging expression. Let

$$\bar{b}_X = (-b_{B1} + b_{B2} + b_{B3} - b_{B4})/4 = (-\bar{b}_{B14} + \bar{b}_{B23})/2 \text{ and}$$
$$\bar{b}_Y = (-b_{B1} - b_{B2} + b_{B3} + b_{B4})/4 = (-\bar{b}_{B12} + \bar{b}_{B34})/2 \tag{24}$$

Then we obtain the final expression for the second order derivative:

$$D^2 UDX_B{}^2 = (\Delta U_R \cdot \bar{b}_R - \Delta U_L \cdot \bar{b}_L) \cdot \bar{b}_X + (\Delta U_U \cdot \bar{b}_U - \Delta U_D \cdot \bar{b}_D) \cdot \bar{b}_Y \qquad (25)$$

The same equation can be obtained from geometrical consideration. Let

$$DUDX_R = \Delta U_R \cdot \bar{b}_R \text{ and } DUDX_L = \Delta U_L \cdot \bar{b}_L \qquad (26)$$

then the first bracketed term in (25) can be written as

$$(DUDX_R - DUDX_L) \cdot \bar{b}_X \qquad (27)$$

which represents the same operation as in the conventional FD formula:

$$\frac{\partial^2 U}{\partial x^2} \approx \left(\frac{U_{j+1} - U_j}{\Delta x_R} - \frac{U_j - U_{j-1}}{\Delta x_L} \right) / \Delta \bar{x} \qquad (28)$$

As in the case of the first order derivative, Eq.(25) consists of two components. First one is the approximation along X direction and the second one along Y direction. Similar formula for y derivative is given below.

$$D^2 UDY_B{}^2 = (\Delta U_R \cdot \bar{c}_R - \Delta U_L \cdot \bar{c}_L) \cdot \bar{c}_X + (\Delta U_U \cdot \bar{c}_U - \Delta U_D \cdot \bar{c}_D) \cdot \bar{c}_Y \qquad (29)$$

where

$$\bar{c}_X = (-c_{B1} + c_{B2} + c_{B3} - c_{B4}) / 4 \text{ and } \bar{c}_Y = (-c_{B1} - c_{B2} + c_{B3} + c_{B4}) / 4 \qquad (30)$$

SUMMARY

The numerical formulas for the first order and second order derivatives for the GFD method were presented, which are rather simple, easy to understand and easy to apply; they can be considered as an extention of the conventional FD formulas.

The motibation behind this version of GFD is the realization that the ADI is possible because of the special structure of the FD grid, in which the node (or grid point) is connected to four nodes, two in the x direction and two in the y direction. The GFD presented here keeps this structure by restricting the element pattern to four element configuration. The modeling flexibility of the GFD method comes from the fact that the gradient is approximated by the surrounding (linear) triangular elements.

REFERENCES

Matsumoto, J., "Mathematical description of the FETEX model based on a new computational method: a simplified finite element method or a generalized finite difference method," Technical Report of the Texas Water Development Board, Austin, Texas,1991.

Matsumoto, J., "Mathematical description of the FETEX model based on a generalized finite difference method" in Vol.I: Numerical Methods in Water resources, *Computational Methods in Water Resources IX*, 1992.

Pinder, G.F. and Gray, W.G., *Finite Element Simulation in Surface and Subsurface Hydrology*, Academic Press, 1977.

Platzman, G.W., "Some response characteristics of finite-element tidal models" *J. of Computational Physics,* vol 40, pp36-63, 1981.

CALCULATION OF VERTICAL VELOCITY IN A THREE-DIMENSIONAL MODEL USING A LEAST SQUARES APPROACH

J.C. MUCCINO[1], W.G. GRAY[1] and M.G.G. FOREMAN[2]

[1]Department of Civil Engineering and Geological Sciences
University of Notre Dame, Notre Dame, IN 46556
USA

[2]Institute of Ocean Sciences
Sidney, British Columbia, V8L-4B2
CANADA

Three-dimensional, finite element, linear, hydrodynamic numerical models generally use a "wave" or Helmholtz equation to solve for free surface elevation, then use the horizontal momentum equation to solve for the horizontal velocities, and finally use the continuity equation to solve for the vertical velocities. The continuity equation is a first order differential equation, and therefore admits only one boundary condition. However, in general, an estimate of vertical velocity is available at both the free surface and the bottom, yielding two boundary conditions and an overdetermined system. This work investigates the use of a least squares approach to solve the continuity equation with both boundary conditions. Preliminary results suggest that the least squares approach conserves mass better than the more traditional finite element approach which solves the vertical derivative of the continuity equation with both boundary conditions.

INTRODUCTION

Finite element models of the three-dimensional shallow water equations are commonly used to gain insight into the behavior of large bodies of water subject to tides and buoyancy forces. Accurate prediction of the vertical component of the current is critical to understanding and managing major fish and zooplankton stocks, as upwelling is responsible for driving nutrients towards the surface from deeper regions. The vertical velocities are found, after the horizontal velocities are calculated, using the continuity equation:

$$\frac{\partial w}{\partial z} = -\nabla \cdot \mathbf{V} \tag{1}$$

where $\mathbf{V}$ is the horizontal velocity with components (u, v), ∇ is the horizontal del operator with components $(\partial/\partial x, \partial/\partial y)$, w is the vertical component of the velocity, and (x, y, z) are the spatial coordinates. However, the continuity equation is a first order equation and therefore admits only one boundary condition. In general, an estimate of the vertical velocity is available at both the free surface and the bottom, leading to an

1105

A. Peters et al. (eds.), Computational Methods in Water Resources X, 1105–1112.

overdetermined system of equations. There is no unique way to solve an overdetermined system of equations, and several approaches have been considered in the literature.

    One approach is to use only one of the boundary conditions in the formulation and use the other as a measure of the error accumulated over the water column (Lynch and Werner, 1991). This approach leads to an accumulation of error at the boundary for which the boundary condition was neglected in the formulation. A second approach is to use the vertical derivative of the continuity equation (henceforth VDC) such that the equation becomes second order and both boundary conditions can be admitted (Naimie and Lynch, 1993):

$$\frac{\partial^2 w}{\partial z^2} + \frac{\partial}{\partial z}(\nabla \cdot \mathbf{V}) = 0 \tag{2}$$

However, with this approach, the original equation of continuity is sacrificed so that both boundary conditions can be used, and the resulting mass conservation properties are questionable.

    The approach considered in this work is a least squares formulation based on the continuity equation (1) and both boundary conditions. In this way, both boundary conditions are retained, as is the original formulation of the continuity equation. This formulation allows nonzero residuals for the boundary conditions and the conservation of mass equation, which is similar to the data assimilation procedure described by Zahel (1991).

THREE-DIMENSIONAL MODEL FRAMEWORK

The framework in which the vertical velocity calculations are tested is a three-dimensional diagnostic model for baroclinic, wind-driven and tidal circulation in coastal seas, as developed by Lynch and Werner (1987). The governing equations are the linearized, harmonic, shallow water equations with hydrostatic and Boussinesq assumptions. These equations are expressed as the continuity equation (1) and the horizontal momentum equation:

$$i\omega \mathbf{V} - \frac{\partial}{\partial z}(N\frac{\partial \mathbf{V}}{\partial z}) + g\nabla \eta + \frac{g}{\rho_0}\int_z^0 \nabla \rho dz = 0 \tag{3}$$

with boundary conditions:

$$N\frac{\partial \mathbf{V}}{\partial z}\bigg|_{z=0} = h\mathbf{\Psi} \tag{4}$$

$$N\frac{\partial \mathbf{V}}{\partial z}\bigg|_{z=-h} = k\mathbf{V} \tag{5}$$

where ω is the radian frequency of the forcing, N is the vertical eddy viscosity, g is gravity, η is the free surface height, ρ_0 is the reference density, ρ is the density, h is the bathymetric depth, $h\Psi$ is atmospheric forcing, k is the linear bottom stress coefficient, and $i = \sqrt{-1}$. These equations are solved for a response which consists of a three-dimensional velocity field (u, v, w) and surface elevation η. In this study, atmospheric forcing is not implemented and the density field is constant, so baroclinic forcing is not implemented. Forcing is introduced at the open boundary as a semi-diurnal tidal signal.

Some manipulation (Lynch and Werner, 1987) of equations (1) and (3) yields a scalar equation in η alone, which can be solved independently from u, v and w. The horizontal components of the velocity can then be obtained from equation (3) without reference to the vertical component since the equations are completely uncoupled in the horizontal and vertical directions. All calculations are performed using the finite element method (Lynch and Werner, 1987).

The domain is a quarter annular harbor, as shown in Figure 1. The bound-

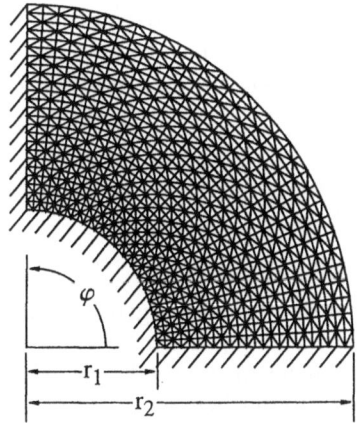

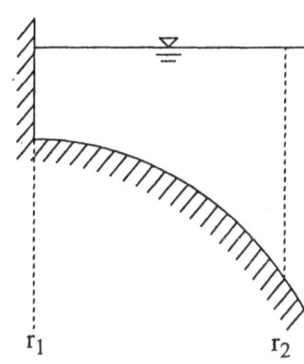

Figure 1a. Section view of quarter annular harbor with opening at r=r₂

Figure 1b. Side view of quarter annular harbor with quadratic bottom.

aries at $r = r_1 = 4 \times 10^4\,\text{m}$, $\varphi = 0$, and $\varphi = \pi/2$ are no flow boundaries. The open boundary is located at $r = r_2 = 5 \times 10^5\,\text{m}$. The bottom of the harbor, h, is quadratic in r and constant in φ, such that at $r = r_1$, $h = 10.0\,\text{m}$, at $r = (r_1 + r_2)/2$, $h = 30.63\,\text{m}$, and at $r = r_2$, $h = 62.50\,\text{m}$. The grid in the horizontal plane was generated with XMGREDIT, a flexible interactive grid generation code developed by Turner and Baptista (1991). It contains 825 nodes and 1536 elements. The three-dimensional mesh is generated by projecting the nodes of the horizontal grid to the bottom of the domain in vertical lines and discretizing the lines into thirty-two vertical elements. Elements within each vertical line are of the same length. The resulting three-dimensional grid is comprised of prismatic elements with perfectly vertical, rectangular sides and triangular top and bottom faces which are, in general, not parallel to each other (Naimie and Lynch, 1993). Further refinement of the horizontal and vertical grid do not change the solution.

IMPLEMENTATION OF LEAST SQUARES PROCEDURE

Horizontal velocities are found, as described above, without reference to the vertical velocities using the horizontal momentum equation. The vertical velocities can now be calculated using these horizontal velocities and the continuity equation.

Straightforward integration of the continuity equation (1) over each layer in the vertical yields one equation for each layer:

$$w|_{z_{i+1}} - w|_{z_i} = - \int_{z_i}^{z_{i+1}} \left(\frac{\partial u}{\partial x} + \frac{\partial v}{\partial y} \right) dz \qquad \text{for } i = 1, n-1 \qquad (6)$$

where n is the number of nodes in the vertical, and $i = 1$ at the bottom and $i = n$ at the free surface. The right side of equation (3) is entirely known. The horizontal derivatives are calculated using standard triangular, linear interpolation and the integral is calculated with the trapezoidal rule. In addition to the $n - 1$ equations arising from (6), the two kinematic conditions at the boundaries are:

$$w|_{z = \eta} = i\omega\eta \qquad (7)$$

$$w|_{z = -h} = -u\frac{\partial h}{\partial x} - v\frac{\partial h}{\partial y} \qquad (8)$$

Equations (6), (7), and (8) result in an overdetermined system of algebraic equations consisting of n unknowns (the vertical velocities at each node), and $n + 1$ equations. A "best fit" solution can be found by minimizing the squares of the $n + 1$ residuals. This approach is commonly known as "least squares," henceforth LSQ.

VERTICAL VELOCITY COMPARISON

The vertical velocities where computed according to equations (2), (7) and (8), using VDC, as well as according to the LSQ procedure described above. Vertical velocities are obtained for the two procedures at $r = 7 \times 10^4$ m, with $k = 0.0001$ m/sec, $N = 0.015$ m^2/sec and $\omega = 1.405 \times 10^{-4}$ sec^{-1} which represents a semi-diurnal tide with a period of 12.42 hours. The forcing at $r = r_2$ has an amplitude of 10 cm and uniform phase. Typical vertical velocity profiles are shown in Figure 2 at one quarter of the tidal cycle (approximately 3.1 hours). It is important to note that although these solutions are converged solutions, the two methods converge to different solutions. Notice that the solution using VDC is smoother than the solution using LSQ which uses the continuity equation in its original form. This might be expected because the second derivative in VDC tends to smooth the solution. The results shown here are representative of results obtained throughout the tidal cycle, for various values of the linear bottom stress coefficient and vertical eddy viscosity, and for different locations within the quarter annular harbor.

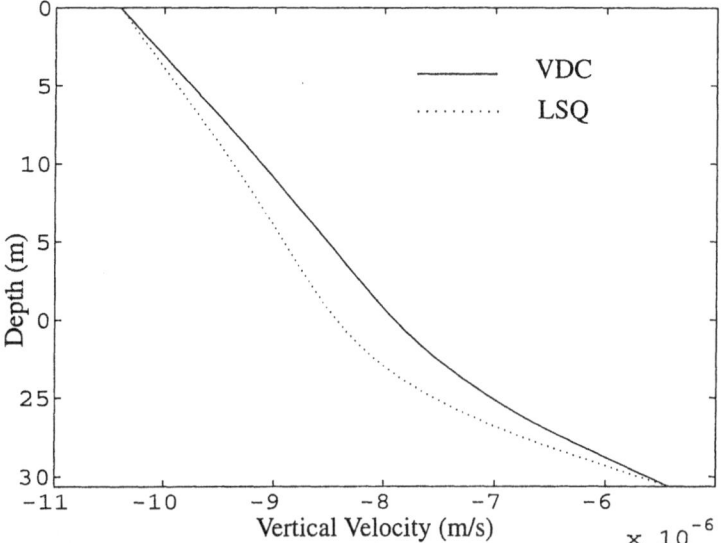

Figure 2. Vertical velocities calculated with VDC and LSQ through the water column at one quarter of the tidal cycle (approximately 3.1 hours).

CONSERVATION OF MASS

It is interesting that the two different methods, VDC and LSQ converge to different solutions, but without further study, there is no evidence that either method is superior. In order to examine the relative accuracy of these two different approaches, a study of their mass conserving properties is performed. For an incompressible fluid, the expression of the principle of conservation of mass for a fixed volume with outer surface Γ is:

$$\int_\Gamma \mathbf{V} \cdot \mathbf{n} d\Gamma = 0 \tag{9}$$

where $\mathbf{n}$ is the normal vector to the surface Γ. In the case of an approximate solution, (9) takes the form:

$$\int_\Gamma \mathbf{V} \cdot \mathbf{n} d\Gamma = R \tag{10}$$

where R is a residual. The residual, R, is examined using both methods with the surface of one three-dimensional finite element as the surface of integration in equation (10). The residual for an element centered at $h = 16.3\,\mathrm{m}$ is plotted versus time in Figure 3. This behavior is typical of elements throughout the water column. The residual obtained by both methods is a sinusoidal wave with a period of 12.42 hours, the period of the forcing. However, the amplitude of the residual obtained with VDC is 0.14 m^3/sec, while the residual obtained with LSQ has an amplitude of 0.01 m^3/sec, more than an or-

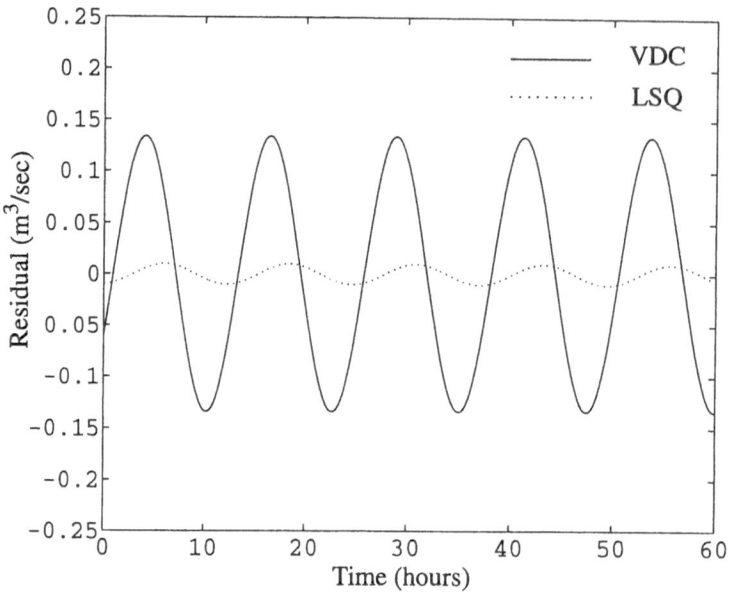

**Figure 3. Residual vs. time calculated with VDC and LSQ
at a depth h = 16.3**

der of magnitude less. In addition, although the two methods produce residuals of the same period, they are out of phase by two hours.

It is also useful to examine the relationship between the residual and the total flow through the element. The residual for the same element considered above is plotted versus total flow in Figure 4. The curve shown represents one full tidal cycle and is repeated for each tidal cycle. These results are representative of elements throughout the water column. A common characteristic of all elements is a relatively constant residual with respect to flux for the LSQ approach while the VDC approach tends to give larger errors for larger fluxes.

Finally, the relationship between the accumulation of volume and time is examined in Figure 5. The accumulation in both cases is a sinusoidal wave with a period of 12.42 hours. The amplitude of the accumulation achieved with VDC is almost $1000m^3$, while the accumulation achieved with LSQ is less than $70m^3$. While LSQ reduced the accumulation by more than an order of magnitude, the total volume of the element is about $4 \times 10^6 m^3$, so the accumulations calculated here are only a small percentage of the total volume. However, the quarter annular harbor is a highly idealized case and some deterioration of results would be expected in a field study, so the improvement shown here is potentially very significant.

CONCLUSIONS

An alternative method to the traditional method of calculation of the vertical velocities is examined. Instead of solving the vertical derivative of the continuity equation with two

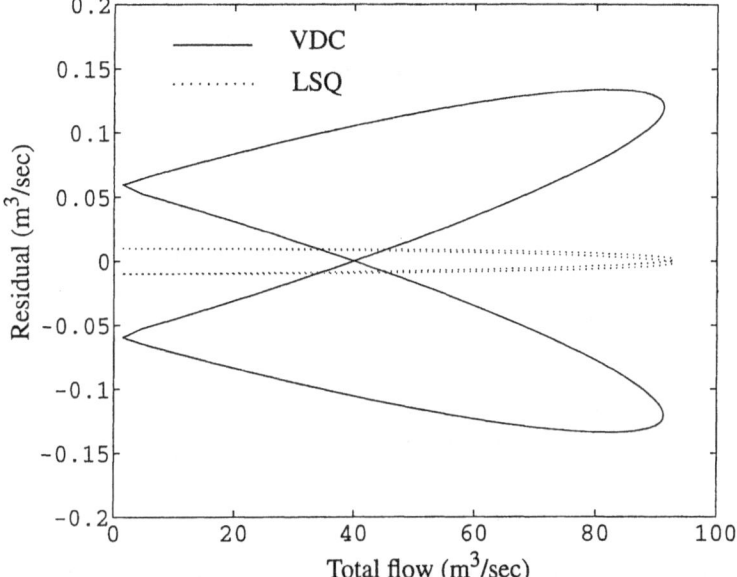

**Figure 4. Residual vs. total flow calculated with VDC and LSQ
at a depth h = 16.3**

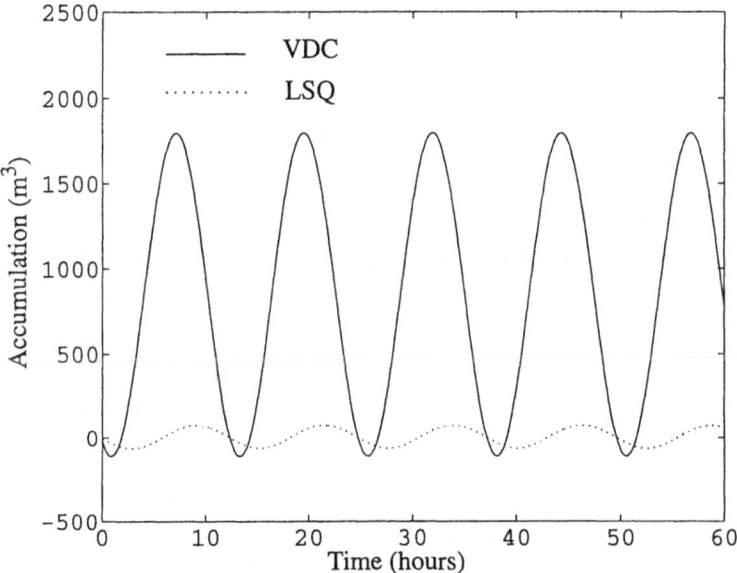

**Figure 5. Accumulation vs. time calculated with VDC and LSQ
at a depth h = 16.3m**

boundary conditions, the continuity equation in its original form is solved with two boundary conditions as an overdetermined system using a least squares approach. The two methods were examined using a quarter annular harbor. The traditional method smooths the vertical velocity profile to a greater extent than the least squares approach. In addition, the least squares approach appears to have better mass conserving properties, as shown by the smaller amplitude of the residuals of the conservation of mass equation in integral form and as shown by a smaller amplitude for accumulation of volume.

These results are preliminary and much work remains to be done. Future studies will include comparison of both methods to an analytic solution in order to more accurately assess the error of each. In addition, the methods will be applied to a field situation west of the Juan de Fuca Strait where vertical velocities are of much interest to government officials formulating fishing regulations.

Further research will also include use of a similar least squares approach to assimilate elevation and horizontal velocity data into two and three-dimensional finite element models.

ACKNOWLEDGEMENTS

We would like to thank D.R. Lynch of Dartmouth College for generously allowing us to use his model, FUNDY5, as a basis for this study. We would also like to thank R.F. Henry of the Institute of Ocean Sciences, Sidney, BC, Canada for his assistance in grid generation. This supported was supported in part by a National Science Foundation Graduate Research Fellowship.

REFERENCES

Lynch, D.R. and Werner, F.E. (1987) "Three-dimensional hydrodynamics on finite elements. Part 1: Linearized Harmonic Model," *International Journal for Numerical Methods in Fluids,* **7,** 871-909.

Lynch, D.R. and Werner, F.E. (1991) "Three-dimensional hydrodynamics on finite elements. Part II: Nonlinear time-stepping model.," *International Journal for Numerical Methods in Fluids,* **12**, 507-533.

Naimie, C.E. and Lynch, D.R. (1993) "Three Dimensional Diagnostic Model for Baroclinic, Wind-driven and Tidal Circulation in Shallow Seas: FUNDY5 User's Manual," Dartmouth College, NH.

Turner, P.J. and Baptista, A.M. (1991) "ACE/Gredit User's Manual: Software for Semiautomatic Generation of Two-Dimensional Finite Element Grids," CCALMR Software Report SDS2(91-2), Oregon Graduate Institute, Beaverton, OR.

Zahel, W. (1991) "Modeling Ocean Tides With and Without Assimilating Data," *Journal of Geophysical Research,* **96**, no. B12, 20,379-20,391.

ALTERNATIVE APPROACHES TO THE WAVE EQUATION IN SHALLOW WATER FLOW MODELLING

SVEN SIGURDSSON
Science Institute, University of Iceland, Dunhaga 3,
IS-109 Reykjavik, Iceland

The use of the wave equation has been advocated in shallow water flow modelling as a tool for stabilizing the numerical solution. We compare three approaches to introducing the wave equation within the context of staggered finite element methods.

1. Derive the appropriate differential equation from the equations of conservation of mass and momentum by first differentiating the mass equation with respect to time. Subsequently discretize the wave and momentum equations in space and time.

2. Discretize the mass and momentum equations in space and then derive the space-discretized wave equation by first differentiating the space-discretized (finite element) mass equation. Finally discretize the wave and momentum equations in time.

3. Discretize the mass and momentum equations in space and the latter also in time by a fully implicit method. Then derive an 'integrated' and space-discretized wave equation from these discretizations. Finally discretize the 'integrated' wave equation in time.

Approaches 1 and 2 lead to identical formulae, the advantage of 2 being that it allows a more 'natural' introduction of boundary conditions, in particular conditions of no normal flow, through the space-discretized (finite element) mass equation.

Approach 3 has the crucial advantage from the point of view of conservation of mass of being based on the mass equation in a non-differentiated form. The linear system that has to be inverted at each timestep is however identical to the one in approaches 1 and 2 and involves in particular a symmetric positive definite matrix.

A. Peters et al. (eds.), Computational Methods in Water Resources X, 1113–1120.
© 1994 Kluwer Academic Publishers. Printed in the Netherlands.

Since adopting approach 3 we have observed in a number of testing simulations that the mass conservation properties have markedly improved, even by allowing the extra stabilizing parameter, β_0 (or G), in the wave equation to be 0. At the same time all other advantages of the wave equation approach seem to be retained.

BASIC EQUATIONS

The finite element scheme to be considered below is based on the following equations describing shallow water flow:

$$\frac{\partial}{\partial t}\eta + \nabla \cdot (h\mathbf{v}) = q_\mathrm{I} \tag{1}$$

$$\frac{\partial}{\partial t}\mathbf{v} + (\mathbf{v} \cdot \nabla)\mathbf{v} = -g\nabla\eta - \beta\mathbf{v} - f\mathbf{k} \times \mathbf{v} + \frac{1}{h}\mathbf{s} - \frac{q_\mathrm{I}}{h}(\mathbf{v} - \mathbf{v}_\mathrm{I}) \tag{2}$$

within a given region Ω, along with boundary conditions on $\partial\Omega$ of either prescribed head or normal flow, as well as initial conditions for η and $\mathbf{v}$.

Here η is surface elevation above mean sea level, $h = d + \eta$ where d is mean water depth of flow, $\mathbf{v}$ is vertically averaged flow velocity, q_I is injected water, g is gravitational acceleration, $\beta = \frac{g}{hc_B^2}|\mathbf{v}|$ where c_B is Chézy bottom friction coefficient, f is the Coriolis parameter, $\mathbf{k}$ is the unit vector in z-direction (vertically upwards), $\mathbf{s}$ is windstress and $\mathbf{v}_\mathrm{I}$ is the velocity of injected water.

The focus of attention in the present paper is, however, the wave equation:

$$\left(\frac{\partial^2}{\partial t^2} + \beta_0\frac{\partial}{\partial t}\right)\eta - g\nabla \cdot (h\nabla\eta) = \nabla \cdot (h(\beta - \beta_0)\mathbf{v}) - f\mathbf{k} \cdot (\nabla \times h\mathbf{v}) \tag{3}$$

$$-\nabla \cdot \mathbf{s} + \nabla \cdot (q_\mathrm{I}(\mathbf{v} - \mathbf{v}_\mathrm{I})) + \nabla \cdot ((h\mathbf{v} \cdot \nabla)\mathbf{v}) - \nabla \cdot (\frac{\partial}{\partial t}\mathbf{v}) + \frac{d}{dt}q_\mathrm{I} + \beta_0 q_\mathrm{I}$$

which follows directly from equations (1) and (2). Here β_0 (also denoted be G in the literature) is an arbitrary constant reference friction factor. The idea that, within the context of finite element approximations, it may be advantageous to advance in time by integrating eqns. (3) (with a suitable choice of β_0) and (2) rather than equations (1) and (2), was put forward by Lynch and Gray [5] in 1979 and has since then been supported by a large number of studies (see e.g. Kolar et al. [4] for references). The advantages of this approach are e.g.:

1. Eqn. (3) is less coupled to eqn. (2) than eqn. (1) is, in particular if we set $\beta_0 \approx \beta$ This may be of importance if we advance in time by first integrating eqn. (3) at each timestep and then integrate eqn. (2).

2. When using an implicit procedure, integrating eqn. (3) will involve the solution of a symmetric positive definite system rather than an unsymmetric system as will be the case with eqn. (1).

Related to the second issue is the fact that codes based on eqn. (3) seem to be more robust and have better spatial stability characteristics, although as we see it, spatial stability is more critically dependent on the proper choice of basis functions.

The wave equation as it stands has, however, at least two potential weaknesses:

1. Normal flow boundary that enter naturally into the formulation of eqn. (1) take on a considerably more complex form when applied to eqn. (3).

2. Eqn. (3) is derived from eqn. (1) in a differentiated form with respect to time. Thus when advancing in time using eqn. (3) rather than eqn. (1) it may prove difficult to retain mass conservation. This has been clearly demonstrated by [3, 4]. While a judicious choice of β_0 will often make it possible to circumvent this difficulty this state of affairs seems rather unsatisfactory.

The main contribution of the present paper is to indicate how one may deal with these weaknesses. The presentation is in terms of a socalled staggered finite element schemes to be reviewed in the next section but we believe that the approach has a more general applicability.

FINITE ELEMENT EQUATIONS

The staggered finite element scheme for shallow water flow was introduced by Sigurdsson et al. [6] in 1988. A fuller and more up to date account may be found in Sigurdsson [7]. Here we only review the scheme to the extent that is required for our discussion of the wave equation.

The staggered finite element scheme is based on triangular elements, using linear approximations for the elevation within the element and constant approximation for the velocity. Thus there is a jump in velocity approximations between elements. While the scheme is based on a mixed Galerkin formulation it is most readily described in terms of the resulting approximation on a single general element $\Delta = ABC$.

We introduce the following notation:
η_i and h_i, $i = A, B, C$ denote the approximate elevation and total depth resp. at the corresponding nodes. $\eta_m = \frac{1}{3}(\eta_A + \eta_B + \eta_C)$, similarily for h_m. $\boldsymbol{\eta} = (\eta_A, \eta_B, \eta_C)^T$. $\mathbf{v}$ denotes the approximate velocity within the element.
$q_{B,i}$ and $q_{I,i}$, $i = A, B, C$ denote the normal flow out through the boundary of the element and prescribed injected flow into the element resp., weighted by the linear basis function at the corresponding nodes. $\mathbf{q}_B = (q_{B,A}, q_{B,B}, q_{B,C})^T$, similarily for $\mathbf{q}_I$. $|\Delta|$ denotes the area of the element and $\mathbf{a}$ the outward normal on the edge opposite node A, whose length is that of the edge, (denoted by $\mathbf{a}^R$ in Sigurdsson [7]) similarily for $\mathbf{b}$ and $\mathbf{c}$. We note, in particular, that the gradient of the linear basis function at node A is $-\frac{1}{2|\Delta|}\mathbf{a}$.

Finally we introduce the following matrices:

$$M = \frac{|\Delta|}{12} \begin{bmatrix} 2 & 1 & 1 \\ 1 & 2 & 1 \\ 1 & 1 & 2 \end{bmatrix} \quad N = -\frac{1}{2}[\mathbf{a}\ \mathbf{b}\ \mathbf{c}]$$

$$K = \frac{1}{4|\Delta|} \begin{bmatrix} \mathbf{a}\cdot\mathbf{a} & \mathbf{a}\cdot\mathbf{b} & \mathbf{a}\cdot\mathbf{c} \\ \mathbf{b}\cdot\mathbf{a} & \mathbf{b}\cdot\mathbf{b} & \mathbf{b}\cdot\mathbf{c} \\ \mathbf{c}\cdot\mathbf{a} & \mathbf{c}\cdot\mathbf{b} & \mathbf{c}\cdot\mathbf{c} \end{bmatrix} = \frac{1}{|\Delta|}N^T N \tag{4}$$

and refer to them as the mass matrix, gradient matrix, and stiffness matrix respectively.

In the staggered finite element scheme, the finite element counterpart of eqn. (1) can then be expressed as:

$$M\frac{d}{dt}\boldsymbol{\eta} + \mathbf{q}_B - h_m N^T \mathbf{v} = \mathbf{q}_I \tag{5}$$

and the counterpart of eqn. (2) as:

$$|\Delta|\frac{d}{dt}\mathbf{v} + (\frac{1}{2}(\mathbf{v}_a + \mathbf{v})\cdot\mathbf{a})(\mathbf{v}_a - \mathbf{v})$$
$$= -gN\boldsymbol{\eta} - |\Delta|\left[\beta\mathbf{v} + f\mathbf{k}\times\mathbf{v} - \frac{1}{h_m}\mathbf{s} + \frac{q_I}{h_m}(\mathbf{v} - \mathbf{v}_I)\right] \tag{6}$$

where $\mathbf{v}_a$ denotes the approximate velocity in the element that the outward normal $\mathbf{a}$ points into and we assume, for the sake of simplicity, to be the only upstream neighbour of the element ABC at a given instance in time.

Finally, we derive the finite element counterpart of eqn. (3) from eqns. (5) and (6) in an exactly analogous way as eqn. (3) is derived from eqns. (1) and (2) and obtain:

$$\left[M\frac{d^2}{dt^2} + \beta_0 M\frac{d}{dt} + h_m g K\right]\boldsymbol{\eta} = -N^T\Big\{h_m(\beta - \beta_0)\mathbf{v} + h_m f(\mathbf{k}\times\mathbf{v})$$
$$-\mathbf{s} + q_I(\mathbf{v} - \mathbf{v}_I) + \frac{h_m}{|\Delta|}(\frac{1}{2}(\mathbf{v}_a + \mathbf{v})\cdot\mathbf{a})(\mathbf{v}_a - \mathbf{v}) - \frac{d}{dt}\eta_m\mathbf{v}\Big\} \tag{7}$$
$$+\frac{d}{dt}(\mathbf{q}_I - \mathbf{q}_B) + \beta_0(\mathbf{q}_I - \mathbf{q}_B).$$

Although eqn. (7) could be derived directly from eqn. (3) using the appropriate Galerkin conditions the derivation above has the advantage of incorporating in a natural way any normal flow boundary conditions through the finite element mass equation by specifying the appropriate value for $\mathbf{q}_B$ on such an outer boundary. (Note that on all inner boundaries the $\mathbf{q}_B$ values will cancel out in the assembly process.)

Thus we have circumvented the first of the weaknesses associated with the wave equation in the preceding section. Furthermore, this derivation ensures that the advective terms are treated in a consistent way in the finite element momentum equation and the wave equation which is shown to be computationally advantageous by Kolar et al. [4]. However, the second weakness of the wave equation, that of retaining mass conservation still remains. It is in an attempt to overcome it, that we now go on to derive the finite element integrated wave equation.

FINITE ELEMENT INTEGRATED WAVE EQUATION

In the following derivation we omit the terms involving injected water from eqns. (5) and (6) and the convective term from eqn. (6) since they enter into the derivation exactly in the same way as the normal boundary flow term in eqn. (5) and the wind stress term in eqn. (6) resp. and may thus be readily substituted into the final result. We let Δt denote the time step and the subscript p refer to that the corresponding values are to be evaluated at the previous time, $t - \Delta t$.

By discretizing eqn. (6) in time using a one-step fully implicit approximation and adding the term $\beta_0 \mathbf{v}$ to both sides we have that:

$$\frac{|\Delta|}{\Delta t}\left((1 + \Delta t \cdot \beta_0)\mathbf{v} - \mathbf{v}_p\right) = -gN\boldsymbol{\eta} - |\Delta|\left[(\beta - \beta_0)\mathbf{v} + f\mathbf{k} \times \mathbf{v} - \frac{1}{h_m}\mathbf{s}\right] \quad (8)$$

Eliminating $\mathbf{v}$ on the left-hand side, substituting it into eqn. (5), using eqn. (4) we derive the finite element integrated wave equation:

$$\left[(\frac{1}{\Delta t} + \beta_0)M\frac{d}{dt} + h_m gK\right]\boldsymbol{\eta} =$$
$$-h_m N^T\left[(\beta - \beta_0)\mathbf{v} + f\mathbf{k} \times \mathbf{v} - \frac{1}{h_m}\mathbf{s} - \frac{1}{\Delta t}\mathbf{v}_p\right] - (\frac{1}{\Delta t} + \beta_0)\mathbf{q}_B \quad (9)$$

We make the following observations on this new form of the finite element wave equation:

1. Eqn. (9) is based on the mass equation in its original form whereas eqn. (7) is based on the mass equation in differentiated form.

2. After a fully implicit timediscretization the assembled global matrix will be exactly the same for eqns. (7) and (9).

3. When advancing forward in time using eqn. (9) only η and $\mathbf{v}$ values from the previous time have to be retained, compared to having to retain η values from two previous times in the case of eqn. (7).

4. As in the case of eqn. (7), the coefficient β_0 has to take the same value within all elements so that the normal boundary flow terms $\mathbf{q}_B$ between elements cancel out in the assembly of the global system.

5. After substituting:
   $\frac{1}{\Delta t}M\left(\frac{d}{dt}\boldsymbol{\eta} - \frac{d}{dt}\boldsymbol{\eta}_p\right)$ for $M\frac{d^2}{dt^2}\boldsymbol{\eta}$, $\frac{1}{\Delta t}(h_m - h_{m,p})\mathbf{v}_p$ for $\frac{d}{dt}\eta_m\mathbf{v}$ and $\frac{1}{\Delta t}(\mathbf{q}_B - \mathbf{q}_{B,p})$

for $\frac{d}{dt}\mathbf{q}_B$ in eqn. (7) the difference between eqns. (9) and (7) amounts to:
$\frac{1}{\Delta t}\left\{ M\frac{d}{dt}\boldsymbol{\eta}_p - h_{m,p}N^T\mathbf{v}_p + \mathbf{q}_{B,p}\right\} = 0$ if the finite element mass equation is satisfied at the previous time.

Consider finally the possibilitiy of rewriting eqn. (8) as:

$$\frac{|\Delta|}{\Delta t}\left((I + \Delta t\begin{bmatrix} \beta & -f \\ f & \beta \end{bmatrix})\mathbf{v} - \mathbf{v}_p\right) = -gN\boldsymbol{\eta} + \frac{|\Delta|}{h_m}\mathbf{s}$$

Eliminating $\mathbf{v}$ on the left hand side, substituting it into eqn. (5), using eqn. (4) and introducing the matrix:

$$L = \frac{1}{2}\begin{bmatrix} 0 & 1 & -1 \\ -1 & 0 & 1 \\ 1 & -1 & 0 \end{bmatrix} = \frac{1}{4|\Delta|}N^T\begin{bmatrix} 0 & 1 \\ -1 & 0 \end{bmatrix}N$$

we obtain:

$$\left[\frac{1}{\Delta t}M\frac{d}{dt} + h_m g(\gamma K + \rho L)\right]\boldsymbol{\eta} = N^T\begin{bmatrix} \gamma & \rho \\ -\rho & \gamma \end{bmatrix}\left\{\mathbf{s} + \frac{h_m}{\Delta t}\mathbf{v}_p\right\} - \frac{1}{\Delta t}\mathbf{q}_B \qquad (10)$$

where

$$\begin{bmatrix} \gamma & \rho \\ -\rho & \gamma \end{bmatrix} = \left(I + \Delta t\begin{bmatrix} \beta & -f \\ f & \beta \end{bmatrix}\right)^{-1} \quad \text{i.e.}$$

$$\gamma = (1 + \Delta t\beta)/D \quad \rho = f/D \quad \text{where } D = \left((1 + \Delta t\beta)^2 + f^2\right)$$

and note that with this formulation the coefficient β may take different values in different elements in the assembly of the global system. Hence we have fully eliminated the velocity terms from the equation (except through the presence of convective terms and the dependence of β and $\mathbf{v}$). This enhanced decoupling from the momentum equation is, however, at the considerable computational price of having replaced the stiffness matrix by a matrix that will vary with time through $\gamma(\beta)$ and include an antisymmetric component through $\rho(f)$.

DISCUSSION

A shallow water code based on finite element eqns. (7) and (6) being advanced one after the other with $\beta_0 \in [10^{-5}, 10^{-4}]$ and fully implicit approximations in time has been used successfully in Iceland over the past years in simulations of both estuaries and shallow lakes. Difficulties linked with unsatisfactory mass conservation in some extreme situations involving eg. flooding and drying led in the spring of 1993 to trial replacements of the finite element wave eqn. (7) by its integrated counterpart (9). The close similarity between the two equations meant that only a few lines of code had to be changed but the improvement in the results was significant. Since the choice $\beta_0 \approx \beta$ in eqn. (9) means that the direct influence of the velocity in the equation is minimized it was assumed that some choice of $\beta_0 > 0$ would still be advisable. Somewhat surprisingly preliminary tests indicated that the choice $\beta_0 = 0$ was in fact to be preferred and that has been the standard setting for all simulations since then. Where comparisons have been made with results obtained by the earlier code there has not been any instance where the new result have been inferior. The possibility of introducing eqn. (10) where the velocity is eliminated more fully from the wave equation has thus as yet not been investigated further.

In order to establish better the mass conservation properties of the new approach the two simulations described in Kolar et al. [4], the first involving an artificial 1-dimensional channel and the second one describing tidal motion in the Bight of Abaco in the Bahamas have been run with the new code based on eqns. (9) and (6) with $\beta_0 = 0$. All parameters, including the timestep size, were identical with those presented in [4] and likewise both local and global mass conservation was investigated comparing the normal flow across internal and external boundaries, integrated over time, with the changes in elevation integrated over the enclosed region. Complete agreement was obtained so that the results are not reproduced here.

It should be noted that in the staggered finite element scheme velocities are not uniquely defined on boundaries between elements. Calculations of normal flow across a boundary is based on solving for q in the finite element mass eqn. (4) on the appropriate boundary and is thus an explicit calculation involving elevation values as well as velocity values in the elements adjacent to the boundary on one side. This, however, is now a well-established procedure for obtaining accurate flow values that applies to more general finite element methods (see e.g. Gresho et al. [2] and references therein].

We finally observe that the simple idea behind the integrated wave equation has recurred in various guises in the lierature. Two unrelated and somewhat arbitrarily chosen examples are the "primitive pseudo wave equation formulation" introduced by Westerink et al. [8] within the framework of harmonic finite element modelling, which amounts to substituting the harmonic form of eqn. (6) into the harmonic form of eqn. (5), and the finite difference shallow water modelling system CYTHERE-ESI described in Benqué et al. [1], where in the propagation step a time discretized form of the momentum equation is substituted into the time discretized mass equation, resulting in a symmetric positive definite system.

ACKNOWLEDGEMENTS

All computations have been carried out by VATNASKIL Consulting Engineers, Reykjavik, using their program package AQUASEA, that implements the staggered finite element scheme. Joannes Westerink, University of Notre Dame, made available a complete datafile for the Bight of Abaco simulation.

References

[1] Benqué , J.P., Cunge, J.A., Feulliet, J., Hauguel, A., and Holly, F.M. (1982)

"New Method for Tidal Current Computation", Jounral of the Waterway, Port, Coastal and Ocean Division, Proceedings of the American Society of Civil Engineers, 108, No. WW3, 396–417.

[2] Gresho, P.M., Lee, R.L., Sani, R.L., Maslanik, M.K., and Eaton, B.E. (1987) "The Consistent Galerkin FEM for Computing Derived Boundary Quantities in Thermal and/or Fluid Problems", Int. J. Num. Meth. in Fluids, 7, 371–394.

[3] Kolar, R.L., Gray, W.G., and Westerink, J.J. (1992) "An Analysis of the Mass Conserving Properties of the Generalized Wave Continuity Equation", in T.F. Russell, R.E. Ewing, C.A. Brebbia, W.G. Gray, G.F. Pinder (eds.) Computational Methods in Water Resources IX, Vol. 2: Mathematical Modeling in Water Resources, Computational Mechanics Publication, Elsevier Applied Science, pp. 537–544.

[4] Kolar, R.L., Westerink, J.J., Cantekin, M.E., and Blain, C.A. (1994) "Aspects of Nonlinear Simulations Using Shallow Water Models Based on the Wave Continuity Equation", to appear in Computers and Fluids.

[5] Lynch, D.R. and Gray, W.G. (1979) "A Wave Equation Model for Finite Element Tidal Computations", Computers and Fluids, 7, 207–228.

[6] Sigurdsson, S., Kjaran, S.P. and Tómasson, G.G. (1988) "A Simple Staggered Finite Element Scheme for Simulation of Shallow Water Free Surface Flows", in M.A. Celia, L.A. Ferrand, C.A. Brebbia, W.G. Gray, G.F. Pinder (eds.) Computational Methods in Water Resources, Vol.: 1 Modeling Surface and Sub-Surface Flows, Elsevier, Computational Mechanics Publications, pp. 329–335.

[7] Sigurdsson, S. (1992) "Treatment of the Convective Term in taggered Finite Element Schemes for Shallow Water Flow" in T.F. Russell, R.E. Ewing, C.A. Brebbia, W.G. Gray, G.F. Pinder (eds.) Computational Methods in Water Resources IX, Vol. 1: Numerical Methods in Water Resources, Computational Mechanics Publications, Elsevier Applied Science, pp. 291–298.

[8] Westerink, J.J., Connor, J.J., and Stolzenbach, K.D. (1987) "A Primitive Pseudo Wave Equation Formulation for Solving the Harmonic Shallow Water Equations", Adv. Water Resources, 10, 188–199.

A FAST METHOD FOR THE COMPUTATION OF FAIRLY LOW AND FAIRLY LONG GRAVITY WAVES

W.A. VAN DER VEEN and F.W. WUBS
Department of Mathematics, University of Groningen
P.O. Box 800, 9700 AV Groningen
The Netherlands

In this paper a system of equations and a numerical model are presented for fairly low and fairly long gravity waves. The equations are derived from an approximate Hamiltonian for gravity waves. This approximation is positive definite which leads to a stable model. The present model is a variant of a model derived by Mooiman. However, by a new derivation the model could be made twice as fast without loss of accuracy.

INTRODUCTION

We consider irrotational nonlinear dispersive gravity waves in an inviscid incompressible fluid. The Boussinesq equations, we are aiming at, describe the propagation of fairly low and fairly long waves and take into account the effect of amplitude and frequency dispersion. Such waves occur in harbors and coastal regions.

The derivation of Boussinesq equations for this type of waves is critical. Very often such equations are not stable for high frequency components. Positive-definite Hamiltonians give canonical equations which are stable for all wave lengths. For this reason we start from the general Hamiltonian formulation of water waves, which has this property. To obtain stable Boussinesq equations, the Hamiltonian is approximated by a simpler form, which is still positive definite (Broer et al. [1, 3]).

Mooiman has derived and implemented a set of Boussinesq equations based on such a Hamiltonian [4, 6]. The solution of the occurring Helmholtz-type equations dominates the run time of the model. Van der Ploeg [7] compared a variety of numerical solution techniques for these equations, which resulted in a speed-up of an order of magnitude. However, this part is still dominating the run time. In this paper, we present equations and a numerical model in which the number of Helmholtz-type equations to be solved per right-hand side evaluation is reduced by a factor two. Moreover, in our case the corresponding matrix is symmetric. This is an additional advantage, because it simplifies the solution process. Numerical experiments show that, as expected, a speed-up of a factor two is attained and the results are as accurate as those of Mooiman's model.

A. Peters et al. (eds.), Computational Methods in Water Resources X, 1121–1128.
© 1994 Kluwer Academic Publishers. Printed in the Netherlands.

THE HAMILTONIAN FOR NONLINEAR GRAVITY WAVES

In this section, we will present the equations for nonlinear dispersive gravity waves and the corresponding Hamiltonian. We consider an incompressible and inviscid fluid. Furthermore, we assume that the flow is irrotational. In this case, the boundary-value problem can be formulated in terms of the velocity potential Φ as follows

$$\Delta\Phi + \Phi_{zz} = 0, \quad -h(x,y) < z < \zeta(x,y,t) \tag{1}$$

$$\Phi_t + \frac{1}{2}(|\nabla\Phi|^2 + \Phi_z^2) + g\zeta = 0, \quad z = \zeta(x,y,t) \tag{2}$$

$$\zeta_t + \nabla\Phi\cdot\nabla\zeta - \Phi_z = 0, \quad z = \zeta(x,y,t) \tag{3}$$

$$\nabla\Phi\cdot\nabla h + \Phi_z = 0, \quad z = -h(x,y) \tag{4}$$

where ∇ and Δ denote the 2-dimensional gradient and Laplace operator, respectively. This notation will be used throughout the paper. Because there is now main flow, we require that $\nabla\Phi$ vanishes as $x^2 + y^2 \to \infty$. Only gradients of Φ are involved in the equations so, in order to make it unique, we also require that Φ vanishes as $x^2 + y^2 \to \infty$. The total energy of the system is given by

$$\mathcal{H} = E_k + E_p \tag{5}$$

where

$$E_k = \frac{1}{2}\rho \int_{-\infty}^{\infty}\int_{-\infty}^{\infty}\int_{-h}^{\zeta}(|\nabla\Phi|^2 + \Phi_z^2)\,dz\,dy\,dx, \quad E_p = \frac{1}{2}\rho\int_{-\infty}^{\infty}\int_{-\infty}^{\infty} g\zeta^2\,dy\,dx$$

Now, the functions $\phi(x,y,t) \equiv \Phi(x,y,\zeta(x,y,t),t)$ and $\zeta(x,y,t)$ fully determine the motion. Given ζ and ϕ, then Φ can be computed by solving for fixed time the boundary value problem given by Equation (1) with boundary conditions (4) and $\Phi(x,y,\zeta) = \phi(x,y)$. Once Φ is known, the time evolution of ϕ and ζ is found from (2) and (3). In view of our previous observation $\mathcal{H}$ is a function of ϕ and ζ and it can be shown [1, 8] that the boundary value problem is equivalent to the Hamiltonian system

$$\rho\frac{\partial\phi}{\partial t} = -H_\zeta, \quad \rho\frac{\partial\zeta}{\partial t} = H_\phi \tag{6}$$

Here H_ζ and H_ϕ are implicitly defined in terms of Gateaux derivatives by

$$(H_\phi, v) = (\frac{d}{ds}\mathcal{H}(\phi + sv, \zeta))_{s=0}, \quad (H_\zeta, v) = (\frac{d}{ds}\mathcal{H}(\phi, \zeta + sv))_{s=0}$$

where v is an arbitrary function and the inner product is given by

$$(v, w) = \int_{-\infty}^{\infty}\int_{-\infty}^{\infty} vw\,dx\,dy \tag{7}$$

In the following section we will present an approximation of the Hamiltonian.

.

APPROXIMATING THE HAMILTONIAN

Before giving an approximation we will write the Hamiltonian in dimensionless form. We assume that all quantities can be written as power series expansions in ε and μ, where ε and μ are defined by $\varepsilon = a/h_0$ and $\mu = (h_0/L)^2$. Here h_0, a and L are characteristic values of the depth, wave amplitude and wavelength, respectively. The parameters ε and μ measure how low and how long a wave is. For fairly low and fairly long waves ε and μ are small with respect to one. The dimensionless variables used are

$$\bar{x} = \frac{x}{L}, \; \bar{y} = \frac{y}{L}, \; \bar{z} = \frac{h_0}{L^2}z, \; \bar{t} = \frac{L}{c_0}t, \bar{h} = \frac{h}{h_0}, \; \bar{\zeta} = \frac{\zeta}{\varepsilon h_0}, \; \bar{\Phi} = \frac{\Phi}{\varepsilon c_0 L}$$

where $c_0 = \sqrt{gh_0}$. Omitting all bars, the Hamiltonian in dimensionless variables reads

$$\mathcal{H} = \frac{1}{2} \int_{-\infty}^{\infty} \int_{-\infty}^{\infty} (\frac{1}{\mu} \int_{-\mu h}^{\varepsilon \mu \zeta} |\nabla \Phi|^2 + \mu \Phi_z^2 \, dz) + \zeta^2 \, dx \, dy \tag{8}$$

The approximation $\mathcal{H}_{app}$ to the Hamiltonian $\mathcal{H}$ should describe the propagation of fairly low and fairly long waves well enough. To this end we require that $\mathcal{H}_{app}$ is a first-order approximation to the Hamiltonian $\mathcal{H}$ given by (8). Hence, $\mathcal{H}_{app}$ should satisfy

$$\mathcal{H}_{app} - \mathcal{H} = O(\varepsilon^2, \varepsilon\mu, \mu^2)$$

To get useful results we assume that ε and μ have the same order of magnitude. Moreover, we require that $\mathcal{H}_{app}$ gives a stable canonical system. Hence, $\mathcal{H}_{app}$ should be a positive definite function of $\nabla \phi$ and ζ.

An approximation $\mathcal{H}_{app}$ to (8) which satisfies these requirements is given by

$$\mathcal{H}_{app} = \frac{1}{2}\rho \int_{-\infty}^{\infty} \int_{-\infty}^{\infty} (\nabla \phi R \nabla \phi + \zeta^2) \, dx \, dy \tag{9}$$

where $R = \frac{1}{\beta}((h + \varepsilon\zeta)^{-1} + \beta\mu A)^{-1} + \frac{\alpha}{\beta}(h + \varepsilon\zeta)$ with $\beta = \alpha + 1$. Here the self-adjoint operator A is defined by

$$Af = -\frac{1}{6}(h\Delta f + \Delta(hf)) + \frac{1}{3}h^{-1}|\nabla h|^2 f$$

with f a sufficient smooth function on which A operates. Recall that an operator is self-adjoint if $(u, Av) = (Au, v)$, where $(\cdot, \cdot)$ signifies the inner product defined in Equation (7). The value of the coefficient α should be nonnegative.

The extra degree of freedom is used to approximate a next term of the linear dispersion relation for a *horizontal* bottom (cf. [3]). This leads to $\alpha = 1/5$. Note that A is a self-adjoint positive operator and from this property, for all positive α, the operator R is positive definite. As a consequence the approximate Hamiltonian is positive definite and the corresponding canonical equations will be stable.

THE CANONCIAL EQUATIONS CORRESPONDING TO $\mathcal{H}_{app}$

Having made an approximation, we can return to the original dimensions and derive the canonical equations. There the Hamiltonian $\mathcal{H}_{app}$ is given by

$$\mathcal{H}_{app} = \frac{1}{2}\rho \int_{-\infty}^{\infty} \int_{-\infty}^{\infty} \nabla\phi R \nabla\phi + g\zeta^2 \, dx \, dy$$

where R is given by

$$R = \frac{1}{\beta}(R_\beta + \alpha(h+\zeta))$$

with R_β

$$R_\beta = [(h+\zeta)^{-1} + \beta A]^{-1}$$

After some algebraic manipulations the canonical equations follow from (6)

$$\phi_t = -\frac{1}{2\beta}|(h+\zeta)^{-1}R_\beta\nabla\phi|^2 - \frac{\alpha}{2\beta}|\nabla\phi|^2 - g\zeta \tag{10}$$

$$\zeta_t = -\text{div}(R\nabla\phi)$$

Note that for the evaluation of the right-hand side, the most time consuming part is the computation of $f = R_\beta\nabla\phi$. It requires in fact the solution of the problem $(h+\zeta)^{-1} + \beta A)f = \nabla\phi$, which is of Helmholtz type. Hence, only one Helmholz-type equation needs to be solved for the right-hand side evaluation.

At this place, it is in order to compare our approximate Hamiltonian to the one proposed by Mooiman based on the work of Broer. That Hamiltonian reads

$$\mathcal{H}_{app} = \frac{1}{2}\rho \int_{-\infty}^{\infty} \int_{-\infty}^{\infty} (h+\zeta)(G\nabla\phi)^2 + g\zeta^2 \, dx \, dy$$

where G is a rational expression like R but now independent of ζ. Of course, $G^*(h+\zeta)G = R + O(\varepsilon\mu, \mu^2)$, where G^* is the adjoint of G. The derivation of the equations is as described in this section. However, instead of R we will find $G^*(h+\zeta)G$ in the continuity equation. The application of this operator requires twice the solution of a Helmholz-type equation.

THE BOUSSINESQ EQUATIONS IN PRIMITIVE VARIABLES

In order to be able to implement the derived equations in Mooiman's model we have to rewrite them in the primitive variables. This can be achieved by differentiating the canonical equation (10) with respect to x and y. We may replace ϕ_x by u because ϕ_x is an $O(\varepsilon\mu)$ approximation to Φ_x at the free surface, likewise for ϕ_y. Further $(Pu)_y$, where

$$P = \check{h}^{-1}R_\beta \tag{11}$$

with $\check{h} = h + \zeta$, may be replaced by $(Pv)_x$, because this is an $O(\mu)$ approximation. In fact this means that P may be replaced by an identity operator, which is allowed because the nonlinear terms in (10) are already $O(\varepsilon)$. Of course this should not be done, because this may destroy the nice stability properties of our approximation. This yields the following equations

$$
\begin{aligned}
\zeta_t &= -\frac{1}{\beta}[(\check{h}Pu + \alpha\check{h}u)_x + (\check{h}Pv + \alpha\check{h}v)_y] \\
u_t &= -\frac{1}{\beta}[(Pu)(Pu)_x + (Pv)(Pu)_y + \alpha[uu_x + vu_y] - \zeta_x \qquad (12) \\
v_t &= -\frac{1}{\beta}[(Pu)(Pv)_x + (Pv)(Pv)_y + \alpha[uv_x + vv_y] - \zeta_y
\end{aligned}
$$

These equations serve as a starting point for our numerical model.

NUMERICAL ASPECTS

In this section we will give attention to the discretization of the Boussinesq model. The most difficult part is the spatial discretization. First, we briefly describe the time discretization. The Boussinesq equations in the previous section can be written as $dw/dt = f(w, t)$ with $w = (\zeta, u, v)^T$. Using the methods of line approach, we obtain a system of ordinary differential equations

$$
\frac{d}{dt}W = F(W, t)
$$

where $W(t)$ is the vector of unknowns at the space grid points. We applied the classical fourth-order Runge Kutta method to this equation This method has a nonempty absolute stability interval along the imaginary axis, which is a prerequisite for numerical integration methods to be applied to wave problems.

Spatial discretization

For the spatial discretization of $f(w, t)$ we use a staggered uniform grid. Let Δx denote the distance between two water-level points. A grid point $(m\Delta x, n\Delta x)$ will be indicated by (m, n) in our formulas.

The boundary consists of closed parts $(u_{m,n} \equiv 0)$ and open parts $(\zeta_{m,n} \equiv f_1(t))$ or $(u_{m,n} \equiv f_2(t))$. For the discretization of the right-hand side of (12) fourth-order discretizations are used, except near the boundaries where first- and second-order discretizations are employed. Higher order approximations are attractive if the solution is smooth, as is the case in these type of problems. For details concerning boundary treatment we refer to [4]. We now focus on the discretization of P. The operator P, see (11) is given by

$$
P = [\check{h} + \beta\check{h}A\check{h}]^{-1}\check{h} = \rho[h(\rho - 1) + h + \beta hAh]^{-1}h
$$

with $\rho = h/\breve{h}$. We want to calculate $q = Pu$. This requires solving the equation

$$[h(\rho - 1) + h + \beta h A h](\frac{q}{\rho}) = hu$$

We must impose boundary conditions on f. We have taken $\frac{\partial q}{\partial n} = 0$ (this is a first-order approximation of the reasonable condition $\frac{\partial u}{\partial n} = 0$).

First we consider the discretization of $h + \beta h A h$. In points that are distant from the boundary we used for second derivatives like u_{xx} the fourth-order discretization

$$u_{xx}(m, n) = \frac{-u_{m+2,n} + 16u_{m+1,n} - 30u_{m,n} + 16u_{m-1,n} - u_{m-2,n}}{12(\Delta x)^2}$$

Near boundary points we used lower order approximations such that the resulting matrix becomes symmetric. This discretization of $h + \beta h A h$ yields a positive-definite matrix M_0 that is nearly an M-matrix. By merely changing the diagonal of this matrix we obtain the discretization of $h(\rho - 1) + h + \beta h A h [= h(\rho - 1) + M_0]$, which will be called M hereafter.

The evaluation of Pu: solving a Helmholtz equation

For calculating $q = Pu$ we must solve the linear equation $M(q/\rho) = hu$. This equation is solved by the conjugate gradient method (CG) using a preconditioning technique. Because of its time dependency M has to be calculated from M_0 (time independent) at the beginning of every right-hand side evaluation. Only M_0 is stored, because M can be calculated by M_0 by merely changing the elements on the main diagonal. We exploit the fact that M is only a small perturbation of M_0; a good preconditioner based on M_0 is also a good preconditioner for M. The preconditioner of M_0 can be built before the time iteration starts.

Two preconditioning techniques are used in conjunction with this algorithm: incomplete Cholesky decomposition and polynomial preconditioning.

The incomplete Cholesky (IC) decomposition used is based on a splitting $M_0 = \tilde{L}\tilde{L}^T - R$ in such a way that all elements of R are in absolute value smaller than a given parameter. For details we refer to [7]. This technique is attractive for sequential computers.

Polynomial preconditioning requires matrix-vector operations only. These operations are very attractive for computations on vector computers. The gain of speed of polynomial preconditioning with respect to incomplete decomposition preconditioning, obtained on vector computers, usually offsets the drawback of a slower convergence. In our case the coefficients of the polynomial are chosen such that $\| 1 - zP(z) \|_2$ is minimized on the interval $[0, 2]$ (see [7]). For solving $M(q/\rho) = hu$ with CG using polynomial preconditioning we used $P_n(\hat{M})$ as a preconditioning for M, where $\hat{M}$ is a symmetrically scaled version of M such that the eigenvalues are in the indicated interval. The scaling is based on M_0.

Experiments

Mooiman's implementation, kindly made available by Delft Hydraulics, has been adapted to the derived Boussinesq equations. Two things have to be checked: the accuracy and the performance. The accuracy of the model is checked in two ways. We compared our results with those found by Mooiman's model. They differed only a few percent. Moreover, Dingemans [2] compared the present method with others and concluded that this new model belongs to the best available.

For the performance a large problem with a spatial grid consisting of 33000 points was available. For a description of this problem we refer to [5]. The new implementation and Mooiman's original implementation have been tested on a 10 Mflop HP720 computer and on the NEC-SX3 vector computer. The results are presented in Table 1. Here old and new refer to Mooiman's model and ours, respectively; ILU/CGS denotes

Table 1: Comparison of computation times

HP720		NEC-SX3	
Implementation	cpu (s)	Implementation	cpu (s)
old with ILU/CGS	80	old with pol.prec (P5/GMRES)	5.2
new with ICCG	32	new with pol.prec (P2/CG)	2.1

the conjugate gradient squared method combined with incomplete LU decomposition as a preconditioner (also here parameter controlled), P5/GMRES means generalised minimal residual method combined with polynomial (of degree 5) preconditioning. In both cases the speed improvement is approximately 2.5. This agrees with our expectations, since the number of systems to be solved is halved and the matrices are symmetric.

Some speed-up may still be possible in both cases by reducing the nine diagonals to five in the matrix used to construct the preconditioner. Another interesting alternative is incomplete decomposition combined with level scheduling which performs well on both scalar and vector computers (for details see [7]).

CONCLUSIONS

In this paper we presented equations that describe the propagation of fairly long and fairly low gravity waves. They were derived in such a way that the number of linear systems to be solved in the numerical model is reduced by a factor two with respect to that of the model of Mooiman. Moreover, in the new model the occurring matrices are symmetric. Numerical experiments showed that the speed-up is proportional to the reduction of the number of linear systems to be solved while the results are at least as good as those obtained by Mooiman's and other models.

Acknowledgement

This work was sponsored by the Stichting Nationale Computerfaciliteiten (National Computing Facilities Foundation, NCF) for the use of supercomputer facilities, with financial support from the Nederlandse Organisatie voor Wetenschappelijk Onderzoek (Netherlands Organization for Scientific Research, NWO).

References

[1] L.J.F. Broer. On the Hamiltonian theory of surface waves. *Appl. Sci. Res.*, 30:430–446, 1974.

[2] M.W. Dingemans. Comparison of computations with Boussinesq-like models and laboratory measurements. MAST-G8M note, in preparation.

[3] E.W.C. van Groesen L.J.F. Broer and J.M.W. Timmers. Stable model equations for long water waves. *Appl. Sci. Res.*, 32:619–636, 1976.

[4] J. Mooiman. Boussinesq equations based on a positive definite Hamiltonian. Technical report, Delft Hydraulics, February 1991. Report Z294.

[5] J. Mooiman. Comparison between measurements and a Boussinesq model for wave deformation by a shoal. Technical report, Delft Hydraulics, September 1991. Report Z294.

[6] J. Mooiman and G.K. Verboom. A new Boussinesq model based on a positive definite Hamiltonian. In T.F. Russell, R.E. Ewing, C.A. Brebbia, W.G. Gray, and G.F. Pinder, editors, *Computational methods in water resources IX, Mathematical methods in water resources*, volume 2, pages 513–527, Southhampton, 1992. Computational mechanics publications. Proceedings of the Ninth International Conference on Computational Methods in Water Resources, Denver.

[7] A. van der Ploeg and F.W. Wubs. Vectorizable preconditioning techniques for solving the boussinesq equations. In Ch. Hirsch, J. Periaux, and W. Kordulla, editors, *Computational fluid dynamics '92*, pages 481–488. Elsevier, 1992.

[8] W.A. van der Veen and F.W. Wubs. A Hamiltonian approach to fairly low and fairly long gravity waves. *submitted for publication*, 1993.

12. FLOW AND TRANSPORT IN RIVERS

NUMERICAL SIMULATION OF FLOW IN RIVER NETWORKS WITH COMPLEX TOPOLOGY

A. A. ALDAMA and J. APARICIO
Mexican Institute of Water Technology
Paseo Cuauhnáhuac 8532, Jiutepec 62550, Mor.
México

Lower river basins are often characterized by channel networks with complex topology and extremely small valley and channel slopes. Flooding in rivers without levees may cause the appeareance of lagoons whose surface changes rapidly with time when the slope in the floodplain is small, thus modifying the length of the interconnected channnels in the river network. A formulation of a model that accounts for the interaction of lagoons and channels with (respectively) time-varying surfaces and lengths is presented in this paper. Furthermore, flood simulation in topologically-complex networks via implicit schemes results in algebraic systems with nonbanded (but sparse) matrices. A very efficient direct method based on the use of numerical Green's functions for the solution of such systems is also presented in this paper. A numerical application of the model is included.

INTRODUCTION

Rivers flowing in lower flat plains usually exhibit complex patterns consisting of interconnecting channel reaches and shallow lagoons which change their surface dimensions rapidly and appreciably. Existing models fail to properly account for the interaction between lagoons with expanding of contracting surfaces and channels with time-dependent lengths. On the other hand, the explicit treatment of interconection nodes in a river network leads to inconvenient stability constraints, whereas their implicit treatment leads to the need of solving large systems of equations with nonbanded matrices. These issues are addressed in this paper.

FUNDAMENTAL EQUATIONS

The basic equations which describe transient, one-dimensional, free surface flow are the Saint-Venant equations (Saint Venant, 1871), i.e., the continuity equation

$$B \frac{\partial H}{\partial t} + \frac{\partial UA}{\partial x} = q \tag{1}$$

A. Peters et al. (eds.), Computational Methods in Water Resources X, 1131–1138.
© 1994 *Kluwer Academic Publishers. Printed in the Netherlands.*

and the momentum equation

$$\frac{\partial U}{\partial t} + U\frac{\partial U}{\partial x} + g\frac{\partial H}{\partial x} + gn^2\frac{U|U|}{R^{4/3}} = 0 \tag{2}$$

where H represents free surface elevation; U, velocity; A, hydraulic area; q, lateral inflow per unit length; R, hydraulic radius; n, Manning roughness coeficient; g, acceleration of gravity; B, free surface width; t, time; and x, distance.

TRANSFORMED EQUATIONS

A river reach conveying transient flow, which is located between expanding or contracting lagoons changes its length with time. The description of the fluid motion in such a reach requires the solution of Eqs. (1) and (2) in a variable domain. The use of fixed grids in the solution of equations of motion in variable domains usually leads to inaccuracies, which in some cases may be quite severe. Following Austria and Aldama (1990), a coordinate transformation strategy is employed here. As shown in Fig. 1, let $x_r = x_r(t)$ and $x_f = x_f(t)$ respectively represent the position of the rear and front of a channel reach whose length changes as a result of expansions or contractions experienced by the free surface of lagoons located at the corresponding boundaries. Thus, let us introduce the following coordinate transformation

$$\xi = \frac{x - x_r(t)}{x_f(t) - x_r(t)} \tag{3}$$

$$\tau = t \tag{4}$$

Evidently, Eq. (3) transforms the time-varying domain $[x_r(t), x_f(t)]$ into the invariant domain $[0,1]$. Employing Eqs. (1) and (2) and the chain rule of differentiation, we get that for any function $f(x,t)$:

$$\frac{\partial f}{\partial x} = \frac{1}{x_f - x_r} \frac{\partial f}{\partial \xi} \tag{5}$$

and

$$\frac{\partial f}{\partial t} = -\frac{1}{x_f - x_r}\left[\xi\frac{dx_f}{dt} + (\xi + 1)\frac{dx_r}{dt}\right]\frac{\partial f}{\partial \xi} + \frac{\partial f}{\partial \tau} \tag{6}$$

Substituting Eqs. (5) and (6) into Eqs. (1) and (2), the following transformed continuity and momentum equations are obtained:

$$B(x_f - x_r)\frac{\partial H}{\partial \tau} - B\xi\frac{\partial H}{\partial \xi}\frac{dx_f}{dt} - B(\xi+1)\frac{\partial H}{\partial \xi}\frac{dx_r}{dt} + \frac{\partial(UA)}{\partial \xi} = q(x_f - x_r) \tag{7}$$

$$(x_f - x_r)\frac{\partial U}{\partial \tau} - \xi\frac{\partial U}{\partial \xi}\frac{dx_f}{dt} - (\xi+1)\frac{\partial U}{\partial \xi}\frac{dx_r}{dt} + U\frac{\partial U}{\partial \xi} + g\frac{\partial H}{\partial \xi} + g(x_f - x_r)n\frac{2U|U|}{R^{4/3}} = 0 \quad (8)$$

NUMERICAL SOLUTION

The advantage of numerically solving Eqs. (7) and (8), rather than the original Eqs. (1) and (2) is that an invariant spatial discretization may be employed in the domain (ξ,τ) and the expansions and contractions of the channel are explicitly accounted for in the equations of motion. For a *single* channel, Eqs. (7) and (8) will be discretized employing the following implicit, fractional step scheme (cf. Yanenko, 1971):

$$B_j^k\frac{H_j^{k+1} - H_j^k}{\Delta\tau} - B_j^k\,\xi_j\,\frac{H_{j+1}^k - H_{j-1}^k}{2\Delta\xi}\frac{x_f^k - x_f^{k-1}}{\Delta t} - B_j^k\left[\xi_j + 1\right]\cdot$$

$$\cdot\frac{H_{j+1}^k - H_{j-1}^k}{2\Delta\xi}\frac{x_r^k - x_r^{k-1}}{\Delta t} + \frac{U_{j+1/2}^{k+1}A_{j+1/2}^k - U_{j-1/2}^{k+1}A_{j-1/2}^k}{\Delta\xi} = 0 \quad (9)$$

and

$$\frac{U_{j+1/2}^{k+1} - U_{j+1/2}^k}{\Delta\tau} - \xi_{j+1/2}\frac{U_{j+3/2}^k - U_{j-1/2}^k}{2\Delta\xi}\frac{x_f^k - x_f^{k-1}}{\Delta t} -$$

$$- \left[\xi_{j+1/2} + 1\right]\frac{U_{j+3/2}^k - U_{j-1/2}^k}{2\Delta\xi}\frac{x_r^k - x_r^{k-1}}{\Delta t} + U_{j+1/2}^{k+1}\frac{U_{j+3/2}^k - U_{j-1/2}^k}{2\Delta\xi} +$$

$$+ g\frac{H_{j+1}^{k+1} - H_j^{k+1}}{\Delta\xi} + gn^2\frac{|U_{j+1/2}^k|}{(R_{j+1/2}^k)^{4/3}}U_{j+1/2}^{k+1} = 0 \quad (10)$$

Subscripts denote spatial position, $\xi_j = j\Delta\xi$, while superscripts denote time level, $\tau^k = k\Delta t$.

From Eq. (10), $U_{j+1/2}^{k+1}$ may be expressed as

$$U_{j+1/2}^{k+1} = \frac{g\theta(H_{j-1}^{k+1} - H_{j+1}^{k+1})}{1 + C_j + F_j} + M_{j+1/2}^k \quad (11)$$

where

$$\theta = \frac{\Delta t}{\Delta x}; \quad C_j = \theta(U_{j+3/2}^k - U_{j-1/2}^k)/2; \quad F_j = gn^2\Delta t\frac{|U_{j+1/2}^k|}{(R_{j+1/2}^k)^{4/3}} \quad (12)$$

and $M_{j+1/2}^k$ depends on values of the discrete flow variables at time level k.

Sustituting Eq. (11) in Eq. (9), a system of algebraic, linear equations is obtained in terms of the vector of unknown free surface elevations, $\{H\}^{k+1}$:

$$[A]^k \{H\}^{k+1} = \{C\}^k \tag{13}$$

where $[A]^k$ and $\{C\}^k$ respectively are a matrix and a vector which depend on values at time level k. A typical equation in system (13) is of the form

$$\alpha_j^k H_{j-1}^{k+1} + \beta_j^k H_j^{k+1} + \alpha_{j+1}^k H_{j+1}^{k+1} = \gamma_j^k \tag{14}$$

where coefficients α, β and γ are known.

For a single channel, Eq. (14) and the proper boundary conditions constitute a tridiagonal system of equations. This kind of system may be solved very efficiently by employing a forward elimination–backward substitution algorithm (Dahlquist and Björck, 1974). However, consider a complex channel network such as that shown in Fig. 2. In such a case, nonzero elements appear outside the three main diagonals of the coefficient matrix (see Fig. 3), a fact that makes the solution procedure considerably less efficient. Thus, the following solution algorithm is proposed.

Let us consider a channel R whose first (rear) and last (front) nodes are shared by other channels. In other words, those nodes are *interconnection nodes*. For such a channel, system (13) may be written as

$$[A_R]^k \{H_R\}^{k+1} + B_{R,r}^k \{H_{R,r}\}^{k+1} + B_{R,f}^k \{H_{R,f}\}^{k+1} = \{C_R\}^k \tag{15}$$

where $[A_R]^k$ is a tridiagonal coefficients matrix $\{H_R\}^{k+1}$ is the vector of unknown water surface elevations *inside* the channel reach, $B_{R,r}^k$ and $B_{R,f}^k$ are scalars, $\{C_R\}^k$ is an independent terms vector, and $\{H_{R,r}\}^{k+1} = \{H_{R,r}^{k+1}, 0, 0, \ldots, 0\}^T$, $\{H_{R,f}\}^{k+1} = \{0, 0, \ldots, 0, H_{R,f}^{k+1}\}^T$, with $H_{R,r}^{k+1}$ and $H_{R,f}^{k+1}$ respectively representing the water surface elevations at the rear and front ends of channel R.

Now let

$$\{H_R\}^{k+1} = \{H_{R,h}\}^{k+1} + \{H_{R,i}\}^{k+1} \tag{16}$$

where $\{H_{R,h}\}^{k+1}$ will be called the *homogeneous solution*, defined by

$$[A_R]^k \{H_{R,h}\}^{k+1} = \{C_R\}^k \tag{17}$$

and $\{H_{R,i}\}^{k+1}$ will be termed the *inhomogeneous solution*. In order to compute $\{H_{R,i}\}^{k+1}$, let us define a *rear numerical Green's function*, $\{G_{R,r}\}^{k+1}$, by

$$[A_R]^k \{G_{R,r}\}^{k+1} = -B_{R,r}^k \{1, 0, 0, \ldots, 0\}^T \tag{18}$$

and a *front numerical Green's function*, $\{G_{R,f}\}^{k+1}$, by

$$[A_R]^k \{G_{R,f}\}^{k+1} = -B_{R,f}^k \{0, 0, \ldots, 0, 1\}^T \tag{19}$$

In other words, $\{G_{R,r}\}^{k+1}$ and $\{G_{R,f}\}^{k+1}$ respectively represent the response of channel R to unit variations in the water surface elevations at its rear and front ends.

From Eqs. (15)-(19) it is evident that

$$\{H_{R,i}\}^{k+1} = H_{R,r}^{k+1} \{G_{R,r}\}^{k+1} + H_{R,f}^{k+1} \{G_{R,f}\}^{k+1} \tag{20}$$

for any channel R.

Eqs. (17), (18) and (19) represent tridiagonal systems and, therefore, the homogeneous solution and the rear and front numerical Green's functions may be very efficiently computed for each channel. Furthermore, equations of the type (16) and (20) for each channel that leaves from or ends at an interconnection node, along with the mass conservation statement for that node, leads to a (relatively) small system of equations in terms of the water surface elevations at the interconnection nodes, which may also be solved very efficiently.

Employing the above procedure, three (small) tridiagonal systems for each channel, and a (small) sparse system for the water surface elevations at the interconnection nodes are solved, instead of solving a large, non-banded system.

NUMERICAL APPLICATION

A numerical application of the above presented techniques is presented in this section. As shown in Fig. 4, a channel network with two common nodes, where lagoons are located, is considered. Channels are rectangular in shape with $B = 20$ m, $S_0 = 0.001$ and $n = 0.025$. As boundary conditions, the hydrograph shown in Figs. 5 and 6 corresponding to section 1, is introduced through each of the upstream ends of channels 0, 1 and 3 and a constant free-surface level $H = 5$ m is mantained at the downstream end of channel 4. Two runs were made to show the effect of the lagoons: In the first one, the lagoons were assumed to have negligible free surface area; while in the second, the lagoons where assumed to possess a truncated cone shape and an initial

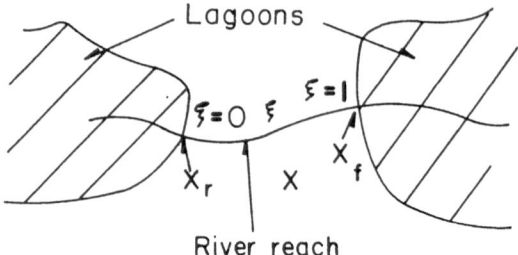

Fig. 1. Time dependent solution domain

Fig. 2. Complex river network

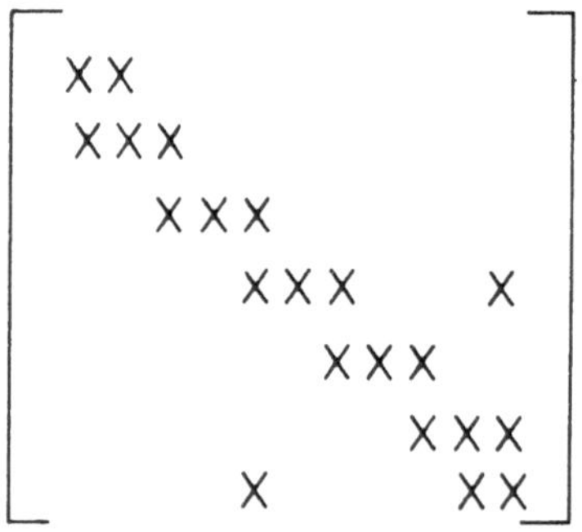

$$\begin{bmatrix} X\,X & & & & & & \\ X\,X\,X & & & & & & \\ & X\,X\,X & & & & & \\ & & X\,X\,X & & & X & \\ & & & X\,X\,X & & & \\ & & & & X\,X\,X & \\ & & X & & & X\,X \end{bmatrix}$$

Fig. 3. Nontridiagonal coefficient
matrix

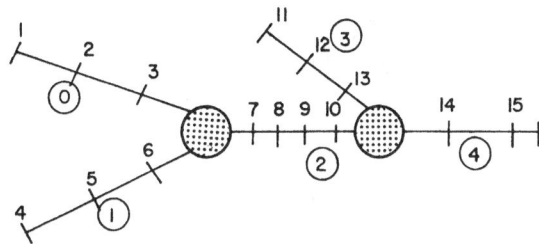

Fig. 4. Channel network for numerical
application

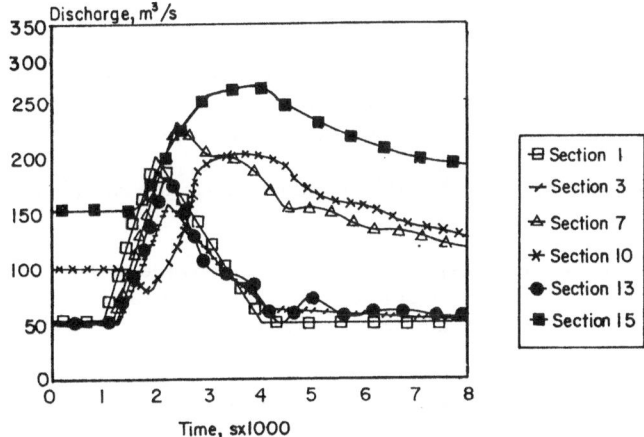

Fig. 5. Flood routing without lagoons

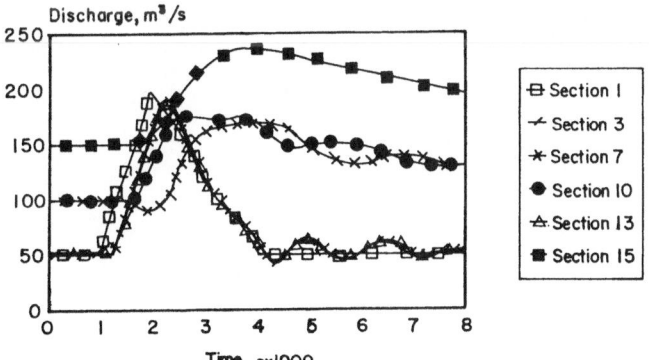

Fig. 6. Flood routing with lagoons

water surface elevation of 5 m. Results in the form of hydrographs at several cross sections are shown in Figs. 5 and 6. Comparing them, the damping effect and the interaction among channels and lagoons can be observed.

CONCLUSIONS

A numerical model for transient flow simulation in river networks with complex topology and floodplain lagoons has been developed. A proper coordinate transformation allows simultaneous, implicit calculation of river channel-lagoons interaction and the use of numerical Green's function makes it possible to efficiently solve resulting algebraic system.

REFERENCES

Austria, P.M. and A. Aldama (1990). "Adaptive mesh scheme for free surface flows with moving boundaries", *Proc.*, VIII Int. Conf. on Computational Methods in Water Resources, Venice, Italy.

Dahlquist, G. and Å. Björck (1974). *Numerical Methods*, Prentice-Hall, New Jersey.

de Saint Venant, Barré (1871). "Théorie du movement non-permanent des eaux avec application aux crues des rivierès et a l'introduction des marées dans leur lit", *Acad. Sci. (Paris) Comptes rendus* V. 73, pp 147-154, 237-240.

Yanenko, N.N. (1971). *The method of fractional steps*, Springer-Verlag, New York.

SURFACE TRANSPORT IN NATURAL RIVER BENDS

G. BJEDOV and J. B. PERRY
Department of Freshman Engineering
Purdue University
West Lafayette, IN 47907 U.S.A.

A two–dimensional discrete element model for uniform circular particles is applied to the problem of surface ice flow in a natural river bend. The trajectories and velocities of individual particles are calculated by integrating Newton's equation of motion. The forces included in the simulation include gravity forces, hydrodynamic drag forces and contact forces between particles which are modeled using a linear viscoelastic contact force model. The particles are allowed to overlap slightly, which is interpreted as a local deformation.

The velocity field in a river bend is obtained using the quasi two–dimensional approach. The growth and decay of the transverse current are obtained by spatial integration of equations of motion along the channel centerline. Special consideration is given to the development of secondary velocity currents.

The importance of several different effects is investigated: upstream concentration, boundary roughness, and existence of the transverse current. The results are intended for use in continuum ice simulation models as local predictors of ice jam occurrence.

INTRODUCTION

The surface transport and accumulation of river ice are important elements of cold regions engineering. In areas of higher latitude, both moving and static ice jams can develop during winter periods. These jams can often cause serious damage. During the 1991/1992 winter season, the capital of Vermont, Montpelier, was completely covered with water for several days due to flooding caused by an ice–jam.

Different mathematical models (e.g. Shen et al. [7], Lal and Shen [6]) have been developed to simulate river conditions during the winter season. These models simulate ice production, transport, and accumulation and growth of ice covers. This approach is adequate if all of the locations of ice jamming are known in advance.

The purpose of the current model is to predict when and where ice jams will occur in a particular section of a river. The model is meant to be used in conjunction with static ice simulation models, to provide the information about the locations of ice jams.

A. Peters et al. (eds.), Computational Methods in Water Resources X, 1139–1146.
© 1994 *Kluwer Academic Publishers. Printed in the Netherlands.*

The model is tested on Muddy Creek, a tributary of the upper Green River in western Wyoming. Muddy Creek was chosen as a test site because extensive data about channel morphology and both longitudinal and transversal velocity profiles is available (Dietrich and Smith [4]). The study reach is 40 m long with channel width ranging from 5.1 m to 6.0 m. At a discharge of about 1.1 m³ /s, the mean depth is 0.5 m. The minimum radius of curvature is 8 m (Figure 1.)

The longitudinal velocity field in the bend is obtained using a stream–tube approach. After the longitudinal velocities are determined, transverse current is obtained by spatial integration of equations of motion along the channel centerline (Chang [3]).

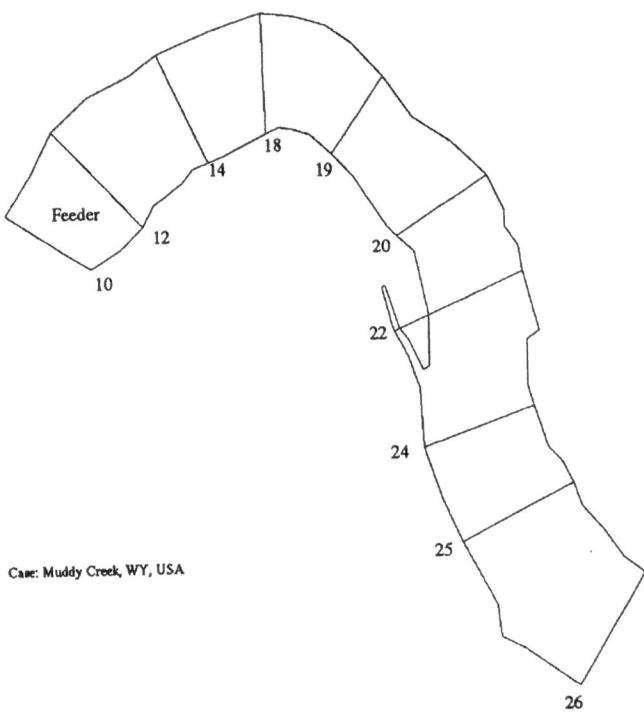

Figure 1. Study reach

DISCRETE ELEMENT METHOD

Ice particles are idealized as uniform circular disks of diameter D, thickness τ and density ρ_i. Newton's equations of motion for particle p are:

$$m\frac{d^2\vec{x}_p}{dt^2} = \tau \sum^{C_p} \vec{F}_{pq} + \vec{G}_p + \vec{H}_p \tag{1}$$

$$I\frac{d\omega_p}{dt} = \tau \sum^{C_p} M_{pq} \tag{2}$$

where $m = \rho_i \tau D^2 \pi / 4$ is the particle mass, $I = \rho_i \tau D^4 \pi / 8$ is the particle moment of inertia, $\vec{x}_p$ is the position vector of the particle centroid, $\vec{v}_p = d\vec{x}_p/dt$ is the velocity vector, ω_p is the angular velocity, C_p is the number of contacts of particle p, $\vec{F}_{pq}$ is the contact force per unit thickness between particles p and q, M_{pq} is the moment per unit thickness between particles p and q, $\vec{G}_p$ is the gravity force acting on particle p, and $\vec{H}_p$ is the hydrodynamic force acting on particle p.

The gravitational force acting on a particle is given by

$$\vec{G}_p = -mg\nabla h(\vec{x}_p) \tag{3}$$

where g is gravitational acceleration, and $h(\vec{x})$ is the water surface elevation field.

The hydrodynamic forces acting on a particle in general include the drag force, the Saffman lift force and the added mass (Basset) force. In the present model, only the drag force is included in the computation. Under turbulent flow conditions, the drag force is proportional to the square of the instantaneous relative velocity between the particle and the fluid. Hence, the hydrodynamic force on particle p is expressed as:

$$\vec{H}_p = \rho_w(C_1 D\tau + C_2 D^2 \pi / 4)(\vec{v}_w(\vec{x}_p) - \vec{v}_p)|\vec{v}_w(\vec{x}_p) - \vec{v}_p| \tag{4}$$

where ρ_w is the water density, C_1 is the drag coefficient on particle perimeter, C_2 is the drag coefficient on the underside of the particle, and $\vec{v}_w(\vec{x})$ is the velocity field at the water surface.

Contact Force Model

The contact force model describes the force-displacement relationship at the point of contact. In the present model a linear viscoelastic contact force model is used.

Particles p and q are in contact if $|\vec{x}_p - \vec{x}_q| < D$. The contact force per unit thickness is expressed as

$$\vec{F}_{pq} = -[(F_n)_{pq}\vec{k}_{pq} + (F_s)_{pq}\vec{t}_{pq}] \tag{5}$$

where $\vec{k}_{pq} = (\vec{x}_p - \vec{x}_q)/|\vec{x}_p - \vec{x}_q|$ is the unit vector pointing from the center of particle p to the center of particle q and $\vec{t}_{pq}$ is perpendicular to $\vec{k}_{pq}$. The contact force per unit thickness in normal direction is modeled as viscoelastic:

$$(F_n)_{pq}^t = (F_n)_{pq}^{t-\Delta t} + K_n \dot{n}_{pq}\Delta t + C_n \dot{n}_{pq} \tag{6}$$

where the superscript indicates current time, Δt is the time step, $\dot{n}_{pq} = (\vec{v}_p - \vec{v}_q) \cdot \vec{k}_{pq}$ is the relative displacement rate in the normal direction, K_n is the effective normal spring stiffness and $C_n = 2\zeta_n\sqrt{K_n m}$ is the normal damping coefficient. The dimensionless damping coefficient ζ_n can be explicitly related to the coefficient of restitution e:

$$\zeta_n = \frac{-\ln e}{\sqrt{\pi^2 + \ln^2 e}} \tag{7}$$

The contact force per unit thickness in the tangential direction is modeled as elastic below the friction limit (given by Mohr-Coulomb law), and frictional at the friction limit:

$$(F_s)^t_{pq} = \min[(F_s)^{t-\Delta t}_{pq} + K_s \dot{q}_{pq} \Delta t \quad ; \quad \mu(F_n)^t_{pq}] \tag{8}$$

where $\dot{q}_{pq} = (\vec{v}_p - \vec{v}_q) \cdot \vec{t}_{pq} + (D/2)(\omega_p + \omega_q)$ is the relative displacement rate in the tangential direction, K_s is the effective tangential spring stiffness and μ is the friction coefficient. The contact moment per unit thickness between particles p and q is $M_{pq} = (D/2)(F_s)_{pq}$.

Boundary Conditions

The present model is designed to simulate the following situation. At time $t = 0$ an ice run coming from the upstream enters the simulated portion of the channel. The objective is to simulate the subsequent flow of ice in the channel and determine whether the steady state flow will develop, and if so, what will be the ice flow rate.

The upstream boundary condition is simulated using the "feeding" scheme developed by Babić, et al. [1]. The part of the channel between the cross-sections 10 and 12 is designated as the "feeder". The number of particles in the feeder N_o is maintained constant throughout the simulation. Particles in the feeder are initially arranged in a regular square packing with a small random deviations. At $t = 0$, particles start flowing from the feeder into the channel. Whenever a particle leaves the feeder and enters the channel, another particle is added into the feeder, thus maintaining the total number of particles N_o in the feeder constant at all times. The particles that leave the feeder are not allowed to return, i.e. there is an artificial one-way "wall" between the feeder and the simulation reach.

The downstream boundary condition is of the "free-end" type. Particles which cross the downstream end of the channel are eliminated from the simulation.

The side boundary conditions (channel banks) are modeled as rough solid walls. The channel banks are "roughened" by placing semi-disks of diameter d along the banks at the specified spacing s. The boundary particles have the same mechanical properties as the flow particles. The interactions between flow particles and wall particles are treated in the same manner as interactions between interior particles.

RESULTS

This study is focused on the relationship between upstream ice concentration and the ice flow rate. The relationship was investigated by varying the concentrations in the feeder from $C_f \approx 0.1$ to $C_f \approx 0.5$. Five different boundary roughnesses were simulated, ranging from $s = 0.25$ m to $s = 0.45$ m.

The time step was equal to $\Delta t = 0.00133$ s. All simulations were carried out for 300,000 time steps, i.e., until $t = 399$ s. The first particles reached the downstream end of the channel after about 200 s. At the end of simulations, the resulting ice flow condition approaches steady state, although the outflow is generally less than the inflow.

During the course of each simulation, the particles crossing the upstream and downstream sections of the channel are counted. The ice flow rates in and out of the channel are obtained as

$$Q_{in} = \frac{dN_{in}}{dt} \qquad ; \qquad Q_{out} = \frac{dN_{out}}{dt} \qquad (9)$$

where N_{in} is the number of particles that entered the simulation reach and N_{out} is the number of particles that exited the simulation reach. The number of particles residing in the channel is simply $N(t) = N_{in}(t) - N_{out}(t)$.

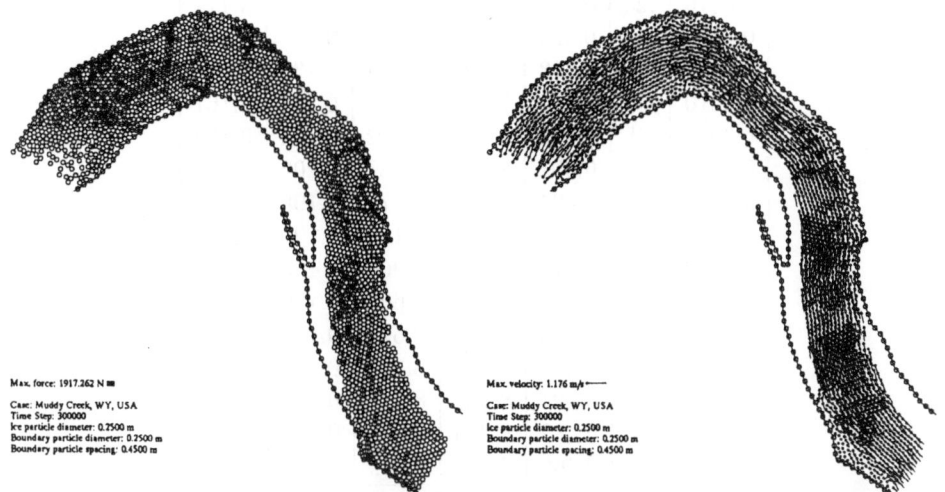

Figure 2. Snapshot of ice configuration, velocities and interaction forces

The average (steady state) values of Q_{in}, Q_{out} and N are obtained by averaging over the last 50,000 steps of the simulation. The average concentration in the channel is obtained as

$$C = N_{av}\frac{D^2\pi/4}{A} \qquad (10)$$

where A is the channel area, N_{av} is the average number of particles in the channel and D is the diameter of the ice particle. The average ice velocities at entrance and exit of the simulation reach are calculated from

$$U_{in} = \frac{Q_{in}D^2\pi/4}{C_f W_{in}} \qquad ; \qquad U_{out} = \frac{Q_{out}D^2\pi/4}{C W_{out}} \qquad (11)$$

where W_{in} is the width of the channel at the entrance of the simulation reach and W_{out} is the width of the channel at the exit of the simulation reach. The results of the simulations are summarized in Table 1.

C_f	s	C	Q_{in}	Q_{out}	U_{in}	U_{out}
0.1003	.25	0.179	3.429	3.353	0.308	0.177
0.1003	.30	0.182	3.338	3.353	0.300	0.174
0.1003	.35	0.180	3.263	3.293	0.293	0.172
0.1003	.40	0.177	3.203	3.353	0.288	0.178
0.1003	.45	0.181	2.947	2.962	0.265	0.155
0.2007	.25	0.321	6.075	5.910	0.273	0.174
0.2007	.30	0.314	5.955	5.865	0.268	0.176
0.2007	.35	0.314	6.015	5.789	0.270	0.174
0.2007	.40	0.317	5.925	5.684	0.266	0.169
0.2007	.45	0.317	5.835	5.444	0.262	0.162
0.3010	.25	0.424	7.835	7.729	0.235	0.172
0.3010	.30	0.420	7.925	7.489	0.237	0.168
0.3010	.35	0.418	7.699	7.594	0.231	0.171
0.3010	.40	0.412	7.534	7.278	0.226	0.167
0.3010	.45	0.416	7.504	6.992	0.225	0.159
0.4014	.25	0.501	8.992	8.481	0.202	0.160
0.4014	.30	0.494	8.827	8.376	0.198	0.160
0.4014	.35	0.489	8.722	8.195	0.196	0.158
0.4014	.40	0.477	8.301	8.060	0.186	0.160
0.4014	.45	0.462	7.955	7.609	0.179	0.156
0.4986	.25	0.506	9.083	8.647	0.164	0.161
0.4986	.30	0.498	8.677	8.331	0.157	0.158
0.4986	.35	0.494	8.662	8.135	0.157	0.155
0.4986	.40	0.482	8.286	8.226	0.150	0.161
0.4986	.45	0.469	7.955	7.639	0.144	0.154

Table 1. Results of DEM simulations.

The influence of secondary current is shown in Figure 3. The results are presented for feeder concentration of $C_f \approx 0.1$ and $C_f \approx 0.5$. Similar results were obtained for other concentrations. The influence of the secondary current increases with ice concentration and boundary roughness. Without the secondary current, the influence of the boundary roughness is minimal, and, as can be seen on in Figure 3, the number of particles in and out of channel remains relatively constant for different boundary particle spacing. The secondary current, however, increases the influence of boundary roughness, resulting in larger differences between the number of particles entering and exiting the channel.

Figure 4 shows a plot of the ice flow rate versus the average concentration. It can be seen that Q increases with C in a nonlinear fashion. For low concentrations $Q_{in} \approx Q_{out}$, but for higher concentrations the actual ice outflow rate begins to deviate from the inflow rate. This is easily understood in terms of force balance: the driving force is the fluid drag and the retarding force is the boundary shear. For low concentrations, the boundary shear force is small compared to the fluid drag, and the ice flow rate is proportional to the concentration. For higher concentrations, the boundary shear increases and the ice flow rate decreases.

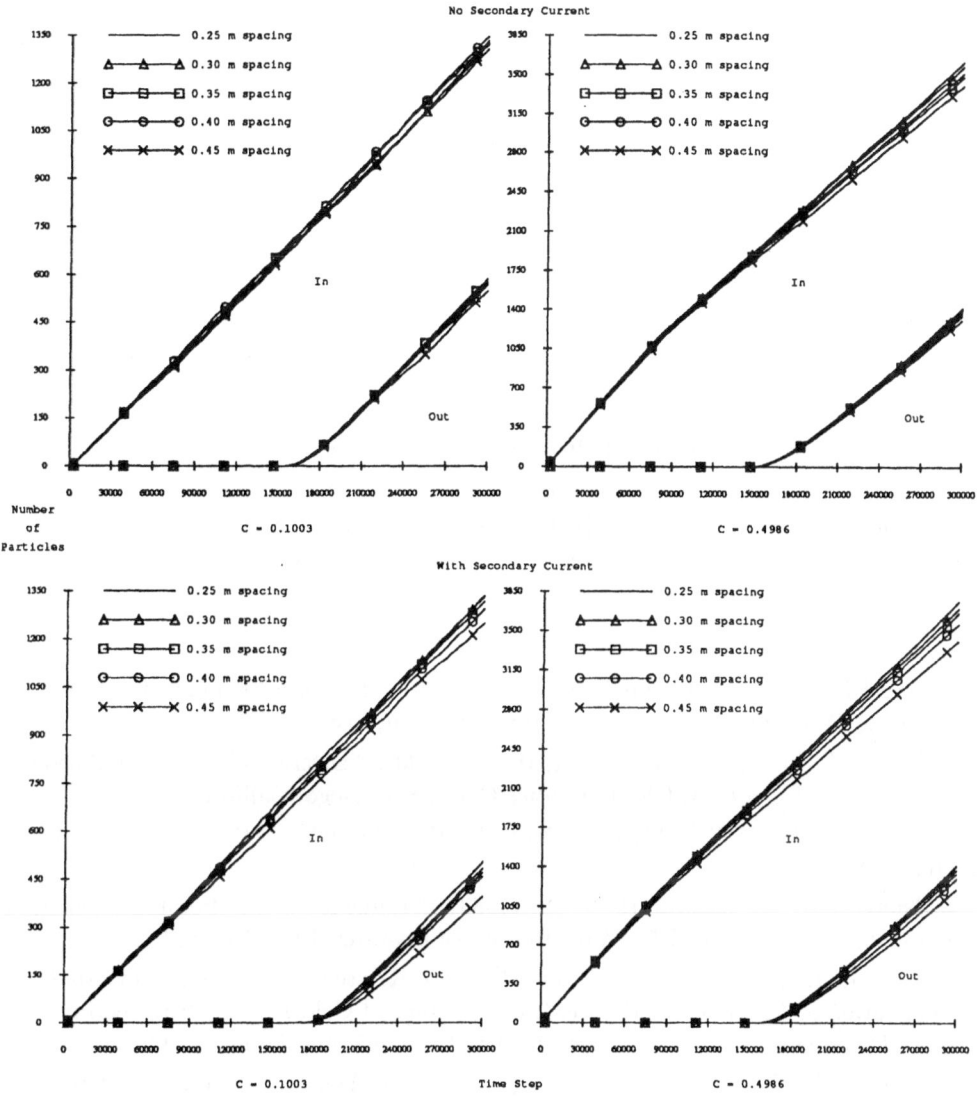

Figure 3. Ice particle flux

The average velocities of ice particles entering and exiting the study reach are also shown in Figure 4. While velocities of the ice particles exiting the reach remain relatively constant and independent of the concentration, the average velocities of the particles leaving the feeder show strong nonlinear dependence on the concentration.

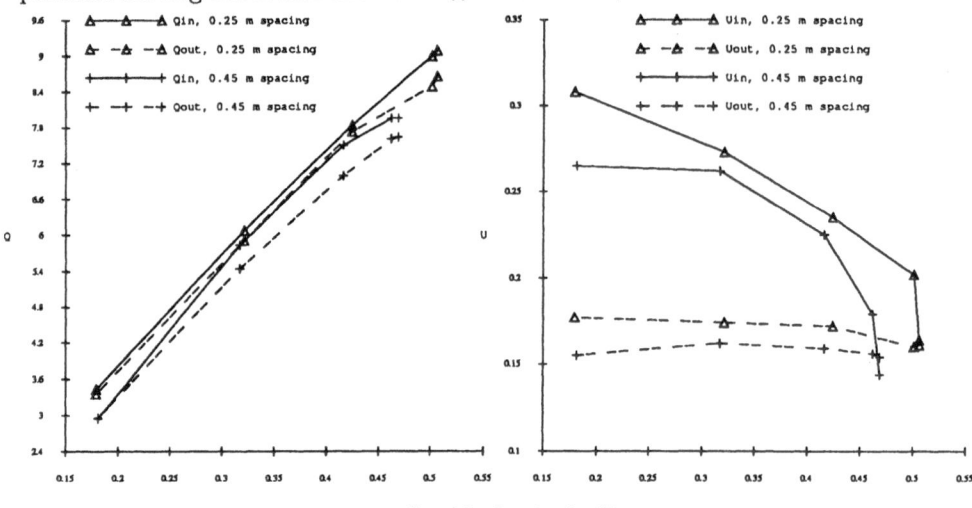

Figure 4. Ice discharges and velocities

Future work will concentrate on the importance of the ice particle size relative to the channel width. Babić et al. [1] have shown that jamming in straight channels is depends on the ratio of the particle size to the channel width.

REFERENCES

1. Babić M., Shen, H. T. and Bjedov, G., (1990). "Discrete Element Simulations of River Ice Transport", Proc. IAHR Ice Symposium, Espoo, Finland

2. Babić M., Shen, H. H. and Shen, H. T., (1990). "Discrete element simulation of river ice jams", Proc. ASCE Hydraulics Conf., San Diego, California.

3. Chang, H. H., (1984). "Regular Meander Path Model", J. Hydraul. Eng. ASCE 110, 1398–1411.

4. Dietrich, W. E. and Smith, J. D., (1983). "Influence of the Point Bar on Flow Through Curved Channels", Water Resources Research 19, 1173–1192.

5. Gerard, R. and Calkins D. J., (1984). "Ice-Related Flood Frequency Analysis: Application of Analytical Estimates", Proceedings of Cold Regions Engineering Specialty Conference, Canadian Society for Civil Engineering, Montreal, Canada.

6. Lal, A. M. W. and Shen, H. T., (1991). "Mathematical Model for River Ice Processes", J. Hydraul. Eng. ASCE 117, 851–867.

7. Shen, H. T., Bjedov, G., Daly, S. F. and Lal, A. M. W., (1991). "Numerical Model for Forecasting Ice Conditions on the Ohio River", U.S. Army Corps of Engineers Cold Regions Research and Engineering Laboratory Report 91–16.

STOCHASTIC DIFFERENTIAL DYNAMIC PROGRAMMING FOR RESERVOIR MANAGEMENT

F. A. EL-AWAR[1], D. G. FONTANE[2], and J. W. LABADIE[2]

1: American Univ. of Beirut, Faculty of Agr. and Food Sci.
850 Third Ave, New York, NY 10022, USA.

2: Colorado State Univ., Civil Engineering Department
Fort Collins, CO 80523, USA.

Dynamic programming (DP) is an optimization technique especially suited for reservoir operation because of its sequential decision abilities. However, one problem is always present whenever DP is used. This problem is the "curse of dimensionality" or the excessive memory and computational time requirements to solve multi-dimensional problems. Due to the multi-dimensional nature of most reservoir operation optimization problems, this DP drawback holds a special significance in this field. One of the most promising techniques to solve the dimensionality problem is differential dynamic programming (DDP).

In this paper, the DDP technique is applied in a stochastic environment. A general stochastic DDP algorithm to maximize hydropower production from multi-reservoir systems is developed for that purpose. The algorithm is developed by extending and improving earlier stochastic DDP approaches. It is tested on a hypothetical four-reservoir hydropower system, and its capabilities as an efficient approach to solve multi-dimensional stochastic problems are shown.

INTRODUCTION

Dynamic programming (DP) is especially suited to optimize sequential decision problems. This characteristic, coupled with a wide variety of its applications, make DP the most powerful technique for reservoir operation optimization. However, due to the highly dimensional nature of most reservoir operation optimization problems, the "curse of dimensionality" or the excessive memory and computational time requirements to solve multi-dimensional problems is frequently encountered in this field.

One of the most promising DP techniques to solve the dimensionality problem is differential dynamic programming (DDP). DDP does not require any discretization of the state and control variables. In abandoning the state and control discretization the memory and computational time requirements grow linearly instead of exponentially with the dimensional expansion of a problem. Mayne (1966), and Jacobson and Mayne

A. Peters et al. (eds.), Computational Methods in Water Resources X, 1147–1155.

(1970) developed DDP as an unconstrained deterministic technique. It was a successive approximation approach which started from an initial control strategy supplied by the user. The objective function was approximated at each stage by a quadratic polynomial through a truncated Taylor series formulation. This objective function was then optimized through a quadratic programming process to produce a new control policy, and the whole procedure was iterated until convergence. Murray and Yakowitz (1979) proposed a constrained DDP formulation. They used their method, which was capable of handling linear constraints, to solve a four-reservoir problem. Both Jacobson and Mayne (1970) and Murray and Yakowitz (1979) methods produced linear feedback control laws as linear functions of the state. Gjelsvic (1982) introduced a discrete time version of the DDP model, which took into account the stochastic nature of the inflows, to operate a multi-reservoir hydropower system. The solution procedure was restricted to the case of hydraulically parallel power stations. Trezos (1986), and Trezos and Yeh (1987,1988) extended another version of DDP to include stochastic inflows. They applied their approach to maximize power production from multi-reservoir hydropower systems. Their methodology was an iterative procedure which transformed the problem's quadratic objective function and linear dynamics into a series of quadratic programming problems. Ouarda (1991) suggested some adjustments to the Trezos and Yeh procedure and compiled his suggestions in one detailed algorithm which he named "Stochastic Differential Dynamic Programming".

PROBLEM SETTING

A typical objective function for maximizing hydropower production of a reservoir system is non-convex and non-linear in both storages and releases of the reservoir. However, its use within a dynamic programming setting transforms it into the standard QP form because it is conditioned on the current estimate of the reservoir storage (or state of the system). An objective function can be formulated as the summation, over the time horizon, of an incremental benefit term which is a non-separable function of reservoir storages and releases. A term representing the value of stored water at the end of the planning horizon can be added if such terminal value is known. In some formulations, a term for the penalty cost of deviating from certain terminal storage target values is added.

In the Trezos and Yeh (1987) algorithm, the benefit term is the product of the multiplication of reservoir release by average head at a particular time stage. The objective function considered in the algorithm is represented by:

$$MAX_{CV} \, E_{RV} \, \{ \sum_{t=1}^{N} HA_t^T . R_t + [S_{N+1} - TARG]^T . PEN . [S_{N+1} - TARG] \} \tag{1}$$

where: CV   = set of control variables,
      RV   = set of random variables,
      E   = expectation operator,
      t   = the time step or stage: 1,2,...,N,

R_t = release vector at time t of a size equal to NR,

HA_t = average head vector at time t of a size equal to the number of reservoirs in the system (NR),

S_{N+1} = storage vector at the end of the planning horizon of size NR,

TARG = storage target value vector at the end of the planning horizon of a size equal to NR,

PEN = a symmetric matrix of order NR representing the penalty cost for deviating from storage target values.

In the formulation of the above objective function, the state is represented by the reservoir storage vector S and the control is represented by the release vector R. The system dynamics is considered as the following mass balance equation:

$$S_{t+1} = S_t + Q_t + SYS \cdot R_t - SP_t \qquad (2)$$

where: t = 1,...,N

S_{t+1} = storage vector at the end of time t,

S_t = storage vector at the beginning of time t,

Q_t = inflow vector during time t,

SYS = system configuration matrix which maps the flow routing in the system,

R_t = release vector during time t,

SP_t = spill vector during time t.

In addition to the mass balance equation, the problem is subject to other constraints on releases, storages, and power production:

$$SLL_t \leq E(S_t) \leq SUL_t \qquad t = 1,...,N \qquad (3)$$

$$RLL_t \leq R_t \leq RUL_t \qquad t = 1,...,N \qquad (4)$$

$$PLL_t \leq E(HA_t^T \cdot R_t) \leq PUL_t \qquad t = 1,...,N \qquad (5)$$

SLL_t, RLL_t, PLL_t, SUL_t, RUL_t, and PUL_t are the allowable lower limits and upper limits of storage, release, and power production at time step t, respectively. Equation (3) is transformed into a release constraint by using the system dynamic equation. The average water head elevation in the reservoir during time t is given by:

$$HA_t = \frac{1}{2}(H_t + H_{t+1}) \qquad (6)$$

H_t being the water head elevation at time t. The reservoir storage and the water head elevation are considered to be related through a continuous piecewise linear function:

$$H_t = A_t \cdot S_t + b_t \qquad t = 1,...,N \qquad (7)$$

where: A_t = diagonal coefficient matrix,

b_t = vector of constant terms.

The reservoir spill is expressed by the equation:

$$SP_t = \alpha_t.S_t + \beta_t.S_{t+1} + \gamma_t.Q_t \qquad t = 1,...,N \tag{8}$$

where: α_t, β_t, γ_t are coefficient matrices obtained from historical reservoir data.

The solution process is initiated with a nominal release policy, and it proceeds recursively in a "backward run" which starts in the last stage N and terminates at the end of stage one. In this backward run, as the procedure iterates, storage estimates calculated in the previous iteration are used to obtain a new open loop release policy for all time stages $t=N,...,1$. This release policy is then considered as another nominal policy to initiate a new iteration of the algorithm. The storage sequence corresponding to this nominal policy is computed in a "forward run" making use of the known state at the beginning of the first stage, the computed releases, and the system dynamic equation.

STOCHASTIC DIFFERENTIAL DYNAMIC PROGRAMMING ALGORITHM

The proposed stochastic DDP procedure uses the hydropower production maximization objective function of equation (1) which is the same as that used by both Trezos and Yeh (1987) and Ouarda (1991). After some mathematical manipulation, and applying the expectation operator to the random variables (reservoir inflows), the optimal value function for time stage t of the problem can be written in the standard QP form:

$$V_t(S_t) = MIN_{R_t} - \{R_t^T.E_t.R_t - F_t^T.R_t + G_t\} \tag{9}$$

where:
E_t = Quadratic matrix coefficient,
F_t^T = Transpose of the cost vector,
G_t = Constant scalar term.

This quadratic programming problem is subject to the same system dynamics of equation (2) and the other constraints on the releases, storages, and power production which are the same as those of equations (3), (4), and (5), respectively. The backward run of the solution process solves the QP problem recursively (starting at the last time stage N) to obtain linear feedback control laws for all stages as follows:
The release R_t is assumed to be a linear function of the storage S_t:

$$R_t(S_t) = X_t + Y_t.S_t \tag{10}$$

where:
X_t is a vector of constant values of dimension NR,
Y_t is a diagonal coefficient matrix of order NR.

Since Y_t is a diagonal matrix, its diagonal elements are transformed into a vector Y'_t of the same order. If, on the other hand, the storage vector S_t is transformed into the principle diagonal of a diagonal matrix S'_t of the same order, then the product $S'_t.Y'_t$ is equal to the product $Y_t.S_t$. Therefore, equation (10) is modified to take the form:

$$R_t(S'_t) = X_t + S'_t.Y'_t \tag{11}$$

These transformations are performed to facilitate the direct implementation of equation (11) in the formula of the optimal value function of equation (9). This direct implementation is necessary to obtain linear control functions for all general cases. Without this implementation the linear control functions cannot be obtained unless the solution is at least on one of the boundaries. Consequently, equation (11) is substituted for the release R_t in the optimal value function, and the optimization is performed directly on the coefficient vectors X_t and Y'_t. Therefore, the linear feedback control laws are obtained in the backward run of the proposed algorithm whether the optimal solution is on the boundaries or not. When the expression of equation (11) is substituted for the release, the optimal value function of equation (9) can be rearranged to take the form:

$$V_t(S'_t) = -\left[[X_t^T, Y'^T_t]\tilde{E}_t\begin{bmatrix} X_t \\ Y'_t \end{bmatrix} - \tilde{F}_t^T\begin{bmatrix} X_t \\ Y'_t \end{bmatrix} + G_t \right] \tag{12}$$

where: $\quad \tilde{E}_t = \begin{bmatrix} E_t & , & E_t S'_t \\ E_t S'_t & , & S'^T_t E_t S'_t \end{bmatrix}$ and $\tilde{F}_t^T = [F_t^T, F_t^T.S'_t]$ $\tag{13}$

The expression of equation (11) is substituted for the release value in the constraints of the QP problem. These constraints can be consolidated into two sets and written into the following system:

$$\begin{bmatrix} I & , & S'_t \\ -I & , & -S'_t \end{bmatrix}\begin{bmatrix} X_t \\ Y'_t \end{bmatrix} \geq \begin{bmatrix} \tilde{d}_1 \\ \tilde{d}_2 \end{bmatrix} \tag{14}$$

$$\text{or} \quad \tilde{C}^T.\begin{bmatrix} X_t \\ Y'_t \end{bmatrix} \geq \tilde{d} \tag{15}$$

The QP problem now is to find the optimal coefficient vectors X_t and Y'_t which minimize equation (9) subject to the system (15). This constrained problem requires a quadratic programming algorithm to be solved. The selected QP algorithm for this procedure is the "active set method" by Fletcher (1971). This algorithm starts the recursive solution with the selection of a nominal release policy from which a set of nominal coefficients X and Y' can be initiated. Then, the nominal storage trajectory for all stages can be calculated and checked for feasibility. The elevation heads, average heads and power production are then computed. At this point, the active set method iterations are initiated. At each step of the QP solution, the decision variables are changed in a search direction that minimizes the objective value. When the Kuhn-Tucker conditions are satisfied and the optimum is attained, the procedure terminates. At convergence, the active set method yields a set of optimal coefficients X_t and Y'_t for all time stages. A linear feedback control law for every time stage t is now obtained. The coefficient vectors X_t and Y'_t as well as the coefficients of the optimal value function are stored in memory to be used later in the forward run of the solution process. This would complete the backward run of the proposed algorithm.

When the backward run is completed, the forward run of the algorithm starts. A non-parametric QP problem is formed at each stage making use of the known starting reservoir storage, the system dynamic equation, and the obtained control laws coefficients. As a result, an open loop optimal control policy for all stages is computed. This policy is considered as a new nominal policy and used to initiate a new iteration of the proposed stochastic DDP algorithm.

APPLICATION

To test the proposed DDP approach for solving multi-dimensional stochastic DP problems, a generalized computer program was developed. This program was coded in PASCAL on a 386 micro-computer. The program initiates the computations by reading the inflows, physical characteristics, and configuration of a general multi-reservoir system along with an initial nominal solution of the problem. The main component of the program is the quadratic programming algorithm (active set method) which is coded to solve the QP problem at each stage of the backward run and the forward run of the solution process. The backward run of the program results in feedback control laws which are linear functions of the state. The end result of the forward run of the program is an open loop optimal release policy for the reservoir system being studied. A flow chart of the program is presented in Figure (1).

The routine is tested on a multi-reservoir system in order to examine its capability in handling multi-dimensional problems. A hypothetical four-reservoir system (Figure 2) has been used for this purpose. The flow routing matrix SYS of this system is a fourth order matrix which represents any release from a reservoir by "-1" and any release into a reservoir (from an upstream reservoir) by "+1". In this case, SYS is given by:

$$ SYS = \begin{bmatrix} -1 & 0 & 0 & 0 \\ 0 & -1 & 0 & 0 \\ 0 & +1 & -1 & 0 \\ +1 & 0 & +1 & -1 \end{bmatrix} $$

This prototypical four-reservoir configuration was first presented by Larson (1968) to illustrate and test the incremental dynamic programming approach, and it has since become a "standard" test configuration system in the literature.

RESULTS AND DISCUSSION

The routine converges in two iterations and yields an objective value of 692348.3 when tested on the hypothetical four-reservoir system. This value is close to the true objective value of this problem which is reported to be 694794.8 by Ouarda (1991). The objective value is found to be virtually insensitive to the initial control policy used to initiate the program. Two initial release policies are used to initiate the problem. The

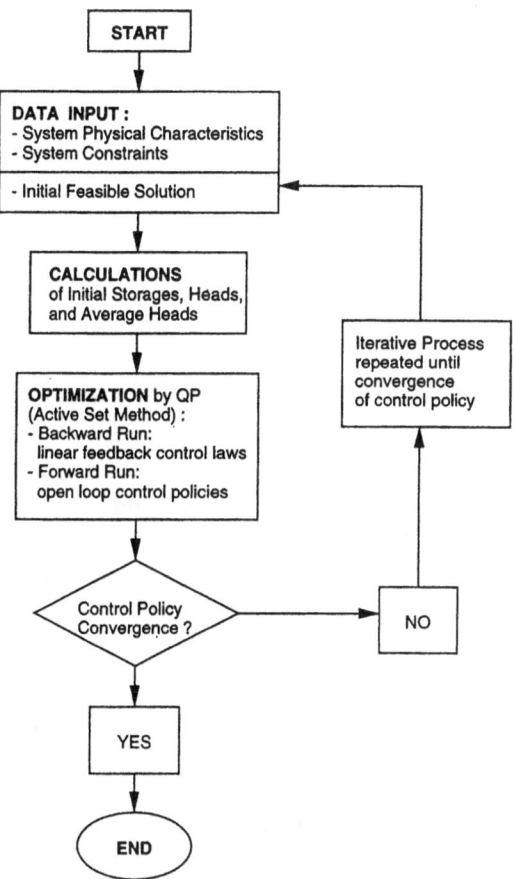

Figure 1 Flow chart of the developed optimization program

first initial policy, in which the total inflows are automatically released at each time period, yields a slightly lower objective value of 689256.3 .

All feasible cases converge in two iterations. The convergence in only two iterations in all cases is a logical result because the objective function is quadratic in both the state and the control of the problem. Consequently, the need to approximate the objective function to a quadratic polynomial through an iterative Taylor series approximation process ceases to exist. It is a known fact that an unconstrained QP problem with an originally quadratic objective function is solved without the need to use any successive approximation iterative procedure. However, in this paper it was decided to allow the computer program to iterate before it reaches the optimal solution. This was done to test whether the QP process with an originally quadratic objective function would still yield an exact solution after the incorporation of the linear constraints and

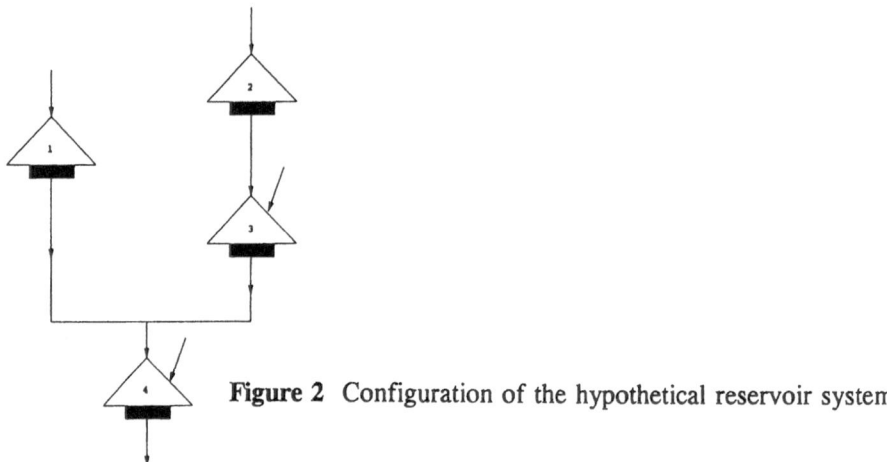

Figure 2 Configuration of the hypothetical reservoir system

stochasticity in the algorithm. The results show that the answer to this query is positive and that the constrained QP approach does not need a successive approximation procedure to converge while optimizing an originally quadratic objective function.

The performance of the proposed stochastic DDP approach as an efficient procedure to solve multi-dimensional problems is clearly illustrated in this case study. The backward run of the algorithm optimizes the coefficients of the linear feedback control laws directly. This means that the dimension of the QP problem is doubled in order to be solved, i.e. in a four-reservoir setting the routine solves an eight dimensional QP problem. In spite of the high dimensionality, the routine requires only about 60 seconds to solve the four-reservoir problem and yield its optimal release policy.

A major difference between the this approach and the one developed by Trezos and Yeh (1987) is the capability of reaching the true optimal objective value of the problem. The Trezos and Yeh approach, when tested by Ouarda (1991), fails regularly to reach the true optimal objective value of the problem setting. The objective value resulting from that approach depends largely on the initial solution used to initiate the problem (Ouarda, 1991). In contrast, the proposed stochastic DDP approach in this research yields a very close objective value to the true optimal solution irrespective of the initial policies used in the case study.

The optimization routine always yields feasible reservoir monthly storages corresponding to the optimal release policies. This is a result of linking the storage and release of every stage in the backward run by obtaining feedback release laws rather than open loop optimal releases. This is another major difference between this proposed stochastic DDP approach and the one developed by Trezos and Yeh (1987) where an entire iteration is completed without any computation of storages. Consequently, sometimes the new nominal release policies generate non-feasible and negative storage values (Ouarda, 1991).

The linkage of the releases and storages in the backward run of the solution process and the building of the stagewise quadratic programming problem in the forward run of the process play a decisive role in leading to this performance. The "direct" linkage of the releases and storages in the backward run, by substituting the release in the objective value expression by a linear function of the storage, leads to obtaining linear feedback control laws in all general cases. This is an improvement which generalizes the stochastic DDP approach. The perturbation approach suggested by Murray and Yakowitz (1979) and Ouarda (1991) works only when the problem would be subject to a set of constraints which includes some equality constraints. With the direct linkage, this pre-condition which limits the capability of the stochastic DDP approach is abolished.

REFERENCES

El-Awar, F. 1993. Multi-Lag Stochastic Differential Dynamic Programming Algorithm for Multi-Reservoir Hydropower Systems. Ph.D. Dissertation, Colorado State Univ, Fort Collins, CO.

Fletcher, R. 1971. A General Quadratic Programming Algorithm. J. Inst. Math. and its Appl., 7: 76-91.

Gjelsvic, A. 1982. Stochastic Seasonal Planning of Multi-Reservoir Hydro-Electric Power Systems by Differential Dynamic Programming. Model. Ident. Contr., 3(3): 131-149.

Jacobson, H. and Q. Mayne. 1970. Differential Dynamic Programming. ed. R. Bellman. Elsevier, New York, NY.

Larson, R. 1968. State Incremental DYNAMIC Programming. Elsevier, New York, NY.

Mayne, Q. 1966. A Second-Order Gradient Method for Determining Optimal Trajectories of Non-Linear Discrete-Time Systems. Intl. J. Control, 3: 85-95.

Murray, D. and S. Yakowitz. 1979. Constrained Differential Dynamic Programming and its Application to Multi-Reservoir Control. Water Resources Research, 15(5): 1017-1027.

Ouarda, T. 1991. Stochastic Optimal Operation of Large Scale Hydropower Systems. Ph.D. Dissertation, Colorado State Univ, Fort Collins, CO.

Trezos, T. and W. Yeh. 1987. Use of Stochastic Dynamic Programming for Reservoir Management. Water Resources Research, 23(6): 983-996.

Trezos, T. and W. Yeh. 1988. Stochastic Dynamic Programming and its Application to Multi-Reservoir Systems. Proceedings of the Third Water Resources Operation and Mgmt Workshop, Colorado State Univ, Fort Collins, CO, June, 1988.

SUBSURFACE SHEAR DISPERSION IN RIVER OIL SPILL MODELLING

J. JÓZSA
VITUKI Consult Rt (Environmental and Water Management Consultants)
Kvassay Jenő út 1, H-1095 Budapest, Hungary

Stochastic Lagrangian techniques are implemented to simulate the subsurface shear dispersion process in oil spill modelling. The Markovian displacement model provides droplet trajectory calculation in three dimensions, making it possible to reproduce the combined effect of buoyancy, velocity shear and turbulent diffusion. Simulations carried out in simple turbulent channel flows show how the droplet size affects the mean longitudinal advection and the rate of spreading of a cloud.

INTRODUCTION

Recently a number of simulation models have been developed to predict the fate of accidental oil spills in surface waters. The models are used either for real-time prediction to assist the clean-up or as scenario tools to analyse the possible impact of hypothetic oil spills on the aquatic environment.

Of the physical processes determining the fate of the spill, the formation of oil-in-water emulsion is an important mode of oil dispersion (see e.g. Hrudney and Kok, 1987, Delvigne and Sweeney, 1988). After the spill has occurred, turbulence and wave action mix the surface oil slick into the water column by breaking it up first into slicklets, then to droplets of various size. The fact that the droplets with different size and density move differently, e.g. larger, more buoyant droplets spend more time in the near-surface layers than smaller ones, results in shear dispersion with droplet fractionation. In spite of its great importance, the above mentioned phenomenon in rivers has remained rather poorly understood. The fundamental problem is certainly the difficulties of investigating the process in the field and reproducing it in lab conditions. Furthermore, when trying to simulate subsurface shear dispersion, fully or at least quasi three-dimensional approach seems to be inevitable due to the complexity of the process.

Most of the oil slick transport models have been developed obviously for sea conditions. There are a number of two-dimensional horizontal models which focus primarily on the surface phenomena. In two dimensions both Eulerian and Lagrangian solution methods are applied. As to three-dimensional modelling, stochastic Lagrangian particle models have recently become extremely popular in simulating transport processes in general, and multiphase oil-in-water systems in particular. Their success is due to the fact that they are free from numerical diffusion and oscillation in the classical sense, moreover, they handle the discrete phase in a very natural way.

A. Peters et al. (eds.), Computational Methods in Water Resources X, 1157–1164.

For rivers the most simplified models are longitudinally one-dimensional in which cross-sectional average values are calculated. Very few river oil spill models are known handling the problem in a more realistic horizontally two-dimensional form (Shen and Yapa, 1988, Józsa, 1989). Three-dimensional models including the direct simulation of the subsurface transport of oil have been developed only for sea conditions. One of the basic works in this field is due to Elliott (1986), who successfully reproduced the tidal and wind induced oil slick transport in the North sea by representing the patch of oil as a distribution of droplets whose buoyancy depended on droplet size, and simulating the turbulence in the water column by a three-dimensional random walk process.

River mixing conditions, however, may be significantly different from the ones found in the sea (Fischer et al., 1979). First of all rivers are shallow waters in which the turbulent currents are highly influenced by the bottom. Flow velocities are usually strong and are induced by gravity rather than by the wind. Wind induced waves are small compared with sea waves and white-caping seldom occurs.

For speeding up the dispersion process, in some cases chemical dispersants have been used (Lewis et al., 1985). Such a treatment of the spills, however, may be accepted only if drinking water abstractions are not affected. In whatever way the oil is mixed into the water column, the simulation of its subsurface fate is of particular importance for assessing its environmental impact.

In the paper the river with oil contamination is considered as an oil-in-water two-phase system in which only the influence of the aquatic phase on the oil phase is taken into account. Attention is paid onto subsurface shear dispersion process only. To simulate the oil transport, stochastic Lagrangian droplet tracking technique is applied. As a first approach, a Markovian displacement model is implemented in which the Fickian law for eddy diffusion holds. This simple version of subsurface transport simulation method needs the advective velocity and the eddy diffusion tensor distribution only. More sophisticated droplet trajectory calculations can be formulated on the basis of Markovian velocity models. These techniques postulate the detailed three-dimensional knowledge of a number of turbulent quantities either from measurements or numerical flow modelling which are, however, far from being available for large stretches of natural rivers. Simulations in simple turbulent channel flows show that the technique reproduces some of the expected main features of the oil shear dispersion process such as the droplet diameter dependent spreading and advection of the droplet cloud in the flow-aligned direction, the fractionation of the droplets in the vertical as well as in the horizontal directions resulting in the skewness of the longitudinal concentration toward the front.

STOCHASTIC LAGRANGIAN MODEL FORMULATION

The stochastic Lagrangian simulation of particle, droplet or bubble dispersion processes is based on handling either the displacement or the velocity of the disperse phase elements as a continuous time Markov process. The most convenient way to establish proper simulation algorythms is using stochastic differential equation theory. First the general three-dimensional formulation of the Markovian displacement model is presented below, then it is implemented to the shear dispersion process of oil droplets in simple turbulent channel flow. Finally, some aspects of applying sophisticated Markovian velocity models are also briefly considered.

Markovian displacement model

Markovian displacement models are in general the stochastic formulation of Fickian diffusion or dispersion processes. The earliest simulations in surface hydraulics are probably due to Bugliarello and Jackson (1964). Starting from deterministic bases in Eulerian reference framework, ommitting droplet coalescence and break-up, the three-dimensional evolution equation of the concentration of an instantaneous point release of droplets with diameter d is

$$\frac{\partial c_d}{\partial t} = -u_i \cdot \frac{\partial c_d}{\partial x_i} - w_d \cdot \frac{\partial c_d}{\partial z} + \frac{\partial}{\partial x_i} \left[D_{nij} \cdot \frac{\partial c_d}{\partial x_j} \right], \quad c_d(x,t_0) = M \cdot \delta(x - x_0) \ ,$$

where x is the Cartesian coordinate vector, x_i is Cartesian space coordinate ($i=1,2,3:=x,y,z$), t is time, u_i is Reynolds-averaged velocity component, c_d is the oil droplet concentration of diameter d, w_d is the rise velocity of droplet with diameter d, D_{nij} is Fickian turbulent diffusion tensor component, M is the released mass. In the equation vertical advection includes the rise velocity of the droplet, which can be estimated in basically two different ways (see e.g. Elliott, 1986), depending upon the critical diameter d_c below

$$d_c = \frac{9.52 \cdot \nu^{2/3}}{g^{1/3} \cdot (1 - \rho_o/\rho_w)^{1/3}} \ ,$$

where ρ_o is the oil density, ρ_w is the water density, ν is the viscosity of water, g is the acceleration due to gravity. For $d < d_c$ (small droplets) the Stokes law

$$w_d = \frac{g \cdot d^2 \cdot (1 - \rho_o/\rho_w)}{18 \cdot \nu} \ ,$$

whereas for $d > d_c$ (large droplets) the Reynolds law applies

$$w_d = \sqrt{(8/3) \cdot g \cdot d \cdot (1 - \rho_o/\rho_w)}$$

In stochastic theory the Eulerian description of advective diffusion processes is based on the Fokker-Planck partial differential equation which governs the evolution of the transition probability density p (Arnold, 1974, van Kampen, 1981). For the above mentioned process it has the form

$$\frac{\partial p}{\partial t} = -\frac{\partial (f_i \cdot p)}{\partial x_i} + \frac{1}{2} \cdot \frac{\partial^2 (b_{ik} \cdot b_{jk} \cdot p)}{\partial x_i \partial x_j} \ , \quad p(x_0, t_0) = p_0 \ ,$$

where f_i is the drift, $b_{ik} \cdot b_{jk}/2$ is stochastic diffusion tensor component. Comparing the deterministic and the stochastic equations shows that the Fokker-Planck equation will satisfy the advective diffusion equation if

$$p(x,t) := \frac{c(x,t)}{M} \, ,$$

$$f_i(x,t) := u_i(x,t) + \frac{\partial D_{nij}(x,t)}{\partial x_j} \, , \quad i=1,2,$$

$$f_i(x,t) := u_i(x,t) + w_d + \frac{\partial D_{nij}(x,t)}{\partial x_j} \, , \quad i=3,$$

$$\frac{1}{2} \cdot b_{ik}(x,t) \cdot b_{jk}(x,t) := D_{nij}(x,t).$$

The same stochastic process can be treated also in Lagrangian framework through the direct description of the droplet trajectories by means of the following Langevin

$$\frac{\partial X_i(t)}{\partial t} = f_i(X,t) + b_{ij}(X,t) \cdot N_j(t) \, , \quad X(t_0) = X_0 \, ,$$

or rather the Itô equation (van Kampen, 1981)

$$dX_i(t) = f_i(X,t) \cdot dt + b_{ij}(X,t) \cdot dW_j(t) \, ,$$

where X_i is the space coordinate as stochastic variable, b_{ij} is the strength of the random fluctuation, $N(t)$ is Gaussian white noise, $W(t)$ is the Wiener-process, and the bold variables stand for three-dimensional vectors.

Taking the turbulent diffusion coefficient in a flow-aligned system in the principal directions, the Lagrangian model has the following vector form:

$$dX(t) = f(X,t) \cdot dt + F^{-1}(X,t) \cdot \begin{bmatrix} \sqrt{2D_I(X,t)}\,dW_I(t) & 0 & 0 \\ 0 & \sqrt{2D_{II}(X,t)}\,dW_{II}(t) & 0 \\ 0 & 0 & \sqrt{2D_{III}(X,t)}\,dW_{III}(t) \end{bmatrix}$$

$$X(t_0) = X_0 \, ,$$

where F is scaling matrix.

The numerical solution of the Lagrangian model can be obtained by standard Monte-Carlo particle technique, which yields the following estimation of the concentration

$$c_d(x,t) \approx \sum_{l=1}^{N_d} m \cdot \delta(x - X_l(t)) = \frac{M}{N_d} \cdot \sum_{l=1}^{N_d} \delta(x - X_l(t)) \; ,$$

where N_d is the number of droplets, m is the droplet mass.

Shear dispersion of oil droplets

In the Fickian scale the above mentioned numerical model comprises all the terms needed for simulating shear dispersion processes. As a simple approximation of more or less unidirectional shallow water flows, let us consider the logarithmic velocity profile of the turbulent boundary layer with parabolic distribution of eddy diffusion, used e.g. by Elder (1959) in his dispersion study. In these conditions the deviation u' of the velocity from the depth-averaged value is

$$u'(\eta) = \left[\frac{u_*}{\kappa} \right] \cdot (1 + \ln \eta) \; ,$$

where $\eta = z/h$, h is the water depth, u_* is the friction velocity, κ is the von Karman's constant. Assuming the Reynolds-analogy, the vertical eddy diffusion D_z distribution is

$$D_z(\eta) = h \cdot u_* \cdot \kappa \cdot \eta \cdot (1 - \eta) \; .$$

For nonbuoyant substances the longitudinal D_x shear dispersion coefficient found by Elder (1959) is

$$D_x = \alpha_x \cdot h \cdot u_* = 5{,}93 \cdot h \cdot u_* \; ,$$

In this hydrodynamic field the Lagrangian droplet tracking model becomes rather simple. Numerical experiments have been carried out in dimensionless form with a number of different droplet diameters. Snapshots of the process for 300 and 100 μm of droplet diameter can be seen in Fig.1 and 2, respectively. Near-surface instantaneous droplet line source was applied, and apart from the droplet cloud, the shape of the longitudinal as well as vertical concentration distribution is also displayed. Although the boundaries have been treated just as reflecting ones (which should be certainly improved), the simulation have shown some important features, as the combined effect of buoyancy, velocity shear and turbulent diffusion. Obviously, the larger the droplets, the more they stay in the near- surface zone. The diameter dependent vertical distribution of the concentration results in diameter specific longitudinal mean velocity and rate of spreading of the droplet cloud. Large particles are drifted faster but spread less efficiently than small ones. For droplets with single diameter the depth-averaged longitudinal concentration distribution is initially rather skewed toward the front, but after sufficient time it becomes first symmetric then Gaussian. The rate of growth of the droplet cloud variance is always smaller than the one for nonbuoyant substances. However, due to the differencies in the mean advection velocity, for a cloud of droplets with a range of

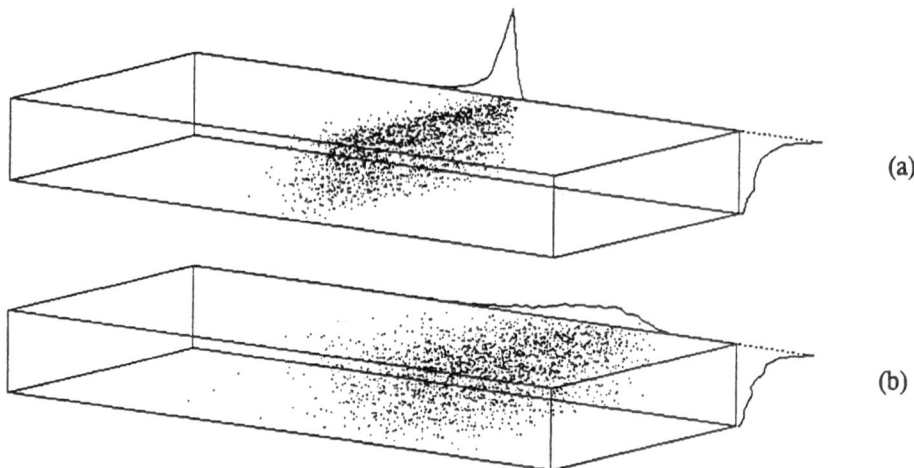

(a)

(b)

Figure 1. Droplet dispersion in turbulent flow with logarithmic velocity difference profile and parabolic eddy diffusion coefficient distribution. $u_*=0.05$m/s, $h=5$m, $d=300\mu$m, droplet number $N_d=3000$, dimensionless time step $\Delta t'=0.05$. (a) - dimensionless time $t'=4$, $\alpha_x=1.92$. (b) - dimensionless time $t'=16$, $\alpha_x=2.68$.

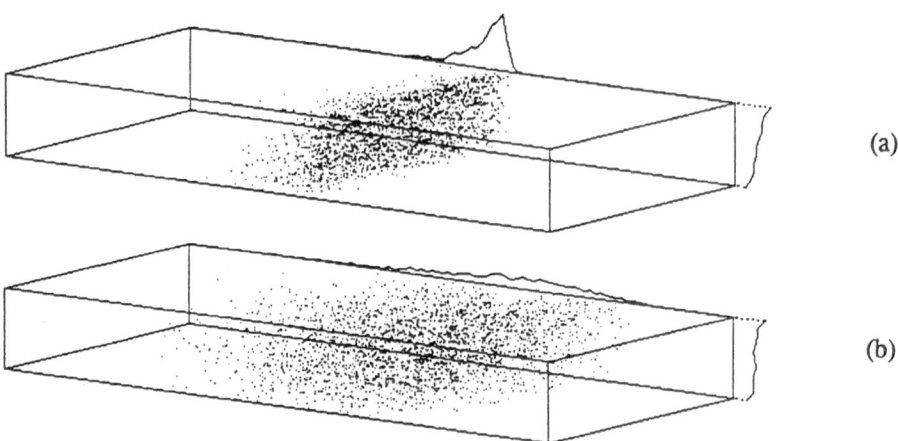

(a)

(b)

Figure 2. Droplet dispersion in turbulent flow with logarithmic velocity difference profile and parabolic eddy diffusion coefficient distribution. $u_*=0.05$m/s, $h=5$m, $d=100\mu$m, droplet number $N_d=3000$, dimensionless time step $\Delta t'=0.05$. (a) - dimensionless time $t'=4$, $\alpha_x=3.78$. (b) - dimensionless time $t'=16$, $\alpha_x=4.98$.

different diameters the longitudinal depth-averaged concentration evolution in general will not follow the Fickian law, thus the conventional depth-averaged advection-dispersion description can not be applied.

Markovian velocity models

The description of turbulent diffusion of droplets below the typical length and time scale of turbulent eddies requires Markovian velocity models not only for the fluid particles (Durbin, 1980), but also for the droplets, which move differently from the ones of the carrying fluid. At this scale the fine component of the droplet motion is treated by momentum balance usually dominated by the drag as well as buoyancy forces. This simulation technique (see e.g. Berlemont et al., 1990) is widely applied in industrial multiphase flow calculations. To obtain realistic trajectories in inhomogeneous, highly anisotropic conditions given either by measurements or advanced turbulence models, the turbulent velocity correlation matrix is used to describe fluid particle trajectories, coupled with discrete particle tracking including the so-called crossing-trajectory effect. However, the difficulties of extensive three-dimensional measurements or modelling of all of the above mentioned turbulence parameters in natural river conditions (Rodi, 1980, Tominaga and Nezu, 1991), furthermore, the influence of other subprocesses not considered here, make it hardly possible to implement realistically this technique in river oil spill models to date. Nevertheless, some numerical experiments are going to be performed to get better insight into e.g. the apparent eddy diffusion coefficients in terms of the droplet size.

CONCLUSIONS

In the paper an attempt has been made toward implementing stochastic Lagrangian techniques to simulate the subsurface shear dispersion fate of river oil spills. The method is based on Markovian droplet displacement description, by which the process can be reproduced numerically at least in a qualitative manner. Reasonable pictures have been obtained in simple turbulent channel flows about the size dependent subsurface advection and spreading of droplet clouds. Unlike for nonbuoyant substances, three-dimensional modelling approach is inevitable even long time after the release. Boundary conditions (like the one developed by Monin (1959) for settling solid particles), droplet size distributions, as well as coalescence and break-up processes, which certainly play an important role in special regions, have not been dealt with in details. These are subjects of future investigation both by experimental and numerical tools.

ACKNOWLEDGEMENTS

This work was part of a project on the "Turbulent mixing of hydrofobic pollutants in natural waters" supported by the National Scientific Research Fund of the Hungarian Academy of Sciences, , Contract No. 715.

REFERENCES

Arnold, L. (1974) Stochastic Differential Equations: Theory and Applications, John Wiley and Sons, New York.

Berlemont, A., Desjonqueres, P., Gouesbet, G. (1990) "Particle Lagrangian simulation in turbulent flows", Int. J. Multiphase Flow 16, 19-34.

Bugliarello, G., Jackson, E.D. (1964) "Random walk study of convective diffusion", ASCE J. Eng. Mech. Div. 94, 49-77.

Delvigne, G.A.L., Sweeney, C.E. (1988) "Natural dispersion of oil", Oil & Chemical Pollution 4, 281-310.

Durbin, P.A. (1980) "A random flight model of inhomogeneous turbulent dispersion", Phys. Fluids 23, 2151-2153.

Elder, J.W. (1959) "The dispersion of marked fluid in turbulent shear flow", J. Fluid Mechanics 5, 544-560.

Elliott, A.J. (1986) "Shear diffusion and the spreading of oil in the surface layers of the North sea", Deutsche Hydrographische Zeitschrift 39, 113-137.

Fischer, H.B., List, E.J., Koh, R.C.Y., Imberger, J., Brooks, N.H. (1979) Mixing in Inland and Coastal Waters, Academic Press, New York.

Hrudney, S.E., Kok, S. (1987) "Environmentally relevant characteristics of oil-in-water emulsions", Proc. Symp. on Oil Pollution in Freshwater, Edmonton, Canada, Pergamon Press, pp. 58-70.

Józsa, J. (1989) "2-D particle model for predicting depth-integrated pollutant and surface oil slick transport in rivers", in Proc. International Conference on Hydraulic and Environmental Modelling of Coastal, Estuarine and River Waters, Bradford, U.K., Gower, pp. 332-340.

van Kampen, N.G. (1981) Stochastic Processes in Physics and Chemistry, North-Holland, Amsterdam.

Lewis, A., Byford, D.C., Laskey, P.R. (1985) "The significance of dispersed oil droplet size in determining dispersant effectiveness under various conditions", Proc. 1985 Oil Spill Conference, API, Washington D.C., pp. 433-440.

Monin, A.S. (1959) "On the boundary condition on the earth surface for diffusing pollution", Advances in Geophysics 6, Academic Press, pp. 435-436.

Rodi, W. (1980) Turbulence Models and their Application in Hydraulics, IAHR Publ., Delft, The Netherlands.

Shen, H.T., Yapa, P.D. (1988) "Oil slick transport in rivers", ASCE J. Hydr. Eng. 114, 529-543.

Tominaga, A., Nezu, I. (1991) "Turbulent structure in compound open-channel flows", ASCE J. Hydr. Eng. 117, 21-41.

MODELLING INPUT-OUTPUT RELATIONS IN CATCHMENTS

H. LANGE and M. HAUHS
BITÖK, University of Bayreuth, 95440 Bayreuth, Germany

Abstract
Different concepts are being used to describe the relation between input and output fluxes of water in natural (forested) catchments. Among the most popular are explicit hydrological models, in which the solution of differential equations (e.g. Richards equation) is sought to determine the internal flowpaths through catchment soils. The parameters of such models are difficult to identify from ecosystem scale observations. On smaller scales, however, they are often unobservable, thus leading to notorious identification problems in ecosystem modelling. Another approach uses transfer functions which are kernels of convolution integrals and give an estimate of residence time distributions.

We describe the connection of inflow and outflow from a different perspective. We consider the catchment (or ecosystem) as being a filter which dampens input signals in a way so that the input signal always has higher complexity than the output. We analyze this filtering by a simple deterministic toy model, which uses a criterion based on the variability of the signal in one soil layer to decide which transfer is made into the next lower layer. It is able to reproduce "realistic" runoff behaviour, and serves as a means for the investigation of complexity measures. Such measures might serve as an abstract tool to assess the abovementioned identification problem directly.

INTRODUCTION

This paper deals with an ecosystem perspective on water flow through catchments. For many years, hydrological modelling mainly was concerned with either rather heuristic approaches connecting input and output with "graphical" procedures (e.g. unit hydrograph, "quick" vs. "slow" flow) or with complicated physically-based models which sought the correct relation between input and output via a differential equation approach. In the latter case, runoff is the final result of a numerical solution of a nonlinear partial differential equation (e.g. Richards equation) which usually requires a lot of knowledge about boundary conditions and internal properties of the catchment soils, like hydraulic conductivity or water content at each point in space for each matrix potential. These parameter functions enter the basic equation, leading to a potentially infinite-dimensional parameter space for

A. Peters et al. (eds.), Computational Methods in Water Resources X, 1165–1172.
© 1994 *Kluwer Academic Publishers. Printed in the Netherlands.*

calibration. Experience with these process-oriented models shows that validation of calibrated parameter functions is hardly ever possible. More specifically, the question of uniqueness of highly nonlinear functions such as hydraulic conductivity is nearly impossible to investigate.

Recent results of soil column experiments place serious doubts on the validity of Richards equation even in such completely controlled situations [Stonestrom and Akstin 1994]. It is therefore unclear up to what extent one searches (in a certainly complicated manner) for solutions of the *correct* equation.

These difficulties show that "application of distributed hydrological models is more an exercise in prophecy than in prediction" [Beven 1993].

One way to overcome these difficulties would be to give up the idea of an unique set of parameters which is to be considered as "optimal" with respect to some measure but to use a continous distribution of parameter values (equifinality assumption). The parameter sets are weighted with a (fuzzy) acceptability measure (likelihood weights) [Beven 1993].

This procedure may be feasible in the case when a computationally simple model such as TOPMODEL is used. Extensive Monte Carlo runs with a finite element numerical model are much more tedious or even impossible. In addition, it is not clear whether this procedure increases the reliability of model extrapolations beyond the observed catchment response (e.g., to climate change).

Alternatively, one may also avoid using physically-based models which resolve internal catchment processes. If the main interest in the hydrological description of a catchment is the rainfall-runoff relationship, as is often the case in practical applications, it is interesting to investigate the question how much parameters are needed for a chosen ("simple") model to give an accurate description of the observed (precipitation and streamflow) data time series. One example for this kind of approach is presented in [Jakeman and Hornberger 1993]. These two authors use a linear transfer-function like model (after a nonlinear transformation of rainfall which is not essential for their purposes) with a variable number of parameters and use statistical measures to identify the number of storages in the catchment needed for a satisfying description of the data. In almost every case, two storages, corresponding to four parameters in their model, are sufficient. This small number shows that the model complexity required is much lower than is usual the case with every process-oriented (e.g. differential equation solving) approach.

ECOSYSTEMS AS INFORMATION DESTROYING UNITS

The findings of [Jakeman and Hornberger 1993] support our hypothesis that ecosystems serve as systems which transform randomly fluctuating input (e.g. rainfall) into output which is much more predictable [Hauhs 1992]. Typical examples of rainfall and runoff from catchments show that the information content of the signals is reduced on the way through the system. The length of the shortest description (number of bits) of the output is smaller than that of the input. Furthermore, the output signal is *redundant* to the input signal: the output can be completely predicted from the input, whereas the input can never be reconstructed from observing the output. We express this hypothesis formally by the equation

$$K(s_{in} \circ s_{out}) = K(s_{in}) \tag{1}$$

where s_{in}, s_{out} are strings which encode the input and output signals in an appropriate (not necessary optimal) way, $\circ$ denotes concatenation and K is the Kolmogorov complexity (see, for example, [Zurek 1989]).

We therefore consider ecosystems as units which gather negentropy from their environment by destroying information from their input signals. Taken in another way, this means that the action of an ecosystem in its environment is that of a filter that dampens arbitrary complex signals to relatively simple ones within given ranges of sensitivity, characteristic for that system. Thus, the formal relationship

$$q_{out} = \mathcal{F} q_{in} \tag{2}$$

where q_{in}, q_{out} are input and output fluxes across the boundary of the ecosystem, and $\mathcal{F}$ is a filter, is invoked by us.

Such filters are rather general mathematical objects; they always imply the desired property of redundancy of the output w.r.t the input. Examples for filters are solutions of Richards equation, the model of [Jakeman and Hornberger 1993] and our model described in the next paragraph.

Output signals which carry information definitely not contained in the associated input would be counterexamples which invalidate eq. (2). We were not able to find such a situation in extensive data sets.

THE MODEL

One of the disadvantages of the Kolmogorov complexity in eq. (1) is that it can be shown that this quantity is not algorithmically computable in a finite number of steps. Thus, in "practical" applications one has to rely on a different concept of complexity.

Note the different meaning of the phrase "complexity" in our approach in comparison with [Jakeman and Hornberger 1993]. There, complexity is defined *via* a given model and is qualitatively determined by the number of parameters required for data description, whereas here complexity refers to the data sets themselves.

Our main purpose is the relationship between input and output of catchments. Resolution or assumptions of internal processes is explicitly excluded here; the ecosystem is considered as a black box connecting precipitation signals with runoff signals.

Exploiting the difficulties discussed above in a *positive* sense, we make two important assumptions. First, there is no *spatial* heterogeneity in the precipitation signal. This is contrary to some detailed theoretical investigations [Waymire et.al. 1984] but rather common in most catchment models (i.e. on small scales). Second, we restrict ourself to a one-dimensional catchment description, as is also often the case (two parallel storages as in [Jakeman and Hornberger 1993] are also *effectively* one-dimensional) but not essential to our conclusions.

In many "simple" approaches to the problem, e.g. unit hydrograph, one separates different types of flow (predominantly quick flow vs. slow flow) which have the common feature of exponential decay after a give input pulse. We therefore include this feature here. Furthermore, the dampening of signals as a function of soil depth - more precisely, as function of the distance from the input zone, which must not be the same - is incorporated in a parametrized way. The flux at location z at time t is given by

$$q(z,t) = q_0 e^{\alpha(t-t_0)(\frac{z}{z_0})^\beta} \tag{3}$$

where q_0 is the precipitation flux incident on the catchment at time t_0, α is a parameter which fixes the time scale of the system (comparable to τ in linebreak [Jakeman and Hornberger 1993]), and β determines the spatial damping. z is measured positive downwards; reasonable values are therefore only $\beta > 0$. z_0 only sets the scale for the spatial coordinate and is therefore no parameter.

We consider a situation where a record of rainfall and runoff at certain sample time points t_i (not necessary equidistant) is given over a certain period of measurement. The soil or catchment region is then subdivided into n layers (which have nothing to do with soil horizons and are not equally spaced). During the interval $]t_i, t_{i+1}[$, the fluxes evolve according to (3). At the sampling points t_i, the following question is answered in each layer: Is the signal in layer k smaller than in layer $k-1$? If so, the signal from layer $k-1$ is taken over to layer k, setting a new t_0 and a new q_0 for layer k. If not so, the evolution (3) is continued in layer k until t_{i+1} is reached and the question is again posed.

This procedure gives a model which is strictly deterministic but simulates the ecosystem behaviour by allowing "soil decisions" in the described way. It has the desired property of damping and allows the definition of a *quantitative measure of complexity*: the number of signal changes ("yes decisions") at a given soil depth z, divided by the total number of data (sampling) points, is the complexity of the time series of fluxes in that region. This complexity measure can easily be evaluated even for the experimental precipitation rates and runoff records.

CONNECTION TO RICHARDS EQUATION

It is possible to find a situation where (3) is a solution of the one-dimensional Richards equation

$$\frac{\partial \Theta}{\partial t} = \frac{\partial}{\partial z}(q - K) \tag{4}$$

if the flux in (3) is considered as Darcian. A unique continous solution is of course impossible to find due to the discontinous jumps at every t_i, but if the solution is restricted to an interval $]t_{i-1}, t_i[$, the following choice leads to a solution of (4) which is given by (3):

$$K = K_0 = \text{const.}$$

$$\Theta(z,t) = \frac{q_0 \beta}{\alpha z}\left[\alpha(t-t_0) + (\frac{z}{z_0})^{-\beta}\right]e^{-\alpha(t-t_0)(\frac{z}{z_0})^\beta} + \Theta_1 \tag{5}$$

$$\Psi(z,t) = \frac{q_0 z_0}{K_0 \beta}[\alpha(t-t_0)]^{-1/\beta}\left(\gamma(\frac{1}{\beta}, 0, \alpha(t-t_0)z_0^{\frac{1-\beta^2}{\beta}}) - \gamma(\frac{1}{\beta}, 0, \alpha(t-t_0)z_0^{-\beta}z^{1/\beta})\right)$$
$$+ \Psi_1$$

where Θ_1 and Ψ_1 are integration constants fixed by the boundary conditions; $\gamma(a, b, c)$ is the generalized incomplete gamma function.

These expressions show that our approach is not incommensurable to the "physical" approach, although the soil properties given in (5) may be rather peculiar. An example for a $\Theta(\Psi)$ curve resulting from (5) is shown in Fig.1. This of course is *not* a predicition of any "real" soil property; it is completely artificial. Another problem with the solution (5) is that the time dependence of Θ and Ψ variables is not identical, i.e. the associated pF-curve at a given point in space would not be stationary, contrary to most approaches.

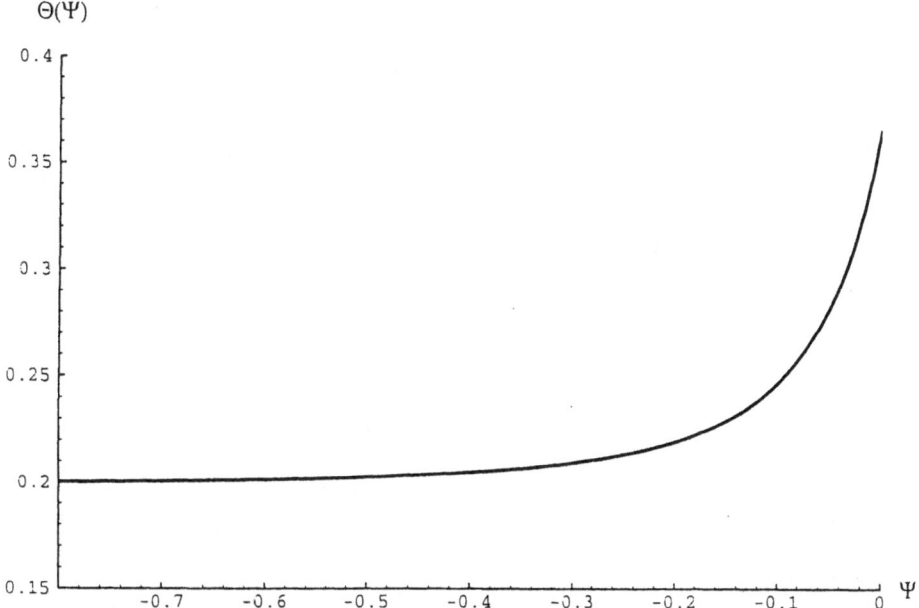

Figure 1: An artificial $\Theta(\Psi)$ curve drawn from (5). Parameters are $q_0 = K_0 = z_0 = \alpha = \beta = 1$ (arb. units), $t_0 = \Psi_1 = 0$ and $\Theta_1 = -1/2$.

AN EXAMPLE

We demonstrate the action of this very simple model by an example. It is the catchment M6 at Maimai, New Zealand, also considered in [Beven 1993]. The data were taken hourly approximately. We have chosen a measurement period during 1991/1992 of length 91 days. The input precipitation has complexity $K_{in} = 0.461$, whereas $K_{out} = 0.311$ (see Fig.2). Compared to other places, input complexity is high but damping is relatively low.

Our model was calibrated against these two complexity values alone. To this end, we fitted α and the number of soil layers, keeping β constant ($= 0.5$) all through. In order to match both input and output complexity, we found $\alpha = 15.6 d^{-1}$ and $n = 3$ as optimal values. The results for the bottom layer are shown in

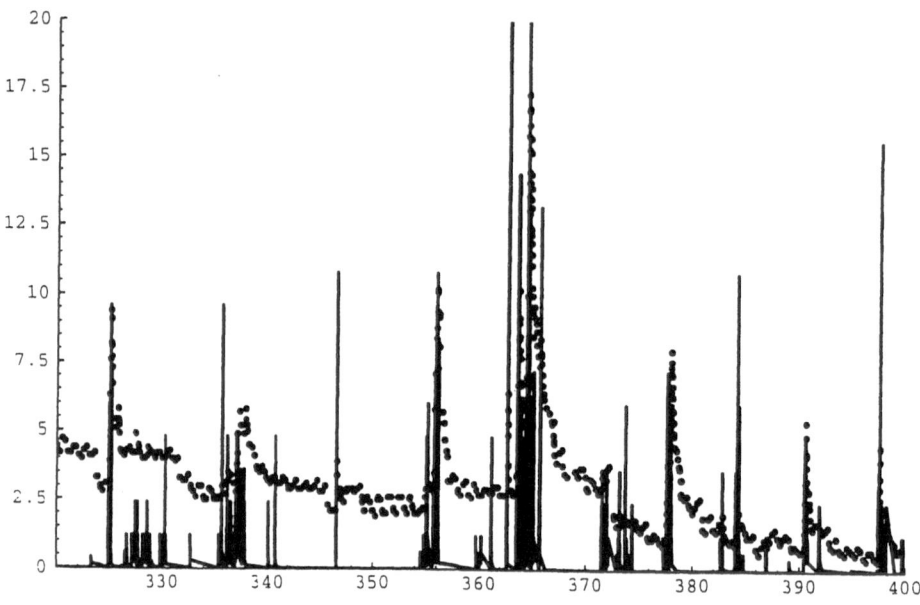

Figure 2: Input (full curve) and Output (dotted) at Maimai catchment vs. day of the year 1991. Units are mm/d for precipitation, l/s for runoff. Note the obvious appearance of base flow in runoff.

Fig.3. Notice that the absolute values of the signals are of no importance here. Base flow could have been included in (3) but was neglected for simplicity. Obviously, the *structure* of the signals (position of maxima etc.) could be reproduced by this simple method.

The efficiency of systems in damping signals determines the number of layers appropriate in this approach. Fig.4 shows the complexity of the Maimai data as function of the number of layers. This curve finally decays to zero but it can easily be seen that most of the damping occurs in the first few layers.

CONCLUSIONS

This paper was concerned with the reconstruction of typical input and output signals from catchments. Emphasis was on the signal complexity, defined in an appropriate way. A rather simple deterministic model with effectively two parameters could qualitatively reproduce the input/output relationship. Ecosystems act as filters which dampen input signals, using the information gained for their self-organization. The complexity of input signals could serve as an abstract measure of the maximum negentropy flux from the environment that ultimately limit any biological process within the ecosystem. The decrease in complexity performed by the system when transforming input into output signals could serve as a measure of efficiency. The M6 catchment would be classified as a relatively unefficient

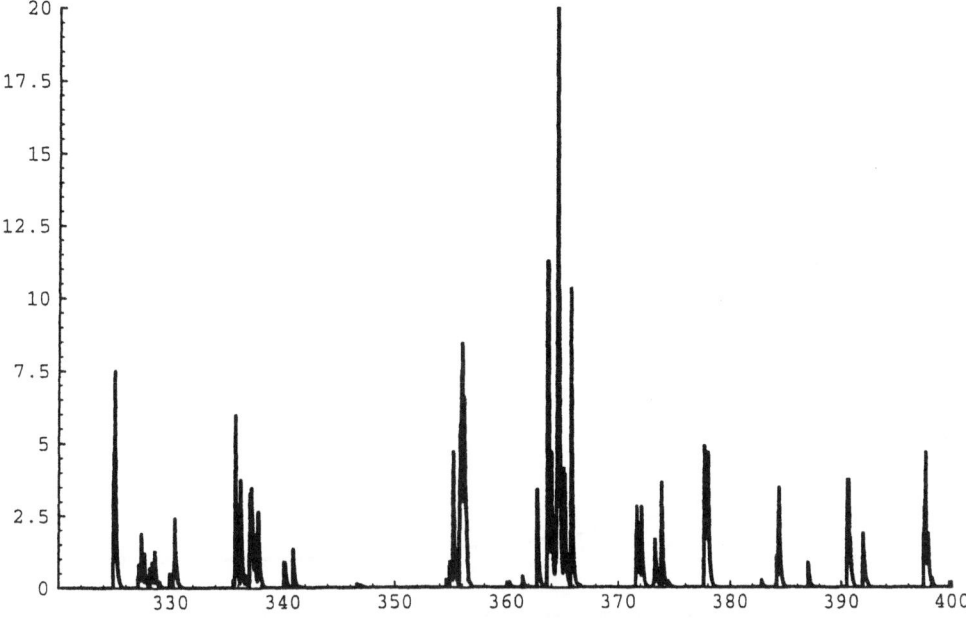

Figure 3: Bottom layer signals from the model vs. day of the year 1991. The parameters are $\alpha = 15.6d^{-1}$, $n = 3$. $\beta(= 0.5)$ was not used for calibration.

one. However, extensive comparisons of such measures with other catchments are needed. Finally, non-redundant output signals may hint to external ecosystem disturbance.

The investigation shows that the description of runoff, given precipitation, does not require soil measurements. Even more important, there is *no unique solution* for the parameter functions inside the soil that reproduces the observed runoff.

References

[Beven 1993] Beven,K: "Prophecy, reality and uncertainty in distributed hydrological modelling." Advances in Water Resources **16**, 41-51.

[Hauhs 1992] Hauhs, M.: "A definition of ecosystems based on information theory." In: Franke, J. and Roeder, A. (eds.), Mathematical Modelling of Ecosystems. J.D. Sauerländers Verlag, Frankurt/Main, p. 83-91.

[Jakeman and Hornberger 1993] Jakeman, A.J. and Hornberger, G.M.: "How much Complexity is warranted in a Rainfall-Runoff Model?" Water Resources Research **29**, 2637-2649.

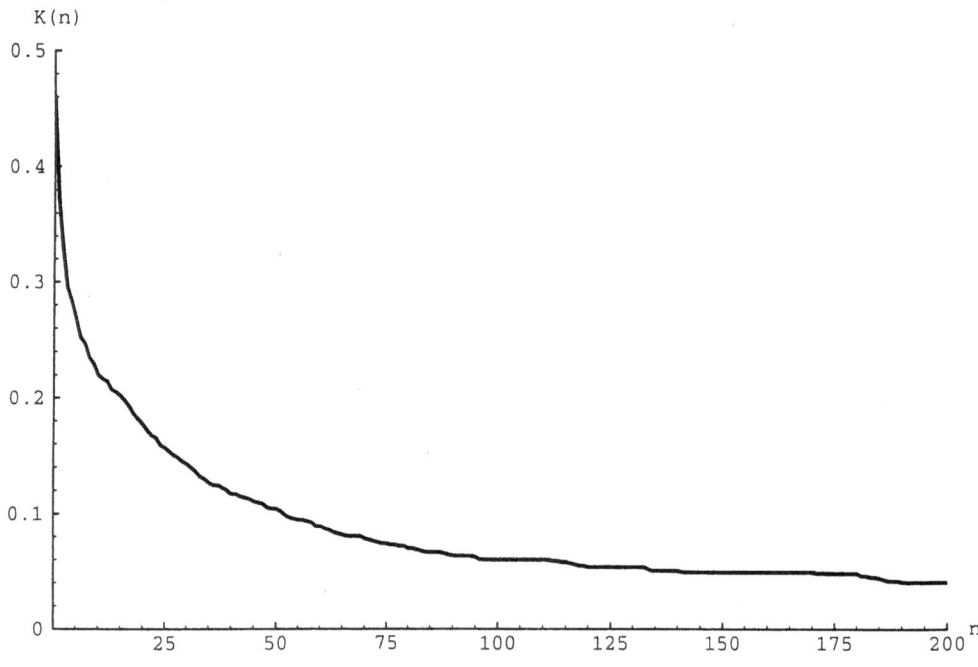

Figure 4: Complexity $K(n)$ as a function of number of layers in the soil. Parameters are as in Fig.3.

[Stonestrom and Akstin 1994] Stonestrom, D.A. and Akstin, K.C.: "Nonmonotonic matric pressure histories during constant flux infiltration into homogeneous profiles." Water Resources Research **30**, 81-91.

[Waymire et.al. 1984] Waymire, E.A., Gupta, V.K. and Rodriguez-Iturbe, I.: "A Spectral Theory of Rainfall Intensity at the Meso-β Scale." Water Resources Research **20**, 1453-1465.

[Zurek 1989] Zurek, W.H.: "Algorithmic randomness and physical entropy." Phys. Rev. A **40**, 4731-4751.

RIVTOX-COMPUTER CODE TO SIMULATE RADIONUCLIDES TRANSPORT IN RIVERS

P.V.TKALICH, M.J.ZHELEZNYAK, G.B.LYASHENKO and A.V.MARINETS
Institute of Mathematical Machines and Systems
Cybernetics Center Ukrainian Academy of Sciences
Pr.Gluskova 42, Kiev, 252207
Ukraine

Computer code RIVTOX is based on the procedures of the numerical solution of the set of one-dimensional equations of unsteady flow in a network of river channels, the equation of bottom deformations, the equations of the radionuclides transport in solute, on suspended sediments and radionuclides depositing in the bottom sediment. The radionuclides and other pollutants' transport in several Ukrainian and West European rivers is simulated based on RIVTOX.

INTRODUCTION

The Commission European Community project to develop the decision support system RODOS for nuclear emergency in Europe includes the elaboration of software for simulation of the radionuclide transport in surface water . RODOS as a real-time on-line system should simulate the short-term and long-term consequences of the accidental radionuclide releases to atmosphere and water (Ehrhardt et al., 1992). The simulated fallout data from atmospheric dispersion modules of RODOS should be used to calculate radionuclide washout from watersheds to river systems. The results of the simulation of the radionuclide concentrations in river water and sediment is used to calculate an internal irradiation dose of population after an accident. The radionuclides intakes via aquatic pathways take place primarily due to consumption of drinking water, fish and products from irrigation farming in the zone affected by accidental contamination. The results of collective dose simulation via aquatic pathways should be used in RODOS in comparison with other sources of irradiation of population to support the post-accidental decision making.

The experience of post-accidental countermeasures in the zones of the Chernobyl radionuclides fallout demonstrates that decision making after the accident could includes combination of countermeasures directed on changes in the water assumption and also measures diminishing the radionuclides flux from the heavy contaminated area through river net (Zheleznyak et al., 1992). The experience obtained in the Cybernetics Center

A. Peters et al. (eds.), Computational Methods in Water Resources X, 1173–1180.

during simulation of radionuclides dispersion in the Ukrainian rivers after the Chernobyl accident on the base of computer code RIVTOX is used to develop code for the rivers' contamination submodel of RODOS (Zheleznyak et al.,1993)

Radionuclides washout from the watersheds is simulated on base of a "box" approach for the long-term scale simulation and a "kinematic wave" approach for the short-term scale simulation in the first stage of the nuclear accident.

The river flow dynamics, sediments and pollutants transport in a river channel is simulated in RIVTOX on the base of set of one-dimensional unsteady equations. The numerical solution of the Saint-Venant equation is based on the implicit finite-difference method (Cunge et al., 1980). In the PC version of the code the "diffusive wave " approximation is used. Taking into account that suspended sediments carry not less than half of the total flux of some kind of radionuclides in a river flow, the accurate simulation of the sediment transport is one of the most important part of the radionuclides concentration modeling in a river net. Rather complicated one-dimensional codes is used now to describe sediment transport in fluvial streams e.g. CHARIMA (Holly et al., 1990), TODAM (Onishi at al, 1981). RIVTOX includes simplified model based on the advection - diffusion equation for suspended sediment transport and "equilibrium concentration" (flow capacity) approach to determine resuspension and sedimentation rates.

To simulate the fate of radionuclides, driven by hydraulic processes the approach similar to the code TODAM is used in RIVTOX. The radionuclides concentration in solute and on suspended sediments are simulated by advection-diffusion equations with explicit description of solute-solid particles exchanges processes including adsorption - desorption and resuspension-deposition. The same processes are described in the equation of radionuclide concentration dynamics in the upper bottom layer, without taking into consideration horizontal transport of radionuclides with bed load. The model validation and parameter identification studies is done on the base of post-Chernobyl and other pollutants spilling studies.

MODEL AND CODE DESCRIPTION

Governing equations

In the long-term scale modeling the "box" approach is used to simulate distributed inflow of the radionuclides from contaminated watersheds to the river net. This parameter is calculated on the base of experimentally defined "washing-out coefficients" (Borzilov et al., 1989) from the total amount of a radionuclides in the fallout on the surface of a sub-watershed ("box"), crossed by the river and on the base of data of the runoff from the watersheds in the specific climatic areas.

For first post-accident period the predictions should be done on the base of day-by-day forecast. For the simulation of this dynamics the area of radioactive fallout under consideration is devided, as in a first approach, on the subwatersheds. The radionuclides concentration data on the watershed surface is received in RIVTOX from the atmospheric dispersion submodel of RODOS or from experimental data. The surface runoff on each catchment is described by "kinematic wave" approach. Washout of the contamination in solute and on sediments are defined by washing-out coefficients (Lyashenko, 1991).

For modeling of the unsteady hydraulic processes in an open stream Saint-Venant equations are used. For plain river these equations may be simplify (Cunge et al.,1980) by neglecting the inertial and acceleration slope terms. In this case the equation of the "diffusive wave" approximation could be written

$$\frac{\partial Q}{\partial t} + \frac{\partial Q}{\partial A}\frac{\partial Q}{\partial x} = \frac{Q}{2BI}\frac{\partial^2 Q}{\partial x^2} + \frac{\partial Q}{\partial A}q \tag{1}$$

where A - cross section area, Q - water discharge, t - time; x- distance along stream, B - width of a stream, I - bed slope, q - flux per unit stream length. This model is used in the PC's-version of RIVTOX.

The aquatic pollutants dispersion submodel describes dynamics of crossectionally averaged values of suspended sediments concentration S, radionuclides concentration in solute C, radionuclides concentration on suspended sediments C^s and radionuclides concentration in bottom deposition C^b by following system of the equations:

$$\frac{\partial S}{\partial t} + \frac{Q}{A}\frac{\partial S}{\partial x} = \frac{1}{A}\frac{\partial}{\partial x}\left(AE\frac{\partial S}{\partial x}\right) - \frac{q^s}{h} + \frac{q^b}{h} + \frac{q}{A}(S_R - S), \tag{2}$$

$$\frac{\partial C}{\partial t} + \frac{Q}{A}\frac{\partial C}{\partial x} = \frac{1}{A}\frac{\partial}{\partial x}\left(AE\frac{\partial C}{\partial x}\right) - \lambda C + \frac{q}{A}(C_R - C) - \atop -a_{1,2}S(K_{ds}C - C^s) - a_{1,3}(K_dC - C^b)\rho_s(1-\varepsilon)Z^*/h \tag{3}$$

$$\frac{\partial C^s}{\partial t} + \frac{Q}{A}\frac{\partial C^s}{\partial x} = \frac{1}{A}\frac{\partial}{\partial x}\left(AE\frac{\partial C^s}{\partial x}\right) + qS_R(C_R^s - C^s)/(AS) - \atop -\lambda C^s + a_{1,2}(K_{ds}C - C^s) + q^b(C^b - C^s)/(hS) \tag{4}$$

$$\frac{\partial C^b}{\partial t} = a_{1,3}\left(K_dC - C^b\right) - q^s\left(C^b - C^s\right)/\left(\rho_s(1-\varepsilon)Z^*\right) - \lambda C^b, \tag{5}$$

$$\frac{\partial Z^*}{\partial t} = \left(q^s - q^b\right)/\left((1-\varepsilon)\rho_s\right). \tag{6}$$

Where h-the river depth; E-the dispersion coefficient (Elder, 1959); q^s and q^b-the vertical fluxes of sediments, i.e.sedimentation and resuspension rates which are calculated as the function of the flow parameters and equilibrium suspended sediments concentration (Bijker, 1968); Z^*-the efficient thickness of the contaminated, upper, bottom deposition layer; S_R, C_R, C^s_R- the concentration in the lateral inflow of suspended sediments and radionuclides in solute and on suspended sediments, respectively. The model parameters are: λ–the decay constant; K_{ds}-the distribution coefficient, $a_{1,2}$-the exchange rate for the system "water-suspended sediment"; K_d, $a_{1,3}$- the same parameters for the system "water - bottom sediments"; ρ_s-the density of the bottom sediments; ε- the porosity of the bottom deposition.

Numerical methods

Numerical solution of "diffusion wave" equation (1) is constructed on the base of splitting method. The procedure allows using the implicit second order in space numerical scheme for advective and diffusive terms separately.

Transport of substances in streams is the combination of different physical processes: advection by the flow, dispersion due to the turbulent diffusion and of the velocity variability in a river crossection. Advection is a dominant process in reservoirs, therefore numerical diffusion of the finite-difference approximation can be higher or the same order of magnitude as a physical diffusion. In RIVTOX this problem is avoided by using the high accuracy numerical method (Tkalich, 1990) to solve the system (2)-(6). Following this method the Hermite cubic interpolation applies for approximation of the advection and diffusion terms for constraining one-step two-points "up-stream" scheme.

Software description

IBM PC's compatible version of RIVTOX includes data base about pattern of a river channel net and channel crossections for considered water basin, distribution of subcatchment areas. The graphical user interface allows an user to perform quick changes of a river net configuration and parameters. The data about temporal and spatial distribution of water discharge and water surface elevation in a river net can be obtained from data base of typical scenarios or simulated by the hydrological submodel.

All submodels: runoff, hydrological and radiological are connected via the data base. Hydrological and radiological submodels use data about the pointed and distributed sources of the contamination along a river channel.

The adaptation of the RIVTOX into the UNIX-based RODOS system (Zheleznyak et al., 1993) is providing with development of the interface with the GIS, atmospheric fallout and fish-chain models.

CODE APPLICATIONS

RIVTOX is used for the simulation of radionuclides' transport in the Pripyat River and Dnieper River after the Chernobyl accident. The results of prediction shows reasonable agreement with observed data. RIVTOX can be used not only for simulation of the radionuclides, but also for other toxic substances after some modification. It has been applied for the simulation of waste release from designed Paper Factory on the Daugava River (Latvia), phenol and formaldehyde accidental spill in the Desna River (1990, Ukraine). Some model validation studies briefly presented below.

Dnestr River study

Dniestr River study was done in 1992 to simulate the river water contamination due to potential threat of the Dubossary Dam's (Moldova) destruction during local military conflict in this region. The comprehensive hydrological information about this region (Sherbak et al., 1978) permits to compare "diffusive wave" model with the observed data from the Dubossary Dam downflow to the river estuary. The example of the simulation on the base of 1989 input data is presented for the Oloneshty crossection 351 km down flow from the Dubossary (Fig. 1)

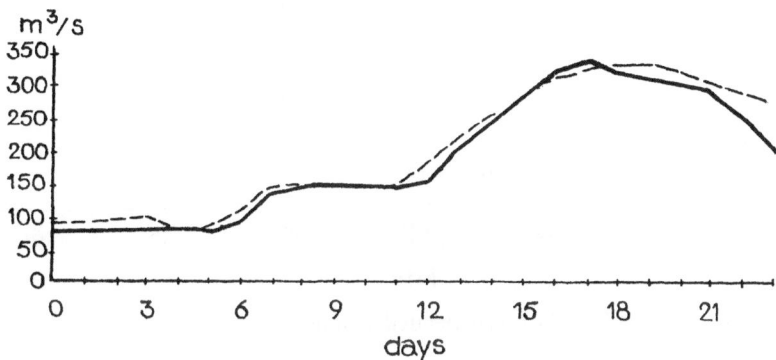

Fig. 1. Simulated (solid line) and measured (dashed line) Dnestr river discharge at Oloneshty 1989 year.

Dnieper River study

The comprehensive experimental studies of the radionuclides transport with different fractions of the suspended sediments have been done in the summer 1993 (Voitsekhovitch, 1993) at the Dnieper River. The measurements from the laboratory sheep of the Ukrainian Hydrometeorological Institute were done on the tributaries, Kiev Reservoir and Kanev Reservoir from June 25 till June 29 1993. Taking into account that water discharge to Kiev Reservoir from Pripyat River and Dnieper River from May 20 till

June 30 changed respectively in the range 650-258 cub.m/sec and 650-342 cub.m/sec, crossectionally averaged flow velocity in Kiev Reservoir were calculated as 5-1 cm/sec. The velocities in this part of Kanev Reservoir (near Kiev) were 0.5-0.1 m/sec.

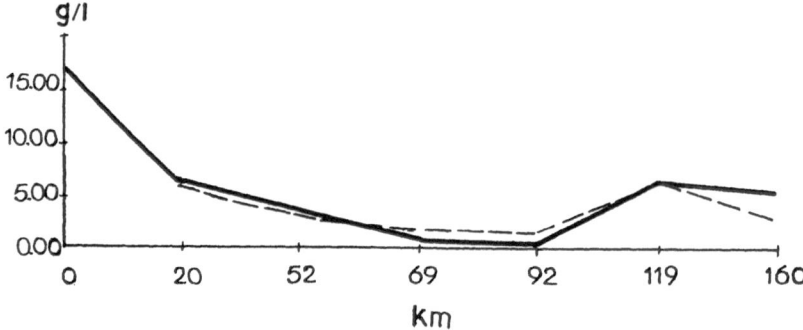

Fig.2. Suspended sediment (size more 40 mkm) concentration distribution in the Kiev and Kanev reservoirs June 25, 1993.

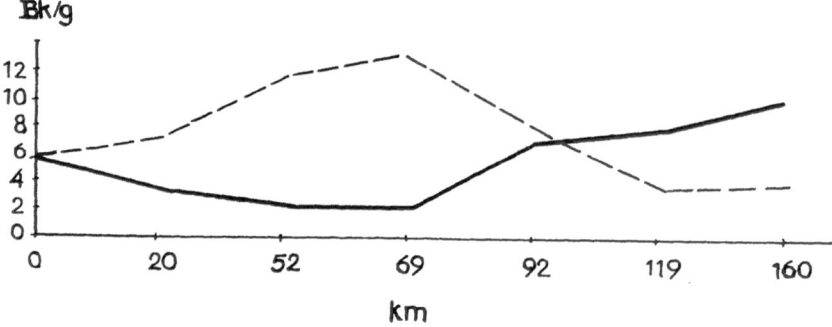

Fig.3. Concentration of Cs-137 on suspended sediments

On the base of these data demonstrated that distribution of the radionuclides in the Kanev and Kiev reservoirs in June 25-29 have been created by contamination, discharged to the Kiev Reservoir from the Dnieper and Pripyat rivers since end of the May. Therefore, to simulate spatial distribution of the radionuclides, it is necessary to start simulation from the middle of May. The results of simulation (solid lines) and observed data (dashed lines) are presented for June 25 on the Fig.2,3. The data distributions are shown from the Pripyat River mouth through the Kiev Reservoir (Kiev HPP -92km) to town Vitachev at the Kanev reservoir (160 km). The comparison of measured an simulated data shows that the main discrepancy of the result is connected with the radionuclides concentration on sediments for the Kiev Reservoir. It is connected with organic matters (algae) that is not taken into consideration during the simulation. The exchange rate parameters were taken $a_{1,2}=1/(100$ day$)$, $a_{1,3}=1/(100$ year$)$.

Sandos accident

On November 1, 1986, a fire broke out in a chemical warehouse located near Basel, Switzerland, on the Rhine River -160 km. The organophosphorus acid ester pesticides has been released into the river. The total mass of the chemicals has been evaluated to be approximately 7000 kg . After 12 h a dam was constructed to capture the runoff water.The results of the model equations (2)-(6) with the following parameters (Schnoor et al.,1992): $\lambda=0.2$/day, $\rho_s=2.6$ kg/l, $K_d=100$ l/kg, $\varepsilon=1$, $K_{ds}=200$ l/kg, $a_{1,2}=1$/day, $a_{1,3}=1/(100$day$)$ are shown in Fig.4.

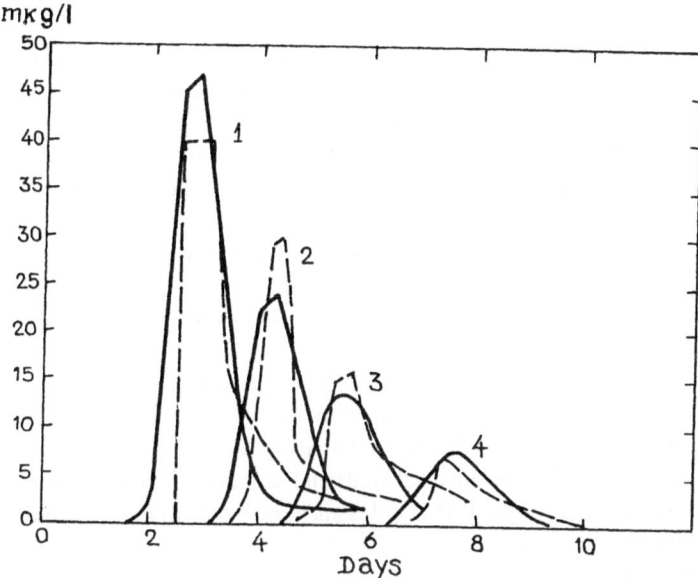

Fig. 4. Simulated concentration of organophosphorus in Rhine after spill in Basel (solid lines) and observed data (dashed lines). *1*-362 km, *2*-496 km, *3*- 640 km, *4*-865 km.

These results demonstrate that the Elder's formula that used to calculate the dispersion coefficient E overestimates its value. A long tail of the concentration simulates contamination of the river bank and the slow dissolution of an oily phase that remained on the bottom of the Rhine River after the accident.

ACKNOWLEDGMENTS

Different aspects of the research have been supported by the Commission European Communities JSP-1 and ESP-3 Projects.

REFERENCES

Borzilov V.A.,Vozhennikov O.I., Dragolubova I.V., Novitski M.A. (1991) "Modelling of removal of radionuclides from soil by rain and thaw water", Vodnye Resursy 3, 103-107.

Bijker E.W. (1968) "Some considerations about scales for coastal models with movable bed", Delft Hydr.Lab. Publ. No 50.

Cunge J.A., Holly F.M. and Verwey A. (1980) Practical Aspects of Computational River Hydraulics, Pitman Publishing Ltd., London.

Ehrhardt J., Paesler-Sauer J., Schuele O., Benz G., Rafat M. and Richter J. (1992) "Development of RODOS, a comprehensive decision support system for nuclear emergencies in Europe", Third International Workshop on Decision-Making Support for Offsite Emergency Management, Schloss Elmau, Bavaria, October 25-30, 4.1.

Elder J.W. (1959) "The dispersion of marked fluid in turbulent shear flow", J.Fluid Mech. 5, 544-560.

Holly F.M., Yang J.C., Schwarz P., Schaefer J., Hsu S.H. and Einhellig R. (1990) "Numerical simulation of unsteady water and sediment movement in multiply connected networks of mobile-bed channels", IIHR Report No.343.

Lyashenko G.B. (1991) "Modeling of the radionuclides migration with surface runoff", in A.A.Morozov (ed), Computer Decision Support Systems in Ecology, Publ. of Cybernetics Institute, Kiev, pp. 36-42.

Onishi Y., Serne J., Arnold E., Cowan C., Thompson F. (1981) "Critical review: radionuclide transport, sediment transport, water quality, mathematical modeling and radionuclide adsorption/desorption mechanism", Report of the Pacific Northwest Laboratory NUREG/CR-1322, Richland.

Sherbak A.V. and Shereshevskiy A.N. (1978) "Investigation of flood characteristics of Dniestr", Annual Report No. 17, Ukrainian Hydrometeorologic Institute, Kiev.

Schnoor J.L., Mossman D.J., Borzilov V.A., Novitsky M.A. and Gerasimenko A.K. (1992) "Mathematical model for chemical spills and distributed source runoff to large rivers", in J.L.Schnoor (ed), Fate of Pesticides and Chemicals in the Environment, A Wiley-Interscience Publ., pp.347-370.

Tkalich P.V. (1990) "Numerical modeling of dissolved contamination's dispersion in water bodies", in A.A.Morozov (ed), System Analysis and Methods of Mathematical Modeling in Ecology, V.Glushkov Institute of Cybernetics, Kiev, pp.62-66.

Voitsekhovitch O.V. (1993) Experimental Study of the Radionuclide-Sediment Interactions in the Pripyat River During the 1993 CEC ESP-3 Project Expedition, Report No. ESP-2/93, Ukrainian Hydrometeorological Institute, Kiev.

Zheleznyak M.J., Demchenko R.I., Khursin S.L., Kuzmenko Y.I., Tkalich P.V. and Vitjuk N.Y. (1992) "Mathematical modeling of radionuclide dispersion in the Pripyat-Dnieper aquatic system after the Chernobyl accident", The Science of the Total Environment 112, 89-114.

Zheleznyak M.J., Tkalich P.V., Marinets A.I. and Lyashenko G.B. (1993) "The computer code for describing the transport of radionuclides in a river system", Report RODOS(D)-RP(93)03, Kernforschungszentrum Karlsruhe, Germany.

ON THE ROLE OF k IN DEPTH-AVERAGED k-ε TURBULENCE MODELLING

T. WENKA
Numerical Models Group
Hydraulics Division, Bundesanstalt für Wasserbau (Federal Waterways Engineering
and Research Institute), 76187 Karlsruhe, Kußmaulstr. 17
FRG

A depth-average calculation procedure is presented for flows in rivers with complex, irregular geometry. The governing flow equations are solved with an efficient finite-volume method employing curvilinear boundary-fitted grids. Horizontal turbulent momentum exchange processes are simulated with a depth-average version of the k-ε turbulence model. The calculation procedure and especially the role of k in the depth-average turbulence model are tested by application to an overbank flow situation in a meandering channel of compound cross-sections. The agreement between calculated and measured velocity field and water surface level is generally good although the calculated and measured k-values do not match well. An interpretation of the discrepancies is given on the basis of the assumptions used within the depth-averaging process of the standard three dimensional k-ε turbulence model.

INTRODUCTION

In the last decade, the tasks of river engineers have been extended by an increased consideration of environmental and landgardening aspects. Due to that change in requirements, rivers are no longer made to be canal-like but have a great geometrical complexity with meanders, irregular boundaries and compound cross-sections. Consequently, research into the understanding of the physical process controlling flow in compound channels as well as the numerical modelling of the momentum transfer between the main channel and its flood plains has been intensified.

Some of the more recent data have been published by Stein (1990), who investigated the flow in a meandering channel with inundated flood plains shown schematically in Fig. 1. In spite of three dimensional flow features, simplifications are necessary in order to get efficient solutions of the problem. In default of empirical and analytical methods to cope in a general way with the flow complexities observed (Fig. 1), more and more flow studies are performed by using 2- or 3-dimensional numerical models. The advances in computer technology and numerical methods allow the efficient use of depth-averaged models based on curvilinear boundary-fitted grids for creating an accurate image of the computational domain.

A. Peters et al. (eds.), Computational Methods in Water Resources X, 1181–1188.

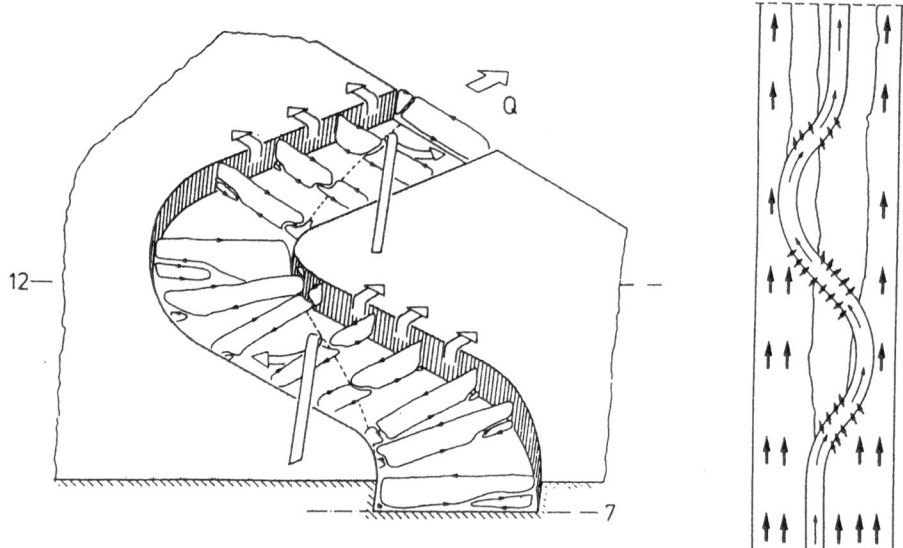

Figure 1: Complex features in overbank flow situations (observed by Stein, 1990)

This paper describes the application of an efficient, flexible depth-average calculation method for river and channel flows to situations with (a) dominant main channel and (b) dominant flood plain flow investigated in laboratory experiments by Stein (1990). The horizontal turbulent momentum exchange processes are simulated with a depth-average version of the well-known k-ε turbulence model. Emphasis is put on the role of k in the turbulence model as well on the diffusion coefficient globally describing the lateral mixing behaviour of the flow.

MATHEMATICAL MODEL

The steady, depth-average flow in rivers is governed by the continuity equation and the momentum equations for the horizontal velocity components, integrated over the local water depth. Using Cartesian velocity components, the resulting equations can be written in common tensor form as given in Table 1 (by putting $\Phi=1$, V_1, V_2).

The bed shear stress components are calculated from the following quadratic friction law:

$$\tau_b^i = \rho c_f V_i V_{res}$$

where the friction coefficient c_f for hydraulically smooth conditions is expressed by

$$c_f = 0.027 \left(\frac{\nu}{Vh}\right)^{1/4}$$

(Schlichting, 1965).

$$\frac{\partial}{\partial x_i}(C_i \Phi + D_{i\Phi}) = J S_\Phi$$

	Φ	C_i	$D_{i\Phi}$	S_Φ
a)	1	U_i	0	0
b)	V_k	U_i	$-\frac{\nu_t}{J} h \left(B^i_j \frac{\partial V_k}{\partial x_j} + \beta^i_j \omega^j_k \right)$	$\frac{1}{\rho} \tau_{b,k} - \frac{h}{\rho J} \frac{\partial}{\partial x_j}\left(P \beta^j_k \right)$
c)	k	U_i	$-\frac{\nu_t h}{\sigma_k J} \left(B^i_j \frac{\partial k}{\partial x_j} \right)$	$\frac{1}{\sqrt{c_f}} \frac{U^3_*}{h} + P_h - \varepsilon$
d)	ε	U_i	$-\frac{\nu_t h}{\sigma_\varepsilon J} \left(B^i_j \frac{\partial \varepsilon}{\partial x_j} \right)$	$c_\varepsilon \Gamma \frac{c_{\varepsilon 2}}{c_f^{3/4}} \sqrt{c_\mu} \frac{U^4_*}{h^2} + c_{\varepsilon 1} \frac{\varepsilon}{k} P_h - c_{\varepsilon 2} \frac{\varepsilon^2}{k}$

$$J = \begin{vmatrix} \frac{\partial y_1}{\partial x_1} & \frac{\partial y_2}{\partial x_1} \\ \frac{\partial y_1}{\partial x_2} & \frac{\partial y_2}{\partial x_2} \end{vmatrix} \qquad \beta^i_j = \text{cofactors of } \frac{\partial y_i}{\partial x_j} \text{ in J} \qquad B^i_j = \beta^i_l \beta^j_l$$

$$U_i = V_j \beta^i_j h \qquad \omega^j_k = \frac{\partial V_j}{\partial x_l} \beta^l_k \qquad P = gH \qquad U_* = \sqrt{\frac{\tau_b}{\rho}}$$

$$P_h = \frac{\nu_t}{J^2} \left(\frac{\partial V_l}{\partial x_n} \beta^n_j + \frac{\partial V_j}{\partial x_m} \beta^m_l \right) \left(\frac{\partial V_l}{\partial x_n} \beta^n_j \right) \qquad \nu_t = c_\mu \frac{k^2}{\varepsilon}$$

$$c_\mu = 0.09 \quad c_{\varepsilon 1} = 1.44 \quad c_{\varepsilon 2} = 1.92 \quad \sigma_k = 1.0 \quad \sigma_\varepsilon = 1.3 \quad c_\Gamma = 3.6$$

Table 1: Complete set of model equations in general curvilinear coordinates

It can be seen from Table 1 that the diffusion term $\mathbf{D}_{i\Phi}$ contains the eddy viscosity

$$\nu_t = c_\mu \frac{k^2}{\varepsilon}$$

which models the horizontal turbulent momentum exchange processes. The eddy viscosity is calculated by means of a depth-average version of the k-ε turbulence model developed by Rastogi and Rodi (1978) involving transport equations for the turbulent kinetic energy k and its dissipation rate ε. It should be mentioned that the constant $c_{\varepsilon \Gamma}$ used corresponds to a value of 0.15 for the dimensionless diffusivity e* recommended by Rastogi and Rodi (1978) for straight wide laboratory flumes. Finally it should be emphasized that the method described here does not account for the dispersion terms resulting from formal integration of the original equations over the water depth.

All equations are solved with an efficient, flexible finite-volume method using cell-centred variable arrangement. Further details on the turbulence model and the numerical method are given in the previous paper of Wenka et al. (1991) and in Wenka (1992).

MODEL APPLICATION

Test Cases provided by IWW of RWTH Aachen

In continuation of main channel / flood plain flow studies in straight channels IWW (Institut für Wasserbau und Wasserwirtschaft) performed further investigations in meandering channels. For the detailed experimental schedule a transparent hydraulic model of 15 m length and 3 m width was built. At selected cross-sections the profiles of the horizontal mean and fluctuating velocity components were measured with LDA in back-scattering technique. The main channel with its rectangular cross-sections of 0.4 m width and 0.1m depth is winding with a 1.8 m mean radius of curvature through the model domain (Fig. 1) and so the flood plain regions on both sides have variable widths.

As part of the above mentioned project, two flow situations with identically smooth bottom roughness of $n = 0.01$ in the main channel and on the flood plains were simulated. Following the experiments of Stein (1990) for the simulation of the dominant flood plain flow (b) with a mean flow depth of $h_{mc} = 0.15$ m in the main channel at inlet and a discharge of $Q = 54.6$ l/s good representation of the flow features by the 2D-model could be expected. For case (a) with dominating main channel flow ($h_{mc} = 0.12$ m at inlet, $Q = 17.0$ l/s), locally prevalent 3D-effects as shown in Fig.1 indicate that the adopted concept of depth-averaging itself is not able to give suitable solutions of the problem. The high quality of experimental data, however, gave reason to examine, how the numerical model was globally able to reflect the observed phenomena and their influence on the main parameters of flow as well as the distribution of the turbulent kinetic energy k.

Computational Details

In the numerical model the lateral profiles of the velocity and, depending on that, the profiles of k and ε were prescribed. For case (a) a constant velocity distribution was supposed. In case (b) different velocities for the flood plains ($V_{fp} = 0.3$ m/s) and the main channel ($V_{mc} = 0.23$ m/s) were prescribed.

High gradients in flow parameters, to be expected near the wall-boundaries and in the range of the banks, required local refinement of the numerical grid (Fig. 2) with its 49 points in lateral and 69 points in streaming direction. Following the embankment lines the three parts of the boundary-orthogonal grid were zonally generated by algebraic methods.

The solution required in both cases 260 iterations to reach convergence, which was declared, when the largest dimensionless residual ε was smaller than 10^{-3}. The calculation took 25 s CPU time on the Siemens SNI S600/20 vector computer of the University of Karlsruhe.

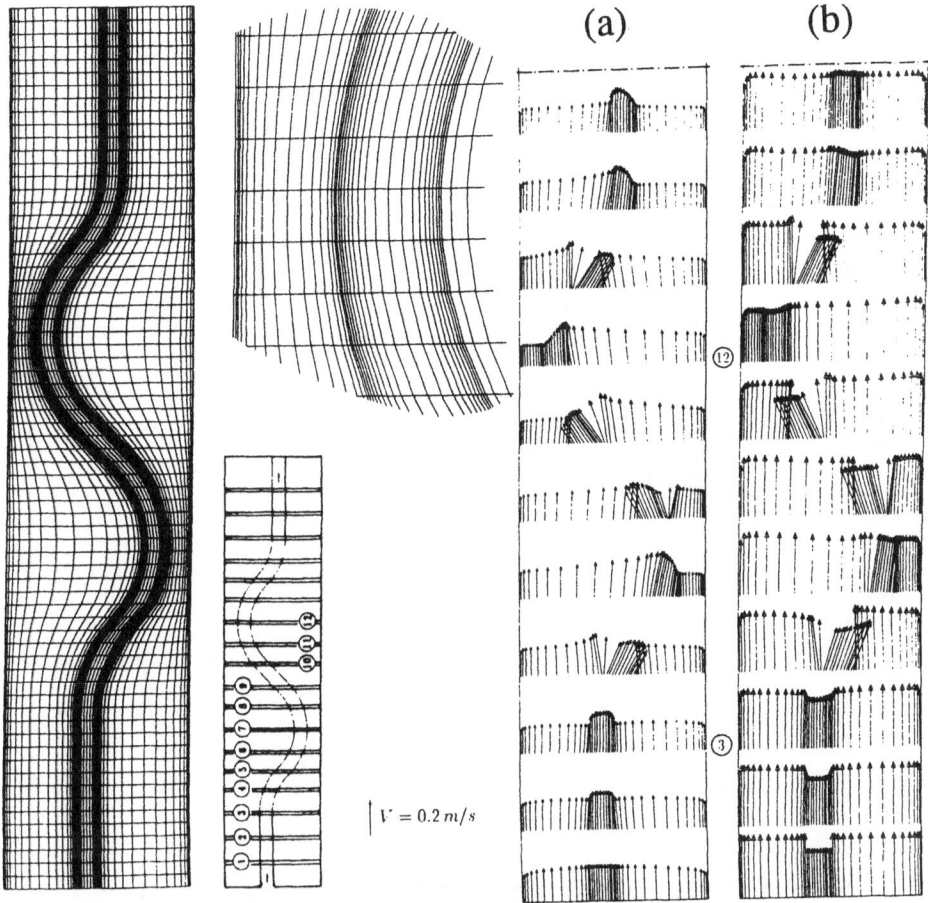

Figure 2: Numerical grid and predicted velocity distribution

Comparison of Predictions and Measurements

For comparison of the turbulent kinetic energy measured in the laboratory model with **k** predicted by the numerical model and in default of measured vertical components of the fluctuating velocity a definition of **k** due to the measurements is used as proposed by Tropea (1982)

$$\mathbf{k} = \frac{3}{4}\left(\overline{\mathbf{v'}_1^{\,2}} + \overline{\mathbf{v'}_2^{\,2}}\right)$$

In spite of that artifical increase of measured **k** the predicted **k** were overestimated by about 85% in case (a) and 170% in case (b). Fig. 3, where the lateral distributions of **k** in a cross-section with strong secondary currents are compared, gives an impression of the discrepancies.

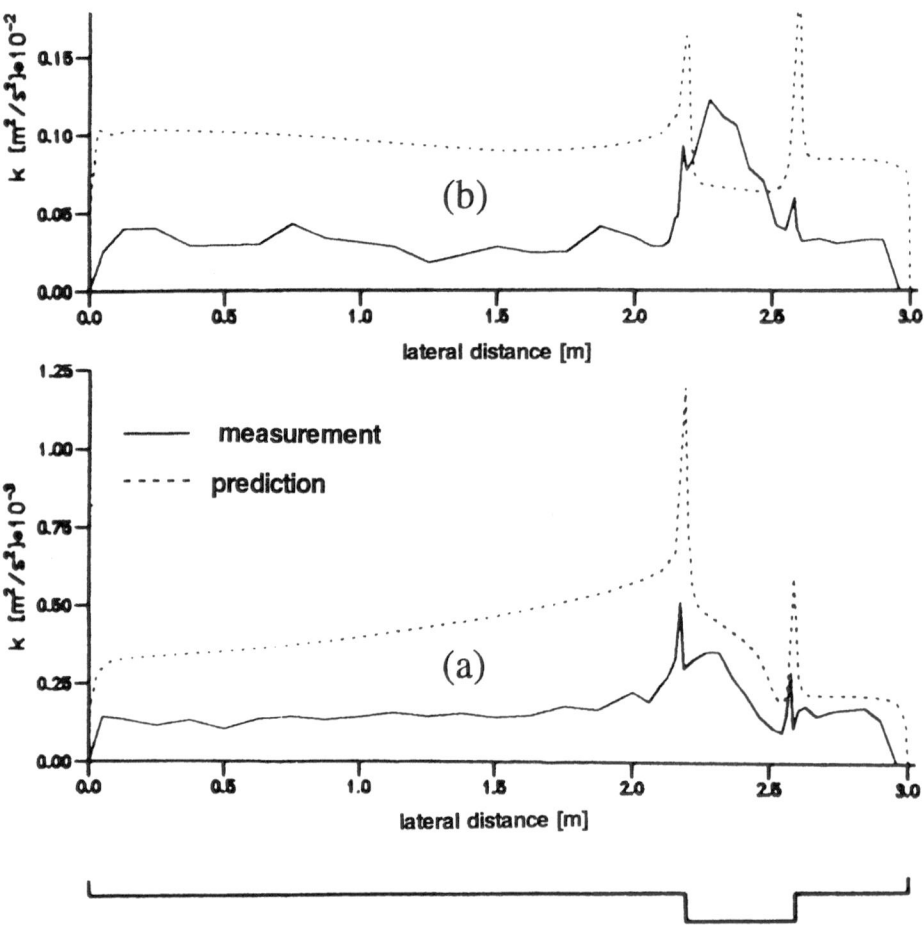

Figure 3: Turbulent kinetic energy distribution in cross-section no. 7

Apart from the global overestimation, in case (a) the predictions agree fairly well with
the experiments. Case (b), however, shows clear discrepancies within the main channel
region. Three dimensional shearing zones originating from curvature-induced
secondary motion and superimposed eddies, induced by interaction of the straight
flood plain and the inclined main channel flow, seem to be responsible for these
discrepancies. The distinctly weaker k-values in the main channel region of the
numerical model, observed only in case (b), are a clear outcome of the simplifications
made by depth-averaging and of the assumptions the standard k-ε turbulence model is
based on, like eddy-viscosity concept and isotropic eddy viscosity (Rodi, 1984). The
fact, that the **k** of the depth-averaged **k-ε** model is not a strictly depth-average quantity
but coarsely defined by direct coupling on roughness-induced vertical turbulence
production as a function of the friction velocity and an empirical depth-mean
diffusivity, seems to be decisive for the global overestimation of **k** in the numerical
model.

The definition of empirical constants of the depth-averaged k-ε model on the proposal of a uniform flow situation leads to the expressions

$$e^{*} = \frac{v_t}{0.5 U_* h} \qquad \text{and} \qquad \varepsilon = \frac{U_*^3}{\sqrt{c_f}\, h}$$

for the dimensionless diffusivity and for the dissipation of **k** respectively. The combination of the definition of v_t as mentioned above with the expressions for e* and ε leads to the following dimensionless relationship between **k** and c_f

$$\frac{k}{U_*^2} = \sqrt{\frac{0.5}{c_\mu}\, e^{*}}\; c_f^{-\frac{1}{4}}$$

According to test case (b) with a mean velocity of $V_{fp} = 0.34$ *m/s*, a mean depth of $h_{fp} = 0.05$ *m* and hydraulically smooth conditions on the flood plains, a mean friction coefficient of $c_f = 0.0024$ is obtained by applicating Schlichting's (1965) formula. A value of $k/U_*^2 = 4.1$ finally results, when the dimensionless diffusivity e* is set to 0.15, standing for laboratory-like mixing characteristics. The value of 4.1 representing **k** in the depth-averaged k-ε model is about twice the depth-mean value of $k/U_*^2 = 2.2$ resulting from measurements in flumes with rectangular cross-sections (Nezu and Nakagawa, 1993). The measured value to be representative for nearly pure 2D-flows is indicating a systematic fault in the numerical model, to be reduced by e* = 0.1.

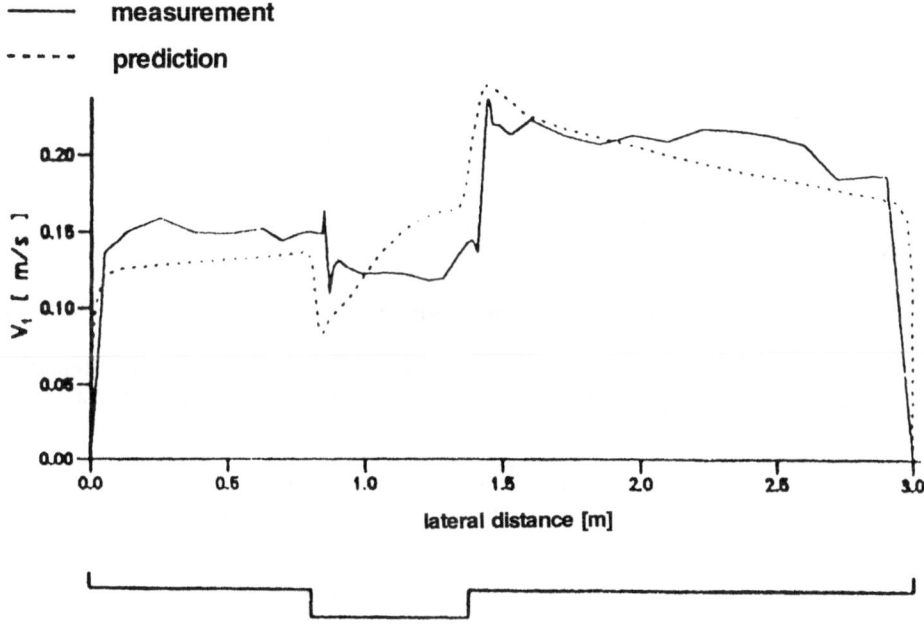

Figure 4: Longitudinal velocity profiles in cross-section no. 10, case (a)

In spite of the observed discrepancies in **k**, the agreement in velocities and water surface elevation between measurement and prediction was fairly good. Fig. 4 shows representatively the lateral profiles of the longitudinal velocity components and indicates, that even in extrem flow situations the numerical method works exactly enough to predict the main parameters of flow suitable for practical engineering purposes.

CONCLUDING REMARKS

Depth-average calculations were presented for two flow situations in a meandering channel with flood plains: a high flood situation with dominant flood-plain flow and an average flood situation with dominating main-channel flow. The governing equations were solved by using an efficient finite-volume method which allows the use of arbitrary boundary-fitted grids and also arbitrary bottom topography. Complex channel flows with locally prevalent 3D-effects could be simulated with the model exactly enough for practical purposes, and generally good agreement with laboratory measurements was obtained for flow velocities and water surface elevation. For the turbulent kinetic energy the foreseen discrepancies between measured and predicted k-values in zones of strong secondary motion were observed. The systematic deviation of measured and predicted k-values observed on the flood plains could be reduced to the simplification adopted in the depth-averaged k-ε model of absorbing all the vertical non-uniformities of flow in production terms related solely to bottom friction.

REFERENCES

Nezu, I., Nakagawa, H. (1993) Turbulence in Open-channel Flows, IAHR Monograph, A.A. Balkema, Rotterdam

Rastogi, A.K., Rodi, W. (1978) "Prediction of Heat and Mass Transfer in Open Channels", ASCE Journal Hydraulics Division, Vol. 104, No. 3

Rodi, W. (1984) Turbulence Models and their Applications in Hydraulics, IAHR Monograph, June

Schlichting. H. (1965) Grenzschicht-Theorie, Verlag G. Braun, Karlsruhe

Stein, C.J. (1990) Mäandrierende Fließgewässer mit überströmten Vorländern, Report No. 76, Institute for Hydraulic Research and Water Resources Development, Aachen Univ. of Technology

Tropea, C. (1982) Die turbulente Stufenströmung in Flachkanälen und offenen Gerinnen, Report of Sonderforschungsbereich 80, Univ. of Karlsruhe, SFB80/E/210

Wenka, T., Valenta, P., and Rodi, W. (1991) "Depth-Average Calculation of Flood Flow in a River with Irregular Geometry", Proc. 24th IAHR Congress, Madrid

Wenka, T. (1992) Numerische Berechnung von Strömungsvorgängen in naturnahen Flußläufen mit einem tiefengemittelten Modell, Dissertation, Univ. of Karlsruhe

MODELLING OF RADIONUCLIDE TRANSPORT IN THE SET OF RIVER RESERVOIRS

M.J. ZHELEZNYAK, YU.I. KUZMENKO, P.V. TKALICH, N.N. DZUBA,
D.S. GOFMAN, I.N. GOLOVANOV, A.V. MARINETS and I.V. MEZHUEVA.
Institute of Mathematical Machines and Systems,
Cybernetics Center Ukrainian Academy of Sciences,
Pr. Glushkova 42, Kiev, 252207, Ukraine

The Chernobyl accident heavily contaminated the Dnieper reservoirs. The set of mathematical models was developed to simulate radionuclide dispersion in the Pripyat-Dnieper river--reservoir system. Box model WATOX is used for seasonal and long term forecasting of radionuclide concentration in the reservoirs. The simulation of the spatial distributions of the radionuclide concentration in the water bodies is done based on the two-dimensional lateral-longitudinal model COASTOX.

INTRODUCTION

The Chernobyl Nuclear Power Plant is situated on the bank of the Pripyat river 20 km from its inflow into the Kiev Reservoir at the Dnieper river. Radionuclides washed out since April 1986 from the heavy contaminated Chernobyl zone are transported downflow by the Dnieper water to the Black Sea through the set of six reservoirs. From the Pripyat River mouth downstream there are Kiev Reservoir - capacity 3.7 cub. km, Kanev Reservoir-2.6, Kremenchug Reservoir-13.5, Dneprodzerzhinsk Reservoir-2.4, Dneprovskoje Reservoir-3.3 and Kakhovskoje Reservoir-18.2 cub. km. Radionuclides are washing out from the watersheds under the impact of snow-melting and rain storm runoff. The larger part of the runoff going through the reservoirs during spring floods from March till June.

The radionuclide fate in the Pripyat river and in the Dnieper river-reservoir system was researched and modeled in Ukraine since May 1986 (Zheleznyak et al., 1992). Modelling of the radionuclide fate in three different phases: in solution, in suspended sediments and in the bottom deposition, is particularly important. Such an approach to radionuclide dispersion simulation has been developed by Onishi et al. (1981) for one-, two- and three-dimensional models Schuckler et al. (1978) for full-mixed box models.

The Pripyat-Dnieper system includes different kinds of water bodies, varying from rivers with their tributaries and floodplains to large reservoirs. The temporal and spatial scales of the processes under consideration may increase by more then two orders of magnitude

A. Peters et al. (eds.), Computational Methods in Water Resources X, 1189–1196.

when one considers the different modelling objects. Such wide range of processes could not be described by one single model. A hierarchy of radionuclide dispersion mathematical models, describing the above mentioned processes, has been developed by consequent averaging of the primitive 3-D equations over the space variables (Zheleznyak et al., 1992). All modells take into account the fate of the radionuclides in three different phases: radionuclides in solution, radionuclides attached to the suspended sediments and radionuclides in the bottom deposition.

The description of adsorption-desorption processes is provided using the distribution coefficient K_d, that is the ratio between contamination on particles and in solution under steady state conditions. The other parameters are transfer rate coefficients $a_{i,j}$, which are determined by the character time of the transfer between i and j phases. The main factors affecting the sediment-contamination interaction may be taken into account when the K_d is considered as a function of water quality, geochemical properties of the sediments, physical-chemical forms of radionuclides, sediment concentration, etc. (Onishi et al. 1979). By this way the description of the physic-chemical behavior of the radionuclides linked with the peculiarities of the Chernobyl fallout has been aggregated. More complicated radionuclide transfer submodels that distinguish the different kinds of physical-chemical forms of radionuclides in the solid phase (exchangeable and non-exchangeable forms) are under development (Borzilov et al., 1991). The latter approach requires detailed experimental data, which as a rule can not be obtained without carefully investigations under site-specific field conditions. Only nowadays the collection of the necessary data sets for Pripyat-Dnieper aquatic system is close to be formed.

For the post-Chernobyl Dnieper river study two models: box model WATOX and two-dimensional lateral-longitudinal model COASTOX have been used directly to predict the radionuclide concentration at water intakes during high spring and rainstorm floods and also to estimate the efficiency of the special hydraulic countermeasures, designed to decrease the rate of radionuclide outflow from the Pripyat river and Kiev Reservoir. Some of the results are presented by Zheleznyak et al. (1992). Some results of recent simulations are considered in the present paper.

MODEL DESCRIPTION

COASTOX model

The COASTOX hydrodynamic submodel contains the shallow water equations

$$\frac{\partial U_j}{\partial t} + U_k \frac{\partial U_j}{\partial x_k} + g \frac{\partial \eta}{\partial x_k} = -\lambda_b U_j |\vec{U}| + \lambda_w W_j |\vec{W}| \tag{1}$$

$$\frac{\partial \eta}{\partial t} + \frac{\partial (hU_k)}{\partial x_k} = R \tag{2}$$

where λ_b-bottom friction coefficient, λ_w-wind friction coefficient, k=1,2; j=1,2; U_j-vertically averaged horizontal flow velocities in the x (j=1) and y (j=2) directions, W-horizontal wind velocities, R-sinks and sources distributed on the free flow surface (precipitation and evaporation).

The depth averaged advection-diffusion equation of the suspended sediment transport is written

$$\frac{\partial hS}{\partial t} + \frac{\partial U_j}{\partial x_k}(hSU_k) = \frac{\partial}{\partial x_k}\left(hE_{ik}\frac{\partial S}{\partial x_i}\right) + \beta w_0(S_* - S) \tag{3}$$

where S and S_* are vertically averaged suspended sediment concentration and equilibrium suspended sediment concentration respectively; β-ratio of the near-bottom suspended sediment concentration to the depth averaged concentration; and E_{ik}-coefficients of the horizontal dispersion (Holly, 1987).

The depth averaged equilibrium suspended sediment concentration S_* is calculated by the Bijker method (Bijker, 1968), and includes the effects of wave bottom shear stress on the magnitude of S_*. The thickness of the upper active bottom deposition layer Z_*, may be described by the equation of bottom deformation

$$\rho_s(1-\varepsilon)\frac{\partial Z_*}{\partial t} = q^s - q^b \tag{4}$$

where the sedimentation and resuspension rates are calculated through the difference of actual and equilibrium suspended sediments concentration (Zheleznyak et al., 1992).

The depth averaged equation of the dilute contamination transport is written

$$\frac{\partial hC}{\partial t} + \frac{\partial}{\partial x_k}(hCU_k) = \frac{\partial}{\partial x_k}\left(hE_{ik}\frac{\partial C}{\partial x_i}\right) - \lambda hC -$$

$$-ha_{1,2}S(K_dC - C^s) - \rho_s(1-\varepsilon)Z_*a_{1,3}(K_dC - C^b) \tag{5}$$

where: C and C^s are depth-averaged concentrations of the radionuclides in solute and in suspended sediments respectively C^b-radionuclide concentration in the bottom deposition averaged over the contaminated layer thickness Z_*.

The transport equation for C^s is as follows

$$\frac{\partial hSC^s}{\partial t} + \frac{\partial}{\partial x_k}(hSC^sU_k) = \frac{\partial}{\partial x_k}\left(hE_{ik}\frac{\partial SC^s}{\partial x_i}\right) -$$

$$-\lambda hSC^s + ha_{1,2}S(K_dC - C^s) + C^bq^b + C^sq^s \tag{6}$$

The contamination of the upper layer of bottom deposition can be described by the equation

$$\frac{\partial Z_*C^b}{\partial t} = Z_*a_{1,3}(K_dC - C^b) - \frac{1}{\rho_s(1-\varepsilon)}(C^bq^b - C^sq^s) \tag{7}$$

The numerical solutions of the hydrodynamic submodel (1)-(2) are obtained by implicit finite-difference method (FDM). For the advection-diffusion equations (3), (5), (6) the fourth order explicit FDM, that is, the modification of the Holly-Preissman scheme (Cunge et al. 1980) is used. The verification of the developed FDM was based on a comparison with some analytical solutions.

WATOX model

Box model WATOX could be derived from the equations of the COASTOX model by their averaging over a compartment bottom area. The resulting set of the ordinary differential equations (Zheleznyak et al., 1992) describes the dynamics of the water volume in a box and the compartmentally averaged value of the suspended sediment concentration, the concentration of the radionuclide in solution, on suspended sediments and in bottom deposition. The results of the prediction of the radionuclide concentration within flood period depend on the operation mode of the reservoirs, i.e. from the changes in the water levels at the HPP dams. The optimization methods are used to provide choices of the reservoir system operation mode under the water quality criteria.

Let ξ_m is the vector of the variables which control water, sediment and radionuclide balance in the boxes (inflow fluxes from the tributaries, outflow to the water intakes, etc.), where m=1,2...M-the number of time intervals. The simplest optimization problem corresponds to the minimization of the target function

$$F(H, C, Q, \xi) \to \min . \tag{8}$$

with the restrictions

$$G_j(H, Q, \xi) \le 0, \qquad j=1, J, \tag{9}$$

where Q_m is the water discharge, H_m is the water elevation, J-the number of the restrictions, G_j-functions for the reservoir. The restrictions take into account the water balance requirements from the different users: Hydropower Plants, Ship navigation companies, Irrigation systems, Fishing and Environmental Institutions. The target function can be rewritten as

$$F(H, C, Q, \xi) = \sum_k p_k f_k, \tag{10}$$

where f_k-optimal criteria for the certain target of the reservoir management; p_k-weight coefficients for each criteria. The WATOX include numerical procedure to solve problem (8)-(10) for 6 certain criteria. From the water quality requirements most important within the used criteria is

$$f_k = \max(C_m^j) \tag{11}$$

It is used to diminish maximum radionuclide concentration in the southern reservoirs by the specific water elevation (discharge) operation mode at hydropower plants.

COASTOX APPLICATIONS

COASTOX code was applied to simulate ^{137}Cs and ^{90}Sr redistribution in the bottom depositions of Dnieper reservoirs during high floods and also to evaluate the efficiency of the dam construction to diminish radionuclides fluxes. One of the most powerful source of ^{90}Sr contamination of the Pripyat water and as result in all down flow Dnieper reservoirs is the fuel particles fallout on the Pripyat river floodplain North-Eastward from the Chernobyl Plant (Laptev and Voitsekhovitch, 1993). Near 10000 Ci of ^{90}Sr are deposited here on the territory 4 km*10 km. The simulation of the potential flooding of this territory has been provided since 1989 on the base of COASTOX computer code (Zheleznyak and Voitsekhovitch, 1991; Zheleznyak et al. 1992). This territory has not been flooded till 1991 because of low spring floods during this period.

The sediment-water interactions have not significant impact on the processes of ^{90}Sr migration in surface water due to the relatively low value of the distribution coefficient K_d for ^{90}Sr. The governing parameters of the model in this case are K_d and the water-bottom exchange rate $a_{1,3}$. For their identification the special laboratory measurements have been provided by Laptev and Voitsekhovich (1993). K_d=50 (l/kg) and $a_{1,3}$=0.0038 (1/day) values have been determined on the base of these data. The simulation of the floodplain flow has demonstrated that the most dangerous situation, causing large increases in radionuclides concentrations is a spring flood with a maximum discharge of 2000 m /sec. The probability of exceeding (PE) this flood magnitude for the area considered of the Pripyat river is 25%. During such spring flood the water covers all parts of the contaminated floodplain, but water depth is less then one for floods with lower PE. It was demonstrated by Zheleznyak and Voitsekhovitch (1991) that during such flooding the concentration of ^{90}Sr would increase from 50 pCi/l at inlet crossection (10 km upflow Chernobyl NPP) to 250-270 pCi/l at outlet crossection (the Yanov Bridge near Chernobyl NPP) due to the interaction with the contamination in the bottom depositions. The Ukrainian Maximum Permissible Level for ^{90}Sr concentration in water is 100 pCi/l. The construction of the special dam around the contaminated area on the left bank of the Pripyat river has been recommended in 1990 on the base of simulated results as the optimal countermeasure to prevent release from floodplain.

In January 1991 before starting of the dam construction ice-jam was formed in the Pripyat river channel between Yanov Bridge and the town of Chernobyl. As a result the Pripyat floodplain near the Chernobyl NPP was covered by water for the first time after the accident. Maximum concentration of ^{90}Sr near the Yanov Bridge increased to the values 240-260 pCi/l. This way the results predicted during simulation of countermeasures effectiveness (Zheleznyak and Voitsekhovitch, 1991) has been confirmed by a wide scale natural experiment. During the flooding the modelling system was use for real-time simulation of the situation on the floodplain on the base of COASTOX model (Fig. 1) and for prediction dynamic ^{137}Cs of ^{90}Sr concentration in the reservoirs and especially near Kiev on the base of WATOX model. Due to dilution and dispersion of contamination in the reservoir the maximum concentration on the way from River Pripyat mouth to Kiev HPP (more than 70 km) diminished from 200 to 30 pCi/l.

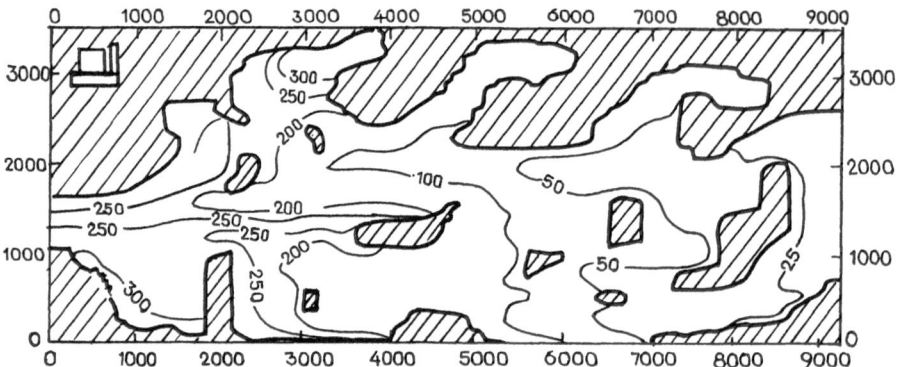

Fig. 1. Simulated distribution of the depth averaged ^{90}Sr concentration (pCi/l) at the Pripyat River floodplain upstream from the Chernobyl NPP (left boundary) during 1991 ice jam . Q=500 cub.m/sec.

The forecasts (confirmed later by direct measurements) have been presented to the Government Commission. It was used to make optimum change in February to the municipal water supply arrangements without having to use water from the River Dnieper. The dam preventing future flooding of the considered part of the Pripyat floodplain has been constructed till 1992 spring flood.

WATOX APPLICATIONS

The predictions of ^{137}Cs and ^{90}Sr concentration in the Dnieper reservoirs during spring flood (March-June) were prepared in February-March each year since the accident by applying the WATOX code. The data of the watersheds contamination density and the averaged values of the radionuclide wash off coefficients were used to predict ^{137}Cs and ^{90}Sr concentrations in the tributaries released into the reservoirs. Since 1986 the level of ^{137}Cs concentration in the Dnieper reservoirs decreases (close to the pre-accident values in the southern reservoirs) due to low magnitude of the spring floods and as a result of the diminishing of ^{137}Cs wash-out coefficient. The coefficient of ^{90}Sr wash out does not diminish in the same manner. Therefore the ^{90}Sr contamination remains most significant problem for high spring floods in the Pripyat River - Dnieper River system. The simulation of processes on the floodplain, considered above, have been supplemented by forecasting of ^{90}Sr dispersion in the Dnieper reservoirs (Zheleznyak et al. 1992).

The optimization module WATOX is used to provide choices of the reservoir operation mode on the base of criteria (11). The optimization efficiency increases with the diminishing of the ratio of spring flood volume to total reservoirs volume. The results for the average spring flood (PE=50%) demonstrate (Fig. 1, 2) that maximum diminishing of the concentration in the reservoirs could be near 25% of the peak value within the hydropower production restrictions. The maximum diminishing (up to 50% of the peak value) could be achieved for low water flood (PE=95%).

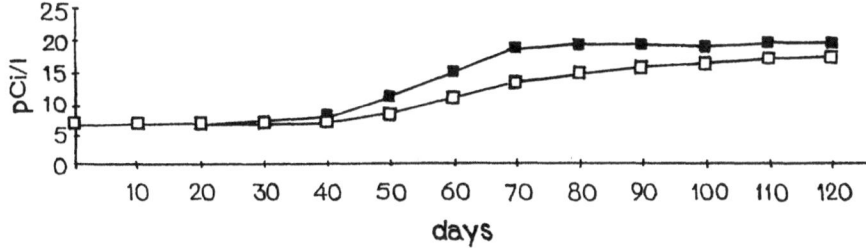

Fig. 2. ^{90}Sr concentration in the Kremenchug Reservoir during 50% PE flood after optimization (—■—) and without it (—□—).

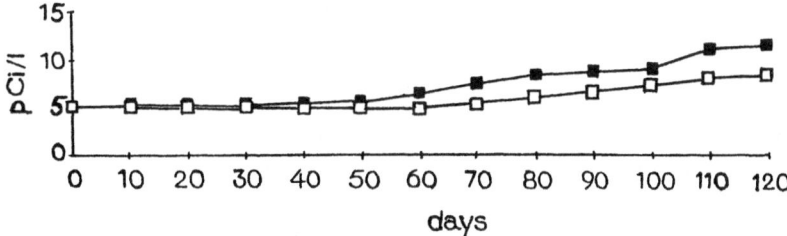

Fig. 3 ^{90}Sr concentration in the Kakhovkaja Reservoir during 50% PE flood after optimization (—■—) and without it (—□—).

To estimate the collective dose for population of Ukraine from the consumption of Dnieper water during 70 years after the accident (till 2057) WATOX code was used in the version based on three months averaged input data. The three month averaged discharge of the Dnieper River, Pripyat River and tributaries to the Dnieper Reservoirs since 1895 were used to create hydrological data base for a long term prediction. The scenario of the worst radiological conditions should be based on the sequences of high runoff years since 1994. The constructed set of the hydrological data used 1970-1992 (high runoff period) and then 1912-1950 (low runoff period) as a "hydrological forecast" for 1994-2057. For the prediction of radionuclides concentration in the tributaries to the Kiev Reservoir the regression relations between concentration and water discharge were constructed based on the experimental data of Laptev and Voitsekhovitch, (1993). These relations were used for simulation of 15 years after the accident (till 2001). It is supposed that the amount of water exchangeable form of ^{90}Sr on the watersheds will remain constant this period, balanced by the leaching from fuel particles and by the decay and the percolation into the soil. For simulation since 2001 the regressions were used with the attenuation coefficient equals to the double decay rate. The simulated results (Fig. 4) demonstrate that the large southern Kakhovka reservoir, damping the seasonal oscillations, will have after some years practically the same level of ^{90}Sr concenration as the Kiev reservoir, where three month averaged concentration will change from 27 pCi/l in the initial period to the 3-5 pCi/l in 2056. These data and data of the better from radiological conditions hydrological scenirious was used to calculate dose for the development of the post-Chernobyl countermeasures in Ukraine.

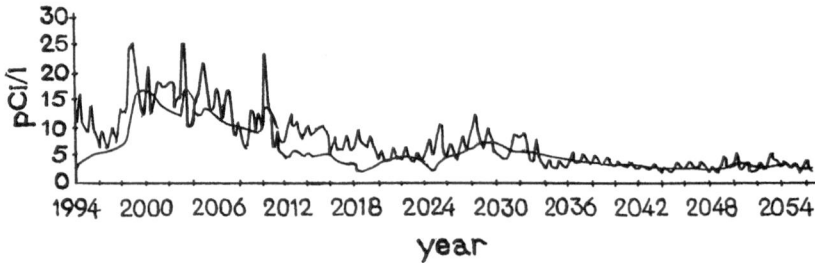

Fig.4. Prediction of Sr concentration in Kiev Reservoir (oscilliting line)and Kakhovskaja Reservoir for the scenario of the worst hydrological conditions.

REFERENCES

Borzilov V.A., Vozhennikov O.I., Dragolubova I.V. and Novitski M.A. (1991) "Modelling of removal of radionuclides from soil by rain and thaw water", Vodnye Resursy 3, 103-107. (in Russian)

Cunge J.A., Holly F.M. and Verwey A. (1980) Practical Aspects of Computational River Hydraulics, Pitman Publishing Ltd., London.

Bijker E.W., (1968) "Some considerations about scales for coastal models with movable bed", Delft Hydr. Lab., Publ. No 50.

Holly F.M., (1987) "Dispersion in rivers and coastal waters. -1.Phisical principles and dispersion equations" Developments in Hydraulic Engineering 3, 1-37.

Laptev O.V. and Voitsekhovitch O.V. (1993) "Experimental investigation of radionuclides liching from floodplain soils of Pripjat river in the conditions of floodplain", Proccidings of Ukrainian Hydrometeorology Institute, 245, 127-143. (in Russian).

Onishi Y. and Trent D.S., (1979) "Mathematical Simulation of Sediment and Radionuclide Transport in Surface Waters" NUREG/CR-1034, Washington D.C.

Schuckler M., Kalckbrenner R. and Bayer A. (1976) "Zukunftige radiologische Belastung durch kerntechnische Anlagen im Einzugsgebiet des Oberrheins" Belastung uber den Wasserweg 2, Conference Dusseldorf Proceedings, ZAED, Eggenstein-Leopoldshafen.

Zheleznyak M.J. and Voitsekhovich O.V. (1991) "Mathematical modellingof radionuclide dispersion in surface waters after the Chernobyl accident to evaluate the effectiveness of water protection measures", CEC, Radiation Protection-53, EUR 13574, 725-748.

Zheleznyak M.J., Demchenko R.I., Khursin S.L., Kuzmenko Yu.I., Tkalich P.V. and Vitjuk N.Ya. (1992) "Mathematical modeling of radionuclide dispersion in the Pripyat-Dnieper aquatic system after the Chernobyl accident", The Science of the Total Environment. 112, 89-114.

13. NAVIER-STOKES EQUATIONS

FINITE ELEMENT CFD ADVANCES FOR ENVIRONMENTAL FLOW MODELING

A. J. BAKER , P. T. WILLIAMS and D. J. CHAFFIN
College of Engineering
University of Tennessee
Knoxville, TN 37996-2030
U.S.A.

The goal is attainment of high quality computational (CFD) solutions to the Reynolds-averaged Navier-Stokes equations for buoyant and turbulent incompressible flows in three-dimensional (3-D) environmental geometries. CFD methods/codes invariably stabilize intrinsic dispersive error mechanisms via artificial diffusion, with resultant potential for compromise of genuine physical diffusion processes. The Taylor weak statement generalization on the classic Galerkin method yields an encompassing theory for comparative analysis. Thereafter, a finite element semi-discretization leads to significant improvement for accurate flowfield prediction. Recent pertinent advances are summarized herein, and performance is documented for incisive verification, benchmark and validation problem statements.

INTRODUCTION

Computational fluid dynamics (CFD) is the maturing science/art of computer-generation of *approximate* solutions to the Reynolds-averaged Navier-Stokes (RaNS) equations. A CFD algorithm can (*at best!*) generate only an *approximate solution,* and intrinsic error mechanisms exist to compromise accuracy, especially artificial diffusion and turbulence closure modeling. Further, in incompressible flows, pressure functions principally as a kinematic enforcement of mass flow conservation. The kinematic issue of incompressibility, i.e., velocity field divergence-freeness, requires very specific CFD modeling, hence represents another significant source of approximation error. Thereupon, any "CFD *algorithm*" transforms the selected, non-linear RaNS partial differential equation (PDE) system into a much larger algebraic system amenable to "*computing.*"

CFD theoretical models have historically employed finite difference (FD) or finite volume (FV) spatial semi-discretizations to produce the computable form. Hence, approximate solution resolution on the scale (Δx) of the

1199

A. Peters et al. (eds.), Computational Methods in Water Resources X, 1199–1208.
© 1994 Kluwer Academic Publishers. Printed in the Netherlands.

generic mesh cell is impossible. This intrinsic failure yields a dominating, dispersive error mechanism, the control of which is central to accuracy and algebraic stability, especially when the mesh is "relatively" coarse. The universal correction strategy has been to augment genuine dissipation phenomena with "artificial diffusion," induced typically via directional (*upwind*) differencing of the fluid convection derivatives, e.g., donor cell, Petrov-Galerkin, "QUICK," flux vector splitting, etc.

A comprehensive CFD theoretical model was reported in 1987 that proved capable of recovering over a dozen historical dissipative methods, Baker and Kim (1987). With recent extensions, this "Taylor weak statement (TWS)" theory has now been extended to produce a *minimal* artificial diffusion CFD theory for application to the RaNS PDE system. This paper highlights this CFD theory for RaNS, and its resultant semi-discretization into computable form via finite elements in a Taylor series-augment Galerkin weak statement. Theoretical performance estimates are summarized, and definitive 2-D and 3-D verification, benchmark and validation results are reported including solution-adaptive remeshing.

THE PROBLEM STATEMENT

Statistical manipulation on the incompressible Navier-Stokes equations yields a "Reynolds-averaged" (PDE) system governing turbulent, thermal essentially-constant density flows. Following nondimensionalization, and identifying the turbulent Reynolds number (Re^t) of the closure model, the RaNS PDE system is

$$\mathcal{L}(\rho_0) = \frac{\partial u_j}{\partial x_j} = 0 \tag{1}$$

$$\mathcal{L}(u_i) = \frac{\partial u_i}{\partial t} + \frac{\partial}{\partial x_j}\left(u_i u_j + P\delta_{ij} - \frac{1}{Re}\left(1 + Re^t\right)\left(\frac{\partial u_i}{\partial x_j} + \frac{\partial u_j}{\partial x_i}\right)\right) - Ar\Theta\delta_{ig} = 0 \tag{2}$$

$$\mathcal{L}(\Theta) = \frac{\partial \Theta}{\partial t} + \frac{\partial}{\partial x_j}\left(u_j\Theta - \frac{1}{Re}\left(\frac{1}{Pr} + \frac{Re^t}{Pr^t}\right)\frac{\partial \Theta}{\partial x_j}\right) - s_\Theta = 0 \tag{3}$$

$$\mathcal{L}(Y_A) = \frac{\partial Y_A}{\partial t} + \frac{\partial}{\partial x_j}\left(u_j Y_A - \frac{1}{Re}\left(\frac{1}{Sc} + \frac{Re^t}{Sc^t}\right)\frac{\partial Y_A}{\partial x_j}\right) - s_A = 0 \tag{4}$$

$$\mathcal{L}(k) = \frac{\partial k}{\partial t} + \frac{\partial}{\partial x_j}\left[u_j k - \frac{1}{Re}\left(1 + \frac{Re^t}{C_k}\right)\frac{\partial k}{\partial x_i}\right] - \frac{Re^t}{Re}\phi + \varepsilon + \frac{Ar}{Re}\frac{Re^t}{Pr^t}\frac{\partial \Theta}{\partial x_j}\delta_{ig} = 0 \tag{5}$$

$$\mathcal{L}(\varepsilon) = \frac{\partial \varepsilon}{\partial t} + \frac{\partial}{\partial x_j}\left[u_j\varepsilon - \frac{1}{Re}\left(1 + \frac{Re^t}{C_\varepsilon}\right)\frac{\partial \varepsilon}{\partial x_i}\right] - C_\varepsilon^1 f_1 \frac{\varepsilon Re^t}{k\,Re}\Phi + C_\varepsilon^2 f_2 \frac{\varepsilon}{k}\varepsilon = 0 \tag{6}$$

The dependent variable (set) in (1)-(3) contains mean-flow velocity vector

$u_i(x_j, t)$, kinematic pressure $P = p/\rho_0 + 2k/3$, where k is turbulence kinetic energy, and potential temperature $\Theta \equiv (T - T_{min})/(T_{max}-T_{min})$. The Boussinesq buoyancy approximation is made in (2) where δ_{ig} is the gravity unit vector. In (4) Y_A is the mass fraction of an inert species, e.g., pollutant, with source term s_A. Finally, (5)-(6) is the TKE two-equation closure model on turbulent kinetic energy (k) and ε, the isotropic dissipation function.

The dimensionless groups appearing in (2)-(6) exert a controlling impact on solution character. The base definitions are

$$ Re \equiv \frac{UL}{\nu}, \quad Gr \equiv \frac{g\, \beta(T_{max}- T_{max})\, L^3}{\nu^2}, \quad Pr \equiv \frac{\rho_0 c_p \nu}{k}, \quad Sc \equiv \frac{\nu}{D_{AB}} \tag{7} $$

where Re is the Reynolds number, Gr is the Grashoff number, Pr is the Prandtl number and Sc is the Schmidt number for binary diffusion with coefficient D_{AB}. A superscript "t" on any variable in (2)-(6) denotes its modeled "turbulent" counterpart, and $Ar=Gr/Re^2$ is the Archimedes number.

The mathematical character of any solution to (1)-(6) depends on these non-dimensional groups, and on the boundary conditions. The second order spatial derivatives appearing in (2)-(6) guarantee each isolated equation is elliptic for any finite Re. In a turbulent simulation, $Re^t>0$ more robustly enforces this ellipticity for Re large. Therefore, all mean-flow variables, and/or their normal derivatives, are required to be given data on the problem statement boundary.

The RaNS system is classically initial-boundary value up to enforcement of the continuity equation (1), which constitutes a differential constraint on solutions to (2). Enforcing of this coupling involves inexact "CFD" theories wherein pressure effects are kinematically "simulated" in an iterative process. Algebraic, pseudo-initial-value and boundary value forms exist; the latter is selected as (Williams, 1993).

$$ \mathscr{L}(\phi) \;\equiv\; -\nabla^2 \phi - \nabla \cdot \widetilde{u} = 0 \quad (8), \qquad P_A^{p+1} = P_n + \phi^{p+1}/\theta\Delta t \tag{9} $$

In (8), the superscript tilde denotes "any" approximation to the true divergence-free velocity field, p is an iteration index and n is the time-step index. Any such CFD-approximate "pressure field" P_A must be replaced at iterative convergence, by solution of the *genuine* pressure Poisson equation, formed via the divergence of (2), as

$$ \mathscr{L}(P) \;=\; \frac{\partial^2 P}{\partial x_i^2} + \frac{\partial}{\partial t}\!\left(\frac{\partial u_i}{\partial x_i}\right) + \frac{\partial^2}{\partial x_i x_j}\!\left(u_i u_j - \frac{1}{Re}\,(1 + Re^t)\!\left(\frac{\partial u_i}{\partial x_j} + \frac{\partial u_j}{\partial x_i}\right)\right) - Ar\frac{\partial \Theta}{\partial x_g} = 0 \tag{10} $$

The third derivatives in (10) are difficult to form to an acceptable discrete order of accuracy. The resolution is achieved via a Galerkin weak statement theory, Williams & Baker (1993), which lowers required differentiability order to first. Since (8) and (10) are elliptic, encompassing boundary conditions are required for well-poseness. For (8), the only known boundary condition is homogeneous Neumann, while a non-homogeneous Newmann condition is available for (10). A closure model for turbulent Reynolds number $Re^t(x,t)$ closes the RaNS system.

WEAK STATEMENT CFD ALGORITHM

A CFD algorithm for (2)-(10) generates approximate solutions parameterized by Re, Gr, Pr, and Re^t and by discrete approximation error. The dependent variable set (the "state variable") is $q(x, t) = \{u_1, u_2, u_3, \Theta, Y_A, k, \varepsilon; \phi, P\}^T$, which at the semi-colon partition satisfies the PDE systems

$$\mathcal{L}(q) = \frac{\partial q}{\partial t} + \frac{\partial}{\partial x_j}\left(f_j - f_j^v\right) - s = 0 \quad (11), \quad \mathcal{L}(q_a) = \nabla^2 q_a - s(q) = 0 \quad (12)$$

In (11), $f_j = f_j(q)$ is the *kinetic* flux vector containing convection and pressure effects, while f_j^v is the *dissipative* flux vector embodiment of natural diffusion (Re,Pr) and the turbulence closure model (Re^t). In (12), the laplacian operates on $q_a = \{\phi,P\}^T$, and $s = s(q)$ is the respective source with dependence on the solution to (11). Hence, (11)-(12) is a coupled non-linear initial-boundary value problem.

A CFD algorithm generates (only) an approximation to the true solution to (11)-(12), via a set of decisions leading to a strictly algebraic system amenable "to computing." As indicated, the dominating (dispersive) error mechanism intrinsic to this process leads to use of "artificial diffusion" mechanisms for stabilizing computations. The Taylor weak statement theory, Baker and Kim (1987), extended to systems yields RaNS system augmentation of (11) as

$$\mathcal{L}^c(q) \equiv \mathcal{L}(q) - \beta \Delta t \frac{\partial}{\partial x_j}\left(A_j A_k \frac{\partial q}{\partial x_k}\right) = 0 \tag{13}$$

where $A_j \equiv \partial f_j/\partial q$, is the kinetic flux vector jacobian (matrix). The replacement of (11) by (13), in the theory produces a tensor invariant, phase selective dispersion error dissipation mechanism, cf., Baker (1983, Ch. 4).

Any approximation $q^N(x, t)$ to the true solution $q(x, t)$ to (11) - (12), hence also (13), in the *continuum* is expressible as

$$q(x, t) \approx q^N(x, t) \equiv \sum_{j=1}^{N} \Psi_j(x) Q_j(t) \tag{14}$$

where $\Psi_j(x)$, $1 \le j \le N$, are *known* functions, selected by the CFD algorithm designer, while $q_j(t)$ is the time-dependent *unknown* expansion coefficient set. Since q^N cannot satisfy (12)-(13) exactly, the *error* associated with q^N can be *extremized*, in the sense of distributions, by rendering it orthogonal to the trial space. This is expressed via the Taylor weak statement (TWS^N) as

$$TWS^N = \int_\Omega \psi_j(x) \, \mathcal{L}^C(q^N) \, d\tau = 0, \qquad 1 \le j \le N \tag{15}$$

which is a "Galerkin" weak statement. If the $\Psi_j(x)$ are not local constants, then upon discretization a finite element (FE) form results. Conversely, if Ψ_j are constants, then a "finite volume" (FV) weak statement form results.

A spatial semi-discretization facilitates the calculus operations required to actually form (15), which *always* produces a matrix ordinary differential equation (ODE) system of the form

$$TWS^h = [M]\frac{d\{Q\}}{dt} + \{R(Q)\} = \{0\} \tag{16}$$

In (16), superscript h denotes a "discretization" Ωh of the domain Ω of (1)-(6) has been invoked, $[M]$ and $\{R(Q)\}$ are global rank square and column matrices respectively, and $\{Q\} = \{Q(t)\}$ is the (unknown) state variable approximation coefficient (set) Q_j at the nodes of the mesh defined by Ω^h.

Any ODE algorithm uses (16) for time derivative evaluation. Selecting the θ-implicit one-step algorithm, and substituting (16) as needed, yields

$$\{FQ\} = [M]\{Q_{n+1} - Q_n\} + \Delta t\big(\theta\{R(Q)\}_{n+1} + (1-\theta)\{R(Q)\}_n\big) = \{0\} \tag{17}$$

which is the terminal matrix algebra equation "for computing," where subscript "n" denotes discrete time level.

The discrete Galerkin weak statement for (12) directly produces the algebraic matrix statements

$$\{FQ_A\} = [D]\{Q_A\} - \{S(Q)\} = \{0\} \tag{18}$$

where square matrix $[D]$ is the discrete laplacian, $\{Q_A\}$ is the nodal array of ϕ^h or P^h, and $\{S(Q)\}$ contains dependence on q^h via the solution $\{Q(t)\}$ from (17).

A matrix solution process for (17)-(18) involves an (any) appropriate linear algebra approximation to the Newton iteration algorithm

$$[JAC]\{\delta Q\}_{n+1}^{p+1} = -\{FQ\}_{n+1}^p \quad (19), \qquad \{Q\}_{n+1}^{p+1} = \{Q\}_n + \sum_{i=0}^{p}\{\delta Q\}_{n+1}^{i+1} \tag{20}$$

where $[JAC]$ is the (Newton) jacobian $\partial\{FQ\}/\partial\{Q\}$ of (17)-(18), easily formed using calculus operations. The range of eligible quasi-Newton iterative procedures use approximations thereto, in concert with preconditioned sparse solvers (e.g., GMRES) preconditioned conjugate gradient (PCG), line-iterative Gauss-Siedel, on block tri-diagonal approximate factorizations.

THEORETICAL, ACCURACY, STABILITY

Weak statement theory confirms the required *fundamental* CFD algorithm decisions to be trial space $\Psi_i(x)$, ODE algorithm, and the dissipative coefficient β. Theoretically, trial space selection is commensurate with the (finite difference) concept of algorithm order of accuracy, hence asymptotic convergence rate under mesh refinement. Linear basis FE (and FV) algorithms are typically third-order accurate, while quadratic/cubic basis FE constructions are higher-order accurate.

Algorithm *stability* is of much greater *practical* significance, since the third-order error mechanism is intrinsically dispersive. In particular, any mesh of "measure Δx" is *incapable* of resolving solution information of wavelength $2\Delta x$, hence a dispersive error mode propagates throughout the mesh, polluting the solution and destablizing the quasi-Newton iterative process.

The "universal" fix is to *dissipate* the oscillations via *numerical diffusion*. For a linear model of the IBV statement (11), a Fourier stability analysis (Chaffin, 1994) determines spectral error in phase speed (ϕ) and amplification factor (g). The exact solution for ϕ and g is unity, i.e., no phase lag or damping is present in the analytical form. Hence, departures from unity, as a function of Fourier mode wavelength $\lambda = n\Delta x$, incisively characterize algorithm error mechanisms and stability. Figure 1 summarizes the comparative error in ϕ and g for FD, FV and various FE CFD algorithms. The horizontal line for g confirms that central FD and all $\beta=0$ FE algorithms exhibit zero dissipation error. The QUICK3 FV algorithm is largely dissipative in the short wavelength region, QUICK5 is a modest improvement, but neither match phase accuracy of the FE TWS^h β-forms. In

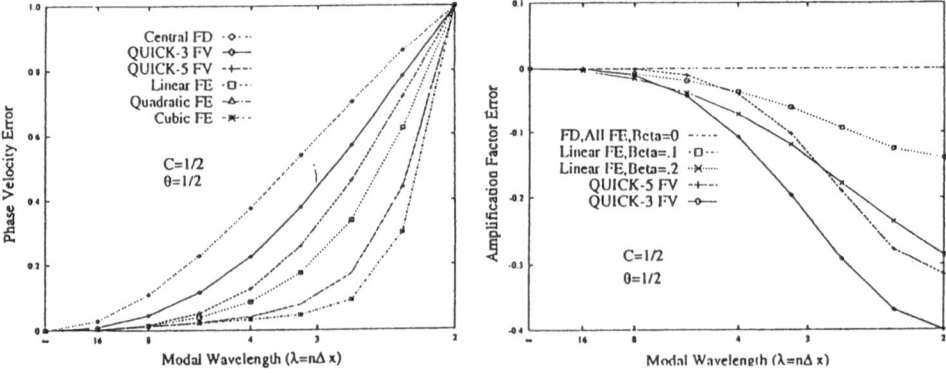

Figure 1. Predicted phase velotiy and application factor spectral error
distributions, FD, FV and FE algorithms.

particular, the phase velocity graph confirms that proceeding from FD through QUICK-FV to linear basis ($k=1$) FE yields improving phase accuracy. Further, the $k=2,3$ FE algorithms are uniform improvements over $k=1$.

RESULTS AND DISCUSSION

An informative, non-dissipative 2-D verification problem for RaNS is the "rotating cone," which constitutes solving the species mass transport equation (4) for infinite Re, i.e., pure unsteady convection. The initial condition is a "cone-shaped" or Gaussian distribution, Fig. 2a). For imposed solid-body rotation velocity, the known analytical solution is exact preservation of the initial condition, as the distribution is convected around the z-plane. Figure 2b)-f) graphs solutions, following a single rotation, for dissipative TWS^h $k = 1,2$ FE basis algorithms, the historical FD algorithm and

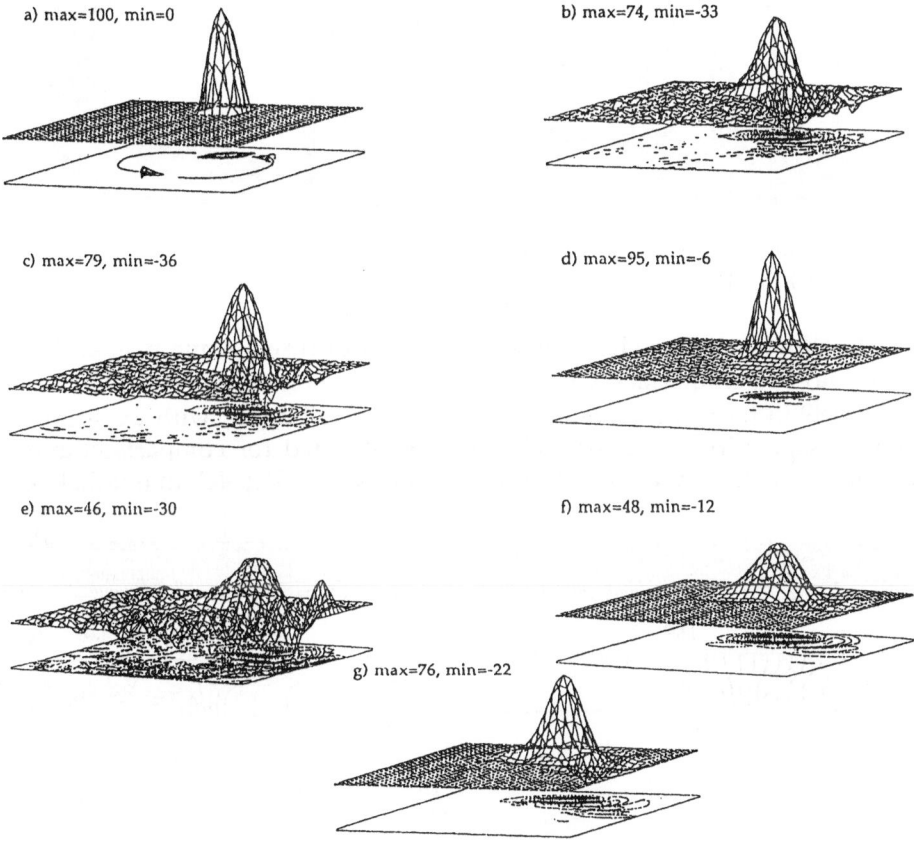

Figure 2. Rotating cone verification, N=1089 nodes, C=1.0, a) IC and exact
solution; after 1 rotation, b) $k=1$ GWS^h, c) $k=2$ GWS^h, d) $k=1$ TWS^h,
$\gamma=1/12$, $\bar{\beta}=1/1024$, e) FD, f) QUICK3 FV, g) QUICK5 FV.

the QUICK3 and QUICK5 "quadratically upwind" dissipative FV algorithms. The non-D time-step (Courant number, C) is unity, and the TWS^h FE solutions are clearly superior, as theoretically predicted, Fig. 1.

Available CFD benchmark tests, i.e., with comparative accepted CFD solutions, are for elementary 2-D geometries. The natural convection thermal cavity, with opposed hot and cold vertical walls, induces a circulatory flowfield parameterized by Ra=GrPr, c.f., de Vahl Davis (1983). The GWS^h non-dissipative, $k=1$ FE algorithm solutions are in excellent agreement, Williams (1993). Figure 3 summarizes the companion GWS^h $k=1$ solutions for a non-square 3-D thermal cavity problem. The illustrated lagrangian particle tracks clearly compare Prandtl number effects.

a)   b)

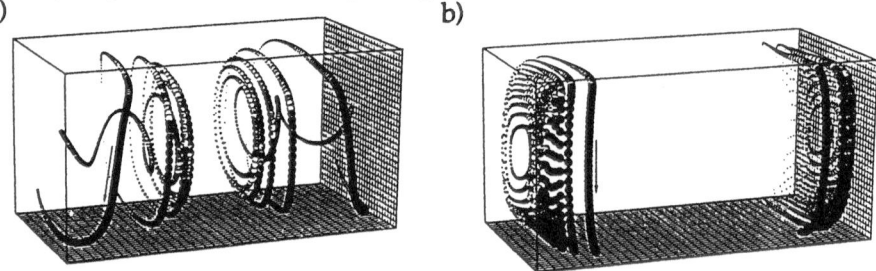

Figure 3. Particle tracks for non-square 3-D thermal cavity, a) Ra=10^4, Pr=0.1,
               b) Ra=10^4, Pr=100.

A 3-D validation is the close-coupled stepwall diffuser, Armaly, et.al. (1983). Figure 4a)-b) summarizes $k=1$ FE GWS^h steady state laminar 2-D solutions for Re=389 in terms of, a) velocity vector $\mathbf{u}^h$ distribution, with experimental primary separation zone reattachment (x_1/S) noted for comparison and, b) pressure. For Re=648, a secondary separation occurs, Fig. 4c), in qualitative

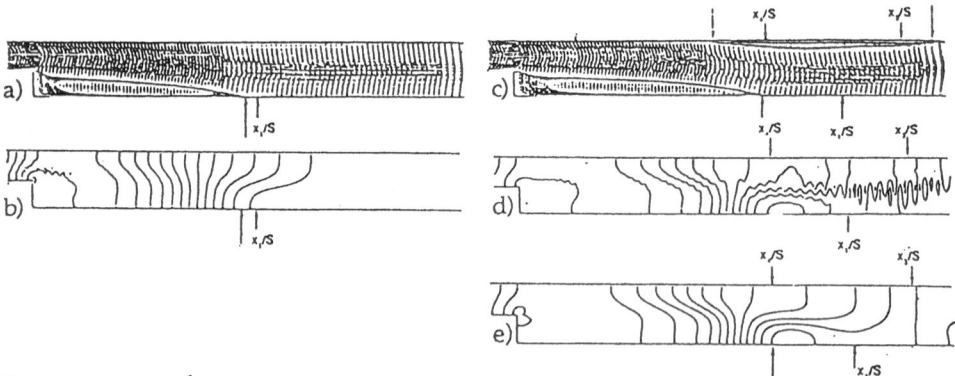

Figure 4. GWS^h solutions, 2-D step-wall diffuser; at Re=390, a) velocity, b)
               pressure; at Re=650, c) velocity, d) pressure, e) TWS^h pressure, $\beta=0.1$.

agreement with the (3-D) experiment. However, the GWS^h solution pressure field is polluted by the dispersive error mode in u^h, Fig. 4d). The TWS^h dissipative algorithm, for β=0.1, does not measurably alter u^h but totally annihilates the GWS^h pressure oscillations, Fig. 4e).

A 3-D validation TWS^h, k=1 FE result has now been obtained for Re≤800, Williams (1993). Figure 5 summarizes solution character at Re=800. Shown left and right are CFD "oil-flow" streamline distributions near the floor and roof confirming a fully 3-D recirculating flowfield. The center graphs show select velocity profiles with surface contours of negative axial velocity, showing the extent of the several recirculation regions. The middle-center graphic confirms CFD prediction of the axial vortex, verified experimentally as the flow transitions from 2-D to 3-D for Re>400.

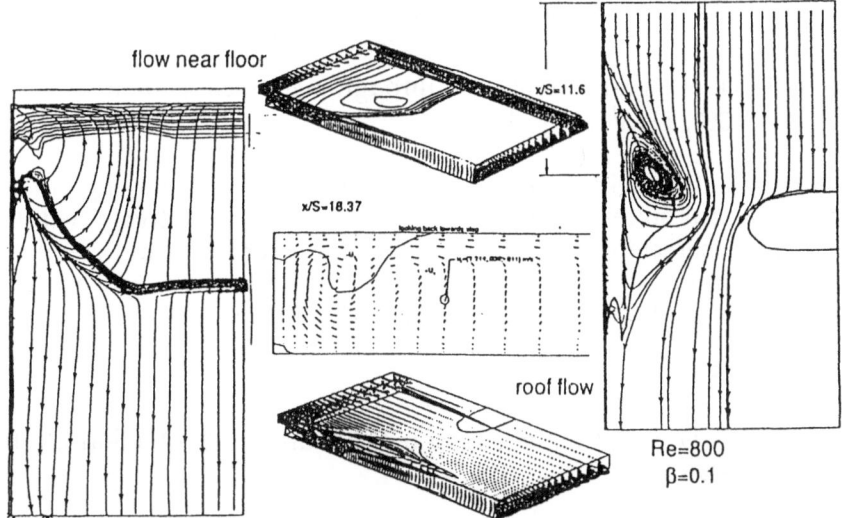

Figure 5. Steady solution summary, 3-D step-wall diffuser benchmark, TWS^h
           k=1 FE algorithm, β-0.1, Re=800.

A lagrangian particle track, Fig. 6, further characterizes three-dimensionality. The symbol size is scaled to vertical elevation above the diffuser floor, to enhance interpretation, and residence time in the main recirculation region is dependent on Re, Williams (1993).

SUMMARY AND CONCLUSIONS

A FE spatial semi-discretization of a Taylor weak statement forms a comprehensive theoretical basis for a robust CFD algorithm/code for thermal

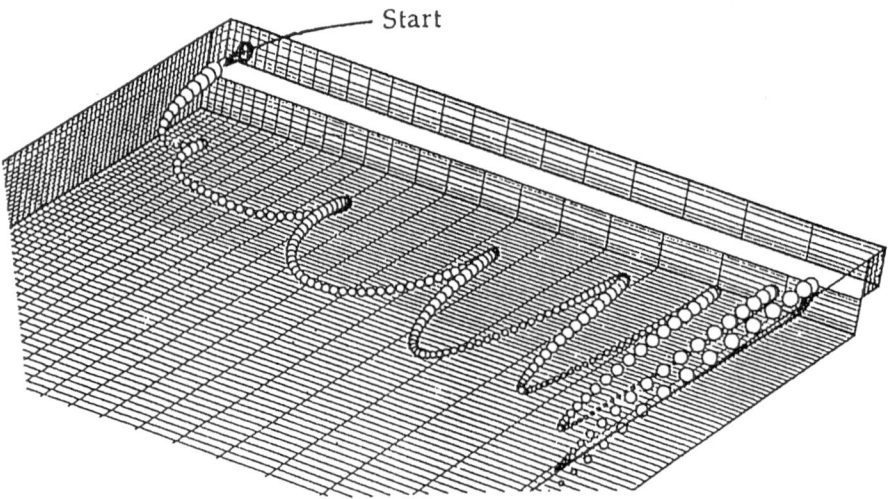

Figure 6. Track of lagrangian particle through main recirculation region,
Re=800, TWS^h algorithm.

incompressible flowfield prediction. The key formulation ingredient of
stability, without excess artificial diffusion, is quantitatively validated.
Mathematical characterizations of asymptotic convergence and stability are
summarized. Summaries of select 2-D and 3-D verification, benchmark and
validation CFD solutions have quantized performance.

REFERENCES

Armally, B.F., F. Durst, J.C.F. Pereira & B. Schonung, (1983) J. Fluid Mech., V.
127, pp. 473-496.

Baker, A.J. (1983) *Finite Element Computational Fluid Mechanics*,
Hemisphere, Washington, D.C.

Baker, A.J. & J.W. Kim (1987), Int. J. Num. Mtd. Fluids, V. 7, pp. 489-520.

Chaffin, D.J. (1994), PhD research in progress, University of Tennessee.

deVahl, Davis (1983), Int. J. Num. Mtd. Fluids, V. 3, pp. 249-264.

Williams, P.T. (1993), Ph.D. dissertation, University of Tennessee.

Williams, P.T. and A.J. Baker (1993), Tech Paper AIAA93-3340.

TIME DEPENDENT INCOMPRESSIBLE FLOW SIMULATIONS WITH FINITE ELEMENTS ON DISTRIBUTED SYSTEMS

H. DANIELS and A. PETERS
IBM Institute for Supercomputing and Applied Mathematics
Vangerowstr. 18
69115 Heidelberg
Germany

We present a very efficient parallel method for the solution of the incompressible time dependent Navier-Stokes equations in arbitrary three dimensional geometries. It uses hexahedral finite elements and the implicit projection-2 algorithm [6], [3]. The data model builds on a parallel domain decomposition variant [9], [7] of the conjugate gradient method [10]. Our implementation is suitable for distributed memory systems. Each subdomain is maped onto one processor. Operations on the subdomains are done in parallel while the interprocessor communication is carried out by passing a small number of messages. The program runs efficiently on fast parallel computers such as the IBM Scalable POWERparallel Systems.

INTRODUCTION

The solution of the Navier-Stokes equations for large Reynolds numbers remains beyond the reach despite remarkable progress in hardware and fast and robust numerical and implementation techniques for computational fluid dynamics (CFD). There is broad consensus that the teraflop speeds required to simulate high Reynolds number flows can be achieved only using massively parallel computer architectures.

Traditional supercomputers employed high-cost bipolar circuits and packing to satisfy the increasing demand of performance. A more recent design approach focuses on the reduction of the price to performance ratio. The effective utilization of a large number of low-cost powerful microprocessors is optimized to obtain highly competitive peak performances. The parallel approach poses a series of tough questions to the designers of CFD codes. Key issues which must be addressed include the minimization of the processor communication and the balancing of the distributed memory.

In this paper we present the parallel implementation of a state-of-the-art scheme for three dimensional, time dependent, incompressible Navier-Stokes simulations on distributed memory parallel computers. The outline of this paper is as follows. Section 2 reviews the basic features of the *projection-2* method [6] which we apply to solve the Navier-Stokes equations. Section 3 presents the domain decomposition variant of the conjugate gradient method [9], [7] underlying our implementation. Section 4 shows benchmark results.

A. Peters et al. (eds.), Computational Methods in Water Resources X, 1209–1216.

THE PROJECTION-2 METHOD

We consider the time dependent Navier-Stokes equations in divergence notation for the primitive variables velocity $(\mathbf{u} = (u, v, w)^T)$ and kinematic pressure $(P = p/\rho_0)$:

$$\frac{\partial \mathbf{u}}{\partial t} + \mathbf{u} \cdot \nabla \mathbf{u} = -\nabla P + \nu \nabla^2 \mathbf{u} + \mathbf{f}, \tag{1}$$

$$\nabla \cdot \mathbf{u} = 0. \tag{2}$$

The parameter ν denotes the kinematic viscosity. The vector $\mathbf{f}$ contains given body forces like buoyancy and surface forces. The system of equations (1) and (2) is defined over a domain Ω with Dirichlet (specified velocities) and Neumann (pseudo-traction) conditions on the boundary $\partial\Omega = \Gamma_1 \cap \Gamma_2$:

$$\mathbf{u} = \mathbf{w} \qquad \text{on } \Gamma_1 \tag{3}$$

$$\nu \frac{\partial u_\eta}{\partial \eta} = F_\eta + P \quad \text{and} \quad \nu \frac{\partial u_\tau}{\partial \eta} = F_\tau \qquad \text{on } \Gamma_2. \tag{4}$$

Here u_η and u_τ are the projections of the velocity onto the tangential plane $(u_\tau = \mathbf{u} \cdot \tau)$ and the normal direction $(u_\eta = \mathbf{u} \cdot \eta)$ at the boundary. Similarly, F_η and F_τ are the normal and tangential components of specified boundary traction. As a prerequisite for the existance of a time dependent solution the initial condition for the velocities $\mathbf{u} = \mathbf{u}_0$ in Ω must fulfill the continuity equation (2) and the normal boundary condition (3) on Γ_1. Details can be found in [6], [3].

Basic steps

The projection-2 method belongs to the more general class of correction methods [8], [2],[11],[12],[6]. Rather than solve the Navier-Stokes equations simultaneously, we decouple and solve the momentum equations (1) and the continuity equation (2) independently. The steps of the projection-2 method are [6]:

1) Given the initial conditions $\mathbf{u}_0$ with $\nabla \cdot \mathbf{u}_0 = 0$ and P_0, solve (1):

$$\frac{\partial \tilde{\mathbf{u}}}{\partial t} + \tilde{\mathbf{u}} \cdot \nabla \tilde{\mathbf{u}} - \nu \nabla^2 \tilde{\mathbf{u}} = \mathbf{f} - \nabla P_0 \tag{5}$$

with (3) and (4) as boundary conditions for the intermediate velocities $\tilde{\mathbf{u}}$, with $\tilde{\mathbf{u}}_0 = \mathbf{u}_0$. They have $\nabla \cdot \tilde{\mathbf{u}} \neq 0$.

2) Therefore project $\tilde{\mathbf{u}}$ to the subspace $\mathbf{v}$ with $\nabla \cdot \mathbf{v} = 0$, i.e. solve the Neumann problem (6) in Ω with no penetration and open outflow boundary conditions (7):

$$(\mathbf{v} - \tilde{\mathbf{u}}) + \nabla \varphi = 0 \quad \text{and} \quad \nabla \cdot \mathbf{v} = 0 \tag{6}$$

$$\eta \cdot \mathbf{v} = \eta \cdot \tilde{\mathbf{u}} \quad \text{on} \quad \Gamma_1 \quad \text{and} \quad \varphi = -\frac{\hat{t}}{2}[F_\eta(\hat{t}) + P_0] \quad \text{on} \quad \Gamma_2. \tag{7}$$

for the unknowns $\mathbf{v}$ and φ. The divergence free velocities $\mathbf{v}$ are approximations to the Navier-Stokes solution $\mathbf{u}$ at the projection time $\hat{t}$. Finally, φ can be used to extrapolate an estimate of the pressure $P(\hat{t})$ (see eq.(15)).

Finite Element Formulation

Gresho [6] proposed the projection-2 method in conjunction with finite elements. Daniels [3] implemented it on vector computers using three dimensional Q1-P0 finite elements in space and weighted finite differences in time. Q1-P0 elements employ trilinear (C^0-continuous) approximation functions for the velocities and elementwise constant (C^{-1}-discontinuous) approximation functions for the pressure. The parallel version presented in this paper is the further development of Daniels' work [3].

The finite element discretization of (1) and (2) yields the semi-discrete equations:

$$M\dot{\mathbf{u}}_i + (D+V)\mathbf{u}_i = -C_i\mathbf{P} + \mathbf{f}_i, \tag{8}$$

$$C_i^T\mathbf{u}_i = \mathbf{g}. \tag{9}$$

The unknowns are the components of the velocity $\mathbf{u}_i \in R^n$ and the pressure $\mathbf{P} \in R^m$. The index i denotes the spacial dimension of the problem, n is the number of grid nodes and m the number of finite elements. The matrices M (mass), D (diffusion), and V (convection) are n by n and $C_i \in R^{m\times n}$ is the gradient matrix. The vectors $\mathbf{f}_i \in R^n$ and $\mathbf{g} \in R^m$ represent the discrete boundary traction and the divergence, respectively. More details about the assembly of the matrix equations (8) and (9) are given in [3].

We apply the projection-2 method (eqs.(5) to (6)) to the discrete system of equations (8) and (9) and use the trapezoidal rule and consistent mass matrix M except for the pressure term in (eqs.(10) to (15)). For other schemes see [3]. We replace the matrix M by the lumped form $L \in R^{n\times n}$, to reduce the computational complexity of inverting a matrix in (13). Hence, the pressure is a lumped quantity and spatially less accurate than the velocities which employ the consistent mass matrix for highly accurate phase speeds. We have [6]:

1) Given $\mathbf{u}_{0i}$ satisfying (9) and $\mathbf{P}_0$ at the begining of a time step, for $i = 1,2,3$ solve the momentum equation i for the intermediate velocity $\tilde{\mathbf{u}}_i$ with $\tilde{\mathbf{u}}_{0i} = \mathbf{u}_{0i}$ from:

$$\left[\frac{2}{\Delta t}M + D + V\right]\tilde{\mathbf{u}}_i = \left[\frac{2}{\Delta t}M - D - V\right]\tilde{\mathbf{u}}_{i0} - ML^{-1}\left(2C_i\mathbf{P}_0 - \tilde{\mathbf{f}}_i - \mathbf{f}_{i0}\right) \tag{10}$$

in case of trapezoidal rule or in case of streamline upwinding solve:

$$\left[\frac{2}{\Delta t}M + \tilde{D}\right]\tilde{\mathbf{u}}_i = \left[\frac{2}{\Delta t}M - \tilde{D} - 2V\right]\tilde{\mathbf{u}}_{i0} - ML^{-1}\left(2C_i\mathbf{P}_0 - \tilde{\mathbf{f}}_i - \mathbf{f}_{i0}\right). \tag{11}$$

Because in (10), (11) the pressure $\mathbf{P}_0$ is taken from the previous time level, body forces $\tilde{\mathbf{f}}_i = \mathbf{f}_{i0}$ are consistent. The systems (10), (11) are solved with a full set of Navier-Stokes boundary conditions (3) and (4). We refer to [3] for details on boundary conditions and streamline upwinding. We recall that the streamline correction alters the diffusion matrix $\tilde{D}$ and thus (11) has a symmectrical left hand side matrix.

2) Project $\tilde{\mathbf{u}}_i$ onto the subspace of discretely divergence-free velocities $\mathbf{v}_i$, i.e. solve the discrete counterpart of (6):

$$\begin{bmatrix} L_1 & 0 & 0 & C_1 \\ 0 & L_2 & 0 & C_2 \\ 0 & 0 & L_3 & C_3 \\ C_1 & C_2 & C_3 & 0 \end{bmatrix} \cdot \begin{bmatrix} \mathbf{v}_1 \\ \mathbf{v}_2 \\ \mathbf{v}_3 \\ \varphi \end{bmatrix} = \begin{bmatrix} L_1\tilde{\mathbf{u}}_1 \\ L_2\tilde{\mathbf{u}}_2 \\ L_3\tilde{\mathbf{u}}_3 \\ 0 \end{bmatrix} \tag{12}$$

where, $\varphi \in R^m$. The indices associated with the lumped mass matrix L suggest that the Dirichlet boundary conditions (3) are built into the system. We rewrite the system (12) in terms of the Shur complement and solve for φ:

$$\left[C_1^T L_1^{-1} C_1 + C_2^T L_2^{-1} C_2 + C_3^T L_3^{-1} C_3 \right] \varphi = C_1^T \tilde{u}_1 + C_2^T \tilde{u}_2 + C_3^T \tilde{u}_3. \qquad (13)$$

The projected velocities are obtained from the expressions:

$$\mathbf{v}_1 = \tilde{u}_1 - L_1^{-1} C_1 \varphi, \; \mathbf{v}_2 = \tilde{u}_2 - L_2^{-1} C_2 \varphi, \; \mathbf{v}_3 = \tilde{u}_3 - L_3^{-1} C_3 \varphi \qquad (14)$$

3) Extrapolate a pressure at time $t + \Delta t$ from

$$\mathbf{P} = \mathbf{P}_0 + 2\varphi/\Delta t, \qquad (15)$$

set $\mathbf{P}_0 = \mathbf{P}, \mathbf{u}_{0i} = \mathbf{v}_i$, advance the time $t = t + \Delta t$ and go back to step 1).

THE PARALLEL DATA MODEL

We introduce a parallel data model for the unstructured finite element projection-2 method based on a parallel domain decomposition implementation of the preconditioned conjugate gradient (PCG) method [5], [9], [7]. The scheme is suitable for parallel, distributed memory computers which communicate by message passing.

The problem domain Ω of m finite elements and n nodes is partitioned into s subdomains Ω_i, where $i = 1, s$, each consisting of m_i elements and n_i nodes. Adjacent meshes overlap at one line of nodes, the interface of two subdomains $\Omega_i \cap \Omega_j$. Elements do not overlap. Each subdomain is maped onto one processor. All data of the subdomain is kept in the local memory during the entire computation. Thus the coefficient matrices of the linear systems of equations (10), (12) and (13) are not assembled globally. Instead, submatrices of dimension n_i by n_i (10) and m_i by m_i (13) are evaluated locally on the s processors.

The following subsection describes the parallel version of the PCG method which we apply to the solution of (10). The version for (13) and further implementation details are presented in [4]. For simplicity we will replace the notations used in (10) with the more general one $A\mathbf{x} = \mathbf{b}$, where A is the coefficient matrix, $\mathbf{b}$ denotes the right hand side vector and $\mathbf{x}$ is the vector of unknowns.

We assume that the linear system $A\mathbf{x} = \mathbf{b}$ is symmetric and positive definite, viz. (11). We assemble on s distributed processors the matrices $A_i \in R^{n_i \times n_i}$, and the right hand side vectors $\mathbf{b}_i \in R^{n_i}$ associated with the subdomains $\Omega_i, i = 1, s$. The matrices A_i and vectors $\mathbf{b}_i$ are *incomplete* in the sense that the coefficients associated with the l_{ij} nodes at the interfaces $\Omega_i \cap \Omega_j$ do not contain the contributions from adjacent subdomains Ω_j. As shown in [7] some steps of the CG-algorithm can be carried out on incomplete vectors [see (31), (32), (36), (37), (32)] due to the associativity property of the operations. Other operations [see (35) and (43)] require that the participating vectors are *complete*, i.e. the coefficients contain contributions from the adjacent domains Ω_j. We have termed the exchange of coefficients and the local completion of the vectors $SWITCH$ operations and represent them by the notation:

$$\mathbf{s}_i = SWITCH_j(\tilde{\mathbf{s}}_i) \qquad (16)$$

where $\mathbf{s}_i$ and $\tilde{\mathbf{s}}_i \in R^{n_i}$ denote a complete and an incomplete vector respectively. A $SWITCH$ operation takes place simultaneously on all s processors and consists of two steps. In the first step the local coefficients corresponding to the l_{ij} nodes at $\Omega_i \cap \Omega_j$ are gathered into an auxiliary vector $\mathbf{v}_i \in R^{l_{ij}}$ and passed on to the processor, which contains the data of subdomain j. The first step is repeated for all neighbors Ω_j of the domain Ω_i, where $1 \le j \le s - 1$:

$$\mathbf{v}_i \xleftarrow[gather]{\Omega_i \cap \Omega_j} \tilde{\mathbf{s}}_i \tag{17}$$

$$\mathbf{v}_i \xrightarrow{send} \Omega_j \tag{18}$$

In the second step the auxiliary vectors $\mathbf{v}_j \in R^{l_{ij}}$ are received from the adjacent subdomains $j, 1 \le j \le s-1$. The coefficients of $\mathbf{v}_j$ are scattered and added to those coefficients of $\mathbf{s}_i \in R^{n_i}$ corresponding to the boundary nodes.

$$\mathbf{v}_j \xleftarrow{recv} \Omega_j \tag{19}$$

$$\mathbf{s}_i = \xleftarrow[scatter-and-add]{\Omega_i \cap \Omega_j} \mathbf{v}_j. \tag{20}$$

With this notation, we represent the diagonal scaled parallel CG-algorithm employed in the solution of (10) by the following pseudo-code. For the initialization we have on each subdomain Ω_i:

$$\tilde{\mathbf{s}}_i = (diag A_i) \tag{21}$$

$$\mathbf{s}_i = SWITCH_j(\tilde{\mathbf{s}}_i) \tag{22}$$

$$(diag A)_i^{-1} = \mathbf{s}_i^{-1} \tag{23}$$

$$\tilde{\mathbf{r}}_i^0 = \tilde{\mathbf{b}}_i - A_i \mathbf{x}_i^0 \tag{24}$$

$$\tilde{\mathbf{s}}_i = (diag A)_i^{-1} \tilde{\mathbf{r}}_i^0 \tag{25}$$

$$\mathbf{s}_i = SWITCH_j(\tilde{\mathbf{s}}_i) \tag{26}$$

$$\mathbf{p}_i^0 = \mathbf{s}_i \tag{27}$$

$$\rho_i^0 = \mathbf{s}_i^T \tilde{\mathbf{r}}_i^0 \quad , \rho_i^0 \xrightarrow{send} HOST \tag{28}$$

$$\rho^0 = \sum_{i=1}^{s} \rho_i^0 \quad , \text{ on the } HOST \tag{29}$$

$$\text{if} \quad \rho^0 \le \varepsilon \quad \text{then} \quad \text{CONVERGED} \tag{30}$$

For $k = 1, limit$ Step 1 Until CONVERGE, Do (31) to (43):

$$\tilde{\mathbf{q}}_i^k = A_i \mathbf{p}_i^k \tag{31}$$

$$\gamma_i^k = \tilde{\mathbf{q}}_i^{kT} \mathbf{p}_i^k \quad , \gamma_i^k \xrightarrow{send} HOST \tag{32}$$

$$\gamma^k = \sum_{i=1}^{s} \gamma_i^k \quad , \text{ on the } HOST \tag{33}$$

$$\alpha^k = \frac{\rho^k}{\gamma^k} \quad , \alpha^k \xrightarrow[broadcast]{i=1,s} \Omega_i \tag{34}$$

$$\mathbf{x}_i^{k+1} = \mathbf{x}_i^k + \alpha^k \mathbf{p}_i^k \tag{35}$$

$$\tilde{\mathbf{r}}_i^{k+1} = \tilde{\mathbf{r}}_i^k - \alpha^k \tilde{\mathbf{q}}_i^k \tag{36}$$

$$\tilde{\mathbf{s}}_i = (diag A)_i^{-1} \tilde{\mathbf{r}}_i^{k+1} \tag{37}$$

$$\mathbf{s}_i = SWITCH_j(\tilde{\mathbf{s}}_i) \tag{38}$$

$$\rho_i^{k+1} = \mathbf{s}_i^T \tilde{\mathbf{r}}_i^{k+1} \quad , \rho_i^{k+1} \rightarrow_{send} \rightarrow HOST \tag{39}$$

$$\rho^{k+1} = \sum_{i=1}^{s} \rho_i^{k+1} \quad , \text{ on the } HOST \tag{40}$$

$$\beta^k = \frac{\rho^{k+1}}{\rho^k} \quad , \beta^k \rightarrow_{broadcast}^{i=1,s} \rightarrow \Omega_i \tag{41}$$

$$\text{if} \quad \rho^{k+1} \le \varepsilon \quad \text{then} \quad \text{CONVERGED} \tag{42}$$

$$\mathbf{p}_i^{k+1} = \mathbf{s}_i + \beta^k \mathbf{p}_i^k \tag{43}$$

The computationally intensive steps (31), (32, 39), (35, 36, 43), (37) and (38) are executed in parallel on s distributed processors for subdomains Ω_i. The global scalar operations (33, 34, 40 to 42) are completed in a control process, which we term $HOST$. The communication between one processor (subdomain Ω_i), its neighbors $1 \le j \le s - 1$ and the $HOST$ is of complexity: a) two times $send$ to and $receive$ from the $HOST$ a scalar value, b) $send$ j interface vectors of length l_{ij} to the neighbors Ω_j .

APPLICATIONS

The first example is a time dependent calculation of three dimensional flow and transport around an object in a numerical wind tunnel. The example is small and can be simulated on a workstation in a couple of hours. The mesh consists of 28,960 3D hexahedral finite elements. The boundary conditions for this time dependent flow and heat transport problem are steady, the initial conditions (IC's) constant. They are:

	Flow BC's	Temperature BC's
Tunnel	slippery ($u_n = 0$ and $u_\tau = $ natural)	adiabatic ($\nabla T = 0 = $ natural)
Inflow	prescribed ($u_n = 1$ and $u_\tau = 0$)	prescribed ($T = 1$)
Outflow	open ($u, v, w = $ natural, $\nabla u_i = 0$)	open ($\nabla T = 0$)
Object	no slip on surface ($u, v, w = 0$)	prescribed ($T = 0$)
IC's:	$\mathbf{u}_0 : u = 0, v = 1, w = 0$	$T_0 : T = 0$

A startup procedure [3] insures that the corrected initial conditions for the flow field fullfil a) the boundary conditions and b) the continuity equation $\nabla \mathbf{u}_0 = 0$. It corrects the ad hoc initial velocities to almost potential flow and produces the corresponding initial pressure. The model parameters, viscosity ν and thermal conductivity $\frac{\lambda}{\rho c}$, are chosen such that the Reynolds number is $Re = 1,000$. Figure 1 shows the flow field in a horizontal and a vertical cut through the mesh at dimensionless time $t = 5.0$. Runs on a RS/6000 cluster for one up to 8 subdivisions show reasonable speed ups with 2 and 3 subdomains when PVM is used with Ethernet for message passing. The application does not scale further, because the ratio between (low) communication bandwith plus network latency and the (high) compute power of the processors is unbalanced for more than 3 subdomains for this small example. Better scalability can be achieved for a large real world application like 3D time dependent

Figure 1: Velocities in a plane at time t=5.0 for numerical wind tunnel.

flow and contaminant transport in a drinking water reservoir. The project and data and the mesh with 241,557 3D hexahedra was provided by the Institut für Wasserbau of the Aachen University of Technology [1]. The benchmark results for mesh subdivisions into 3,4,6,7 and 8 subdomains are shown in Table 1. Obviously we achieve high scalability for the large problem even with the low speed Ethernet for the communication. The load balancing is not perfect, because the mesh is unstructured.

A very important feature of the parallel data model is the scalability of the feasible problem size (mesh resolution) with the growth of the distributed memory. One workstation cannot handle the reservoir job, because the local memory (256 MB per node) is too small.

N_{mesh}	3	4	6	7	8
$CPU/\Delta t[s]$	464-484	426-468	284-310	185-214	161-208
$elapsed/\Delta t[s]$	505-554	502-525	345-375	242-263	238-258
$efficiency$	98.3%	93.3%	88.3%	85.6%	87.4%

Tab.1: Benchmark for reservoir with PVM 2.4.2 and Ethernet on IBM 9076 SP1
(8 nodes RS/6000 Model 370 with 256 MB each and 121.1 MFLOPS each for SPECfp92)

CONCLUSIONS

We combined finite elements and projection-2 - already efficient on sequential computers [3] - with a domain decomposition variant of the conjugate gradient method and designed a code which exploits parallel computers with distributed memory. Each subdomain is assigned to one processor and all data of that subdomain is kept in the local memory. This allows the parallel assembly of subsystems of linear equations without communication. In the subsequent solution of the linear systems, almost all operations can be carried out in parallel while few interface coefficients are exchanged between adjacent subdomains. An important advantage of the model is the fact that it allows the computation of very large problems. A key issue is the scalability to many parallel processors. We have used up to 8 powerful processors with slow Ethernet connection. With regard to further development we expect a boost of the scalability by more than one order of magnitude by a) using the IBM High Performance Switch in the SP1 with MPI instead of Ethernet and PVM b) optimizing the current implementation of the parallel PCG solver with regard to better preconditioning and communication reduction.

References

[1] Bergen, O., *Numerische Simulation der Strömung und des Transports wasserlöslicher Stoffe in Seen und Talsperren am Beispiel des Vorbeckens der Möhnetalsperre*, Thesis, Inst. f. Wasserbau, RWTH Aachen, 1993.

[2] Chorin, A.J., *Numerical Solution of the Navier-Stokes Equations*, Math. Comp., **22**, 1968, 745-763.

[3] Daniels, H., *PASTIS-3D Finite Element Projection Algorithm Solver for Transient Incompressible Flow Simulations - Manual*, **UCRL-MA-111833**, LLNL, Livermore, CA, 1992.

[4] Daniels, H. and A. Peters, *Solving Large Incompressible Time-Dependent Flow Problems on Scalable Parallel Systems*, prep. for Int. J. Num. Meth. Fluids, IBM TR **75.94**, 1994.

[5] Dryja, M., *A finite element capacitance method for elliptic problems on regions partitioned into subdomains*, Numer. Math., **44**, 1984, 153-168.

[6] Gresho, P.M., *On the theory of semi-implicit projection methods for viscous incompressible flow and its implementation via a finite element method that also introduces a nearly-consistent mass matrix, Part 1: Theory*, Int. J. Num. Meth. Fluids, **11**, 1990, 587-620.

[7] Haase, G. and U. Langer, *Parallelisierung und Vorkonditionierung des CG-Verfahrens durch Gebietszerlegung*, Num. Algebra auf Transputersystemen, Teubner, May 1993

[8] Harlow, F.H. and J.E. Welch, *Numerical Calculation of Time-Dependent Viscous Incompressible Flow of Fluids with Free Surface*, Physics of Fluids, **8**, No. 12, 1965, 2182-2189.

[9] Keyes, D.E. and W.D. Gropp, *A comparison of domain decompositions techniques for elliptic partial differential equations and their parallel implementation*, SIAM J. SCI. STAT. COMPUT., **8**, No. 2, 1987, 166-202.

[10] Peters, A., *Non-symmetric CG-like schemes and the finite element solution of the advection-dispersion equation*, Int. J. Num. Meth. Fluids, **17**, 1993, 955-974.

[11] Témam, R., *Sur l'approximation de la solution des équation de Navier-Stokes par la méthode des pas fractionnaires (I)*, Arch. Rat. Mech. Anal., **32**, 1969, 135-153.

[12] Van Kan, J., *A Second-Order Accurate Pressure Correction Scheme for Viscous Incompressible Flow*, SIAM J. Sci. Comp., **7**, No. 3, 1986, 870-891.

A FINITE ELEMENT FORMULATION FOR LARGE EDDY SIMULATIONS

C. FORKEL, J. BIRKHÖLZER and G. ROUVÉ
Institute for Hydraulic Engineering and Water Resources Management
University of Technology, RWTH Aachen
Mies-van-der-Rohe-Straße 1, Aachen, 52056
Germany

In this paper a finite element formulation for large eddy turbulent flow simulations is presented. The subgrid scale models are based on the easier Smagorinsky (1963) model and for more complex flow situations on the model proposed by Germano et. al. (1991). A new spatial filtering procedure is introduced which fits to the finite element interpolation functions. The large eddy models were implemented into an existing code for laminar flow and transport (Daniels, 1992). Experiences with the new code have not been yet obtained but will be presented on the conference.

INTRODUCTION

Most of the turbulence models for flow simulations like k-ε models need a lot of constants which are hardly well known for natural flows. Large eddy simulations which need only one constant might therefore be a better approach to this problem. But still large eddy simulations have mostly been applied with finite differences or spectral methods. Both of these methods have difficulties in discretizing irregular geometries which normally exist in natural flow systems. So a finite element formulation for large eddy simulations seems to be the best way to model turbulent flow in nature if there is sufficient computer capacity and a need for more accurate results.

As it is not possible to use an arbitrary combination of the different parts of a large eddy simulation model (like spatial approximation, filtering approach, filter and subgrid scale model), it is the aim of this paper to present a compatible combination of all different parts of a finite element large eddy simulation model.

MATHEMATICAL FORMULATION

The idea of large eddy turbulent flow simulations is to resolve the major vortices directly and to model only the minor turbulent structures with a turbulence model. As the minor turbulent structures are nearly isotropic and very similar to each other a very simple turbulence model can be used for these small eddies. Formally the distinction between resolved and not resolved flow field can be obtained by filtering the Navier-Stokes equations, which leads to following equations

A. Peters et al. (eds.), Computational Methods in Water Resources X, 1217–1224.

$$\frac{\partial \overline{v_i}}{\partial x_i} = 0 \tag{1}$$

$$\frac{\partial \overline{v_i}}{\partial t} + \frac{\partial \overline{v_i v_j}}{\partial x_j} = -\frac{\partial \left(\overline{P} + \frac{2}{3} k \right)}{\partial x_i} + \nu_0 \frac{\partial^2 \overline{v_i}}{\partial x_j^2} \tag{2}$$

Here v_i is the velocity vector, P the kinematic pressure, k the kinetic energy and ν_0 the kinematic viscosity. The overbar denotes the filtered quantities. A second filtering of the convective term in eq. (2) leads to

$$\frac{\partial \overline{v_i}}{\partial t} + \overline{v_j} \frac{\partial \overline{v_i}}{\partial x_j} = -\frac{\partial \left(\overline{P} + \frac{2}{3} k \right)}{\partial x_i} + \nu_0 \frac{\partial^2 \overline{v_i}}{\partial x_j^2} - \frac{\partial \tau_{ij}}{\partial x_j} \tag{3}$$

with

$$\tau_{ij} = L_{ij} + C_{ij} + Q_{ij}, \tag{4}$$

the Leonard stresses

$$L_{ij} = \overline{\overline{v_i} \, \overline{v_j}} - \overline{v_i} \, \overline{v_j}, \tag{5}$$

the cross terms

$$C_{ij} = \overline{\overline{v_i} v_j{}'} + \overline{v_i{}' \overline{v_j}}, \tag{6}$$

and the subgrid scale stresses

$$Q_{ij} = \overline{v_i{}' v_j{}'} + \frac{2}{3} k \delta_{ij}. \tag{7}$$

It is now possible to model all stresses τ_{ij} together with one turbulence model or to model only the Q_{ij} and to approximate the L_{ij} and C_{ij} by Taylor series (Leonard, 1974; Aldama, 1990). Another approach is to model the C_{ij} and Q_{ij} and to calculate the L_{ij} by explicitly filtering this term (Ferziger, 1977). Of course a Taylor approximation of the L_{ij} and C_{ij} terms is the most accurate technique. But as the resulting error in using a turbulence model for all τ_{ij} stresses is only of second order (Rogallo/Moin, 1984) and the numerical model we use is only second order accurate itself (Daniels, 1992), there is no need for a detailed evaluation of the L_{ij} and C_{ij} terms. Therefore we use a turbulence model for all τ_{ij} stresses, for example a Smagorinsky (1963) turbulence model

$$\tau_{ij} = -2\nu_t \overline{S_{ij}}, \tag{8}$$

with the eddy viscosity

$$\nu_t = \left[C_s\Delta\right]^2 \cdot \left(\overline{S_{mn}}\,\overline{S_{mn}}\right)^{\frac{1}{2}} \tag{9}$$

and the filtered deformation tensor

$$\overline{S_{ij}} = \frac{1}{2} \cdot \left(\frac{\partial \overline{v_i}}{\partial x_j} + \frac{\partial \overline{v_j}}{\partial x_i}\right). \tag{10}$$

C_s is a model constant (~0,2) and Δ the filter width which can be determined for non equidistant grids from

$$\Delta = \sqrt{\frac{\Delta_x^2 + \Delta_y^2 + \Delta_z^2}{3}}. \tag{11}$$

The Smagorinsky model is very often used today despite the fact that it is not accurate in modeling the exact stresses. But as it correctly simulates the global energy drain from the large scales to the small scales (Piomelli et al., 1987), it is preferred in our calculations against other models like the mixed model (Bardina et. al, 1980) or the two part model by Schumann (1975). The Smagorinsky model does not give accurate results in the near wall region. Therefore it is better to use approximate boundary conditions instead of simulating this region with the Smagorinsky model. A good overview over possible boundary conditions is given by Piomelli et al. (1987). Another promising model was developed by Germano et al. (1991) and improved by Lilly (1992). They also use a subgrid scale model of the Smagorinsky type but calculate a variable C_s by applying a coarser "test filter" to the equations. The variation of C_s gives the possibility to account for the backscatter near the wall and to simulate also laminar flow regions.

COMBINATION OF FILTER, FILTERING PROCEDURE AND SUBGRID SCALE MODEL FOR FINITE ELEMENT METHODS

The previous chapter gives an overview over different parts of a large eddy simulation model. These different parts can not be used in any arbitrary combination but depend on each other. Table 1 gives some combination examples from the literature. The numerical method used in the PASTIS (Projection Algorithm Solver for Transient Incompressible Flow Simulations) code leads to a filtering approach where all stress terms are modeled with one subgrid scale model (see also previous chapter). For ease of computation a Smagorinsky model was implemented first, a Germano model will be implemented after an extensive testing of the Smagorinsky model.

Table 1: Possible combinations of numerical methods, filtering approach,
filters and subgrid scale models

author	numerical method	filtering approach	filter	turbulence model
PIOMELLI et al. (1987)	spectral methods	explicit filtering of Leonard, modeling of cross terms	Gaussian ---------- Fourier	mixed model ----------------- Smagorinsky
SCHUMANN (1975)	finite differences	volume averaging method	box	Schumann
LILLY (1967)	finite differences	modeling of Leonard and cross terms	box	Smagorinsky
ALDAMA (1990)	quadratic finite elements (1D)	spatial and temporal approximation for Leonard and cross terms, three-scale approach	Gaussian	Smagorinsky
GERMANO et.al., 1991; LILLY, 1992	spectral methods	modeling of Leonard and cross terms by variation of C_s	Fourier	Smagorinsky

Not yet mentioned is the choice of the filter. The filter should be related to the numerical method (like finite differences - box filter, spectral methods - Fourier filter, for definition of the filters see Aldama, 1990). The spatial approximation functions in the PASTIS-code are linear for the velocities and the transport variables and elementwise constant for the pressure. Therefore different filters are used for the pressure and the node variables (only spatial filtering is applied here). The pressure is filtered by taking an element volume weighted arithmetic average (similar to a box filter) of the adjoining element pressure values. For the node variables (velocities and scalar transport), the following filter is applied (1D-formulation, see also Fig 1)

$$h(x - x') = \frac{1}{\Delta} \begin{cases} 1 + \frac{x - x'}{\Delta} & \text{for} \quad -\Delta \leq x - x' < 0 \\ 1 - \frac{x - x'}{\Delta} & \text{for} \quad 0 \leq x - x' \leq \Delta \end{cases}. \tag{12}$$

In Fourier space this filter gives (see also Fig. 2)

$$\hat{h}(k) = \frac{4}{\Delta^2 k^2}(1 - \cos \Delta k). \tag{13}$$

h is the filter function and $\hat{h}$ its Fourier transform, Δk is the dimensionless wave number. As it is shown in Fig. 2, the triangle-filter gives a strong weighting of the lower wave numbers. The Fourier transform is a damped sinusoid without amplitude changes. A disadvantage is the fact that certain wave numbers are totally supressed (for example $\Delta k = 2\pi$).

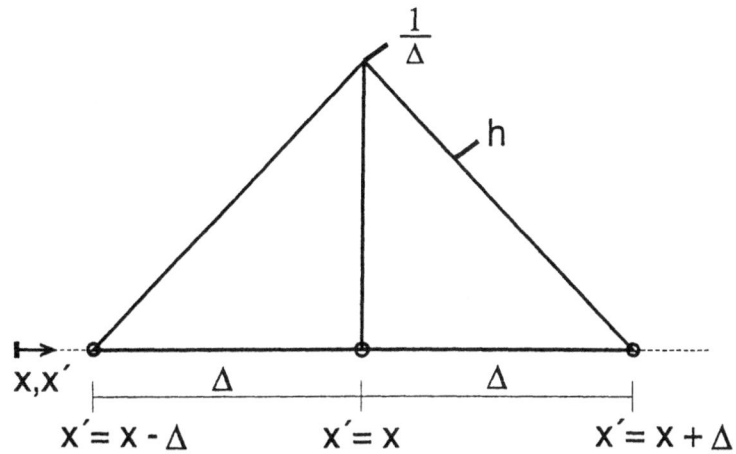

Fig. 1: "Triangle"-filter for node variables in the finite element method

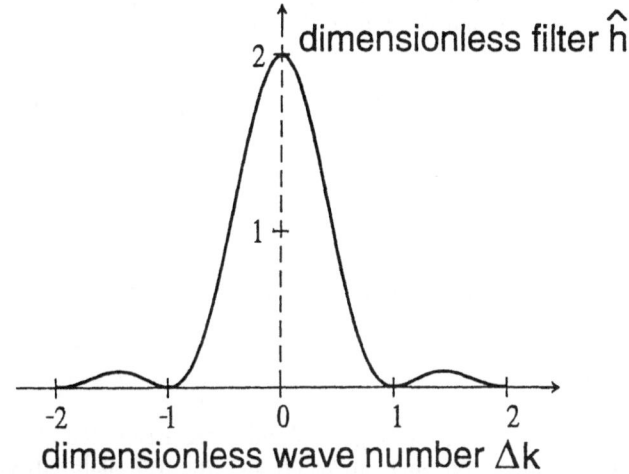

Fig. 2: "Triangle"-filter in Fourier space

In the Eqs. (11) and (12) the filter width Δ is regarded as a constant. In unstructured finite element meshes the filter width may change from node to node. In this case it can be calculated as the nodal average $\overline{\Delta}$ of all adjoining node connections.

$$\overline{\Delta}_j = \left(\frac{1}{n_{k_j}} \sum_{i=1}^{n_{k_j}} (\Delta_k)_{ij}^2 \right)^{1/2} \tag{14}$$

with n_{k_j}:   connectivity of node j

and $\Delta_{k_{ij}}$:   distance of node i adjoined to node j

NUMERICAL MODEL

The new large eddy formulation was set up on the existing, very efficient finite element code PASTIS-3D for laminar flow and transport. PASTIS-3D solves the transient incompressible Navier-Stokes equations time accurately in two and three space dimensions. Finite elements of arbitrary shape and with unstructured mesh topology are used to approximate spatial operators. The element types have mixed C^0 and C^{-1} approximation functions. In 3D the trilinear velocity/constant pressure hexahedron and in 2D the bilinear velocity/constant pressure quadrilateral isoparametric element is used. Because of the arbitrary shape of the elements all integrations on the element level are obtained by full 2x2x2 Gauss point integration in the general case. Time-accurate solutions are obtained by using the "projection 2" algorithm recently described and analyzed for stability and convergence behaviour by Gresho (1990).

The projection 2 algorithm takes the following discrete form in PASTIS-3D with the time discretization by Leismann/Frind (1989) (for details of the numerical model and its implementation see Daniels (1992)):

0) Given the initial divergence free nodal vector v_0 with $C^T v_0 = 0$ (C is the assembled global divergence matrix), and the initial element vector P_0; i.e. those values at the beginning of a new time step

1) Solve three systems of equations for the three components of the intermediate velocity vector $\tilde{v}$ with $\tilde{v}_0 = v_0$

$$\left(\frac{1}{\Delta t}M + D + \Theta_{BTD}D_{BTD}\right)_i \tilde{v}_i =$$
$$\left(\frac{1}{\Delta t}M - (1 - \Theta_{BTD})D_{BTD} - V\right)_i v_{0i} - MM_{L_i}^{-1}\left(C_i P_0 - \tilde{F}_i\right), \qquad (15)$$

i=1, 2, 3, in Ω

where Δt is the time step size, M, D, D_{BTD}, V and F are assembled global matrices for Galerkin (consistent) mass, physical diffusion and subgrid scale stresses, balancing tensor diffusivity, convection and traction boundary conditions, respectively. M_L is the assembled lumped mass vector. Θ_{BTD} is the time weighting factor for the balancing tensor diffusivity.

2) Project the intermediate velocity $\tilde{v}$ to the divergence-free sub-space, i.e. solve for the vector φ from

$$\left(C^T M_{L_i}^{-1} C\right)\varphi = C^T \tilde{v} \quad \text{in } \Omega \qquad (16)$$

3) Compute the final, divergence-free velocities v from

$$v = \tilde{v} - M_L^{-1} C \varphi \qquad \text{in } \Omega \qquad (17)$$

4) Update the pressure

$$P = P_0 + 2\varphi / \Delta t \qquad \text{in } \Omega \tag{18}$$

5) Report and optionally output the nodal velocity vectors v and the element vector P, then set $P_0 = P$, $v_0 = v$ in Ω

6) Go back to step 1

In addition to the incompressible Navier-Stokes equations (now with filtered variables) the term

$$\frac{\partial}{\partial x_j}\left(\tau_{ij}\right) = -\frac{\partial}{\partial x_j}\left[C_S^2 \Delta^2 \left(\overline{S_{mn} S_{mn}}\right)^{1/2} \left(\frac{\partial \overline{v_i}}{\partial x_j} + \frac{\partial \overline{v_j}}{\partial x_i}\right)\right] \tag{19}$$

has to be implemented in the existing code. The term $(S_{mn} S_{mn})^{1/2}$ is the absolute value of the deformation tensor. The deformation tensor is calculated with the velocities $\overline{v}$ of the old time step. The filter width Δ is a nodal value, C_S is constant in the whole domain. Applying the Gauss theorem to the eqs. (3) and (14) leads to the term

$$-C_S^2 \Delta^2 \left(\overline{S_{mn} S_{mn}}\right)^{1/2} \frac{\partial^2 \overline{v_i}}{\partial x_j^2},$$

which is of a form very similar to the viscous stress term. Therefore the new stress term is built into the global matrix together with the viscous stress term. The nodal values of the diffusion matrix are now multiplied with the term

$$a = v_0 + C_S^2 \Delta^2 \left(\overline{S_{mn} S_{mn}}\right)^{1/2}. \tag{20}$$

COMMENTS ON THE VERIFICATION AND APPLICATION OF THE MODEL

The previously described model will be tested first in 2D-space in the comparison with direct numerical simulations (DNS) of channel flow. After this extensive testing of the new filter and the connected subgrid scale models, the combination with the best results will be verified with threedimensional large eddy simulations in comparison to the DNS by Kim et al (1987) and the DNS of a curved channel by Moser/Moin (1987). We intend to present these results on the conference.

ACKNOWLEDGEMENTS

This work is supported by the DFG (Deutsche Forschungsgemeinschaft) under Grant Ro-365/44-2.

REFERENCES

Aldama, Alvaro A. (1990) Filtering Techniques for Turbulent Flow Simulation, Lecture Notes in Engineering 56, Ed. C.A. Brebbia and S.A. Orszag, Springer Verlag, Heidelberg, ISBN 3-540-52137-2

Bardina, J., Ferziger, J.H. and Reynolds, W.C. (1980) "Improved Subgrid Scale Models for Large Eddy Simulation", AIAA 13th Fluid and Plasma Conference, July 14-16, Snowmass, Colorado

Daniels, Helmut (1992) PASTIS-3D, Finite Elemente Projection Algorithm Solver for Transient Incompressible Flow Simulations, UCRL-MA-111833; Lawrence Livermore National Laboratory, California, USA

Ferziger, J.H. (1977) "Large Scale Eddy Numerical Simulation of Turbulence", AIAA Journal, 15, 1261

Germano, M., Piomelli, U., Moin, P. and Cabot, W.H. (1991) "A Dynamic Subgrid-Scale Eddy Viscosity Model", Phys. Fluids A, Vol. 3, 1760-1765

Gresho; P.M. (1990) "On the Theory of Semi-Implicit Projection Methods for Viscous Incompressible Flow and Its Implementation via a Finite Element Method that also Introduces a Nearly Consistent Mass Matrix, Part 1", Int. J. Num. Meth. in Fluids, Vol. 11, Wiley and Sons Ltd, 587-620

Kim, J., Moin, P. and Moser, R. (1987) "Turbulence Statistics in Fully Developed Channel Flow at Low Reynolds Number", J. Fluid Mech., Vol. 177, 133-166

Leismann, H.M. and Frind, E.O. (1989) "A Symmetric-Matrix Time Integration Scheme for the Efficient Solution of Advection-Dispersion Problems", Water Resources Research, Vol. 25, No. 6, 275-280

Leonard, (1974) "Energy Cascades in Large-Eddy Simulation of Turbulent Fluid Flows", Adv. in Geophysics, 18 A, 237-248

Lilly, D.K. (1967) "The Representation of Small-Scale Turbulence in Numerical Simulation Experiments", Proc. of the IBM Sci. Comp. Symp. Environm. Sci., IBM Data Process. Div., White Plains, N.Y., 195-220

Lilly, D.K. (1992) "A Proposed Modification of the Germano Subgrid-Scale Closure Method, Phys. Fluids A 3", 1760

Moser, R. D. and P. Moin (1987) "Effect of curvature in wall bounded turbulent flows"; Report TF-20, Dept. Mech. Eng., Stanford University, USA

Piomelli, U., Ferziger and J.H.; Moin, P. (1987) "Models for Large Eddy Simulations of Turbulent Channel Flows Including Transpiration", Report TF-32, Thermosciences Division, Dep. of Mech. Eng., Stanford University, USA

Rogallo, R.S. and Moin, P. (1984) "Numerical Simulation of Turbulent Flows", Ann. Rev. Fluid Mech., 99-137

Schumann, U. (1975) "Subgrid Scale Model for Finite Difference Simulations of Turbulent Flows in Plane Channels and Annuli", J. of Computational Phys., Vol. 18, 376-404

Smagorinsky, J. (1963) "General Circulation Experiments with the Primitive Equations, I. The Basic Experiment", Monthly Weather Review 91, 99-104

AN EQUAL-ORDER APPROXIMATE PROJECTION FEM

P.M. GRESHO, S.T. CHAN, and M. CHRISTON
Physical Sciences Department
Lawrence Livermore National Laboratory
P.O. Box 808, L-262
Livermore, California 94551, USA

In this short paper we introduce and test a technique for solving the incompressible Navier-Stokes equations using bilinear basis functions for velocity and pressure.

INTRODUCTION

The equations of principal interest are

$$\partial \underline{u}/\partial t + \underline{u} \cdot \nabla \underline{u} + \nabla P = v\nabla^2 \underline{u} \text{ and } \nabla \cdot \underline{u} = 0 \quad \text{in} \quad \Omega, \tag{1}, (2)$$

$$\text{with } \underline{u} = \underline{w}(t) \text{ on } \Gamma \text{ and } \int_\Gamma \underline{n} \cdot \underline{w} = 0 , \tag{3}$$

$$\text{and } \underline{u} = \underline{u}_0 \text{ with } \nabla \cdot \underline{u}_0 = 0 \text{ in } \Omega \text{ at } t = 0 \quad \text{and with } \underline{n} \cdot \underline{u}_0 = \underline{n} \cdot \underline{w}(0) \text{ on } \Gamma. \tag{4}$$

Since these "primitive equations" are usually considered as too difficult to solve as stated (fully-coupled), the following pressure Poisson equation (PPE) formulation is of much interest: Using $\partial(\nabla \cdot \underline{u})/\partial t = 0$ in (1) leads to the PPE:

$$\nabla^2 P = \nabla \cdot (v\nabla^2 \underline{u} - \underline{u} \cdot \nabla \underline{u}) \equiv \nabla \cdot \underline{a} \quad \text{in} \quad \Omega , \partial P/\partial n = \underline{n} \cdot \left(v\nabla^2 \underline{u} - \underline{u} \cdot \nabla \underline{u} - \frac{\partial \underline{w}}{\partial t} \right) \text{ on } \Gamma. \tag{5}, (6)$$

The PPE (derived) formulation is (1) and (3)–(6): i.e., (2) is omitted because it is implied [See, e.g., Gresho (1991a)].

The weak (Galerkin) form of the PPE formulation is

$$\int \underline{v} \cdot \left(\frac{\partial \underline{u}}{\partial t} + \underline{u} \cdot \nabla \underline{u} \right) + v(\nabla \underline{u})^T : \nabla \underline{v} - P\nabla \cdot \underline{v} = 0 \quad \forall \underline{v} \in \underline{H}_O^1 \tag{7}$$

$$\text{and } \int \nabla \varphi \cdot \nabla P = \int \underline{a} \cdot \nabla \varphi - \int_\Gamma \varphi \underline{n} \cdot \partial \underline{w}/\partial t \quad \forall \varphi \in H_0^1. \tag{8}$$

Remarks:
1. The above PPE formulation assures that $\nabla \cdot \underline{u} = 0$ in a very weak (and vague) sense in that *if* (7) and (8) $\Rightarrow$ (1) and (3)–(6) then $\nabla \cdot \underline{u} = 0$. I.e., *if* our weak solution is also a classical solution, then $\underline{u}$ will be (strongly) div-free. These are big if's.

A. Peters et al. (eds.), Computational Methods in Water Resources X, 1225–1232.
© 1994 Kluwer Academic Publishers. Printed in the Netherlands.

2. The PPE formulation has more solutions than the $\underline{u} - P$ formulation—and the extras are spurious, a particular example being that $\nabla \cdot \underline{u}_0$ need not vanish in order that PPE solutions exist—*any* $\underline{u}_0$ 'works'. [See Gresho (1992) for others.]

If we discretize (1)–(4) via the GFEM, we arrive at the following Index 2 DAE system:

$$M\dot{u} + N(u)u + CP = Ku + f, \ u(0) = u_0 \text{ and } C^T u = g \ , \ \text{with} \ C^T u_0 = g_0 \qquad (9), (10)$$

where f and g correspond to BC data. Note that (9) also corresponds to the finite-dimensional version of (7). Converting this to a lower index (easier) problem is achieved by inserting $\dot{u}$ into the *time-differentiated* version of (10): the resulting Index 1 DAE system is (9) and

$$\left(C^T M^{-1} C\right) P = C^T M^{-1} \left(Ku + f - N(u)u\right) - \dot{g}. \qquad (11)$$

As with the PPE formulation, the Index 1 formulation uncouples the pressure from the acceleration (its "raison d´ etre"). Also as with the continuous formulation, violation of $C^T u_0 = g_0$ results in an ill-posed Index 2 problem but the Index 1 version—having more solutions—is not ill-posed. It is just not *right;* the resulting discrete velocity field will not satisfy $C^T u = g$. [It *will* satisfy $C^T u = g + C^T u_0 - g_0$; see Gresho (1991b).].

For comparison with what follows below, we now introduce the projection matrix, $\wp$:

$$\wp \equiv I - M^{-1} C \left(C^T M^{-1} C\right)^{-1} C^T, \qquad (12)$$

which is a 'formal' construction only. $\wp$ has the following properties:
1. $\wp^2 = \wp$; it satisfies the *definition* of a projection.
2.   If w is an arbitrary velocity field, then $\wp w$ is its projection to the (discretely) div-free subspace; i.e. $u \equiv \wp w$ satisfies $C^T u = 0$. In fact, $C^T \wp \equiv 0$.
3.   The eigenvalues of $\wp$ are either zero or 1, and its norm is unity. All discretely divergence-free vector fields are eigenvectors of $\wp$ with eigenvalue unity.

In terms of $\wp$, the Index 1 problem can be 'condensed' to an Index 0 problem (i.e., to a set of ODE's):

$$\dot{u} = \wp \left[M^{-1}(Ku + f - N(u)u)\right] + M^{-1} C \left(C^T M^{-1} C\right)^{-1} \dot{g} \ , \qquad (13)$$

which clearly satisfies $C^T \dot{u} = \dot{g}$; (only) an initially div-free velocity field will remain div-free.

Suppose though that we use instead (7) and (8) to generate our GFEM equations? The result is (9) and

$$LP = h \ , \ \text{where} \ L_{ij} \equiv \int \nabla \varphi_i \cdot \nabla \varphi_j \qquad (14), (15)$$

is the 'conventional' GFEM Laplacian and φ_k is the bilinear basis function associated with node k. The RHS vector is

$$h_i \equiv \int \nabla \varphi_i \cdot \left(\nu \nabla^2 \underline{u}^h - \underline{u}^h \cdot \nabla \underline{u}^h\right) - \int_\Gamma \varphi_i \underline{n} \cdot \ \partial \underline{w}/\partial t \qquad (16)$$

from the finite-dimensional form of (8). Clearly some remedial action is required if C^0 basis functions are to be employed (for which $\nabla^2 \underline{u}^h$ is not well-defined); for one type of response, see Sohn and Heinrich (1990). For a 'rationalization', note that $\nu \int \nabla \varphi \cdot \nabla^2 \underline{u} = \nu \int \varphi \underline{n} \cdot \nabla^2 \underline{u}$ when $\nabla \cdot \underline{u} = 0$ and thus the *omission* of the viscous term in (16) can *only* 'affect things' near Γ—an especially valid approximation for large Reynolds number (small ν). (See Hassanzadeh *et al.* 1994) In our formulation, an approximation to (14)–(16) is actually only required at $t = 0$—to estimate the *initial* pressure field; for $t > 0$, the pressure is determined by the (approximate) projection method—described below.

We, and many others, have generated codes based on the Index 1 formulation of (9) and (11) using the Q_1P_0 element; i.e., the quadrilateral element with bilinear basis functions (2D version) for velocity and piecewise-constant basis functions for P. But, with the exception of our ad hoc procedure in Gresho and Chan (1990), all Q_1P_0 Index 1 codes 'required' the also ad hoc (and more deleterious) approximation of *mass lumping* (because M^{-1} is otherwise dense), which introduces a *serious loss* of *accuracy* for the bilinear element when the flow is advection-dominated (vortex shedding, for example). Additional bad features of this element are:

1. It suffers from the "Bent-Element Blues"—(see Gresho and Leone #71); the CP approximation to ∇P is not very accurate when the elements are bent pretty hard (are far from rectangular).
2. The staggered mesh 'bookkeeping' is not fun.
3. The Laplacian matrix $C^T M_L^{-1} C$ is 'awkward' and is suspected to converge 'too slowly' when iterative solvers are used. (M_L is the lumped version of M.)
4. It fails the LBB test! I.e., the velocity and pressure spaces are not quite compatible.
5. The (necessary) integration by parts of the ∇P term causes 'problems' in 'flow-through' domains wherein a homogeneous normal natural boundary condition $\left(\nu \partial u_n/\partial n - P = 0\right)$ is employed and a significant 'body force' is present [see Sani and Gresho (1994)].

THEORY

So, to get back 'in vogue', we now describe *our* attempt at creating a *stabilized* equal-order element; viz Q_1Q_1, which overcomes at least flaws (1), (2), and (5) above. But (3) and (4) still 'shoot it down' *and* mass lumping still seems required. To implement a consistent mass Q_1Q_1 element *and* to pass 'LBB' (which Q_1Q_1 fails in spades), we invoke an "approximate projection" [Almgren *et al.* (1993), and Dvinsky and Dukowicz (1993)] in much the same way that many before us have done, viz, we replace the 'bad' Laplacian in (11) by the good (conventional GFEM) Laplacian, (15). Actually, the procedure is closer to the following:

1. Replace $-P\nabla \cdot \underline{v}$ in (7) by $-\underline{v} \cdot \nabla P$

   which implies $\nu \partial \underline{u}/\partial n = \underline{0}$ as an OBC which is generally an ill-posed boundary condition (see Sani and Gresho #137). But we get away with this crime because we will no longer require a 'stringent' version (even discretely) of $\nabla \cdot \underline{u} = 0$.

2. Generate the following Index 2 DAE's:

   $$M\dot{u} + N(u)u + GP = Ku + f, \ u(0) = u_0 \text{ and } Du = g, \ Du_0 = g_0 \qquad (17)\text{--}(20)$$

   where no longer are G and D transposes of each other.

3. Solve the DAE's via the following approximate projection method: Given u_n and P_n,

   (*i*) Solve $M\frac{(\tilde{u}_{n+1}-u_n)}{\Delta t} + N(u_n)u_n + GP_n = \frac{1}{2}K(\tilde{u}_{n+1} + u_n) + f_n$ for $\tilde{u}_{n+1}$. $\qquad (21)$

   (*ii*) Project $\tilde{u}_{n+1}$ to the approximately discretely div-free subspace as follows:

       (1) Solve $\Delta tL(P_{n+1} - P_n) = D(\tilde{u}_{n+1} - u_n) - (g_{n+1} - g_n)$ for ΔP. $\qquad (22)$

       (2) Compute the projected velocity from $u_{n+1} = \tilde{u}_{n+1} - \Delta t \, M_L^{-1} G(P_{n+1} - P_n)$ $\qquad (23)$

       (3) Update P and go to (*i*)

The approximate projection above is derived as follows: Given $\tilde{u}_{n+1}$, solve

$$M_L \frac{(u_{n+1} - \tilde{u}_{n+1})}{\Delta t} + G(P_{n+1} - P_n) = 0 \text{ and } Du_{n+1} = g_{n+1} + (Du_n - g_n), \qquad (24)\text{--}(25)$$

which is called 'projecting the difference'—a required 'trick' obtained from J. Bell (LLNL, personal communication). Note that, because of the next trick, below, $Du = g$ is never quite achieved in this method.

These equations imply the following 'PPE' for the pressure update:

$$\Delta t \left(DM_L^{-1}G\right)\left(P_{n+1} - P_n\right) = D\left(\tilde{u}_{n+1} - u_n\right) - \left(g_{n+1} - g_n\right), \tag{26}$$

which is replaced by (22) $\left(DM_L^{-1}G \to L\right)$ and is the final "trick" referred to above—a replacement made necessary because $DM_L^{-1}G$ has too many spurious pressure modes (fails LBB) and made 'interesting' because iterative solvers should like L better (i.e., require fewer iterations).

Remarks:

If $Du_n - g_n$ is omitted from the RHS of (22), the results are disappointing in that (at least) a steady pressure cannot be attained even when the velocity becomes steady.

Now comes the 'hard part'—analyzing the resulting algorithm to show that it is actually *useful*. And we admit up front that we have not yet been totally successful. But, being more like engineers than mathematicians (mathematical engineers?), we tested it in the computational laboratory anyway. (And it works. This happens often with engineers—but definitely not always.) If we study (21), (22), and (23) for $\Delta t \to 0$ we get, with $F_n \equiv f_n + Ku_n - N(u_n)u_n - GP_n$,

(i) $\tilde{u}_{n+1} = u_n + \Delta t \, M^{-1} F_n + O\left(\Delta t^2\right)$

(ii) $L\left(P_{n+1} - P_n\right) = DM^{-1}F_n - \dot{g}_n + O(\Delta t)$

(iii) $u_{n+1} = \tilde{u}_{n+1} - \Delta t M_L^{-1} \, GL^{-1}\left\{DM^{-1}F_n - \dot{g}_n\right\} + O\left(\Delta t^2\right)$

$= u_n + \Delta t\left[\wp_a M^{-1}F_n + M_L^{-1}GL^{-1}\dot{g}_n\right] + O\left(\Delta t^2\right)$

(iv) $Du_{n+1} = Du_n + \Delta t\left[EL\left(P_{n+1} - P_n\right) + \dot{g}_n\right] + O\left(\Delta t^2\right)$

(v) Thus $\Delta t \to 0 \Rightarrow \dot{u} = \wp_a M^{-1}\left[f + Ku - N(u)u - GP\right] + M_L^{-1}GL^{-1}\dot{g}$

(vi) $D\dot{u} = D\wp_a M^{-1}F + DM_L^{-1}GL^{-1}\dot{g} = EDM^{-1}\left[f + Ku - N(u)u - GP\right] + \left(DM_L^{-1}G\right)L^{-1}\dot{g}$

and the *problem* is to *reconcile* (iv) with (vi); the former $\Rightarrow D\dot{u} = EL \cdot \lim_{\Delta t \to 0} \left(P_{n+1} - P_n\right) + \dot{g}$ and the two are only in accord if $P_{n+1} - P_n \to 0$ as $\Delta t \to 0$, *and if*

$$\left(DM^{-1}G\right)P = DM^{-1}\left[f + Ku - N(u)u\right] - \dot{g} , \tag{27}$$

which is just the PPE that we'd *like* the pressure to obey! And if this is true, then (v) above yields

$$\dot{u} = M^{-1}\left[f + Ku - GP - N(u)u\right] , \tag{28}$$

also the desired result. (Note the mysterious disappearance of the L-matrix.) *But*—we have not been able to show that $P_{n+1} - P_n = O(\Delta t)$. In fact, somewhat the converse *seems* to be true; viz, starting from the initial pressure, P_0, obtained from

$$LP_0 = DM^{-1}\left[f_0 + Ku_0 - N(u_0)u_0\right] - \dot{g}_0 , \tag{29}$$

proceeding through a few steps, and generalizing via induction leads to

$$L\left(P_{n+1} - P_n\right) = E^{n+1}LP_0 + O(\Delta t) \text{ where } E \equiv I - DM_L^{-1}GL^{-1} \text{ and to} \tag{30}$$

$$Du_{n+1} = Du_n + \Delta t\left[E^{n+2}LP_0 + \dot{g}_n\right] + O\left(\Delta t^2\right) . \tag{31}$$

Thus, at least initially, $P_{n+1} - P_n = O(1)$ in Δt and, for large n, we recover to $P_{n+1} - P_n = O(\Delta t)$ iff $E^n \to 0$. But since that E has eigenvectors with unit eigenvalues whenever the initial vector has any non-zero projection onto the null space of G, E^n may not $\to 0$. It seems that *only* if LP_0 (and $\therefore P_0$) has no spurious pressure node components can we assert that the algorithm can 'recover'—and even then the early (small n) results are likely to be poor...

But the resulting code seems to perform much better than these gloomy results would indicate—as we shall show.

NUMERICAL RESULTS

We have successfully (for the most part) compared $Q_1 Q_1$ against our workhorse $Q_1 P_0$ code for lid-driven cavities and vortex shedding past circular cylinders. Here we show a sample of results for flow past an airfoil at $Re = 10^4$ (based on chord, c) at a small $\left(1.2°\right)$ angle of attack. The particular airfoil, "a NACA 16 thickness form with maximum thickness $t_0/c = 8.84\%$ and maximum camber 2.576% with a beveled anti-singing trailing edge"—E. H. Lurie (1993), was tested in a *water* tunnel (*incompressible* flow!) at MIT's Ocean Engineering Department as part of a Navy/ARPA program on unsteady fluid dynamics. But our laminar flow simulations were not meant to be compared with their measured data at $Re > 10^6$ because we do not yet have a believable turbulence model. Maybe later.

We designed a 'truth' mesh of ~40,000 elements and a test mesh of ~6400 elements. The fine mesh was run at $Re = 10^3$, 10^4, 10^5, and 10^6 using $Q_1 P_0$ and our 'projection 2' algorithm (Gresho and Chan 1990). The results were not believable at $Re = 10^6$ (chaotic), semi-so at $Re = 10^5$ (nearly periodic), and believable at $Re \leq 10^4$. $Re = 10^3$ (only) produced a steady-state result. The coarse mesh for $Re = 10^3$ was then run with $Q_1 P_0$ and the new $Q_1 Q_1$; all three results were acceptably close to each other.

We now focus on the $Re = 10^4$ case and compare $Q_1 P_0$ on the two meshes with $Q_1 Q_1$ on the coarse mesh. Qualitatively, the flow is one of periodic vortex formation and shedding of vortices at the trailing edge. Nowhere else is there any interesting dynamics. Fig. 1 shows snapshots of vectors, streamfunction, and vorticity from the coarse mesh (vectors are interpolated via a coarser-yet mesh graphics package.). Fig. 2 shows time histories, starting from potential flow, of the x-component of velocity at a node just above ($\sim .002c$) the trailing edge—and the pressure there, as well as the RMS-norm of the discrete divergence for $Q_1 Q_1$. The agreement of $Q_1 Q_1$ with $Q_1 P_0$ on the coarse mesh is quite close—the range of u's during vortex shedding being $\sim -.07$ to $+.16$ for $Q_1 P_0$ (1c) and $\sim -.08$ to $+.15$ for $Q_1 Q_1$ (1e). (The fine mesh results (1a) show $\sim -.10$ to $+.20$, somewhat stronger, but still reasonably close). The corresponding pressure histories are (Dank Gott!) of a similar quality; the coarse mesh results ranged from $\sim -.17$ to $-.31$ for $Q_1 P_0$ (1d) and $\sim -.16$ to $-.27$ for $Q_1 Q_1$ (1f), compared to the 'true' results (1b) of $\sim -.15$ to $-.33$. (In all cases, the inlet velocity is 1.0, as is that on the top and bottom boundaries (tow tank BC's). The OBC's were: 'natural'; i.e., $\upsilon \, \partial u/\partial x - P = 0 = \upsilon \, \partial v/\partial x$ for $Q_1 P_0$ and $\upsilon \, \partial u/\partial x = \upsilon \, \partial v/\partial y = P = 0$ for $Q_1 Q_1$, the latter being imposed as an *essential* BC in the PPE.).

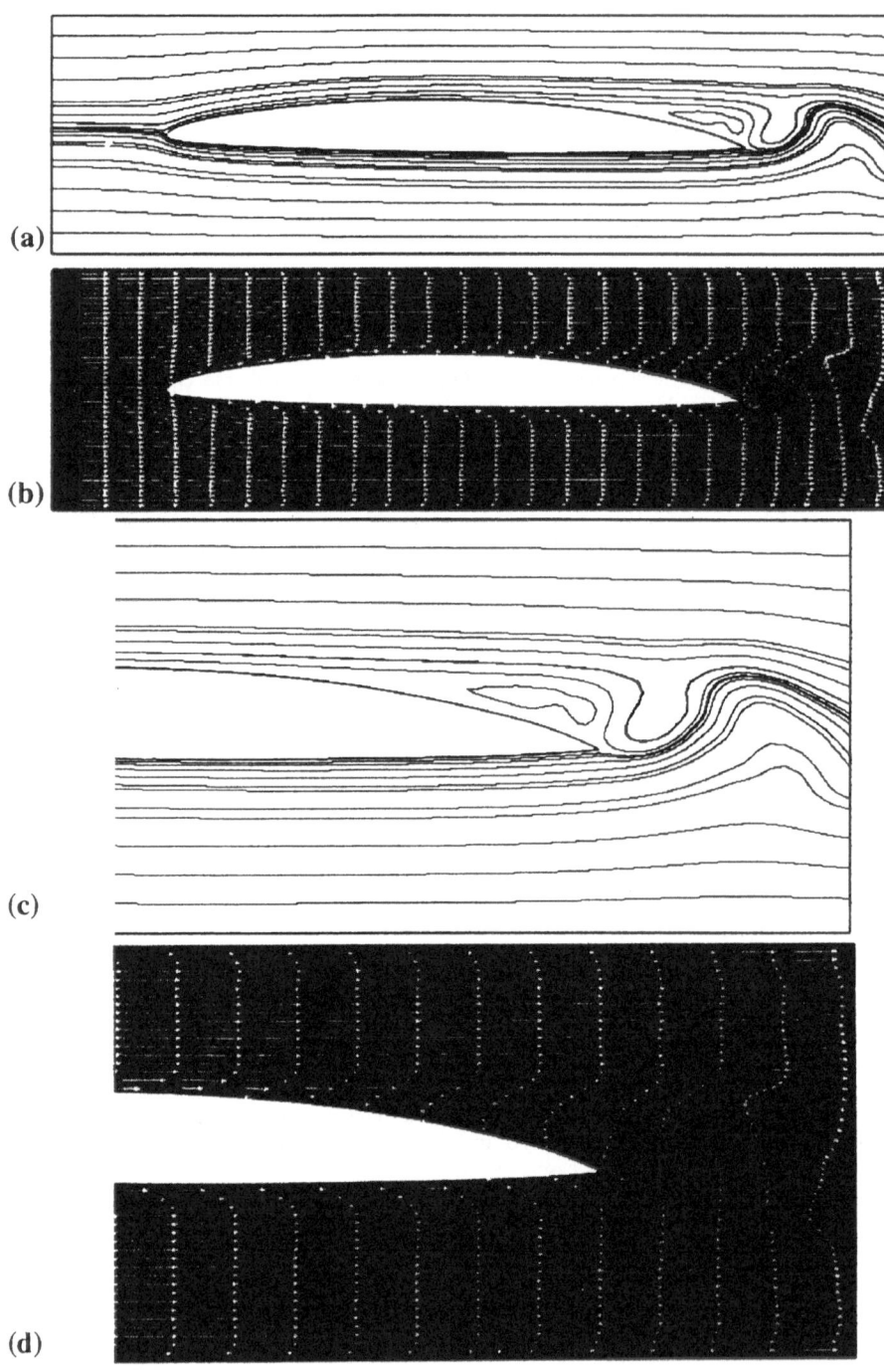

Figure 1.

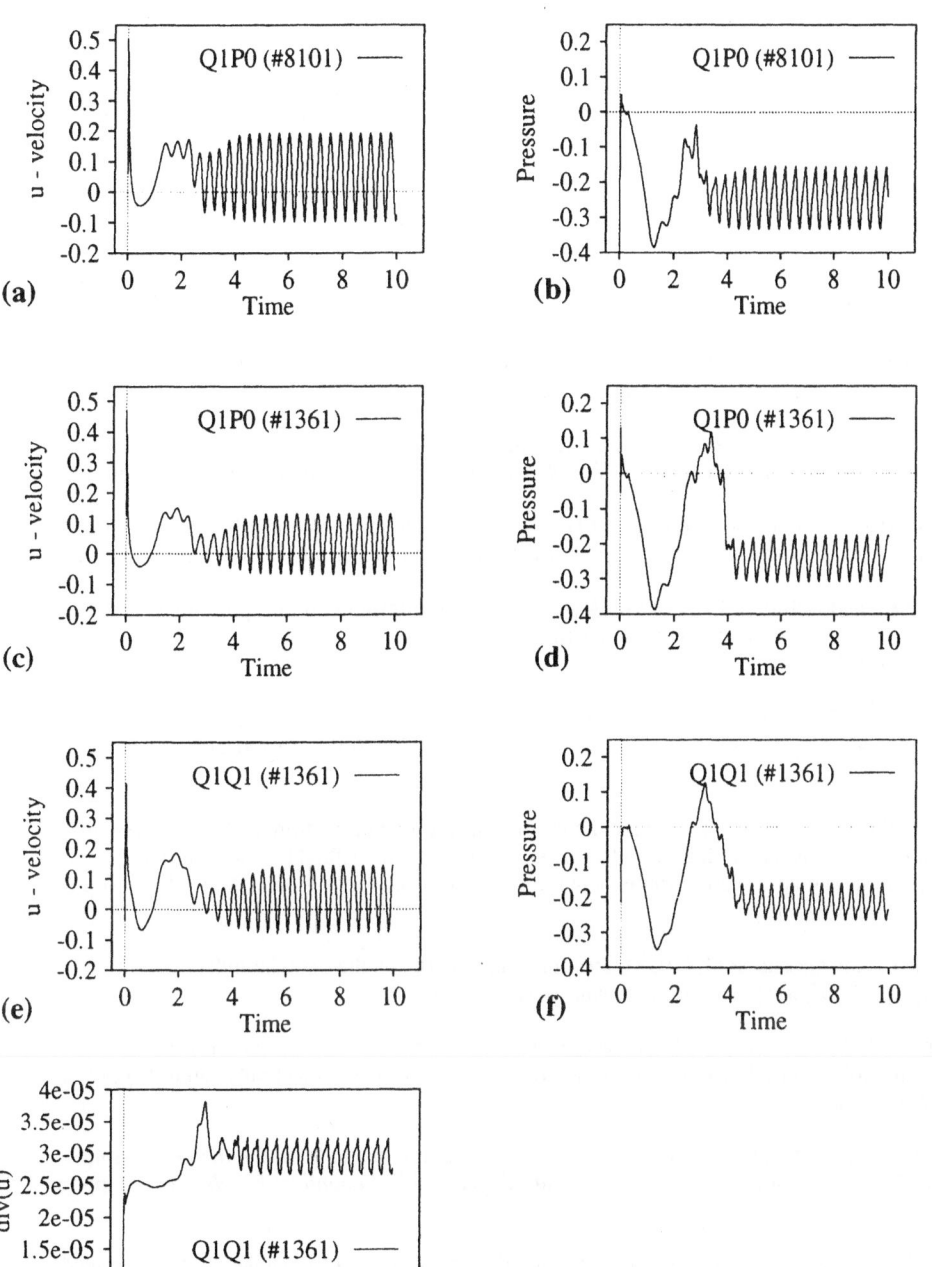

Figure 2:   Time-histories.

CONCLUSIONS

While interesting and viable (apparently—in spite of the shortcomings in our analysis), the new $Q_1 Q_1$ has not yet displaced 'old faithful' ($Q_1 P_0$), largely because the PPE solution—via DSCG—was little cheaper (10–15% fewer iterations). Perhaps it would look better using multi-grid.

For a simpler application of this $Q_1 Q_1$ stabilization technique using forward Euler time marching, see Gresho and Chan (1994).

REFERENCES

Gresho, P.M. (1991a) "Incompressible fluid dynamics: Some fundamental formulation issues", *Annu. Rev. Fluid Mech.*, Vol. **23**, pp. 413-53

Gresho, P.M. (1992) "Some interesting issues in incompressible fluid dynamics, both in the continuum and in numerical simulation", *Advances in Applied Mechanics*, Vol. **28**, pp. 46–133.

Gresho, P.M. (1991b) "Some current CFD issues relevant to the incompressible Navier-Stokes equations", Computer Methods in Applied Mechanics and Engineering, Vol. 87, pp. 201–252.

Sohn, J.L., and Heinrich, J.C. (1990) "A poisson equation formulation for pressure calculations in penalty finite element models for viscous incompressible flows", *Int. J. Num. Meth. Fluids*, Vol. **30**, pp. 349–361.

Hassanzadeh, S., Sonnad, V., and Foresti, S. (1993) "Finite element implementation of boundary conditions for the pressure poisson equation of incompressible flow", to appear in *Int. J. Num. Meth. Fluids* (1994).

Gresho, P.M., and Chan, S.T. "On the theory of semi-implicit projection methods for viscous incompressible flow and its implementation via a finite element method that also introduces a nearly consistent mass matrix. Part 2: Implementation", *Int. J. Num. Meth. Fluids*, Vol. **11**, pp. 621–659.

Sani, R.L., and Gresho, P.M. (1993) "Resume and remarks on the open boundary condition minisymposium", to appear in *Int. J. Num. Meth. Fluids*.

Almgren, A.S., Bell, J.B., and Szymczak, W.G. (1993) "A numerical method for the incompressible Navier-Stokes equations based on an approximate projection", submitted to *SIAM J. of Sci. Comp.*

Dvinsky, A.S., and Dukowicz, J.K. (1992) "Null-space-free methods for the incompressible Navier-Stokes equations on non-staggered curvilinear grids", *Computers Fluids*, Vol. **22**, pp. 685–696.

Lurie, E.H. (1993) "Unsteady response of a two-dimensional hydrofoil subject to high reduced frequency gust loading", Ocean Engineering's Report No. 93–5.

Gresho, P.M., and Chan, S.T. (1994) "Another stabilized incompressible Navier-Stokes solver using explicit time-marching", in Proceedings, Second US-Japan Symposium on Finite Element Methods in Large-Scale CFD, Chuo University, 14–16 March 1994.

AN ARBITRARY LAGRANGIAN–EULERIAN FINITE ELEMENT METHOD FOR FLUID–STRUCTURE INTERACTION PROBLEM

Akira MARUOKA, Akira ANJU and Mutsuto KAWAHARA
Department of Civil Engineering
Chuo University
1–13–27 Kasuga Bunkyou-ku Tokyo 112
Japan

1. INTRODUCTION

This paper presents a finite element analysis of a fluid–structure interaction problem, in which the fluid is treated as incompressible viscous flow, and a structure is idealized as a rigid body supported by elastic springs. The arbitrary Lagrangian–Eulerian (ALE) method[1, 2] is employed to solve the flow field around the structure, and the fractional step method[2, 3] is adopted for the time integration. For the numerical example, the present method is applied to the flow analysis around the oscillating rectangular cylinder.

2. BASIC EQUATIONS

2.1 Basic equations for fluid in the ALE description

The Navier–Stokes equation and the incompressible continuity equation for fluid in the ALE description are expressed as:

$$\rho\frac{\partial u_i}{\partial t} + \rho(u_j - w_j)u_{i,j} + p_{,i} - \mu(u_{i,j} + u_{j,i})_{,j} = 0 \tag{1}$$

$$u_{i,i} = 0 \tag{2}$$

where u_i is velocity of fluid particle, w_i is mesh velocity, p is pressure, ρ is density and μ is viscosity coefficient. The mesh velocity w_i can be chosen arbitrary in the ALE description.

2.2 Equation of motion for structure

The structure surrounded by the fluid is assumed to be written as follows.

(a) The structure is a single rigid body.

(b) The motion of the structure is described by the three degrees of freedom, which are translational displacements X and Y in x and y directions, and rotational displacement θ defined at the center of rotation G.

(c) The structure is supported by the three elastic springs and the three dashpots, and each degree of freedom has only one spring and one dashpot that are uncoupled from the other degrees of freedom.

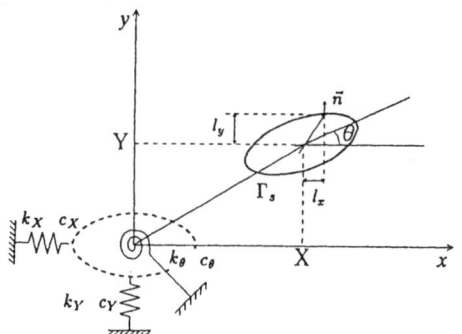

Figure 1 Model of structure

A. Peters et al. (eds.), Computational Methods in Water Resources X, 1233–1238.

Thus, the equation of motion for the structure can be written as:

$$M_{ij}\ddot{X}_j + C_{ij}\dot{X}_j + K_{ij}X_j = F_i \tag{3}$$

$$X_i = \{X, Y, \theta\} \qquad F_i = \{D, L, M\}$$

where M_{ij}, C_{ij} and K_{ij} are mass matrix, damping matrix and stiffness matrix. Each matrix is a diagonal matrix with constant coefficients. X_i and F_i are displacement and fluid force. In this paper, the conventional linear acceleration method[4] is used to solve eq.(3).

D, L and M are calculated as:

$$D = \int_{\Gamma_s} (\sigma_{xj}n_j)d\Gamma \qquad L = \int_{\Gamma_s} (\sigma_{yj}n_j)d\Gamma \qquad M = \int_{\Gamma_s} \{-l_y(\sigma_{xj}n_j) + l_x(\sigma_{yj}n_j)\}d\Gamma \tag{4}$$

where Γ_s is boundary around the structure, n_i is unit normal vector to the boundary Γ_s, l_i is distance between the point on the boundary Γ_s to the center of rotation G, and σ_{ij} is stress tensor as:

$$\sigma_{ij} = -p\delta_{ij} + \tau_{ij} \qquad (\tau_{ij} = \mu(u_{i,j} + u_{j,i})) \tag{5}$$

where δ_{ij} is Kronecker's delta function.

3. FRACTIONAL STEP FINITE ELEMENT METHOD

For the numerical integration in time, eqs.(1),(2) can be discretized at nth time point as:

$$\rho\left(\frac{u_i^{n+1} - u_i^n}{\Delta t}\right) + \rho(u_j^n - w_j^n)u_{i,j}^n + p_{,i}^{n+1} - \mu(u_{i,j}^n + u_{j,i}^n)_{,j} = 0 \tag{6}$$

$$u_{,i}^{n+1} = 0 \tag{7}$$

where Δt means time increment.

In this paper, the velocity correction method[3] that is one of the fractional step method is applied to eqs.(6),(7). The algolithm of the velocity correction method can be summarized as follows.

1. The intermediate velocity $\tilde{u}_i^{n+1}$ is calculated by the momentum equation which subtracts the pressure p^{n+1} from eq.(6).

$$\rho\tilde{u}_i^{n+1} = \rho u_i^n - \Delta t \left(\rho(u_j^n - w_j^n)u_{i,j}^n - \mu(u_{i,j}^n + u_{j,i}^n)_{,j}\right) \tag{8}$$

2. The pressure p_i^{n+1} is calculated by the pressure poisson equation.

$$p_{,ii}^{n+1} = \frac{\rho}{\Delta t}\tilde{u}_{,i}^{n+1} \tag{9}$$

3. The velocity u_i^{n+1} is calculated by the velocity correction equation.

$$\rho u_i^{n+1} = \rho\tilde{u}_i^{n+1} - \Delta t p_{,i}^{n+1} \tag{10}$$

4. $n \leftarrow n + 1$ and go to step 1.

Employing the conventional Galerkin method for the space discretization, the finite element equations can be obtained as:

$$\rho\bar{M}_{\alpha\beta}^{n+1}\tilde{u}_{\beta i}^{n+1} = \rho\bar{M}_{\alpha\beta}^n u_{\beta i}^n - \Delta t \left(\rho K_{\alpha\beta\gamma i}^n(u_{\beta j}^n - w_{\beta j}^n)u_{\gamma i}^n + \mu S_{\alpha i\beta j}^n u_{\beta j}^n - \Sigma_{\alpha i}^n\right) \tag{11}$$

$$A_{\alpha i \beta i}^{n+1} p_\beta^{n+1} = -\frac{\rho}{\Delta t} H_{\alpha \beta i}^{n+1} \tilde{u}_{\beta i}^{n+1} + \Lambda_{\alpha i}^{n+1} \tag{12}$$

$$\rho \bar{M}_{\alpha \beta}^{n+1} u_{\beta i}^{n+1} = \rho \bar{M}_{\alpha \beta}^{n+1} \tilde{u}_{\beta i}^{n+1} - \Delta t H_{\alpha \beta i}^{n+1} p_\beta^{n+1} \tag{13}$$

in which $\bar{M}_{\alpha \beta}$ is lummped matrix of $M_{\alpha \beta}$, and coeffcent matrices mean:

$$M_{\alpha \beta} = \int \Phi_\alpha \Phi_\beta d\Omega \quad \cdot K_{\alpha \beta \gamma i} = \int \Phi_\alpha \Phi_\beta \Phi_{\gamma,i} d\Omega \quad S_{\alpha i \beta j} = \int \Phi_{\alpha,k} \Phi_{\beta,k} d\Omega \, \delta_{ij} + \int \Phi_{\alpha,j} \Phi_{\beta,i} d\Omega$$

$$H_{\alpha \beta i} = \int \Phi_\alpha \Phi_{\beta,i} d\Omega \quad A_{\alpha i \beta i} = \int \Phi_{\alpha,i} \Phi_{\beta,i} d\Omega \quad \Sigma_{\alpha i} = \int \Phi_\alpha \tau_{ij} n_j d\Gamma \quad \Lambda_{\alpha i} = \int \Phi_\alpha p_{,i} n_i d\Gamma$$

where Φ_α is the interpolation function.

4. ALGORITHM

The computational algorithm of the fluid–structure interaction problem from the n step to the $n+1$ step can be described as follows.

1. The right hand side of eq.(11) is calculated using u_i^n and w_i^n.
2. The displacement X_i^{n+1} is calculated using eq.(3) based on the fluid force F_i^n.
3. The nodal point is moved from x_i^n to x_i^{n+1}.
4. Using the new location of nodal point, the mesh velocity w_i^{n+1} is calculated as:

$$w_i^{n+1} = \frac{x_i^{n+1} - x_i^n}{\Delta t} \tag{14}$$

5. Each coefficient matrix is calculated for the nodal point x_i^{n+1}.
6. The intermediate velocity $\tilde{u}_i^{n+1}$ is calculated using the right hand side of eq.(11).
7. The pressure p_i^{n+1} is calculated using eq.(12).
8. The velocity u_i^{n+1} is calculated using eq.(13).
9. The fluid force F_i^{n+1} is calculated using eq.(4).
10. $n \leftarrow n+1$ and go to step 1.

5. NUMERICAL EXAMPLE

For the numerical example, the flow around the oscillating rectangular cylinder is calculated. The cylinder oscillates only perpendicular direction to the flow. Figure 2 shows the calculated domain, boundary conditions and physical parameters. The same parameters used in a wind tunnel test[5] are chosen. To express the phenomenon, non-dimensional parameters in table 1 are used. The ALE domains are denoted the domains Ω_1 and Ω_2, and the Eulerian domain(fixed domain) is denoted the domain Ω_3. In the ALE domain Ω_1, the mesh moves with the structure as one united body. Figure 3 shows the mesh deformation pattern when the cylinder displacement is D. Total numbers of nodes and elements are 4340 and 8400. The linear triangular element is employed for all velocity, mesh velocity and pressure. For the numerical stability, the BTD(Balancing Tensor Diffudivity) technique[6] is introduced. The calculated results are shown in Figures. 4–6. Figure 4 shows the time history of displacement Y/D. When the non–dimensional velocity V is equal to 5.5, it can be clearly seen that amplitudes of the cylinder come to wider and wider. Figure 5 shows the stream line and pressure distribution at the time of lift coefficient $C_L = 0$ and $V = 5.5$. The relation between the maximum amplitude Y/D and the non-dimensional velocity V is presented in Figure 6. The tendency of calculated results is nearly equal to that of the wind tunnel test[5].

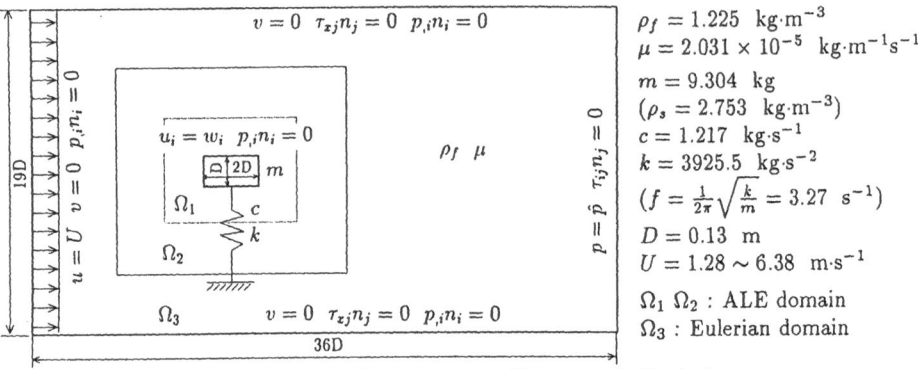

Figure 2 Calculated domain, boundary conditions and physical parameters

Table 1 Non–dimensional parameters

Density ratio	$R = 224.7$
Damping coefficient	$h = 0.00318$
Non–dimensional velocity	$V = 3.0 \sim 15.0$
Reynolds number	$Re = 1.0 \times 10^4 \sim 5.0 \times 10^4$
Non–dimensional time increment	$\Delta T = 0.01$

$$R = \frac{\rho_s}{\rho_f} \qquad h = \frac{c}{2\sqrt{km}} \qquad V = \frac{U}{fD} \qquad Re = \frac{\rho U D}{\mu} \qquad \Delta T = \frac{\Delta t U}{D}$$

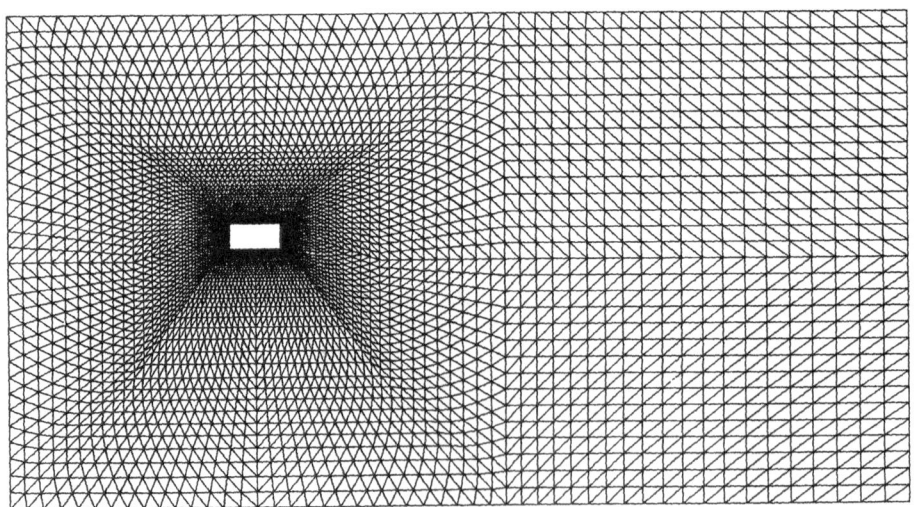

Figure 3 Mesh deformation pattern

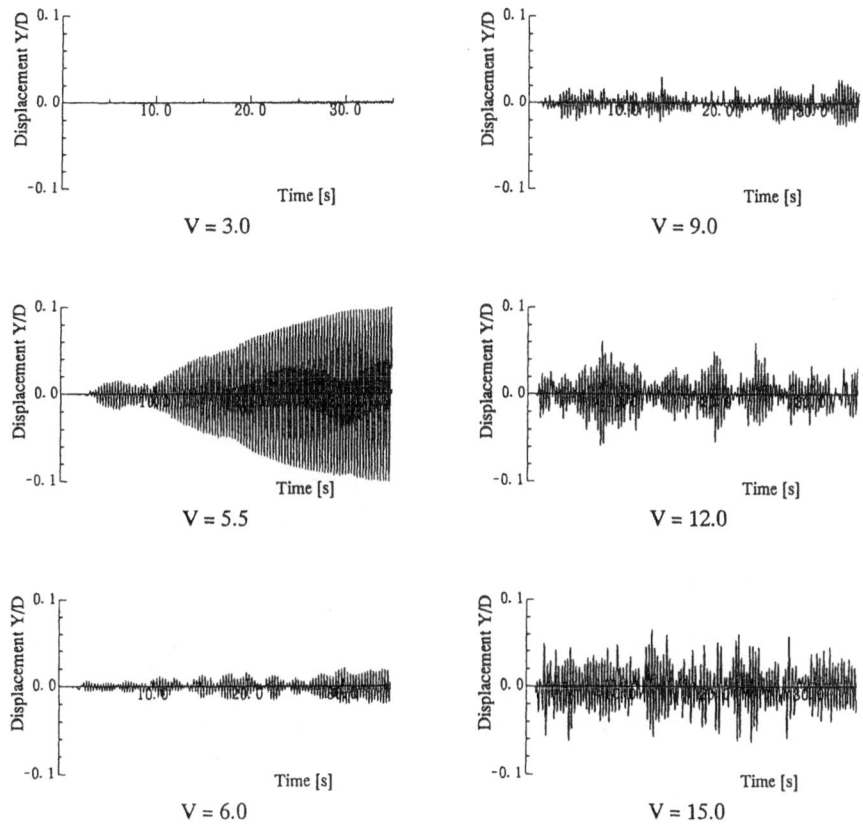

Figure 4 Time history of displacement Y/D

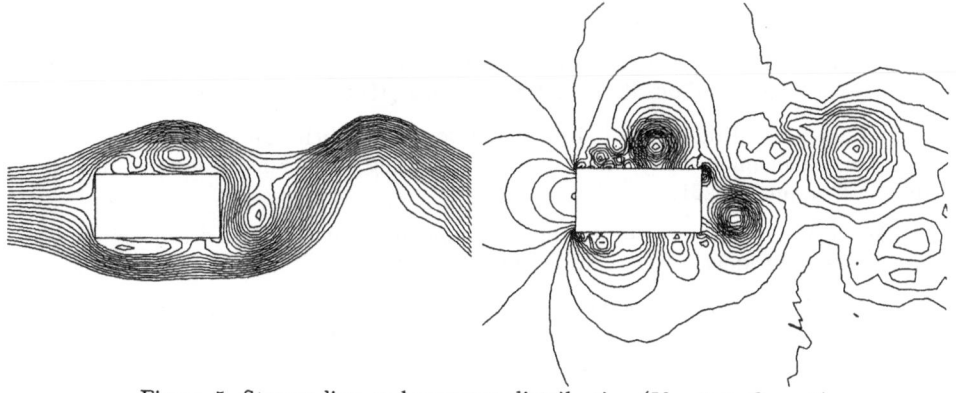

Figure 5 Stream line and pressure distribution ($V = 5.5$, $C_L = 0$)

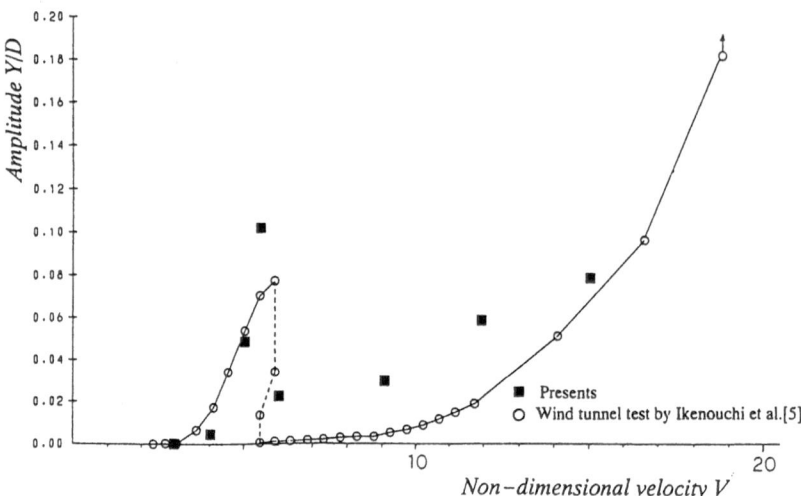

Figure 6 Relation between maximum amplitude Y/D and non-dimensional velocity V

6. CONCLUSION

In this paper, an ALE finite element analysis using the fractional step method has been proposed for the fluid–structure interaction problem. For the numerical example, the flow around the oscillating rectangular cylinder is calculated. In this paper, the technique in which the analysis domain is divided into Ω_1, Ω_2 and Ω_3 is employed. This technique not only reduces the computer core storage but also avoids the large element distortion.

References

[1] T. Nomura and T.J.R. Hughes, An arbitrary Lagrangian–Eulerian finite element method for interaction of fluid and a rigid body, Comput. Meths. Appl. Mech. Engng. 95 (1992), pp.115–138.

[2] V. Palanisamy and M. Kawahara, A factional step arbitrary Lagrangian–Eulerian finite element method for free surface density flows, Comp. Fluid. Dyn. vol.1 (1993), pp.57–77.

[3] M. Shimura and M. Kawahara, Two dimensional finite element flow analysis using the velocity correction method, Proc. of JSCE No.398 (1988), pp.255–263.

[4] M. Kawahara, H. Hirano, and T. Kodama, Two–step explicit finite element method for high Reynolds number flow passed through oscillating body, Finite Elements in Fluids vol.5 (1984), pp.227–261.

[5] M. Ikenouchi et al., Fundamental investigation on wind resistant design of bridges, Mitsui Zosen Technical Review No.116 (1982), pp.31–38.

[6] P. M. Gresho et al., A Modified finite element method for the solving the time-dependent incompressible Navier–Stokes equations, Part 1 & 2, Int. J. Num. Meth. Fluids, Vol.4 (1984), pp.557–598 & pp.619–640

ACCURATE TIME DISCRETIZATION SCHEMES FOR COMPUTING NONSTATIONARY INCOMPRESSIBLE FLUID FLOW

R. RANNACHER
Institute for Applied Mathematics
University of Heidelberg
INF 293, D-69120 Heidelberg

The long-time simulation of viscous flows requires apart of a stable space discretization and an efficient elliptic solver particularly an accurate time-stepping procedure. In this note, we discuss the pros and cons of two methods, the fractional step θ-scheme and a second-order projection method, in combination with a finite element discretization in space.

THE NAVIER-STOKES PROBLEM

We consider the incompressible Navier-Stokes problem

$$\mathbf{v}_t - \nu\Delta\mathbf{v} + \mathbf{v}\cdot\nabla\mathbf{v} + \nabla p = \mathbf{f}, \quad \nabla\cdot\mathbf{v} = 0 \quad \text{in } \Omega\times[0,T), \tag{1a}$$

$$\mathbf{v}_{|t=0} = \mathbf{v}^0, \quad \mathbf{v}_{|\Gamma_0} = 0, \quad \mathbf{v}_{|\Gamma_{in}} = \mathbf{v}_{in}, \quad \nu\partial_n\mathbf{v} - p\mathbf{n}_{|\Gamma_{out}} = 0, \tag{1b}$$

in a bounded domain $\Omega \subset \mathbf{R}^d$ (d=2 or 3), where Γ_0, Γ_{in} and Γ_{out} are the rigid, the inflow and the outflow part of the boundary, respectively, and ν is the kinematic viscosity. Here and below, quantities related to the velocity vector are written with bold-face type. Assuming the characteristic length and speed to be normalized to one, the Reynolds number is $Re \equiv 1/\nu$. We are interested in laminar flows related to values $Re \approx 10^1\text{-}10^5$.

SPATIAL DISCRETIZATION

We consider a finite element (FE) discretization in space of problem (1) based on its standard variational formulation in the primary variables $\{v,p\}$. On a finite mesh T_h (triangulation, etc.) covering the domain Ω, with variable element width h, let H_h and L_h be spaces of piecewise polynomial ansatz

1239

A. Peters et al. (eds.), Computational Methods in Water Resources X, 1239–1246.
© 1994 Kluwer Academic Publishers. Printed in the Netherlands.

functions for the velocity and pressure, respectively. The spaces H_h may be "nonconforming", i.e., the discrete velocities are required to be continuous across the interelement boundaries and to vanish along the rigid boundary only in an approximate sense (see the first example below). The discrete analogues of (2) read: Find $v_h \in H_h + v_{in,h}$ and $p_h \in L_h$, such that

$$(v_{h,t}, \phi) + v(\nabla v_h, \nabla \phi) + (v_h \cdot \nabla v_h, \phi) - (p_h, \nabla \cdot \phi) = (f, \phi) \quad \forall \, \phi \in H_h , \qquad (2a)$$

$$(\chi, \nabla \cdot v_h) = 0 \quad \forall \, \chi \in L_h , \qquad (2b)$$

where $(\cdot, \cdot) = (\cdot, \cdot)_\Omega$ denotes the inner product of $L^2(\Omega)^d$, $v_{in,h}$ is an approximation to v_{in}, and the nabla operator ∇ is to be understood in the "piecewise" sense (with respect to T_h) if H_h is nonconforming.

Many suitable pairs of velocity/pressure spaces H_h / L_h have been proposed in the literature (see, e.g., [3]). Below, two particularly simple examples of quadrilateral elements will be described which have satisfactory approximation properties and are applicable in two as well as three space dimensions. For simplicity, we specify them only for the 2-dimensional case.

1) In the first example piecewise "rotated" bilinear shape functions are used for the velocity (on the reference element) spanned by $\{1, x, y, x^2 - y^2\}$, and piecewise constants for the pressure. The nodal values are the mean values of the velocity over the element edges and the mean values of the pressure over the elements. This "nonconforming" element is the natural quadrilateral analogue of the well-known triangular Stokes element of Crouzeix/Raviart (see [3]). A corresponding convergence analysis is given in [12] and very promising computational results are reported in [14].

2) In the second example continuous bilinear shape functions are used for both the velocity and the pressure where the nodal values are the respective function values at the vertices of the mesh. This originally unstable "conforming" element is stabilized by adding certain least-squares terms in the continuity equation (2b) (pressure stabilization of Hughes, et al., [6]),

$$(\chi, \nabla \cdot v_h) + \frac{\sigma}{v} \sum_{K \in T_h, \Gamma \subset \partial K} \left\{ h_K^2 (\nabla \chi, \nabla p_h)_K + h_\Gamma ([\chi]_\Gamma, [p_h]_\Gamma)_\Gamma \right\} = c_h(v_h; \chi) ,$$

$$c_h(v_h; \chi) = \frac{\sigma'}{v} \sum_{K \in T_h} h_K^2 (\nabla \chi, f + v \Delta v_h - v_h \cdot \nabla v_h)_K ,$$

where $[\chi]_\Gamma$ denotes the jump of χ over the edge $\Gamma \subset \partial K$, $h_K = \mathrm{diam}(K)$ and

$h_\Gamma = \text{diam}(\Gamma)$. This modification is fully consistent as (for $\sigma=\sigma'$) the stabilization terms cancel out if the solution $\{v,p\}$ of problem (1) is inserted. Normally, one takes $\sigma \approx 0.1$ and $\sigma'=\sigma$ or simply $\sigma'=0$.

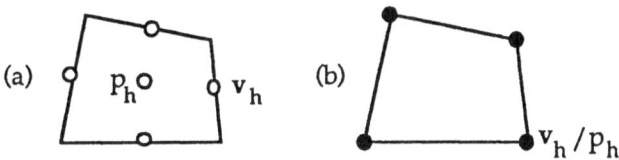

Figure 1. Nodal points of a) the nonconforming "rotated" $\widetilde{Q}_1/Q_0$- and b) the conforming (stabilized) Q_1/Q_1-Stokes element

These quadrilateral elements admit simple upwind strategies which lead to matrices with certain M-matrix properties. Further, efficient multigrid solvers are available which work satisfactorily over the whole range of relevant Reynolds numbers, $1 \le \text{Re} < 10^5$ (see [14] and [1]).

Using the same symbols v_h and p_h also for the coefficient vectors corresponding to the nodal bases of the spaces H_h and L_h, the discrete problem (3) can be written as a differential-algebraic system

$$M_h \dot{v}_h + A_h v_h + N(v_h)v_h + B_h p_h = g, \tag{3a}$$

$$B_h^T v_h + \sigma C_h p_h = \sigma' c_h(v_h). \tag{3b}$$

Here, M_h is the mass, A_h the diffusion, B_h the gradient and $N_h(\cdot)$ the transport matrix, while C_h and $c_h(\cdot)$ are the Neumann-like matrix and the correction term corresponding to a least-squares pressure stabilization (see example 2). The load vector g contains also the contribution from the nonhomogeneous inflow condition. For dominant transport, the matrix $N_h(\cdot)$ may also include some upwind mechanism (see [14] and [1]).

TIME-DISCRETIZATION

System (3) is highly stiff, with a stiffness ratio of the order $O(vh^{-2})$ essentially depending on the smallest mesh-size h_{min}. Further stiffening is caused by the pressure-velocity coupling (3b). Therefore, in the choice of time-stepping methods for solving this system, one is limited to implicit schemes. In the past, explicit time-stepping schemes were commonly used in non-stationary flow calculations mainly for managing the transition to steady-state limits. Because of the severe stability problems inherent in this approach (for moderately sized Reynolds numbers and non-uniformly refined meshes), the required very small time-steps prohibited the accurate solution of really time-dependent flows. Since implicit methods became feasible thanks to more efficient linear system solvers, the schemes most frequently used were either the first-order backward Euler scheme or the second-order Crank-Nicolson scheme. These two methods belong to the class of "one-step θ-schemes", which, dropping the subscript h, read as follows (k = time step):

$$[M+\theta k\overline{A}^{n+1}]v^{n+1}+Bp^{n+1} = [M-(1-\theta)kA^n]v^n+\theta kg^{n+1}+(1-\theta)kg^n, \qquad (4a)$$

$$B^T v^{n+1}+\sigma Cp^{n+1} = \sigma'\overline{c}^{n+1}, \qquad (4b)$$

where $A^m = A+N(v^m)$, $g^m = g(t_m)$, $\overline{A}^m = A+N(\overline{v}^m)$ and $\overline{c}^m = c(\overline{v}^m)$, with some predicted value $\overline{v}^m \approx v^m$ obtained, e.g., by linear extrapolation from the preceding time levels: $\overline{v}^m = 2v^{m-1}-v^{m-2}$. The Crank-Nicolson scheme (i.e., $\theta=1/2$) occasionally suffers from unexpected instabilities because of its only weak damping properties (not *strongly* A-stable). For a discussion of this aspect see [9] and [10]. This defect can in principle be cured by an adaptive step-size reduction, but this may necessitate the use of an unreasonably small time step thereby increasing the computational cost.

Alternative schemes of higher order are based on the (diagonally) implicit Runge-Kutta formulae or the backward differencing multi-step formulae, being well-known from ODE literature. These schemes, however, have not yet found wide applications in flow computations, mainly because of their higher complexity and storage requirements compared with the traditional Crank-Nicolson scheme. There is still another method which seems to have the potential to excel in this competition, the so-called "fractional step θ-scheme" (FS_θ scheme). Applied to the usual linear model equation $y' = \lambda y$, this 3-level one-step method derives from a third-order rational approximation of the exponential,

$$R_\theta(\lambda) \equiv \frac{(1+\alpha\theta'\lambda)(1+\beta\theta\lambda)^2}{(1-\alpha\theta\lambda)^2(1-\beta\theta'\lambda)} = e^\lambda + O(|\lambda|^3), \ \lambda < 0 , \qquad (5)$$

where $\theta = 1-\sqrt{2}/2 = 0.292893...$, $\theta'=1-2\theta$, and $\alpha\in(0.5,1]$, $\beta=1-\alpha$. This choice of parameters insures the second-order accuracy and, additionally, the strong A-stability of the scheme,

$$\overline{\lim}_{\lambda \to \infty} |R_\theta(\lambda)| = \beta/\alpha < 1. \tag{6}$$

For $\alpha = (1-2\theta)/(1-\theta) = 0.585786...$, one has $\alpha\theta = \beta\theta'$, which may be useful in solving linear autonomous problems. Further, for Re $\lambda \to 0$, the factor $R_\theta(\lambda)$ becomes nearly one (e.g., $R_\theta(0.8i) \approx 0.9998$), which means that the resulting time-stepping scheme contains only very little numerical dissipation. Applied to the Navier-Stokes system (6) the FS_θ scheme reads:

$$[M+\alpha\theta k\overline{A}^{n+\theta}]v^{n+\theta}+Bp^{n+\theta} = [M-\beta\theta kA^n]v^n+\theta kg^n, \tag{7a}$$

$$B^T v^{n+\theta}+\sigma Cp^{n+\theta} = \sigma'\overline{c}^{n+\theta}, \tag{7b}$$

$$[M+\beta\theta' k\overline{A}^{n+1-\theta}]v^{n+1-\theta}+Bp^{n+1-\theta} = [M-\alpha\theta' kA^{n+\theta}]v^{n+\theta}+\theta' kg^{n+1-\theta}, \tag{8a}$$

$$B^T v^{n+1-\theta}+\sigma Cp^{n+1-\theta} = \sigma'\overline{c}^{n+1-\theta}, \tag{8b}$$

$$[M+\alpha\theta k\overline{A}^{n+1}]v^{n+1}+Bp^{n+1} = [M-\beta\theta kA^{n+1-\theta}]v^{n+1-\theta}+\theta kg^{n+1}, \tag{9a}$$

$$B^T v^{n+1}+\sigma Cp^{n+1} = \sigma'\overline{c}^{n+1}. \tag{9b}$$

The use of the FS_θ scheme for solving the Navier-Stokes equations was first proposed by Glowinski, et al., [2], in the form of an operator-splitting scheme separating the two problems "nonlinearity" and "incompressibility" within each cycle $t_n \to t_{n+\theta} \to t_{n+1-\theta} \to t_{n+1}$. This aspect may not be relevant if efficient quasi-Stokes solvers are available, but the FS_θ scheme is also attractive as a mere time-stepping method. It possesses the full smoothing property, which is important in the case of rough initial or boundary data, and it contains only very little numerical dissipation, which is crucial in the computation of non-enforced temporal oscillations in a flow. Furthermore, one cycle of length $(2\theta+\theta')k=3k$ provides nearly the same accuracy as three steps of the Crank-Nicolson scheme with total step-length $3k$ (for more details see [14] and [11]). A complete convergence analysis of the FS_θ scheme for the Navier-Stokes problem has recently been given in [7] including an optimal-order error estimate $\|v_h^n-v(\cdot,t_n)\| \le C(h^2+k^2)$.

For solving the linear Stokes-like systems arising in each substep of the FS_θ scheme, efficient and robust multigrid methods are available (see [14] and [1]). This is particularly important in long-time simulations when the time step has to be taken rather large. For illustration, Figures 1 shows relative streamline plots for a flow around an inclined ellipse at Re=500 (van Kármán vortex shedding) developing from time t=7 to t=50, computed with a) the backward Euler scheme, b) the Crank-Nicolson scheme, and c) the FS_θ scheme, all with time step length k=0.33 .

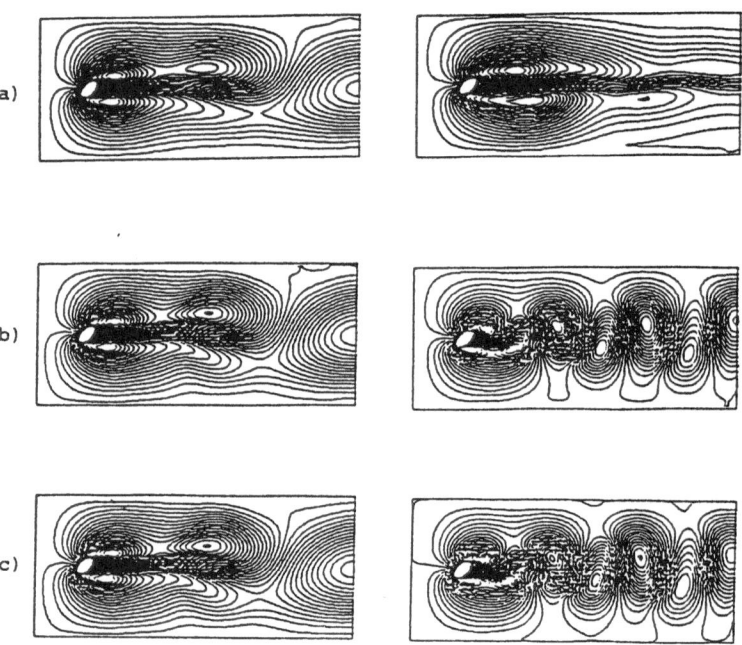

Figure 2. Relative streamlines of van Kármán vortex shedding, with Re=500, at times t = 7 and t = 50 (from [14], [10]).

If a quasi-Stokes solver is not available or if it is not efficient enough, one may separate the incompressibility constraint from the momentum equation through splitting approach as proposed by A. Chorin (see [4] and [13]). Chorin's original method was only of first order accurate, but later also second-order projection methods have been devised (see [4] and [13]). One of these schemes reads in our algebraic notation as follows (see, e.g., [10] and [8]):

$$[M+\tfrac{k}{2}\overline{A}^{n+1}]\tilde{u}^{n+1}+Bp^n = [M-\tfrac{k}{2}A^n]u^n+kg^{n+1/2} \tag{11a}$$

$$B^T M^{-1} Bp^{n+1} = \tfrac{1}{k}B^T \tilde{v}^{n+1}, \quad v^{n+1} = \tilde{v}^{n+1}-kM^{-1}Bp^{n+1}, \quad p^{n+1} = p^n + p^{n+1}, \tag{11b}$$

where again $A^m = A+N(v^m)$, $g^m = g(t_m)$, $\overline{A}^m = A+N(\overline{v}^m)$ and $\overline{c}^m = c(\overline{v}^m)$. This scheme requires in the first step the solution of a convection-diffusion system (alike the multi-dimensional Burgers equations) and in the second step the solution of a scalar Neuman-type problem for the pressure correction. In this way the coupling of velocity and pressure is broken within the time-stepping and only efficient Poisson- and Neumann solvers are needed. This approach appears very attractive and is widely used in engineering practice (see [4]). A theoretical analysis has been given in [13] and [8].

There are various versions of the projection method according to the construction of the "Neumann-matrix" $C \equiv B^T M^{-1} B$. In Chorin's original method, the matrix $C \equiv C_N$ was derived by discretizing the "Neumann operator" Δ_N (Laplace operator with homogeneous Neumann boundary conditions) by some finite difference scheme. In the context of the stable FE discretization (3) a more natural choice would be to actually use the form $C \equiv B_0^T M_0^{-1} B_0$, where the matrices B_0 and M_0 are generated within the FE ansatz $H_h \times L_h$ observing the no-slip ("zero" Dirichlet) boundary conditions for the velocity basis functions. Alternatively, one could work with $C \equiv B_s^T M_s^{-1} B_s$ or with $C = B_f^T M_f^{-1} B_f$, where the indices s and f refer to the use of either "slip" boundary conditions, $n \cdot v_{|\partial\Omega} = 0$, or no boundary restrictions at all ("free") for the velocity. The case $C \equiv B_s^T M_s^{-1} B_s$ is related to a "mixed" discretization of the Neumann operator. These different choices show various interesting effects on the stability and accuracy of the resulting schemes which will be discussed in more detail in a forthcoming paper.

REFERENCES

[1] Becker, R., and Rannacher, R., Finite element methods for the Stokes and Navier-Stokes equations on highly stretched grids, Proc. 10th GAMM-Seminar, Kiel, January 14-16, 1994, Vieweg, to appear.

[2] Bristeau, M.O., Glowinski, R., and J. Periaux, Numerical methods for the Navier-Stokes equations: Applications to the simulation of compressible and incompressible viscous flows, in Computer Physics Report 1987.

[3]   Girault, V., and P.A. Raviart, Finite Element Methods for Navier-Stokes Equations, Springer, Berlin-Heidelberg 1986.

[4]   Gresho,P.M., and Chan, S.T., On the theory of semi-implicit projection methods for viscous incompressible flow and its implementation via a finite element method that also introduces a nearly consistent mass matrix, Part 1: Theory, Part 2: Implementation, Int.J.Numer.meth. in Fluids 11, 587-620, 621-659 (1990).

[5]   Heywood, J.G., and R. Rannacher, Finite element approximation of the nonstationary Navier-Stokes problem, Part 1, SIAM J.Numer.Anal. 19, 275-311 (1982), Part 2, ibidem, 23, 750-777 (1986), Part 3, ibidem, 25, 489-512 (1988), Part 4, ibidem, 27, 353-384 (1990).

[6]   Hughes, T.J.R., Franca, L.P., and M. Balestra, A new finite element formulation for computational fluid mechanics: V. Circumventing the Babuska-Brezzi condition: A stable Petrov-Galerkin formulation of the Stokes problem accommodating equal order interpolation, Comp.Meth. Appl.Mech.Eng. 59, 85-99 (1986).

[7]   Müller-Urbaniak, S., Eine Analyse des Zwischenschritt-θ-Verfahrens zur Lösung der instationären Navier-Stokes-Gleichungen, PhD Thesis, University of Heidelberg, 1993.

[8]   Prohl, A., and Rannacher, R., An analysis of first- and second-order projection methods for the nonstationary incompressible Navier-Stokes equations, to appear.

[9]   Rannacher, R., Numerical analysis of nonstationary fluid flow. A survey, in "Applications of Mathematics in Industry and Technology" (V.C.Boffi and H.Neunzert, eds.), pp. 34-53, B.G. Teubner, Stuttgart 1989.

[10]   Rannacher, R., On the numerical solution of the incompressible Navier-Stokes equations, Survey Lecture on the GAMM-Conference 1992, Leipzig, Z.Angew.Math.Mech. 73, 203-216 (1993).

[11]   Müller-Urbaniak, S., Rannacher, R., and, Turek, S., Implicit time discretization of the incompressible Navier-Stokes equations, to appear.

[12]   Rannacher, R., and Turek, S., A simple nonconforming quadrilateral Stokes element, Numer.Meth.Part.Diff.Equ. 8, 97-111 (1992).

[13]   Shen, J., On error estimates of some higher order projection and penalty-projection methods for Navier-Stokes equations, Numer.Math. 62, 49-73 (1992).

[14]   Turek, S., Tools for simulating nonstationary incompressible flow via discretely divergence-free finite element models, Int.J. Numer.Meth. in Fluids 18, 71-105 (1994).

TWO-DIMENSIONAL FLOW SIMULATION AND MASS TRANSPORT FOR RECTANGULAR TANKS

H.E. SCHULZ[1] and P. KREBS[2]
1)Department of Hydraulics and Sanitary Engineering, University of Sao Paulo, 13560/250, PO Box 359, Sao Carlos, Sao Paulo, Brazil.
2)Institut for Hydromechanics, University of Karlsruhe, Kaiserstrasse 12, 76128, PO Box 6980, Karlsruhe,Germany.

Flows in rectangular tanks are calculated with the aid of numerical tools and the k-ε model. Numerical results for some flow properties are presented and compared with experimental data from other sources. The present study is concerned with mass transport in turbulent flows, more specifically, with the mass concentration evolution of a tracer at the outlet of the tank in the form of Flow-Through Curves (FTC). Considerations about the injection time of the tracer are made in view of analytical solutions for the case of isotropic and homogeneous turbulence.

INTRODUCTION

Tracer transport in flows in rectangular tanks is a matter of interest for engineering sciences. One example is the study of settling basin characteristics using Flow- Through Curves (FTC). The tracer transport is associated with the fluid flow, which is generally turbulent and does not have proper analytical solutions. On the other hand, experimental studies may become too expensive. Therefore, the use of numerical tools appears to be the best way to predict some characteristics of settling basins or other flows in complex geometries. FTC may also be simulated through numerical analyses. However, the injection time of the tracer, a parameter needed to perform the simulation, is generally not given or chosen as a fixed proportion of the theoretical residence time, a procedure still without clear interpretation (Stamou et al. 1989). Some trends of FTC may be estimated based on analytical solutions of simple flows. In this study the effect of the injection time on the FTC is studied through such analytical solutions. An approach to evaluate a priori a proper range for the injection time to be used in numerical calculations is then proposed.

Analytical Approximation

The tank shape used in this study is sketched in Fig.1, which is similar to that studied by Stamou and Adams (1988), also discussed in Adams and Rodi (1990). The ideal case taken to evaluate some trends of the FTC is shown in Fig.2, where homogeneous and isotropic turbulence is assumed. In the figures,

A. Peters et al. (eds.), Computational Methods in Water Resources X, 1247–1254.

C is the tracer concentration; U, the mean velocity of the flow; L, the tank length and l is the length of a volume source of tracer. Other parameters of interest are the eddy diffusivity, D_T, the theoretical residence time, $T_O = L/U$, and the time associated with l, $t_O = l/U$, used here as the injection time. We are thus using the instantaneous volume source solution as an approximation to the pulse injection problem. The governing equation for the ideal case is the one-dimensional convection-diffusion equation, with constant eddy diffusivity. The solution for the FTC in section 2 of Fig. 2 is given by Eq.1 (Carslaw and Jaeger, 1947):

$$\overline{C} = \frac{L}{2l}\left\{ erf\left[\frac{1-2\left(\dfrac{L-Ut}{l}\right)}{\sqrt{\dfrac{16D_T t}{l^2}}}\right] + erf\left[\frac{1+2\left(\dfrac{L-Ut}{l}\right)}{\sqrt{\dfrac{16D_T t}{l^2}}}\right]\right\} \tag{1}$$

$\overline{C}$ is a nondimensional concentration, defined as the ratio between the observed concentration and the fictitious concentration obtained dividing the injected mass over the volume between sections 1 and 2.

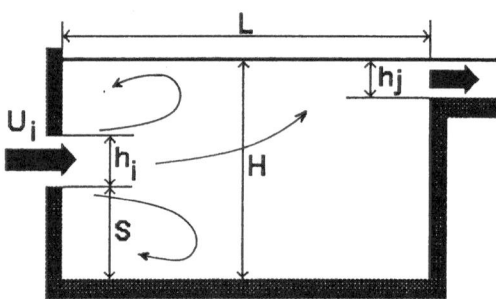

Figure 1: Geometry of the tank. H = water depth, h_i and h_f = inlet and outlet dimensions, S = inlet height, L = length and U_i = velocity of the flow at the inlet of the basin

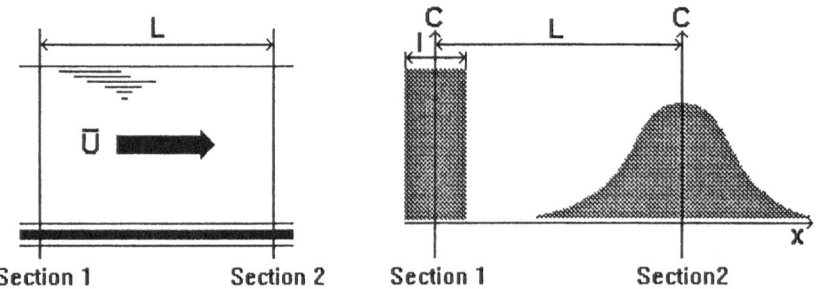

Figure 2: Ideal flow situation. Homogeneous and isotropic turbulence between two sections of a channel. $\overline{U}$ is the mean flow velocity.

The peak concentration occurs when $t = T_0 = L/U$, and is given by Eq.2.

$$\overline{C_{peak}} = \frac{L}{Ut_0} \, erf \sqrt{\frac{U^3 \, t_0^2}{16D_T \, L}}$$ (2)

The FTC are commonly associated with instantaneous sources, that is, situations for which $t_0 \rightarrow 0$. This leads also to the maximum value of the peak, $\overline{C_{max}}$, as:

$$\overline{C_{max}} = \sqrt{\frac{LU}{4\pi D_T}}$$ (3)

Equation 2 shows that $\overline{C_{peak}}$ decreases when t_0 is increased. As was mentioned previously, although the FTC are commonly associated with the situation $t_0 \rightarrow 0$, it is necessary to adopt proper injection times to perform the calculations. Expanding Eq.2 in a power series and using third order approximation, a theoretical injection time, t_0, may be defined to be (valid for $t_0 \ll T_0$):

$$t_0 = \sqrt{48\alpha \frac{D_T \, L}{U^3}} \qquad \text{or} \qquad \frac{t_0}{T_0} = \sqrt{48\alpha \frac{D_T}{LU}}$$ (4)

α is the ratio between the calculated and the maximum peak concentrations. It is necessarily a small value, because we are interested to maintain the calculated solution near the instantaneous source solution. Thus, the criterion to define t_0 is to obtain small reductions of the concentration peak. In this paper we compare FTC simulated for different injection times, to verify the reduction of the peaks. Further, we compare injection times obtained using Eq.4 and those which produce small peak concentration reductions in numerical calculations.

Flow equations

The flows are assumed to be two-dimensional. Continuity and momentum equations (Eq.5 to 7) together with k-ε equations (Eq.8 to 11) were solved using the FAST 2D program from the Institute of Hydromechanics of the University of Karlsruhe, Germany (Zhu, 1991). The finite volume discretization is used in this program. The boundary conditions at solid walls involve the use of log-profiles. The free surface is treated as a symmetry plane. The velocity at the entrance is assumed to be constant over the inlet area. The turbulent kinetic energy per unit mass, k, is defined at the inlet as $k_{inlet} = 0.001 \, (U_{inlet})^2$. The energy dissipation rate per unit mass, ε, is defined at the inlet as $\varepsilon_{inlet} = 9 \, (k_{inlet})^2 /(10000 \, v)$, where v is the kinematic viscosity of the fluid.

$$\frac{\partial U}{\partial x} + \frac{\partial V}{\partial y} = 0$$ (5)

$$\frac{\partial U^2}{\partial x}+\frac{\partial VU}{\partial y}=-\frac{1}{\rho}\frac{\partial P}{\partial x}+\frac{\partial}{\partial x}\left(v_T\frac{\partial U}{\partial x}\right)+\frac{\partial}{\partial y}\left(v_T\frac{\partial U}{\partial y}\right)+\frac{\partial}{\partial y}\left(v_T\frac{\partial V}{\partial x}\right)+\frac{\partial}{\partial x}\left(v_T\frac{\partial U}{\partial x}\right) \qquad (6)$$

$$\frac{\partial UV}{\partial x}+\frac{\partial V^2}{\partial y}=-\frac{1}{\rho}\frac{\partial P}{\partial y}+\frac{\partial}{\partial x}\left(v_T\frac{\partial V}{\partial x}\right)+\frac{\partial}{\partial y}\left(v_T\frac{\partial V}{\partial y}\right)+\frac{\partial}{\partial x}\left(v_T\frac{\partial U}{\partial y}\right)+\frac{\partial}{\partial y}\left(v_T\frac{\partial V}{\partial y}\right) \qquad (7)$$

$$\frac{\partial Uk}{\partial x}+\frac{\partial Vk}{\partial y}=\frac{\partial}{\partial x}\left(\frac{v_T}{\sigma_k}\frac{\partial k}{\partial x}\right)+\frac{\partial}{\partial y}\left(\frac{v_T}{\sigma_k}\frac{\partial k}{\partial y}\right)+\mathrm{Pr}-\varepsilon \qquad (8)$$

$$\frac{\partial U\varepsilon}{\partial x}+\frac{\partial V\varepsilon}{\partial y}=\frac{\partial}{\partial x}\left(\frac{v_T}{\sigma_\varepsilon}\frac{\partial\varepsilon}{\partial x}\right)+\frac{\partial}{\partial y}\left(\frac{v_T}{\sigma_\varepsilon}\frac{\partial\varepsilon}{\partial y}\right)+c_1\mathrm{Pr}\frac{\varepsilon}{k}-c_2\frac{\varepsilon^2}{k} \qquad (9)$$

$$\mathrm{Pr}=v_T\left[2\left(\frac{\partial U}{\partial x}\right)^2+2\left(\frac{\partial V}{\partial y}\right)^2+\left(\frac{\partial U}{\partial x}+\frac{\partial V}{\partial y}\right)^2\right] \qquad (10)$$

$$v_T=c_\mu\frac{k^2}{\varepsilon} \qquad (11)$$

U and V are the velocities in the horizontal and vertical directions, respectively. P is the pressure, ρ is the fluid density, Pr is the production of turbulent energy and v_T is the eddy viscosity. The constants of the k-ε model were taken from the literature (Rodi, 1980), so that σ_μ=0.09, c_1=1,44, c_2=1,92, σ_ε =1,22 and σ_k =1.0. The HLPA scheme is used for the flow calculations, together with the SIMPLEC algorithm for the pressure correction equation. Six flow situations are calculated using the flow conditions of Stamou and Adams (1988). Table 1 presents the main parameters necessary to perform the calculations.

Table 1: Parameters used to perform the numerical calculations in the present study

Case	L (cm)	H (cm)	S (cm)	hi (cm)	hf (cm)	Ui (cm/s)	To (s)	Ph	Pv
1	250	11.0	9.0	2.0	1.14	12.5	110	139	75
2	250	13.0	9.0	2.0	1.14	12.5	130	139	97
3	100	17.0	9.0	2.0	1.14	12.5	68	85	128
4	175	17.0	9.0	2.0	1.14	12.5	119	113	128
5	250	17.0	9.0	2.0	1.14	12.5	170	139	128
6	250	29.0	9.0	2.0	1.14	12.5	290	139	150

Ph and Pv are, respectively, the number of points of the numerical cells in the horizontal and vertical directions. Proper grids were adjusted for each flow situation, with variable mesh sizes in both the horizontal and vertical directions.

Mass transfer equation

Equation 12 was solved together with the flow equations, also using the FAST

2D programm (Zhu, 1991). In this study we present FTC obtained for the outlet of the tanks, calculated as the temporal evolution of the mean tracer concentration at this position. The convection schemes HYBRID, SOUCUP, HLPA, SMART and QUICK are compared in Fig.3. In this study, the HYBRID scheme furnishes results closer to the experimental observations of Stamou and Adams (1988), being thus used in the further calculations. D_T is calculated as $D_T = v_T/\sigma_C$, with $\sigma_C = 0.7$.

$$\frac{\partial C}{\partial t} + \frac{\partial UC}{\partial x} + \frac{\partial VC}{\partial y} = \frac{\partial}{\partial x}\left(D_T \frac{\partial C}{\partial x}\right) + \frac{\partial}{\partial y}\left(D_T \frac{\partial C}{\partial y}\right) \qquad (12)$$

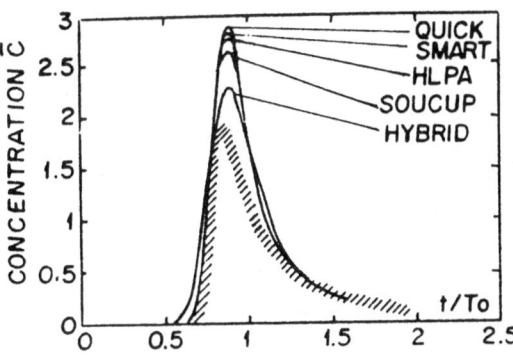

Figure 3: Comparison between the five convection schemes used. (t_0 / T_0 is 9%) ////// = Stamou and Adams (1988) data.

Numerical results

As sketched in Fig.1, recirculation zones form near the bottom and the free surface. Figures 4 a, b and c show the comparison between measured and calculated values of areas and lengths of such cells. As can be seen, a good agreement was obtained, showing that the flow simulations are of good quality. An example of the distribution of the eddy viscosity, needed to calculate the eddy diffusivity, is shown in Fig.5. This example corresponds to case 1 of Table 1. The effect of the injection time over the FTC, also calculated for this case, is shown in Fig.6. The values of t_0 / T_0 are 4.5%, 9.0% and 27%, which correspond to $\overline{C_{peak}}$ values of 2.33, 2.30 and 2.02. The peaks maintain a nearly constant value for small t_0 (4.5 and 9.0% T_0), but clearly decrease when t_0 is increased. This fact is in agreement with Eq.2 and 3, obtained for the ideal case of Fig. 2.

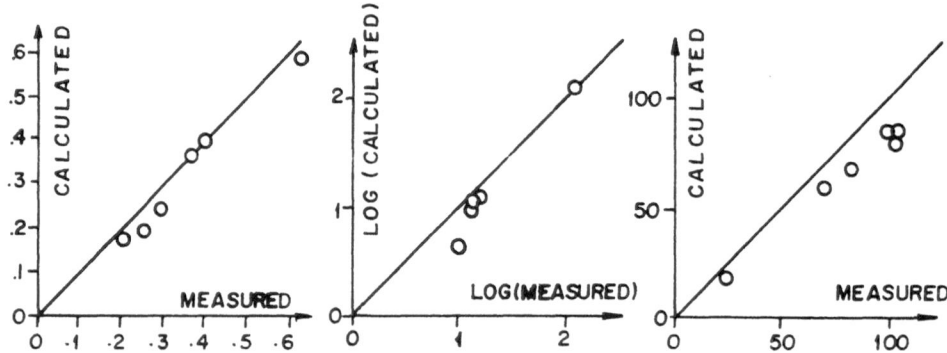

Figure 4: Comparisons between measured and calculated: (a) cell areas (cm^2); (b) cell longitudinal lengths at the surface (log cm); (c) cell longitudinal lengths at the bottom (cm). (Data from Stamou and Adams,1988)

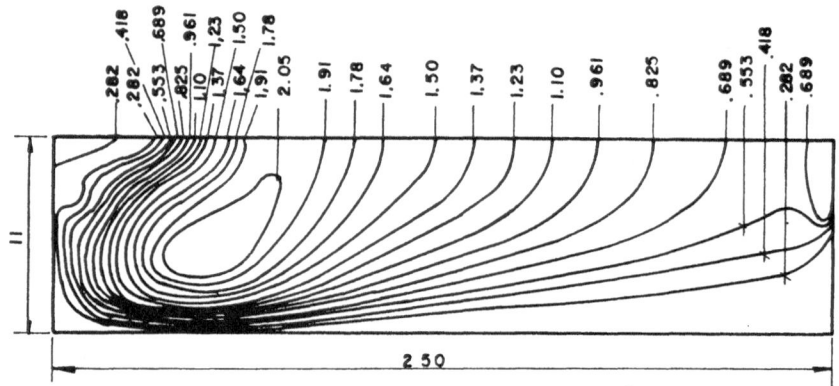

Figure 5: Eddy viscosity distribution for case 1 of Table 1. The values shown in the figure are multiplied by 10^4. Values in m^2/s.

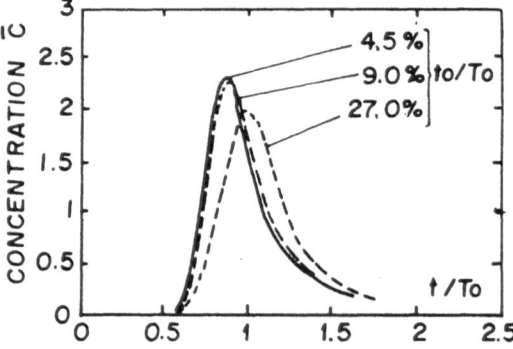

Figure 6: FTC for case 1 using three injection times: 4.5, 9.0 and 27.0% T_0

To use Eq. 2, 3 and 4 for the flows described in Table 1, a proper D_T must be used. The following mean value $\overline{D_T}$ is defined:

$$\overline{D_T} = \frac{\sum_i D_T^i A^i}{\sum_i A^i} = \frac{\overline{v_T}}{\sigma_c} \quad where \quad \overline{v_T} = \frac{\sum_i v_T^i A^i}{\sum_i A^i} \tag{13}$$

A^i are the areas of the cells which correspond to D^i or v^i. Figure 8 shows the comparison between the values of $\overline{C_{peak}/C_{max}}$ calculated through simulation and using Eq.2 and 3. Although the values calculated for the same situation using both procedures are different, the shape of the obtained curves is similar. This points to the possibility to use Eq.4 to predict a proper range for t_O also for numerical calculations.

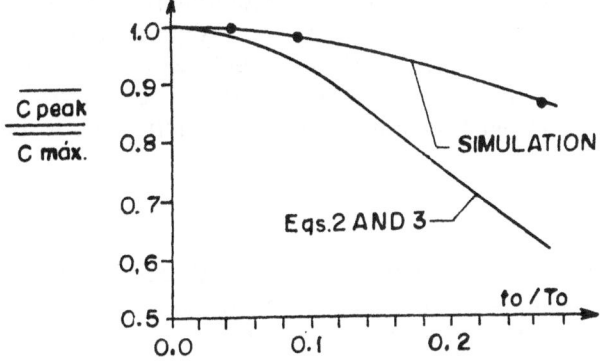

Figure 7: Comparison between the curves of $\overline{C_{peak}}$ calculated through numerical simulation and Eq.2. Case 1 of Table 1. $\overline{D_T}$ = 1.4 10^{-4} m^2/s.

For α=0.01 and 0.07, we obtain (case 1 of Table 1):

$$t_0^{(\alpha=0.01)} = 3.8s \quad or \quad \frac{t_0}{T_0}^{(\alpha=0.01)} = 3.4\% \tag{14}$$

$$t_0^{(\alpha=0.07)} = 10s \quad or \quad \frac{t_0}{T_0}^{(\alpha=0.07)} = 9.1\% \tag{15}$$

The results obtained for the ideal case show that if we desire to work with a peak reduction less than 1% (α=0.01), we need to choose t_O / T_O less than 3.4%. If we accept a peak reduction of 7% (α=0.07), we may choose t_O / T_O to be less than 9.1%. As the changes of the concentration peaks for real flows are lower than those of the ideal case, we may also use these values as upper limits of t_O/T_O for numerical simulations in rectangular geometries. As an example, for case 1 of Table 1 we have obtained, through numerical simulation, a reduction of about 1.5% for t_O / T_O = 9.0%, which is lower than the 7% suggested by Eq.15. In the present analysis only the peak concentration reduction was observed. As mentioned previously, small values of α must be chosen to maintain the solution close to the situation $t_O \rightarrow 0$. High values of t_O translate the FTC in time (Fig. 6), which, of course, is not desirable.

CONCLUSIONS

Flows in rectangular geometries were simulated to reproduce experimental situations found in literature. The Flow-Through Curves for these flows were calculated and the effect of the injection time over the obtained curves was analysed. It was observed that the value of the concentration peak decreases when the injection time is increased. On the other hand, for small times the peaks mantain a nearly constant value. The same behaviour was observed for an ideal case with available analytical solution. An equation which permits to calculate the injection time for the ideal case was furnished. The injection time is expressed as a function of an acceptable concentration peak reduction. Flows in rectangular geometries, like those studied here, produce smaller changes on the peak concentrations in comparison with the ideal case. As a consequence, the same equation for t_O may be used in numerical procedures as an *a priori* evaluation of the upper limit for the range of injection times to be used in the calculations.

AKNOWLEDGEMENTS

The authors are indebted to Professors Fazal H. Chaudhry and Wolgang Rodi, as well as to the colleagues of the Institute of Hydromechanics of the University of Karlsruhe, Tobias Buchal and Gerhard Bosch.

REFERENCES

Adams, E.W. and Rodi, W. (1990) "Modeling flow and mixing in sedimentation tanks", Journal of Hydraulic Engineering, Vol.116, No.7, july, pp.895-913.

Carslaw, H.S. and Jaeger, J.C. (1947) Conduction of Heat in Solids, Oxford University Press, London.

Rodi, W. (1980) Turbulence Models and their Application in Hydraulics-a State of Art Review, International Hydraulics Research Association, Delft.

Stamou, A.I and Adams, E.W. (1988) "Study of the hydraulic behaviour of a model settling tank using flow-through-curves and flow patterns", Technical Report SFB 210/E/36, March, University of Karlsruhe.

Stamou, A.I., Adams, E.W. and Rodi, W. (1989) "Numerical modeling of flow and settling in primary rectangular clarifiers", Journal of Hydraulics Research, Vol.27, No.5, pp. 665-681.

Zhu, J. (1991) Fast-2D: A Computer Program for Numerical Simulation of Two-Dimensional Incompressible Flows with Complex Boundaries, Institute for Hydromechanics, University of Karlsruhe, Germany.

Tools for Predicting Incompressible Flows via Nonconforming Finite Element Models

Stefan Turek
Institut für Angewandte Mathematik, Universität Heidelberg
Im Neuenheimer Feld 294, D-69120 Heidelberg
Germany

Introduction

We consider the nonstationary incompressible Navier–Stokes equations,

$$\mathbf{u}_t - \nu \Delta \mathbf{u} + (\mathbf{u} \cdot \nabla)\mathbf{u} + \nabla p = \mathbf{f}, \ \nabla \cdot \mathbf{u} = 0, \text{ in } \Omega \times (0, T),$$

for a given force $\mathbf{f}$ and viscosity ν, with prescribed boundary values on $\partial\Omega$ and an initial condition at $t = 0$. The variables $\mathbf{u}$ and p describe the velocity and the pressure of a viscous incompressible flow in a bounded region $\Omega \subset \mathbf{R}^n, n = 2, 3$. These fundamental equations are of interest to both more theoretical scientists like mathematicians or physicists, and more applied ones like engineers or industrial users. What are the theoretical aspects needed to develop an algorithm to solve these equations? Our mathematical research led us to examine the following problems:

Efficient treatment of the nonlinearity, error estimates for spatial and time discretizations, construction of fast linear solvers, modelling of boundary conditions, and many others. It is very important not only to solve these problems in a theoretical way, but also to see the results in "real life", via computer simulations. It is not sufficient to prove very good convergence rates if the corresponding numerical amount is too large. For instance, some methods of (theoretically) first order accuracy may yield better results than certain second order methods, since the asymptotic range implicit in the proof is (nearly) never reached. Therefore, also, numerical experiences have a large influence on our theoretical work, and should not be neglected. On the other hand, engineers are also interested in an efficient solution method for these equations. The simulation of some complex fluid structures, the development of new models for "real" problems and, not to forget, the saving of money, by doing a computer simulation rather than an expensive laboratory test, these are all wishes, which appear every day. Now, our aim is to develop and really to implement a solution method, which satisfies both parties, this means we would like to construct a so called "Black Box solver". This term is often used, and in many cases its meaning is not clear, so we

A. Peters et al. (eds.), Computational Methods in Water Resources X, 1255–1262.

must define it. What we cannot offer is a "black universe method", we only present a method which attempts to give the user a "fast", "robust" and "accurate" solution. Then, the problem is to determine whether the computed "solution" is physically relevant. But our method, by itself, does not necessarily produce the "right solution", satisfying a prescribed accuracy corresponding to the real flow.

In this sense, we will develop a **FRA**-method (Fast–Robust–Accurate), with an emphasis on the term method, as opposed to algorithm or solver. So, first we have to explain the key words "fast", "robust" and "accurate".

Our implemented algorithm shall produce the results in a "short" amount of time, working on a workstation. We prefer this class of machine, because at today's prices (nearly) every mathematical institute can afford such computers. Additionally, these machines are very fast, complete as well as robust, and the future trend indicates that these will only improve. By fast we mean that we want to compute a fully developed nonstationary flow in hours (2D) or days (3D), and this with an accuracy, which is essentially determined by the available RAM memory. Another aspect of the term "fast" involves being able to change and to easily implement new ideas by adding subroutines to the existing code. Hence, we need a very modular programming structure and a very clever data management. The basis of all our programs is the finite element package **FEAT** ([1]), which was developed by our group in Heidelberg. This basis enables several people to work and to coorporate in the same field. Finally, another important aspect of "fast" is the fast graphical presentation of the computed data. Most workstations have very fine graphic features and enable, for example, one to compute and visualize at the same time. Hence, we also have to provide the corresponding graphical algorithms.

If we say our method is stable, we think to several aspects. First, the implemented algorithm shall be independent of given "data", like the shape of the domain (not convex, not polygonal, several boundary components), the mesh (not equidistant, refined in some parts), and the size of force, viscosity and boundary values. So by "robust" we mean that our finite element discretization works for (nearly) all meshes, that our linear solvers (almost) always produce the same convergence rates, and that the nonlinear convergence rates are not becoming too bad for higher Reynolds numbers. What we do not claim is that our algorithm works completely independently of all these "parameters", since, for instance, it is clear that the stationary Navier-Stokes solver cannot converge for higher Reynolds numbers, if there is no stationary solution, and it is also clear that the time step size is dependent on the physical flow structure. The aim is to develop a method, which requires only defining the domain, the right hand side, the boundary values and the viscosity, and then, the algorithm attempts to produce results in an "optimal" way. We think, the proposed method is a step in this direction. Another aspect is the robustness of the implementation. The program should be organized in a very modular structure in order to simplify and to control possible errors when making changes to the program. The worst case is to have a code with about 30,000 lines, and to change a little part resulting in errors in other routines. This often costs much more time than many simulation processes with the (correctly) modified algorithm. We also emphasize that "robustness" includes the aspect, as to whether the algorithm is analysable, and in a certain sense "deterministic" for the user. Therefore, for instance, variational formulations and discretizations by finite elements are used, because this allows, at least for some simpler model problems, a strong analysis, and can be helpfull to explain some "funny" results of the implemented code. So, only well known "standard methods" are used, like stable finite element pairs for the discretization, fast multigrid algorithms, fixed point defect correction methods for the nonlinearity, and well known time discretization schemes for the nonstationary processes. Finally, also our graphical support should work (in a robust way) independently of the above mentioned "data", something which is not fulfilled by some existing graphic packages.

The question of accuracy is the most critical and difficult point of our method. As mentioned above we cannot present a "black universe method", which tells the user the quality of the computed solutions compared with the exact ones. What we can present is a "solution" with the "best accuracy" reachable for a given machine and time. Here, one has to consider that the more accurate methods are sometimes not the fastest ones. Therefore, only very simple finite elements (of second order), upwind schemes for the convective parts and time and space discretization schemes of second order are used, since then corresponding solution methods which are very fast can be constructed (divergence–free finite elements, multigrid methods). For the grid construction process the best method would be a fully adaptive method. One reason we do not use this is because at this moment we do not have the right tool, and on the other hand our "semi–adaptive" method leads to a very regular data structure, which gives perhaps not the absolutely most accurate results, but leads to a very efficient multigrid solution method.

Now, we want to introduce the three main parts of our "FRA Black Box" method: Input of all needed "data/Preprocessing part", Visualization/Postprocessing part and the fully modular finite element solvers.

Input/Preprocessing part

The perfect method would be to define only the computational domain (perhaps in a graphical way), the input data (like force term, viscosity and boundary values), and then have the algorithm produce the most accurate solution, depending on the available memory and proposed time limit. Unfortunately, at this time, we are unable to accomplish this goal. At present our realized method needs, beside the description of the domain (by parametrization or given boundary points), also a "coarse" grid, which can be already slightly adapted to the expected solution (for example after small test calculations, or after comparing with existing theoretical and practical results). Then, this "coarse" grid is systematically refined (corresponding to the

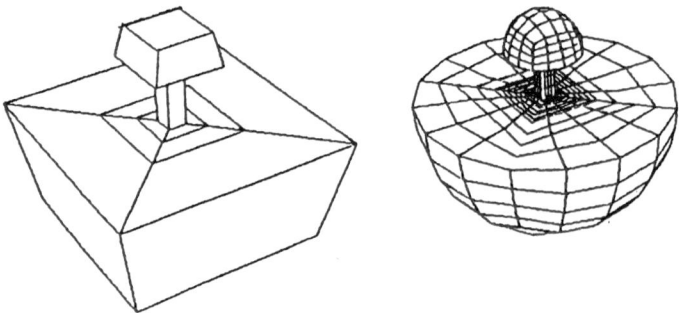

Figure 1: Coarse "semi–adaptive" grid and two times refined grid

available memory), and the computation may begin. For nonstationary problems, one still has to define the time step size, because (presently) we have not implemented an adaptive control. But, since our implicit time discretization scheme is unconditionally stable, this parameter can be chosen depending solely on physical reasons, and not on (purely) mathematical stability conditions. These non–self adaptive methods do not guarantee the most accurate "solution", but do give a nearly optimal "solution" in a short amount of time.

Visualization/Postprocessing part

We use standard graphic packages like MOVIE.BYU (for 2D) and AVS for 3D applications. Additionally, we developed some modules for nonstationary flows, for example a fully interactive particle tracing with the corresponding data structure. These "movies" can also be transfered directly from screen to video tape. We think, this nonstationary postprocess modul is the most important one, since only with its aid one is really able to examine nonstationary flows, because these graphical results differ essentially from the streamline or vector plot snapshots (see Figure 2).

The fully modular finite element solvers

They are based on the finite element package **FEAT** (in FORTRAN 77) with pseudodynamic memory management and basic finite element tools for grid generation, stiffness matrices, right hand sides and boundary components, and a corresponding **BLAS**–package with basic linear algebra tools and linear solvers. The most important components of our finite element code are the following. We delay a more detailed description to the subsequent papers in the reference list.

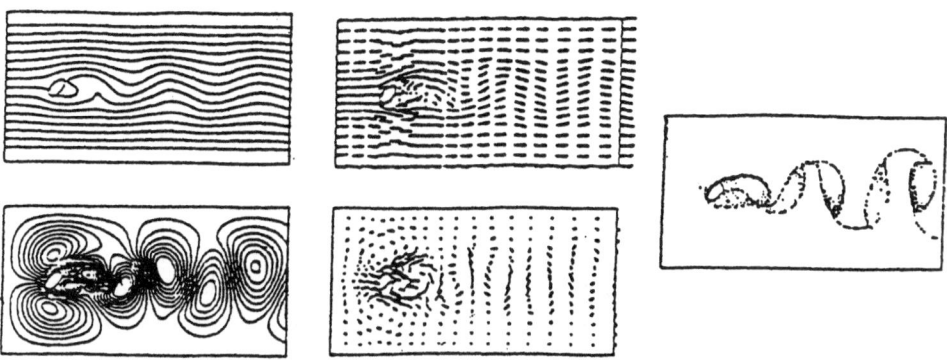

Figure 2: Different visualization techniques for 2D flow around an ellipse

We use nonconforming "rotated bi/trilinear" finite element spaces for the velocity (generated by $\langle x^2 - y^2, x^2 - z^2, x, y, 1\rangle$) at the midpoints of the edges/areas of the triangulation, and a piecewise constant pressure approximation. In this case an explicit construction of the discretely divergence–free subspaces satisfying $\int_T \nabla \cdot \mathbf{v}_h^d \, dx = 0$ for all elements T is possible. Then, the pressure is eliminated, and can be calculated in a postprocess by a marching process. This pair of finite element spaces fulfills the LBB–condition, and guarantees h^2/h approximation properties. The amount of storage is about 750 Byte per cell, independent of the used mesh.

Corresponding to these special finite elements we developed a suited multigrid algorithm with "makro-elementwise full interpolation". After a careful implementation this multigrid method with simple Jacobi/Gauss–Seidel/ILU–iteration in the divergence–free case or a block Gauss–Seidel method in the primitive formulation for smoothing, an adaptive step length control for the correction and the "full" interpolation leads to a very efficient and robust Stokes–solver with a numerical effort which is comparable to scalar Poisson–solvers with conforming quadratic finite elements.

For an efficient treatment of the Oseen–equations we use for the convective parts

$$b(\mathbf{u}, \mathbf{v}, \mathbf{w}) := \int_\Omega \mathbf{u}_i D_i \mathbf{v}_j \mathbf{w}_j \, dx$$

a central or an upstream discretization with lumping regions Λ_l around each midpoint B_l (N number of midpoints)

$$\sum_{l=1}^N \sum_{k \in \Lambda_l} \int_{\Gamma_{lk}} \mathbf{u}\mathbf{n}^{lk} d\gamma \, (1 - \lambda_{lk}(\mathbf{u})) \, (\mathbf{v}_j(B_k) - \mathbf{v}_j(B_l)) \, \mathbf{w}_j(B_l)$$

Here, we have several choices for λ_{lk} (with "local Re number" $x := \dfrac{1}{\nu} \displaystyle\int_{\Gamma_{lk}} \mathbf{u} n^{lk} d\gamma$):

$$\lambda_{lk}(\mathbf{u}) := \left\{ \begin{array}{ll} 1 & \text{if } x \geq 0 \\ 0 & \text{else} \end{array} \right\} \quad \text{or} \quad \lambda_{lk}(\mathbf{u}) := \left\{ \begin{array}{ll} \frac{\frac{1}{2}+\alpha x}{1+\alpha x} & \text{if } x \geq 0 \\[2mm] \frac{1}{2(1-\alpha x)} & \text{else} \end{array} \right\}.$$

Choice 1 (equivalent to "$\alpha = \infty$") leads to a very stable discretization scheme with M-matrices, but only of first order accuracy, while the second one (also called Samarski-scheme) has better approximation properties, but no complete theorie in 2D is known.

We solve the steady nonlinear Navier–Stokes problems

$$A(\mathbf{u})\mathbf{u} := -\nu \Delta \mathbf{u} + (\mathbf{u} \cdot \nabla)\mathbf{u} + \nabla p = \mathbf{f}$$

by a fixed point–defect correction–method with adaptive step length control:

$$\mathbf{u}^{n+1} = \mathbf{u}^n + \omega^n [\tilde{A}(\mathbf{u}^n)]^{-1}(\mathbf{f} - A(\mathbf{u}^n)\mathbf{u}^n).$$

This means, in each step an Oseen equation with unknown $\mathbf{u}$ and known coefficient function $\tilde{\mathbf{u}}$ must be solved:

$$-\nu \Delta \mathbf{u} + (\tilde{\mathbf{u}} \cdot \nabla)\mathbf{u} + \nabla p = \text{right hand side}, \quad \nabla \cdot \mathbf{u} = 0.$$

This can be done if we use for $\tilde{A}(\mathbf{u}^n)$ one of our proposed upwind–schemes, since the resulting matrices are similiar to M–matrices, and, therefore, can be easily inverted by our proposed multigrid–solver. The "optimal" value for ω^n can be calculated by a simple matrix–vector multiplication.

For solving the unsteady Navier–Stokes equations we use the fractional step θ-scheme: Choosing $\theta \in (0,1)$, $\theta' = 1 - 2\theta$, and $\alpha \in [0.5,1]$, $\beta = 1 - \alpha$, the time step $t_n \to t_{n+1}$ is split into three substeps as follows:

$$[I + \alpha\theta\Delta t A^{n+\theta}] \ \mathbf{u}^{n+\theta} \quad = \quad [I - \beta\theta\Delta t A^n]\mathbf{u}^n + \theta\Delta t \mathbf{f}^n$$

$$[I + \beta\theta'\Delta t A^{n+1-\theta}] \ \mathbf{u}^{n+1-\theta} \quad = \quad [I - \alpha\theta'\Delta t A^{n+\theta}]\mathbf{u}^{n+\theta} + \theta'\Delta t \mathbf{f}^{n+1-\theta}$$

$$[I + \alpha\theta\Delta t A^{n+1}] \ \mathbf{u}^{n+1} \quad = \quad [I - \beta\theta\Delta t A^{n+1-\theta}]\mathbf{u}^{n+1-\theta} + \theta\Delta t \mathbf{f}^{n+1}.$$

For the special choice $\theta = 1 - \sqrt{2}/2$ this scheme is of second order and strongly A-stable. For $\alpha = (1-2\theta)/(1-\theta)$, $\beta = \theta/(1-\theta)$ the coefficient matrices are the same in all substeps. Therefore we think that this scheme is able to combine the advantages of the implicit Euler and Crank-Nicolson–scheme, without more additional numerical amount. This scheme applied to the Navier-Stokes equations leads to the following

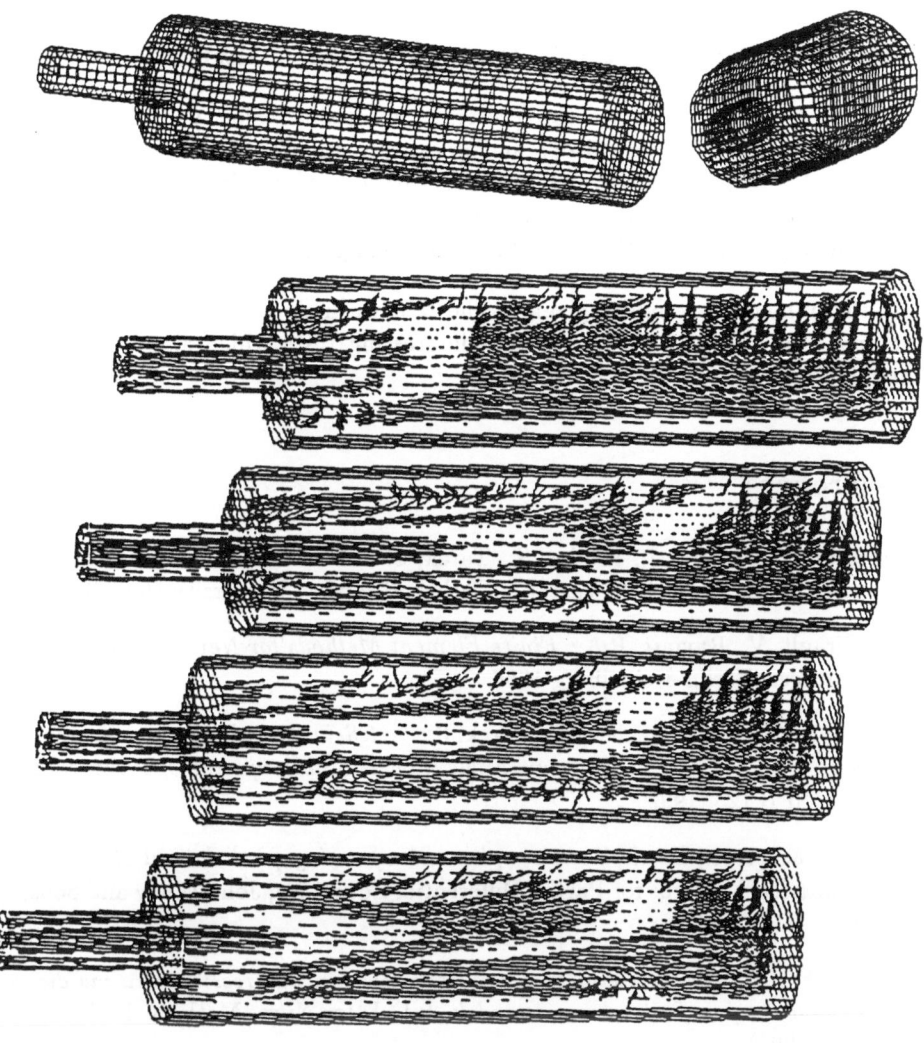

Figure 3: Vector plot representation at different times of a flow in a 3D pipe with five slits on the top. The flow is driven by time dependent pressure drop boundary conditions. The Re number is of order $O(10^2)$.

type of problems (M mass matrix)

$$[M + \alpha \Delta t\, A(\mathbf{u}^{n+1})]\mathbf{u}^{n+1} = [M - (1 - \alpha)\Delta t\, A(\mathbf{u}^n)]\mathbf{u}^n + \alpha \Delta t\, \mathbf{f}^{n+1} + (1 - \alpha)\Delta t\, \mathbf{f}^n\,.$$

As boundary conditions we can prescribe nonsteady velocity profiles, nonsteady fluxes, nonsteady pressure drops and nonsteady tangential velocities (for cylinder rotations). Additionally we can calculate with nonsteady forces or start from rest. At "free" boundaries (this means where we don't know what to prescribe, for instance at outflows), we use our Neumann-like "do nothing" conditions (see [3]).

By the numerical examples given in the talk we hope to show that our proposed components work well, even in really interesting tests which also have "real life" applications.

References

[1] Blum, H., Harig, J., Müller, S., Turek, S.: **FEAT2D**. *Finite element analysis tools. User Manual. Release 1.3*, Technical report, University Heidelberg, 1992

[2] Girault, V., Raviart, P.A.: *Finite Element Methods for Navier–Stokes equations*, Springer, Berlin–Heidelberg 1986

[3] Heywood, J., Rannacher, R., Turek, S.: *Artificial boundaries and flux and pressure conditions for the incompressible Navier–Stokes equations*, Technical Report 681, SFB 123, 1992

[4] Rannacher, R., Turek, S.: *A simple nonconforming quadrilateral Stokes element*, Numerical Methods for Partial Differential Equations, John Wiley and Sons, Inc., 8, 97–111 (1992)

[5] Turek, S.: *Tools for simulating non–stationary incompressible flow via discretely divergence–free finite element models*, Int. J. for Num. Meth. in Fluids, 18, 71–107 (1994)

ASPECTS OF NON-SMOOTHNESS IN FLOW COMPUTATIONS

P. WESSELING, P. VAN BEEK and R.R.P. VAN NOOYEN
Faculty of Technical Mathematics and Informatics
Delft University of Technology
Mekelweg 4, 2628 CD Delft
The Netherlands

We discuss some numerical consequences of non-smoothness of the pressure gradient in the incompressible Navier-Stokes equations in general coordinates, and non-smoothness of grid and diffusion coefficient for interface problems.

DISCRETIZATION OF GRADIENTS ON NON-SMOOTH BOUNDARY-FITTED GRIDS

An example where non-smoothness requires special measures in flow computations is the following. Suppose one wishes to solve the incompressible Navier-Stokes equations in general curvilinear boundary-fitted coordinates on a staggered grid, generalizing the classical "marker and cell" method [7] from Cartesian to general coordinates. Let $x = x(\xi)$ be the coordinate mapping, with x Cartesian coordinates in the two-dimensional domain Ω and $\xi \in G = (0,1) \times (0,1)$. G is divided uniformly into cells. The mapping is generated numerically; $x = x(\xi)$ is given in cell vertices and extended by bilinear interpolation. The situation is sketched in figure 1.

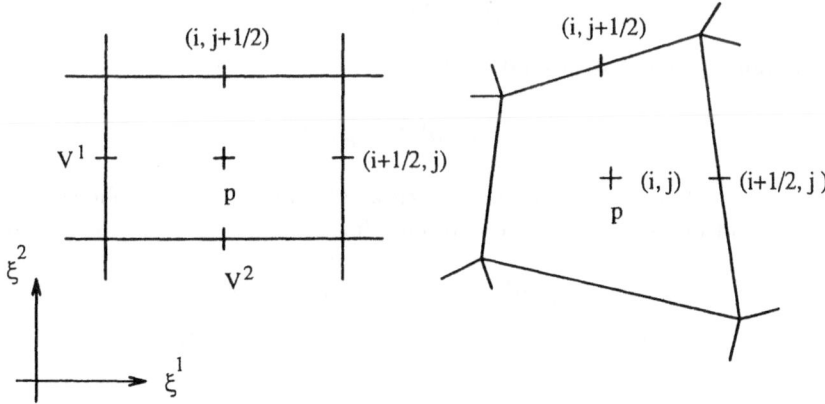

Figure 1: Cell in computational domain G and its image in the physical domain Ω

A. Peters et al. (eds.), Computational Methods in Water Resources X, 1263–1271.
© 1994 Kluwer Academic Publishers. Printed in the Netherlands.

The discrete unknowns are the pressure p, located in cell centers, and the normal mass fluxes V^1, V^2 located in cell face centers. In order to obtain a discretization, in general coordinates, the tensor formulation of the Navier-Stokes equations may be discretized on the grid in G. This is done in [11], [21], [13], [23], [16], [15], [14], [12] for example. Here we will consider the discretization of the pressure gradient. Omitting all other terms, the momentum equation in general coordinates becomes, using the summation convention,

$$\frac{\partial V^\alpha}{\partial t} + \frac{\partial \sqrt{g} g^{\alpha\beta} p}{\partial \xi^\beta} + \sqrt{g} \left\{ \begin{matrix} \alpha \\ \beta\gamma \end{matrix} \right\} g^{\beta\gamma} p = 0 \tag{1}$$

where

$$V^\alpha = \sqrt{g} a^{(\alpha)} \cdot u, \qquad g^{\alpha\beta} = a^{(\alpha)} \cdot a^{(\beta)}, \qquad \sqrt{g} = a^1_{(1)} a^2_{(2)} - a^2_{(1)} a^1_{(2)},$$
$$\left\{ \begin{matrix} \alpha \\ \beta\gamma \end{matrix} \right\} = a^{(\alpha)} \cdot \frac{\partial a_{(\beta)}}{\partial \xi^\gamma} \tag{2}$$

and

$$a^{(\alpha)} = \nabla \xi^\alpha, \quad a_{(\alpha)} = \partial x / \partial \xi^\alpha \tag{3}$$

A straightforward way to discretise (1) is to compute $\alpha_{(\alpha)}$ by finite differences, $a^{(\alpha)}$ from $a^{(\alpha)} \cdot a_{(\beta)} = \delta^a_\beta$ and taking finite differences in (1). However, the resulting discretization is not exact if the gradient $\nabla p = 0$. The underlying reason is discontinuities in the quantities defined in (2) across cell boundaries. As a result large errors may result on non-uniform grids, as in the case of the isobar pattern shown in figure 2(b).

A systematic way to obtain a coordinate-invariant discretization which is exact for $\nabla p = 0$ on arbitrary (structured) grids is as follows. In vector notation we have

$$\frac{\partial u}{\partial t} + \nabla p = 0 \tag{4}$$

Taking the inner product with $\sqrt{g} a^{(\alpha)}$ gives

$$\frac{\partial V^\alpha}{\partial t} + \sqrt{g} a^{(\alpha)} \cdot \nabla p = 0 \tag{5}$$

Instead of writing this in a tensor form such as (1) we proceed as follows. On the staggered grid of figure 1 we have to compute, for example, $dV^1_{i+1/2,j}/dt$. We have

$$\sqrt{g} a^{(1)} = (a^2_{(2)}, -a^1_{(2)})^T \tag{6}$$

which can be evaluated exactly in $(i + 1/2, j)$. Furthermore, we may write

$$p|^{i+1,j}_{ij} = \int_{ij}^{i+1,j} \nabla p \cdot dx \cong \nabla p_{i+1/2,j} \cdot \tilde{a}_{(1)} \Delta \xi^1 \tag{7}$$

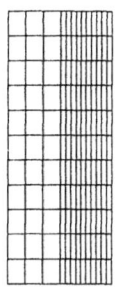 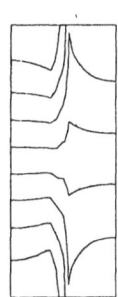 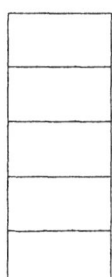

Figure 2: Poiseuille flow; (a): the grid; (b): wrong isobars; (c): correct isobars.

$$\frac{1}{4}\{p|_{i,j-1}^{i,j+1} + p|_{i+1,j-1}^{i+1,j+1}\} = \frac{1}{4}\{\int_{i,j-1}^{i,j+1} + \int_{i+1,j-1}^{i+1,j+1}\}\Delta p \cdot d\boldsymbol{x} \cong \nabla p_{i+1/2,j} \cdot \tilde{\boldsymbol{a}}_{(2)}\Delta \xi^2 \qquad (8)$$

where

$$\tilde{\boldsymbol{a}}_{(1)} \equiv \boldsymbol{a}_{(1)i+1/2,j} \equiv \frac{1}{\Delta \xi^1}\boldsymbol{x}|_{ij}^{i+1,j} \qquad (9)$$

$$\tilde{\boldsymbol{a}}_{(2)} \equiv \boldsymbol{a}_{(2)i+1/2,j} \equiv \frac{1}{4\Delta \xi^2}\{\boldsymbol{x}|_{i,j-1}^{i,j+1} + \boldsymbol{x}|_{i+1,j-1}^{i+1,j+1}\} \qquad (10)$$

Note that (7) and (8) are exact for ∇p constant. We can solve for ∇p from (7) and (8) as follows. Define

$$\tilde{\boldsymbol{a}}^{(1)} \equiv \boldsymbol{a}^{(1)}_{i+1/2,j} \equiv (\tilde{a}^2_{(2)}, -\tilde{a}^1_{(2)})^T/\sqrt{\tilde{g}} \qquad (11)$$

$$\tilde{\boldsymbol{a}}^{(2)} \equiv \boldsymbol{a}^{(2)}_{i+1/2,j} \equiv (-a^2_{(1)}, \tilde{a}^1_{(1)})^T/\sqrt{\tilde{g}} \qquad (12)$$

$$\sqrt{\tilde{g}} \equiv \sqrt{\tilde{g}}_{i+1/2,j} \equiv \tilde{a}^1_{(1)}\tilde{a}^2_{(2)} - \tilde{a}^2_{(1)}\tilde{a}^1_{(2)} \qquad (13)$$

Then $\tilde{\boldsymbol{a}}^{(\alpha)} \cdot \tilde{\boldsymbol{a}}_{(\beta)} = \delta^\alpha_\beta$, and the solution of (7), (8) can be written as

$$\nabla p_{i+1/2,j} \cong \frac{1}{\Delta \xi^1}\tilde{\boldsymbol{a}}^{(1)}p|_{ij}^{i+1,j} + \frac{1}{4\Delta \xi^2}\tilde{\boldsymbol{a}}^{(2)}\{p|_{i,j-1}^{i,j+1} + p|_{i+1,j-1}^{i+1,j+1}\} \qquad (14)$$

In a similar way an approximation to $\nabla p_{i,j+1/2}$ can be obtained. Substitution of (6) and (14) in (5) gives a spatial discretization which is exact if $\nabla p = constant$, regardless of the smoothness of the grid, and which results in the correct isobars of figure 2. Generalization to three dimensions is straightforward.

DISRETIZATION OF AN INTERFACE PROBLEM

Consider the following equation:

$$- div\,(D\nabla\varphi) = \sigma \qquad (15)$$

This equation plays an important role in the theory of flow in porous media, and closely resembles the pressure equation in the IMPES (implicit pressure, explicit saturation) model in reservoir engineering. Often the domain contains interfaces across which D has large jumps. When this is the case we call (15) an interface problem.

Suppose the domain has a complicated shape, and (15) is solved with a boundary-fitted grid, such as shown in figure 1. We assume that φ is located in cell-centers, and that interfaces consist only of cell-faces. Integration of (15) over a cell in a finite volume discretization procedure leads to integrals over cell faces, such as

$$\int_{i+1/2,j-1/2}^{i+1/2,j+1/2} D\nabla\varphi ds \cong (D\nabla\varphi) \cdot (\sqrt{g}\boldsymbol{a}^{(1)})_{i+1/2,j}\Delta\xi^2 \qquad (16)$$

with $\sqrt{g}\boldsymbol{a}^{(1)}$ given by (6). This is exact for $D\nabla\varphi = constant$. Everything hinges on the approximation of $(D\nabla\varphi)_{i+1/2,j}$. The fact that D may be discontinuous at $(i+1/2,j)$ has to be taken into account, and we would like to obtain zero error for $D\nabla\varphi$ constant on non-smooth grids. This may be achieved in a way quite similar to what we just did for the pressure term in the Navier-Stokes equations.

Assuming D to be constant in each cell, we may write

$$\varphi|_{ij}^{i+1,j} = \int_{ij}^{i+1,j} \nabla\varphi \cdot d\boldsymbol{x} \cong (D\nabla\varphi)_{i+1/2,j} \cdot \int_{ij}^{i+1,j} \frac{1}{D} d\boldsymbol{x} \qquad (17)$$

which is obviously exact for $D\nabla\varphi$ constant. We have, exactly for D piecewise constant and the mapping $\boldsymbol{x} = \boldsymbol{x}(\boldsymbol{\xi})$ piecewise bilinear,

$$\boldsymbol{b}_{(1)} \equiv \int_{ij}^{i+1,j} \frac{1}{D} d\boldsymbol{x} = \{(\boldsymbol{a}_{(1)}/D)_{ij} + (\boldsymbol{a}_{(1)}/D)_{i+1,j}\}\frac{\Delta\xi^1}{2} \qquad (18)$$

Next, we choose the following two integration paths:

$$\begin{aligned} C1:&\ (i+1,j-1) \to (i+1,j) \to (i-1,j) \to (i-1,j+1) \\ C2:&\ (i-1,j-1) \to (i-1,j) \to (i+1,j) \to (i+1,j+1) \end{aligned} \qquad (19)$$

Adding the integrals of $\nabla\varphi$ over C_1 and C_2 gives

$$\frac{1}{4}\{\varphi|_{i,j-1}^{i,j+1} + \varphi|_{i+1,j-1}^{i+1,j+1}\} \cong (D\nabla\varphi)_{i+1/2,j} \cdot \boldsymbol{b}_{(2)} \tag{20}$$

with

$$\boldsymbol{b}_{(2)} = \frac{1}{4}\{\int\limits_{C_1} + \int\limits_{C_2}\}\frac{1}{D}d\boldsymbol{x} \tag{21}$$

We have for piecewise constant D

$$\begin{aligned}
\boldsymbol{b}_{(2)} &= \frac{1}{D_{i,j-1}}(\boldsymbol{x}_{i,j-1/2} - \boldsymbol{x}_{i,j-1}) + \frac{1}{D_{ij}}(\boldsymbol{x}_{i,j+1/2} - \boldsymbol{x}_{i,j-1/2}) \\
&+ \frac{1}{D_{i,j+1}}(\boldsymbol{x}_{i,j+1} - \boldsymbol{x}_{i,j-1/2}) + \frac{1}{D_{i+1,j-1}}(\boldsymbol{x}_{i+1,j-1/2} - \boldsymbol{x}_{i+1,j-1}) \\
&+ \frac{1}{D_{i+1,j}}(\boldsymbol{x}_{i+1,j+1/2} - \boldsymbol{x}_{i+1,j-1/2}) + \frac{1}{D_{i+1,j+1}}(\boldsymbol{x}_{i+1,j+1} - \boldsymbol{x}_{i+1,j-1/2})
\end{aligned} \tag{22}$$

Defining

$$\begin{aligned}
\boldsymbol{b}^{(1)} &\equiv (b_{(2)}^2, -b_{(2)}^1)^T/B, \quad \boldsymbol{b}^2 \equiv (-b_{(1)}^2, b_{(1)}^1)^T/B \\
B &\equiv b_{(1)}^1 b_{(2)}^2 - b_{(1)}^2 b_{(2)}^1
\end{aligned} \tag{23}$$

we find the solution of (17) and (21) to be

$$(D\nabla\varphi)_{i+1/2,j} = \boldsymbol{b}^{(1)}\varphi|_{ij}^{i+1,j} + \boldsymbol{b}^{(2)}\{\varphi|_{i,j-1}^{i,j+1} + \varphi|_{i+1,j-1}^{i+1,j+1}\} \tag{24}$$

The treatment of $(D\nabla\varphi)_{i,j+1/2}$ is similar. Extension to three dimensions is straight-forward, but somewhat laborious.

MULTIGRID FOR INTERFACE PROBLEMS

Next, we consider the solution of the discrete system obtained by discretization of the interface problem by multigrid methods. Because only this case has been considered in the literature, and because the principles involved can also be discussed in the context of this case, we consider the Cartesian case only, i.e. $\boldsymbol{x} = \boldsymbol{x}(\boldsymbol{\xi})$. The remaining non-smoothness is that of D; this has consequences for the application of multigrid, which we will discuss.

Let us consider finite volume discretization on the fine grid of figure 3. The grid is uniform with mesh-size h in both directions. The resulting finite volume discretization is called cell-centered or block-centered. Domain boundaries are located at cell faces. The discretization discussed before leads to the following stencil for the discretization matrix A:

$$[A] = \begin{bmatrix} & -w_{i,j+1/2} & \\ -w_{i-1/2,j} & \Sigma & -w_{i+1/2,j} \\ & -w_{i,j-1/2} & \end{bmatrix} \tag{25}$$

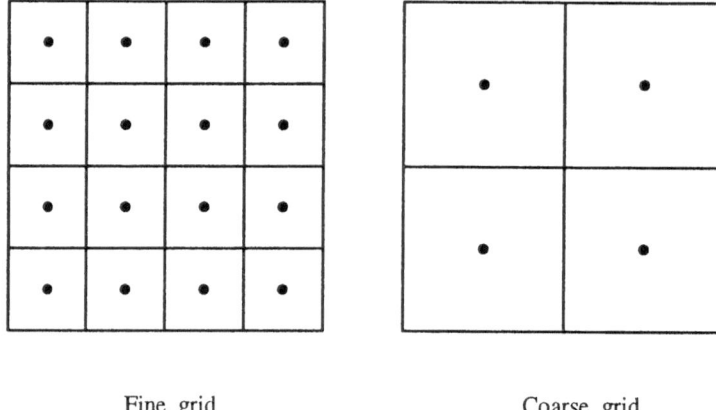

Fine grid   Coarse grid

Figure 3: Fine and coarse grid for cell-centered multigrid

where

$$w_{i+1/2,j} = 1/(\frac{1}{2D_{ij}} + \frac{1}{2D_{i+1,j}}) \tag{26}$$

and a similar formula for $w_{i,j+1/2}$; Σ is the negative sum of the surrounding coefficients. A derivation of (26) from first principles may be found, for example, in [22].

When using multigrid to efficiently solve the resulting system of equations, it is natural to use cell centered multigrid, in which coarse grids are constructed by taking unions of fine grid cells, as illustrated in figure 3. The alternative, more widely used, is vertex-centered multigrid, in which coarse grids are constructed by deleting grid points where the numerical approximation is defined. This is more natural than cell-centered multigrid for vertex-centered discretizations, and is illustrated in figure 4. For interface problems, there is a huge difference between cell-centered and vertex-centered multigrid. In the cell-centered case, standard multigrid works with the usual interpolatory transfer operators between the grids. Some theoretical backing of this is given in [20]; details and numerical experiments may be found in [19], [9] and [10]. In the vertex-centered case, however, standard multigrid does not work. As explained in [1] and [8], the transfer operators have to be chosen in a more complicated way, in dependence on [A], at a price of extra computing time and memory. This approach has been further developed in [6], [2], [3], [4] and [5]. The cell-centered approach has been further developed in [17] and [18]. Both approaches can be made to work efficiently with $O(N)$ complexity.

The reason why standard vertex-centered multigrid does not work for interface problems is that $grad\ \varphi$ is discontinuous where D is discontinuous, so that prolongation by linear interpolation is inaccurate. In the cell-centered case prolongation may be done by piecewise constant interpolation, which does not suffer from discontinuities

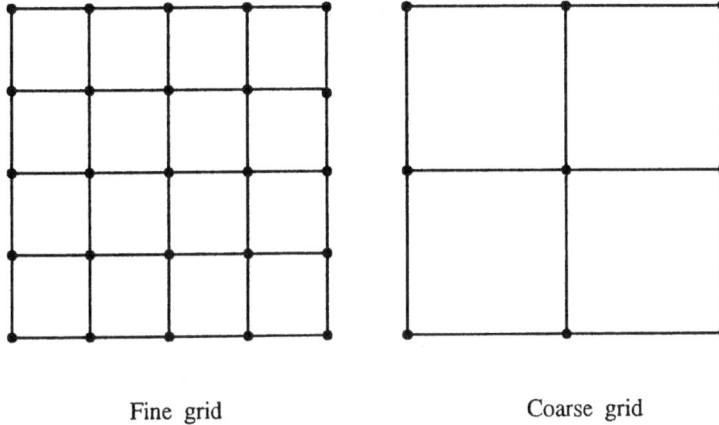

Fine grid   Coarse grid

Figure 4: Fine and coarse grid for vertex-centered multigrid

in *grad* φ.

References

[1] R.E. Alcouffe, A. Brandt, J.E. Dendy Jr., and J.W. Painter. The multigrid method for diffusion equations with strongly discontinuous coefficients. *SIAM J. Sci. Stat. Comp.*, 2:430–454, 1981.

[2] J.E. Dendy Jr. Black box multigrid. *J. Comp. Phys.*, 48:366–386, 1982.

[3] J.E. Dendy Jr. Black box multigrid for nonsymmetric problems. *Appl. Math. Comp.*, 13:261–284, 1983.

[4] J.E. Dendy Jr. Black box multigrid for systems. *Appl. Math. Comp.*, 19:57–74, 1986.

[5] J.E. Dendy Jr. Two multigrid methods for three-dimensional problems with discontinuous and anisotropic coefficients. *SIAM J. Sci. Stat. Comp.*, 8:673–685, 1987.

[6] J.E. Dendy, Jr. and J.M. Hyman. Multi-grid and ICCG for problems with interfaces. In M.H. Schultz, editor, *Elliptic problem solvers*, pages 247–253, New York, 1981. Academic Press.

[7] F.H. Harlow and J.E. Welch. Numerical calculation of time-dependent viscous incompressible flow of fluid with a free surface. *The Physics of Fluids*, 8:2182–2189, 1965.

[8] R. Kettler and J.A. Meijerink. A multigrid method and a combined multigrid-conjugate gradient method for elliptic problems with strongly discontinuous coefficients in general domains. Technical Report Shell Publ. 604, KSEPL, Rijswijk, The Netherlands, 1981.

[9] M. Khalil. *Analysis of Linear Multigrid Methods for Elliptic Differential Equations with Discontinuous and Anisotropic Coefficients*. PhD thesis, Delft University of Technology, Delft, The Netherlands, 1989.

[10] M. Khalil and P.Wesseling. Vertex-centered and cell-centered multigrid for interface problems. *J. Comp. Phys.*, 98:1–10, 1992.

[11] A.E. Mynett, P. Wesseling, A. Segal, and C.G.M. Kassels. The ISNaS incompressible Navier-Stokes solver: invariant discretization. *Applied Scientific Research*, 48:175–191, 1991.

[12] C.W. Oosterlee. *Robust multigrid methods for the steady and unsteady incompressible Navier-Stokes equations in general coordinates*. PhD thesis, Delft University of Technology, The Netherlands, 1993.

[13] C.W. Oosterlee and P. Wesseling. A multigrid method for an invariant formulation of the incompressible Navier-Stokes equations in general co-ordinates. *Communications in Applied Numerical Methods*, 8:721–734, 1992.

[14] C.W. Oosterlee and P. Wesseling. A robust multigrid method for a discretization of the incompressible Navier-Stokes equations in general coordinates. *Impact Comp. Science Engng.*, 5:128–151, 1993.

[15] C.W. Oosterlee, P. Wesseling, A. Segal, and E. Brakkee. Benchmark solutions for the incompressible Navier-Stokes equations in general co-ordinates on staggered grids. *Int. J. Num. Meth. Fluids*, 17:301–321, 1993.

[16] A. Segal, P. Wesseling, J. Van Kan, C.W. Oosterlee, and C.G.M. Kassels. Invariant discretization of the incompressible Navier-Stokes equations in boundary fitted co-ordinates. *Int. J. Num. Meth. Fluids*, 15:411–426, 1992.

[17] G.J. Shaw and P.I. Crumpton. Discretization and multigrid soution of elliptic equations with strongly discontinuous coefficients. Report 92/16, Oxford University Computing Laboratory, 1992.

[18] R. Teigland and B.G. Ersland. Comparison of two cell-centered multigrid schemes for problems with discontinuous coefficients. IBM Report 91/6, Bergen Scientific Centre, 1991. ISBN 82-415-0081-5.

[19] P. Wesseling. Cell-centered multigrid for interface problems. *J. Comp. Phys.*, 79:85–91, 1988.

[20] P. Wesseling. Two remarks on multigrid methods. In W. Hackbusch, editor, *Robust multi-grid methods*, pages 209–216, Braunschweg, 1988. Vieweg. Notes on Numerical Fluid Mechanics 23.

[21] P. Wesseling. Large scale modeling in computational fluid dynamics. In E.F. Deprettere and A.-J. van der Veen, editors, *Algorithms and parallel VLSI architectures, Volume A: Tutorials*, pages 277–308, Amsterdam, 1991. Elsevier.

[22] P. Wesseling. *An introduction to multigrid methods*. John Wiley & Sons, Chichester, 1992.

[23] P. Wesseling, A. Segal, J.J.I.M. van Kan, C.W. Oosterlee, and C.G.M. Kassels. Finite volume discretization of the incompressible Navier-Stokes equations in general coordinates on staggered grids. *Comp. Fluid Dynamics Journal*, 1:27–33, 1992.

14. COASTAL FLOW

DOMAIN AND GRID SENSITIVITY STUDIES FOR HURRICANE STORM SURGE PREDICTIONS

C. A. BLAIN[1], J. J. WESTERINK[1] and R. A. LUETTICH, JR.[2]

[1]Department of Civil Engineering and Geological Sciences,
University of Notre Dame, Notre Dame, IN 46556
[2]University of North Carolina at Chapel Hill, Institute of Marine Sciences
3431 Arendell St., Morehead City, NC 28557

ABSTRACT

The formulation of a numerical storm surge model directly affects the predicted physics of the storm surge response. Both the size and discretization of the computational domain are studied to quantify their influence on storm surge generation in a coastal region. Results of a domain size sensitivity study demonstrate that the largest domain examined, which covers the Gulf of Mexico, adjacent basins, and extends out into the deep Atlantic ocean, best represents the storm surge response associated with hurricane forcings. An assessment of the error in the storm surge predictions associated with various grid discretizations suggests an optimal grid structure which has significant refinement in near shore regions but allows grid spacing up to the spatial scale of the hurricane in deep waters.

INTRODUCTION

Numerical modeling is an increasingly important tool used for understanding the hydrodynamic behavior of the coastal ocean. Of particular concern is the lack of studies documenting the convergence of storm surge predictions with regard to grid spacing and domain size. This paper presents the findings of two sensitivity studies which investigate the influence of domain size and grid structure on modeled storm surge response. In the first study, an actual hurricane that made landfall on the Florida shelf in the Gulf of Mexico is applied over three domain sizes allowing comparisons between storm surge elevations computed over each domain. In the second study, errors in predicted storm surge generated by a synthetic hurricane are calculated over regular and variably graded grids. Assessments of these errors indicate regions of maximum error and the influence of grid spacing.

HYDRODYNAMIC MODEL DESCRIPTION

The hydrodynamic computations were performed using the finite element model, ADCIRC-2DDI (Luettich et al., 1992), which uses the vertically averaged equations of

1275

A. Peters et al. (eds.), Computational Methods in Water Resources X, 1275–1282.

mass and momentum conservation, subject to the hydrostatic pressure approximation. Using the standard quadratic parameterization for bottom stress and neglecting baroclinic terms as well as lateral diffusion/dispersion leads to the following set of conservation statements in primitive, non-conservative form expressed in a spherical coordinate system (Kolar et al., 1993):

$$\frac{\partial \zeta}{\partial t} + \frac{1}{R\cos\phi}\left[\frac{\partial UH}{\partial \lambda} + \frac{\partial (VH\cos\phi)}{\partial \phi}\right] = 0 \tag{1}$$

$$\frac{\partial U}{\partial t} + \frac{1}{R\cos\phi}U\frac{\partial U}{\partial \lambda} + \frac{1}{R}V\frac{\partial U}{\partial \phi} - \left[\frac{\tan\phi}{R}U + f\right]V =$$

$$-\frac{1}{R\cos\phi}\frac{\partial}{\partial \lambda}\left[\frac{P_s}{\rho_0} + g(\zeta - \eta)\right] + \frac{\tau_{s\lambda}}{\rho_0 H} - \tau_* U \tag{2}$$

$$\frac{\partial V}{\partial t} + \frac{1}{R\cos\phi}U\frac{\partial V}{\partial \lambda} + \frac{1}{R}V\frac{\partial V}{\partial \phi} + \left[\frac{\tan\phi}{R}U + f\right]U =$$

$$-\frac{1}{R}\frac{\partial}{\partial \phi}\left[\frac{P_s}{\rho_0} + g(\zeta - \eta)\right] + \frac{\tau_{s\phi}}{\rho_0 H} - \tau_* V \tag{3}$$

where t represents time, λ, ϕ are degrees longitude (east of Greenwich positive) and degrees latitude (north of the equator positive), ζ is the free surface elevation relative to the geoid, U, V are the depth averaged horizontal velocities, R is the radius of the Earth, $H = \zeta + h$ is the total water column depth, h is the bathymetric depth relative to the geoid, $f = 2\Omega\sin\phi$ is the Coriolis parameter, Ω is the angular speed of the Earth, p_s is the atmospheric pressure at the free surface, g is the acceleration due to gravity, η is the effective Newtonian equilibrium tide potential, ρ_0 is the reference density of water, $\tau_{s\lambda}, \tau_{s\phi}$ are the applied free surface stresses, and τ_* is given by the expression $C_f(U^2 + V^2)^{1/2}/H$ where C_f equals the bottom friction coefficient.

Equations (1)-(3) are reformulated into a generalized wave continuity equation (GWCE) and are subsequently discretized using the finite element (FE) method (Lynch and Gray, 1979; Kinnmark, 1984). It is noted that wetting and/or drying of elements is not currently accommodated.

METEORLOGICAL FORCING

Computations of hurricane wind stress and pressure fields are carried out using a modified form of the HURWIN wind model (Cardone et al., 1992). An exponential pressure law is used to generate a circularly symmetric pressure field situated at the low pressure center of the storm. Wind speeds are then obtained through a solution of the equations of horizontal motion which have been vertically averaged through the depth of the planetary boundary layer. These wind speeds are then converted to surface wind stresses using a quadratic drag law proposed by Garratt (1977).

DOMAIN SIZE SENSITIVITY STUDY

Three domain sizes, shown in Figure 1 are defined to clearly demonstrate the relationship between the domain size and open boundary elevation specification. Characteristics pertinent to each domain are summarized in Table 1. The smallest domain considered, the Florida Coast domain pictured in Figure 1a, is a semi-circular basin similar to the one used by the National Weather Service in conjunction with the SLOSH storm surge model (Jelesnianski et al., 1992). Much of the Florida Coast domain lies on the continental shelf at depths less than 130m. The second domain, the Gulf of Mexico domain shown in Figure 1b, includes the entire Gulf of Mexico and is similar to domains used by other investigators for storm surge modeling studies in the Gulf of Mexico (e.g. Bunpapong et al., 1985). The final and largest domain is the Eastcoast domain shown in Figure 1c which has been previously used by Westerink et al. (1993) to study tides in the western North Atlantic. The Eastcoast domain encompasses the western North Atlantic ocean, the Caribbean Sea, and the Gulf of Mexico and has been constructed with a single deep ocean boundary.

Table 1: Characteristics of the Model Domains

Domain	Area (km^2)	Max Depth (m)	Discretization		Grid Size (km)	
			Nodes	Elements	Max	Min
Florida coast	5.07x10^4	1094	1451	2326	32.5	0.5
Gulf of Mexico	1.41x10^6	3781	6325	11441	50.0	0.5
Eastcoast	8.35x10^6	7765	23711	41709	105.0	0.5

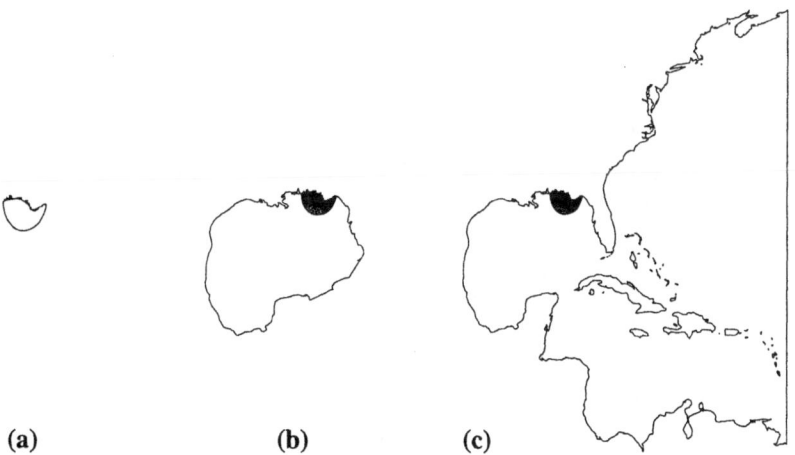

(a)   (b)   (c)

Figure 1: Boundaries of the (a) Florida Coast domain, (b) Gulf of Mexico domain, and (c) Eastcoast domain.

A series of simulations is conducted using the historical storm Hurricane Kate (1985) as the meteorological forcing. Since all domains have identical discretizations over corresponding regions and simulations were conducted using identical model parameters and wind and pressure forcing, differences between the model responses are due solely to the domain size and/or the boundary condition specification. The computed storm surge hydrograph at Apalachicola along the Florida coast is shown in Figure 2a using a still water boundary condition and in Figure 2b using an inverted barometer

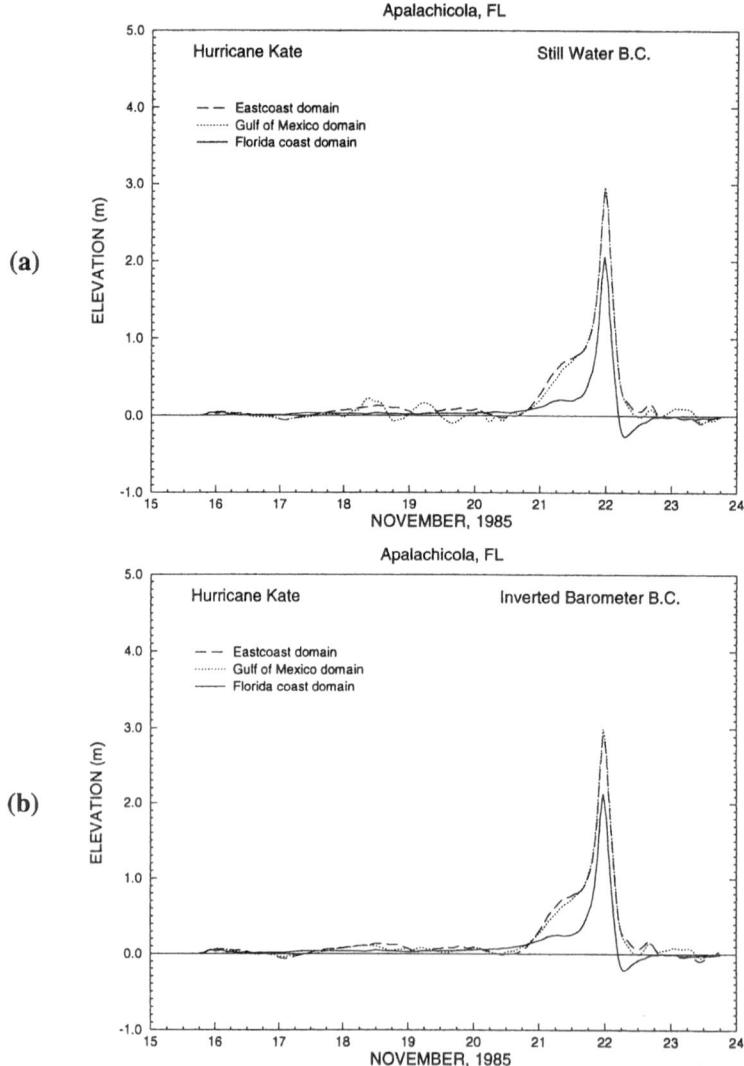

Figure 2: Computed storm surge for Hurricane Kate using a still water (a) and an inverted barometer (b) open boundary condition at Apalachicola, Florida.

boundary condition. The storm surge profile at Apalachicola is representative of conditions on the right-hand side of the hurricane, east of the hurricane landfall region.

The storm surge hydrographs in Figure 2 clearly illustrate that a model domain such as the Florida Coast domain which is situated largely on the continental shelf and whose domain size is limited relative to the size of the storm significantly underestimates the primary storm surge response. Substantial storm surge occurs in the vicinity of open ocean boundaries of such domains when water is pushed up on the shelf by hurricane winds. While the peak surge for the Gulf of Mexico and Eastcoast domains closely correspond, Figures 2a and 2b demonstrate that different oscillatory patterns before and after the peak storm surge are predicted for the Gulf of Mexico and Eastcoast domains. These oscillatory patterns are associated with resonant modes in the Gulf of Mexico basin and relate to the surge forerunner. While the predicted modes in the Gulf of Mexico domain appear to be very sensitive to the boundary conditions specified, modes in the Eastcoast domain solution exhibit no sensitivity to boundary condition specification.

The Eastcoast domain which includes the western North Atlantic ocean, the Gulf of Mexico, and the Caribbean Sea leads to convergent computations of both the primary storm surge and the surge forerunner. The inclusion of contiguous basins allows the proper set up of basin resonant modes and facilitates the realistic propagation of storm surge throughout the domain onto the continental shelf where development of the storm surge is most critical. The main advantage of the Eastcoast domain is that the open boundaries lie within the deep Atlantic ocean and are far from the intricate processes occurring on the continental shelf and within the Gulf of Mexico basin in response to the storm. A complete analysis of the influence of domain size on storm surge prediction including hydrographs at a number of stations is given by Blain et al., 1993.

GRID SENSITIVITY STUDY

Three rectangular grids of dimension 2500 km x 3000 km are constructed to correspond to the areal extent of the Eastcoast domain. A land boundary representing the coastline lies along the leftmost 3000 km length of these rectangular grids and open ocean conditions apply at the remaining grid boundaries. These grids, named G01, G02, and G03, have regular nodal spacings of 50 km, 25 km, and 12.5 km, respectively. A synthetic hurricane, H011, serves as the meteorological forcing for the hydrodynamic model. Hurricane H011 moves along a path perpendicular to the coastline and eventually makes landfall at 192 hours. Simulation of the storm surge generated by hurricane H011 is carried out over grids G01, G02, and G03. Storm surge predictions computed over the most finely discretized grid, G03, are considered the "truth" solution and will be used to assess errors in the storm surge computations performed over the remaining grids.

Maximum overprediction (positive) and underprediction (negative) errors in the storm surge computed over grids G01 and G02 are shown in Figures 3a and 3b. The absolute error is calculated simply as the difference between computed storm surge values and the "true" storm surge elevations. The relative error normalizes the absolute error with respect to the highest surge elevation over the entire domain at the specific point in time being considered. Ocean bathymetry at the locations of the maximum error is also represented in Figure 3. Observe in Figures 3a and 3b the large increase in the

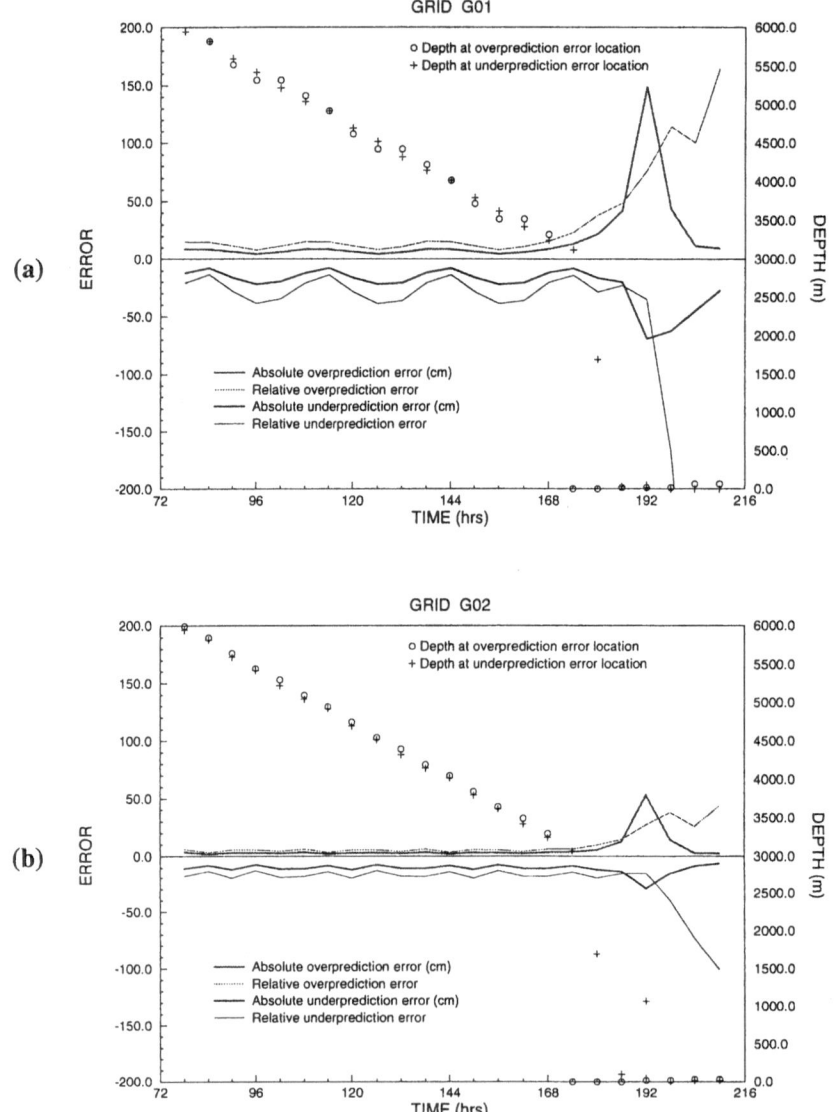

Figure 3: Maximum errors in the storm surge computed over grid G01 (a), grid G02 (b), and grid VG01 (c).

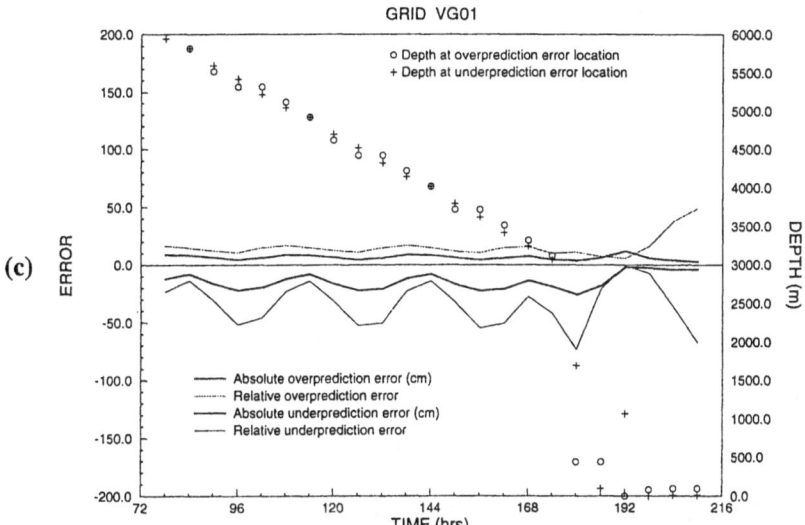

Figure 3c: Maximum errors in the storm surge computed over grid VG01.

relative error which coincides with a decreasing absolute error and a declining peak surge as the storm makes landfall. This behavior of the relative error indicates that processes which dissipate the peak surge are much more rapid than those which dissipate the errors in the storm surge. A comparison of the error magnitudes between Figures 3a and 3b demonstrates that halving the nodal spacing leads to a reduction of the error in deep waters by a factor of two and a threefold reduction of the errors generated on the continental shelf. The peak errors over grids G01 and G02 are concentrated on the continental shelf nearest the shoreline suggesting that a variably graded mesh may provide an optimal grid structure for minimization of the storm surge prediction errors.

A variably graded grid, VG01, is constructed such that the grid spacing ranges from a maximum of 50 km over the deep ocean and including regions up to 50 km offshore to a minimum of 12.5 km near the coastline. Maximum error values for the grid VG01 are presented in Figure 3c. Errors in storm surge elevations over the deep ocean are the same as those computed over grid G01 due to an identical grid spacing of 50 km. However, errors over the continental shelf are less than those presented for grid g02 by a factor of four. In general, the maximum errors computed over grid VG01 are relatively uniform in time and space and are significantly less than those computed over either grids G01 or G02.

CONCLUSIONS

These sensitivity studies demonstrate that both the size and discretization of the computational domain influence storm surge generation in the coastal region. To minimize the effect of cross shelf boundary specification and properly represent the resonant characteristics within the Gulf of Mexico, a very large domain encompassing the Gulf of

Mexico, adjacent basins, and extending out into the deep ocean is most appropriate for accurate storm surge prediction. A variable grid structure which has extensive refinement in the near shore region leads to low uniform errors throughout the domain. Since discretization errors can never be completely eliminated, uniform errors over a mesh are desired with the acceptable magnitude of these errors representing a balance between computational effort and required accuracy. The variable grid structure utilized here yields the smallest errors with the least computational effort particularly when implemented in conjunction with a large domain such as the Eastcoast domain.

REFERENCES

Blain, C.A., Westerink, J.J., Luettich, R.A. (1993) "The Influence of Domain Size on the Response Characteristics of a Hurricane Storm Surge Model", *Journal of Geophysical Research,* in review.

Bunpapong, M., Reid, R.O. and Whitaker, R.E. (1985) "An Investigation of Hurricane-Induced Forerunner Surge in the Gulf of Mexico", *Coastal Engineering Research Center*, U.S. Army Engineers, Technical Report CERC-85-5.

Cardone, V.J., Greenwood, C.V. and Greenwood, J.A. (1992) "Unified Program for the Specification of Hurricane Boundary Layer Winds Over Surfaces of Specified Roughness", *Coastal Engineering Research Center*, U.S. Army Engineers, C.R.-CERC-92-1.

Garratt, J.R. (1977) "Review of Drag Coefficients Over Oceans and Continents", *Mon. Wea. Rev., 105*, 915-929.

Jelesnianski, C.P.,Chen, J. and Shaffer, W.A. (1992), "SLOSH: Sea, Lake, and Overland Surges from Hurricanes", *NOAA Technical Report NWS 48.*

Kinnmark, I.P.E. (1984) "The Shallow Water Wave Equations: Formulation, Analysis and Application", Ph.D. Dissertation, Department of Civil Engineering, Princeton University.

Kolar, R.L., Gray, W. G., Westerink, J.J. and Luettich, R.A. (1993) "Shallow Water Modeling in Spherical Coordinates: Equation Formulation, Numerical Implementation, and Application", *J. Hydraul. Res.,* in press.

Luettich, R.A., Westerink, J.J., and Scheffner, N.W., (1992) "ADCIRC: An Advanced Three-Dimensional Circulation Model for Shelves, Coasts and Estuaries, Report 1: Theory and Methodology of ADCIRC-2DDI and ADCIRC-3DL", *Tech. Report DRP-92-6,* Department of the Army.

Lynch, D.R. and Gray, W.G. (1979) "A Wave Equation Model for Finite Element Tidal Computations", *Comp. Fluids, 7,* 207-228.

Westerink, J.J., Luettich, R.A. and Muccino, J.C. (1993), "Modeling Tides in the Western North Atlantic Using Unstructured Graded Grids", *Tellus,* in press.

Finite Element Simulation of Turbidity Current with Internal Hydraulic Jump

Sung–Uk Choi and Marcelo H. Garcia
Department of Civil Engineering
University of Illinois at Urbana–Champaign
Urbana, Illinois 61801, USA

ABSTRACT

A turbidity current passing through a submarine canyon changes its flow regime from supercritical to subcritical when it meets a submarine fan which has a much lower slope. Therefore, a discontinuity in the flow profiles, called internal hydraulic jump, is expected to occur near the vicinity of the canyon mouth where the slope changes. A numerical model developed for unsteady turbidity currents is applied to simulate the internal hydraulic jump at the canyon–fan region. The computational algorithm is based on the Petrov–Galerkin finite element method. Artificial viscosity is introduced to remove the oscillations at the front of the jump, making it possible to simulate the development of an internal hydraulic jump in a turbidity current through a change in flow regime. The steady state equations are integrated numerically with specified boundary conditions, and the results are compared with profiles computed with the unsteady model and laboratory observations.

INTRODUCTION

Turbidity currents are sediment–driven density currents, whose driving force comes from the density difference due to suspended sediment. Turbidity currents occur in lakes, reservoirs, and oceans. They interact with the surrounding environment by entraining clear water from above into the flow, depositing suspended sediment onto the bed, and entraining bed sediment into suspension. The balance between sediment erosion and deposition determines the life expectancy of a given current.

In the ocean, turbidity currents are known as the architects of many submarine canyons. The change in channel slope at the canyon–fan transition induces the formation of an internal hydraulic jump. For turbidity currents, an internal hydraulic jump has great importance, because it indicates the transition from an overall erosive environment upstream to an overall depositional environment downstream. Garcia and Parker (1989) observed the internal hydraulic jump at a canyon–fan transition by generating continuous turbidity currents in laboratory experiments. They found that the thickness of the sediment deposited from the suspended load decreased roughly exponentially in the downstream direction. The existence of internal hydraulic jump at this region was also foreseen from numerical computations by Akiyama and Stefan (1986). They integrated numerically the

A. Peters et al. (eds.), Computational Methods in Water Resources X, 1283–1290.
© 1994 *Kluwer Academic Publishers. Printed in the Netherlands.*

steady state governing equations for turbidity currents to obtain the profiles, however, they could not proceed with the computations beyond the point where the hydraulic jump took place.

Except for the steady state computations of Garcia (1993), no numerical attempts to simulate the hydraulic jump associated with turbidity currents can be found. A few attempts to capture the shock numerically have been done in hydraulic engineering applications. Dispersive higher–order schemes such as MacCormack scheme are well known to reproduce hydraulic jumps well (Garcia et al., 1986 ; Jimenez et al., 1988). Recently, Gharangik and Chaudhry (1991) applied a scheme which is second–order accurate in time and fourth–order accurate in space, to the St. Venant and Boussinesq equations. They took advantage of the fact that the equations are hyperbolic regardless of Froude number value by using unsteady flow techniques to solve the steady problem. The computed profiles were compared with measured data for various Froude numbers, and good agreement was observed. In terms of finite element methods, Katopodes (1984a,b) applied the dissipative Galerkin method to the St. Venant equations, and showed that the scheme has a shock capturing ability. Thompson (1990) obtained a two–dimensional profile of a hydraulic jump by using the FESWMS–2DH software. However, the computed profile did not exhibit a sharp increase in the water surface profile at the front of the jump.

Choi and Garcia (1994) developed a numerical model for one–dimensional, unsteady turbidity currents, which is based on the dissipative Galerkin finite element method. Using this model, the simulation of an internal hydraulic jump near a transition region is investigated herein. The difficulty of the problem lies in obtaining a jump profile that is steep and is also free from numerical oscillations. The easiest way may be to refine meshes near discontinuities in advance, but this requires prior knowledge about the location of discontinuities. Otherwise, an adaptive mesh refinement technique should be employed. Here, an artificial viscosity is added to the original formulations to avoid oscillatory solutions instead of changing the grid systems. For comparison with the results of the numerical simulation, the experimental data of Garcia (1993) are used. Such experiments were performed in a flume that is 30 cm wide, 78 cm deep, about 12 m long. The channel consisted of a 5 m long model canyon with a slope of 0.08 followed by a 6.6 m horizontal model fan. At the end of the fan, there was a free fall which provided an appropriate boundary condition. Turbidity currents driven by sediment particles of different sizes were generated with such facility.

LAYER–AVERAGED BALANCE EQS. FOR TURBIDITY CURRENTS

The spatial development of an unsteady, one–dimensional turbidity current can be described by the following approximate set of layer–averaged equations (Parker et al., 1986):

$$\frac{\partial h}{\partial t} + \frac{\partial uh}{\partial x} = e_w u \tag{1}$$

$$\frac{\partial uh}{\partial t} + \frac{\partial u^2 h}{\partial x} = -\frac{1}{2} Rg \frac{\partial ch^2}{\partial x} + RgchS - u_*^2 \tag{2}$$

$$\frac{\partial ch}{\partial t} + \frac{\partial uch}{\partial x} = v_s(e_s - r_o c) \tag{3}$$

where the dependent variables are, h = the current depth; u = depth–averaged velocity; and c = the depth–averaged volumetric concentration. Also, e_w = water entrainment coefficient; v_s = sediment fall velocity; e_s = sediment entrainment coefficient; r_o = shape factor; u_* = bed shear velocity; R = submerged specific gravity which has a value of 1.65 for quartz; and S is the bottom slope, taken to be much smaller than unity.

In the above relations, eq.(1) represents the fluid mass balance, where $e_w u$ denotes the rate of clear water entrainment from the upper layer into the underflow. Conservation of streamwise momentum is given by eq.(2), where on the right hand side the first term represents the effect of the streamwise pressure gradient caused by the suspended sediment, the second term represents the current driving force due to the downslope action of gravity on the sediment–water mixture, and the third term represents flow resistance due to bottom friction. The mass balance of suspended sediment is given by eq.(3), where the right hand side represents the difference between the rate of sediment entrainment into suspension (i.e., $v_s e_s$) and the rate of sediment deposition onto the bed (i.e., $v_s c_b$).

The governing equations can be written in conservative form as follows:

$$\frac{\partial Y}{\partial t} + A \frac{\partial Y}{\partial x} = b \tag{4}$$

where $Y = \{h, q, p\}^T$, $q = uh$, $p = ch$, and

$$b = \begin{Bmatrix} e_w u \\ RgchS - u_*^2 \\ v_s(e_s - r_o c) \end{Bmatrix}, \qquad A = \begin{bmatrix} 0 & 1 & 0 \\ \frac{1}{2}Rgch - u^2 & 2u & \frac{1}{2}Rgh \\ -uc & c & u \end{bmatrix} \tag{5),(6}$$

For water entrainment from the upper layer, the following relationship proposed by Parker et al. (1987) is used,

$$e_w = \frac{0.075}{\left(1 + 718 \, Ri^{2.4}\right)^{0.5}} \tag{7}$$

where $Ri = (gRhc)/u^2$ is the bulk Richardson number. No sediment entrainment into suspension, i.e., $e_s = 0$ is assumed, and a value of the bed friction coefficient c_D ($\equiv u_*^2/u^2$) = 0.01 is maintained throughout the computations. The near bed volumetric sediment concentration c_b at an elevation equal to five percent of the layer thickness h is represented as $r_o c$ in eq.(3). A constant value of $r_o \approx 2.0$ is used herein (Parker et al., 1987).

INTERNAL HYDRAULIC JUMP

When a turbidity current reaches the mouth of the canyon and proceeds across the lower slope of the channel, an initial discontinuity will take place within the channel. Depending on the amount of entrainment, the internal hydraulic jump will either remain within the channel or will migrate up the channel into the canyon, before becoming stationary, and achieving a steady state (Komar, 1971; Garcia, 1993). As in the case of surface flow, a relationship for the ratio of the depths, after (h_2) and before (h_1) the jump is given by :

$$\frac{h_2}{h_1} = \frac{1}{2}[\sqrt{1 + 8\ Ri_1^{-1}} - 1]$$ (8)

where Ri_1 denotes the bulk Richardson number evaluated before jump. The underlying assumptions in deriving the above equation are ; (1) the fan is rectangular, and horizontal, (2) pressure distribution is hydrostatic, (3) bed friction in the jump region, and water entrainment from upper layer due to jump are negligible, and (4) the rate of sediment entrainment equals the rate of sediment deposition within the jump (i.e., $v_s e_s = v_s c_b$).

DISSIPATIVE GALERKIN FINITE ELEMENT

It is well known that the standard Galerkin method when applied to convection dominated flows, yields oscillatory solutions, which come from the symmetry of the weighting function. A small variation in the symmetric linear test function N to give more weight in the upwind direction than in the downwind direction leads to a dissipative Galerkin scheme, better known as the "Petrov–Galerkin" scheme. This scheme has a selective damping property that removes only spurious high frequency waves behind the head while maintaining overall accuracy at the same level. The computational algorithm applied here is the same as that used by Katopodes (1984a) for open–channel flows, with the weighting function:

$$N_* = N + \varepsilon\ A^T \frac{\partial N}{\partial x}$$ (9)

where A^T is the transpose of the convection matrix A. Notice that the weighting is dependent on A, and the level of upwinding can be controlled by a parameter ε. Using this weighting function, we can express the weighted residual form of eq.(4) as:

$$\sum_{n=1}^{n_e} \int_L N_*^T (\frac{\partial Y}{\partial t} + A \frac{\partial Y}{\partial x} - b)\ dx = 0$$ (10)

where n_e = total number of elements, and L = the distance between two nodes.

The Newton–Raphson technique is applied to solve the nonlinear finite element equation given by eq.(10). For the time integration, the partial derivatives with respect to time can be substituted as variables themselves at the same time level by using a second–order finite difference scheme such as :

$$(\frac{\partial h}{\partial t})^{n+1} = ah^{n+1} + \beta^{n+1}$$ (11)

where n represents time level, $a = 1/(\lambda \Delta t)$, and $\beta^{n+1} = ah^n + (\partial h/\partial t)^n$. As a weighting factor between two time levels, $\lambda = 0.5$ is used. The optimal level of upwinding in eq.(9) can be found analytically with homogeneous, linear governing equations (Choi and Garcia, 1994), and is controlled by a parameter ε given by :

$$\varepsilon = \frac{L}{\sqrt{15}} \frac{1}{u + \sqrt{Rgch}} \tag{12}$$

which is a function of element size and flow parameters.

The dissipative higher–order schemes produce spurious high–frequency oscillations near the shock in the solution. Artificial viscosity is effective in removing oscillatory waves near the regions of sharp gradients while smooth regions are undisturbed. The computed variables, for example f_i^{k+1}, can be modified according to the following equation (Jameson et al., 1981 ; Gharangik et al., 1991) :

$$f_i^{k+1} = f_i^{k+1} + \xi_{i+1/2}(f_{i+1}^{k+1} - f_i^{k+1}) - \xi_{i-1/2}(f_i^{k+1} - f_{i-1}^{k+1}) \tag{13}$$

where

$$\xi_i = \frac{|h_{i+1} - 2h_i + h_{i-1}|}{|h_{i+1}| + 2|h_i| + |h_{i-1}|} \tag{14}$$

and

$$\xi_{i+1/2} = \varkappa \frac{\Delta x}{\Delta t} \max(\xi_{i+1}, \xi_i) \tag{15}$$

Here, $\varkappa$ is a parameter to control the amount of dissipation.

STEADY STATE MODELING

The steady state form of governing equations, eq.(1)~eq.(3), can be rewritten as (Parker et al., 1986) :

$$\frac{dh}{dx} = \frac{- R_i S + e_w(2 - \frac{1}{2}R_i) + c_D + \frac{1}{2}\frac{v_s}{U} r_o R_i(\frac{\psi_e}{\psi} - 1)}{1 - R_i} \tag{16}$$

$$\frac{h}{U} \frac{dU}{dx} = \frac{R_i S - e_w(1 + \frac{1}{2}R_i) - c_D - \frac{1}{2}\frac{v_s}{U} r_o R_i(\frac{\psi_e}{\psi} - 1)}{1 - R_i} \tag{17}$$

$$\frac{h}{\psi} \frac{d\psi}{dx} = \frac{v_s}{U} r_o(\frac{\psi_e}{\psi} - 1) \tag{18}$$

where $\psi = uch$ denotes the volumetric sediment transport rate per unit width, and $\psi_e = e_s hu/r_o$ means the equilibrium value of Ψ when neither erosion nor deposition occurs.

If there is no change in flow regime within the domain, the above set of ordinary differential equations can be integrated by a numerical integration technique such as the fourth–order Runge–Kutta method with prescribed boundary conditions. The profiles can be sought in the downstream direction for a supercritical flow case ($Ri<1$), and in the upstream direction for a subcritical flow ($Ri>1$).

For the problem at hand including a critical section within the domain, the following strategy can be used. First, assume that the hydraulic jump occurs at a point near the break in slope, and then obtain the supercritical flow profile up to the assumed critical section. If we assume no water entrainment and no sediment entrainment into suspension after the hydraulic jump, we have a constant discharge q_j. Then, an analytical expression for ψ can be obtained as follows (Garcia, 1993) :

$$\psi = \psi_j \, exp(- v_s r_o x' / q_j) \tag{19}$$

where ψ_j is the value of ψ at the point of the hydraulic jump, and x' is the distance from the jump. Now, it becomes possible to route the depth profile by integrating eq.(16) with the downstream boundary condition of the critical depth at the free fall, $h_c = q_j/B_f^{1/3}$, where $B_f = gR\psi$ denotes the buoyancy flux at the free fall estimated with the help of eq.(19). The velocities can be also estimated by dividing the constant q_j by h. The computations may be carried out continuously until the height of the current obtained by the backward routing coincides with the current depth predicted by the forward routing and eq.(8).

APPLICATIONS

Simulation results of an internal hydraulic jump associated with a turbidity current, having upstream boundary conditions, $h_o = 3.0$ (cm), $u_o = 8.3$ (cm/s), and $c_o = 2.48 \times 10^{-3}$, are shown in FIG.1. The sediment used in generating the current has a mean particle diameter, $D_{sg} = 4$ μ, and a fall velocity of 0.0016 cm/s. The initial Courant number, Cr_o [$\equiv (u_o + \sqrt{Rgc_oh_o}) \, \Delta t/\Delta x$] is made approximately equal to unity by choosing $\Delta x = 0.1$ cm and $\Delta t = 1.0$ sec. In FIG.1, the solid lines are the results of the unsteady finite element simulation after a steady state has been reached, and the dashed lines are the results from the steady–state modeling. Excellent agreement between the two computed results were found in terms of the location of the hydraulic jump, and the flow depth after the jump. This proves the numerical method is capable of capturing the shock. Also in this figure, the symbols denote current depths computed with the help of measured velocity profiles (Garcia, 1993). General agreement between computed and observed results is good in the supercritical region and subcritical region far away from the jump. However, in the transition region, the predicted jump is much more abrupt than that observed in the laboratory.

The development in time of a hydraulic jump is shown in FIG.2. It can be seen that the front of the internal hydraulic jump migrates in the upstream direction to a point in the proximity of the break in slope as time goes on. The strength of the shock increases until a steady–state has been reached.

The profiles of a steady–state turbidity current in the proximity of the jump are shown in FIG.3 for different values of the parameter $\varkappa$ in the artificial viscosity. The selective removal of high frequency waves in the solutions near the critical section by using artificial viscosity is clear. Without artificial viscosity, i.e., $\varkappa = 0$, the spurious oscillations prevent convergence to the right solution. As the value of $\varkappa$ increases, the smoothing effect become more apparent. The value of $\varkappa$ beyond 3 does not seem to affect the solution much. Herein, $\varkappa = 3$ was used, and the hydraulic jump was captured within one or two elements.

CONCLUSIONS

A finite element model developed for unsteady turbidity currents was applied to simulate the internal hydraulic jump at the canyon–fan transition region. Unsteady equations were solved for a steady problem. The computational results were compared with solutions of steady state model as well as with observed data. Through the numerical experiments, the shock capturing capability of the numerical scheme was demonstrated. Also, the

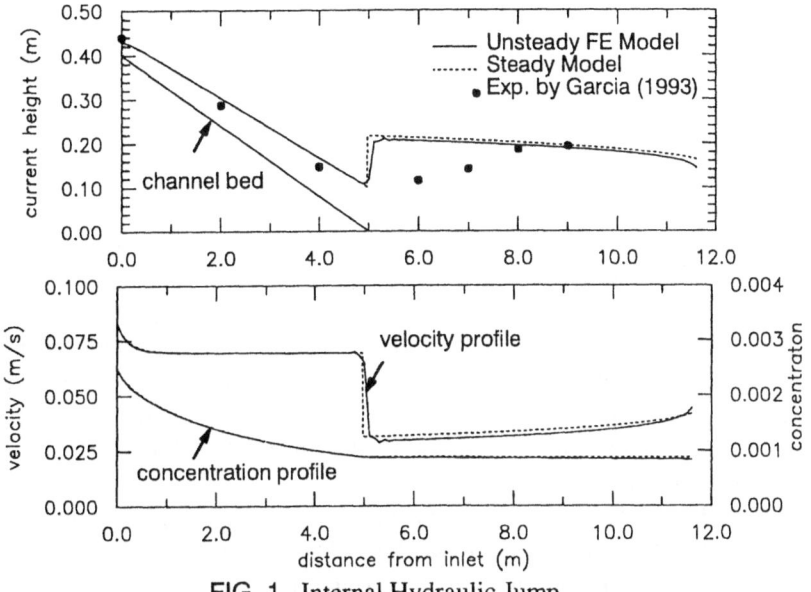

FIG. 1. Internal Hydraulic Jump

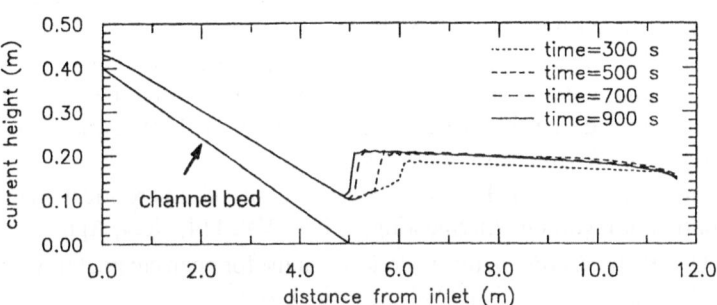

FIG. 2. Process of an Internal Hydraulic Jump Development

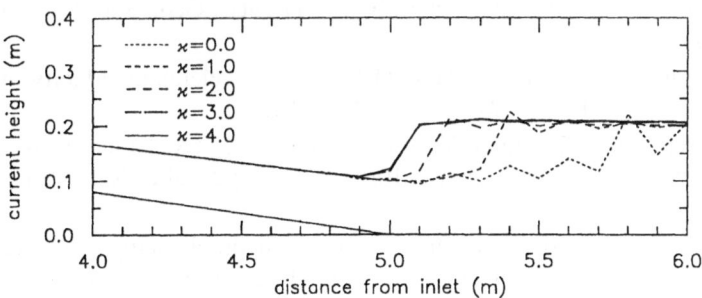

FIG. 3. Effect of Control Parameter $\varkappa$ in Artificial Viscosity

numerical experiments revealed a number of deficiencies in representing the physics involved in the internal hydraulic jump.

Acknowledgements

The support of the Marine Geology and Geophysics program of the U.S. Office of Naval Research is gratefully acknowledged.

REFERENCES

Akiyama, J. and Stefan, H. (1986) "Prediction of turbidity currents in reservoirs and coastal regions", Third International Symposium on River Sedimentation, University of Mississippi, Miss., pp. 1295–1305.

Choi, S.–U. and Garcia, M. (1994) "One–Dimensional Turbidity Current Modeling with Finite Element Methods", submitted to Journal of Hydraulic Engineering, ASCE.

Garcia, R. and Kahawita, R.A. (1986) "Numerical solution of the St. Venant equations with the MacCormack finite–difference scheme", International Journal for Numerical Methods in Fluids, Vol.6, pp.259–274.

Garcia, M. and Parker, G. (1989) "Experiments on hydraulic jumps in turbidity current near a canyon–fan transition", Science, Vol. 245, pp. 393–396.

Garcia, M. (1993) "Hydraulic jumps in sediment–driven bottom current", Journal of Hydraulic Engineering, ASCE, Vol. 119, No.10, Oct.

Gharangik, A.M. and Chaudhry, M.H. (1991) "Numerical simulation of hydraulic jump", Journal of Hydraulic Engineering, ASCE, Vol. 117, No.9, Sep.

Jameson, A., Schmidt, W., and Turkel, E. (1981) "Numerical simulation of the Euler equations by finite volume methods using Runge–Kutta time–stepping schemes", AIAA 14th Fluid and Plasma Dynamics Conf.: AIAA 81–1259, American Institute of Aeronautics and Astronautics, Palo Alto, Calif.

Jimenez, O.F. and Chaudhry, M.H. (1988) "Computation of supercritical free–surface flows", Journal of Hydraulic Engineering, ASCE, Vol. 114, No.4, Apr.

Katopodes, N.D. (1984a) "A dissipative Galerkin scheme for open channel flow", Journal of Hydraulic Engineering, ASCE, Vol. 110, No.4, Apr.

Katopodes, N.D. (1984b) "Two–dimensional surges and shocks in open channels", Journal of Hydraulic Engineering, ASCE, Vol. 110, No.7, June.

Komar, P.D. (1971) "Hydraulic jumps in turbidity currents", Geol. Soc. Am. Bull., Vol. 82, pp. 1477–1488.

Parker, G., Fukushima, Y., and Pantin, H.M. (1986) "Self accelerating turbidity currents", Journal of Fluid Mechanics, Vol. 171, pp. 145–181.

Parker, G., Garcia, M., Fukushima, Y., and Yu,W. (1987) "Experiments on turbidity currents over an erodible bed", Journal of Hydraulic Research, Vol.25, pp.123–147.

Thompson, J. (1990) "Numerical Modeling of irregular hydraulic jumps", in H.H. Chang and J.C. Hill (eds.), Proceedings of the 1990 Hydraulic Engineering Conference, ASCE, pp.749–754.

PREDICTIVE FINITE ELEMENT MODELLING OF STRATIFIED TIDAL FLOW

A. HAQUE and J. BERLAMONT
Laboratory of Hydraulics
K.U. Leuven
de Croylaan 2, B-3001 Heverlee
Belgium

ABSTRACT

A predictive numerical model in a 2D vertical plane has been developed. Finite elements have been used for the spatial discretization, and a three level semi-implicit scheme has been used for the time discretization. Using various levels of turbulence closures, the model has been applied in a tidal flume experiment performed at the Delft Hydraulics. Predictive abilities of the mixing length model, two equation turbulence model and algebraic stress/flux model have been compared. The model has been applied to study the behaviour of Reynolds stresses and buoyancy effects in a stratified tidal medium.

INTRODUCTION

Modelling of a stratified flow in a tidal medium requires a vertical resolution of the flow, density and turbulence fields. In this flow situation, turbulent transport processes have to be defined accurately for a predictive model. A mixing length model is a simplified representation of these turbulent transport processes. Most of the predictive numerical models developed so far in this flow situation used the mixing length model as a turbulence closure (Bloss et al., 1988; Boericke and Hogan, 1977). Because of the difficulties of specifying the velocity scale and the length scale accurately, mixing length models involve locally adjustable empirical constants for a particular flow situation. These assumptions introduce uncertainty on the predictive ability of a mixing length model. A stratified tidal flow may be considered as 'complex' flows because they posses some additional physical features due to the presence of buoyancy force which are absent from the classical thin shear layer. It is thus unlikely that a simplified representation of the turbulence field by a mixing length type turbulence closure and a reduction function type buoyancy correction will represent all the details of this type of flow. On the other hand, most of the features of a stratified tidal flow can be represented in their exact form in a higher order turbulence closure. Within the limit of computational economy, the algebraic stress/flux model is the most appropriate choice as a turbulence closure for a predictive numerical model.

A. Peters et al. (eds.), Computational Methods in Water Resources X, 1291–1298.

MATHEMATIFCAL BASIS

The starting point to describe the physics of a stratified tidal flow where stratification is mainly due to the salinity is the conservation law of mass, momentum and salt. In a 2D vertical plane, the equations are (Bloss et al., 1988):

$$\frac{\partial}{\partial x}(BU) + \frac{\partial}{\partial z}(BW) = 0 \tag{1}$$

$$B\frac{\partial U}{\partial t} + BU\frac{\partial U}{\partial x} + BW\frac{\partial U}{\partial z} = -Bg\frac{\partial \zeta}{\partial x} - Bg\alpha\frac{\partial}{\partial x}\int_z^\zeta S dz + B\frac{\partial}{\partial x}(-\overline{uu}) + B\frac{\partial}{\partial z}(-\overline{uw}) \tag{2}$$

$$B\frac{\partial S}{\partial t} + BU\frac{\partial S}{\partial x} + BW\frac{\partial S}{\partial z} = B\frac{\partial}{\partial x}(-\overline{us}) + B\frac{\partial}{\partial z}(-\overline{ws}) \tag{3}$$

where x coordinate has been measured positive eastwards and z coordinate has been measured positive upwards, u is the turbulent mean longitudinal velocity, w is the turbulent mean vertical velocity, u and w are the corresponding fluctuating parts, $\overline{uu}$ and $\overline{uw}$ are the Reynolds stresses, B is the width, ρ is the saltwater density, ρ_0 is the freshwater density, g is the acceleration due to gravity, s is the turbulent mean salt concentration, $\overline{us}$ and $\overline{ws}$ are the Reynolds fluxes. In equations (1) to (3) it has been assumed that the longitudinal velocity components are several orders of magnitude larger than the vertical components. Also, the influence of variable density has been considered only in the buoyancy term. The relation between the density and concentration can be expressed by the following equation of state :

$$\rho = \rho_0(1 + \alpha S) \tag{4}$$

where α is the constant relating the density and concentration. The surface elevation has been calculated by vertically integrating equation (1) and using the kinematic free surface condition. The resulting equation is :

$$B\frac{\partial \zeta}{\partial t} + \int_{-h_b}^\zeta B\frac{\partial U}{\partial x} dz = 0 \tag{5}$$

where ζ is the surface elevation above the datum and h_b is the bottom elevation. Due to the hydrostatic pressure assumption, the vertical velocity has been calculated from the mass conservation and bottom kinematic condition as:

$$BW(z) = BUS_0 - \int_{-h_b}^z B\frac{\partial U}{\partial z} dz \tag{6}$$

where s_0 is the bottom slope.

THE TURBULENCE CLOSURE

The Reynolds stresses and fluxes appearing in equations (2) and (3) have to be defined by a turbulence closure. In the mixing length model, the turbulent transport of momentum and mass are expressed by the gradients of mean quantities, and by a mixing length, where the mixing length is defined as a purely geometric parameter (Bloss et al., 1988). The stratification effect is represented by the local gradient Richardson number by an empirical reduction function (ASCE task committee, 1988).

The empiricism involved in the mixing length model can be overcome in the two equation turbulence model, where the velocity and length scale of turbulence are expressed by transport equations (Rodi, 1980). These transport equations describe the production/dissipation and transport of fluctuating quantities by advection/diffusion. The stratification effects (stable or unstable) come in an exact form and no empirical representation is required. The isotropic eddy viscosity/diffusivity is determined from the Kolmogorov-Prandtl relation (Rodi, 1980).

Although the two equation turbulence model is quite successful in different flow situations, its main drawback are the assumption of isotropic eddy viscosity/diffusivity and one velocity scale for all the fluctuating quantities. Measurements in stably stratified steady open channel flow (Komori et al., 1983) showed different behaviours for Reynolds shear stress/normal stresses and longitudinal/vertical Reynolds fluxes. Experiments also showed a much higher longitudinal Reynolds flux than the vertical and an increased ratio of them with the increased stability. Instead of a constant value of the turbulent Prandtl number, this value was found to vary with stability (Arya, 1975). The Kolmogorov-Prandtl constant was determined in the non-buoyant shear layer where stratification effect was not considered.

Considering these facts, the second order closure level of turbulence is particularly attractive in a stratified medium where both the direct and the indirect effects of stratification can be represented naturally with the governing equations. The exact equations for the Reynolds stresses and fluxes can be derived from the Navier-Stokes equations (Hinze, 1975). The governing equations can be derived after introducing certain model assumptions to close the unknown correlations (Launder et al., 1975; Gibson and Launder, 1978). The differential transport equations for the Reynolds stresses and fluxes can be simplified into algebraic expressions by eliminating the gradients of the dependent variables. This requires the assumption of neglecting the rate of change and the total transport of Reynolds stresses and fluxes, together with the local equilibrium assumtion. In this way, most of the basic features of the physics of the system can be represented in a more simplified manner. After solving the algebraic relations algeraically, the expressions for the individual Reynolds stresses and fluxes have been obtained. If the finite element method is used for the spatial discretization, the Reynolds stresses and fluxes can be eliminated at the element level. So, the total number of equations to be solved remains the same which is required to solve the turbulence closure with a two equation turbulence model.

THE NUMERICAL MODEL

Finite elements have been used for the spatial discretization of the governing equations and a three level semi-implicit time integration scheme has been used for the time discretization. The resulting 2D vertcal plane model equations after implementing this time scheme have been given by Haque and Berlamont (1993). A Fourier analysis of the time scheme used for the finite element discretization of the long-wave equations shows that the scheme is unconditionally stable for $0.5 \leq \theta \leq 1$, and selectively damps the shortest wavelengths responsible for the nodal oscillations. This time scheme has been used for an explicit coupling of the flow, density and turbulence fields. The element equations have been formed by applying the Galerkin weighted residual method. The resulting global equations have been solved sequentially.

MODEL APPLICATIONS

The model has been applied in a tidal flume experiment performed in the tidal flume of Delft Hydraulics under the E.C Large Installation Plan (LIP1). The flume is 130m long and 1m wide. The sea has been modelled by a basin with a surface area of 120m^2, where a time varying water level has been introduced to generate a tidal current. The vertical density distribution in the model sea has been controlled by a fresh water skimmer and a jet-mixing system. The parameters that have been measured in the experiment are the velocity, salinity and water level at various positions in the flume. In the experiment, a mean water depth of 25cm has been used for a 5cm tidal range and 650s tidal period. A 10 l/s fresh water discharge has been used to maintain the maximum density difference between the sea water and the fresh water at 20 kg/m^3.

The 2D vertical plane finite element model has been applied to predict the flow and the density fields of the tidal flume experiment using various levels of turbulence closures. For the mixing length model, a parabolic distribution of the mixing length has been assumed (Bloss et al., 1988). The stratification effect has been represented by the widely used Munk-Anderson relation (ASCE task committee, 1988). The longitudinal salt transport has been modelled by using a parameterization of the longitudinal diffusion. The form for the longitudinal diffusion coefficient that has been used here is similar to the form used by Boericke and Hogan (1977).

The actual experience with the application of the mixing length model is that the numerical model is very sensitive to the distribution of eddy viscosity and diffusivity. The numerical model responds very quickly to any local instability of the eddy viscosity/diffiusivity. This may happen if the local velocity gradient becomes zero and/or a strong stratification arises. This local instability has been avoided by applying a best-fit smoothing to the eddy viscosity/diffusivity at each time step. When an unstable stratification develops, turbulent kinetic energy is produced from the turbulent buoyancy flux and turbulent mixing is enhanced. There is no mechanism in the reduction function used to treat this, although some empirical modification is possible. Instead, in this case, the density field has been re-calculated with the computed maximum eddy diffusivity, giving the same effect of an enhanced turbulent mixing. To improve the predictive ability of the mixing length model, it has been found that a problem dependent description of the mixing length distribution is much more important than to parameterize the stratification effect. In fact, the sensitivity of the stratification reduction function on the computed flow and density fields depends on the mixing length distribution. The longitudinal distribution of salt and its intrusion length also depends on the mixing length distribution.

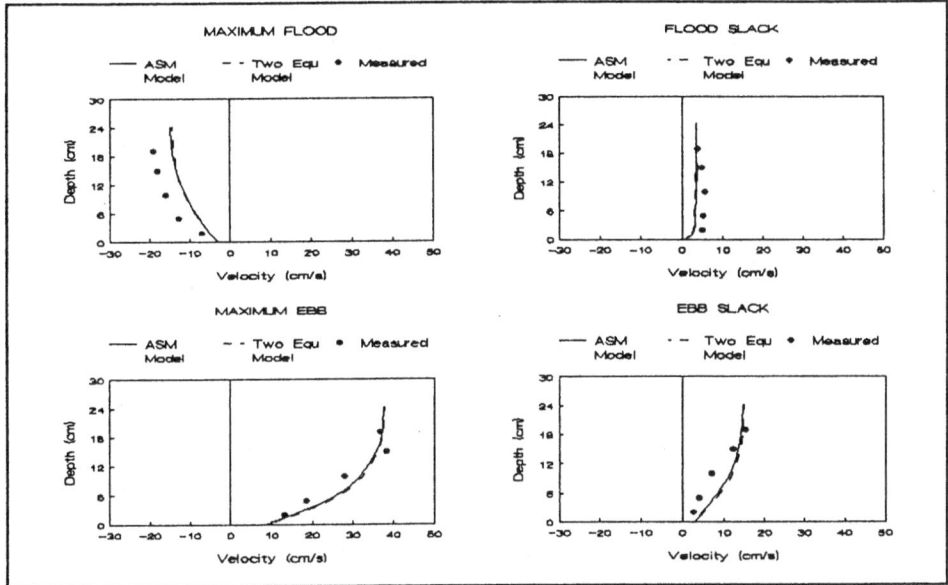

FIG.1 Comparison of predicted velocity profiles at four characteristic times between the two equation turbulence model and the ASM model.

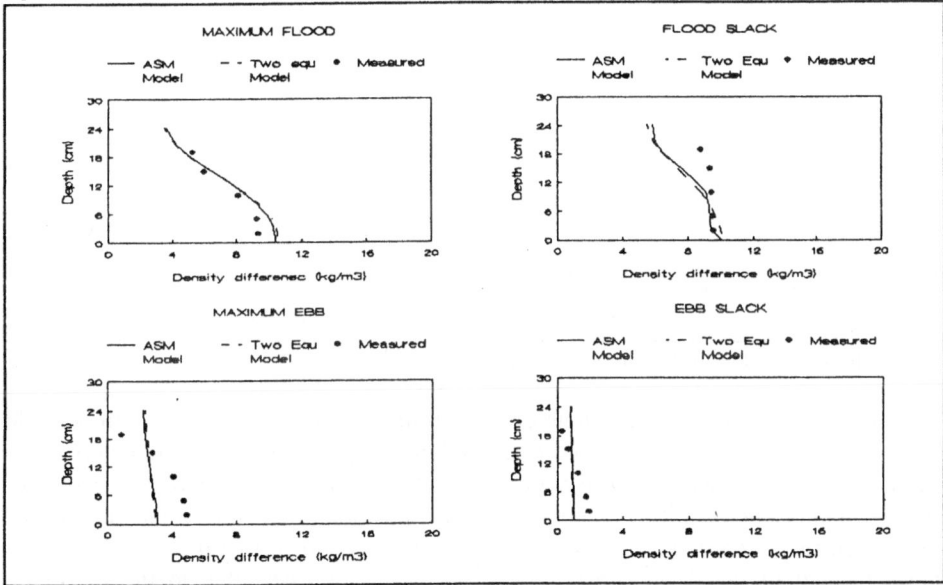

FIG.2 Comparison of predicted density difference profiles at four characteristic times between the two equation turbulence model and the ASM model.

Predictive ability of the numerical model has been found to improve when higher order turbulence models are used. The turbulence models have been applied without any coefficient tuning. The comparison of the flow and density fields prediction with the two equation turbulence model and the algebraic stress/flux model (ASM) at a section 12m from the model sea boundary are shown in Figures 1 and 2. Prediction with the ASM model is only marginally better than the prediction with the two equation model. This feature was also noticed by Leschziner and Rodi (1983) in predicting a coaxial heated water discharge experiment and by Belhassan et al. (1991) in predicting a low speed variable density flow. Both the ASM and the two equation turbulence model show a weaker stratification compared to the measurements at the time of maximum ebb. This may be due to a significant 3D effect at the peak tidal instants which the model can not predict.

REYNOLDS STRESSES AND BUOYANCY EFFECTS

Reynolds stresses have been computed with the ASM model. The relative magnitude of three Reynolds stress components $\overline{uu}$, $\overline{ww}$ and $\overline{uw}$ at the time of maximum flood and ebb at a section 24m from the model sea boundary is shown in Figure 3. For both instants, the normal stress component $\overline{uu}$ is greater than $\overline{ww}$ and $\overline{uw}$. This order of relative magnitude is similar to what was found for a homogeneous steady symmetric channel flow experiment (Launder et al., 1975).

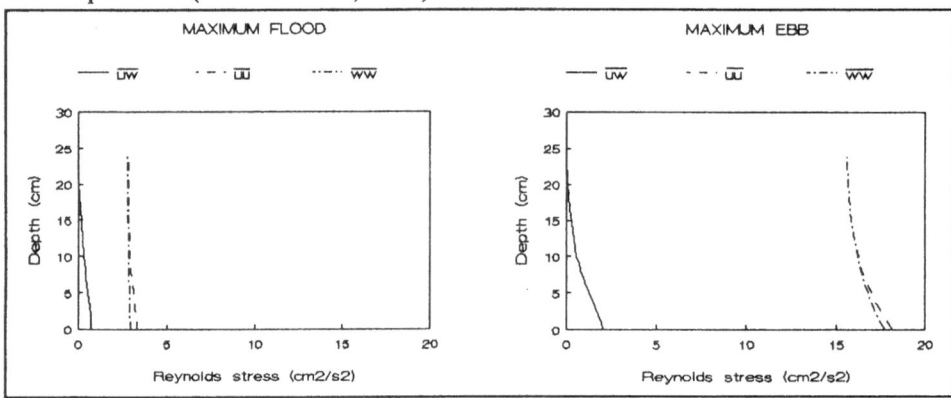

FIG.3 Three Reynolds stress components at the peak tidal instants.

To study the effect of buoyancy alone on the turbulent transport of momentum, a model run has been made in a nonstratified tidal medium keeping the other parameters of the stratified medium unchanged. The computed vertical profiles of Reynolds stresses in a stratified and in a nonstratified homogeneous tidal medium at the time of maximum flood and at a section 3.4m from the model sea boundary is shown in Figure 4. The figure shows that all of the three stress components are suppressed due to the buoyancy. Maximum suppression of the Reynolds stresses occur at some distance above the bottom, where the buoyancy effect is expected to be dominant. Among the two normal Reynolds stress components $\overline{uu}$ and $\overline{ww}$, which are comparable in magnitude, the Reynolds stress $\overline{ww}$ experiences higher reduction than the Reynolds stress $\overline{uu}$.

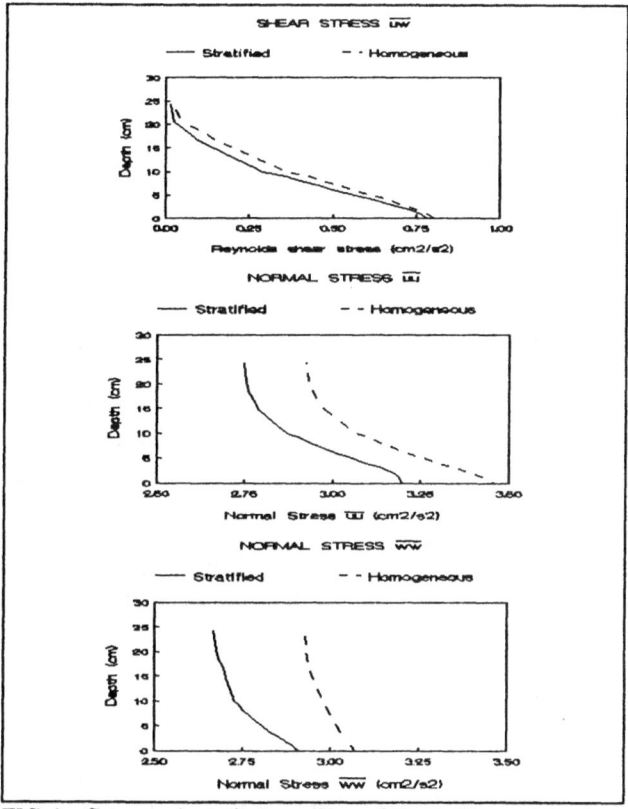

The spatial variation of the turbulent Prandtl number, which is the ratio of eddy viscosity to diffusivity, is shown in Figure 5 at the time of maximum flood. The figure shows that the turbulent Prandtl number has a higher value near the model sea boundary where the buoyancy effect is high compared to the values near the model fresh water boundary where the buoyancy effect is low. This shows the increase of the turbulent Prandtl number with the increase of stratification, a behaviour also experimentally observed for stable stratified free shear flow (Gibson and Launder, 1978). An increase of turbulent Prandtl number indicates a suppression of the vertical mixing.

FIG.4   Computed vertical profiles of Reynolds stresses in a stratified and in a nonstratified homogeneous tidal medium.

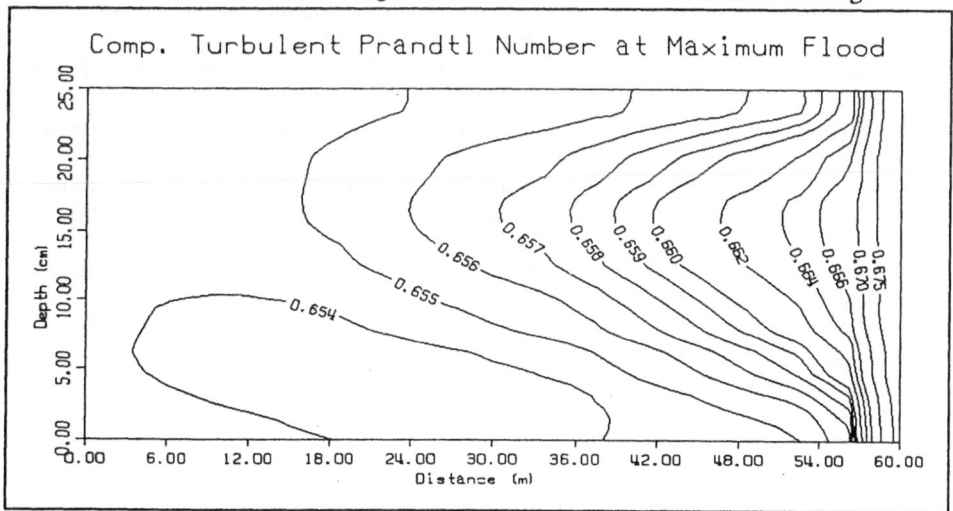

FIG.5   Computed contour of turbulent Prandtl number at the time of maximum flood.

CONCLUSIONS

A predictive model in a stratified tidal flow requires an accurate description of the turbulent transport processes. Using finite elements in space and a three level semi-implicit scheme in time, a predictive numerical model in a 2D vertical plane has been developed. The mixing length model has a limited ability to predict the flow and density fields of a stratified tidal flow. The predictive ability of higher order turbulence models is superior than the predictive ability of the mixing length model. The flow and density fields prediction with the algebraic stress/flux model is only slightly better than the prediction with the two equation turbulence model. Among the three Reynolds stress components $\overline{uu}$, $\overline{ww}$ and $\overline{uw}$, the Reynolds stress $\overline{uu}$ is greater than the Reynolds stresses $\overline{ww}$ and $\overline{uw}$ at the peak tidal instants. All of these Reynolds stress components are suppressed due to buoyancy. An increased stratification causes an increase of the turbulent Prandtl number, and results a suppression of vertical mixing in a stratified tidal medium.

ACKNOWLEDGEMENT

The tidal flume experiment has been carried out at the Delft Hydraulics under the E.C. Large Installation Plan (LIP1). Assistance of the technical personal at Delft Hydraulics is gratefully acknowledged.

REFERENCES

[1] Arya, S.P.S. : *Buoyancy effects in a horizontal flat-plat boundary layer*, J. Fluid Mechanics, Vol. 68, 321-343, 1975.
[2] Belhassan, M., Flamain, D. Gabillard, M. and Viollet, P.L. : *Numerical modelling and experimental study of low speed variable density flow*, XXIV IAHR Congress, Madrid, 1991.
[3] Bloss, S., Lehfeldt, R. and Patterson, J.C. : *Modelling turbulent transport in stratified estuary*, J. of Hydraulic Engineering, ASCE, Vol. 114(9), 1115-1133, 1988.
[4] Boericke, R.R. and Hogan, J.M. : *An X-Z hydraulic/thermal model for estuaries*, J. Hydr. Div.,ASCE, Vol. 103(1), 19-37, 1977.
[5] Gibson, M.M. and Launder, B.E. : *Ground effects on pressure fluctuations in the atmospheric boundary layer*, J. Fluid Mechanics, Vol. 86, 491-511, 1978.
[6] Haque, A. and Berlamont, J. : *A finite element model for density induced flow*, Proc. of the 1st Int. conf. on Hydro-Science and Engg., Washington, Sam S.Y. Wang ed., Part A(I), 1993.
[7] Hinze, J. : *Turbulence*, McGraw Hill Book Co., New York, 1975.
[8] Komori, S., Ueda, H., Ogino, F. and Mizushina, T. : *Turbulence structure in stably stratified open channel flow*, J. Fluid Mechanics, Vol.130, 13-26, 1983.
[9] Launder, B.E., Reece, G.J. and Rodi, W. : *Progress in the development of a Reynolds-stress turbulence closure*, J. Fluid Mechanics, Vol.68, 537-566, 1975.
[10] Leschziner, M.A. and Rodi, W. : *Calculation of coaxial heated water discharge*, J. of Hydraulic Engineering, ASCE, Vol. 109(10), 1380-1385, 1983.
[11] Rodi, W. : *Turbulence models and their application in hydraulics, A state of the art review*, IAHR, Delft, The Netherlands, 1980.
[12] Task Committee on Turbulence Models in Hydraulic Computations, ASCE : *Turbulence modelling of surface water flow and transport*, J.Hydr.Engg., ASCE,Vol.114(9),970-1073, 1988.

THE EFFECTS OF HORIZONTAL DENSITY GRADIENT ON THE FLOW IN THE SCHELDT ESTUARY

JABBARI E. and BERLAMONT J.
Hydraulics Laboratory
Catholic University of Leuven
de Croylaan 2, 3001 Heverlee, Belgium

The Scheldt estuary is a well-mixed to partially stratified estuary. Although this estuary can be considered as vertically homogeneous, most of the time of a tidal cycle, it shows significant salinity variations in the longitudinal direction. At the mouth in the Netherlands, salinity is about 36ppt, and at Prosperpolder in Belgium, the average salinity is about 17ppt. So far, hydrodynamic modelling of the Scheldt has been performed with the assumption of a constant density over the whole estuary. The baroclinic forcing due to the horizontal density gradients have been neglected. This study was carried out to investigate the effects of the horizontal density gradients on the flow in a 16km reach of the western Scheldt using a barotropic model. Baroclinic effects on momentum were represented by the density gradient terms in the momentum equations. The equation of state was used to approximate the density gradients by the salinity gradients. The conservation equation for salt was solved coupled with the hydrodynamic equations. A finite element method was used to solve the system of equations. A full implicit scheme was used for temporal discretization.

INTRODUCTION

The flow condition in an estuary depends on the relative strength of the forces due to the surface gradients, the density gradients, and the Coriolis effects. The force due to the density differences is often ignored in modelling the flow in an estuary assuming that relative to the force due to the surface gradients its effects are not considerable. By illustrating their importance in the James estuary, however, Pritchard demonstrated that these longitudinal density gradients can not always be ignored in a hydrodynamic simulation. The longitudinal density gradients induced force, generally, acts in the direction of the force due to the surface gradients during a landward flow situation and opposes this force during a seaward flow situation due to the fact that generally the density decreases in the landward direction.

To demonstrate the effect of this force on the flow in the Scheldt estuary a two-dimensional depth-integrated hydrodynamic model was extended to include this force and the model was used to simulate the flow on 4th of October 1990. Extensive information was available for hydrodynamic boundary conditions from a data-collection campaign. For salinities, however, a one dimensional analysis of salt motion over a longer reach

1299

A. Peters et al. (eds.), Computational Methods in Water Resources X, 1299–1306.
© 1994 Kluwer Academic Publishers. Printed in the Netherlands.

of the estuary had to be performed. The results of this analysis at the boundaries of the region of interest were taken as salinity boundary conditions.

The results of the simulations with two models, with and without baroclinic effects, were compared. Comparison showed that horizontal density variations contribute significantly to the momentum fluxes in the hydrodynamics computation.

GOVERNING EQUATIONS

Assuming that the only force that, actually, balances the gravity in the momentum equation in the vertical direction is the pressure force (Praagman 1979), this equation reduces to

$$-g - \frac{1}{\rho}\frac{\partial p}{\partial z} = 0 \tag{1}$$

Integration of this equation between the water surface and any level gives the pressure at that level as

$$p = \rho_0 g(\xi - z) + g\int_z^\xi (\rho - \rho_0)d\theta + p_a \tag{2}$$

where p = pressure, ρ = density, g = acceleration due to the gravity, ξ = the distance between the horizontal reference level (x,y) and water surface, θ = water depth relative to the level z, ρ_0 = density of fresh water, p_a = atmospheric pressure. Variation of the atmospheric pressure is small compared to the variation of the hydrostatic pressure. The variation of pressure over the horizontal plane is illustrated by the definition sketches in figure (1). From equation (2) one finds in two horizontal directions

$$\begin{aligned}
\frac{\partial p}{\partial x} &= \bar{\rho} g\frac{\partial \xi}{\partial x} + (\xi - z)g\frac{\partial \bar{\rho}}{\partial x} \\
\frac{\partial p}{\partial y} &= \bar{\rho} g\frac{\partial \xi}{\partial y} + (\xi - z)g\frac{\partial \bar{\rho}}{\partial y}
\end{aligned} \tag{3}$$

The first terms in the right side of the above equations gives the difference in pressure resulting from surface changes. These terms remain constant over the depth. The second terms result from the differences in pressure caused by the longitudinal gradients of density. The values of these terms increase linearly with the increase of the distance from the surface and would be maximum at the bed of the estuary. The difference in density, as far as this study is concerned, is due to the difference in salinity in space.

Substitution of equation (3) for the pressure gradients in the depth-averaged momentum equations in the horizontal directions gives

$$\begin{aligned}
\frac{\partial u}{\partial t} + u\frac{\partial u}{\partial x} + v\frac{\partial u}{\partial y} &= fv - g\frac{\partial \xi}{\partial x} - \frac{gh}{2\rho}\frac{\partial \rho}{\partial x} \\
&+ v(\frac{\partial^2 u}{\partial x^2} + \frac{\partial^2 u}{\partial y^2}) + \frac{\rho_a \gamma u_w}{\rho h}(u_w^2 + v_w^2)^{\frac{1}{2}} - \frac{gn^2 u(u^2 + v^2)^{\frac{1}{2}}}{h^{\frac{1}{3}}}
\end{aligned} \tag{4}$$

$$\frac{\partial v}{\partial t}+u\frac{\partial v}{\partial x}+v\frac{\partial v}{\partial y} = -fu-g\frac{\partial \xi}{\partial y}-\frac{gh}{2\rho}\frac{\partial \rho}{\partial y}$$

$$+\nu(\frac{\partial^2 v}{\partial x^2}+\frac{\partial^2 v}{\partial y^2})+\frac{\rho_a \gamma \nu_w}{\rho h}(u_w^2+v_w^2)^{\frac{1}{2}}-\frac{gn^2 v(u^2+v^2)^{\frac{1}{2}}}{h^{\frac{1}{3}}} \tag{5}$$

where u, and $v=$ depth-averaged tidal velocities in the x and y directions respectively, $h=$ total water depth, $n=$ Manning friction coefficient, $\gamma=$ empirical wind stress coefficient, $\rho_a=$ air density, u_w, $v_w=$ the components of wind speed, $\nu=$ horizontal eddy viscosity coefficient, $f=$ Coriolis parameter.

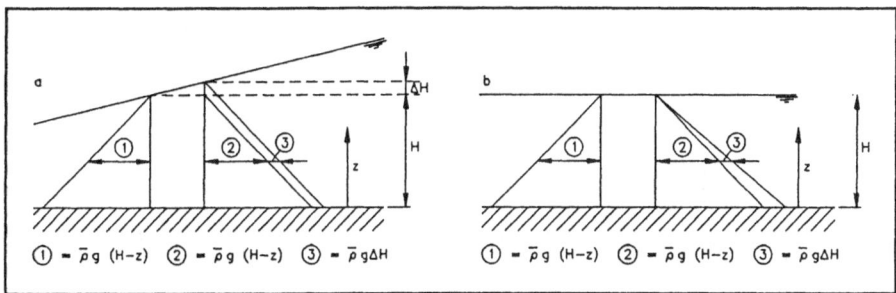

Figure 1. *Pressure gradient (a) due to the surface gradient,*
(b) due to the density gradient (after Abraham, 1976)

The density of estuarine water in the momentum equations may be replaced through the equation of state by salinity, temperature, and pressure. A simple relation between the density and the above variable does not exist (Massie 1982). However, for salinities up to 30ppt and temperatures between $0°$ and $20°$, the equation of state is effectively independent of temperature and may be approximated by (Perrel, Karelse 1982)

$$\rho = \rho_0 (1 + 0.75s) \tag{6}$$

The depth-averaged mass conservation equation for salt reads

$$\frac{\partial s}{\partial t}+u\frac{\partial s}{\partial x}+v\frac{\partial s}{\partial y} = \frac{1}{h}\frac{\partial}{\partial x}(D_{xx}h\frac{\partial s}{\partial x}+D_{xy}h\frac{\partial s}{\partial y})+\frac{1}{h}\frac{\partial}{\partial y}(D_{yx}h\frac{\partial s}{\partial y}+D_{yy}h\frac{\partial s}{\partial y}) \tag{7}$$

in which $s=$ salt concentration and $\begin{bmatrix} D_{xx} & D_{xy} \\ D_{yx} & D_{yy} \end{bmatrix}=$ dispersion tensor. The continuity equation in depth-averaged form is written as

$$\frac{\partial \xi}{\partial t} + \frac{\partial(uh)}{\partial x} + \frac{\partial(vh)}{\partial y} = 0 \tag{8}$$

The above equations complete the set of equations for the unknowns u, v, h, and s. These equations are to be solved over the whole domain and over the time under consideration. As an approximate solution method, finite elements are used for spatial discretization and an implicit scheme is used for temporal discretization.

FINITE ELEMENT IMPLEMENTATION

The Galerkin's weighted residual method is used to solve the governing equations. Two types of elements are used to subdivide the domain of interest: quadrilateral and triangular elements. Velocities and salinities are distributed over the elements by quadratic shape functions and water-depths by linear shape functions. Consequently each side of the elements should contain a node at the middle (Fig. 2).

The equations are multiplied by the appropriate weighting functions which are the same as the shape functions in the Galerkin scheme, momentum and salt balance equations by quadratic and fluid continuity equation by linear functions. The equations are integrated

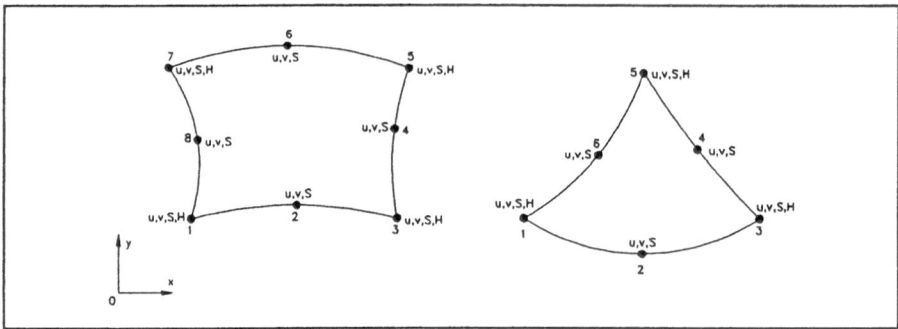

Figure 2. Parabolic four and three sides elements with the description of unknowns

over the horizontal plane. The second order derivatives (diffusion terms) are reduced to the first order by partial integration or Green-Gauss theorem. The Matrix form of the equations for a single node at the corner of an element will be

$$
\begin{bmatrix}
C_{11} & C_{12} & C_{13} & C_{14} \\
C_{21} & C_{22} & C_{23} & C_{24} \\
C_{31} & C_{32} & C_{33} & C_{34} \\
C_{41} & C_{42} & C_{43} & C_{44}
\end{bmatrix}
\begin{bmatrix} u \\ v \\ S \\ h \end{bmatrix}
=
\begin{bmatrix} f_1 \\ f_2 \\ f_3 \\ f_4 \end{bmatrix}
\tag{9}
$$

For a node at the middle of an element the forth equation of the matrix is omitted.

APPLICATION: THE SCHELDT ESTUARY

The western Scheldt estuary is located partly in Belgium and partly in the Netherlands (51° North, 4° east). The water level in this estuary is determined by the tidal variations in the North sea (Fig. 3). At about 200km from the mouth of the estuary, however, a barrier prevents the tidal variations to influence the flow upstream. Fresh water flow into the estuary is very small as compared to the tidal inflow and outflow. The average fresh water discharge during the months of September, October and November 1990 has been recorded as 33, 46, 58m³/s respectively.

The saline water intrusion into this estuary extends up to Rupelmonde where the river Rupel joins Scheldt. The average longitudinal salinity gradient over a spring tide was about 3.6×10^{-4}‰/m. Figure 4 shows the variations of water-level and cross-sectional average salinity at Prosperpolder (Fig. 3), during the spring tide of October 1990. The time variations of salinity correspond to the tidal variations. Examination of depth-averaged salinity at several locations in a cross-section shows that there is a lateral gradient of salinity across the cross-section. salinity is higher in high velocity areas (deeper parts of the channel) and is lower in low velocity areas. This contributes to higher dispersion of intruded salt from the North sea by causing a lateral circulation of water. However, the cross-sectional dispersion of the saline water is caused, mainly, by the vertical and transverse shears due to the vertical and transverse velocity gradients.

At Zandvliet cross-section (Fig. 3), measurements were carried out on 4th of October 1990, at

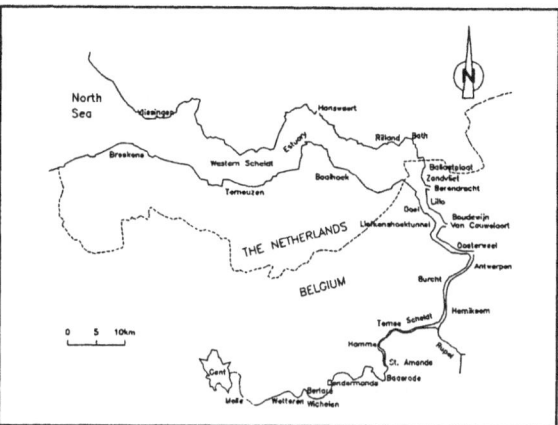

Figure 3. Location of western Scheldt estuary

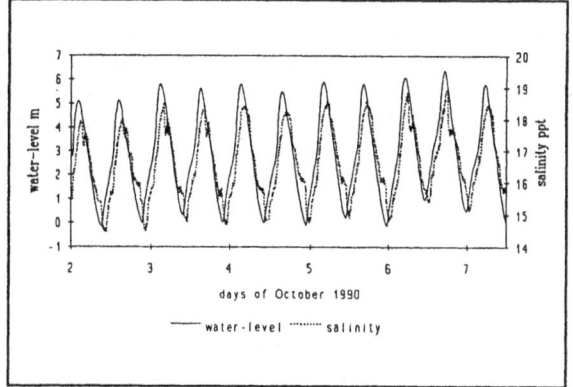

Figure 4. Water-level and salinity versus time, spring tide, October 1990

five points for several parameters including velocity and salinity. The point numbers in the next sections refer to the number of these points. The study reach extended some 8km upstream of this cross-section and about 8km downstream of it. Water-level variations were recorded at these two cross-sections, upstream and downstream, at two permanent gagging stations.

BOUNDARY CONDITIONS

At both ends of the domain, the recorded surface elevation versus time with an interval of 10 minutes were available and were imposed as boundary conditions. Generally, the computation would be more stable if at one end time variations of velocity were specified. Since no velocity at the boundaries was available, a weak condition for velocity was imposed at both seaward and landward boundaries. For salt concentration, no data were available at the boundaries of the region of interest. Salinity had been recorded at cross-sections upstream and downstream of the domain with 6 minutes interval. A one dimensional modelling was performed over a reach between these cross-sections setting the average tidal variations of cross-sectional velocity and water-depth over the five years prior to the simulation date as flow field. These water-depth and velocity variations were compared with the results of a barotropic simulation over the region of interest and proved to be accurate enough for the purpose of this study. The one dimensional model was calibrated for appropriate dispersion coefficient by comparing the results with the recorded salt concentration at a cross-section in the middle of the reach. The salinity variations for the boundaries of the region of interest were taken from the results of this one dimensional modelling. Salinity variations at both ends of the domain were imposed in the final simulation. Simulation, however, was performed imposing a weak condition at the outflow boundary of the region (upstream during flood and downstream during ebb). The model failed to reflect the specified salinity variations at the other boundary.

RESULTS OF THE SIMULATIONS

Flow and transport simulations were performed with both models, without and with the effects of the horizontal density gradients with the same boundary conditions. The models were run with a time-step of 10 minutes. A 7.5 m^2/s was taken as the coefficient of diffusion-like processes. An appropriate steady-state solution was achieved by driving the barotropic model for 4 hours prototype time. So, reliable results could be obtained for a tidal cycle by running the model for two tidal cycles (approximately 25 hours prototype time). However, since a transport process was involved in the new model, the simulations were performed for three tidal cycles. Modelling the salt motion with a transport model uncoupled from the hydrodynamic model required over eight tidal cycles simulation to achieve an appropriate steady-state solution. Comparison between the results of the two models showed that the differences between the surface elevations were not considerable. The difference between the velocity variations, however, were significant. The r.m.s. tidal velocity for point 2 of the measurement stations, from the first model (barotropic model) was 0.541m/s and from the second model was 0.515m/s. Figure 5 shows the difference between the velocity variations from two models for points 1 and 2. The calculated discharge from the results of two models are shown on figure

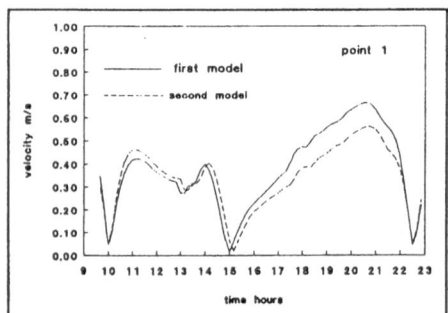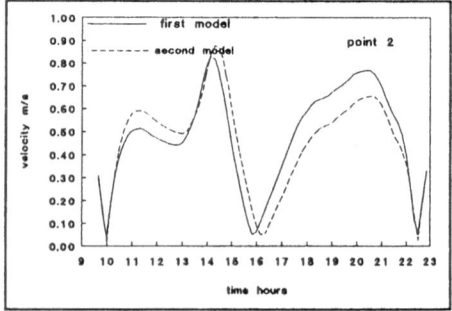

Figure 5. Difference between velocities from two models

6. The first and second model, in the figures, refer to the model without the effects of density gradients induced force and the model with the effects of this force respectively. The figures indicate a longer duration of the landward flow and a shorter duration of the seaward flow from the results of the second model.

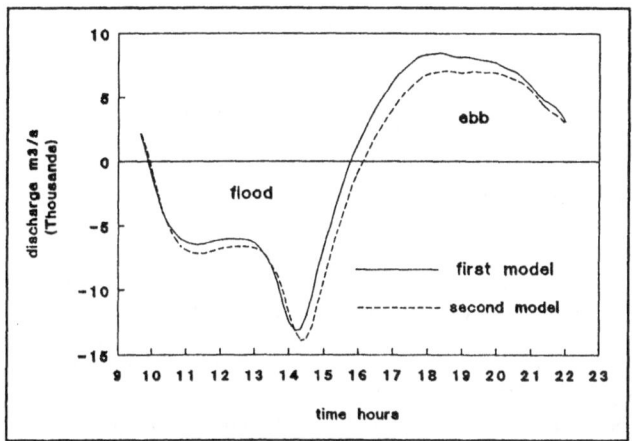

Figure 6. difference between discharges from two models

CONCLUSIONS

A two-dimensional depth-averaged model was used to simulate the flow in a reach of the Scheldt estuary. Due to the presence of the horizontal density gradients this model was extended to incorporate the effects of this gradients on the velocities by including the corresponding terms in the momentum equations and coupling the salt balance equation with the hydrodynamic equations for the salinities in conjunction with an equation of state for the density variations over the horizontal plane. The effects of the horizontal density variations on the surface elevation was not considerable. It was, however, shown that the horizontal density variations contribute significantly to the momentum fluxes determined in the hydrodynamic computations by different velocity variations. The computed discharge variations at the measurement cross-section showed that ignoring the effects of baroclinic forces will result in overestimation or underestimation of the discharge. The lack of an accurate initial salinity distribution required a transport model uncoupled with the hydrodynamic model to be run for over 8 tidal cycles for a reliable steady state solution. In simulation with the baroclinic model, however, the salinity results were reproducible at the third tidal cycle.

ACKNOWLEDGEMENT

The first author wishes to express his sincere thanks to the International Marine and Dredging Consultants (IMDC), Antwerp, in particular to Mr. Marc Sas, for the implementation of the original version of the computer program and for providing the necessary information and data for the application part of this study.

REFERENCES

Chen Ching L., and Lee Kwang K. (1991), "Great lakes river-estuary hydrodynamic finite element model", Journal of the Hydraulics Division, ASCE, vol. 117, HY11, 1531-1550.
Dyer K. R. (1973), "Estuaries: A physical introduction", John Wiley.
Fischer Hugo B. (1968), "Dispersion prediction in natural channels", Journal of the Sanitary Engineering Division, ASCE, Vol. 94 SA8, 927-943.
Fischer Hugo B. (1972), Mass transport mechanism in partially stratified estuaries, Journal of Fluid Mechanics, Vol. 53, Part 4, 671-687.
King I. P., Norton W. R., and Iceman K. R. (1975), "A finite element solution for two-dimensional stratified flow problems", Finite elements in fluids, 1, R. H. Gallagher, J. T. Oden, C. Taylor, and O. C. Zienkiewicz, eds., John Wiley and Sons, London, U.K, 133-156.
King, I. P. (1990), "Modelling of flow in estuaries using combinations of one- and two-dimensional finite elements", Hydrosoft, vol. 3, No. 3, 108-119.
McAnally, W. H., Jr., Letter, J. V., Jr., Stewart, J. P., Thomas, W. A., and Brogdon, N. J., Jr. (1984), "Colombia river hybrid modelling system", Journal of the Hydraulics Division, ASCE, vol. 110, HY3., 301-311.
Massie W. W. (1982), "Coastal engineering", Delft University of Technology, 16-21.
Partridge P. W., and Brebbia C. A. (1976), "Quadratic finite elements in shallow water problems", Journal of the Hydraulics Division, ASCE, vol. 102, HY9, 1299-1313.
Perrels P.A.J., Karelse M. (1982), "A two dimensional lateral averaged model for salt intrusion in estuaries", Delft Hydraulics Laboratory, publication No. 262.
Abraham G. (1976), "Density currents due to the differences in salinity", Rijkswaterstaat Communications, Salt Distribution in Estuaries, No. 26, 17-28.
Rijkwaterstaat, Dienst Getijdewateren, 1991, Waterkwaliteitsmodel Westerschelde, SAWES-nota 91.01 (in Dutch).
Simmons Henry B. (1955), Some effects of upland discharge on estuarine hydraulics, Journal of the Hydraulics Division, ASCE, Vol. 81, HY9, 1-21.
Smith, L. H., and Cheng, R. T. (1987), "Tidal and tidally averaged circulation characteristics of Suisun bay, California", Water Resources Research, Vol. 23, No. 1, 143-155.
Zienkiewicz O. C. (1977), "The finite element method in engineering science", McGraw Hill, London, U.K.

THE LOW-FREQUENCY OSCILLATIONS OF A THIN, ELASTIC INCLUSION IN A NONHOMOGENEOUS COASTAL FLOW.

V. M. KHARIK
Spencer Laboratory
Department of Mechanical Engineering
University of Delaware
Newark, DE 19716

A generation of the small-amplitude wave motion in a nonhomogeneous coastal flow is considered in a time-dependent setting. In the present approach the essential effects of the stratification and rotation of a fluid are included. Such problems arise in a study of the radiation and propagation of the gravity-inertial waves in subsurface hydrodynamics, as well as in the offshore structure engineering.

The complexity of a conventional boundary integral method is moderated via the symmetry properties of the hydrodynamic potentials. Some recent developments in the time-dependent potential theory, which are stimulated by the growing applicability of the boundary element methods (Brebbia, 1981) are analyzed. Such analysis provides the explicit solution to a problem of small-amplitude oscillations of an elastic inclusion with singular points in the flow. A variety of the pressure distribution on the surface of the considered thin inclusion is examined.

A comprehensive analysis of the amplitudes of the generated waves is presented for the large time intervals. The results are shown to be in a good correspondence with those obtained by a different method for a similar problem (Gabov, Sveshnikov, 1990). Similarities with other problems (Kharik, 1989)[3] are also discussed.

INTRODUCTION.

A modelling of the dynamics of an offshore structure can be quite complicated, especially when the essential effects of the stratification and rotation of a fluid are included. The singular points in an offshore structure geometry usually increase the complexity of such problems. These singularities may also lead to the instability of the numerical scheme used. Since the domain of a problem under consideration in most geophysical applications is often infinite, such numerical methods as finite differences or finite elements are not easily employed, if they are used at all. The unboundedness of a domain not only reduces the number of choices of numerical techniques, but also requires a selection of an appropriate analytical method for a problem of wave generation in such domains (Kharik, Kozhevnikov, 1991).

A linear mathematical model involving an unbounded domain is more manage-

A. Peters et al. (eds.), Computational Methods in Water Resources X, 1307–1314.
© 1994 Kluwer Academic Publishers. Printed in the Netherlands.

able if one utilizes the boundary element methods (Brebbia, 1981), which are closely connected with the potential theory methods (Hsiao, Wendland). A variety of techniques in the potential theory involve potential-type integral operators, the kernels of which inherit the complexity of a problem under consideration.

To avoid some of the aforementioned difficulties we discuss a number of applications of a boundary integral method (Kharik, 1992), which is based on the reciprocal symmetry properties of hydrodynamic potentials (Kharik,1993)[7] with respect to the differential operators involved in a generalized Newmann boundary condition. The symmetry properties of the potential-type integral operators were investigated first for the Gabov potentials (Kharik, 1989[3]; Kharik, Pletner, 1990). Later on, a similar analysis was carried out for a generalization of those potentials (Kharik, 1993)[7] and used to solve a number of problems of the transient fluid dynamics. Such results are a consequence of the development of an approach emphasizing the symmetry analysis in the time-dependent potential theory (Kharik, 1989)[3], as well as in the structure of some Sobolev-type equations (Kharik, 1993)[9,10].

STATEMENT OF THE PROBLEM.

We consider small-amplitude, time-dependent oscillations of a thin, elastic inclusion Γ flowing with a rotating stratified non-Boussinesq coastal flow in a Cartesian system (x_1, x_2, x_3) moving with the inclusion. The dissipative effects are assumed to be negligible. We also require that the incompressibility condition: $div\vec{V} = 0$, is satisfied by the velocity vector $\vec{V}$ of a fluid particle, $\vec{V} = (v_1, v_2, v_3)$. The angular velocity of the flow rotation equals to $f/2$, where f is the inertial frequency. The rotation of the fluid can be considered as one around the axis Ox_3, as the linear size of an unbounded coastal flow is much smaller than the Earth radius. It is well known, that the fluid stratification occurring mainly along axis Ox_3 due to gravity is analogous to the density distribution along axis Ox_1 (or Ox_2) caused by the rotation of a fluid (Kharik, 1993)[7]. An axially symmetric inertial stratification and the gravitational stratification along the axis of rotation creates a symmetry in the structure of the coastal flow dynamic system.

The aforementioned symmetry can be exploited to derive a symmetric mathematical model for small wave motions in the considered flow (Kharik, 1993)[7] :

$$[\frac{\partial^2}{\partial t^2} + N^2(x_3)](\rho_0 v_3)_{x_1} - [\frac{\partial^2}{\partial t^2} + f^2](\rho_0 v_1)_{x_3} - f p_{x_2 x_3} = 0 \tag{1}$$

where ρ_0 is the fluid density in an unperturbed stationary state, p is the dynamic pressure, and $N(x_3)$ is the Brunt-Vaisala frequency ($N^2 = -g\rho_0'/\rho_0(x_3), g$ is the acceleration due to gravity). Note that term $f p_{x_2 x_3}$, as well as $\vec{V}$, depends upon x_1 and x_2. However, that term is zero in the case of small-amplitude, plane-parallel oscillations of a thin, elastic inclusion Γ which are independent from x_2, where $\Gamma := \{(x_1, x_3) : x_1 = x_1(s), x_3 = x_3(s), s \in [0,l]\}$. Considering an exponential stratification, as far as the buoyancy effects are concerned, we have $N^2 = 2g\beta$, where β is a parameter of stratification. Introducing the following weighted stream function $U(x_1, x_3; t)$: $\rho_0(x_3)v_1 = U_{x_3}$, $\rho_0(x_3)v_3 = -U_{x_1}$; we derive a symmetric form of Sobolev-type equation

$$[\frac{\partial^2}{\partial t^2} + N^2]\frac{\partial^2 U}{\partial x_1^2} + [\frac{\partial^2}{\partial t^2} + f^2]\frac{\partial^2 U}{\partial x_3^2} = 0. \tag{2}$$

describing small-amplitude, plane-parallel wave motions in a rotating, exponentially stratified, non-Boussinesq coastal flow. This symmetric equation has important similarities with some other partial differential equations of mathematical physics (Kharik, 1993)[7]. A numerical integration of the Eq. (2) in the region of R^2 exterior to Γ can be unstable due to the presence of two singular end points of the curve Γ.

Let us orient the inclusion Γ by setting its sides Γ^+ and Γ^- in the following manner. We denote the tangent and normal vectors at a point $x(s) = (x_1(s), x_3(s))$ of the curve Γ by $\vec{\tau}_s$ and $\vec{n}_s = (n_1, n_3)$, respectively. If one will rotate $\vec{\tau}_s$ for $\pi/2$ counterclockwise, one will obtain $\vec{n}_s$. The side of the inclusion Γ corresponding to the positive direction of $\vec{n}_s$ will be denoted by Γ^+, and the opposite side by Γ^-. The smoothness of Γ is defined by the condition: $\Gamma \in A^{(2,\lambda)}$.

We assume that there were no oscillations in the rotating, stratified coastal flow before time $t = 0$. Thus we have the initial conditions of the form:

$$U(x,0) = U_t(x,0) = 0, \quad U(x,t) := U(x_1, x_3, t). \tag{3}$$

After time $t = 0$, the pressure distributions on two sides $\Gamma^\pm$ of the curve Γ are prescribed. They are different, in general. Mathematically, it is equivalent to the prescription of the generalized Neumann boundary conditions on $\Gamma^\pm$ (Gabov, Sveshnikov, 1990). We introduce a symmetric form of such conditions:

$$N_{tx}U(x,t) := (\frac{\partial^2}{\partial t^2} + N^2)\frac{\partial U}{\partial n_1} + (\frac{\partial^2}{\partial t^2} + f^2)\frac{\partial U}{\partial n_3} = \varphi_\pm(s,t),$$

for $x = x(s) \in \Gamma^\pm$, where $\partial/\partial n_i := cos(\vec{n}, x_i)\partial/\partial x_i, i = 1,3$; and functions $\varphi_\pm(s,t)$ satisfy the following compatibility condition:

$$\int_\Gamma (\varphi_+(s,t) - \varphi_-(s,t))ds = 0, \quad for \ \ t > 0.$$

Since we consider small-amplitude time-dependent oscillations of the inclusion Γ in a nonhomogeneous non-Boussinesq flow, which is unbounded, we need to set the conditions of regularity at infinity for function $U(x,t)$:

$$|\partial_t^k U| < A_k(t)/|x|, \quad |\partial_t^k \partial_{x_i} U| < B_k(t)/|x|^2, \tag{4}$$

where $\partial_t^k := \partial^k/\partial t^k, k = 0, 1, 2; \partial_{x_i} := \partial/\partial x_i; |x| = (x_1^2 + x_3^2)^{1/2}; A_k(t)$ and $B_k(t) \in C^{(0)}[0, \infty)$. The presence of singular points in the unbounded region $R^2 \setminus \Gamma$ may cause singularities of the function $U(x,t)$ or its gradient in the neighborhoods of the end points of curve Γ. An analysis of the possible character of such singularities suggests (Gabov, Sveshnikov, 1990) that the function $U(x,t)$ and its derivative U_t are bounded. Other derivatives of that function have the following behavior:

$$|\nabla U|, \ |\nabla U_{tt}| \sim O(r_{1,2}^{-1/2}), \tag{5}$$

where $r_{1,2}$ are the distances to the ends of the curve Γ.

THE CLASSICAL SOLUTION AND ITS PROPERTIES.

We shall look for a classical solution to the formulated above problem, which will be referred to as the Problem Σ, in the form of a linear combination of two hydrodynamic potentials. The integral operators of those potentials have properties analogous to those of the single- and double-layer potentials, and the form given by the following expressions (Kharik, 1993)[7] :

$$V[\nu](x,t) = \int_0^t \int_\Gamma (\ln|x - y(s)|\nu_\tau(s,t;\tau) + \frac{\nu(s,t-\tau;t)}{\tau}[1 - \cos\frac{\tau|x-y(s)|_*}{|x-y(s)|}])dsd\tau,$$

$$W[\mu](x,t) = \int_0^t \int_\Gamma \{K(x,s)\mu_\tau(s,t;\tau) - \mu(s,t-\tau;t)F[K(x,s);\tau]\}dsd\tau,$$

where the potential densities $\nu(x,t;\tau)$ and $\mu(x,t;\tau)$ depending on a positive parameter τ are to be determined, $y(s) = (y_1(s), y_3(s)) \in \Gamma$,

$$F(\xi,\tau) = \int_0^\xi \sin\{\tau\bar{\theta}^{1/2}\}/\bar{\theta}^{1/2}d\theta, \ \ \bar{\theta}(\theta) = N^2\sin^2\theta + f^2\cos^2\theta;$$

and K(x,s) is the kernel of the angle potential defined by the expressions (Gabov, Sveshnikov, 1990):

$$\cos K(x,s) = (x_1 - y_1(s))/|x - y(s)|, \ \ \sin K(x,s) = (x_3 - y_3(s))/|x - y(s)|.$$

Notice that the Gabov potentials (Kharik, 1989)[3] may be derived from the aforementioned integral operators for a special type of density functions. Later on, to assure the single-valuedness of $W[\bar{\mu}](x;t)$, we shall require:

$$(1,\mu)_{L_2(\Gamma)} = 0, \ \ \bar{\mu}(s,t;\tau) = \int_0^\tau (t-\tau)\mu(s,\tau)d\tau, \tag{6}$$

where $\tau \in [0,t]$ and $\mu(s,t) = \mu(s,t;t)$.

Using Holder spaces, we define the following set of functions depending on parameter s, where $s \in [0,l]$:

$$C_r^{(0,h)}(\Gamma) := \{f(s) : r(s)f(s) \in C^{(0,h)}(\Gamma)\}$$

where $1 > h > 0$ and $r = r(s) := |x(s) - x(0)|^{1/2}|x(s) - x(l)|^{1/2}$. Applying the results of (Kharik, 1989)[3], one may prove the following lemma:

Lemma 1. If $\nu(s,t), \mu(s,t) \in C^{(2)}[0, \infty; C_r^{(0,h)}(\Gamma)]$, then function

$$U(x,t) = V[\bar{\nu}](x,t) + W[\bar{\mu}](x,t)$$

satisfies Eq. (2), the initial conditions (3), the conditions of regularity (4), and conditions (5) in the neighborhoods of the end points of the inclusion Γ.

The definition of function $U(x,t)$ and the boundary conditions can be employed to derive a system of singular integro-differential equations for the density functions $\nu(s,t)$ and $\mu(s,t)$ (Kharik, 1989)[3]. To simplify the complex structure of that system, we utilize the results of the Lemma 1 and Kharik (1993)[7], and prove that the potentials V and W possess the reciprocal symmetry properties with respect to two differential operators ∂_{τ_s} and N_{tx}:

Lemma 2. If $\bar{\mu}(s,t) \in C^{(0)}[0, \infty; C^{(0,h)}(\Gamma)), \nu(s,t) \in C^{(2)}[0, \infty; C_r^{(0,h)}(\Gamma))$, functions $\mu(s,t)$ and $\nu(s,t)$ satisfy (16) , then:

$$\lim_{x \to x(s) \in \Gamma^\pm} \frac{\partial}{\partial \tau_s} V[\bar{\mu}](x,t) = \lim_{x \to x(s) \in \Gamma^\pm} N_{tx} W[\bar{\mu}](x;t),$$

$$\lim_{x \to x(s) \in \Gamma^\pm} \frac{\partial}{\partial \tau_s} W_\pm[\bar{\nu}_d](s,t) = - \lim_{x \to x(s) \in \Gamma^\pm} N_{tx} V[\bar{\nu}](x;t)$$

where $s \in (0,l), \bar{\nu}_d(s,t) = (\partial_t^2 + N^2)(\partial_t^2 + f^2)\bar{\nu}(s,t)$ and $\bar{\bar{\mu}}$ (or $\bar{\nu}$) is defined by

$$\bar{\bar{\mu}}(s,t;\tau) = \int_0^\tau (t - \tau)\bar{\mu}(s,\tau)d\tau.$$

Note, it is known (Kharik, Pletner, 1990) that the Gabov potentials possess similar properties of less symmetric form for a smaller set of the density functions. Using Lemma 1 and Lemma 2, as well as the results of Kharik (1989)[3], one may prove the following statement:

Theorem. If $\nu(s,t), \mu(s,t)$ belong to $C^{(2)}[0, T; C_r^{(0,h)}(\Gamma)]$, satisfy condition (8) and $\varphi_\pm(s,t)$ belong to $C^{(2)}[0, \infty; C^{(0,h)}(\Gamma)]$ and satisfy the compatibility conditions, then the function $U(x,t)$ is the unique explicit solution to the Problem Σ for $\Gamma_0 := \{(x_1, x_3) : x_1 = s \cos \varphi, x_3 = s \sin \varphi; s \in [-1,1], \varphi \in (0, \pi/2)\}$, and

$$\nu(s,t) = \frac{1}{2\pi}(I - NJ_{Nt}*)(I - fJ_{ft}*)(\varphi_+(s,t) - \varphi_-(s,t)),$$

$$\mu(s,t) = \frac{(2\pi)^{-1}}{(1 - s^2)^{1/2}} \int_{-1}^1 \frac{(1 - \xi^2)^{1/2}}{\xi - s}[\varphi_+(s,t) + \varphi_-(s,t)]d\xi,$$

where the convolution-type operator $J_{\alpha t}$ is defined by $J_{\alpha t} * m(t) := \int_0^t J_1(\alpha(t - \tau))m(\tau)d\tau, h > 0$ and ξ is a dummy integration variable. In the case of a curve Γ, solution $U(x,t)$ exists with the density $\mu(s,t)$ defined implicitly.

Notice, a numerical analysis of the derived solution $U(x,t)$ is not nearly as troublesome as that of the Eq. (2) in the unbounded region with singularities. To the point, to obtain good accuracy of the calculations in a finite region with singular points (Kharik, Dubov, 1983), one has to find an optimal configuration of appropriate finite elements.

ANALYSIS OF THE RESULTS.

The Laplace transformation can be applied to analyze the asymptotic behavior (as $t \to \infty$) of the wave motion amplitudes caused by the plane-parallel oscillations of the inclusion Γ_0 in a non-Boussinesq flow. In this section we shall examined a set of the pressure distributions having a property: functions $\varphi_\pm(s,t)$ are not zero for $t < T, T > 0$, and zero for $t > T$. Following a technique discussed by Kharik (1989)[3] for a Boussinesq fluid, one may show that the amplitude of the generated gravity-inertial waves, which are described by the two components of the function $U(x,t)$, satisfies the following estimate:

$$\|\hat{T}^n U(x,t_0)\|_{C^{(0)}(R^2)} < C/n^{1/2}, \quad as$$

$$\|\hat{T}^n V[\bar{\nu}](x,t_0)\|_{C^{(0)}(R^2)} < C/n^{1/2}, \quad \|\hat{T}^n W[\bar{\mu}](x,t_0)\|_{C^{(0)}(R^2)} < C/n^{1/2},$$

for large enough n, where $\hat{T}U(x,t_0) := U(x, t_0 + T)$, $t_0, T > 0$. As a result, we have derived the estimates for all given pressures on the surface of the inclusion Γ_0, such that $\varphi_\pm(s,t) \in C^{(2)}[0, T; C^{(0,h)}]$ and $\varphi_\pm(s,t) = 0$ for $t > T$. A similar problem with the Dirichlet conditions is shown to have a solution satisfying analogous estimates (Gabov, Sveshnikov, 1990).

Now we also examine time-harmonic oscillations of a linear inclusion Γ, e.g. the pressure distributions are given by functions $\varphi_\pm(s,t) = exp(-i\omega_\pm t)g_\pm(s)$, where $g_\pm(s) \in C^{(0,h)}[-1,1]$ and $\omega_\pm < f$ (or $\omega_\pm > N$, $N > f$). The symmetry of the field $R^2 \setminus \Gamma_0$ geometry and the linearity of the problem under consideration allow us to present function $U(x,t)$ as the sum $U_+(x,t) + U_-(x,t)$, each satisfying the boundary conditions with $\varphi_+(s,t)$ and $\varphi_-(s,t)$ respectively. Using the method of stationary phase (Fedoryuk, 1987), we show that the aforementioned oscillations of the inclusion Γ_0 cause the following, so called, limiting amplitudes $\bar{U}_\pm(x,t)$:

$$\lim_{n \to \infty} \hat{T}^n U_\pm(x,t_0)exp(i\omega_\pm t_0) = \bar{U}_\pm(x), \quad where \tag{7}$$

$$\bar{U}_\pm(x) = A \int_{-1}^{1} \{g_\pm(s) \ln[(x_1 - s \cos\varphi)^2 + (x_3 - s \sin\varphi)^2/(1 - k_l^2)] +$$

$$+\mu_\pm(s)(1 - k_h^2)^{1/2}(N^2 - \omega_\pm^2)^{-1}\mathrm{Arctan}[\tan K(x,s) \times (1 - k_l^2)^{-1/2}]\}ds,$$

A=const, and the coefficients k_l and k_h are defined by

$$k_l^2 := \frac{N^2 - f^2}{N^2 - \omega_\pm^2}, \quad k_h^2 := \frac{N^2 - f^2}{\omega_\pm^2 - f^2}, \quad k_h^2 = (1 - \frac{1}{k_l^2})^{-1}.$$

Notice that in practical computations, number n is finite, so the equation (7) will be satisfied only approximately. The direct calculations show that the limiting amplitudes $\bar{U}_\pm(x)$ satisfy an elliptic equation:

$$(1 - k_h^2)\frac{\partial^2 \bar{U}_\pm(x)}{\partial x_1^2} + \frac{\partial^2 \bar{U}_\pm(x)}{\partial x_3^2} = 0.$$

Note that $k_h \to 1$, as the frequency of oscillations $\omega_\pm$ approaches the Brunt-Vaisala frequency N from above, and $k_l \to 1$, as $\omega_\pm \to f$ from below. In the case when $\omega_\pm \in (f, N)$, the above equation changes its type and becomes a hyperbolic one, as the coefficients $k_{h,l}$ pass their critical values. Analogous transition can be observed in a transonic flow. The usefulness of the aforementioned coefficients can be also demonstrated by using calculations of Appleby and Crighton (1987) for a stationary case, when $f = 0$.

It is necessary to mention that the Problem Σ may serve as a mathematical model not only for the wave motion in a coastal flow in the Gulf of Mexico, say, but also for such motions in a non-coastal flow rotating simply due to the Earth rotation (Greenspan, 1968; Howard, Siegmann, 1969). Moreover, a problem of a generation of, so called, gravity-gyroscopic waves in a rotating, stratified fluid can be formulated similarly to the Problem Σ (Kharik, 1989)[3]. This model can be also partially used to analyze some other wave phenomena (Kharik, 1989)[16]. One may find other details on the applications of a variety of the hydrodynamic potentials in the work of Ladyzhenskaya (1968).

Acknowledgement. This work was begun, when the author was supported at Moscow State University by National Aerodynamic Research Center under contract #40/90. The author is indebted to Professor L. Schwartz and Unidel Professor Emeritus I. Greenfield of University of Delaware for their generous moral support, as well as to I-Harima Heavy Industries Co. of Japan for the financial assistance.

REFERENCES.

1. Brebbia, C. (1981) "Boundary Element Methods", Springer-Verlag, New York.
2. Gabov, S. and Sveshnikov, A. (1990) "Linear Problems of the Transient Internal Waves Theory", Nauka, Moscow (in Russian).
3. Kharik, V. (1989) "Some Mathematical Models in Hydrodynamics", Washington and Lee University, Lexington.
4. Kharik, V. and Kozhevnikov, V. (1991) "Modelling of Nonlinear Problems of a Compressible Shear Flow Over an Obstacle", Report #04.01.91, Department of Mathematical Physics, Moscow State University, Moscow, Russia.
5. Hsiao, G. and Wendland, W. (1994) "Variational Methods for Boundary Integral Equations and Mathematical Foundation of Boundary Element Methods", Springer-Verlag (to appear).
6. Kharik, V. (1992) "A Boundary Integral Method Based on the Symmetry Properties of the Hydrodynamic Potentials: Sobolev-Type Equations", Abstrs. of the

Midwest and Southeastern Atlantic Second Joint Regional Conference on Differential Equations, November 13-15, Lexington, USA, p. 12.

7. Kharik, V. (1993) "Generation of Internal Waves in a Rotating, Stratified Fluid: Some Symmetry Properties", J. Math. Physics 34 (1), p. 206.

8. Kharik, V. and Pletner, U. (1990) "The Problem of Gravity-Gyroscopic Waves, Which are Excited by the Oscillations of a Curve", J. Math. Physics 31 (5), p. 1280.

9. Kharik, V. (1993) "The Homogenized Spaces in the Theory of Rotating Stratified Flows (Part I: A Generalized Neumann Problem)", Abstrs. of SIAM Conference on Mathematical and Computational Issues in the Geosciences, April 19-21, Houston, USA, p. A23.

10. Kharik, V. (1993) "The Homogenized Spaces in the Theory of Rotating Stratified Flows (Part II: A Non-Boussinesq Flow)", Abstrs. of SIAM 1993 Annual Meeting, July 12-14, Philadelphia, USA, p. A56.

11. Kharik, V. and Dubov, A. (1983) "A Geophysical Application of the Optimal Triangulation of a Polygon With a Finite Number of Corners", Report #07.02.83, Engineering Geodesy Division, Polytechnic Institute, Chernivtsi, Ukraine.

12. Fedoryuk, M. (1987) "Asymptotics, Integrals and Series", Nauka, Moscow.

13. Appleby, J. and Crighton, D. (1987) "Internal Gravity Waves Generated by Oscillations of a Sphere", J. Fluid Mechanics 183, p. 439.

14. Greenspan, H. (1968) "The Theory of Rotating Fluids", Cambridge University Press, Cambridge.

15. Howard, L. and Siegmann, W. (1969) "On the Initial Value Problem for Rotating Stratified Flow", Stud. Appl. Math. 48, p. 153.

16. Kharik, V. (1989) "Physics and Mathematics Around Us", Journal of Science (Washington and Lee University) 5, p. 7.

17. Ladyzhenskaya, O. (1968) "The Mathematical Theory of Viscous Incompressible Flow", Gordon and Breach, New York.

A MAPPING PROCEDURE FOR CHESAPEAKE BAY DATA

NEERCHAL K. NAGARAJ and RONALD W. SHAFER
Department of Mathematics and Statistics
University of Maryland Baltimore County
Baltimore, MD 21228 USA
and
United States Environmental Protection Agency,
401 M Street
Washington, DC 20460 USA

United States Environmental Protection Agency and the offices of the States surrounding the Chesapeake Bay are conducting an assessment of the Chesapeake Bay to determine the effectiveness of the forty percent nutrient reduction strategy for supporting benthic community restoration goals. One important need of this effort is to develop methods for graphically representing the various information, including developing a map of benthic conditions for the Chesapeake Bay. Maps will provide a way to examine and correlate the Benthic Index, dissolved oxygen, and contaminant data from the Bay. In this article, a method of mapping based on multiple regression techniques and kriging is presented. Because of the noisy nature of the data, some robust and resistant techniques are incorporated into the mapping procedure.

INTRODUCTION

For the past two decades, environmental decision makers were concerned mainly with identifying the appropriate abatement strategies needed by industry and the government to meet national ambient standards for clean water and air. Their decision making process was principally (and still is) focused on what are the best available control technologies. These determinations are the basis for the control limits set in natonal environmental regulations. Consequently, the process to develop regulations for national ambient standards is primarily an engineering and legal decision and does not require comprehensive status and trends information about environmental, ecological, or natural resources.

However, the luxury of needing only limited environmental information to implement national and regional problems is changing. Environmental issues are becoming

A. Peters et al. (eds.), Computational Methods in Water Resources X, 1315–1322.
© 1994 *Kluwer Academic Publishers. Printed in the Netherlands.*

more complex, but fortunately our understanding of environmental problems and their interrelationships is increasing. Presently the United States Environmental Protection Agency (USEPA) is addressing a broader set of environmental issues than national ambient standards, which will require more environmental and ecological data, different types of data, and additional analytical methods beyond what is being used today.

USEPA's top priorities are: to implement a pollution prevention program; to ensure ecological resources are managed properly; to reduce public exposure to toxic chemicals; and to address the potential global impacts associated with environmental stressors. These issues demand a multi-media, holistic analysis of the problems. To do this, the problem must be looked at in the context of its location and therefore a geographic approach is required. Temporal and spatial data that accurately represents an area is needed to identify stressors, to link those stressors to environmental impacts, and to evaluate government and public responses to the problems. Accurate and reliable maps, displaying environmental information, will become an important tool in environmental decision making.

BENTHIC INDEX DEVELOPMENT

USEPA's initiative for the Chesapeake Bay, in partnership with the states bordering the Bay, was established in 1984. The goal of the initiative is the restoration (abundance, health, and diversity) of living resources to the Chesapeake Bay by reducing nutrient loadings, reducing toxic chemical impacts, and enhancing habitats. USEPA's Chesapeake Bay Program Office is responsible for implementing this initiative and has established an extensive monitoring program that includes the traditional water chemistry sampling. Also, importantly the monitoring program collects data on living resources to measure their progress towards meeting Chesapeake Bay restoration goals.

Sampling of benthic invertebrate assemblages have been an integral part of the Chesapeake Bay monitoring program due to their ecological importance and their value as biological indicators. The condition of benthic assemblages is a measure of the ecological health of the Chesapeake Bay, including the effects of multiple types of environmental stresses. Nevertheless, regional-scale assessment of ecological status and trends using benthic assemblages are limited by the fact that benthic assemblages are strongly influenced by naturally variable habitat elements, such as salinity, sediment type, and depth. Also, different state agencies and USEPA programs use different sampling methodologies, limiting the ability to integrate data into an unified assessment. To circumvent these limitations , USEPA has standardized benthic data from several different monitoring programs into a single database, and from that database developed a Restoration Goals Benthic Index that identifies whether benthic restoration goals are being met.

Index development was conducted independently for eight habitat classes defined by salinity and sediment type. This was done to ensure that natural differences in the benthic communities related to these habitat factors did not confound interpretation of the Index. Restoration Goals include metrics from each of the following five categories: benthic biodiversity, abundance and biomass, life history strategy, activity beneath surface, and feeding guild.

The validation results indicate that the preliminary Restoration Goals Index is effective in distinguishing between sites of high quality and those of lower quality in six of the seven habitats for which data were available for goal development. The only habitat class for which the restoration goals index did not validate well was the tidal freshwater. No samples were taken for low mesohaline sand, due to lack of reference sites in this class.

Assessments using benthic monitoring data have been extremely useful for characterizing changes in environmental conditions at individual sites over time, or for relating the condition of sites to pollution loadings and sources. The full potential of these assessments, however, for addressing larger management questions, such as what is the overall condition of the Bay, or how does the condition of the eastern shore compare to that of the western shore tributaries, has not been realized. The development of spatial statistical methods is necessary for geographical mapping of benthic invertebrate condition in the Chesapeake Bay.

THE PROCEDURE

We developed a mapping procedure that interpolates Benthic Index values over the mainstem of the Chesapeake Bay. This procedure will be used to address the assessment question of how much acreage in the Chesapeake Bay has degraded biotic condition. Data from the State of Maryland's long term benthic monitoring was program used to developed this procedure.

The long term benthic monitoring data base contains Benthic Index values for the various locations of the Bay shown in Figure 4b. For a detailed description of the sampling design and the methods used in long term benthic monitoring, the reader is referred to Ranasinghe et al.(1992). Here, we consider the data from the mainstem of the Bay only. The location of 231 Benthic Index values used in this article is shown in Figure 4b. The objective is to obtain a colored map of the Bay, where color will show the benthic health at any given location. (Only grey-scale versions are presented in this article. Colored versions are available for on-screen display.) The colored maps will facilitate correlational studies between Benthic Index and the other stressors. Furthermore, as described in the previous sections, they will also provide effective management tools.

Since, data is available only at the sampled locations, we need to interpolate the values for every other grid location prior to mapping. The method of interpolation used here is a variant of universal kriging. Conceptually, universal kriging decomposes the data as:

$$\text{Data} \; = \; \text{Large Scale Variation} \; + \; \text{Small Scale Variation}$$

Large scale variation can be explained by spatial trend that depend on predictor variables. Spatial correlation among neighbors and measurement error constitute the smaller scale variation. Large scale variation is usually captured by fitting a regression model for the data using predictor variables. Spatial statistical tools, such as semivariograms and kriging are used to deal with the smaller scale variations. This combination is known as universal kriging. In the case of Benthic Index we use depth as the primary predictor variable in addition to latitude and longitude. Figure 1 and 2 illustrate the relationship between Benthic Index and depth. Figure 1 is a straightforward plot of Benthic Index against the bottom depth at the location. The most significant feature of the data apparent from Figure 1 is the large amount of noise. Figure 2 plots the predicted probability of bad benthos at a given location against the bottom depth. A technique called logistic regression is used in obtaining these predicted probabilities. It is clear that deep locations usually correspond to bad benthos. These Figures indicate that bottom depth is a potentially useful predictor of Benthic Index. Furthermore, the bottom depth data is available for a 1km x 1km grid throughout the mainstem of the Bay.

Thus, latitude, longitude, and the bottom depth form our set of predictors for the Benthic Index. Let $Z(s)$, $Lt(s)$, $Ln(s)$ and $D(s)$ denote the Benthic Index value, latitude, longitude, and bottom depth, at any arbitrary spatial location s, respectively. The spatial location s is given by the latitude and longitude of the location. Let $s_1, s_2, \ldots, s_n$ be the sampled locations, where Benthic Index values are available. Then the decomposition of the data mentioned above can be written formally as

$$\mathbf{Z} = \mathbf{X}\beta + \epsilon,$$

where

$$\mathbf{Z} = \begin{pmatrix} Z(s_1) \\ Z(s_2) \\ \cdot \\ \cdot \\ Z(s_n) \end{pmatrix}, \quad \mathbf{X} = \begin{pmatrix} 1 & Lt(s_1) & Ln(s_1) & D(s_1) \\ 1 & Lt(s_2) & Ln(s_2) & D(s_2) \\ \cdot & \cdot & \cdot & \cdot \\ \cdot & \cdot & \cdot & \cdot \\ 1 & Lt(s_n) & Ln(s_n) & D(s_n) \end{pmatrix},$$

$\beta = (\beta_0, \beta_1, \beta_2, \beta_3)'$, and $\epsilon = (\epsilon(s_1), \ldots, \epsilon(s_n))'$ is a n-dimensional random vector of errors with a mean 0 and a variance-covariance matrix Σ. Elements of the matrix Σ model the correlation among the spatial neighbors. Thus, if one assumes that there is no spatial correlation among neighbors, Σ will be a diagonal matrix. In the spatial

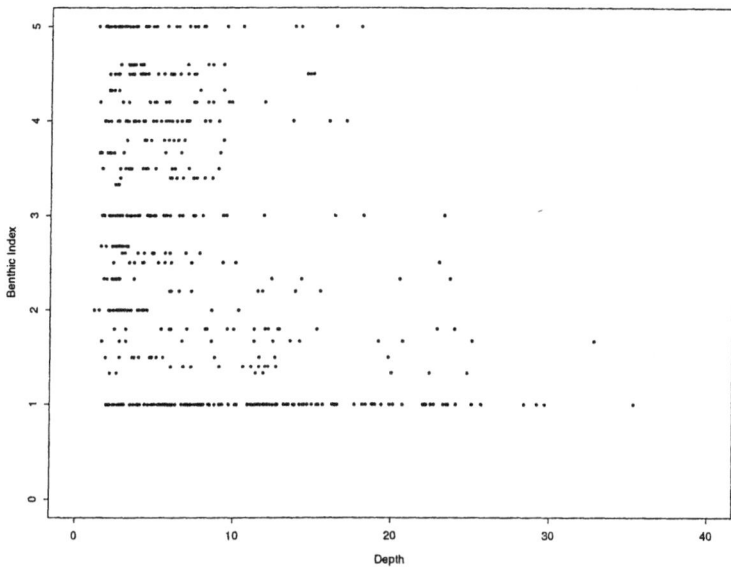

Figure 1: Relationship of Benthic Index and Bottom Depth

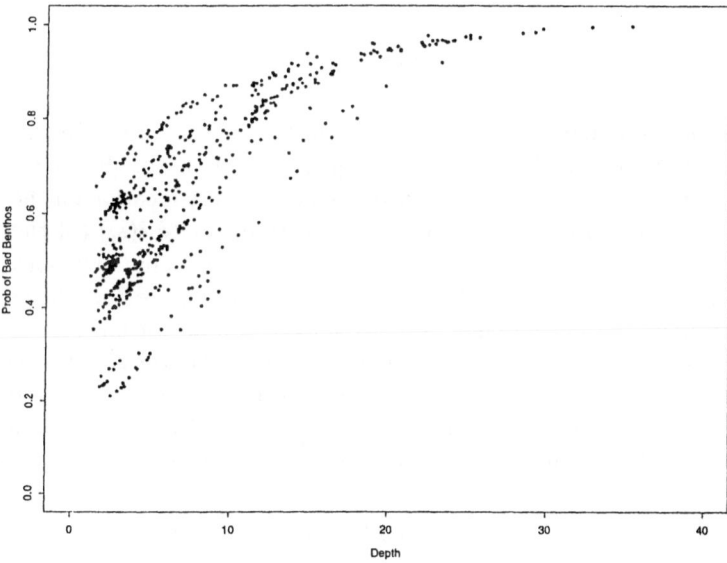

Figure 2: Relationship of Probablity of Bad Benthos and Bottom Depth

statistics literature, it is more common to describe spatial correlation in terms of a semivariogram. For a stationary error process $\epsilon(s)$, semivariogram is defined as

$$\gamma(h) = \frac{1}{2}\mathrm{Var}[\epsilon(s+h) - \epsilon(s)].$$

For a stationary error process

$$\gamma(h) = \mathrm{Var}[\epsilon(s)] - \mathrm{Cov}[\epsilon(s+h), \epsilon(s)].$$

Thus, for any stationary process, estimating the correlation among the neighbors is equivalent to the estimation of the semivariograms. The mapping procedure outlined in this article has, broadly, three steps. For the purposes of the description, suppose that the sampled locations are on a regular grid as in Figures 3a - 3d.

Figure 3: Mapping Procedure

In the first step, the parameters of the regression model are estimated. In this step the spatial correlation among the neighbors is ignored, and the predictions which capture the large scale variation of the data is obtained for every location on the Bay. At the sampled locations (Figure 3a), one arrives at an estimate $\hat{\epsilon}(s_i)$ (of the true error $\epsilon(s_i)$) by comparing the predicted value with the observed value. The locations where these error estimates are available are denoted by $\otimes$ in Figure 3b. The second step consists of kriging the estimated error $\hat{\epsilon}(s_i)$'s to obtain an estimated error for every grid point in the map. Prior to kriging, one has to estimate the semivariogram, and fit a model to it in terms of the distance. Considering the noisy nature of the data (see Figures 1 and 2) we shall use the robust estimate of the semivariogram as described in Cressie (1991). The robust estimator of the semivariogram is given by

$$2\hat{\gamma}(h) = \left\{ \frac{1}{|N(h)|} \sum | \hat{\epsilon}(s_i) - \hat{\epsilon}(s_j) |^{\frac{1}{2}} \right\}^4 / (0.457 + 0.494/ | N(h) |),$$

where $| N(h) |=$ the number of pairs in $\{(s_i, s_j) : \mathrm{dist}(s_i, s_j) = h, i, j = 1, \cdots, n\}$. In our application, a linear model for the semivariogram gave an adequate fit. Thus, for any pair of grid locations s and t in the map, an estimate $\gamma(s-t)$ of the semivariogram can be obtained. The kriged value of the error at an arbitrary location on the map is given by

$$\tilde{\epsilon}(s) = \sum_{i=1}^{n} \lambda(s_i)\hat{\epsilon}(s_i),$$

where $\hat{\Gamma} = ((\hat{\gamma}(s_i - s_j)))$ is an $n \times n$ matrix, $\hat{\gamma} = (\hat{\gamma}(s - s_1), \ldots, \hat{\gamma}(s - s_n))'$ is an n dimensional column vector, and $(\lambda(s_i))$'s are components of

$$\lambda = (\lambda_1, \cdots, \lambda_n)' = \hat{\Gamma}^{-1}\left(\hat{\gamma} + 1\frac{(1 - 1'\hat{\Gamma}^{-1}\hat{\gamma})}{1'\hat{\Gamma}^{-1}1}\right).$$

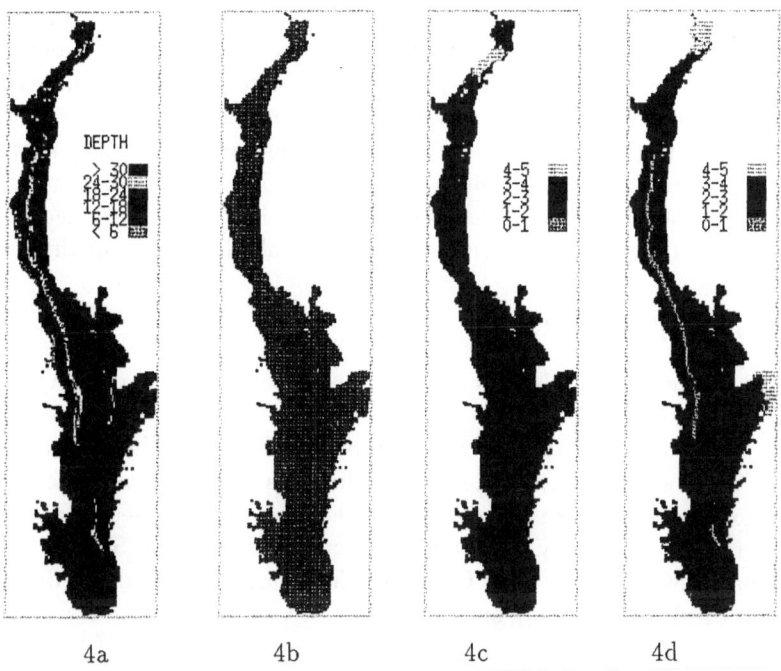

4a	4b	4c	4d

Figure 4: Benthic index maps for the Chesapeake Bay

4a: Bottom depth(in feet) contours for the mainstem of Chesapeake Bay
4b: Sampling locations for long term benthic monitoring in mainstem of the Bay
4c: Benthic index maps obtained by ordinary kriging
4d: Benthic index maps obtained by universal Kriging

Thus by kriging the residuals at the locations denoted by '+' in Figure 3c, we obtain estimated errors $\tilde{\epsilon}(s)$'s for all locations, denoted by circles in 3d. Final value to be mapped is obtained by adding the regression predictions from Figure 3b to the kriged errors at the corresponding locations in Figure 3d.

The mapping procedure was implemented for the Benthic Index data on the Chesapeake Bay and resulting map is given in Figure 4d. Figure 4b shows locations on the mainstem of the Bay, where the benthic samples were collected. Thus, actual Benthic Index value available for these locations. Figure 4c shows the map one would obtain if kriging were implemented without making use of the regression model involving the site characterizing variable. That is, the map in Figure 4c is the result of ordinary kriging. Because of the sparse data, maps based on ordinary kriging has large patches of same color. Note that in Figure 4c, many regions close to shore line are shaded same as the mid-channel. This map seem to ignore the general relationship between Benthic Index and bottom depth (See Figures 1 and 2). Figure 4a shows the depth contours of the Bay. Note that the map in Figure 4d, the one obtained by the procedure outlined in this article, follows the general pattern of the bottom depth and shows some local variation. Referring back to the our procedure, the general pattern of the map is captured in the regression model where as the local variations are modeled by the semivariogram during the kriging. Next steps will be to verify the interpolation accuracy of the mapping procedure, using data collected by a different agency. If the verification process proves encouraging, we will expand the benthic condition map to include Chesapeake Bay tributaries, and will initiate efforts to develop maps that characterize Chesapeake Bay water quality conditions.

ACKNOWLEDGEMENTS

The work of Neerchal K. Nagaraj was supported by a Cooperative Agreement between the Department of Mathematics and Statistics, University of Maryalnd Baltimore County and the Statistics and Computing Branch of United States Environmental Protection Agency. Any opinions, findings, and conclusions or recommendations expressed here are those of the authors and do not necessarily reflect the views of the above Agency. The authors would like to thank Steve Weisberg and Ana Ranasinghe of Versar Inc., Columbia, Maryland, for insightful discussions on long term benthic monitoring and for the use of Chesapeake Bay benthic data.

REFERENCES

Cressie, N. (1991) Statistics for Spatial Data, John Wiley and Sons, New York.

Ranasinghe, J. A., Scott, L. C., and Newport, R. (1992) "Long-term benthic monitoring and assessment program for the Maryland Portion of the Bay, Jul 1984-Dec 1991", Report prepared for the Maryland Department of the Environment and the Maryland Department of Natural Resources by Versar, Inc. Columbia, MD.

MESHING REQUIREMENTS FOR LARGE SCALE COASTAL OCEAN TIDAL MODELS

J. J. WESTERINK[1], R. A. LUETTICH, JR.[2] and S. C. HAGEN[1]
[1]Department of Civil Engineering and Geological Sciences,
University of Notre Dame, Notre Dame, IN 46556
[2]University of North Carolina at Chapel Hill, Institute of Marine Sciences
3431 Arendell St., Morehead City, NC 28557

INTRODUCTION

In our studies of tidal circulation on the continental shelves of the western North Atlantic ocean and Gulf of Mexico, we have found it advantageous to define a domain which encompasses a large expanse of the deep ocean in addition to the continental margin regions of interest. In fact, our Western North Atlantic Tidal (WNAT) model domain, shown in Figure 1, encompasses a significant portion of the Atlantic ocean, as well as the entire Gulf of Mexico and Caribbean Sea and covers an area of more than 8.347×10^6 km^2. The advantages of this domain are related to the geometry and location of the open ocean boundary which significantly simplify the task of boundary condition specification (Blain et al, 1994; Westerink et al., 1994).

It is clear that due to the large variability in bathymetry, coastline features and the resulting hydrodynamic response, unstructured meshes with graded grid spacing will allow the domain to be optimally discretized and will result in the most accurate solutions. Often the criterion to establish grids for tidal computations is based on the one-dimensional, linear, frictionless, constant topography wavelength to grid size ratio computed as:

$$\frac{\lambda}{\Delta x} = \frac{\sqrt{gh}}{\Delta x} T \qquad (1)$$

(where g = gravitational constant, h = water depth and T = tidal period). We have found that this criterion does not in of itself lead to unstructured graded grids which perform satisfactorily. Among the shortcomings of this criterion are its inability to identify the two dimensional structure of the waves associated with features such as amphidromes and with circulation forced by two dimensional topographic and/or coastline features. In these cases the actual wavenumber content of the response is much greater than that predicted by the one dimensional criterion and we require additional mesh resolution. Another significant shortcoming of the one dimensional criterion is that it is based on uniform waves propagating in constant depth water. Therefore it does not take into consideration the rate of change of wavelength or the associated significant gradients in response which occur as the waves propagate over steep topographic changes. Nonetheless, as is shown in Figure 1, very steep topographic changes are a wide spread feature of the WNAT model domain and occur in the vicinity of the continental shelf break and the continental slope.

A. Peters et al. (eds.), Computational Methods in Water Resources X, 1323–1330.
© 1994 Kluwer Academic Publishers. Printed in the Netherlands.

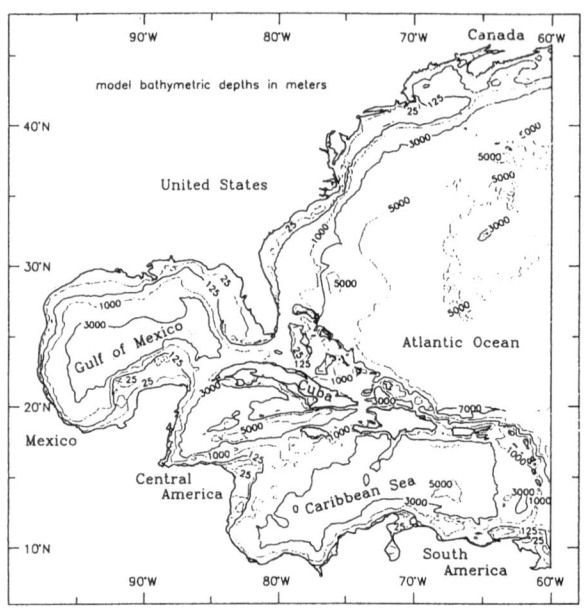

Figure 1: The WNAT model domain including bathymetry (in meters).

The focus of this paper is to examine the mesh resolution requirements in the vicinity of the continental shelf break. Westerink et al. (1992) look at this question using numerical experiments in the context of a one dimensional idealized slice of ocean and found that the $\lambda/\Delta x$ criterion had to be significantly exceeded in the vicinity of the continental shelf break and slope in order to obtain highly accurate unstructured graded grid solutions which matched uniform fine grid solutions. In this paper, we extend that study by developing a formal truncation error analysis of the governing equations as well as implementing a series of numerical experiments which examine the influence of mesh structure using the two dimensional WNAT model domain.

The results presented in this paper are based on ADCIRC-2DDI, a two dimensional depth integrated finite element model which solves the fully nonlinear shallow water equations formulated in generalized wave-continuity equation (GWCE) form. ADCIRC-2DDI applies linear finite elements to interpolate elevation, velocity and depth. A description of the numerical aspects and implementation of ADCIRC-2DDI is given by Luettich et al. (1992) and Kolar et al. (1994a,b)

TRUNCATION ERROR ANALYSIS

In order to identify regions which require a high level of localized mesh refinement, we performed a local truncation error analysis of the one dimensional linearized form of the harmonic shallow water equations in GWCE form. Utilizing Taylor series expansions in the generic discrete nodal momentum and GWCE equations leads to expressions for nodal truncation error. Developing a very fine mesh ($\Delta x = 1$ km) truth solution allows us to estimate the spatial derivatives of elevation and velocity in the expressions for the local truncation er-

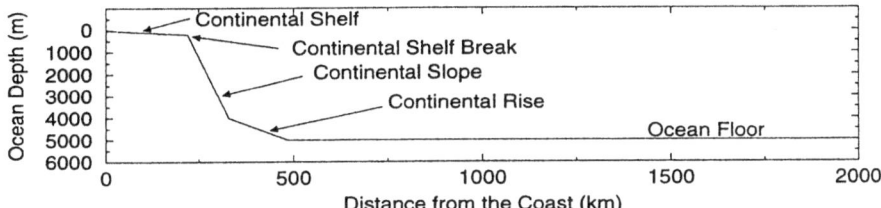

Figure 2: One dimensional idealized bathymetry for truncation error analysis.

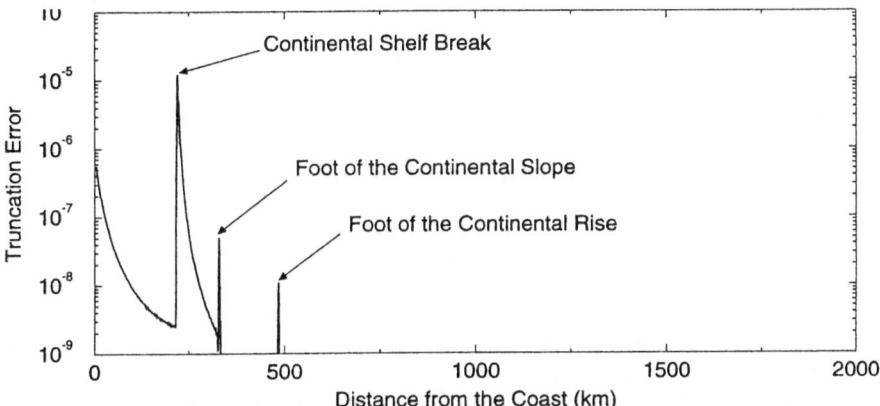

Figure 3: Local truncation error for the one dimensional momentum equation.

rors. Applying this analysis for a typical slice of our WNAT model topography, shown in Figure 2, leads to the localized truncation error in the momentum equation (with up to second order terms and an assumed uniform mesh) shown in Figure 3. We note that the predominant local truncation errors occur near the coast on the continental shelf and near the continental shelf break and slope. The local truncation errors for the GWCE are very similar to those of the momentum equation. Since a high local truncation error will adversely affect the accuracy of the entire solution, we want to reduce the local truncation error in these areas by providing a much higher level of mesh refinement very near the coast and in the vicinity of the shelf break than on most of the shelf or in the deep ocean. We note that the $\lambda/\Delta x$ criterion does not indicate the need for a locally high level of mesh refinement in the vicinity of the shelf break.

UNIFORM AND GRADED GRIDS OF THE WNAT MODEL DOMAIN

We now examine the influence of mesh resolution in the vicinity of the continental shelf break by comparing responses obtained using a uniform and two unstructured graded grids (grids SS4, T1 and T2) to a "truth" response computed using a very fine uniform grid (SSS5). The characteristics of these four grids are summarized in Table 1. For simplicity and to ensure identical boundaries, all grids approximate the WNAT model using a coarse boundary

Grid SS4 uniformly discretizes the domain with a resolution of approximately $12' \times 12'$. Bathymetry is interpolated onto the grid using the ETOPO5 data base with a minimum defined depth of between 3 m (for the Gulf of Mexico and Caribbean Sea) and 7 m (for the northeast U.S. and Canadian coasts) to avoid drying of elements during maximum ebb. Grid SSS5 is derived from grid SS4 by splitting each element into four smaller triangular elements effectively doubling the resolution to $6' \times 6'$. Bathymetry from the ETOPO5 data base was interpolated onto grid SSS5 and minimum depth values were again set. Responses from grid SSS5 are used as our "truth" solution against which other solutions are compared.

Table 1: Grid Characteristics

Grid	Nodes	Grid Structure	Grid Size		Resolution	
			(degrees/min)	(km)	$(\frac{\lambda}{\Delta x})_{max}$	$(\frac{\lambda}{\Delta x})_{min}$
SS4	24255	uniform	12′	19	704	11.3
SSS5	95999	uniform	6′	9	1322	22.6
T1	11712	graded	5′→0.8°	7→70	1184	16.0
T2	28889	graded	5′→0.8°	7→70	1013	16.0

Grid T1, shown in Figure 4a, is an unstructured graded grid with element sizes varying between 5′ and 0.8° and is based on a target minimum $\lambda/\Delta x$ criterion of between 25 and 150. The higher $\lambda/\Delta x$ criterion was predominantly applied in deep water regions where two-dimensional flow features require additional resolution. The actual $\lambda/\Delta x$ values achieved for the T1 grid were somewhat higher in some regions due to adjacent element size ratio and element skewness restrictions (i.e. elements should not change in size too rapidly). We note that grid T1 does not provide a very high level of mesh refinement over the shelf break. Bathymetry for grid T1 was obtained by interpolating directly from grid SSS5 depths.

Grid T2, shown in Figure 4b, was developed from grid T1 by providing additional refinement over the shelf break region (typically 5′ resolution between 100m and 500m depth) and the continental slope (typically 10′ resolution between 500m and 4000m). This results in a grid with a very high level of resolution in the vicinity of the shelf break and slope with $\lambda/\Delta x$ ratios ranging between 250 to 500. Again bathymetry was interpolated onto grid T2 from grid SSS5.

All four grids were identically forced using an M_2 tide on the open ocean boundary as well as an M_2 tidal potential function within the interior domain (Westerink et al., 1994). The fully nonlinear form of the shallow water equation was used and the nonlinear bottom friction coefficient was set to $C_f = 0.003$ throughout the entire domain. The simulations were run for 55 days in order for all start up transients to dissipate and to reach a dynamic steady state. Harmonic analysis of the stationary elevation responses provides elevation amplitudes and phases at all nodes and establishes a basis for intercomparing the results obtained with these four grids.

We now compare elevation amplitude and phase responses for grids SS4, T1, and T2 to our "truth" solution obtained with grid SSS5. This is accomplished by interpolating the amplitude and phase responses from grids SS4, T1, and T2 onto grid SSS5 and then subtracting off the nodal response values obtained with grid SSS5. We then summarize these

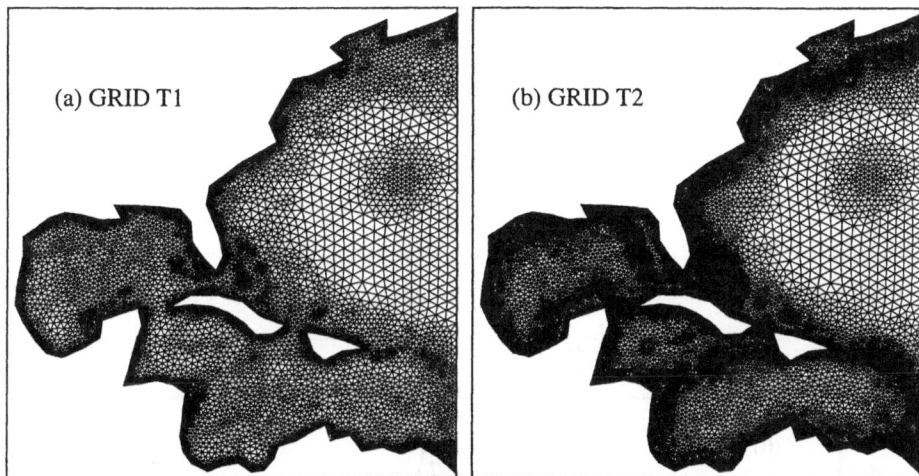

Figure 4: (a) Grid T1 and (b) Grid T2

error distributions by defining the fraction of the domain area which has levels of under- or overprediction which exceed a set limit. This results in the cumulative error distribution curves for absolute elevation amplitude error, relative amplitude error (the absolute elevation errors are normalized with the "truth" amplitude values) and elevation phase error shown in Figure 5. We note that a domain median under- or overprediction error corresponds to a cumulative fraction approximately equal to 2.5×10^{-1} while the L_∞ or maximum domain errors correspond to the end points of the curves. Furthermore we note that a cumulative area fraction of 5×10^{-6} (0.0005% of the WNAT domain) equals the area of one finite element in the SSS5 grid (~ 28 km^2). Therefore the lower portion of the under- and overprediction curves correspond to very small fractions of the WNAT model domain.

We first examine the cumulative area error distribution of the absolute elevation amplitude, shown in Figure 5a. For grid SS4 we note that over 97.6% of the domain underprediction errors do not exceed 1 cm and over 98.9% of the domain overprediction errors are less than 1 cm. Regions with greater than 1 cm error are typically located adjacent to the coastline. The extreme underprediction error equals 16 cm and the extreme overprediction error equals 7.7 cm. Both these extreme errors occur in the Gulf of Maine, a region with high M_2 amplitude values and gradients. We note that the underprediction errors are in general more severe than the overprediction errors. For grid T1, the absolute elevation amplitude errors are significantly greater than for grid SS4 over almost the entire domain. In fact for this grid only 91.8% of the domain underprediction errors do not exceed 1 cm and only 80.9% of the domain overprediction errors are less than 1 cm. Regions with amplitude errors exceeding 1 cm now extend over the entire Atlantic and Florida shelves. Furthermore while the extreme underprediction error is less than for grid SS4, the extreme overprediction error is about the same as for grid SS4. The extreme underprediction error now occurs near an amphidrome located off of Haiti while the extreme overprediction error is still located in the Gulf of Maine. For grid T2, the error levels are slightly less than for grid SS4 over most of the domain. Like grid SS4, error levels in excess of 1 cm are again restricted to regions very near the Atlantic coast. However the extreme underprediction errors have been dramat-

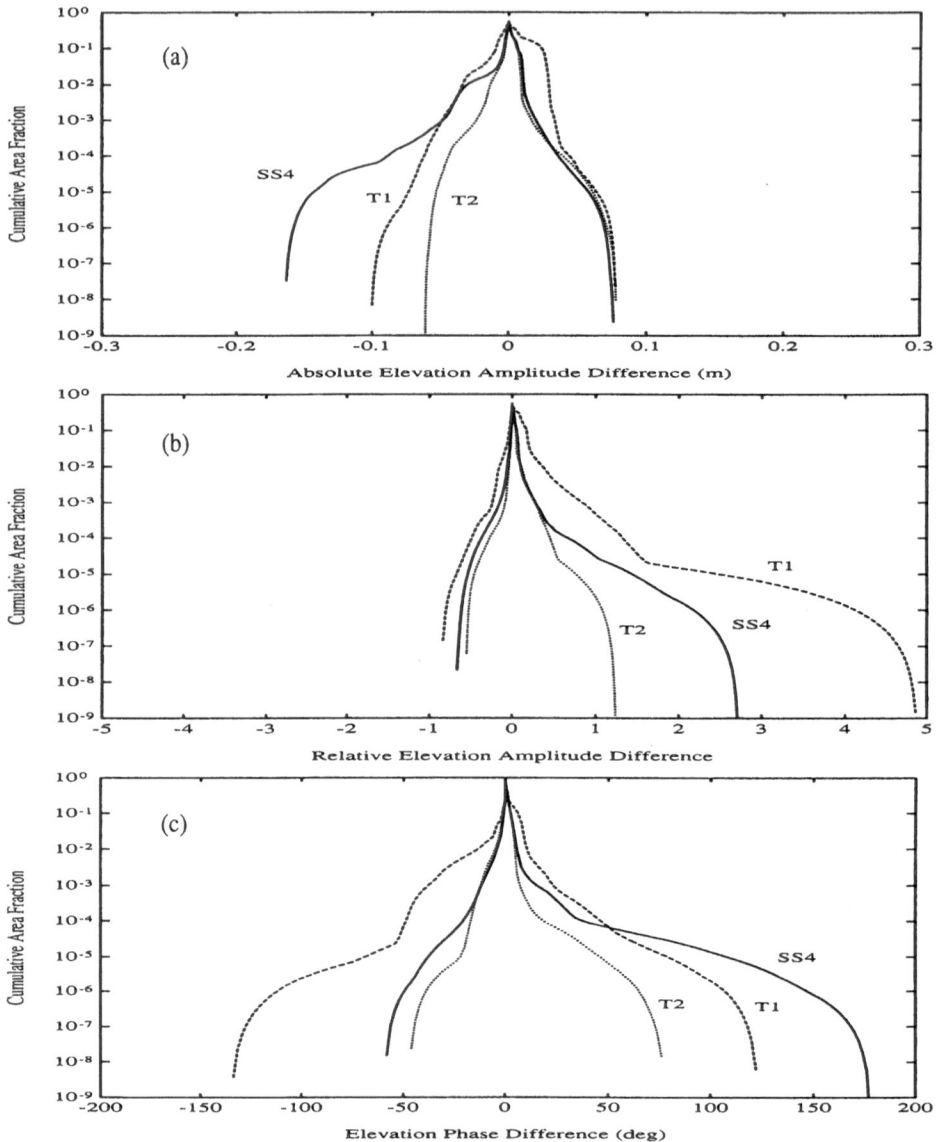

Figure 5: Cumulative area fraction curves.

ically reduced as compared to grid SS4. Furthermore the amplitude error levels of this unstructured graded grid with a high level of resolution over the shelf break are dramatically better than for grid T1 over the entire domain. Finally we note that there is no bias in the under and overprediction errors with these curves being much more symmetrical about the zero amplitude error axis than results for grids SS4 or T1.

We now examine the cumulative area distribution of the relative elevation amplitude error, shown in Figure 5b. We note that the patterns and especially relative positioning of the three plotted curves is somewhat different for the relative amplitude error as compared to the absolute amplitude error (since the relative errors tend to weigh errors in regions where the actual amplitude is small). For grid SS4 over 98.5% of the domain underprediction errors don't exceed a factor of -0.05 and 89.0% of the overprediction errors don't exceed a factor of 0.05. These 0.05 factor error levels are located predominantly in the Gulf of Mexico and Caribbean Sea in the vicinity of amphidromes. The extreme underprediction error equals a factor of -0.68 and the extreme overprediction error equals a factor of 2.73. Both these extreme errors are located within amphidromes off of Haiti. We note that the very low amplitude values within amphidromes tend to bias the absolute error there, resulting in large relative errors. For grid T1, the relative amplitude errors are always significantly greater than for grid SS4 over the entire domain. The 0.05 factor error levels now extend over most of the Gulf and Caribbean. For grid T2, the relative amplitude errors are always less than for grid SS4 and dramatically less than for grid T1. The 0.05 factor error levels extend over smaller parts of the Gulf and Caribbean than grid SS4 and are again located in the vicinity of amphidromes. The extreme overprediction error as compared to grid SS4 has in particular dropped to a factor of 1.24 and is still located near an amphidrome off of Haiti.

Finally we examine the cumulative area distribution of the phase error, shown in Figure 5c. For grid SS4, over 99.3% of the domain underprediction in phases do not exceed $-5°$ and over 97.6% of the domain overpredictions in phase do not exceed $5°$. The $5°$ phase error regions are located primarily within the Gulf and Caribbean and are typically in the vicinity of amphidromes. Amphidromes have associated high gradients in phase. Extreme errors are significant with an extreme underprediction in phase of $-59°$ occurring in the Caribbean and an extreme overprediction in phase of $178°$ occurring in the Gulf on the Florida shelf. Both these extreme errors occur very near amphidromic points. For grid T1, phase errors are generally significantly greater than for the SS4 grid except for the extreme overprediction errors. Phase errors exceed $5°$ in large portions of the Gulf and Caribbean. Finally the T2 grid results in phase errors which are always less than grid SS4 and significantly less than grid T1. Again the $5°$ phase error regions are limited to regions near amphidromes in the Gulf and Caribbean as was the case for grid SS4. The extreme overprediction error has significantly dropped as compared to grid SS4 and now equals $79°$ and is located in the Caribbean adjacent to an amphidrome.

CONCLUSIONS

Grids based on a minimum $\lambda/\Delta x$ criterion do not generally lead to a satisfactorily converged solution. This is related to the two-dimensional structure of the tides as well as the high local truncation errors caused by rapid changes in topography. The influence of changes in topography in the vicinity of the shelf break was clearly demonstrated using a truncation error analysis. Furthermore, numerical experiments using the WNAT domain with a coarse boundary clearly demonstrated the need for additional mesh refinement over the shelf break. The unstructured graded grid T2, with a high level of refinement over the shelf break, performed dramatically better than the unstructured graded grid T1, essentially the same as grid T2 except without the mesh refinement over the shelf break. Furthermore grid T2 performed better than the uniform grid SS4 which has approximately the same number of nodes as grid T2. Extreme errors in particular were substantially less.

The regions where errors remained the highest in grid T2 were almost entirely located either very near the Atlantic coast (absolute elevation amplitude errors) or near amphidromes (relative elevation amplitude and phase errors). Both regions near the coast and near amphidromic points often have high gradients in amplitude and phase associated with them. Based on examining contour maps of error distributions for the simulations presented in this paper as well as others, we conclude that grid T2 requires some additional mesh refinement on the shelf in shallow waters (these are the most underresolved in our T2 grid with $\lambda/\Delta x = 25$). This additional mesh refinement will provide more resolution in the higher elevation gradient regions that occur adjacent to the coast. In addition it will improve the representation of the tides as they propagate on the shelf which will in turn improve the location of the amphidromic points (even if they are located off the shelf) and thus reduce these errors as well.

ACKNOWLEDGEMENTS

This research was funded under contract DACW-39-90-K-0021 with the U.S. Army Engineers Waterways Experiment Station. We also thank Antonio Baptista of OGI for allowing us to use the Xmgredit software to generate the grids presented in this paper.

REFERENCES

Blain, C.A., Westerink, J.J., Luettich, R.A. (1994) "The Influence of Domain Size on the Response Characteristics of a Hurricane Storm Surge Model", *J. of Geophys. Res.,* To appear.

Kolar, R.L., Gray, W. G., Westerink, J.J. and Luettich, R.A. (1994a) "Shallow Water Modeling in Spherical Coordinates: Equation Formulation, Numerical Implementation, and Application", *J. Hydraul. Res.,* In press.

Kolar, R.L., Westerink, J.J., Cantekin, M.E. and Blain, C.A. (1994b) "Aspects of Non-linear Simulations Using Shallow Water Models Based on the Wave Continuity Equation", *Computers and Fluids,* **23**, 3, 523-538.

Luettich, R.A., Westerink, J.J., and Scheffner, N.W., (1992) "ADCIRC: An Advanced Three-Dimensional Circulation Model for Shelves, Coasts and Estuaries, Report 1: Theory and Methodology of ADCIRC-2DDI and ADCIRC-3DL", *Tech. Report DRP-92-6,* Department of the Army.

Westerink, J.J., Muccino, J.C. and Luettich, R.A. (1992), "Resolution Requirements for a Tidal Model of the Western North Atlantic and Gulf of Mexico", *Comp. Methods in Water Res. IX, Vol. 2,* T.F. Russell et al., Eds., 667-674, Elsevier Applied Science, London.

Westerink, J.J., Luettich, R.A. and Muccino, J.C. (1994), "Modeling Tides in the Western North Atlantic Using Unstructured Graded Grids", *Tellus,* In press.

NUMERICAL MODELLING OF STORM SURGES ALONG THE BELGIAN COAST

C.S. YU, M. MARCUS AND J. MONBALIU
K.U.Leuven, Laboratory of Hydraulics
de Croylaan 2, B-3001 Heverlee
Belgium

The application of the North-West European Continental Shelf Model (*mu-CSM*) for the computation of storm surges along the Belgian coast has been investigated. The surface winds were deduced from the atmospheric pressure field. The commonly applied empirical formulae for describing this relationship has been modified to fit the local wind measurements at Ostend. Tidal levels recorded at Ostend have been used for calibrating the drag coefficient at the air-sea interface. The inclusion of the barometer effect along the open boundary of the model was found to be necessary. The simulations of two types of storms are presented and discussed.

INTRODUCTION

From September till April, major displacement of water, apart from the basic tidal motion, of the entire North Sea and its associated shelf seas occurs due to the winds generated by the frequent build up of large depression systems. The large fluctuation of the atmospheric forcing on the sea surface often introduces large sea surface variations in the shallow coastal areas along the southern part of the North Sea. Besides flooding of the coastal low lands and danger to the navigation, the storm surges may cause large amounts of coastal erosion due to the accompanying strong currents. numerical models can not only provide accurate simulation of surge levels and currents but also information for estimating and understanding the coastal sediment transport and morphological changes. Regional models, e.g. Duun-Christensen [1971], Banks [1974] and Prandl [1975], were used in the 70's. Later models were improved by using finer grids, e.g. Davies and Flather [1978], and by expanding the model area, e.g. [Heaps, 1983]. Nowadays, storm surge prediction models operate on routine basis providing on-line accurate predictions of the flood levels.

Atmospheric forcing was introduced [Marcus,1993] to a fine-grid, 2.5' (latitude) by 5'(longitude), two-dimensional depth averaged numerical model (*mu-CSM*), constructed and calibrated to simulate tidal flows in the north-west European continental shelf seas [Yu et al., 1990]. From the formulations to derive surface winds from atmospheric pressure, the formula proposed by Duun-Christensen [1975] was chosen and

A. Peters et al. (eds.), Computational Methods in Water Resources X, 1331–1338.
© *1994 Kluwer Academic Publishers. Printed in the Netherlands.*

calibrated to the wind measured at the Belgian Coast (Ostend). For the air-sea drag, a relationship as suggested by Wu [1982] was tuned to correspond with surge levels observed at Ostend. The prescribed tidal levels at the open boundary (composed by 8 astronomical tides), were corrected for the inverse barometer effect.

Two different types of storms have been simulated. One (January 7 to 13, 1993) has been used for calibration, the other (November 8 to 13, 1992) as verification.

METEOROLOGICAL DATA

The weather forecasts used here were obtained from the United Kingdom Meteorological Office (UKMO). The forecasts are provided on a spherical coordinate system with a grid resolution of 1.25 degree, situated between 35°W to 35°E and 32.5°N to 75°N. Weather predictions were given twice daily at 00:00 GMT and at 12:00 GMT. Information included are grid values of atmospheric pressure, and surface winds at a six-hour interval. Since the atmospheric pressure gradient has been included in the model computation and the interpolation of pressure is much easier, it was decided to derive the surface winds, rather than using the provided surface winds. By doing so, the interpolation procedure has to be performed for pressures only. Moreover, it allows for calibration of winds to correspond with local measurements.

PREDICTION OF SURFACE WINDS

The surface wind is the actual wind speed acting (observed or evaluated) near the surface (usually taken at 10 meters above the mean-sea-level), whereas the geostrophic wind is a good approximation to the actual wind in smooth flow aloft in the atmosphere. The components of the geostrophic winds on a spherical coordinate grid are calculated as:

$$W_{g\chi} = -\frac{1}{\rho_a f R}\frac{\partial P_a}{\partial \phi} \qquad ; \qquad W_{g\phi} = +\frac{1}{\rho_a f R\cos\phi}\frac{\partial P_a}{\partial \chi} \qquad (1)$$

where χ, ϕ are coordinates in longitude and latitude; $W_{g\chi}$, $W_{g\phi}$ are the components of the geostrophic wind; ρ_a is the air density (= 1.25 kg m^{-3}); f is the Coriolis parameter (=$2\omega\sin\phi$); ω is the angular speed of Earth's rotation (= 0.73 10^{-4} rad/s); R is the radius of the Earth (= 6367 km); and P_a is the atmospheric pressure.

Inspired on the work of Duun-Christensen [1975], Hasse [1974] and Verboom et al. [1987], the following formula for deriving surface winds (W) from geostrophic winds (Wg) was found to give good agreement with the observed winds at Ostend during the January storm:

$$W = 6.82 \ (\ 0.56 \ W_g + 2.4 \ )^{0.5} - 15.4 \tag{2}$$

Defining a constant cross isobar angle δ (= 18 degrees) to approximate the frictional effect, the components of the surface wind can be expressed as:

$$W_\chi = W \ W_g^{-1} \ (\ W_{g\chi}\cos\delta - W_{g\phi}\cos\delta \ )$$

$$W_\phi = W \ W_g^{-1} \ (\ W_{g\chi}\cos\delta + W_{g\phi}\cos\delta \ ) \tag{3}$$

AIR-SEA DRAG COEFFICIENT

The relationship between the sea surface stresses, τ_s, and the wind is usually described by a quadratic law :

$$\tau_s = C_D \ \rho_a \ W \ |W| \tag{4}$$

where C_D is an air-sea drag coefficient empirically defined. Many formulations suggest a linear relationship between the drag coefficient and the surface wind (Smith and Banke [1975], Wu [1982], ...). Good agreement with local observations during the January storm was found with the following calibrated expression:

$$10^3 \ C_D = 1.29 + 0.066W \qquad W \leq 30 \ ms^{-1}$$

$$= 3.27 \qquad W > 30 \ ms^{-1} \tag{5}$$

OPEN BOUNDARY CONDITION

Along the open boundary, the water levels are prescribed as a function of time. The total elevation at each point on the open boundary is given by :

$$\zeta_B = \zeta_B(T) + \zeta_B(M) \tag{6}$$

where $\zeta_B(T)$ is the part due to tidal motion, and $\zeta_B(M)$ that due to meteorological influence (inverse barometer effect). The water elevations due to the tidal motion are composed by 8 tidal constituents, i.e. $O_1, P_1, Q_1, K_1, N_2, M_2, S_2$ and K_2[Yu et al., 1990], as

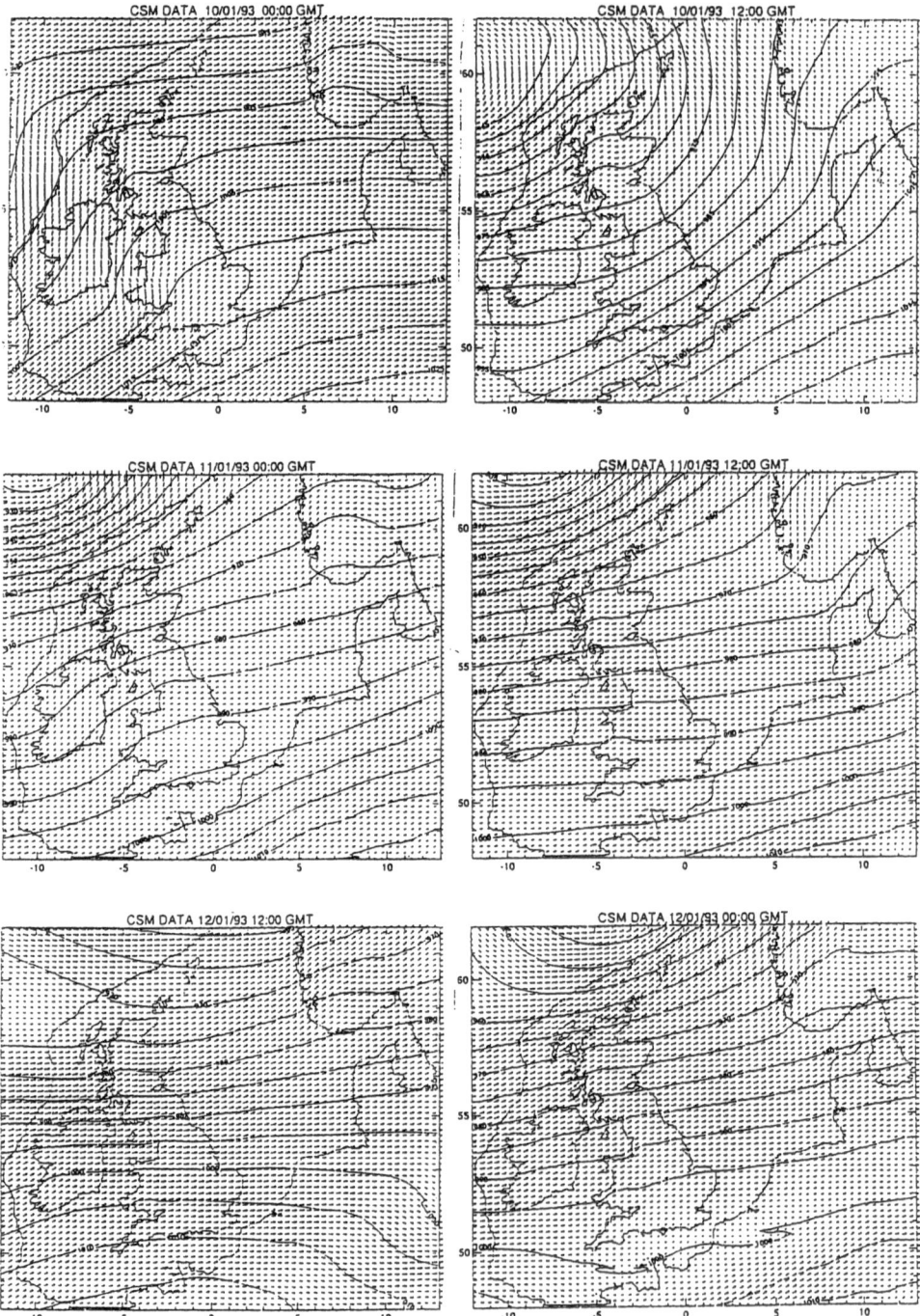

Figure 1: Weather charts of the January storm (1993).

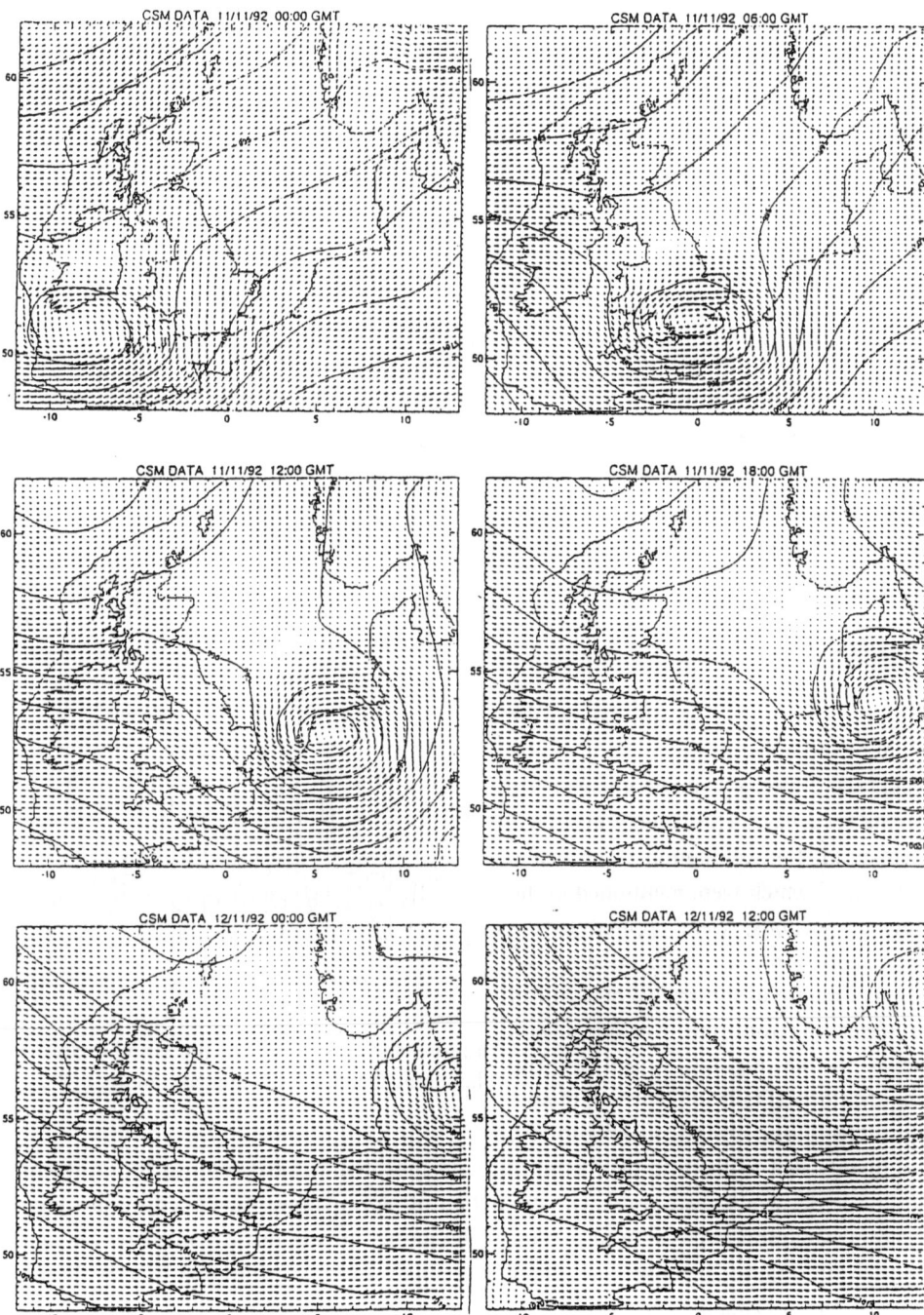

Figure 2: Weather charts of the November storm (1992).

$$\zeta_B(T) = \sum_{n=1}^{8} A_n f_n \cos(w_n t - G_n + (V_0 + u)_n)$$ (7)

where t is time; A_n is the amplitude of the nth-tidal component; f_n is the node factor of the nth-tidal component; w_n is the frequency of the nth-tidal component; G_n is the phase angle of the nth-tidal component; $(V_0 + u)_n$ is the equilibrium argument of the nth-tidal component.

The change in sea surface elevation $\zeta_B(M)$ due to the meteorological influence is given by the hydrostatic law as

$$\zeta_B(M) = (\overline{P} - P_a) / \rho\, g$$ (8)

where $\overline{P}$ is the mean atmospheric pressure over the modelled area, here taken as 1013 [mbar].

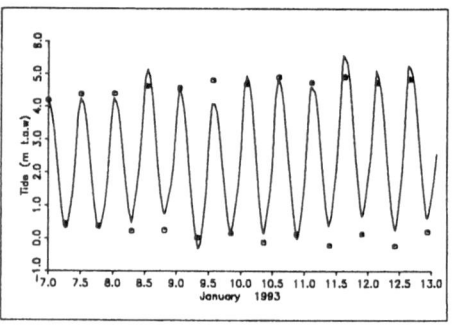

Figure 3: Comparison of the observed (—) surge levels with tides (⊖) at Ostend.

SIMULATIONS

For every storm period simulation, the initial conditions were generated by running the model without atmospheric forcing for 3 days. The January event has been used for calibration. The same parameters, which were mentioned in the previous sections, were applied to the November storm as verification. The different nature of theses two storms can be observed from the weather charts shown in the Figures 1 and 2. The January storm (Figure 1), causing almost one meter surge at Ostend (Figure 3),

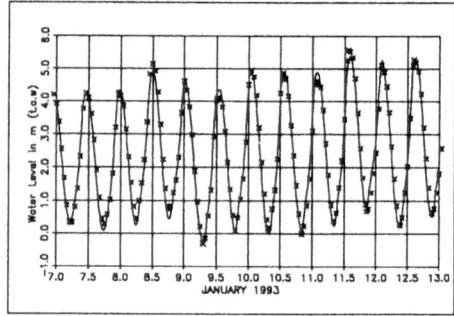

Figure 4: Comparison of the simulated (—) levels at Ostend with measurements (∗).

was mainly due to the low depressions around the Shetland Islands. In the southern North Sea the winds were mild. The prediction of storm surges generated by this type of storm would be impossible to predict if the model area is not extended far enough to catch it. Figure 4 shows the good agreement of the calibration run with the observations at Ostend.

The November storm caused almost the same amount of surge in Ostend, see

Table 1 : Comparison of the computed (A) and the observed (B) water levels at Ostend.

A = 0.980 B + 0.018   for the Period of 8 - 13 November 1992

A = 1.006 B - 0.057   for the Period of 7 - 13 January 1993

Figure 5. The surge generated by this type of storms is mainly due to local weather fluctuations. The depression center travelled through the southern North Sea (Figure 2) and the associated strong winds caused major water displacement on November 11, 1992. The parameters calibrated by the January event were used to simulate this storm period. Model results at Ostend show very good agreement with the surge levels observed (Figure 6). The general quality of the model can be summarized in Table 1.

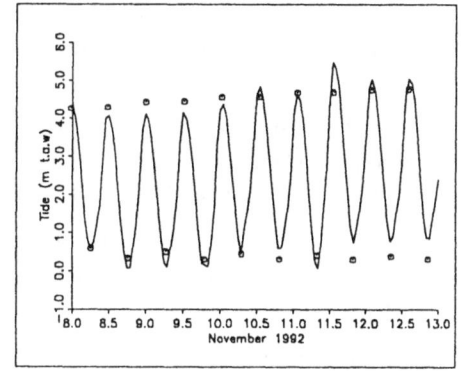

Figure 5: Comparison of the observed (—) surge levels at Ostend with tides (⊖).

CONCLUDING REMARKS

The storms surge model reproduced observed water level variations along the Belgian coast during the two moderate storm surge periods considered in this study, implying accurate reproduction of the tide and good representation of the meteorological data input.

      The calibration of the model was a two step procedure. First the derivation of the surface winds from the atmospheric pressure charts was adjusted. Then the air-sea drag formulation was tuned. Although the verification exercise was found to be quite satisfactory, it was too limited to be conclusive. Possible tuning of the bottom friction, and further verification, using more and different storms and involving observations at

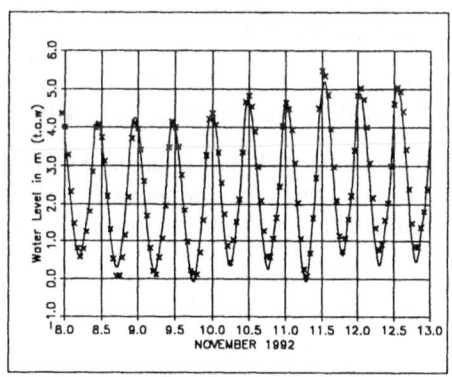

Figure 6: Comparison of the simulated (—) surge levels at Ostend with observation (*).

more stations, might prove to be necessary for the final calibration of the model.

ACKNOWLEDGEMENT

The "*mu-CSM*" was developed in close collaboration with MUMM (Management Unit of Mathematical Models for the North Sea and Scheldt River). Their support and their help in obtaining the weather charts is greatly appreciated. The wind data and the water levels in Ostend were provided by the Coastal Service in Ostend (DDK) of the Ministry of the Flemish Community.

REFERENCES

Banks, J. E. (1974) "A mathematical model of a river-shallow sea system used to investigate tide, surge and their interaction in the Thames-Southern North Sea region", Phil. Trans. R. Soc. London, A 275, 567-609.

Davies, A. M., and R. A. Flather (1978) "Application of numerical models of the North West European Continental Shelf and the North Sea to the computation of storm surges of November to December 1973", Dt. Hydr. Z., A., No. 14, 72 pp.

Duun-Christensen, J. T. (1975) "The representation of the surface pressure field in a two-dimensional hydrodynamic numeric model for the North Sea, the Skagerrak and the Kattegat", Dt. hydrogr. Z., **28**, 97-116.

Hasse, L. (1974) "On the surface to geostrophic wind relationship at sea and the stability dependence of the resistance law", Beit. Phys. Atmos., **47**, 45-55.

Heaps, N. S. (1983) "Storm Surges, 1967-1982" Geophys. J. R. astr. Soc., **74**, 331-376.

Prandl, D. (1975) "Storm surges in the southern North Sea and River Thames" Proc. R. Soc. A, **344**, 509-539.

Smith, S.D. and E.G. Banke (1975) "Variation of the sea surface drag coefficient with wind speed" Quart. J. R. Met. Soc., **101**, pp. 665-673.

Verboom, G.K., R.P. van Dijk and J.G. de Ronde (1987) Een model van het Europese Kontinental Plat voor windopzet and waterkwaliteitsberekeningen, Delft, DELFT HYDRAULICS and Rijkswaterstaat Dienst Getijwateren, Report Z9600/GWAO-87.021.

Wu, J. (1982) "Wind stress coefficient over sea surface from breeze to hurricane" JGR, Vol. 87, No. C12, 9704-9706.

Yu, C.S., A. Vermunicht, M. Rosso, M. Fettweis and J. Berlamont (1990) Numerical simulation of long waves on the North-West European Continental Shelf, Part 2: Model set-up and calibration, Report to the Ministry of Public Health and Environment, Belgium, Ref.BH/88/28.

15. SEDIMENT TRANSPORT

MODELLING OF SEDIMENTATION AND RADIONUCLIDES DEPOSITION IN A BOTTOM TRAP

R.I. DEMCHENKO, M.J. ZHELEZNYAK and L.A. KOZIY
Institute of Mathematical Machines and Systems,
Cybernetics Center Ukrainian Academy of Sciences,
Pr.Glushkova 42, Kiev, 252207,
Ukraine

The problem of suspended sediment and radionuclide transport in river flow is studied on the base of two-dimensional (width averaged) model. The model includes submodels of velocity distribution, submodel of suspended sediment transport and submodel of radionuclide transport in dilute and with suspended sediments. The verification of the velocity distribution and sediment transport submodels has been done using the Delft Hydraulics flume experimental data for the dredged trenches. The model was applied to simulate processes in the bottom traps in the Pripyat river channel created to settle down the contaminated suspended sediments after the Chernobyl accident.

INTRODUCTION

The radionuclides released into rivers and coastal zone adhere to the sediment, with some exchange between sediment and water. The suspended sediment transport in river flow is main carrier of radionuclides with high adsorption possibility - e.g. isotopes of plutonium, ruthenium-106, cobalt-90. As for cesium-137, that is one of the most toxic radionuclide released during the accidents from a nuclear installation, approximately near half of its amount in river flow is transported by the sediments.

The processes of secondary contamination of water bodies and water self purifications from such kind of pollutants take place under the influence of particle-water exchanges. Comparative influence of sedimentation-erosion processes and physicochemical exchange processes on the fate of pollutants in the bottom sediments could be analyzed on the base of bottom traps studies. The trench, quarries and other place of the increasing depth play a role of a trap for contaminated sediments in a fluvial flow. Such kinds of the special traps for radionuclides attached to sediments have been constructed in the Pripyat River after the Chernobyl accident. This study was initiated by the problem to estimate efficient configuration of such traps for water protection measures after the nuclear accident. Two-dimensional vertical-longitudinal model VERTOX has been developed to describe such kind of processes in river channels (Zheleznyak et al., 1992). Velocity distribution in a trap is simulated on the base of simplified approach, which

A. Peters et al. (eds.), Computational Methods in Water Resources X, 1341–1348.

includes the formula definition of the eddy viscosity coefficient. The VERTOX development is continued now to incorporate in it the k-ε model and the Reynolds stress transport model.

The dredged trenches sedimentation in a channel has been studied in a lot of experimental and modeling investigations (e.g. Rijn, 1981, 1984). Their results were used to develop VERTOX's suspended sediment transport submodel (Demchenko and Zheleznyak, 1990).

The VERTOX's submodel of radionuclide transport governed by the advection-diffusion equation describes the radionuclide concentration in solute, the concentration in the suspended sediments and the concentration in the bottom deposition. The structure of the submodel is close to the model SERATRA developed in the Pacific Northwest Laboratories (Onishi and Wise, 1979). The main processes of the radionuclide exchange between radionuclides in solute and in solid phase are described as adsorption-desorption and sedimentation-resuspension processes. The analysis of the comparative effects of these mechanisms on the radionuclide transport by river flow is the main objective of the paper.

MODEL DESCRIPTION

The model VERTOX considered in the present work belongs to the hierarchy of radionuclide dispersion models developed in the Cybernetics Center. (Zheleznyak et al., 1992). The model has been derived from 3-D models of hydraulic and pollutant transport processes by averaging their equations over the flow width. The continuity and momentum equations can then be written as:

$$\frac{\partial u}{\partial t} + u\frac{\partial u}{\partial x} + w\frac{\partial u}{\partial z} = -g\frac{\partial \eta}{\partial x} + K\frac{\partial^2 u}{\partial x^2} + \frac{1}{b}\frac{\partial}{\partial z}\left(bv\frac{\partial u}{\partial z}\right) -$$
$$-\frac{1}{b}\left(\tau_{xz}^{B_2}\frac{\partial B_2}{\partial z} - \tau_{xz}^{B_1}\frac{\partial B_1}{\partial z}\right) - \frac{1}{b}\left(\tau_{xx}^{B_2}\frac{\partial B_2}{\partial x} - \tau_{xx}^{B_1}\frac{\partial B_1}{\partial x}\right) + \frac{1}{b}\left(\tau_{xy}^{B_2} - \tau_{xy}^{B_1}\right) = 0 \tag{1}$$

$$\frac{\partial(bu)}{\partial x} + \frac{\partial(bw)}{\partial z} = 0 \tag{2}$$

where u, w are the width-averaged velocity in longitudinal and vertical direction x and z respectively, η is the water surface elevation above the undisturbed level, K is the width averaged turbulent diffusion coefficient for longitudinal direction, v is the width-averaged vertical turbulent diffusion coefficient, g is the acceleration of gravity, b = B_1-B_2 is the width of a channel, y=B_1(x,z) and y=B_2(x,z) discribe surfaces of the left and

right rigid boundaries of a channel respectively, $\tau_{xz} = v\frac{\partial u}{\partial z}$, $\tau_{xx} = K\frac{\partial u}{\partial x}$, $\tau_{xy} = K\frac{\partial u}{\partial y}$ are

shear stresses, superscripts [B1], [B2] point out that these values are taken at the respective boundaries.

The hydrostatic approximation is used for the vertical momentum equation. The vertical turbulent diffusion coefficient v is calculated through the Prandtl relation and the Montgomery formula for the turbulent length scale l (Montgomery, 1943):

$$v = l^2 \left| \frac{\partial u}{\partial z} \right|, \qquad\qquad l = \frac{\chi}{H}(z + H + z_H)(-z + \eta + z_\eta) \qquad (3)$$

where χ = von Karman constant, H = undisturbed water depth, $h = H + \eta$, z_H, z_η are the parameters of bottom roughness and water surface roughness respectively. The efficiency of this approach to simulate the vertical structure of velocity field has been demonstrated for different kinds of open flow, including currents in the presence of a wind stress (Voltsinger et al., 1989).

The bottom boundary condition is
$z = -H + z_H$: $\qquad\qquad\qquad\qquad u = 0, \quad w = 0 \qquad\qquad\qquad\qquad (4)$
At the free surface $z = \eta$ we use the kinematic boundary condition

$$w = \frac{\partial \eta}{\partial t} + u \frac{\partial \eta}{\partial x} \qquad (5)$$

and put the shear stress equal to zero.

Taking into account the aforementioned boundary conditions (4), (5) the vertical velocity component may be expressed as

$$w = -\frac{1}{b} \int_{-H}^{z} \frac{\partial(bu)}{\partial x} dz \qquad (6)$$

The integral of the continuity equation over the depth yields

$$b \frac{\partial \eta}{\partial t} + \frac{\partial}{\partial x} b \int_{-H}^{\eta} u\, dz = 0 \qquad (7)$$

Further we shall omit terms that content the derivatives of the width b, that are supposed to be lower order of magnitude than other terms of the equation for the flows under consideration.

The advection-diffusion equation governing the transport of suspended sediment is

$$\frac{\partial S}{\partial t} + u \frac{\partial S}{\partial x} + (w - w_0) \frac{\partial S}{\partial z} = K \frac{\partial^2 S}{\partial x^2} + \frac{\partial}{\partial z}\left(v \frac{\partial S}{\partial z}\right) \qquad (8)$$

where S is the sediment concentration, w_0 is the settling velocity of the particles.

The vertical flux of the sediments through the free surface should be zero. The corresponding boundary condition is

$z = \eta$: $\qquad\qquad\qquad (w - w_0)S = v \frac{\partial S}{\partial z} \qquad\qquad\qquad (9)$

The vertical flux of suspended sediments at the bottom is accepted to be equal to the difference of the resuspension and sedimentation rates. This assumption yields the following formulation of boundary condition:

$z=-H+a$:
$$v\frac{\partial S}{\partial z} + w_0 S = q^s - q^b \tag{10}$$

where a is the level of boundary between suspended sediment and bottom sediment transport, q^s is the sedimentation rate, q^b is the resuspension rate. It is supposed that in the case of fine noncohesive sediments this rates may be estimated by the equations

$$q^s = \begin{cases} w_0(S_0 - S_*), & S_0 > S_* \\ 0, & S_0 < S_* \end{cases}, \qquad q^b = \begin{cases} 0, & S_0 > S_* \\ E_r w_0(S_* - S_0), & S_0 < S_* \end{cases} \tag{11}$$

where E_r is the erodibility coefficient, S_0 is the actual sediment concentration S at the bottom level $z=-H + a$, $S*$ is the near-bottom equilibrium sediment concentration that corresponds to the sediment capacity of the steady and uniform flow with the same local parameters. Rijn (1984) method is used to calculate $S*$ as function of the bottom shear stress, bottom roughness, sediment grain diameter and settling velocity.

The submodel of radionuclide transport describes the radionuclide concentration in solute C, the concentration in the suspended sediments C^s and the concentration in the bottom deposition C^b. The exchanges between these variables have described as adsorption-desorption and sedimentation - resuspension processes. The governing equations are similar to ones used by Onishi and Wise (1979). The transport equation of dissolved pollutant may be written as:

$$\frac{\partial C}{\partial t} + u\frac{\partial C}{\partial x} + w\frac{\partial C}{\partial z} = K\frac{\partial^2 C}{\partial x^2} + \frac{\partial}{\partial z}\left(v\frac{\partial C}{\partial z}\right) - \lambda C - a_{1,2}S(K_d C - C^s) \tag{12}$$

where K_d - the distribution coefficient in solid particle-water system in geochemical equilibrium conditions, $K_d = C^s/C$ when $t \rightarrow \infty$ in hydraulic steady state conditions; $a_{1,2}$ is the rate of water-suspended sediment exchange, λ is the decay constant for the radionuclide under consideration.

The boundary condition at the free surface $z = \eta$ is

$$v\frac{\partial C}{\partial z} = wC \tag{13}$$

The diffusion flux into the bottom is taken as

$z=-H+a$:
$$v\frac{\partial C}{\partial z} = \rho_s(1-\varepsilon)Z_* a_{1,3}(K_d C - C^b) \tag{14}$$

where ε is the bottom porosity, $Z*$ is the efficient thickness of the contaminated upper bottom layer, $a_{1,3}$ is the rate of water-bottom exchange, ρ_s is the sediment density. The value of the distribution coefficient K_d is assumed to be the same for the suspended and bottom sediments.

The particulate-contaminant transport, i.e. radionuclide transport by suspended sediments is described by the equation

$$\frac{\partial SC^s}{\partial t} + u\frac{\partial SC^s}{\partial x} + (w - w_0)\frac{\partial SC^s}{\partial z} = K\frac{\partial^2 SC^s}{\partial x^2} + \frac{\partial}{\partial z}\left(v\frac{\partial SC^s}{\partial z}\right) -$$
$$- \lambda SC^s + a_{1,2}S(K_d C - C^s) \tag{15}$$

The boundary conditions describing zero flux of the C through the water surface and its flux equals to the difference of amount of eroded and deposited on the bottom pollutants may be written as

$z=\eta:$
$$(w - w_0)SC^s - v\frac{\partial(SC^s)}{\partial z} = 0 \tag{16}$$

$z=-H+a:$
$$w_0 SC^s + v\frac{\partial(SC^s)}{\partial z} = C^s q^s - C^b q^b \tag{17}$$

The thickness Z_* of the upper layer of the contaminated bottom deposition is governed by the equation of the bottom deformation.

$$\rho_S(1-\varepsilon)\frac{\partial Z_*}{\partial t} = q^s - q^b \tag{18}$$

Radionuclide concentration in the upper bottom layer is described by the equation:

$$\frac{\partial(Z_* C^b)}{\partial t} = a_{1,3}Z_*(K_d C - C^b) - \frac{1}{\rho_S(1-\varepsilon)}(C^b q^b - C^s q^s) \tag{19}$$

The boundary conditions in the inflow section $x=x_0$ were determined through the depth averaged value of flow parameters $\bar{f}$. The modifications of the free propagation conditions used at outflow boundary.

The numerical solution of the equations (1) - (17) is obtained by the explicit-implicit finite-difference method similar to one presented in (Voltsinger et al., 1989). The transformation of the variables conforms the flow area into the rectangle where the equations are approximated on the vertically inhomogeneous grid.

MODEL APPLICATIONS

The laboratory flume experimental data for dredged trenches (Rijn, 1981) have been used for the verification of the velocity field and sediment transport submodels of the VERTOX code. The comparison of the simulated flow velocity and suspended sediment concentration with the experimental data demonstrates overall reasonable agreement between them. The larger discrepancies take place at the corner sections of the bottom profile. The approximation error of the description of real flow field in these local zones has not large influence on the simulation of overall sediment deposition in the trench.

The follows parameters were used during simulation: $w_0 = 0.013$m/sec , particle diameter of bed material $d_{50} = 0.00016$m, $\bar{u}(x_0, t) = 0.5$m/sec, h= 4.6m, $z_H = 0.0008$m.

The model application for the Pripyat River radionuclide transport is considered here for the trap-bottom quarry, constructed in 1986 at village Otashev downflow Chernobyl NPP (Fig.2).

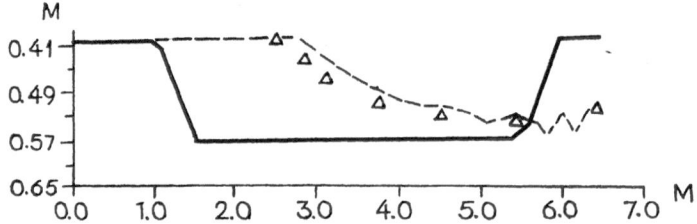

Fig. 1 Measured (triangles) and computed (solid line) bed level profile after 7.5 hours of the experiment

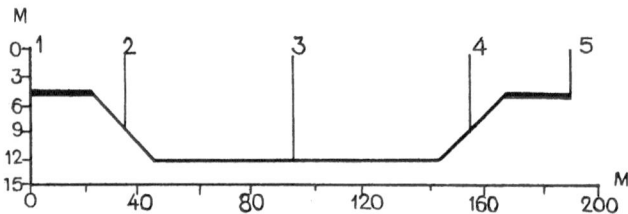

Fig. 2 Designed profile of the Otashev trap

To diminish simulating time the analysis of the adsorption-desorption processes was provided with the overestimated value of the parameters $a_{1,2}$ and especially $a_{1,3}$:

$a_{1,2}=a_{1,3}=3\frac{1}{day}$. The reasonable value of $a_{1,3}$ for cesium-137 is near 1/year. Simulations were done for $K_d=5000l/kg$, $\overline{C}(x_0,t)=100pci/l$, $\overline{C^S}(x_0,t)=10C_e^b$, $C^b(x,t_0)=10^{-4}C_e^b$, where $C_e^b=\overline{C}K_d$. Temporal dynamics of the bottom contamination (Fig.3) demonstrates that radionuclide concentration increases in the deposition zone and decreases in the resuspension zone of the trap.

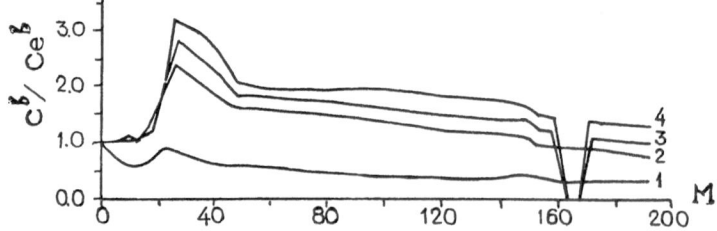

Fig. 3. Dimensionless radionuclide concentration in the bottom deposition ($d_{50}=0.005cm$, $\overline{u}_0=0.2m/sec$). $t_1=1day$, $t_2=5days$, $t_3=7days$, $t_4=10days$.

The vertical profiles of dissolved contaminant concentration C (solid line) and concentration with suspended sediment, $P=SC^s$ (dashed line) show large difference of the distributions in the sedimentation zone (section 2) and resuspension zone (section 4).

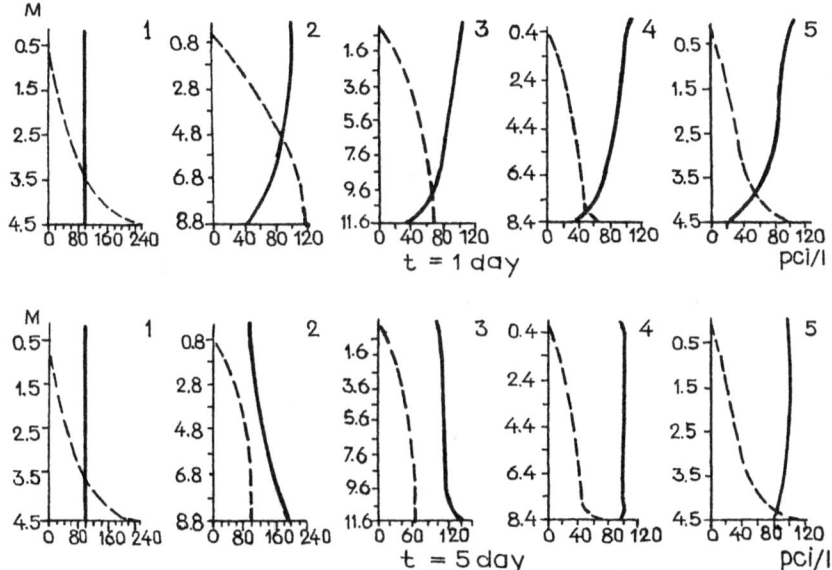

Fig. 4

The efficiency of the bottom trap to decrease radionuclide flux in the river is determined by the ratio N=R/F, where $R=\rho_s(1-\varepsilon)\int_{x_0}^{x_L}[C^bZ_*(x,t)-C^bZ_*(x,t_0)]dx$ is the integrated amount of the radionuclide stored in the bottom upper layer Z_* within time $(t-t_0)$, $F=(\overline{S}\,\overline{C}^s+\overline{C})\overline{u}(x_0,t)(t-t_0)$ is the amount of the radionuclides transported through the inlet boundary within the same time. This ratio can vary in different hydrological seasons.

Table 1. Calculated ratio N.

Sediment size	N	
(cm)	$\overline{u}_0=0.2$	$\overline{u}_0=0.8$
0.0005	0.01	4 10^{-6}
0.0025	0.06	9 10^{-6}
0.005	0.15	2 10^{-5}
0.01	---	8 10^{-4}

The results for the averaged velocitiy magnitudes typical for low water period - 0.2m/sec and 0.8m/sec for flood period (Table 1) demonstrate that only small part of radionuclide flux could be depositid in the trap.

The data observed in the Pripyat River (Voitcekhovich et al., 1989) are consistent with the simulated results.

CONCLUSIONS

The simulated results demonstrate that sedimentation processes have lager influence on the river water's selfpurification from radionuclide than direct physicochemical exchange with the bottom. The rate of the radionuclide deposition in the bottom layer abruptly decreases with the diminishing of the typical sediment grain size and grouth up of flow velocity. Such kind of a trap could not diminish considrable total radionuclide discharge Their influence could be important mainly for coarse particles that could be a part of the destroyed nuclear fuel

REFERENCES

Demchenko, R. and Zheleznyak, M. (1990) "Numerical modelling of the vertical sediment distribution for unhomogeneous depth", in A. Morozov and M. Zheleznyak (eds.), System analysis and methods of the mathematical modelling in the ecology, Cybernetics Center, Kiev, pp. 66-72 (in Russia).

Montgomery, R. (1943) "Generalization for cylinders of Prandtl's linear assumption for mixing length", Ann. N.Y. Acad. Sci. 44, 89-103.

Onishi, Y. and Wise, S. (1979) Mathematical model, SERATRA, for sediment - contaminant transport in rivers and its application to pesticide transport in Four mile and Wolf Greeks in Iowa, Pacific Northwest Laboratories, Richland, Washington.

Rijn, L. van (1981) "Model for sedimentation predictions", in Proc. XIX Congress of International Association for Hydraulic Research Sedimentation Engineering for Rivers and Reservoirs, New Delhi, pp. 321-328.

Rijn, L. van (1984) "Sediment transport, Part II: Suspended load Transport", J. Hydraulic Engineering 110, 1613-1641.

Voitcekhovich, O., Kanivets, V. and Shereshevsky, A. (1989) "On the effectiveness of intersection of radionuclides by river - bed open cut of Pripyat River", in E.Ignatenko (ed.), Chernobyl - 1988, vol. 6, Gidrometeoizdat, Moscow, pp. 46-63 (in Russian).

Voltsinger, N., Klevanniy, K. and Pelinnovskiy, E. (1989) Long-wave dynamics of coastal areas, Gidrometeoizdat, Leningrad (in Russian).

Zheleznyak, M., Demchenko, R., Khursin, S., Kuzmenko, Y., Tkalich, P. and Vitiuk, N. (1992) "Mathematical modeling of radionuclide dispersion in the Pripyat - Dnieper aquatic system after the Chernobyl accident", The Science of the Total Environment 112, 89-114.

FLOW AND SEDIMENT TRANSPORT MODELLING IN WESTERN SCHELDT

JABBARI E. and BERLAMONT J.
Hydraulics Laboratory
Catholic University of Leuven
de Croylaan 2, 3001 Heverlee, Belgium
SAS M.
International Marine and Dredging Consultants
Wilrijkstraat 37, 2140 Antwerp, Belgium

For the study of the construction of a container terminal on the north bank of the western Scheldt in Antwerp (Belgium) an extensive measurement campaign was carried out on 27th of september 1990 during a neap tide and on 4th of october during a spring tide. The measurement data were used to calibrate a set of mathematical models. The hydrodynamics model was calibrated for the roughness coefficient to provide the transport model the necessary flow field and the necessary information about bottom shear stress. For the modelling of sediment transport three different approaches were used (Englund-Hansen, Acker-White, and van Rijn). The results were compared with the measured values and some modifications were implemented.

INTRODUCTION

The western Scheldt is the main navigational access to the port of Antwerp from the North sea. The huge shipping activities between the mouth and the port of Antwerp requires a minimum depth of 13.3m below T.A.W. (Local Chart Datum) to be maintained in the navigation channel. As a consequence , constant dredging is carried out to remove the sediment which is trapped in the estuary by the residual landward circulation of water along the bottom of the channel. This costly operation requires a better understanding of the interaction between sediment and tidal flows.
Most of the developed formulae to determine the amount of sediment transported by rivers were calibrated under ideal steady uniform flow conditions. In addition, the complexity of the transport mechanism of granular material causes the model not to be reliable in all cases. Three approaches were proved to be satisfactory to simulate sediment transport by tidal flows in the Eastern Scheldt. These were implemented in an existing transport model. This study was undertaken to evaluate how good the selected formulae perform in the oscillatory tidal currents.

DESCRIPTION OF THE MODELS

In the hydrodynamics model, RMA-2V, (King, 1990) the depth-integrated equations of mass continuity and momentum conservation are solved using finite element as mathematical tool. Manning's formula is used for friction losses and eddy viscosity coefficients are used to define turbulent losses. The transport model, STUDH , (Mc Anally et al., 1984) solves the advection-diffusion equation for sediment.

A. Peters et al. (eds.), Computational Methods in Water Resources X, 1349–1356.

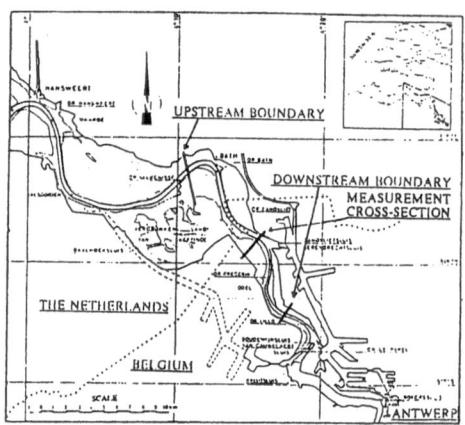

Figure 1. Location of the Scheldt estuary and the study area

The approaches of Englund-Hansen, Acker-White, and van Rijn were employed to define the sink-source term. Time derivatives are replaced by finite difference approximations using a fully implicit scheme. Galerkin weighted residual method is used for solution of the governing equations.

GOVERNING EQUATIONS

The vertically averaged conservation equations in Cartesian coordinates are the continuity equation

$$\frac{\partial \xi}{\partial t} + \frac{\partial (hu)}{\partial x} + \frac{\partial (hv)}{\partial y} = 0 \tag{1}$$

the momentum equations assuming a hydrostatic pressure distribution and the Boussinesq approximation

$$\frac{\partial u}{\partial t} + u\frac{\partial u}{\partial x} + v\frac{\partial u}{\partial y} = fv - g\frac{\partial \xi}{\partial x} + \frac{1}{\rho h}(\tau_x^s - \tau_x^b) + v\nabla^2 u \tag{2}$$

$$\frac{\partial v}{\partial t} + u\frac{\partial v}{\partial x} + v\frac{\partial v}{\partial y} = -fu - g\frac{\partial \xi}{\partial y} + \frac{1}{\rho h}(\tau_y^s - \tau_y^b) + v\nabla^2 v \tag{3}$$

and the conservation equation for the sediment

$$\frac{\partial C}{\partial t} + u\frac{\partial C}{\partial x} + v\frac{\partial C}{\partial y} = \frac{1}{h}[\frac{\partial}{\partial x}(hD_x\frac{\partial C}{\partial x}) + \frac{\partial}{\partial y}(hD_y\frac{\partial C}{\partial y})] + S \tag{4}$$

where u, and v = depth-averaged tidal velocities in the x and y directions respectively, ξ = elevation of the water surface above the reference level, h = total water depth, ρ = density of water, $[\tau_x^b, \tau_y^b] = \rho g n^2 [u, v](u^2+v^2)^{1/2}/h^{1/3} = x$ and y components of the bottom stress, n = Manning friction coefficient, $[\tau_x^s, \tau_y^s] = \phi\rho_a w^2 [\sin\theta, \cos\theta] = x$ and y components of wind induced stress, ϕ = wind stress coefficient, ρ_a = air density, w = wind speed, θ = wind direction, v = horizontal eddy viscosity coefficient, f = Coriolis parameter, c = sand concentration, D_x, and D_y = horizontal dispersion coefficients.

ADAPTATION TIME FOR A NON-EQUILIBRIUM SITUATION

Most of the sediment transport formulae have been developed to predict the equilibrium sediment transport rate in a uniform, steady current field. In an estuary the flow condition, however, changes in magnitude and direction with time. The adaptation of the local sediment transport to the changed flow condition does not proceed fast in comparison with the period concerned. The rate with which flow moves from a non-equilibrium condition to an equilibrium one, depends on different parameters for different interaction between fluid and bed. In the equation of mass conservation the source or sink term is defined as $S = (C_{equ} - C)/t_a$, where C_{equ} = equilibrium concentration and t_a = adaptation time. For the adaptation time the Corps of Engineers(Mc Anally, et al., 1984) gives $t_a = (C_d\ h)/w$ and $t_a = (C_e\ h)/\bar{U}$ for deposition and erosion respectively, where C_d = coefficient for deposition, C_e = coefficient for erosion, w = fall velocity and $\bar{U}$ = depth-averaged velocity.

Examination of the results of several simulations with constant values for the other parameters and considering the measured values of concentration gave 6. and 20. as the best values for C_d and C_e respectively.

BED SOURCE TERM

Three different formulae used to calculate the equilibrium concentration were Englund-Hansen, Acker-White, and van Rijn. Englund-Hansen's equation (1967) expresses the total load q_{sb} in terms of a friction factor f, a dimensionless sediment discharge ϕ, and a dimensionless bed shear stress θ as

$$f\phi = 0.1\theta \tag{5}$$

in which

$$f = \frac{2u_*}{\bar{U}^2} \quad , \quad \theta = \frac{\tau_*}{v(s-1)d} \quad , \quad \phi = \frac{q_T}{\sqrt{(s-1)gd^3}} \tag{6}$$

where u_* = shear velocity, $\bar{U}$ = depth average velocity, τ_* = bed shear stress, v = kinematic viscosity, $s = \rho_s/\rho$, d = mean fall diameter and q_T = the volumetric sediment discharge per unit width.

The general function of Acker and White (1973) is based on the stream power concept and dimensional considerations, and is expressed as

$$G_{gr} = C\left(\frac{F_{gr}}{A} - 1\right)^m \tag{7}$$

where

$$G_{gr} = \frac{ch}{sd}\left(\frac{u_*}{U}\right)^n \quad , \quad F_{gr} = \frac{u_*^n}{\sqrt{gd(s-1)}}\left(\frac{\bar{U}}{5.75\log[10h/d]}\right)^{-n} \tag{8}$$

in which c = dimensionless concentration by mass. The coefficients A, C, m and n are given in terms of a dimensionless grain size

$$D_{gr} = d\left(\frac{g(s-1)}{v^2}\right)^{1/3} \tag{9}$$

In van Rijn's sediment transport formula (1984a,b) bed load and suspended load are calculated separately. The dimensionless particle diameter of Acker and White , D_{gr}, is used and a transport stage parameter is defined as

$$T = \frac{(u'_*)^2 - (u_{*,cr})^2}{(u_{*,cr})^2} \tag{10}$$

in which, u'_* = bed shear velocity related to grains by assuming a logarithmic velocity profile and taking $3d_{90}$ as roughness length, and $u_{*,cr}$ = critical shear velocity according to Shields. The bed load is given as

$$\frac{q_b}{\sqrt{(s-1)g}\ d^{1.5}} = 0.053\frac{T^{2.1}}{D_{gr}^{0.3}} \tag{11}$$

where q_b = volumetric sediment discharge per unit width and for suspended load

$$q_s = F\bar{U}hc_a \tag{12}$$

in which F and c_a are respectively a dimensionless parameter and the reference concentration defined as

$$F = \frac{[\frac{a}{h}]^{z'} - [\frac{a}{h}]^{1.2}}{[1 - \frac{a}{h}]^{z'}(1.2 - z')} \quad , \quad c_a = 0.015\frac{D_{50}}{a}\frac{T^{1.5}}{D_{*}^{0.3}} \tag{13}$$

where a = the reference level and z' = the Rouse suspension number modified for the influence of the sediment particles on the turbulence structure of the fluid.

STUDY AREA AND COMPUTATIONAL NETWORK

The part of Scheldt under study starts at Liefkenshoek downstream Antwerp and extends about 16Km to Bath in the Netherlands before reaching the North sea (Figure 1). The fresh water discharge has no influence on the intensity of the flow in this part of the estuary. The water-level and flows in the reach are determined by tidal water-levels in the south-eastern part of the North sea. The cross-sectional area of the river during a spring tide varies between 6400 to 11200m² at Liefkenshoek and between 14000 to 26000 m² at Bath.

The finite element mesh was constructed using a 1:5000 scale map. The map was based upon a bathymetry survey made during July-September 1990. The resulting finite element mesh consisted of 674 elements and 2215 nodes. The elements were generally aligned along the bed elevation contour-lines.

IN SITU MEASUREMENTS

During a thorough-tide measurements at spring and neap tide velocity, temperature, sand and mud concentration were measured from five anchored vessels across the river Scheldt in front of the future container terminal and from two supplementary vessels located upstream and downstream of the cross-section.

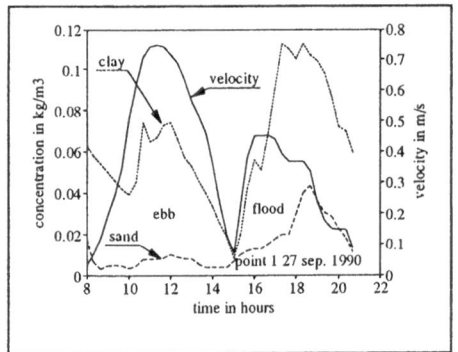 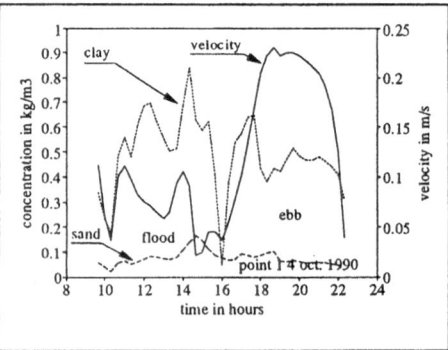

Figure 2. Measurement results during neap tide (left) and spring tide (right)

The Ultrasonic Doppler Scatterometer was used to monitor the vertical variation of the sand concentration by taking 240 concentration readings during 120 seconds. The results of the measurement are shown in figure 2 for a typical location for depth-averaged values.

BOUNDARY CONDITIONS AND SIMULATIONS

For hydrodynamic simulation water-levels were imposed as a function of time at both ends of the domain and velocities were initially set to zero at all nodes. Simulation was performed for three similar tidal cycles. The results of the third cycle were used for sediment transport runs. The model was calibrated for roughness coefficients. Comparison with the measured results showed that higher Manning values had to be used for the flood for both spring tide and neap tide simulations.

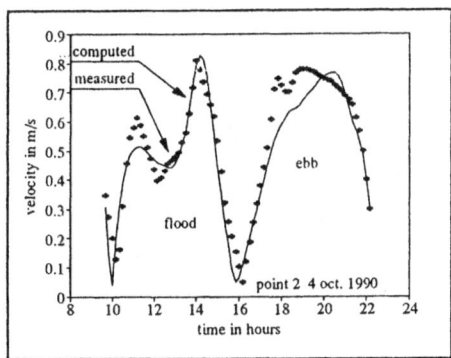 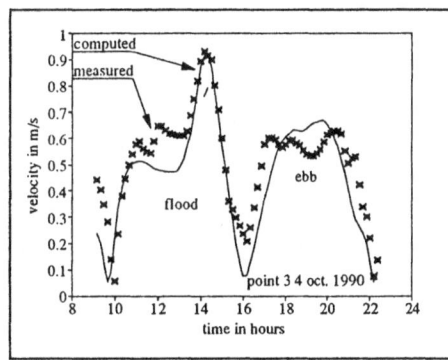

Figure 3. comparison of measured and computed velocities during spring tide

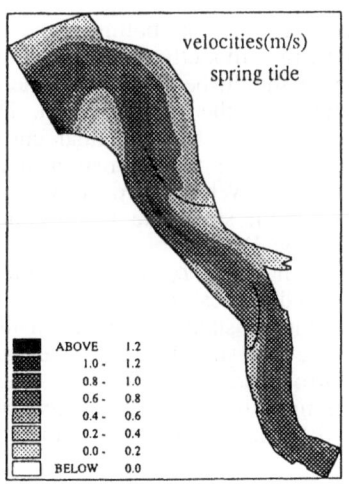 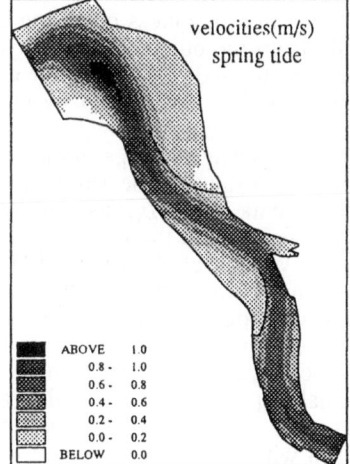

Figure 4. contours of velocity, before high water(left), and after high water(right)

For sand transport simulation, the concentration was specified on the upstream boundary during ebb and at the downstream during flood. A weak condition was imposed at the other end. As initial condition, concentrations were set to the values at the specified boundary condition. All the simulations were performed for 5 similar tidal cycles.

DISCUSSION

With all three transport formulae, it was necessary to relate the shear stress or the shear velocity to the depth-averaged velocity. This relation was of particular importance for the Acker-White and Englund-Hansen formulae. As stated earlier the calibration of the hydrodynamics model required higher values for the Manning coefficient for the flood period. The computations showed little sensitivity to the roughness when unit discharge was imposed as upstream boundary condition.

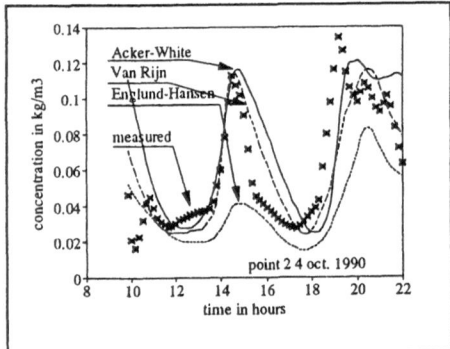 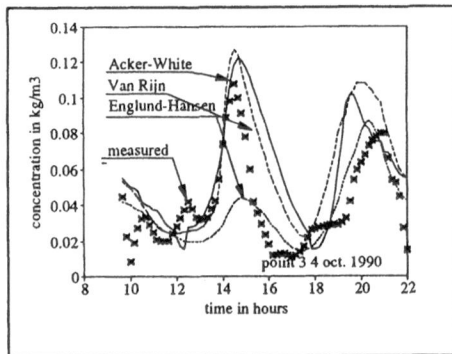

Figure 5. Time history of concentration for spring tide

This means that the roughness condition and consequently the bottom shear stress may differ greatly from the outcome of the hydrodynamics modelling. So, the bottom shear stress to be used for sediment transport simulation should be calculated independent of the hydrodynamics simulation and by consideration of the bed condition. This would be achieved by assuming a logarithmic velocity profile for the tidal currents. The source of error in using the logarithmic velocity profile is the estimation of the roughness length. The results of the simulation with Acker-White were in good agreement with measured data, with D_{35} as representative grain size. The total sediment load, however, was higher than the measured values (table 1).

In the original version of the Englund-Hansen formulation D_{50} has been introduced as grain size. Comparison of the results with recorded values showed that D_{35} was more suitable. .16 was found to be the best value for the constant (Eq. 5) rather than .1.

The van Rijn equation was not sensitive to the particle size. It was, however, sensitive to the reference level, a (Eq. 13), which was introduced by van Rijn as .01h. In fact, in his original paper he has taken this as a minimum value for the reference level. Simulations were carried out with coefficients different from .01 and the best results were obtained with .001h as reference level.

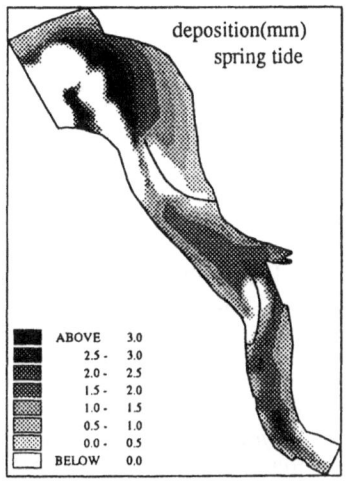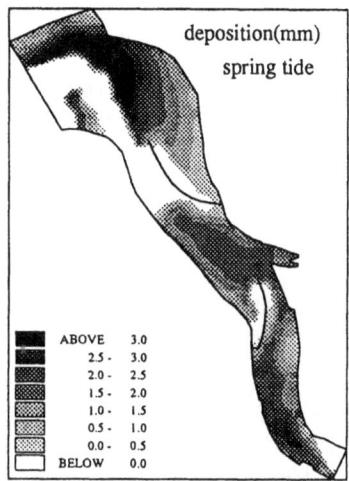

Figure 6. deposition after 1 tidal cycle Acker-White(left), van Rijn(right)

Table 1 shows the sediment transport load per tidal cycle for spring tide and neap tide. Comparison of the total sediment load transported through the measurement cross-section with calculated values from the measured concentrations and velocities showed that calibrated van Rijn and Englund-Hansen formulae give very close estimations of the total load. Acker-White overestimated the total load.

Table 1. total transport in tons/tidal cycle

	Landward		Seaward		Net transport	
	spring tide	neap tide	spring tide	neap tide	spring tide	neap tide
Acker-White	8352.	3070.	10377.	2084.	2025.	986.
Englund-Hansen	5527.	2142.	8293.	2047.	2766.	95.
van Rijn	5612.	2376.	8699.	1917.	3088.	459.
Measurement	6223.	2206.	7699.	1258.	1476.	948.

CONCLUSIONS

In calibration of the hydrodynamics model different values of the Manning coefficient had to be used for ebb and flood. In this regard, it was argued that the roughness coefficient which is indicated by flow simulation may be unsuitable for the calculation of the shear stress as the most important factor for the transport capacity of the flow.

The results of the transport simulation with the Acker-White formula were in good agreement with measured concentrations. The total transport load through the measurement cross-section, however, were overestimated.

The Englund-Hansen formula was used with D_{35} rather than D_{50} as grain diameter and a constant `.16` was taken instead of `.1`. The calculated concentrations are still lower than the measured values, but the total transport load was in good agreement with measurements.

Van Rijn's approach was not sensitive to the grain diameter. It was modified by taking a smaller value for the reference level. With this changes both concentrations and the transport load were in good agreement with the measurements.

REFERENCES

Acker, P., and White, R. (1973), "Sediment transport: New approach and analysis", Journal of the Hydraulics Division, ASCE, vol. 99, HY11, 2041-2060.

Einstein, H. A. (1950), "The bed load function for sediment transportation in open channel flows", Technical Bulletin No. 1026, Department of Agriculture, Washington D.C..

Englund, F., Hansen E. (1967), "A monograph on sediment transport in alluvial stream", Teknisk Forlag, Copenhagen, Denmark, 62p.

IMDC-WLB (1993), "Containerkaai Noord, hydraulisch en sedimentologisch onderzoek", Deelraport 8., Hydraulisch en Morfologisch Onderzoek - Stormtij.

Heathershaw, A.D. (1981), "Comparison of measured and predicted sediment transport rates in tidal currents", Marine Geology, 42, 75-104.

King, I. P. (1990), "Modelling of flow in estuaries using combinations of one- and two-dimensional finite elements", Hydrosoft, vol. 3, No. 3, 108-119.

McAnally, W. H., Jr., Letter, J. V., Jr., Stewart, J. P., Thomas, W. A., and Brogdon, N. J., Jr. (1984), "Colombia river hybrid modelling system", Journal of the Hydraulics Division, ASCE, vol. 110, HY3, 301-311.

Rijn, L. C. van (1984a), "Sediment transport, Part I: Bed load transport", Journal of the Hydraulics Division, ASCE, vol. 110, HY10, 1431-1456.

Rijn, L. C. van (1984b), "Sediment transport, Part II: Suspended load transport", Journal of the Hydraulics Division, ASCE, vol. 110, HY11, 1613-1641.

Smith, L. H., and Cheng, R. T. (1987), "tidal and tidally averaged circulation characteristics of Suisun bay, California", Water Resources Research, Vol. 23, No. 1, 143-155.

Task Committee for Preparation of Sediment Manual (1971), "Sediment transport formulas", Journal of the Hydraulics Division, ASCE, vol. 97, HY4, 523-567.

MATHEMATICAL MODELLING OF RESERVOIR SEDIMENTATION: APPLICATION TO SEFID - RUD RESERVOIR, IRAN

A. SHARGHI and J. BERLAMONT
Laboratory of Hydraulics
Catholic University of Leuven(K.U.L.)
de Croylaan 2, B-3001 Heverlee, Belgium

ABSTRACT

Reservoir sedimentation creates major problems in the planning, design and operation of water resource systems. The useful life time of a reservoir can be reduced rapidly because of high sediment yields from drainage basins, and the number of reservoirs deteriorated due to complete settling up with sediments is continuously increasing.

An elegant method for recovering lost storage, which has been applied successfully for reservoirs in several countries, such as Russia, China, India and Iran is known as "Flushing". The flushing technique is simply the flushing of deposited sediments with water available in the reservoir by opening the bottom outlets and consequently lowering the water level abruptly. In order to plan flushing operations and to predict the efficiency, one should be able to predict the magnitude and distribution of sediment deposits in the reservoir. It is therefore important to develop a mathematical model for simulating both velocity fields and sediment movements.

A mathematical model has been developed which describes the unsteady flow of water and entrainment of sediments in a reservoir. The compound flow model approach with jet equations are used to predict the magnitude and lateral distribution of the sediments in the reservoir. The differential equations for simulating the gradually varied unsteady flow, the continuity and momentum equations for sediment laden water and the continuity equation for the sediment, are solved by a finite difference scheme.

The mathematical model has been applied to the Sefid - Rud Reservoir in Iran. The model results show a reasonable agreement with the observed bed profiles both for flushing and accumulation.

INTRODUCTION

The phenomenon of reservoir sedimentation poses serious problems in the planning, design and operating of water resource systems.The useful life time of a reservoir can reduce rapidly because of high sediment yields from drainage basins. In order to fight reservoir sedimentation, two basic possibilities are available: one is to prevent sediment from entering the reservoir and the other is allowing it to enter and to extract it afterwards. One of the methods for recovering lost storage without using external energy,

A. Peters et al. (eds.), Computational Methods in Water Resources X, 1357–1364.

which has been applied for reservoirs is known as flushing. Usually by the end of the irrigation period, some water remains in the reservoir and this water can be used to flush out sediment which has been deposited during the year.

Prediction of the sediment distribution pattern in reservoirs is a complex task and depends on many interrelated factors, such as: size and texture of the sediment particles, seasonal variations in river and sediment flow, size and shape of the reservoir and its operation schedule. The mathematical model offer the engineer a technique to explore the interrelationship between the important variables of the process and it also a good opportunity to study long-term system responses.

GOVERNING EQUATIONS

The basic equations which govern the flow are differential equations for simulating the gradually varied unsteady flow in a natural alluvial channel:

Momentum equation for sediment - laden water:

$$\frac{\partial(Q\rho)}{\partial t} + \frac{\partial(\rho\beta Q V)}{\partial x} = -gA\frac{\partial(\rho y)}{\partial x} + \rho g A(S_x - S_f) + \rho q_l V_l \tag{1}$$

Continuity equation for sediment - laden water:

$$\frac{\partial Q}{\partial x} + \frac{\partial A}{\partial t} + \frac{\partial A_d}{\partial t} - q_l = 0 \tag{2}$$

Continuity equation for sediment:

$$\frac{\partial Q_s}{\partial x} + (1 - \lambda)\frac{\partial A_d}{\partial t} + \frac{\partial(AC)}{\partial t} - q_{sl} = 0 \tag{3}$$

Sediment transport relation:

$$Q_s = mV^n \tag{4}$$

in which Q = the discharge of sediment - laden water; ρ = the density of the sediment - laden water given by $\rho = \rho_w(1 - C) + \rho_s C$; ρ_w = the water density, ρ_s = the sediment particle density; v = the mean flow velocity; c = the sediment concentration in the cross section on a volume basis; g = the gravitational acceleration; β = the momentum correction factor for velocity distribution; A = the cross-sectional area of the channel; y = the flow depth; S_x = the bed slope; s_f = the friction slope given by the Manning equation = $Q^2 n^2 / A^2 R^{4/3}$; q_l = the lateral inflow of sediment - laden water into the stream; v_l = the velocity component of the lateral flow in the x - direction; A_d = the volume of deposition or erosion of sediment on unit length of channel bed; Q_s = the total sediment load in units of volume per unit time; λ = porosity of sediment in bed layer($0 \leq \lambda \leq 1$); q_{sl} = the lateral sediment inflow into the stream; and m and n = constants that depend on the sediment and flow characteristics.

Equations (1) - (3) are a set of nonlinear first - order partial differential equations of the hyperbolic type. They contain two independent variables x and t and three dependent variables, Q, y, A_d; the remaining terms are functions of the three basic unknowns. Equation (4) is an empirical relation based on laboratory and field measurements under steady - uniform flow conditions, and it is considered a good first approximation for moderate changes in the total sediment load.

The compound flow model approach developed by Dass (7,8) and modified by Lopez (2,8) to solve the governing equations is modified and used to route the flow of water and sediment through the river - reservoir system. The river - reservoir system in this study is shown in Figure 1. The model considers a reservoir formed by a dam which has been constructed on a river course. The river is presented by a single channel; one - dimensional flow phenomena are predominant, whereas a set of multiple channels is used to simulate the river and the flood plains in the reservoir. This development is based on dividing a nonuniform cross section into a set of subsections in such a way that each subsection can be treated as uniform within itself. Therefore, the basic equations (1) - (3) are solved by first treating the whole river reach as a single channel and then solving the sediment continuity equation by considering a multiple channel approach. In this way the variations of bed deviations with time for each individual subsection can be determined. The following assumptions based on the compound flow model approach procedure have been used in the mathematical model: (1) the term $\partial A_d/\partial t$ is much smaller than $\partial A/\partial t$; and (2) within a short time period the change in channel geometry due to sediment routing is not significant, so that the three basic equations can be solved simultaneously. The calculation procedure is then summarized as follows. First equations (1) and (2) are solved for two unknowns Q and y at each cross section in the reach. The next step is to determine the value of these unknowns for each individual stream by using the equation: $Q = Q_m(K_m/K)$, in which Q_m and K_m are the water discharge and conveyance for any subchannel at the known cross section (such as mth subsection) and Q and K are the total discharge and conveyance of a cross section. Then the flow continuity equation, equation (1), is rewritten as follows at each individual mth stream to solve the lateral flow transfer:

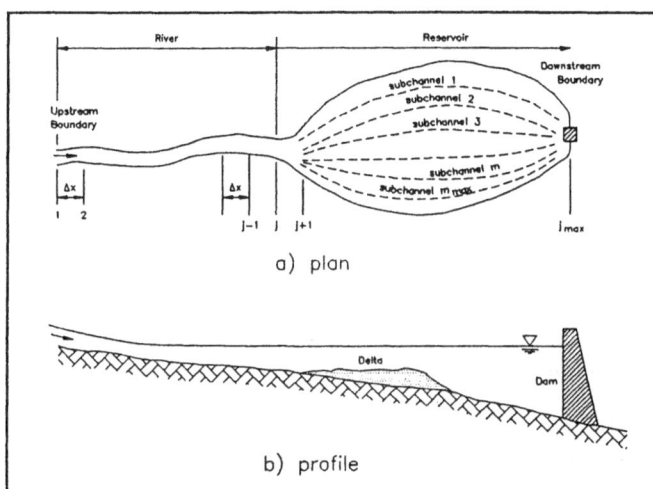

FIG. 1 Schematic representation of the river - reservoir system

a) plan

b) profile

$$\frac{\partial Q_m}{\partial x} + \frac{\partial A_m}{\partial t} - q_m + q_{m-1} = 0 \qquad (5)$$

Noting that Q_m and A_m are known from the flow continuity calculations and that q_{m-1} for $m = 1$ is specified in advance, the only unknown is q_m. Finally the sediment continuity equation for each individual stream which is to be solved for the variation in bed elevations will be introduced as follows:

$$\frac{\partial Q_{s_m}}{\partial x} + (1 - \lambda)_m \frac{\partial A_{d_m}}{\partial t} + \frac{\partial (A_m C_m)}{\partial t} - q_{s_m} + q_{s_{m-1}} = 0 \qquad (6)$$

The entering of the river into reservoir is comparable to a submerged plane jet flow discharge from a slot of width equal to the width of the river. The two dimensional equations for velocities in a submerged turbulent jet discharging into a fluid of the same density have been given by Albertson and others (1,4,8). With reference to Figure 2, they distinguish two basic zones in their theory. The first zone is called the zone of flow establishment, in which the fluid undergoes lateral diffusion and deceleration

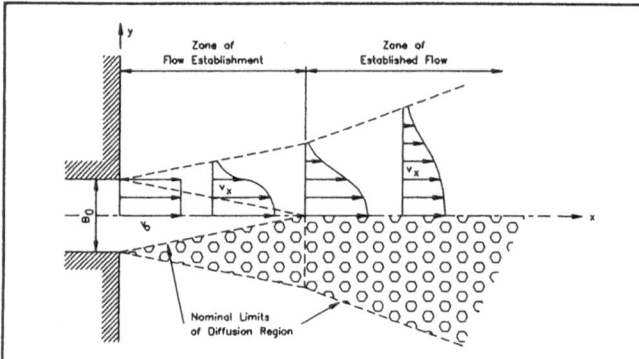

FIG. 2 Schematic representation of jet diffusion

as a direct result of the turbulence generated at the border of a submerged jet. The constant - velocity core of the jet will steadily decrease in lateral extent. The limit for this initial zone of flow establishment (which is almost equal to five times the width of the river) is achieved when the mixing region has reached the centre line of the jet. The second zone is the zone of established flow, in which the diffusion process continues without essential change in character (8,9).

The velocity field in the no diffusion zone:

$$V_{(x,y)} = V_0 \tag{7}$$

The velocity field in the zone of flow establishment:

$$V_{(x,y)} = V_0 \exp\left[-42.084\left(0.096 + \frac{y - \frac{B_0}{2}}{x}\right)^2\right] \tag{8}$$

The velocity field in the zone of established flow:

$$V_{(x,y)} = V_0 \sqrt{\frac{B_0}{x}} \exp\left[0.812 - 41.58\frac{y^2}{x^2}\right] \tag{9}$$

In which v_0 = the mean velocity at the mouth of the river; B_0 = the river width; and x and y = the longitudinal and transverse coordinates depicted in Figure 2.

For simulation of the flushing technique in this study, close to the dam, the velocity field is approximated by the following expression:

$$V_{(x,y)} = U_0 \sqrt{\frac{D}{x}} \exp\left[0.812 - 41.58\frac{y^2}{x^2}\right] \tag{10}$$

Where v_0 = the velocity of the orifice = $c_v \sqrt{2gh}$; D = the bottom outlet diameter; and c_v = the velocity coefficient. From the integration of the velocity distribution equations, correction factors for the water discharge distribution at each section are obtained and incorporated in the compound flow model. The distribution of sediment in the reservoir highly depends on the shape of the basin. If the reservoir is regular in shape, deposits will normally spread almost uniformly along the axis. On the other hand, if the shape of the reservoir is irregular, the sediment deposition might be distributed in irregular patterns. In narrow reservoirs, the incoming flow spreads evenly across the pool and therefore the jet effect is not significant but if the stream enters a wide reservoir, the flow tends to act like a jet with a considerable effect.

In this mathematical model, nonuniform grain size distribution is treated by considering different size fractions separately and by calculating the transport capacities for each grain size. Then, the total variation in bed elevations can be obtained by superimposing the results obtained from the sediment routing for each sediment fraction. In this way the sorting process and its effect in the delta formation can be investigated.

NUMERICAL SCHEME

Equations (1) - (3) are a set of nonlinear hyperbolic equations, and closed - form solutions are available only for idealized cases. Therefore, they are solved by numerical schemes. In this model, an implicit finite difference scheme developed by Preissmann (6,9) is used. According to this scheme, a variable, say f, and its derivatives are discretized as follows:

$$f(x,t) \approx \frac{1}{2}\left[\theta(\Delta f_{i+1} + \Delta f_i) + f_{i+1}^j + f_i^j\right] \tag{11}$$

$$\frac{\partial f}{\partial x} \approx \frac{1}{\Delta x}\left[\theta(\Delta f_{i+1} - \Delta f_i) + f_{i+1}^j - f_i^j\right] \tag{12}$$

$$\frac{\partial f}{\partial t} \approx \frac{1}{2\Delta t}(\Delta f_{i+1} + \Delta f_i) \tag{13}$$

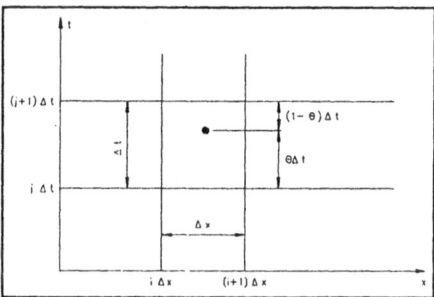

FIG. 3 Finite difference scheme

In which f represents Q, A, T, etc.; $\Delta f_i = f_i^{j+1} - f_i^j$; Δx and Δt = the length and time increments, respectively; i =node along x-axis; j =node along t-axis (Figure 3); and θ =a weighting factor. All the variables are known at all nodes of the network on the time line t^j. The unknown values of the variables on time line t^{j+1} can be found by solving the system of linear algebraic equations formulated from substitution of the finite-difference approximations, equations (11), (12), and (13), into the basic governing equations with prescribed initial and boundary conditions by the double sweep method (5,8). Once the values of water elevation and discharge at a new time step are known, the bed level z^{j+1} was computed by solving the continuity equation for sediment (3) with an explicit finite difference scheme.

MODEL APPLICATION

The model was applied to the Sefid - Rud Reservoir located at the town of Manjil, 260 km north of Tehran (IRAN) to check if the model is suitable to simulate with reasonable accuracy the distribution of the sediment in the reservoir. The study area (Fig. 4) in this mathematical model extends from the Sefid - Rud Dam itself up to 8.43 km upstream of the dam. Based on the topographic and hydraulic characteristics, the system was divided into 3 subchannels. The data available from the periods 1963 to 1970 and 1982 - 1983 (flushing period) were selected for the calibration of this mathematical model in case of sediment deposition and flushing technique respectively. After the model had been calibrated, the data from the periods 1970 to 1976 and 1983 - 1984 (flushing period) were used for verification of the model. The boundary used in this model consisted of the following conditions. At the upstream boundary condition a discharge hydrograph and sediment discharge hydrograph were used. The downstream boundary condition was a stage hydrograph in case of sediment deposition and an orifice equation (as a discharge being a function of water elevation variations) for the case of flushing. The sediment discharge hydrograph used in the model was determined from the power relationship between the sediment, size fractions of the sediment and the mean velocity. The following procedures were carried out to properly adjust or calibrate the required parameters of the model: 1) Manning's coefficients and the water and bed elevations in the system are initially estimated, based on the available data and visual observation. 2) Then Manning's coefficients were adjusted through the assumption of a fixed bed condition without sediment movement, as well as the reconstitution of observed water elevation variations at different sections along the river under study. As there is no doubt that the sediment distribution in the reservoir will effect the water elevations, further adjustment will be needed after completion of step 4.

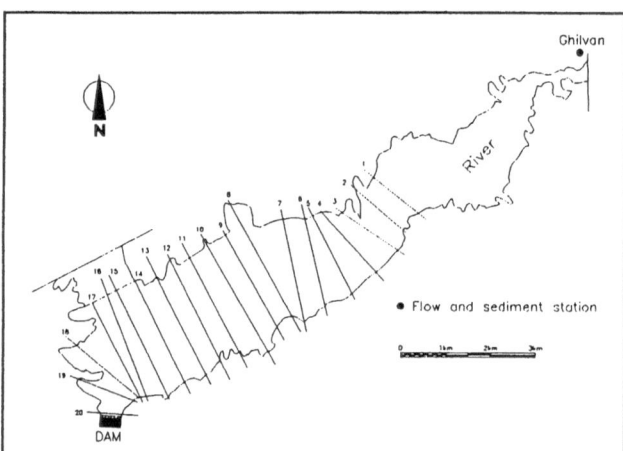

FIG. 4 Schematic representation of the Sefid - Rud Reservoir showing the cross sections

3) The next step is to release the sediment inflow and to adjust the sediment parameters m and n in the sediment transport equations by comparing the sediment distribution patterns with observed bed profiles. 4) The distribution of sediment patterns at the different sections and subsections in the system was further adjusted by proper selection of the jet width constant "α" at sections where jet corrections were required.

The bed profiles computed by the mathematical model, together with the average bed elevations observed at different cross sections along the reservoir under study, both for accumulation and flushing, are shown in Figure 5 and 6 respectively.

RESULTS AND DISCUSSION

As far as sediment deposition is concerned, according to Figure 5, aggradation took place almost uniformly in all sections after six years of simulation. The predicted and observed bed elevations in subchannel 1 are lower than the bed elevations in subchannels 2 and 3 at the downstream sections. This is due to the fact that the reservoir is connected with a small river, as indicated in Figure 4. The lateral inflow and sediment discharge which enter the reservoir from sections 14 to 17 of subchannel 1 have also been accounted for in the model. On the basis of this phenomenon, it can be said that most of the sediment deposition took place on the main channel (subchannel 2) and subchannel 3. This appears to be in agreement with field observations. The results also showed that the computed averaged bed profiles are lower than the observed bed profiles in the farthest downstream sections, which is due to the fact that the density currents were not considered in this mathematical model. Satisfactory results of the mathematical model for flushing suggest that the model is able to predict the magnitude and lateral distribution of the sediments in the Sefid - Rud Reservoir. Figure 6 shows the changes in bed profiles which have taken place at different cross sections along the reservoir by applying flushing during year 1983 - 1984. According to this figure, the model predicts higher bed elevations in sections 7 to 12 of the main channel (subchannel 2) as compared to observed bed elevations.

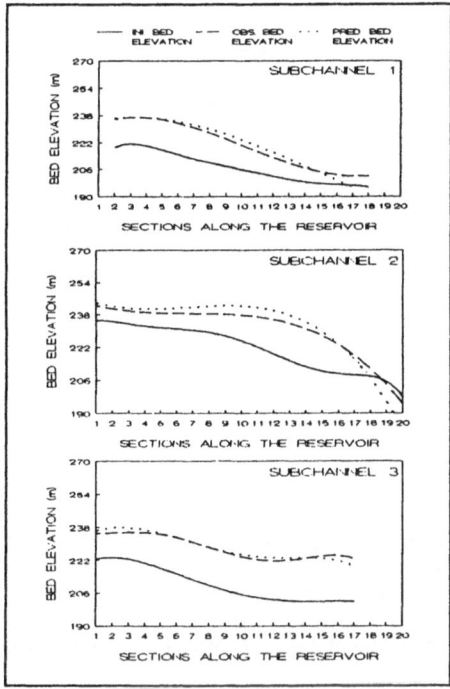

FIG. 5 Variation of the bed elevations after six years of simulation (1970 - 1976)

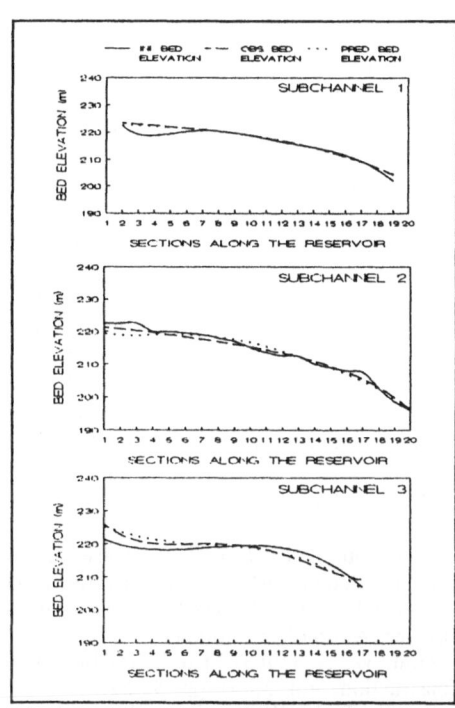

FIG. 6 Variation in bed elevations after applying the flushing technique (1983 - 1984)

The reason for this difference is the sudden high rain fall which occurred during the flushing period. This sudden high rain fall, which has not been taken into account in this model simulation, causes the bed to be washed and moved in the downstream direction in those sections. Figure 6 also indicates that there are no big changes in the bed profiles of subchannel 1. Instead some sediment accumulation and slight degradation have occurred. In subchannel 3 on the other hand, some sediment accumulation took place in sections 1 to 8; in section 9 and consecutive sections downstream degradation has occurred. These phenomena are due to low water elevation in the reservoir and to the velocity in lateral channels being lower than in the main channel (subchannel 2). Finally, Figure 6 suggests that during the flushing simulation in the Sefid - Rud Reservoir, sediment was washed away mostly from the main channel (subchannel 2) and subchannel 3. This phenomenon appears to be in good agreement with the observations. The results obtained by the mathematical model also show that the bed material is becoming coarser with time in the upstream sections, and finer in the downstream sections. This shows that the model is able to predict with good accuracy the sediment sorting process.

CONCLUSIONS

A mathematical model is presented for studying the unsteady flow of water and entrainment of sediment in a reservoir. The model considers a reservoir formed by a dam which has been constructed on a river course. The river is presented by a single channel, whereas a set of multiple channels is used to simulate the reservoir. The compound flow model approach, together with a two dimensional jet theory, is used to predict the magnitude and lateral distribution of the sediment in the reservoir. A simple equation for the sediment discharge as a power function of flow velocity is used.

The model was applied to the Sefid - Rud Reservoir in Iran to determine the feasibility of the model. The computed bed profiles were compared with the measured bed profiles. The agreement between them is satisfactory, for both flushing and accumulation. The results also showed that the compound model approach, together with a two dimensional jet theory, are highly useful to predict the magnitude and lateral distribution of the sediment in the reservoir. The model may well be used by incorporating seasonal variations in the flow of water and sediment for studying the long-term responses in the river-reservoir system. Also it can be applied as a predictive tool for reservoir sedimentation. Satisfactory results of the model for flushing suggest that the model can also be used to develop operational methods by which the maximum amount of sediment can be flushed with the available amount of water in the reservoir.

REFERENCES

1. Albertson, M. L., Dai, Y. B., Jensen, R. A., and Rouse, H. (1950) "Diffusion of submerged jets", Transaction, ASCE, Vol. 115, pp. 2041-2060.
2. Annandale, G. W. (1987) "Reservoir sedimentation", Developments in water science, Elsevier, Amsterdam, Holland.
3. Graf, W. H. (1983) "The hydraulics of reservoir sedimentation", J. Water Power & Dam Construction, Vol. 35(4), pp. 45-52.
4. French, R. H. (1985) "Open channel hydraulics", McGraw - Hill Book Company.
5. Liggett, A. J., and Cunge, J. A. (1975) "Numerical methods of solution of the unsteady flow equations", Vol. 1 (Mahmood, ed.), Fort Collins, Colorado, pp. 89-182.
6. Sharghi, A., and Berlamont, J. (1990) "Reservoir sedimentation", 3rd Iranian Congress of Civil Engineering, Shiraz, Iran.
7. Sharghi, A., and Berlamont, J. (1994) "Mathematical model for reservoir sediment-ation", 2nd International conference on river flood hydraulics, York, England (in press).
8. Sharghi, A. (1994) "Reservoir sedimentation", Ph.D thesis, Hydraulics Laboratory, Catholic University of Leuven, Belgium.
9. Sloff, C. J. (1991) "Reservoir sedimentation", A Literature Survey, Report No. 91-2, Faculty of Civil Engineering, Delft University of Technology, Delft, Holland.
10. Yalin, M. S. (1992) "River mechanics", Pergamon Press Ltd, Headington Hill Hall, Oxford, England.

MODELLING POSSIBLE IMPACTS OF RIVER SEDIMENTS ON A WATERED MEADOW ECOSYSTEM

K.ULBRICH[1], G.HANSCHMANN[2] and E.WEIßBRODT[3]
UFZ-Centre for Environmental Research Leipzig-Halle Ltd.
Departments of Ecological Modelling[1], Analytical Chemistry[2],
Remediation[3]
Permoserstr.15, 04318 Leipzig
Germany

A 2-dimensional mathematical model was developed to investigate the influence of remobilized river sediments on a meadow ecosystem. In the north of the city of Leipzig (Germany) a riverine meadow, threatened by drying up, is planned to be flooded with water from the river *Weiße Elster*. However, the pollutants, contained in the water, present a danger for a lot of species living in the meadow. A great part of the pollutants is adsorbed to resuspended river sediments and will be deposited together with them in the meadow. The composition of the meadow sediments depends on the flow regime of the watering and on the specific kind of the introduced river sediments. It is necessary to investigate the so emerging pollution patterns to estimate their impacts on biological processes in the meadow ecosystem. The mathematical model describes the settling behaviour of essential fractions of particles. Results derived from special calculations are consistent with measured data for the iron transport by the sediment.

INTRODUCTION

The meadows at the river *Weiße Elster* in the north of the city of Leipzig are threatened by drying up. The ground water level has come down to a level, where the species living there are endangered. A possible solution of the problem is the flooding of the meadows with river water. However, an additional danger is caused by the sediments settled in the river. There are placed about 1,5 million tons of sediments in the area of Leipzig. The greatest part of them is located in a high tide basin. The thickness of the sediments deposited there reaches nearly 2 meters (Stottmeister and Weißbrodt,1993).

As a consequence of the input of sewages by the carbochemical industry for 3 decades they are loaded with polyaromatic hydrocarbons (PAH), phenols, heavy metals and other pollutants (see table 1). The threat for the ecosystems results both from the mechanical mobilization and from chemical remobilization processes, leading to increased toxic effects. Examples are autoxydation reactions of phenoles with synthesis of high molecular products and the synthesis of readily soluble heavy metal compounds from not readily soluble sulfides.

A. Peters et al. (eds.), Computational Methods in Water Resources X, 1365–1372.
© 1994 Kluwer Academic Publishers. Printed in the Netherlands.

Most of the pollutants are bound at solid particles while the poral water is much less polluted (Stottmeister and Weißbrodt,1993). Therefore it is of importance to consider the dispersal of those particles when estimating the dangers resulting from pollution.

Table 1 Contents of heavy metals and some other important organics in the sediments of the river Weiße Elster (high tide basin)

element or substance	contents,ppm 0-50 cm deep
Cd	14
Pb	204
Zn	3200
Cu	294
Ni	250
Fe	58900
Mn	1086
Cr	593
As	29
PAH and Naphthalines	7,2
Phenoles	4,8

The sediments are continuously resuspended and carried down-stream by the current. The amount of the suspended matter depends on the water flow regime. When the wetlands are watered by the river, the particles enter the meadow and deposit there. The chemical analyses show that the concentration of heavy metals (copper, lead, chromium, iron) is significantly decreasing with the distance from the *Weiße Elster* river. The same decrease is found with the concentrations of nitrate, sulphate and of organic toxicants (Hanschmann &al,1993). Due to high biological activity in the watered area in summer the concentration of total organic matter (mainly humic and fulvic acids) is increased. The formation of patterns is caused also by the differences between the particle fractions (gravel, clay, silt, organical materials, e.g. soot). Due to differences in their physico-chemical properties different substances are bound. Silt particles bind mainly organic chemicals (Umlauf and Bierl,1987). For clay particles the ionic adsorption of metals is typically. Chemical load effects specific properties of the sediments in relation to their ability to biological degradation and to toxicity. To understand the correlations between biological processes and temporal-spatial patterns generated by the deposition of river sediments mathematical methods are helpful. Thereby a first step is the description of the

Thereby a first step is the description of the generated chemical and physical patterns (Ulbrich et al.,1993). They depend on the watering regime and the content and composition of the river sediments. At the same time it is necessary to investigate the fate of some special substances. Taking as an example the ditch system in the meadows described above, we explored the dispersal of the river sediments using experimental data of iron content.

THE MODEL

The mathematical model was developed to describe spatial-temporal patterns in the meadow, which are important for biological processes, occuring in the ecosystem. A lot of physical, chemical and biological interactions can be observed. Because it is impossible to consider the whole complexity in a model, we concentrated on the sediment transport by the water flow for a first approach. We regarded following processes: advection, sedimentation, diffusion and erosion (see fig.1). According to the non-stationary character of the processes a time-dependent model was developed.

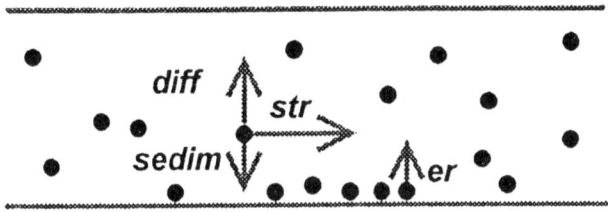

fig.1 Processes, describing sediment transport
str advection, sedim sedimentation, er erosion, diff diffusion

Model structure

The model is based on a discretisation of advection-diffusion equations, which are commonly used to describe the transport of suspended particles (Brüggemann,1991 and Müller,1991). The water body was subdivided into segments with a length of e.g. 5m and vertical layers with a height of e.g. 0,1 m. Cells formed by this way are regarded as homogeneous areas, which are the structural basis of the model (see fig.2). The cells are interacting through the processes named in fig.1.

cell   water layers

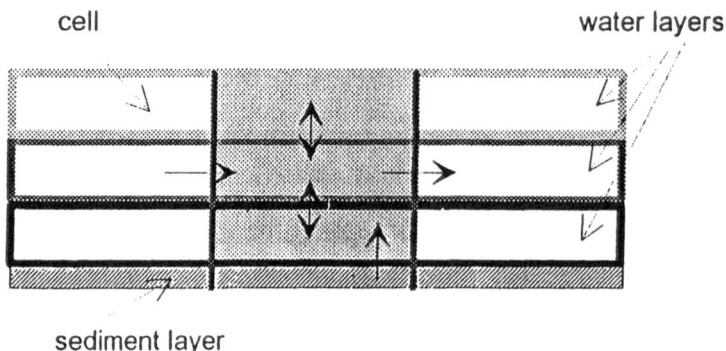

sediment layer

Fig.2 Basic scheme of the model structure, interactions between the cells

The dynamics of the sediment layer are defined by the interactions of settling and resuspending particles. In order to describe the processes mentioned above simple submodels are developed.

Submodels

For each particle fraction the spatial-temporal distribution was determined according to the following equations:
for the sediment

$$sd(t,x) = SEDIM(t-1,x,y_k) - ER(t-1,x)$$

for suspended particles

$$p(t,x,y) = STR(t-1,x,y) + SED(t-1,x,y) + DIFF(t-1,x,y) + ER(t-1,x,y)$$

where p is the particle mass in a cell, sd the sediment mass in the river segment and t the time. The submodel terms are described by STR (advection term), SED (sedimentation term), DIFF (diffusion term) and ER (erosion term).
The advection term is calculated by means of the advection rate:

$$STR(t,x,y) = str(t,x-1,y) * p(t,x-1,y) - str(t,x,y) * p(t,x,y)$$

with

$$str(t,x,y) = \frac{\Delta t * w(t,x,y)}{\Delta l}$$

where str(t,x,y) is the advection rate, Δt the time step, w the flow velocity and Δl is the segment length.
The sedimentation term is calculated by

$$SEDIM(t,x,y) = sedim(t,x,y-1) * p(t,x,y-1) - sedim(t,x,y) * p(t,x,y)$$

$$sedim(t,x,y) = \frac{\Delta t * w_s}{\Delta h}$$

where sed(t,x,y) is the sedimentation rate, w_s the settling rate, Δh the height of the vertical layer.

Due to turbulent diffusion, particles move upwards in the layers above. The submodel of diffusion looks as follows:

$$DIFF(t,x,y) = \Delta t * (diff(t,x,y) * (p(t,x,y+1) - 2 * p(t,x,y) + p(t,x,y-1))$$

where diff(t,x,y) is the diffusion rate.

Describing the erosion one has to consider the sediment layer:

$$ER(t,x) = \Delta t * er(t,x) * sd(t,x)$$

where er(t,x) is the erosion rate.
where al is the active layer.

Model calculations

The following assumptions are made:
1. Four particle fractions are described (fractions 1,2,3 and 4).
2. Each particle fraction can be characterized by a mean settling velocity.
For lack of measured data for w_s, we assumed values between $1*10^{-3}$ and $1*10^{-5}$ $m*s^{-1}$.
3.Aggregation of particles does not occur, i.e. the settling behaviour of the particles is assumed to be constant.
4.Impacts of seasons (e.g. summer) are considered by changing the settling velocities. This assumption is based on the observations of Greiser (1988), conclusing that biological activity has a considerable influence on settling velocity.

For model calculations following scenarios are choosen:
A) silent water flow and B) highwater. For each fraction the sediment mass is calculated.

Silent water flow is characterized by a flow velocity of 3 $cm*s^{-1}$ and a content of suspended matter in the input stream of 10 $mg*l^{-1}$. Calculations are carried out for 2 different settling velocities, differing by the factor 1.6 (see fig.3 and 4). The increasing of w_s may be caused by biological activity in spring.

High water is described by increased both flow velocity and content of suspended matter in the input stream (see fig.5). The values applied here for w_s and p_0 are 15 $cm*s^{-1}$ and

200 mg*l⁻¹ respectively.

Figure 3 shows the formation of sediment layers in the meadow for scenario A. The river water is entering the meadow ditch system at x=0. The settling velocities of the 4 fractions are assumed to be typical for winter (no biological activities).

The fractions are settling in a various manner, forming sediments with different compositions. Fraction 1 with $w_s=1*10^{-3}m*s^{-1}$ is predominating in the sediment of the first 100 m, fraction 2 with $w_s=5*10^{-4}m*s^{-1}$ is prevailing from 100 m until 320 m.

Fraction 3, which is characterized by $w_s=1*10^{-4}cm*s^{-1}$, represents the main part of the sediment, settling behind 320m. Fraction 4 with a very low settling velocity ($w_s=1*10^{-6}m*s^{-1}$) is containing in all sediments by small parts.

The pattern showed in figure 3 was received after 10^{-4} time steps. Increasing of the calculation time does not change the pattern, i.e. it is a stationary one.

Figure 4 shows the influence of increased settling velocity on the sedimentation pattern. It is assumed, that w_s is increased by the factor 1.6 due to biological activity in spring. In comparison to figure 3 the sedimentation pattern does not differ considerably.

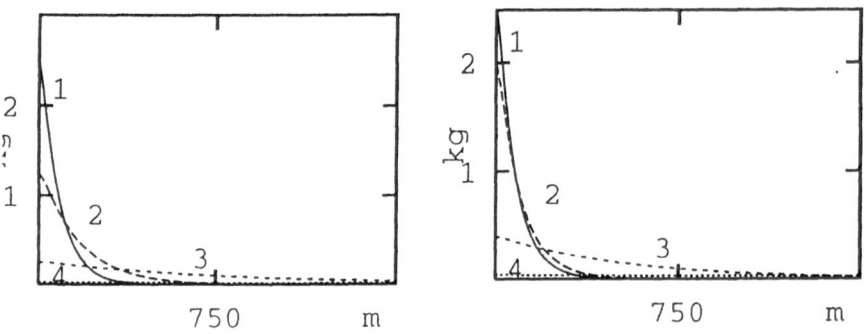

Fig.3 Silent water, winter   Fig.4 Silent water,spring

Results obtained by simulating scenario B are represented on figure 5. The influence of high water on the sedimentation pattern is clearly shown. The composition of the sediments is crucial another one than for silent water (fig.3).

The transport of substances bound to the sediment was investigated for the scenario A at the example of iron-containing sediments (see fig.6). Experimental values have been received filtering the water samples taken from several sites along the flooding route through a glass-fibre filter (Sartorius, 1000μm). The content of heavy metals in the sediments deposited on the filters was determined by X-ray fluorescence analysis (Hanschmann et al.,1993). Figure 6 shows the consistence of experimental data and calculated values for $w_s = 1*10^{-3} m*s^{-1}$.

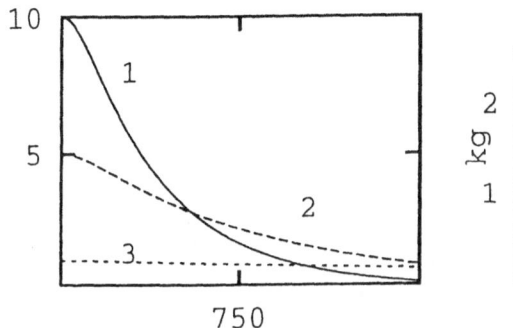
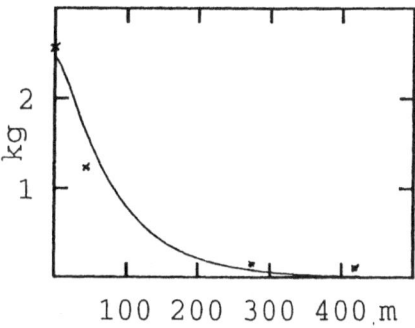

Fig.5 High water

Fig.6 Iron transported by sediment points are measured data, line calculated with $w_s = 1*10^{-3} cm*s^{-1}$

DISCUSSION and CONCLUSIONS

The knowledge of the distribution and chemical loadings of different sediment fractions is necessary to estimate possible impacts of river water, used for flooding, on biological processes occuring in the riverine meadow ecosystem. Sediment distribution depends mainly on the watering regime and sediment properties (fig.3 and 5). The influence of increased settling velocity (e.g. due to biological activities) is relatively small (fig.4).

We compared the results of several model calculations with measured data concerning several important pollution components. This yields information about the settling behaviour of defined substances. For instance, the settling velocity of iron-containing sediments was calculated as about $1*10^{-3} m*s^{-1}$. The impact of the pollution pattern on the ecosystem after that can be described by developing models including correlations between quality and quantity of substances, entering the meadow system by sediments, and population dynamics (e.g. of microorganisms).

Sediment distribution in a riverine meadow, flooded by river water, can be described by means of a 2-dimensional time-dependent mathematical model.

For given values of settling velocities and sediment composition, realistic pollution pattern could be derived.

Modelling pollution patterns is a first step in modelling biological processes in the meadow.

LITERATURE

Brüggemann,R.,Trapp,S.,Matthies,M.(1991) "Behaviour assessment of a volatile chemical in the Rhine river",Environmental Toxicology and Chemistry,vol.10,1097-1103

Greiser,N. (1988) "Zur Dynamik von Schwebstoffen und ihren biologischen Komponenten", Diss.,Hamburger Küstenforschung, Heft 45

Hanschmann,G.,Geyer,W.,Ziegler,M.,Morgenstern,P.,Mattusch,J.(1993) "Tracer Applications of Optical Spectroscopy to Water Quality",Final report of the summercamp 1993,UFZ-Centre for Environmental Research Leipzig-Halle Ltd.

Müller,A.,Grabemann,I.,Krohn,J.,Kunze,B.,Lobmeyr,M.,(1991) "Modellierung des Transportes von Wasserinhaltsstoffen in Fließgewässern unter besonderer Berücksichtigung des Schwebstoffs, Tagung der AGF vom 28.bis 29.11.91 im Wissenschaftszentrum Bonn-Bad Godesberg", Bonn-Bad Godesberg: AGF,1991.20-24

Stottmeister,U. and Weißbrodt,E. in cooperation with Pürschmann,J., Martius,S., Kuschk,P.,Kopinke,F.-D.,Remmler,M.and the Department for Analytical chemistry (1993) "Gefährdungen der Gewässer im Raum Leipzig durch Abwasserlasten der braunkohlenverarbeitenden Industrie",BMFT-Förderkennzeichen 0 3394 19

Ulbrich,K.,Weißbrodt,E.,Marsula,R.,Jeltsch,F.(1993) "Modellierung ökotoxikologicsher Belastungsmuster in Fließgewässern anhand von Experimentaluntersuchungen an der Weißen Elster", Verhandlungen der Gesellschaft für Ökologie,Band 23

Umlauf,G.and Bierl,R.(1987) "Distribution of Organic Micropollutants in Different Size Fractions of Sediment and Suspended Solid Particles of the River Rotmain",Z.Wasser-Abwasser-Forschung 20,203-209

16. ALGEBRAIC METHODS

A MULTIGRID-BASED SOLVER FOR TRANSIENT GROUNDWATER FLOWS USING CELL-CENTERED DIFFERENCES

Myron B. Allen[1] and Zhonghe Wang[2]
Departments of Mathematics[1] and Geology and Geophysics[2]
University of Wyoming, Laramie, WY 82071, U.S.A.

Cell-centered finite-difference methods enjoy a close relationship with mixed finite-element methods, especially in the context of the groundwater flow equation. Both methods have the attractive feature that they generate accurate velocities and heads in steady-state flows. In addition, it is possible to formulate multigrid-based solvers that allow for computationally efficient solutions to the algebraic equations that arise, even in the presence of strongly heterogeneous hydraulic conductivities. We examine the extensions of these ideas to transient groundwater flows. In particular, we consider numerical experiments that compare cell-centered differences and mixed finite elements and that show the ability of the former scheme to produce reasonable results in the presence of discontinuous conductivity fields.

1. INTRODUCTION

The equation

$$S \, \partial_t p - \nabla \cdot (K \nabla p) = f \tag{1}$$

governs the flow of groundwater in a saturated, horizontal, two-dimensional aquifer Ω. Here, $p(\mathbf{x}, t)$ denotes the hydraulic head; $S(\mathbf{x})$ is the specific storage; $K(\mathbf{x})$ is the saturated hydraulic conductivity, and $f(\mathbf{x}, t)$ represents sources. We assume that $p = 0$ on $\partial\Omega$. Equation (1) arises from the mass balance,

$$S \, \partial_t p + \nabla \cdot \mathbf{v} = f, \tag{2}$$

when we substitute for the groundwater velocity $\mathbf{v}(\mathbf{x}, t) = (u(\mathbf{x}, t), v(\mathbf{x}, t)$ using Darcy's law:

$$K^{-1}\mathbf{v} + \nabla p = 0. \tag{3}$$

We examine two closely related discretizations of Equation (1) or, equivalently, Equations (2) and (3). The first discretization uses a mixed finite-element method to discretize in space and employs a two-level iterative procedure to solve the resulting linear system at each time step. The inner iteration in this procedure uses a multigrid algorithm. The second discretization uses cell-centered finite differences in space and

A. Peters et al. (eds.), Computational Methods in Water Resources X, 1375–1382.

has the computational advantage that it requires only a one-level iterative procedure at each time step. We confirm, through a set of numerical experiments, that the two discretizations have comparable orders of accuracy and that they yield good numerical results in heterogeneous aquifers exhibiting sharp changes in hydraulic conductivity.

Our work rests on several previous efforts. The basic theory behind the iterative solution scheme for the lowest-order mixed finite-element method appears in the context of steady groundwater flow in Allen, Ewing, and Lu (1992). A subsequent paper (Allen and Curran, 1992) investigates the parallelism of this algorithm on distributed-memory computers. Shen (1992) establishes a close relationship between the lowest-order mixed method and the cell-centered finite-difference method, proving that the two techniques yield comparable errors in the steady problem. Beckie et al. (1993) examine a multigrid-based algorithm for the steady version of equation (1), using a higher-order mixed method.

In what follows, Section 2 reviews the mixed finite-element method and its algebraic relationship to cell-centered finite difference. Section 3 presents the results of numerical tests, and Section 4 summarizes our conclusions.

2. REVIEW OF THE NUMERICAL METHODS

We first review the mixed finite-element method. Consider a rectangular grid having mesh size h on $\Omega = (0,1) \times (0,1)$. We seek an approximate solution of the form $(p_h, \mathbf{v}_h)$. Here, the approximate head p_h belongs to the trial space Q_h of functions that are piecewise constant on the grid, and the approximate velocity $\mathbf{v}_h$ belongs to W_h, the space of functions whose first coordinates are piecewise linear in x and piecewise constant in y and whose second coordinates are piecewise constant in x and piecewise linear in y. Corresponding to Equations (2) and (3) is the following variational system:

$$\int_\Omega K^{-1}\mathbf{v}_h \cdot \mathbf{w} - \int_\Omega p_h \nabla \cdot \mathbf{w} = 0 \qquad \forall \mathbf{w} \in W_h,$$

$$\int_\Omega qS\,\partial_t p_h + \int_\Omega q\,\nabla \cdot \mathbf{v}_h = \int_\Omega qf \qquad \forall q \in Q_h. \tag{4}$$

Using standard bases for Q_h and W_h (Douglas, Ewing, and Wheeler, 1983) in these equations and adopting a variably weighted approximation to the time derivative yields a linear system having the following structure:

$$\begin{bmatrix} \theta A & \theta N \\ \theta N^\top & \Delta t^{-1}M \end{bmatrix} \begin{bmatrix} U \\ P \end{bmatrix}^{n+1} = \begin{bmatrix} G_1 \\ G_2 \end{bmatrix}^n, \tag{5}$$

where

$$\begin{bmatrix} G_1 \\ G_2 \end{bmatrix}^n = \begin{bmatrix} 0 \\ F \end{bmatrix}^{n+\theta} + \begin{bmatrix} (\theta-1)A & (\theta-1)N \\ (\theta-1)N^\top & \Delta t^{-1}M \end{bmatrix} \begin{bmatrix} U \\ P \end{bmatrix}^n.$$

Here, the vector U contains approximate x- and y-velocities, associated with cell edges perpendicular to the x- and y-directions, respectively. The vector P contains values of head, associated with cell centers. Under proper lexicographic ordering of unknowns, the block matrix A, which contains information about the hydraulic conductivity, is tridiagonal. The block matrix N is a bidiagonal differencing matrix, and M is a diagonal storage matrix.

There are many approaches to solving systems like this (Ewing et al., 1989). One, based on the idea of eliminating velocities in favor of heads, extends the two-level scheme presented in Allen, Ewing, and Lu (1992). Let $D_1 = \text{diag}(A)$. Then solve as follows:

1. $G^{n+1,m} \leftarrow G_2^n - N^\mathsf{T} D_1^{-1}[G_1^n + (D_1 - A)U^{n+1,m}]$.

2. Solve $(\Delta t^{-1}M - N^\mathsf{T} D_1^{-1}N)P^{n+1,m+1} = G^{n+1,m}$ for $P^{n+1,m+1}$.

3. $U^{n+1,m+1} \leftarrow D_1^{-1}[G_1^n + (D_1 - A)U^{n+1,m}] - D_1^{-1}N P^{n+1,m+1}$.

4. If convergence fails, $m \leftarrow m + 1$.

Lagging off-diagonal terms in A in this way allows us to solve a pentadiagonal matrix equation for head in step 2, instead of solving a system involving the full matrix $\Delta t^{-1}M - N^\mathsf{T} A^{-1}N$. For the pentadiagonal system, we use several cycles of a highly parallelizable multigrid algorithm (Allen and Curran, 1992). In the steady analog of Equation (1), this algorithm, with the choice $D_1 = \text{diag}(A)$, exhibits good convergence properties even in the presence of fine spatial grids and heterogeneous coefficients K (Allen, Ewing, and Lu, 1992). In the transient problem, the matrix $\Delta t^{-1}M$ enhances the diagonal dominance of the system, thereby improving the conditioning with respect to the steady problem.

In the corresponding cell-centered difference scheme, we associate an approximate head value $p_{i-1/2,j-1/2}$ with the center $(x_{i-1/2}, y_{j-1/2})$ of each cell $[x_{i-1}, x_i] \times [y_{j-1}, y_j]$ formed by the grid. We associate a value $u_{i,j-1/2}$ with each cell edge parallel to the y-axis and a value $v_{i-1/2,j}$ with each cell edge parallel to the x-axis. On a uniform grid, this association leads to the difference approximations

$$S_{i-1/2,j-1/2}^{n+\theta}\left(p_{i-1/2,j-1/2}^{n+1} - p_{i-1/2,j-1/2}^n\right)$$

$$+\frac{\Delta t}{h}\left(u_{i,j-1/2}^{n+\theta} - u_{i-1,j-1/2}^{n+\theta} + v_{i-1/2,j}^{n+\theta} - v_{i-1/2,j-1}^{n+\theta}\right) = 0,$$

$$K_{i,j-1/2}^{-1}v_{i,j-1/2}^{n+\theta} + \frac{1}{h}\left(p_{i+1/2,j-1/2}^{n+\theta} - p_{i-1/2,j-1/2}^{n+\theta}\right) = 0, \qquad (6)$$

$$K_{i-1/2,j}^{-1}v_{i-1/2,j}^{n+\theta} + \frac{1}{h}\left(p_{i-1/2,j+1/2}^{n+\theta} - p_{i-1/2,j-1/2}^{n+\theta}\right) = 0.$$

For the conductivities $K_{i,j-1/2}$ and $K_{i-1/2,j}$ associated with cell edges we take the harmonic averages of conductivities in the adjacent cells.

The system (5) has the equivalent form

$$
\begin{bmatrix} \theta D_2 & \theta N \\ \theta N^\mathsf{T} & \Delta t^{-1} M \end{bmatrix} \begin{bmatrix} U \\ P \end{bmatrix}^{n+1} = \begin{bmatrix} G_1 \\ G_2 \end{bmatrix}^n ,
\tag{7}
$$

where

$$
\begin{bmatrix} G_1 \\ G_2 \end{bmatrix}^n = \begin{bmatrix} 0 \\ F \end{bmatrix}^{n+\theta} + \begin{bmatrix} (\theta-1)D_2 & (\theta-1)N \\ (\theta-1)N^\mathsf{T} & \Delta t^{-1} M \end{bmatrix} \begin{bmatrix} U \\ P \end{bmatrix}^n .
$$

This form is similar to the system (5), except that in place of the tridiagonal block A we have the diagonal block D_2 that results when we lump A by assigning its row sums to the diagonal entries.

The computational advantage of the cell-centered scheme (7) lies in the fact that we can dispense with the outer iteration used in the algorithm for the mixed finite-element method. In particular, at each time step, we need only apply a multigrid solver once, to step 2 in the following sequence:

1. $G^{n+1} \leftarrow G_2^n - D_2^{-1} G_1^n$.

2. Solve $(\Delta t^{-1} M - N^\mathsf{T} D_2^{-1} N) P^{n+1} = G^{n+1}$ for P^{n+1}.

3. $U^{n+1} \leftarrow D_2^{-1}(G_1^n - N P^{n+1})$.

Moreover, as Shen (1992) demonstrates, we expect no loss in order of accuracy with respect to the lowest-order mixed finite-element method. In particular, both heads and velocities obey $\mathcal{O}(h)$ global error estimates, and the $\mathcal{O}(h^2)$ superconvergence results of Ewing, Lazarov, and Wang (1991) hold for the nodal values of both unknowns.

3. NUMERICAL EXPERIMENTS

We turn now to a set of numerical experiments illustrating the effectiveness of the cell-centered scheme. All experiments involve the following initial-boundary-value problem on $\Omega = (0,1) \times (0,1)$:

$$
S\, \partial_t p - \nabla \cdot (K \nabla p) = f, \quad \text{for} \quad (\mathbf{x}, t) \in \Omega \times (0, \infty);
$$

$$
p(\mathbf{x}, t) = 0 \quad \text{for} \quad \mathbf{x} \in \partial\Omega; \qquad p(\mathbf{x}, 0) = 0, \quad \text{for} \quad \mathbf{x} \in \Omega.
$$

We take $S(x,y) = \frac{1}{5}$ and $f(x,y) = 10^3 \delta(x - \frac{2}{3}, y - \frac{2}{3})$. We set $\theta = 1$ throughout. In the first experiment, the conductivity field has a diagonal discontinuity:

$$
K(x,y) = \begin{cases} 1, & \text{if} \quad x+y \leq 1, \\ 10^{-3}, & \text{if} \quad x+y < 1. \end{cases}
$$

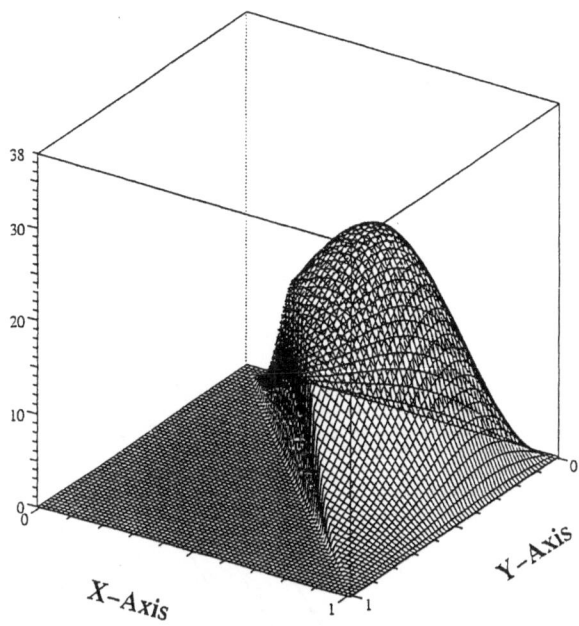

Figure 1. Cell-centered difference solution at $t = 0.1$ for a diagonally discontinuous conductivity field on $(0,1) \times (0,1)$, with $h = 2^{-6}$.

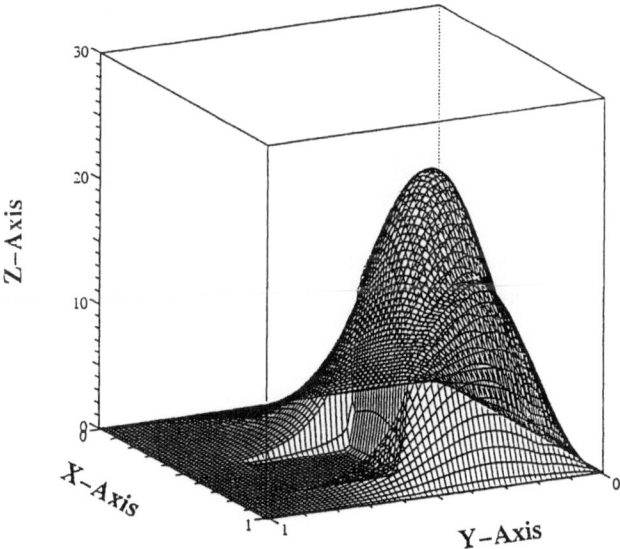

Figure 2. Cell-centered difference solution at $t = 0.1$ for a piecewise constant conductivity field on $(0,1) \times (0,1)$, with $h = 2^{-6}$.

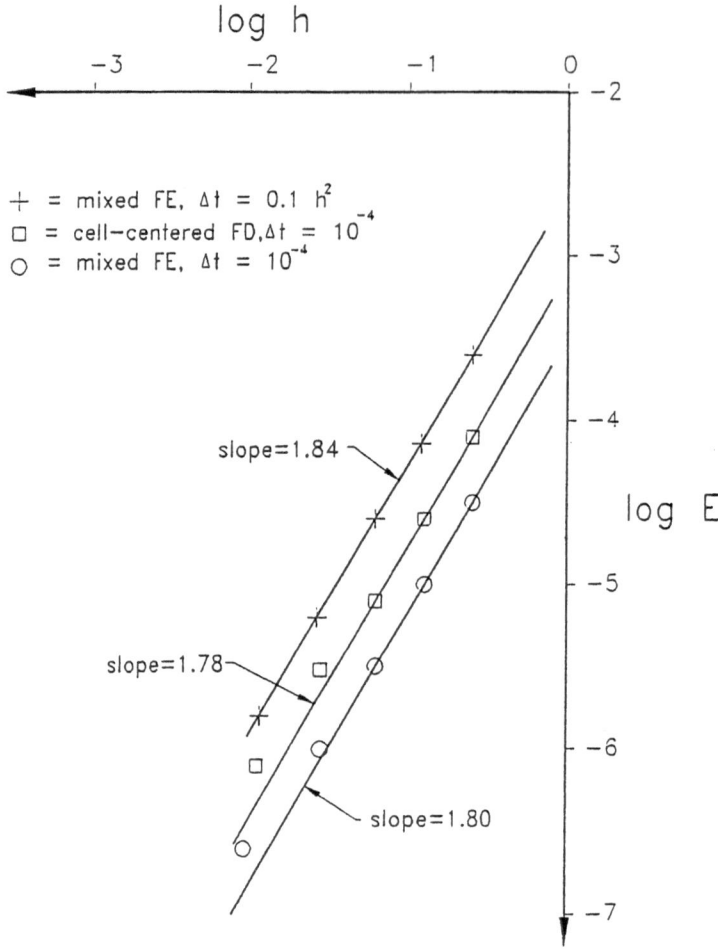

Figure 3. Spatial convergence plots for the lowest-order mixed finite-element method and the cell-centered difference scheme.

Figure 1 shows a perspective plot of the numerical solution at $t = 0.1$, computed using $h = 2^{-6}$ and $\Delta t = 10^{-4}$.

In the second experiment, the conductivity field has an irregular, piecewise constant form. Defining the regions $\Omega_1 = (0,1) \times (0, \frac{1}{8})$ and $\Omega_2 = (0.5, 0.7) \times (0.5, 0.7)$, we take

$$K(x,y) = \begin{cases} 10, & \text{if } (x,y) \in \Omega_1, \\ 10^{-2}, & \text{if } (x,y) \in \Omega_2, \\ 1, & \text{otherwise.} \end{cases}$$

Figure 2 shows a perspective plot of the numerical solution at $t = 0.1$, again computed using $h = 2^{-6}$ and $\Delta t = 10^{-4}$.

Our last experiment compares the numerical convergence rates of the cell-centered difference scheme and the mixed finite-element scheme as $h \to 0$. For this purpose, we solve the model problem with $S = \frac{1}{5}$, $K(x,y) = 1$, and $f(x,y) = 2t \sin(\pi x)[\pi^2 y^2 (1 - y) + 2] + 2Sy(1 - y)\sin(\pi x)$, computing the maximum nodal error

$$E_h = \max |p(x_{i-1/2}, y_{j-1/2}) - p_h(x_{i-1/2}, y_{j-1/2})|$$

for various mesh sizes h. Figure 3 plots $\log E_h$ versus $\log h$ for the two discretizations with $\Delta t = 10^{-4}$, along with corresponding results for the mixed method with $\Delta t = h^2/10$. These plots corroborate the theoretical convergence results cited earlier.

4. CONCLUSIONS

The cell-centered difference approximation is closely related to the lowest-order mixed finite-element method, not only in its grid structure but also in its ability to deliver accurate solutions to the groundwater flow equation. In addition, the cell-centered difference approach allows for more efficient use of multigrid solution schemes at each time step. Numerical experiments confirm the ability of the cell-centered difference scheme to produce good numerical solutions in discontinuous conductivity fields and show good agreement with theoretical convergence results.

ACKNOWLEDGMENT

The U.S. National Science Foundation supported this work in part through grant number EHR-910-8774.

REFERENCES

Allen, M.B., and Curran, M.C. (1992), "A multigrid-based solver for mixed finite-element approximations to groundwater flow," in *Computational Methods in Water Resources IX, Volume I: Numerical Methods in Water Resources,* ed. by T.F. Russell et al., Elsevier Applied Science Publishers, London, 579-585.

Allen, M.B., Ewing, R.E., and Lu, P. (1992), "Well conditioned iterative schemes for mixed finite-element models of porous-media flow," *SIAM Jour. Sci. Stat. Comp.* *13*, 794-814.

Beckie, R., Wood, E.F., and Aldama, A.A. (1993), "Mixed finite element simulation of saturated groundwater flow using a multigrid accelerated domain decomposition technique," *Water Resour. Res. 29:9*, 3145-3157.

Douglas, J., Ewing, R.E., and Wheeler, M.F., "The approximation of the pressure by a mixed method in the simulation of miscible displacement," *R.A.I.R.O. Analyse Numérique 17*, 17-33.

Ewing, R.E., Lazarov, R.D., Lu, P., and Vassilevski, P.S. (1989), "Preconditioning indefinite systems arising from mixed finite element discretization of second-order elliptic systems," *Proceedings of the Conference on Preconditioned Conjugate Gradient Methods*, Nijmegen, Netherlands, June 15-17.

Ewing, R.E., Lazarov, R.D., and Wang, J. (1991), "Superconvergence of the velocities along the Gauss lines in mixed finite element methods," *SIAM J. Numer. Anal. 26:4*, 1015-1029

Shen, J. (1992) *Mixed Finite Element Methods: Analysis and Computational Aspects*, PhD dissertation, University of Wyoming, Laramie, Wyoming.

SPARSE GRID AND EXTRAPOLATION METHODS FOR PARABOLIC PROBLEMS

R. BALDER , U. RÜDE*, S. SCHNEIDER AND C. ZENGER

INSTITUT FÜR INFORMATIK
TECHNISCHE UNIVERSITÄT MÜNCHEN
D-80290 MÜNCHEN
GERMANY
E-MAIL: RUEDE@INFORMATIK.TU-MUENCHEN.DE

Abstract. Sparse grids are a recently introduced new technique for discretizing partial differential equations having a very favorable complexity in the number of unknowns for higher dimensional problems. Therefore, sparse grids are especially attractive for instationary equations when time is treated as an additional dimension. The paper will introduce the *sparse grid finite element technique* and the *sparse grid combination technique* which can be interpreted as a *multivariate extrapolation method*. The concepts are closely related to the multilevel principle so that multigrid methods and multilevel preconditioning strategies are the natural solvers. Thus the overall solution process has optimal complexity. Furthermore, the combination technique is easily parallelizable and applicable to nonlinear problems, like the Richardson equation. Besides an introduction of the algorithms with their basic analysis we will present numerical tests for a suite of characteristic model problems.

1. Introduction. The efficient solution of the parabolic equations arising from the simulation of flows in porous media requires a combination of advanced numerical techniques. Besides efficient algebraic solvers, accurate discretization techniques play a central role. In recent research for elliptic equations, a new class of discretizations, the so-called *sparse grids*, have been developed. It is the main goal of this paper to explore the suitability of these methods for time dependent problems.

Real life physical problems require the simulation of three space and one time dimension, so that conventional discretization techniques based on implicit time steps produce a sequence of very large systems of linear equations. Typically in each of N time steps a system with $O(N^3)$ unknowns must be solved, where N is the number of grid points along each coordinate direction. Though these systems can be solved with work proportional to the number of unknowns by using advanced multilevel algorithms, the overall work is of order $O(N^4)$. Thus the work increases quickly with the accuracy demand.

Sparse grids provide methods to discretize partial differential equations on grids with only $O(N \log_2(N)^{(d-1)})$ unknowns, where d is the dimension. Clearly, for higher dimensional problems the savings can be dramatic. On the other hand, sparse grids require a certain regularity of the solution and result in a slight deterioration of accuracy compared to the conventional full grid solution. In the case of multilinear approximation, it can be shown that the deterioration in accuracy is a factor of $(\log_2(N)^{(d-1)})$ as compared with the accuracy of the full grid.

---

* present adress: Fachbereich Mathematik, Reichenhainer Str. 41, TU Chemnitz, D-09009 Chemnitz

A. Peters et al. (eds.), Computational Methods in Water Resources X, 1383–1391.

In present research, sparse grids have been explored in depth for elliptic problems. For a two dimensional situation typical sparse grids are shown in Fig. 1. To obtain

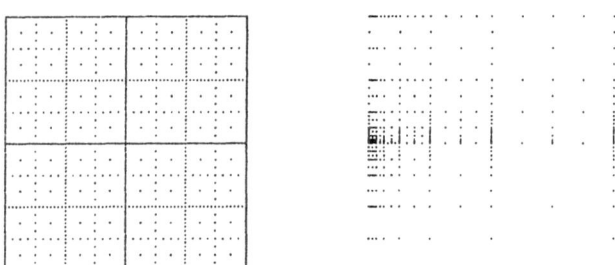

Fig. 1. *Regular and adaptive sparse grids*

sparse grid solutions, two basic approaches can be used. The first is an extrapolation-like *combination technique*, introduced by Griebel, Schneider, and Zenger [4]. We will discuss the combination technique and its extension to time-dependent problems in Section 2. Alternatively, the so-called *sparse grid finite element technique* can be used, as introduced by Zenger [11] and Bungartz [3]. The main advantage of this technique is that it can be combined easily with adaptive strategies, to obtain grids like the one depicted on the right half of Fig. 1. Possibilities to extend these results to time-dependent problems will be discussed in Section 4.

The extrapolation character of the combination technique can be generalized to a full *multivariate extrapolation*. In contrast to regular Richardson extrapolation for partial differential equations (see Marchuk and Shaidurov [6]), the mesh size is varied separately in each coordinate direction, so that the complexity advantages of sparse grids are retained for higher dimensional grids. This idea is also known as splitting extrapolation, see Schüller and Qun [5]. For elliptic problems, a suite of numerical experiments comparing all these different approaches is presented in Rüde [8].

2. Combination extrapolation. In this chapter we develop the combination technique for the heat equation on $\Omega = (0,1) \times (0,1)$ as an illustrating example.

$$(1) \qquad\qquad L\,u = u_t - \Delta u = f(t,x,y) \text{ in } \mathbb{R}^+ \times \Omega$$

with Dirichlet boundary conditions

$$(2) \qquad\qquad u(t,x,y) = g(t,x,y) \text{ on } \Sigma = \mathbb{R}^+ \times \partial\Omega$$

and initial values

$$(3) \qquad\qquad u(0,x,y) = u_0(x,y) \text{ in } \Omega.$$

As a model problem let us consider the problem (1) with right hand side,

$$f(t,x,y) = \left[\frac{\pi}{2}\cos(\frac{\pi}{2}t) + 2\pi^2\sin(\frac{\pi}{2}t)\right]\sin(\pi x)\sin(\pi y)\,,$$

Dirichlet boundary conditions and initial values choosen such that the exact solution is

$$(4) \qquad u(t,x,y) = \sin(\frac{\pi}{2}t)\sin(\pi x)\sin(\pi y).$$

Assume that we need to compute the solution for $t = 1.0$. To solve this equation by means of numerical finite difference methods, we define grids G^{τ,h_x,h_y} by

$$G^{\tau,h_x,h_y} = \{(i\,\tau, j\,h_x, k\,h_y) \mid 0 \le i \le N_\tau\,,\ 0 \le j \le N_{h_x}\,,\ 0 \le k \le N_{h_y}\} \subset [0,1] \times \Omega\,,$$

where $\tau = \frac{1}{N_\tau}$, $h_x = \frac{1}{N_{h_x}}$ and $h_y = \frac{1}{N_{h_y}}$. Depending on the context, u^{τ,h_x,h_y} is defined pointwise, or denotes a multilinear interpolant. The discrete Dirichlet boundary conditions and initial values are given by

$$u^{\tau,h_x,h_y} = g^{\tau,h_x,h_y} \text{ in } G^{\tau,h_x,h_y} \cap \Sigma\,, u^{\tau,h_x,h_y} = u_0^{h_x,h_y} \text{ on } G^{\tau,h_x,h_y} \cap (t=0) \times \Omega\,,$$

f^{τ,h_x,h_y}, g^{τ,h_x,h_y} and $u_0^{h_x,h_y}$ are suitable restriction of the functions f, g and u_0. We discretize L by the Crank-Nicholson scheme in time and central 5−point differences in space and obtain the discrete version of (1)

$$L_{\tau,h_x,h_y}\,u^{\tau,h_x,h_y} = f^{\tau,h_x,h_y} \text{ in } G^{\tau,h_x,h_y} \cap \Omega.$$

Note that possibly $h_x \ne h_y$, so that the operator L_{τ,h_x,h_y} may be anisotropic in x and y. Each time step requires the solution of an elliptic problem that we solve using the program MDG9V, developed at the Centre of Mathematics and Computer Science (CWI) in Amsterdam By de´Zeeuw, see [10]. MDG9V is robust for discontinuous coefficients and dominating convection terms. In our context it is especially useful, because it handles even strongly anisotropic grids. Now we introduce the combination technique in (quasi) 3 dimensions. We will study the main ideas on the simplest example. We calculate the solutions to the following grids

$$u^{\tau,h_x,h_y} \text{ on } G^{\tau,h_x,h_y}\,,\quad u^{\frac{\tau}{2},h_x,h_y} \text{ on } G^{\frac{\tau}{2},h_x,h_y}\,,\quad u^{\tau,\frac{h_x}{2},h_y} \text{ on } G^{\tau,\frac{h_x}{2},h_y} \text{ and } u^{\tau,h_x,\frac{h_y}{2}} \text{ on } G^{\tau,h_x,\frac{h_y}{2}}.$$

From these solutions we define a linear combination by

$$(5) u_{3+1+0}^{combination} = \beta^{\tau,h_x,h_y} u^{\tau,h_x,h_y} + \beta^{\frac{\tau}{2},h_x,h_y} u^{\frac{\tau}{2},h_x,h_y} + \beta^{\tau,\frac{h_x}{2},h_y} u^{\tau,\frac{h_x}{2},h_y} + \beta^{\tau,h_x,\frac{h_y}{2}} u^{\tau,h_x,\frac{h_y}{2}}\,.$$

For the combination technique we choose the weights

$$\beta^{\tau,h_x,h_y} = -2.0\,,\quad \beta^{\frac{\tau}{2},h_x,h_y} = \beta^{\tau,\frac{h_x}{2},h_y} = \beta^{\tau,h_x,\frac{h_y}{2}} = 1.0\,.$$

Assuming the existence of an error splitting of the form

$$e^{\tau,h_x,h_y} = u - u^{\tau,h_x,h_y} = e_\tau(\tau^2) + e_{h_x}(h_x^2) + e_{h_y}(h_y^2) + R(\tau, h_x, h_y)$$

it is easy to show, that

$$u_{3+1+0}^{combination} = u + e_\tau\left[(\frac{\tau}{2})^2\right] + e_{h_x}\left[(\frac{h_x}{2})^2\right] + e_{h_y}\left[(\frac{h_y}{2})^2\right] + \tilde{R}(\tau, h_x, h_y)\,.$$

TABLE 1
Errors of combination, extrapolation, and full grid solution

	$u_{3+1+0}^{combination}$	$u_{3+1+0}^{extrapolation}$	$u^{\frac{\tau}{2},\frac{h_x}{2},\frac{h_y}{2}}$
number of unknowns	660	660	648
l_2 error	$1.00\ 10^{-2}$	$2.92\ 10^{-3}$	$0.96\ 10^{-2}$
AP error	$1.78\ 10^{-2}$	$5.84\ 10^{-3}$	$1.92\ 10^{-2}$

Except for the remainder term, this has the same accuracy as $u^{\frac{\tau}{2},\frac{h_x}{2},\frac{h_y}{2}}$ on $G^{\frac{\tau}{2},\frac{h_x}{2},\frac{h_y}{2}}$. Now we focus on the number of unknowns, needed to compute the solution, including boundary points. We denote by $u_{M+N+P}^{combination}$ a linear combination of M grids with $\tau h_x h_y = K$, N grids with $\tau h_x h_y = \frac{K}{2}$ and P grids with $\tau h_x h_y = \frac{K}{4}$, where K is a constant concerning the combination technique. The numbers M, N and P are given by the number of different grids that fulfill the conditions above. To calculate $u_{3+1+0}^{combination}$ requires asymptotically a total of $\frac{7}{8}N_\tau(N_{h_x}+1)(N_{h_y}+1)$ unknowns, whereas the solution $u^{\frac{\tau}{2},\frac{h_x}{2},\frac{h_y}{2}}$ needs $N_\tau(N_{h_x}+1)(N_{h_y}+1)$ unknowns. As we can see, the number of unknowns is reduced by 12.5 percent and therefore the combination solution is cheaper. The corresponding error functions at $t = 1$ for $\tau = \frac{1}{4}$, $h_x = \frac{1}{4}$ and $h_y = \frac{1}{4}$ are shown in Fig. 2. This, and the numerical errors in Table 1, confirm that the error is roughly of

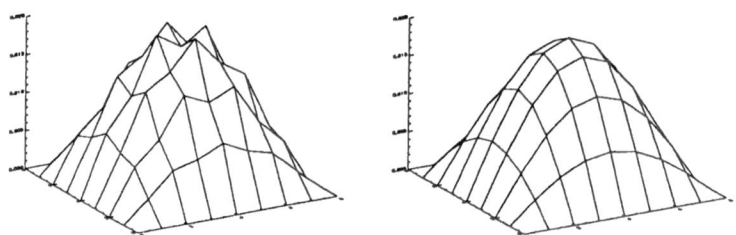

FIG. 2. *Combination error in comparison with the full grid error, point by point*

the same magnitude. Here, the AP error be defined as

$$\text{AP error}(u^{\tau,h_x,h_y}, u) = u^{\tau,h_x,h_y}(1.0, 0.5, 0.5) - u(1.0, 0.5, 0.5),$$

and the l_2 error as

$$l_2\ \text{error}(u^{\tau,h_x,h_y}, u) = \sqrt{h_x h_y \sum_{G^{\tau,h_x,h_y}} (u^{\tau,h_x,h_y} - u)^2}.$$

In our numerical experiments, the AP error is roughly proportional to the l_2 error. For that reason, we use the easier computable AP error for the following comparisons of the different methods.

A 12.5% saving may not yet seem worthwhile, however the combination technique can be generalized easily. In the same manner as described above, we calculate other combinations referred to as $3+1+0$, $6+3+1$, $10+6+1$,.... . These are depicted in Fig. 3, where we show the accuracy versus the total number of unknowns involved in the computation process. The top linie $(+)$ shows the full grid solution, the lines below $(*,$ $\cdot, \Diamond, ...$) the solutions $u_{3+1+0}^{combination}$, $u_{6+3+1}^{combination}$, $u_{10+6+3}^{combination}$, ..., respectively. Clearly, the

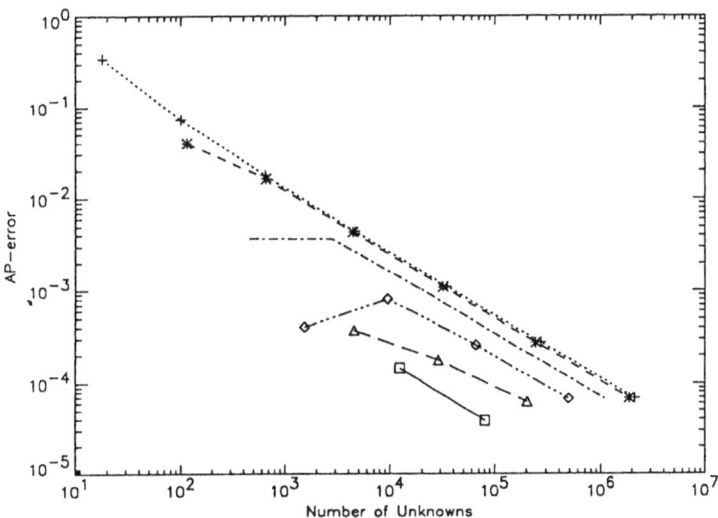

FIG. 3. *Time-Space-Combination*

cost effectivities of the combination technique increases when a larger number of grids is used. In Table 2, we present the corresponding numerical values, where q denotes the asymptotically behavior of the number of unknowns in relation to the number of unknowns of the corresponding full grid.

3. **Multivariate Extrapolation.** We again use the model problem (2) - (4) of Section 2. This time it is our aim to completely eliminate all error terms of second order. The multivariate extrapolation process is described in more detail by Rüde [8]. It is based on the following expansion

$$u = u^{\tau,h_x,h_y} + T\tau^2 + Xh_x^2 + Yh_y^2 + \eta^{\tau,h_x,h_y}$$

where T, X, Y are functions independent of τ, h_x, h_y and the remainder term

$$\eta^{\tau,h_x,h_y} = O(\tau^4) + O(\tau^2 h_x^2) + ... + O(h_y^4) \ .$$

The existence of this expansion is proved in the Diplomarbeit of Schneider [9]. Now we can compute a linear combination similar to (5):

$$u_{3+1+0}^{extrapolation} = \gamma^{\tau,h_x,h_y} u^{\tau,h_x,h_y} + \gamma^{\frac{\tau}{2},h_x,h_y} u^{\frac{\tau}{2},h_x,h_y} + \gamma^{\tau,\frac{h_x}{2},h_y} u^{\tau,\frac{h_x}{2},h_y} + \gamma^{\tau,h_x,\frac{h_y}{2}} u^{\tau,h_x,\frac{h_y}{2}} \ ,$$

TABLE 2

Number of unknowns in relation to the full grid

combination technique	q
full grid	1.000
$3 + 1 + 0$	0.875
$6 + 3 + 1$	0.516
$10 + 6 + 3$	0.215
$15 + 10 + 6$	0.084
$21 + 15 + 10$	0.030
$28 + 21 + 15$	0.010
$36 + 28 + 21$	0.003

but this time we use the weights

$$\gamma^{\tau,h_x,h_y} = -3.0 \, , \gamma^{\frac{\tau}{2},h_x,h_y} = \gamma^{\tau,\frac{h_x}{2},h_y} = \gamma^{\tau,h_x,\frac{h_y}{2}} = \frac{4}{3} \, .$$

Clearly,

$$u_{3+1+0}^{extrapolation} = u + \tilde{\eta}^{\tau,h_x,h_y} \, ,$$

which may be interpreted as follows: We have constructed from a low order discretization scheme, here second order, by means of extrapolation a high order discretization
scheme, here fourth order, with not more work than needed to calculate $u_{3+1+0}^{combination}$.
The error values are shown in the corresponding colum of Table 1. More complicated
extrapolations schemes will be discussed in [9].
Similar to Fig. 3, Fig. 4 presents an accuracy versus work graph for various methods.
The dotted line gives again the conventional full grid solution for reference. the un-
broken line marked with boxes ($\square$) represents $u_{21+15+10}^{combination}$. The line without any marks
summarizes the results for the multivariate extrapolation technique introduced in this
section. The line marked with ($\times$) gives the results for a sparse grid finite element
method that will be discussed in detail in the following Section. Note that the extrapolation technique constitutes a full 4th order method and is thus asymptotically more
efficient than any combination method with a fixed number of grids.

4. Sparse grid finite elements. We will now examine the properties of sparse
grids in a Finite Element Galerkin approach. We consider the stationary version of (1),

$$\Delta u = f \qquad \text{in } \Omega = (0,1)^d$$

with boundary conditions $u = g(x)$ on $\partial\Omega$ for dimensions $d = 2, 3$. Our goal is finding
an approximate solution in a space, spanned by multilinear hierarchical basis functions,
that contribute essentially to the solution. A priori estimates lead to the regular sparse
grids we know from the combination technique. For smooth solutions, this is absolutely

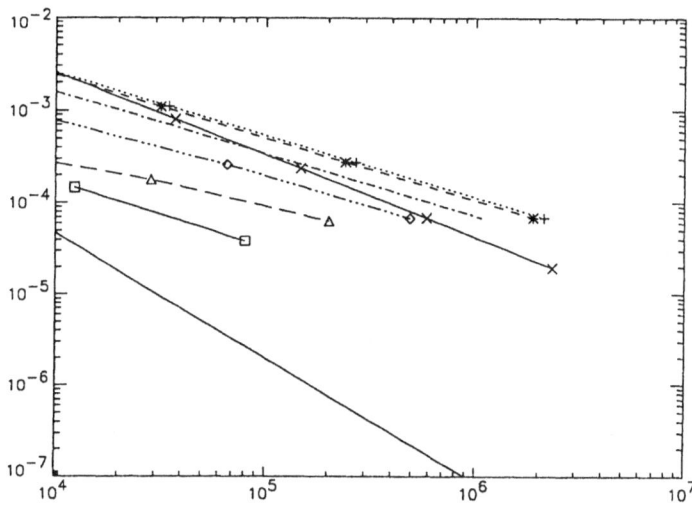

FIG. 4. *Time-Space-Combination, Time-Space-Extrapolation and Sparse Grid Finite Elements*

sufficient. If we expect non smooth solutions, the adaptive approach is superior, where we do not use a priori estimations but computed results and a posteriori estimates to construct the grid. See Fig. 1 for a regular and an adaptive sparse grid. Note, that we do not need to change our data structure or our algorithms to treat the adaptive case. Such problems have been studied intensively by Bungartz [3]. Numerical experiments support the assumption, that the L_∞ error estimations for interpolation are also valid for the solution of boundary value problems.

For parabolic problems we now present a time stepping method, using sparse grids only for the space dimensions and piecewise constant basis functions in time. The latter corresponds to a backward Euler scheme, and leads from $u_t - \Delta u = f$ to

$$(6) \qquad -\Delta u(x,t) + \frac{1}{\tau}u(x,t) = f + \frac{1}{\tau}u(x,t-\tau),$$

which is a Helmholtz equation, and can be solved with the techniques discussed by Balder and Zenger [1]. The results for regular sparse grids are not yet competitive with the results from the combination technique because we compute in each time step the same resolution in the space dimensions. On the other hand, with a mixed technique, if we combine solutions of (6) with different mesh sizes in space and time, we obtain comparable results. Note that our time discretization is only of first order. Therefore, we have to combine solutions where the mesh sizes in time direction differ by factors of 4. Formally, we define a combined solution based on sparse grids in space by

$$u_c = \sum_{i=0}^{n} u_{2^i h_x, 4^{n-i}\tau} - \sum_{i=0}^{n-1} u_{2^{i+1} h_x, 4^{n-i}\tau}.$$

$u_{h_x,\tau}$ is a solution with finest mesh size h_x in the space dimensions, and step size τ in time. A comparison of this method with the other methods presented in this paper is included in Fig. 4.

If we use a hierarchical basis for piecewise constant functions, (see Figure 5,) the

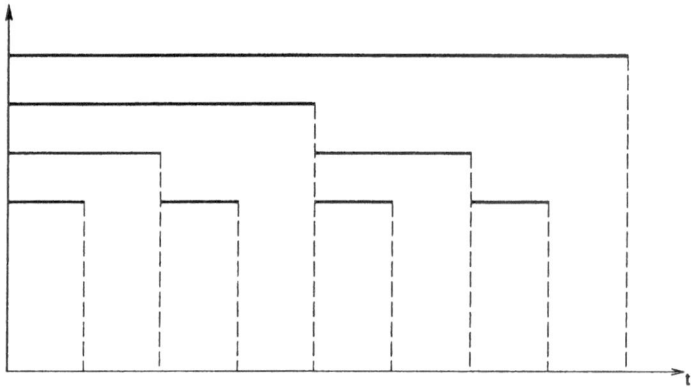

FIG. 5. *piecewise constant hierarchical basis*

same ideas as in the Elliptic case lead to a direct Finite Element approach on regular or adaptive sparse grids for parabolic problems with different resolutions in space for different time steps. A computer implementation for this strategy is planned for the near future.

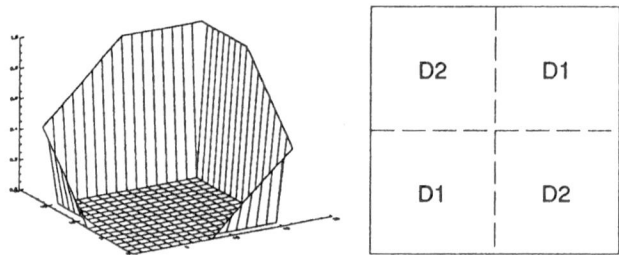

FIG. 6. *Boundary Conditions and Piecewise constant diffusion coefficient on rectangular subdomains*

5. Extensions. In this section we present results for a problem with a structural singularity, similar to Rüde and Zenger [7], which usually causes bad convergence of the solution. Let us consider a generalization of (1)

$$\tilde{L}\, u = u_t - \nabla(D(x,y) \cdot \nabla u) = f(t,x,y) \text{ in } \mathbb{R}^+ \times \Omega$$

with $f(t,x,y) = 0.0$, initial values $u_0 = 0.0$. The Dirichlet boundary conditions are constant in time corresponding to Fig. 6. Let $D(x,y)$ be piecewise constant on rectangular subdomains of Ω, as shown in Fig. 6. We choose $D1 = 10.0$ and $D2 = 1.0$. The

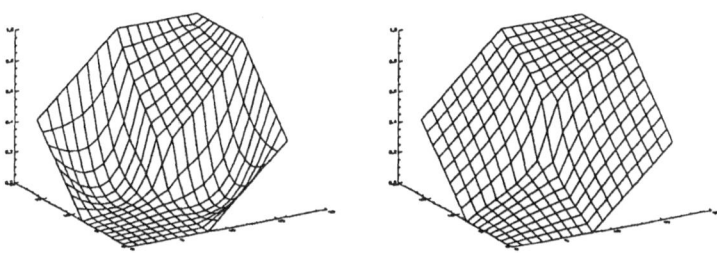

FIG. 7. *Timeevolution at t=0.01 and t=1.0*

solutions we get by using a $3 + 1 + 0$ extrapolation technique, at time $t = 0.01$ and $t = 1.0$ are shown in Fig. 7. The (almost) stationary solution at $t = 1.0$ nicely shows the singular solution behavior. As in the stationary case, the pollution effect of the singularity which usually causes pour global convergence can be compensated by either an appropriate extrapolation scheme, see Blum and Rannacher [2], or by correcting the discretization locally as in Rüde and Zenger [7]. An error analysis and intensive numerical tests are contained in Schneider [9].

REFERENCES

[1] R. BALDER, C. ZENGER, *The d-dimensional Helmholtz equation on sparse Grids* , Technische Universität München, SFB-Bericht Nr. 342/21/92 A, (1992).

[2] H. BLUM, R. RANNACHER, *Extrapolation Techniques for Reducing the Pollution Effect of Reentrant Corners in the Finite Element Method*, Numer. Math. 52, (1988), pp. 539-564.

[3] H.-J. BUNGARTZ, *Dünne Gitter und deren Anwendung bei der adaptiven Lösung der dreidimensionalen Poisson-Gleichung*, Technische Universität München, Dissertation, (1992).

[4] M. GRIEBEL, M. SCHNEIDER UND C. ZENGER, *A combination technique for the solution of sparse grid problems*, in Proceedings of the IMACS International Symposium on Iterative Methods in Linear Algebra, Brüssel, 2.4.–4.4. 1991, P. de Groen und R. Beauwens, Hrsg., Elsevier, Amsterdam, 1992 (auch: Technische Universität München, SFB-Bericht Nr. 342/19/90 A).

[5] A. SCHÜLLER, QUN LIN, *Efficient High Order Algorithms for Elliptic Boundary Value Problems Combining Full Multigrid Techniques and Extrapolation Methods*, Arbeitspapiere der GMD Nr. 192 (1985).

[6] G.I. MARCHUK, V.V. SHAIDUROV, *Difference Methods and their Extrapolations*, Springer, New York (1983).

[7] U. RÜDE, CHR. ZENGER, *On the Treatment of Singularities in the Multigrid Method*, Lecture Notes in Mathematics, Multigrid Methods II, Proceedings, Cologne (1985), pp. 261-271.

[8] U. RÜDE, *Extrapolation and Related Techniques for Solving Elliptic Equations*, Technische Universität München, TUM-I9135) (1991).

[9] S. SCHNEIDER, *Extrapolations- und Dünngittermethoden für parabolische Gleichungen*, TU-München, Institut für Informatik, Diplomarbeit (in preparation) (1994).

[10] P.M. DE ZEEUW, *Matrix-dependent prolongations and restrictions in a black box multigrid solver*, Journal of Computational and Applied Mathematics 33, page 1 - 27 (1990).

[11] C. ZENGER, *Sparse grids*, in: Parallel Algorithms for Partial Differential Equations: Proceedings of the Sixth GAMM-Seminar, Kiel, 19.1.–21.1. 1990, Notes on Numerical Fluid Mechanics (Vol. 31), W. Hackbusch, Hrsg., Vieweg, Braunschweig, 1991 (also: Technische Universität München, SFB-Bericht Nr. 342/18/90 A).

ON THE NUMERICAL SOLUTION OF THE EQUATION OF SATURATED/UNSATURATED FLOW IN POROUS MEDIA

JÜRGEN FUHRMANN
fuhrmann@iaas-berlin.d400.de
Institut für Angewandte Analysis und Stochastik
D-10117 Berlin, Mohrenstraße 39
Germany

The solution of Richards' equation of saturated-unsaturated water flow in porous media using implicit time discretization, Newton's method for the nonlinear problems and preconditioned iterative methods for the linear problems is described. A central feature is the multigrid preconditioned BICGstab solver for the linear problems which is implemented for rectangular meshes. On unstructured meshes, the same code can be used, but with an ILU preconditioner for the linear problems. The correctness of the space discretization is verified by a numerical experiment. A test example shows the performance of the method in dependence of the parameters of the problem. The paper continues the investigation of this problem documentend in [Kna87, Fuh93].

THE PROBLEM.

Consider Richards' equation of saturated-unsaturated water flow in porous media [Ric31]

$$
\begin{aligned}
\partial_t \theta(x, \psi(x,t)) + \nabla \cdot q(x,t) &= f(x,t) & \text{in } \Omega \times [0,T] \\
-q(x,t) \cdot \nu &= g(x,t) & \text{on } \Gamma_N \times [0,T] \\
\psi(x,t) &= \psi_0(x,t) & \text{on } \Gamma_D \times [0,T]
\end{aligned}
$$

together with Darcy's law

$$
q(x,t) = -K(x, \psi(x,t))(\nabla \psi(x,t) - \gamma)
$$

where $\Omega \subset \boldsymbol{R}^d$, $d = 2,3$ is a domain with $\partial\Omega = \Gamma_D \cap \Gamma_N$. The meaning of the variables is the following:

A. Peters et al. (eds.), Computational Methods in Water Resources X, 1393–1400.

x	$[m]$	space variable
t	$[s]$	time variable
$\psi(x,t)$	$[m]$	capillary pressure
$\theta(x,\psi)$	$[m^3/m^3]$	volumetric water content
$K(x,\psi(x,t))$	$[m^2/s]$	hydraulic conductivity
$q(x,t)$	$[m/s]$	specific water flux density
$f(x,t)$	$[(m^3/m^3)/s]$	volumetric source/sink density
γ	1	gravity

Using the variable substitution $u(x,t) := \psi(x,t) - \gamma x$ and replacing $K(x, u + \gamma x)$ and $\theta(x, u + \gamma x)$ by $K(x,u)$ and by $\theta(x,u)$, respectively, we exclude the gravity and arrive at the equation

$$\partial_t \theta(x,u) - \nabla \cdot (K(x,u)\nabla u) = f(x,t) \quad \text{in } \Omega$$
$$(K(x,u)\nabla u) \cdot \nu = g(x,t) \quad \text{on } \Gamma_N$$
$$u = u_0(x,t) \quad \text{on } \Gamma_D.$$

For the nonlinearities, we use a regularized van Genuchten ansatz [vG80], other curves can be used, too.

DISCRETIZATION AND NUMERICAL SOLUTION.

Let $0 = t_0 < t_1 < \cdots < t_N = T$ and $\tau^n = t^n - t^{n-1}$. Set $u^n(x) = u(x, t^n)$. For the time discretization, we use implicit Euler's method and arrive at the time discrete problem at $t = t_n$:

$$\frac{\theta(x, u^n(x)) - \theta(x, u^{n-1}(x))}{\tau^n} -$$
$$\nabla \cdot (K(x, u^n(x))\nabla u^n(x)) = f(x, t^n) \quad \text{in } \Omega$$
$$(K(x, u^n(x))\nabla u^n(x)) \cdot \nu = g(x, t^n) \quad \text{on } \Gamma_N$$
$$u^n(x) = u_0(x, t^n) \quad \text{on } \Gamma_D$$

We implement a simple heuristic time step control scheme: Hold the value $\|u^n - u^{n-1}\|_\infty$ within a given range defined by the parameter u_{step} by slightly changing the next time step size assuming a linear dependence between step size and control norm. We reject updates if they are too large or if Newton's method didn't converge and lower the timestep.

The solution of the time dependent problem demands then the ability to solve stationary problems of the form

$$\theta(x,u) - \nabla \cdot (\tau K(x,u)\nabla u) = F(x) \quad \text{in } \Omega$$
$$(\tau K(x,u)\nabla u) \cdot \nu = G(x) \quad \text{on } \Gamma_N$$
$$u = U_0(x) \quad \text{on } \Gamma_D.$$

with

$$
\begin{aligned}
F(x) &= \tau f(x, t^n) + \theta(x, u^{n-1}(x)) \\
G(x) &= g(x, t^n) \\
U(x) &= u_0(x, t^n).
\end{aligned}
$$

We discretize this problem using a finite volume discretization with arithmetical coefficient averaging (which can be seen also as a FE method with quadrature).

To solve this discrete nonlinear problem, we use Newton's method with line search features and affine invariant residual calculation. The lumping of the Jacobi matrix which will be discussed below and the desire to have time steps as large as the approximation allows demands the implementation of a damping scheme characterized by the damping parameters δ_i:

$$
u_h^{(i)} = u_h^{(i-1)} - \delta^i (\tilde{A}'_{u_h^{(i-1)}})^{-1} (A_h(u_h^{(i-1)}) - F_h).
$$

After [DH91] we use an affine invariant measurement of the residual at the cost of the solution of one more linear system per Newton step, which allows a far more stable and time step independent control of Newton's method:

$$
r^i = \|(\tilde{A}'_{u_h^{(i-1)}})^{-1} (A_h(u_h^{(i)}) - F_h)\|_2.
$$

The damping strategy looks as follows. Choose $q > 1$, $0 < \delta_{\min} \le \delta^0 \le 1$.

- $r^i < r^{i-1} \Rightarrow \delta^{i+1} = \max(\delta^i q, 1)$.
- $r^i > r^{i-1} \Rightarrow \delta^i = \min(\delta^i/q, \delta_{\min})$ und calculation of a new $u_h^{(i)}$, accept $u_h^{(i)}$ if $\delta_{\min}$ is reached.

We try to iterate as close as possible to roundoff error to ensure discrete mass conservation. The construction of δ_i ensures that we can reach quadratic convergence if there is no lumping.

To be able to solve the linear problems which in general have matrices which are neither symmetric, nor M - matrices, we lump the linear operators to obtain at least the M-property by adding all the positive off-diagonal entries onto the main diagonal. The linear problems are solved iteratively using preconditioned CG-like methods. In the case of unstructured meshes, an ILU preconditioner can be used. An algebraic multigrid method for this case is under consideration. In the case of rectangular grids we use a semi-algebraic multigrid preconditioner [Fuh92, FG93] which is capable to handle the asymmetries and coefficient jumps occuring in the linear problems. Here, some more features of this multigrid code are listed:

- 2D and 3D problems on rectangular and toroidal grids
- dimension independent implementation
- theoretical interpretation for nonsymmetric problems with matrices of the form $A = CE$, C symmetric and positive definite, and E an positive diagonal

matrix, but shows good performance for wider classes of problems provided, the matrices have the M-property.

- operator dependent intergrid transfer operators
- coarse grid operator generation by harmonical averaging fine grid operator coefficients
- "semi-algebraic": grid structure and coarse grid nodes are fixed, all the other components are created using only the fine grid matrix without relying on a nested FEM structure
- can be used as preconditioner for CG, BICGstab etc.
- ILU smoothing
- Chebyshev polynomials, ILU preconditioned CG or BICGstab for coarsest grid solution

The whole code is organized so that the computaional domain is given as a rectangular mesh. If it is necessary to work on more complicated geometries, a mesh generator [Sch93] can be used to cut out some part of it. In this case the grid is considered a unstructured, so currently, no multigrid is available.

A NUMERICAL EXPERIMENT.

To test the discretization we obtain a 1D test problem with exact solution: Given $\Omega = (\xi_0, \xi_1)$ and $d(u) = D'(u) > 0$, consider

$$-(d(u)u')' = (D(u))'' = 0$$

with $u(\xi_0) = u_0$, $u(\xi_1) = u_1$. Then for the solution $\hat{u}(\xi)$ one has

$$D(\hat{u}(\xi)) = \bar{D}(\xi) = D(u_0) + \frac{D(u_1) - D(u_0)}{\xi_1 - \xi_0}(\xi - \xi_0)$$

Thus,

$$\hat{u}(\xi) = D^{-1}(\bar{D}(\xi))$$

can be yielded numerically by Newton's method. Assume $0 \in (\xi_0, \xi_1)$, $D(0) = 0$ and $u_0 < 0$. Then we can adjust u_1 to ensure $\hat{u}(0) = 0$:

$$u_1 = D^{-1}(K(u_0)\frac{\xi_1}{\xi_0})$$

Now, given d_0, d_1, ε, choose

$$d(u) = \begin{cases} d_0 & , \quad u < -\varepsilon \\ d_1 & , \quad u > \varepsilon \\ \text{cubic spline} \quad S_3(u) & , \quad \text{else} \end{cases}$$

Below we see a plot of $d(u)$ for different values of ε:

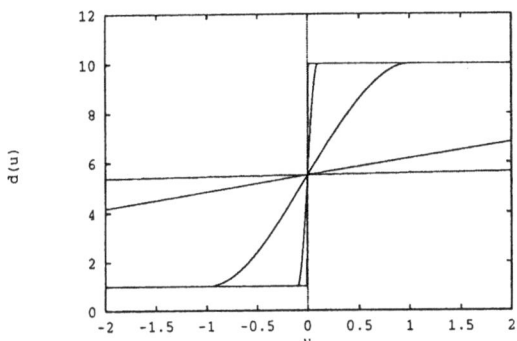

Assume $S_3(u) = S_4'(u)$. Then

$$D(u) = \begin{cases} S_4(-\varepsilon) + d_0(u + \varepsilon) & , \ u < -\varepsilon \\ S_4(\varepsilon) + d_1(u - \varepsilon) & , \ u > \varepsilon \\ S_4(u) & , \ \text{else} \end{cases}$$

Assume $S_3'(u) = S_2(u)$. Then

$$\max d'(u) = S_2(0) = \frac{3(b - a)}{4\varepsilon} = O(\varepsilon^{-1}) \quad (\varepsilon \to 0).$$

Now, we "blow up" the problem to the 2D case. Let $0 \in \Omega \subset R^2$.
Let for given β and $(x, y) \in \Omega$ $\xi_\beta(x, y) = x \cos \beta + y \sin \beta$ and $\xi_0 = \min_{\partial\Omega} \xi_\beta(x, y)$,
$\xi_1 = \max_{\partial\Omega} \xi_\beta(x, y)$ Let $U_0 = \hat{u}(\xi_\beta(x, y))$ Consider

$$-\nabla(d(u)\nabla u) = 0 \ \text{ in } \Omega$$
$$u = U_0 \ \text{ in } \partial\Omega$$

For this problem we have the exact solution

$$\hat{u}_\beta(x, y) = \hat{u}(\xi_\beta(x, y)).$$

Now we are able to make the following experiment. Let $\Omega = (-1, 1) \times (-1, 1)$ with $\beta = \pi/4$, $u_0 = -1$. Solve the 2D test problem with different values of ε and different values of the discretization parameter h. We consider the error on the different refinement stages and estimate the approximation order in dependece of ε. Assume $\|u_h - u_{\text{exact}}\|_\infty = Ch^\alpha$. Given two grids with mesh sizes h_1 and h_2 and measured discretization errors e_1 and e_2, we can estimate

$$\alpha \approx \log_{h_1/h_2} \frac{e_1}{e_2}.$$

Below we see a plot of the average of α over three refinement stages $32 \times 32, 64 \times 64, 128 \times 128$ depending on ε

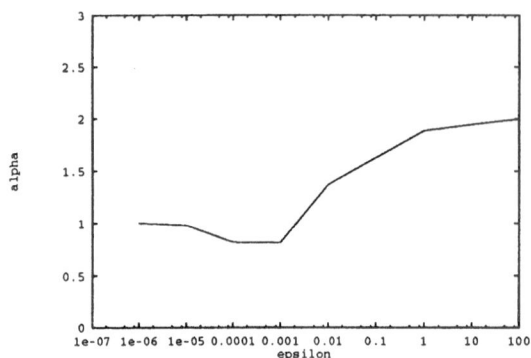

It comes out what one would expect: for smooth coefficients (with respect to a fixed h) we have an estimated approximation of order 2 which deteriorates to 1 as the jump becomes sharper. The worst approximation we have when h interplays with ε. For $h \to 0$ independently of ε we can expect $O(h^2)$ convergence. Unfortunately, the real calculations cannot rely on this assumption and we have to expect the worst convergence order estimated.

AN INFILTRATION EXAMPLE.

Let $\Omega = [0,6] \times [-3,0]$ be discretized by a 33×33 rectangular mesh. Assume Dirichlet boundary conditions

$$\begin{aligned} \psi &= 0.2 \text{ on } [0,1] \times 0, T < 1/2d \\ \psi &= 2 - x \text{ on } 2 \times [-3,-2] \end{aligned}$$

and as an initial solution the hydrostatic equilibrium with

$$\psi = -2 \text{ on } [0,1] \times 0$$

Use the van Genuchten data for "fine to medium sand" [SR85] ($\alpha = 2.5$, $n = 4.875$, $\theta_s = 0.362$, $\theta_r = 0.084$, $K_s = 0.0000125$) or "Beit Netova clay" [vG80] ($\alpha = 0.169$, $n = 1.169$, $\theta_s = 0.45$, $\theta_r = 0.00$, $K_s = 9.25 \cdot 10^{-9}$), respectively. Below we see the behaviour of the capillary pressure ψ at $(0.0, -1.5)$ and the average contraction rates of Newton's method for two different time step control values u_{step}. The time scale is logarithmic.

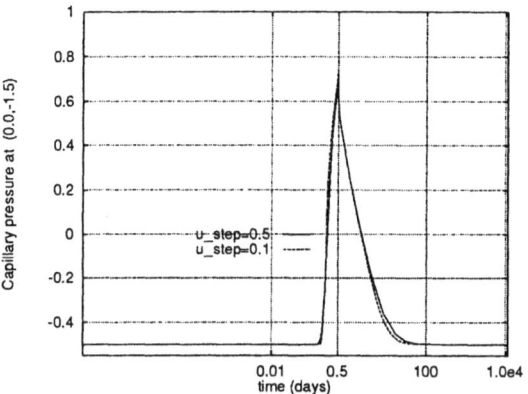

Behaviour of the capillary pressure for sand

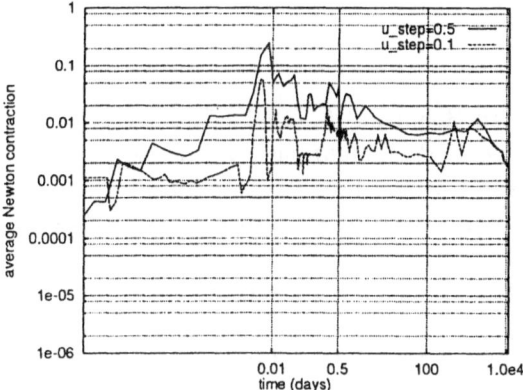

Behaviour of Newton's method for sand

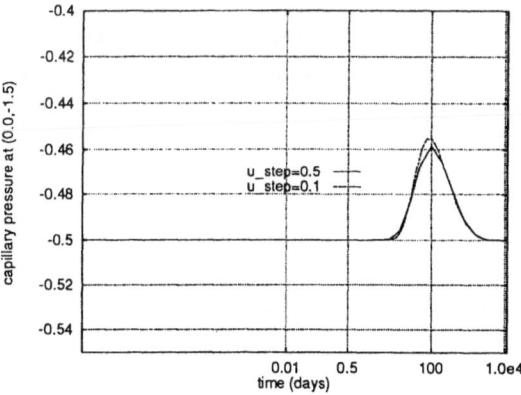

Behaviour of the capillary pressure for clay

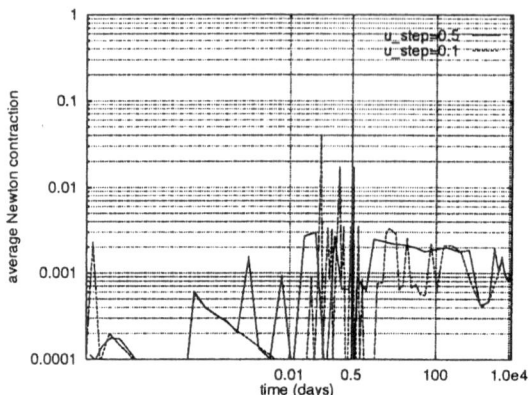

Behaviour of Newton's method for clay

REFERENCES

[DH91] Deuflhard, P. and Hohmann, A. (1991), "Numerische Mathematik - eine algorithmisch orientierte Einführung", de Gruyter, Berlin, New York.

[FG93] Fuhrmann, J. and Gärtner, K.(1993) "On matrix data structures and the stability of multigrid algorithms", Proceedings of EMG'93, Amsterdam, submitted.

[Fuh92] Fuhrmann, J. (1992), "On the convergence of algebraically defined multigrid methods", preprint no.3, IAAS Berlin

[Fuh93] Fuhrmann, J. (1993) "Calculation of saturated-unsaturated flow in porous media with a Newton-multigrid method", in GAMM-Seminar on Multigrid Methods, Gosen, September 1992, S. Hengst, ed., Berlin, IAAS Berlin. Report No. 5.

[Kna87] Knabner, P. (1987), "Finite element solution of saturated- unsaturated flow through porous media", in Large scale scientific computing, P. Deuflhard and B. Engquist, eds., vol. 7 of Progress in Scientific Computing, Birkhäuser, Boston.

[Ric31] Richards, L. (1931), "Capillary conduction of liquids through porous mediums", Physics, vol. 1, pp. 318–333.

[Sch93] I. Schmelzer, I. (1993), "3D anisotropic grid generation with intersection-based geometry interface", IMA, Minneapolis, 1993, preprint no. 1180.

[SR85] D.Stephens ,D. and Rehfeldt, K. (1985), " Evaluation of closed form analytical models in a fine sand", Soil Sci. Soc. Amer. J., vol. 49, pp. 12–19.

[vG80] van Genuchten, M. (1980), "A closed-form equation for predicting the hydraulic conductivity of unsaturated soils", Soil Sci. Soc. Amer. J., vol. 44 ,pp. 892–898.

ON THE COMBINATION OF ITERATIONS

WOLFGANG HACKBUSCH
Institut für Informatik und Praktische Mathematik
Christian-Albrechts-Universität zu Kiel
D-24098 Kiel, Germany

In the following, we consider different types of combined iterations and discuss their convergence and efficiency.

1 INTRODUCTION

Often, different iterations are combined. An example demonstrating the arising problems is the following one. The discrete nonlinear Navier-Stokes equation can be solved by the Newton method. Within one Newton step one has to the solve the linearised problem. This is a perturbation of the Stokes problem. If a Stokes solver is available, one may approximate the linearised problem by k steps of a Stokes solver. The Stokes solver may be an k-step iterative process involving the inversion of the Poisson problem. Finally, the Poisson solver may consist of k steps of some elementary iteration, e.g., an SSOR method. The resulting iterative process is rather costly: One Newton step involves k^3 SSOR steps, whereas the convergence speed is limited by the worst speed of the involved iterations.

The last mentioned fact (the bottleneck principle) is another drawback of the combination of iterations: Often, the overall iterative process is slow if one of the involved iterations is so.

Here, the usual preconditioning of, e.g., the conjugate gradient method by a basic iteration is not understood as a combination of different iterative methods. If a basic iteration is defined, it may performed as simple iteration, in a semi-iterative manner (e.g., Chebyshev method), or embedded into a conjugate gradient process (cf. §7 and §9 in [H]). In any case, the spectral properties of the basic iteration determine the arising convergence behaviour.

1401

A. Peters et al. (eds.), Computational Methods in Water Resources X, 1401–1408.
© 1994 *Kluwer Academic Publishers. Printed in the Netherlands.*

§2 discusses the product iteration. In §3 we investigate the composed iteration. Due to page limitations, we cannot study the respective questions for subdomain or subspace iterations. Other types of hybrid methods are discussed, e.g., in [B].

2 PRODUCT ITERATIONS

Let the system $Ax = b$ of linear equations be given. A one-step iterative method $x^m \mapsto x^{m+1} = \Phi(x^m, b)$ is characterised by the function Φ. The iteration is linear, if Φ is linear in x and b, i.e., if there are matrices M and N such that

(2.1) $\Phi(x, b) = Mx + Nb$ (M: iteration matrix).

We call (2.1) or (2.1') the first normal form of the method:

(2.1') $x^{m+1} := M x^m + Nb.$

The consistency of the method requires the relation $M + NA = I$ (cf., §3.2.2 in [H]). Using $N = (I - M) A^{-1}$ in (2.1'), one obtains the second normal form

(2.2) $x^{m+1} := x^m - N(A x^m - b).$

The matrix N will be called the «matrix of the second normal form of Φ». Often, the iteration is written in the third normal form

(2.3) $W(x^m - x^{m+1}) = A x^m - b.$

The matrix W of the third normal form is also called the preconditioner. Equation (2.3) should be understood in the algorithmic form: Solve $W \delta = A x^m - b$ and define $x^{m+1} := x^m - \delta$. Obviously,

(2.4) $N = W^{-1}$

holds. The m-th iterate $x^m = x^m(x^0, b)$ starting with x^0 and depending on the right-hand side b of the system $Ax = b$ has the explicit representation

(2.5) $x^m(x^0, b) = M^m x^0 + \sum_{k=0}^{m-1} M^k N b$ for $m \geqslant 0.$

Given two iterative methods Φ and Ψ, one defines the product iteration $\Phi \circ \Psi$ by $x^{m+1} = (\Phi \circ \Psi)(x^m, b) := \Phi(\Psi(x^m, b), b).$

The iteration matrices of Φ, Ψ, and $\Phi \circ \Psi$ are related by $M_{\Phi \circ \Psi} = M_\Phi M_\Psi$. Furthermore, $N_{\Phi \circ \Psi} = N_\Phi + N_\Psi - N_\Phi A N_\Psi$ holds (cf. §3.2.7 in [H]). In contrast to consistency, the convergence of the factors Φ and Ψ is not inherited by product methods. On the other hand, products of divergent methods may converge.

The bestknown example of an product iteration may be the symmetric over-relaxation method (SSOR); here, a foreward SOR step Φ is combined with backward SOR step Ψ. The purpose of combination is not the improvement of the convergence rate but the fact that $M_{\Phi \circ \Psi}$ has a real spectrum (cf., §4.8.5 in [H]).

The alternating direction iterative method (ADI) is an example, where the product of iterations produces a very fast product iteration, at least in the commutative case (cf., §7.5 in [H]).

The multi-grid iteration is a product two quite different iterations: the smoothing iteration Φ and the coarse-grid iteration Ψ. The good convergence of the multi-grid iteration is due to the complementary properties of both steps. Note that both Φ and Ψ may be divergent (cf. [M] and §10 of [H]).

A recent example of a multiple product iteration is the method of filtering decompositions (cf. Wittum [W] or §10.9.4 in [H]).

The given examples show clearly that the bottleneck problem mentioned in the introduction is not valid for product iterations.

By definition, product iterations are to be performed sequentially. This may be regarded as disadvantage with respect to applications to parallel computers. However, when replacing product iterations (interpreted as multiplicative subspace methods) by the additive counterpart (e.g., BPX instead of the standard V-cycle), one must expect a loss in the convergence speed.

3 SECONDARY ITERATIONS

When constructing iterations, one is often led to formula (2.2) with $N = B^{-1}$:

(3.1) $\qquad x^{m+1} := x^m - B^{-1}(A x^m - b),$

or (2.3) with $W = B$ which is not so easy to invert directly. As a simple example consider, e.g., a blockwise Jacobi method where B is a block-diagonal. In the case of three spatial dimensions, this block may correspond to a plane and involve a five-point formula. An obvious idea is to approximate $B^{-1}(A x^m - b)$ by an iterative method solving equations of the form $B \delta = c$. The iteration for solving the auxiliary problem $B \delta = c$ is called the secondary iteration and leads to the following composed iteration:

....

(3.2)   composed iteration Φ_k:

....

(3.2a) $x^m \mapsto c := Ax^m - b$,

(3.2b)   perform the secondary iteration for solving $B\delta = c$:

(3.2b$_1$)   set the starting iterate $\delta^0 := 0$,

(3.2b$_2$)   perform the (semi-)iteration $\delta^0 \mapsto \delta^1 \mapsto \ldots \mapsto \delta^k$.

(3.2c) $x^{m+1} := x^m - \delta^k$.

....

The larger k is, the better the sequences x^m from (3.1) and (3.2a–c) coincide. On the other hand, one would like to choose k as small as possible, since the amount of work per (outer) iteration step increases with k. The iteration matrix of (3.1) is

(3.3) $M_A = I - B^{-1}A$.

For solving the auxiliary equation $B\delta = c$, we apply the secondary iteration Φ_B:

(3.4) $\delta^{m+1} = \delta^m - C^{-1}(B\delta^m - c) = M_B \delta^m + N_B c$

with the iteration matrix $M_B = I - C^{-1}B$. The composed iteration Φ_k defined for fixed $k \geqslant 0$ by (3.2a–c) is a linear and consistent iteration $\Phi = \Phi_k$ for the solution of $Ax = b$. By (2.5), its iteration matrix equals

(3.5a) $M_k = I - \sum\limits_{q=0}^{k-1} M_B^q N_B A$ (M_B, N_B from (3.4)).

If, in addition, Φ_B is consistent, (3.5a) simplifies to

(3.5b) $M_k = M_A + M_B^k B^{-1}A$ (M_A from (3.3)).

The matrix of the second normal form (2.2) is

(3.5c) $N_k = (I - M_B^k)B^{-1}$.

If M_B does not have an eigenvalue λ with $\lambda^k = 1$ (divergence of Φ_k), the matrix of the third normal form (2.3) can be written as

(3.5d) $W_k = B(I - M_B^k)^{-1}$.

The representation (3.5b) permits interpretation of the iteration matrix M_k as a perturbation of the iteration matrix M_A. The contraction number of Φ_k with respect to the spectral norm and, if B or A are positive definite, with respect to the norms $\|x\|_B = \|B^{1/2}x\|_2$ and $\|x\|_A = \|A^{1/2}x\|_2$ are

(3.6a) $\|M_k\|_2 \leqslant \|M_A\|_2 + \|M_B\|_2^k \|B^{-1}A\|_2$,

(3.6b) $\|M_k\|_B \leqslant \|M_A\|_B + \|M_B\|_B^k \|B^{-1/2}AB^{-1/2}\|_2$ (if $B > 0$),

(3.6c) $\| M_k \|_A \leqslant \| M_A \|_A + \| M_B \|_A^k \| A^{1/2} B^{-1} A^{1/2} \|_2$ (if $A > 0$).

Knowledge of the spectral radius $\rho(M_B)$ is not sufficient for analysing the secondary iteration, because the spectral radius describes the convergence only asymptotically, whereas here we need precise upper bounds after the fixed number of k iteration steps.

The factor $\| B^{-1} A \|_2$ in (3.6a) is bounded by

(3.6d) $1 - \| M_A \|_2 \leqslant \| B^{-1} A \|_2 \leqslant 1 + \| M_A \|_2$.

The conclusions that one can draw from (3.6a-c) are the subject of the next remark, which uses the term «effective amount of work». Let Cn be the number of arithmetical operations involved in one step of the iteration Φ (n: dimension of the sparse system), while $\rho(M)$ is the convergence rate. Then $-C / \log \rho(M)$ is called the effective amount of work of Φ (notation: Eff(Φ)). It measures the work needed for an error improvement by the factor $1/e$.

Remark 3.1. Assume that one of the expressions $\| M_A \|_2$, $\| M_A \|_B$, $\| M_A \|_A$ together with the corresponding quantity $\| M_B \|_2$, $\| M_B \|_B$, $\| M_B \|_A$ is smaller than 1. Then, the composed method Φ_k converges for sufficiently large k.
One should choose k sufficiently large, so that the right-hand side of (3.6a) has a size comparable with $\| M_A \|_2$, e.g., $\frac{1}{2}(1 + \| M_A \|_2)$ (similarly for (3.6b-c)). If $\| M_A \|_2 \leqslant \zeta < 1$ (ζ independent of h) and $\| M_B \|_2 = 1 - O(h^\beta)$ ($\beta > 0$), inequality $\| M_k \|_2 \leqslant (1 + \zeta)/2$ can be achieved with $k = O(h^{-\beta})$. In this case, the effective amount of work for Φ_k is also of the order $Eff(\Phi_k) = O(h^{-\beta})$. If, however, $\| M_A \|_2 = 1 - O(h^\alpha)$ ($\alpha > 0$), inequality (3.6a) admits only the unfavourable estimate $Eff(\Phi_k) = O(h^{-\alpha - \beta})$.

In particular, (3.6a-c) yields no statement that could guarantee the convergence of Φ_k for small k (but compare [F] for a particular situation). Since, according to (3.6d), the factor $\| B^{-1} A \|_2$ attains at best the value ≈ 1, one needs at least $k = O(h^{-\beta})$ iterations to make the right-hand side of (3.6a) smaller than 1.

Since Φ_k is again a linear and consistent iteration, Φ_k may be used as a basic iteration, e.g., of a semi-iteration. Another situation arises, if k is not fixed but determined by means of some stopping criterion or a semi-iteration is applied as a secondary process. In these cases, Φ_k is nonlinear; hence, the suitability of Φ_k as the basic iteration of a semi-iterative method is questionable. For a discussion of this problem, we refer to Golub-Overton [G].

In the following, we assume that the matrices A, B, C satisfy $A = A^H$, $B > 0$, $C > 0$. In this case, we call Φ_A and Φ_B symmetric. A necessary condition for the convergence of Φ_B is $0 < B < 2C$. Under this assumption, the composed iteration Φ_k defined in (3.2) is also symmetric for all $k \in \mathbb{N}$. The symmetry is, e.g., required for the application of the conjugate gradient method. In particular, the matrix

$$(3.7) \qquad W_k = B^{1/2}[I - (I - B^{1/2}C^{-1}B^{1/2})^k]^{-1}B^{1/2}$$

of the third normal form of Φ_k is positive definite: $W_k > 0$.

It is not true that the convergence of Φ_A and Φ_B implies that of Φ_k, but convergence can always be achieved for a suitable damping. In the following, we assume the inequalities

$$(3.8a) \qquad \gamma B \leqslant A \leqslant \Gamma B \qquad\qquad \text{with } 0 < \gamma \leqslant \Gamma,$$
$$(3.8b) \qquad \delta C \leqslant B \leqslant \Delta C \qquad\qquad \text{with } 0 < \delta \leqslant \Delta.$$

The spectrum of $B^{1/2}C^{-1}B^{1/2}$ lies in $[\delta, \Delta]$, i.e., $\sigma(I - B^{1/2}C^{-1}B^{1/2}) \subset$ $\subset [1 - \Delta, 1 - \delta]$. The spectrum of $I - (I - B^{1/2}C^{-1}B^{1/2})^k$ is contained in the interval $[\underline{\beta}, \bar{\beta}]$ with

$$(3.8c_1) \qquad \underline{\beta} := \begin{cases} 1 - (1 - \delta)^k & \text{for odd } k \\ 1 - \max\{(1 - \Delta)^k, (1 - \delta)^k\} & \text{for even } k \end{cases},$$

$$(3.8c_2) \qquad \bar{\beta} := \begin{cases} 1 - (1 - \Delta)^k & \text{for odd } k \text{ or } \Delta < 1 \\ 1 & \text{for even } k \text{ and } \Delta \geqslant 1 \end{cases}.$$

(3.7) proves that $\underline{\beta} W_k \leqslant B \leqslant \bar{\beta} W_k$. By means of (3.8a), one obtains

Lemma 3.2. The inclusions (3.8a,b) prove (3.9) for W_k from (3.7):

$$(3.9) \qquad \gamma_k W_k \leqslant A \leqslant \Gamma_k W_k \quad \text{with } \gamma_k := \gamma \underline{\beta}, \ \Gamma_k := \Gamma \bar{\beta} \ (\underline{\beta}, \bar{\beta} \text{ from (3.8c)}).$$

Let δ, Δ, γ, Γ be the optimal bounds in (3.8a,b). Then (3.10) holds:

$$(3.10) \qquad \varkappa(W_k^{-1}A) \leqslant \frac{\Gamma_k}{\gamma_k} = \frac{\Gamma}{\gamma}\frac{\bar{\beta}}{\underline{\beta}} = \frac{\bar{\beta}}{\underline{\beta}}\varkappa(B^{-1}A).$$

Analysing the iteration Φ_B separately, we obtain the optimal damping parameter

$$(3.11a) \qquad \Theta_B = 2/(\delta + \Delta).$$

The matrix of the third normal form of the *damped* iteration is $\Theta_B^{-1}C$ instead of C and leads to the bounds $\delta\Theta_B$ and $\Delta\Theta_B$ instead of δ and Δ. This scaling changes the ratio $\bar{\beta}/\underline{\beta}$. Next, we discuss the factor Θ_B minimising the condition number (3.10).

For even k, the parameter Θ_B from (3.11a) yields the optimal condition number (3.10). For odd k, however, the minimum of $\varkappa(W_k^{-1}A)$ is attained by a value of Θ_B in the open interval

(3.11b) $1/\Delta < \Theta_B < 2/(\delta+\Delta)$.

For $k=1$, $\varkappa(W_1^{-1}A) \leqslant \varkappa(B^{-1}A)\varkappa(C^{-1}B)$ holds independently of Θ_B. For $k=3$, the optimal value is $\Theta_B = 3/[\Delta+\delta+\sqrt{\Delta(\Delta-\delta)+\delta^2}]$.

An important question concerns the number k of secondary iterations for which one obtains an effective amount of work as favourable as possible. A trivial statement is the following: The effective amount of work $Eff(\Phi_k)$ is minimal for finite k, because $Eff(\Phi_k) = O(k)$ as $k \to \infty$.

We assume that $\varkappa := \varkappa(C^{-1}B) = \Delta/\delta \gg 1$ holds for the condition number corresponding to the method Φ_B. Furthermore, let Φ_B already be optimally damped, i.e., $\Theta_B = 2/(\delta+\Delta)=1$ (cf. (3.11a)). Then

$$-(1-\Delta) = 1-\delta = (\varkappa-1)/(\varkappa+1) = 1-\tfrac{1}{\varkappa}+O(\varkappa^{-2})$$

proves

(3.12a) $(1-\delta)^k = 1-\dfrac{k}{\varkappa} + O((k/\varkappa)^2), \quad (1-\Delta)^k = (-1)^k(1-\delta)^k$.

From (3.8c$_{1,2}$), one obtains the following expansion for $k \leqslant \varkappa$:

(3.12b) $\dfrac{\bar{\beta}}{\underline{\beta}} = \begin{cases} \varkappa/k + O(1) & \text{for odd } k, \\ \varkappa/2k + O(1) & \text{for even } k. \end{cases}$

First, we consider the case in which Φ_k serves as the (stationary) iterative method. Then the convergence rate

(3.12c) $\rho(\Phi_k) = \dfrac{\varkappa(W_k^{-1}A)-1}{\varkappa(W_k^{-1}A)+1} \approx 1-2/\varkappa(W_k^{-1}A) = 1-2\dfrac{\gamma}{\Gamma}\dfrac{\underline{\beta}}{\bar{\beta}} \approx 1-2\alpha k$

can be shown for optimal damping with $\alpha = \gamma/(\varkappa\Gamma)$ for odd and $\alpha = 2\gamma/(\varkappa\Gamma)$ for even k. From $-\log\rho(\Phi_k) \approx 2\alpha k$ we obtain

(3.12d) $Eff(\Phi_k) \approx \dfrac{C'+kC''}{2\alpha k} = (\tfrac{1}{k}C'+C'')/(2\alpha)$.

The effective amount of work of the iterative method Φ_k initially decreases with k until for $k \approx \varkappa$ the asymptotical representations (3.12b,c) lose their validity. Because of the better value $\bar{\beta}/\underline{\beta}$ for even k, one should prefer even numbers k.

A different situation arises when the (symmetric) iteration Φ_k is used as the basic iteration of the Chebyshev method, since then the asymptotical rate is given by $c = (1-\sqrt{\varkappa})/(1+\sqrt{\varkappa})$ with $\varkappa = \varkappa(W_k^{-1}A)$ instead of (3.12c). For the semi-iterative Chebyshev method, the effective work equals

$$Eff_{\text{semiiterative}}(\Phi) = -(C_\Phi + \text{const})/\log c,$$

where C_Φ corresponds to the work taken by one iteration step of Φ. For Φ_k this number becomes $C' + kC''$. The expansion

(3.13) $$Eff_{\text{semiiterative}}(\Phi_k) \approx \tfrac{1}{2}[C' + kC'' + \text{const}]\left(\frac{\Gamma}{\gamma}\frac{\bar{\beta}}{\underline{\beta}}\right)^{1/2} \approx$$
$$\approx \tfrac{1}{2}[C' + kC'' + \text{const}]/\sqrt{\alpha k}$$

holds with the same α as in (3.12c).

The semi-iterative effective amount of work (3.13) becomes minimal for the even number k next to the value $k_0 = [C' + \text{const}]/C''$. Since $k_0 < 3$ is realistic, $k = 2$ is the optimum.

Numerical examples for the application of secondary iterations can be found, e.g., in §8.4.1 and §8.4.6 of [H].

A particular case of a composite iteration is again the multi-grid iteration in contrast to the two-grid iteration. The coarse-grid equation is not solved exactly, but by one (V-cycle) or two (W-cycle) steps of the same procedure at the lower level. Since $k = 1, 2$, the total amount of work is not increased to much.

REFERENCES

[B] Brezinski, C. and Redivo-Zaglia, M. (1994) "Hybrid procedures for solving linear systems". Numer. Math. 67, 1-19

[F] Frommer, A. and Szyld, D. B. (1992) "H-Splittings and two-stage iterative methods". Numer. Math. 63, 345-356

[G] Golub, G. H. and Overton, M. L. (1988) "The convergence of inexact Chebyshev and Richardson iterative methods for solving linear systems". Numer. Math. 53, 571-593

[H] Hackbusch, W. (1994) Iterative Solution of Large Sparse Systems of Equations, Springer-Verlag, Berlin - Iterative Lösung großer schwachbesetzter Gleichungssysteme, 2. deutsche Auflage, Teubner, Stuttgart

[M] Hackbusch, W. (1985) Multi-grid methods and applications. Springer-Verlag, Berlin

[W] Wittum, G. (1992) Filternde Zerlegungen: Schnelle Löser für große Gleichungssysteme (Filtering decomposition, Fast solvers for large systems of equations). Teubner Skripten zur Numerik, Teubner, Stuttgart

PRECONDITIONING AND SOLVING LINEAR SYSTEMS FOR THE COMPUTATION OF FREE SURFACE FLOWS

J.-M. HERVOUET*, J.-M. JANIN**, C. MOULIN*
* Fluvial hydraulics section
** Maritime hydraulics section
Laboratoire National d'Hydraulique
Direction des Etudes et Recherches, Electricité de France
6 QUAI WATIER, 78401 CHATOU Cedex
FRANCE

ABSTRACT:

Linear systems stemming from De Saint-Venant Equations are described and the efficiency of different iterative techniques and preconditioners to solve these systems are assessed. The comparison of two different formulations of shallow water equations solved by the computer programme TELEMAC-2D leads to a specific preconditioning when the unknowns are the depth and the velocity.

INTRODUCTION:

Finite Element models are now widely used for the computation of free surface flows and prove to be very flexible tools. At the "Laboratoire National d'Hydraulique" the software system TELEMAC, based on Element by Element techniques, has been dedicated to several kinds of equations occurring in free surface hydraulics: shallow water equations in 2D or 3D, Serre equations, and mild slope equations, along with water quality and sedimentological modules.

Among the Finite Element models, those with implicit schemes offer an extra flexibility: the choice of the time-step is not theoretically limited by a stability criterion. The price to pay is the need to solve linear systems. As direct solvers would not be practicable with large meshes, one must resort to iterative techniques such as conjugate gradients. In the very active field of iterative techniques, many different methods may be used and a wealth of new ones is published every year. The behaviour of those methods is highly dependent on the conditioning, on the equations solved, and, sadly enough, on each particular application. A good choice may lead to a considerable speed up and dramatic changes in the efficiency of numerical schemes. To find the best fitting technique, one has actually little else to do than test them all !

A. Peters et al. (eds.), Computational Methods in Water Resources X, 1409–1416.
© 1994 Kluwer Academic Publishers. Printed in the Netherlands.

A series of iterative solvers and preconditioners have thus been implemented and compared in the framework of the TELEMAC system. After some preliminary explanations of the kind of linear system we have to solve shallow water equations, the efficiency of various methods will be discussed.

LINEAR SYSTEMS IN TELEMAC-2D:

Two different options are available in TELEMAC 2D: the h-u option where the unknowns are the two components of velocity and the depth, and the c-u option where the depth h is replaced by c, the celerity of shallow water waves.

In the first option, the Saint-Venant equations read:

$$\frac{\partial h}{\partial t} + \vec{u} \cdot \overrightarrow{\mathrm{grad}}(h) + h \, \mathrm{div}(\vec{u}) = 0$$

$$\frac{\partial u}{\partial t} + \vec{u} \cdot \overrightarrow{\mathrm{grad}}(u) + g \frac{\partial h}{\partial x} - \mathrm{div}(\,v\,\overrightarrow{\mathrm{grad}}(u)\,) = S_x - g \frac{\partial Z_f}{\partial x}$$

$$\frac{\partial v}{\partial t} + \vec{u} \cdot \overrightarrow{\mathrm{grad}}(v) + g \frac{\partial h}{\partial y} - \mathrm{div}(\,v\,\overrightarrow{\mathrm{grad}}(v)\,) = S_y - g \frac{\partial Z_f}{\partial y}$$

With the latter option, the equations become:

$$\frac{\partial c}{\partial t} + \vec{u} \cdot \overrightarrow{\mathrm{grad}}(c) + \frac{c}{2} \, \mathrm{div}(\vec{u}) = 0$$

$$\frac{\partial u}{\partial t} + \vec{u} \cdot \overrightarrow{\mathrm{grad}}(u) + 2\,c \frac{\partial c}{\partial x} - \mathrm{div}(\,v\,\overrightarrow{\mathrm{grad}}(u)\,) = S_x - g \frac{\partial Z_f}{\partial x}$$

$$\frac{\partial v}{\partial t} + \vec{u} \cdot \overrightarrow{\mathrm{grad}}(v) + 2\,c \frac{\partial c}{\partial y} - \mathrm{div}(\,v\,\overrightarrow{\mathrm{grad}}(v)\,) = S_y - g \frac{\partial Z_f}{\partial y}$$

with the following notation:

u, v	:	velocity components.
h	:	water depth.
c	:	celerity of shallow water waves.
z_f	:	bottom level.
S_x, S_y	:	source terms.
g	:	gravity acceleration.
v	:	dispersion coefficient.

The discretisation of the equations in the h-u option leads to a linear system $A\,X = B$, with:

$$A = \begin{pmatrix} M1 & B_x & B_y \\ -C_x^T & M2 & 0 \\ -C_y^T & 0 & M3 \end{pmatrix} \quad \text{and} \quad X = \begin{pmatrix} H \\ U \\ V \end{pmatrix}$$

With the c-u option, the matrix A and unknown vector X are changed:

$$A = \begin{pmatrix} M1 & B_x & B_y \\ -B_x^T & M2 & 0 \\ -B_y^T & 0 & M3 \end{pmatrix} \quad \text{and} \quad X = \begin{pmatrix} C \\ U \\ V \end{pmatrix}$$

H, U, V, and C are vectors containing respectively the unknown depths, components of velocity and celerities.

$M1$, $M2$, $M3$, B_x, B_y, C_x, C_y are the following square matrices:

$$M1 = \frac{2}{\theta_u\,DT} M$$

where M is the mass matrix, θ_u the implicitation coefficient on the velocity and DT the time-step.

$$M2 = M3 = \frac{1}{2\,\theta_c\,DT} M + \frac{1}{2\,\theta_c} D$$

where D is the diffusion matrix and θ_c the implicitation coefficient on the celerity or on the depth. B_x and B_y are gradient-like matrices of the form:

$$B_x(i,j) = -\int_\Omega \psi_j \frac{\partial}{\partial x}(c^n\,\psi_i)\,d\Omega \quad \text{and} \quad B_y(i,j) = -\int_\Omega \psi_j \frac{\partial}{\partial y}(c^n\,\psi_i)\,d\Omega$$

Ψ_i and Ψ_j are the basis and c^n is the celerity at time t^n.

C_x and C_y are of the form:

$$C_x(i,j) = -\int_\Omega \psi_j \frac{\partial}{\partial x}(\psi_i)\,d\Omega \quad \text{and} \quad C_y(i,j) = -\int_\Omega \psi_j \frac{\partial}{\partial y}(\psi_i)\,d\Omega$$

The linear systems stemming from these two options of the Saint-Venant equations are not positive definite, but with the c-u formulation they can be put in a symmetric form (with negative terms on the diagonal). In the systems given above, it has been assumed that the advection terms have been treated in a fractional step with the characteristics method and appear only in the right-hand side terms. When other techniques are used to deal with advection terms, such as S.U.P.G. (see [9] for an application in TELEMAC), new matrices must be added to M1, M2 and M3 and spoil their symmetry, but all the rest remains unchanged.

ITERATIVE TECHNIQUES AND PRECONDITIONING:

The following methods have been tested to solve our linear systems in both options:

Iterative solvers: Conjugate Gradient, Conjugate Residuals, Normal Equation (i.e. solving $^tA \, A \, X = {}^tA \, B$ instead of $A \, X = B$), Minimal Error, Bi-conjugate Gradient Stabilised (BICGSTAB, see ref. [8]), Generalised Minimum Residual (GMRES, ref. [7])

Preconditioners: Diagonal scaling, Bloc-Diagonal preconditioning (ref. [6]), Crout Preconditioning (ref. [5]), Gauss-Seidel element-by-element (GS-EBE, ref. [5]).

Iterative solvers:

A common characteristic of all the iterative solvers is their sensitivity to the time-step. The higher the time-step, the higher the number of iterations. Even when using unconditionally stable schemes, this fact leads to the existence of an optimum time-step to simulate a given period of time, e.g. a tide.

The conjugate Gradient method has been rapidly discarded. Probably because it theoretically works only with positive definite systems, it can only deal with very easy test-cases but fails to converge in more complex situations. The Conjugate Residual method should also suffer from the same drawbacks but sometimes works fairly well, as in the tidal application presented below.

Depending on the test cases, the two best methods for solving our linear systems are the Normal Equation and GMRES. The Minimal Error generally behaves like the Normal Equation, but is often slightly less efficient. Surprisingly BICGSTAB did not appear to be better than the Normal Equation.

The Normal Equation and GMRES are thus the currently used methods in the TELEMAC system. The first one remains the better in very difficult cases such as the computation of a tide in the Dunkerque harbour with Courant numbers greater than 50. On the other hand, GMRES is by far the most efficient when computing flows with tidal flats or when using S.U.P.G. in the continuity equation. In GMRES, the dimension of the Krylov space is an additional parameter that has to be adjusted; in TELEMAC values ranging from 3 to 6 are generally enough and the tuning seems to have little importance.

Preconditioning:

All the methods were tested with a diagonal scaling. More sophisticated techniques such as Block-diagonal preconditioning, Crout preconditioning and GS-EBE have been tried in various combinations but gave no additional speed; however, they proved to be efficient in TELEMAC with other sets of equations such as diffusion equations or with different systems (ref. [6]).

A specific preconditioning for shallow water equations:

It appears clearly by comparison with the c-u formulation that the h-u formulation leads to badly conditioned systems. This may be due to the loss of symmetry and also to the fact that the linear system is not dimensionless. With high values of depth, for example in maritime situations, GMRES is slowed down and the Normal Equation fails to converge. On the contrary, in the same case, both methods are efficient in the c-u formulation. This striking difference between the formulations c-u and h-u led to a new preconditioning of the h-u formulation; the change of variables leading to the unknown c is done in the h-u option after discretisation, at the level of the linear system:

First the value of c for each point (denoted $c^n(i)$) is taken from the previous time-step at time t^n or from the initial conditions. In case of drying zones, this value of c taken for preconditioning may be limited to avoid divisions by zero. Then, in the momentum and in the continuity equations, the unknown $h^{n+1}(i)$, at time t^{n+1}, is replaced by $h^{n+1}(i)\dfrac{g}{2c^n(i)}$, g being the gravity acceleration. At last the continuity equation is multiplied by $\dfrac{2g}{c^n(i)}$. The whole operation is formally equivalent to a diagonal preconditioning: the matrix A of the system is replaced by D1 A D2, where the matrices D1 and D2 are diagonals function of c^n.

With this new preconditioning, the linear systems of the two formulations becomes broadly equivalent and the Normal Equation does converge. There is also an increase in speed of about 30% for GMRES.

APPLICATIONS:

We show in figures 1 and 2 two applications: one in maritime hydraulics and one in fluvial hydraulics.

Computation of tides: Coupling this new preconditioning with a diagonal scaling and using GMRES, an implicit scheme appears to be very advantageous: the computation of a tide, with time steps of 150 s (i.e. about 300 time steps), on a 9500 elements mesh covering Western Europe (see attached figure) takes 35 s on a CRAY C90 and 17 mn on a HP 9000 (series 750) workstation. The accuracy on the tidal range is about 10 cm, without resorting to parameter estimation for bottom friction. The computation of very long sequences of tides for the study of lagrangian drifts or sedimentological processes is thus now possible.

Computation of a thermal plume in a river, with dry zones: the mesh is made of 11250 elements and the time-step is 10 s. 1000 iterations on a HP workstation takes 9071 s. It appears that solvers, and accuracy, are more sensitive to a Courant number based on the velocity than on celerity. Optimal time-steps are thus lower for computation in rivers. When a steady state is obtained, different scenarios of thermal outlets may be tested by solving only the temperature equation. The computation of 1000 time steps takes then 2000 s.

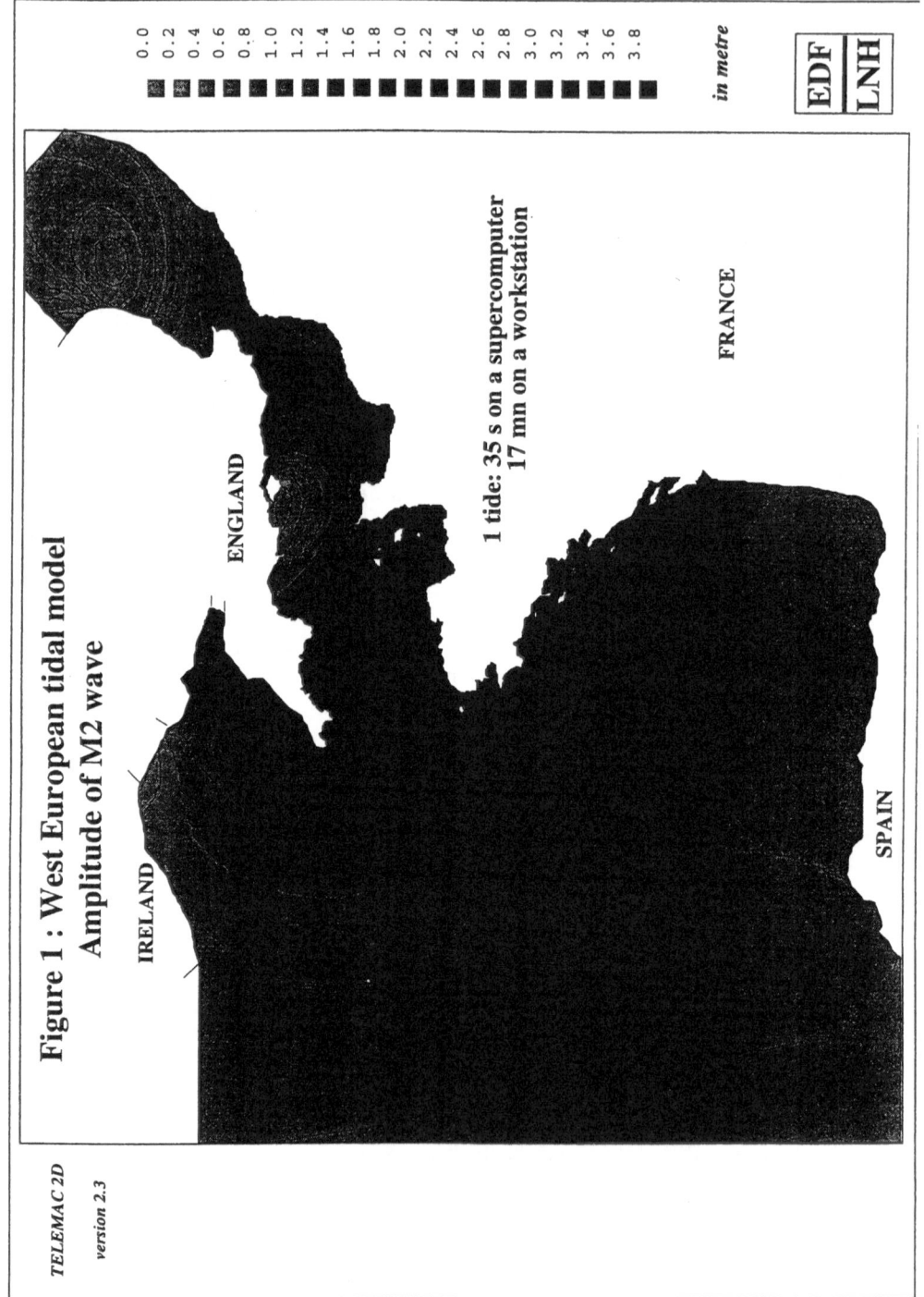

**Figure 1 : West European tidal model
Amplitude of M2 wave**

TELEMAC 2D

version 2.3

IRELAND

ENGLAND

FRANCE

SPAIN

1 tide: 35 s on a supercomputer
17 mn on a workstation

0.0
0.2
0.4
0.6
0.8
1.0
1.2
1.4
1.6
1.8
2.0
2.2
2.4
2.6
2.8
3.0
3.2
3.4
3.6
3.8

in metre

EDF
LNH

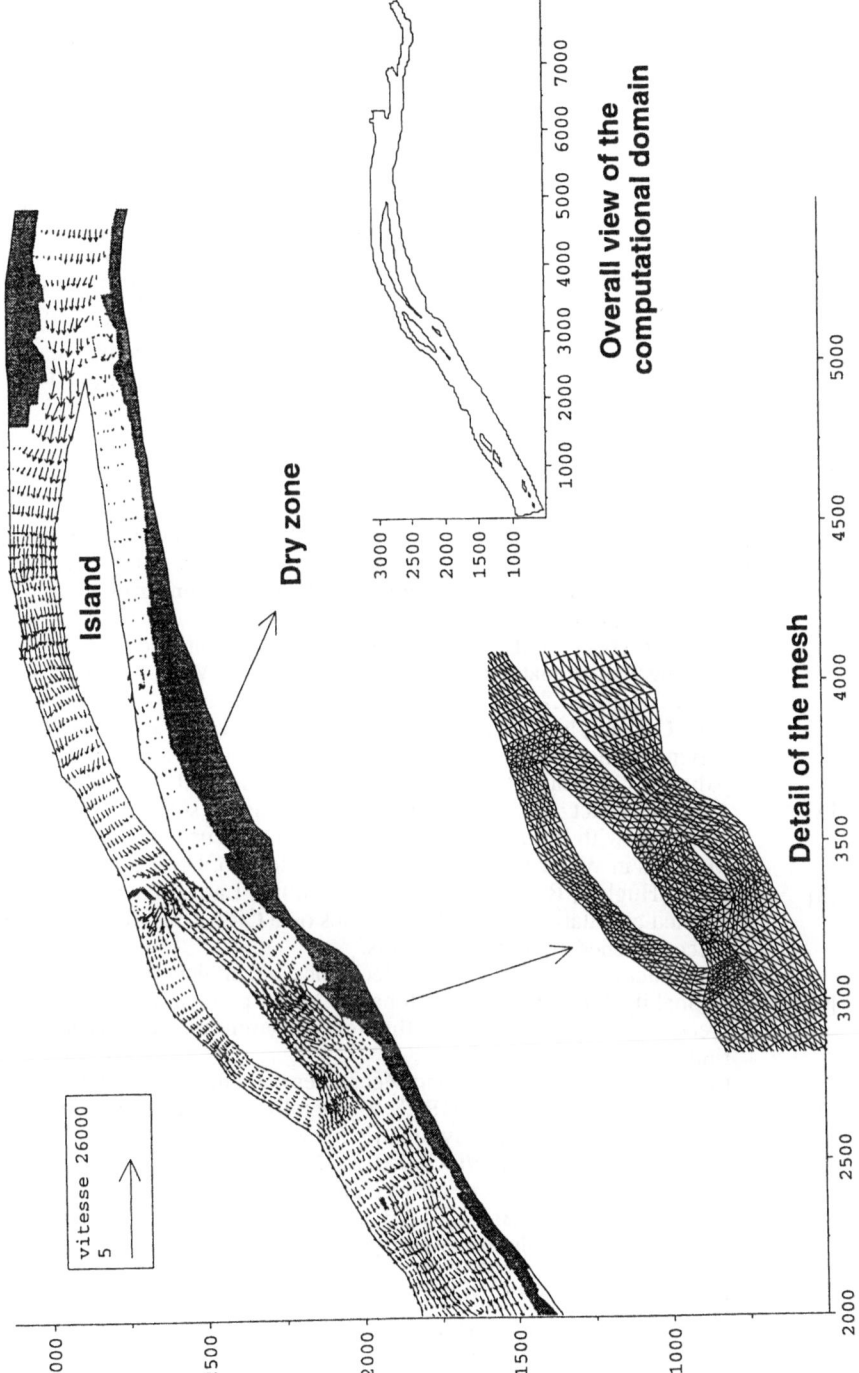

FIGURE 2: VELOCITY FIELD IN THE RIVER LOIRE

The following table gives the total computational time of the two applications given here. All the solvers have been tested with the combination of a diagonal preconditioning and the specific one described in this paper.

	4 tides 1200 time steps 9500 elements time step: 150 s	River with dry zones 1000 time steps 11250 elements time step: 10 s
Conjugate Gradient	no convergence	no convergence
Conjugate Residual	149 s (CRAY C90)	11557 s (HP)
Normal Equation	151 s (CRAY C90)	9071 s (HP)
Minimal Error	150 s (CRAY C90)	15864 s (HP)
BICGSTAB	159 s (CRAY C90)	no convergence
GMRES	139 s (CRAY C90)	9920 s (HP)

BIBLIOGRAPHY:

[1]   O. Daubert, J.-M. Hervouet and A. Jami: Description of some numerical tools for solving incompressible turbulent and free surface flows. International Journal for Numerical Methods in Engineering. Vol. 27,3-20(1989).

[2]   J.-M. Hervouet : TELEMAC: a fully vectorised finite element software for shallow water equations. Computer Methods and Water Resources. Rabat, Morocco.7-11 Oct. 1991.

[3]   J.-C. Galland, N. Goutal, J.-M. Hervouet: A new numerical model for solving shallow water equations. Advances in Water Resources. Vol. 14 n°3 June 1991 pp.138-148.

[4]   J.-M. Hervouet : Element by Element methods for solving shallow water equations with F.E.M.". IX International Conference on Computational Methods in Water Resources. Denver, Colorado, USA. June 9-12,1992.

[5]   T.J.R. Hughes, R.M. Ferencz, J.O. Hallquist, Large-scale vectorized implicit calculations in solid mechanics on a CRAY X-MP/48 utilizing EBE preconditioned conjugate gradients. Computer Methods in Applied Mechanics and engineering 61(1987) 215-248.

[6]   F. Shakib, T.J.R. Hughes, Z. Johan: A multi-element group preconditioned GMRES algorithm for nonsymmetric systems arising in finite element analysis. Computer Methods in Applied Mechanics and engineering 75(1989) 415-456.

[7]   Y. Saad, M.H. Schultz: GMRES: a generalized minimum residual algorithm for solving nonsymmetric linear systems. Research report YALEU/DCS/RR-254. Department of Computer Science, Yale University, 1983.

[8]   H.A. van der Vorst: Bi-CGSTAB: A fast and smoothly converging variant of Bi-CG for the solution of nonsymmetric linear systems. Pre-print, Utrecht. 1990.

[9]   J.-M. Hervouet, C. Moulin: Nouveaux schémas de convection de TELEMAC-2D: apport de la méthode SUPG. EDF Report HE43/93.27

THREE-DIMENSIONAL D4Z RENUMBERING FOR ITERATIVE SOLUTION OF GROUND-WATER FLOW AND TRANSPORT EQUATIONS

KENNETH L. KIPP
Water Resources Division, U.S. Geological Survey
P.O. Box 25046, M.S. 413, Denver, CO 80225-0046
U.S.A.

THOMAS F. RUSSELL and JAMES S. OTTO
Department of Mathematics, University of Colorado at Denver
P.O. Box 173364, Campus Box 170, Denver, CO 80217-3364
U.S.A.

D4 zigzag (D4Z), previously introduced in two space dimensions, is a renumbering scheme for incomplete LU preconditionings of conjugate-gradient-like iterative solvers for linear systems arising from five-point (seven-point in three dimensions) stencils. Compared to other orderings, it reduces sensitivity of convergence to the sequence of coordinate directions for renumbering in anisotropic media. This paper extends D4Z to three dimensions and demonstrates its insensitive behavior on two test problems.

INTRODUCTION

With increasing frequency, practical applications of models of ground-water flow and transport involve many thousands of discrete finite-difference or finite-element un-knowns. In two and especially in three space dimensions, for reasons of computing time and storage, it is often infeasible to solve directly the systems of linear equations that determine these unknowns (or nonlinear iterates of them). Thus, robust, efficient iterative solvers are of considerable and increasing importance. In practice, where there may not be time or expertise to fine-tune algorithm parameters, a robust approach that behaves in an average manner, avoiding worst-case scenarios, is of significant interest.

Due to anisotropies in subsurface formations and to aspect ratios in grid cells, discrete flow coefficients commonly favor one coordinate direction over others. It is well-known that this affects the convergence efficiency of many iterative methods; for example, line successive overrelaxation is motivated by this. However, anisotropies and aspect ratios may be quite heterogeneous, causing methods based on the choice of a preferred direction to be unreliable. Historically, modelers of subsurface flows have found conjugate-gradient-like iterative solvers such as Orthomin, preconditioned by incomplete LU (ILU) decompositions (Meijerink and van der Vorst, 1977), to be

A. Peters et al. (eds.), Computational Methods in Water Resources X, 1417–1424.

more dependable. In this context, directional sensitivity arises more subtly in the preconditioner.

The idea of ILU preconditioning is to apply the conjugate-gradient-like algorithm to the system $\mathbf{MA}x = \mathbf{M}b$, where $\mathbf{A}x = b$ is the discrete system to be solved, $\mathbf{M} = (\mathbf{LU})^{-1} \approx \mathbf{A}^{-1}$, and $\mathbf{L}$ and $\mathbf{U}$ are approximate lower- and upper-triangular factors of $\mathbf{A}$. Since $\mathbf{MA}$ is in some sense close to the identity matrix, the preconditioned iteration should converge more rapidly than the original. Another useful device is a red-black reduction of the original 2-cyclic system, assuming a standard seven-point finite-difference stencil as in the HST3D code (Kipp, 1987) used in this study. This procedure assigns colors in a checkerboard fashion, then numbers all red nodes first, resulting in the block matrix

$$\mathbf{A} = \begin{bmatrix} \mathbf{D}_R & \mathbf{A}_{RB} \\ \mathbf{A}_{BR} & \mathbf{D}_B \end{bmatrix},$$

where $\mathbf{D}_R$ and $\mathbf{D}_B$ are diagonal matrices. Elimination yields the "black matrix"

$$\mathbf{R} = \mathbf{D}_B - \mathbf{A}_{BR}\mathbf{D}_R^{-1}\mathbf{A}_{RB},$$

a matrix with half as many unknowns and enhanced diagonal dominance. These features generally result in still faster preconditioned iterative convergence. This paper considers ILU preconditioning of $\mathbf{R}$ with no added fill, i.e., the nonzero diagonals of $\mathbf{L}$ and $\mathbf{U}$ are limited to those of $\mathbf{R}$, corresponding to ICCG(1,1) in the notation of Meijerink and van der Vorst (1977). The basic iterative procedure is Orthomin(s), in which Orthomin is restarted after every $s + 1$ iterations, i.e., a maximum of s search directions are saved.

ORDERINGS

Without preconditioning, the ordering of the unknowns of $\mathbf{R}$ would not affect the Orthomin iterates. Different orderings yield different ILU factors and hence different preconditioned algorithms. For the reduced matrix $\mathbf{R}$, the simplest ordering, which we denote by RB (red-black), numbers the black nodes in natural order. This was first considered for iterative subsurface-flow computations by Tan and Letkeman (1982). Natural ordering has a clear directional bias, so the alternate diagonal (D4) ordering of the black nodes was proposed as an alternative (Behie and Forsyth, 1984; Eisenstat et al., 1988). In a previous paper (Kipp et al., 1992), we have found that D4 exhibits the same sensitivity to direction as RB in two dimensions. That is, traversing the same D4 diagonal lines in the opposite direction has the same effect as exchanging the primary and secondary directions in the RB ordering.

This can be intuitively understood by examining the (in two dimensions) nine-diagonal $\mathbf{R}$ under RB or D4 ordering, assuming significant anisotropy. Depending on directional choices, the largest off-diagonal elements will be far from the main diagonal (faster convergence) or close to it (slower convergence), in a similar manner for both methods. For details, see Kipp et al. (1992), where the two-dimensional alternating

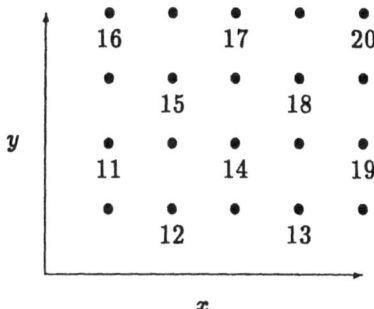

Figure 1: Two-dimensional D4Z ordering for black nodes of 5 × 4 grid.

diagonal zigzag (D4Z) scheme was proposed. This approach reverses direction from one diagonal to the next (see Figure 1), resulting in about the same proximity of large off-diagonal elements to the main diagonal irrespective of directional choices (one could reverse the direction of traverse of all diagonal lines and still have a zigzag pattern). For model problems, the expected insensitivity of D4Z was observed in iteration counts and in the condition numbers and eigenvalue distributions of preconditioned reduced matrices. D4Z iteration counts were near the average of those for different RB and D4 directional choices in cases with at least slight compressibility. D4Z was closer to the slower RB and D4 choices in an incompressible case.

D4Z IN THREE DIMENSIONS

If i, j, k denote coordinate indices, the black nodes in three dimensions are those for which $i + j + k$ is even; denote it by 2ℓ. Diagonal planes are obtained by a constant value $2, 3, 4, \ldots$, of ℓ. Associated with each plane (value of ℓ) is one of the designations kji, jik, ikj in cyclic order. For example, kji for $\ell = 2$, jik for $\ell = 3$, ikj for $\ell = 4$, kji again for $\ell = 5$, and so on. It remains to describe how each plane is ordered; consider jik for $\ell = 3$ as an illustration. The first index, j, is primary, so the nodes of plane $i + j + k = 6$ are ordered in lines along which j is constant, starting at the maximum value, 4, of j and proceeding in descending order through $j = 3, 2, 1$. The next index, i, is secondary, so it is secondary in lines $j = 4$ and $j = 2$, tertiary in lines $j = 3$ and $j = 1$. Thus, in lines $j = 4$ and $j = 2$, i starts at its maximum and decreases, while in lines $j = 3$ and $j = 1$, k does the same. The resulting zigzag ordering of the nodes (i, j, k) in the plane is shown in Figure 2.

The preceding paragraph describes three possible D4Z orderings of the three-dimensional grid: zyx, yxz, xzy, in which the first plane $\ell = 2$ is designated kji, jik, ikj, respectively (thus we used zyx in the example). Three others (xyz, yzx, zxy) use the designations ijk, jki, kij in an analogous fashion. The same total of six permutations is reached for RB and D4 by choosing a primary, secondary, and tertiary direction in

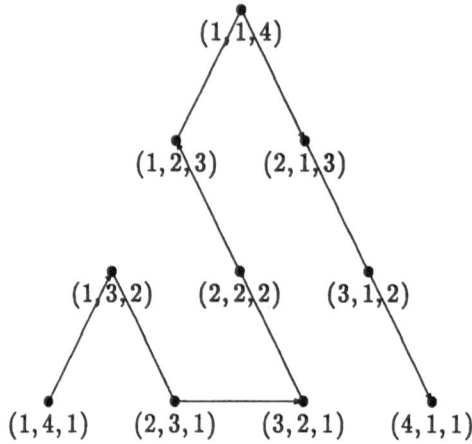

Figure 2: *jik* zigzag ordering of nodes in plane $i + j + k = 6$

all possible ways.

TEST PROBLEMS

Problem 1 discretizes a 2 m × 0.2 m × 2 m region (vertical slab) with a uniform 11 × 3 × 11 point-centered grid (Figure 3 shows boundary conditions). The porosity is 10%, with directional permeabilities 10^{-8}, 10^{-8}, and 10^{-10} m². At time zero, with initial hydrostatic equilibrium of fresh water, a pressure increase of 5386 Pa is imposed at the short horizontal column $x = z = 0$, while atmospheric pressure is maintained at $x = z = 2$ m, causing inflow of saline water (density 10% greater than that of the resident fresh water) at $x = z = 0$ and outflow at $x = z = 2$. As time progresses, a plume of dense saline water invades the region. Incompressible and compressible (water compressibility 5×10^{-10} Pa^{-1}, matrix 10^{-8} Pa^{-1}) versions were run. The difference discretizations were centered in both space and time, covering one second in five equal time steps comprising ten linear solutions (two outer Picard iterations per step). The (nonsymmetric) flow and transport equations were solved sequentially; only flow results (totals over the ten linear solutions) are presented because the transport solution was always obtained in one or two Orthomin iterations. The solver was Orthomin(4), with convergence upon reduction of the L^2 norm of the residual by a factor of 10^{-7}.

Problem 2 is a field-scale simulation (see Huyakorn *et al.*, 1987, for further details). The region is 14400 ft × 9600 ft × 200 ft with a uniform 62 × 19 × 11 point-centered grid (12,958 nodes). Porosity is 20%; horizontal and vertical hydraulic conductivities are 2000 ft/day and 200 ft/day, respectively. The problem calculates the steady-state response of the aquifer to two pumping wells, respectively screened in the upper and

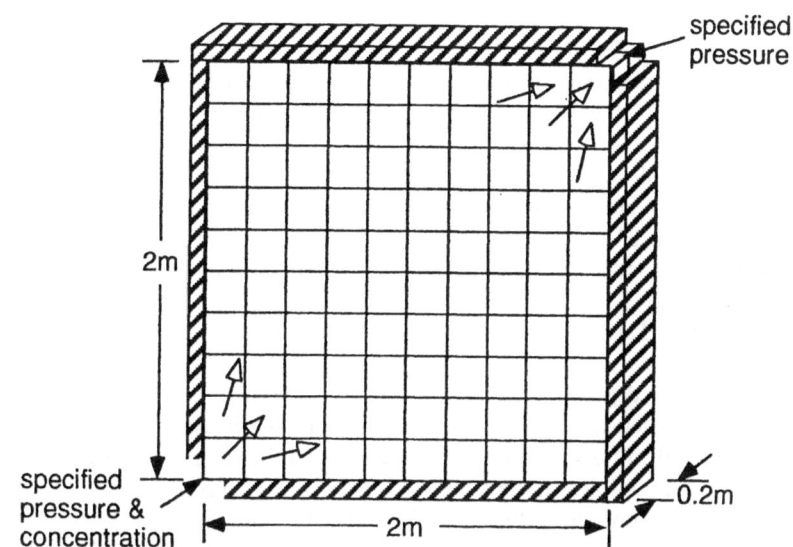

Figure 3: Region and boundary conditions for Problem 1

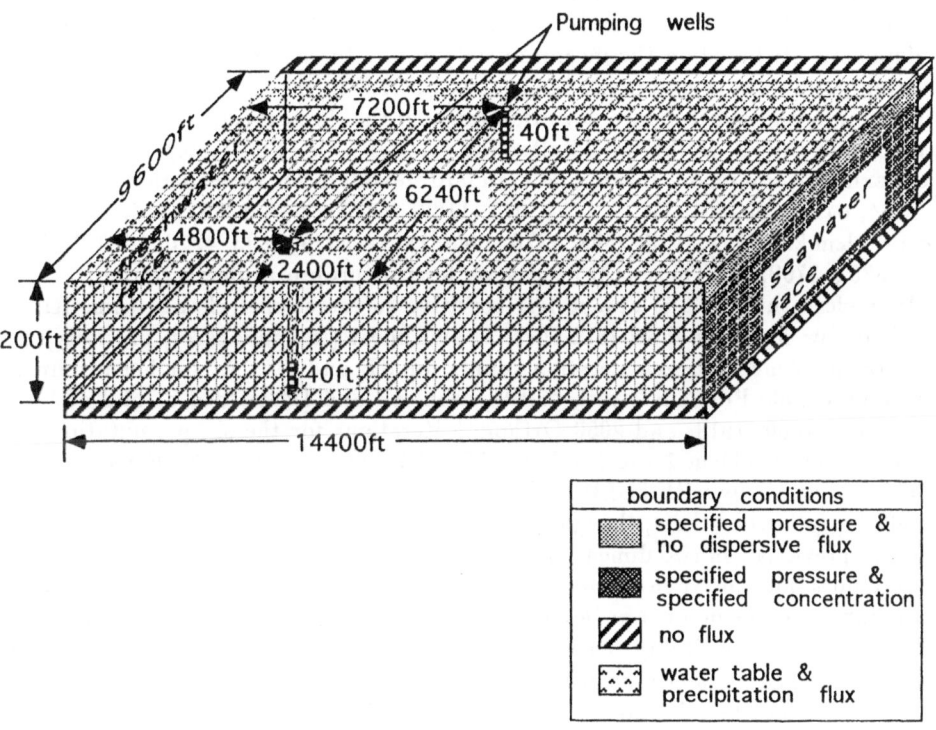

Figure 4: Region and boundary conditions for Problem 2

lower 40 ft and withdrawing water at 0.5×10^6 and 1.0×10^6 ft^3/day. The bottom, north, and south boundaries are impermeable; the west boundary is fresh water with horizontally linear pressure; the east is seawater on the lower 180 ft with no dispersive flux on the upper 20 ft; the top is uniformly recharged at 1 ft/year (see Figure 4). Initial conditions are hydrostatic pressure (uniform horizontally) and fresh water. Again incompressible and compressible (water 3.5×10^{-6} psi^{-1}, matrix 7×10^{-6} psi^{-1}) versions were run. Differencing was upstream in space and backward in time over 25,000 days in 31 time steps (62 outer Picard iterations). Again Orthomin(4) was used, this time with tolerance 10^{-6}.

RESULTS AND DISCUSSION

Results are presented graphically in Figures 5 and 6 for Problems 1 and 2, respectively. In both problems, modified ILU (MILU) preconditioning of the flow equation, where discarded fill-in terms were added to the main diagonal, was tried in the incompressible and compressible versions and was much more efficient than ILU, by factors of 2 to 5. This agrees with earlier experience (Behie and Forsyth, 1984; Eisenstat *et al.*, 1988). The figures present total iteration counts, over all Picard iterations for all time steps, for the incompressible (MILU only, with all three numbering methods) and compressible (MILU only, with RB and D4Z) versions of both problems (MILU was not run with D4 in the compressible case because these results could be expected to duplicate RB based on the incompressible data). In each case, counts for the six ordering permutations determine the representations in the figures. The maximum, minimum, mean, and standard deviation range of the six numbers are depicted. ILU data are not shown, as this would alter the vertical scale. For Problem 1 with ILU, the maximum, minimum, mean, and standard deviation for each method were: RB — 1764, 641, 1182, 561; D4 — 1764, 641, 1182, 561; D4Z — 1380, 1134, 1273, 103. For Problem 2: RB — 5928, 2893, 4370, 1449; D4 — 5928, 2893, 4370, 1449; D4Z — 3784, 3751, 3769, 11. These ILU results are for the incompressible cases only.

It is clear that D4Z is always less sensitive to directional change than RB or D4. The raw numbers show that there is an almost (in some cases, precisely) exact correspondence between iteration counts for RB and D4. As further examples, in incompressible Problem 2 with MILU, both RB and D4 yielded a total of 1110, 1415, 1692, 1880, 1919, and 2050 Orthomin iterations for the six permutations; in incompressible Problem 1 the numbers differed by no more than 2. For the model problem 1, especially with MILU, D4Z is at some disadvantage relative to the average efficiency of the others, though compressibility mitigates this to some extent as observed previously in two dimensions. This makes it very interesting that the disadvantage does not appear in the more practical field problem 2, where D4Z performs in a reliable average manner as intended, and better than average in one case (ILU), even without compressibility. Further study is needed to understand the reasons for this, but it appears that the somewhat pessimistic behavior observed in some model examples may not be relevant in practical cases. The present serial implementation may be parallelizable by incorporating zigzag ordering in recently developed parallel

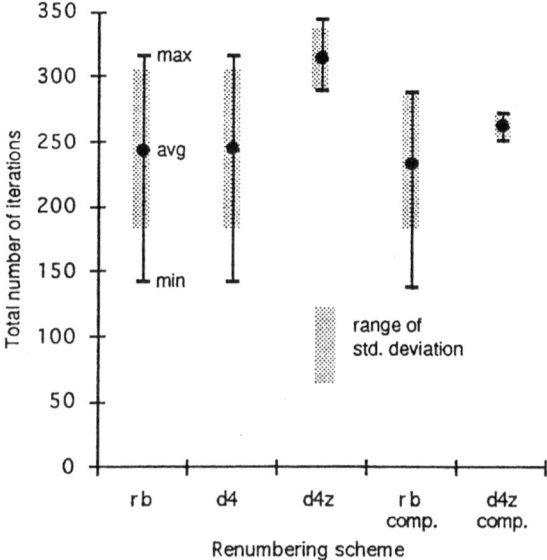

Figure 5: Distribution of iteration counts for directional permutations (Problem 1)

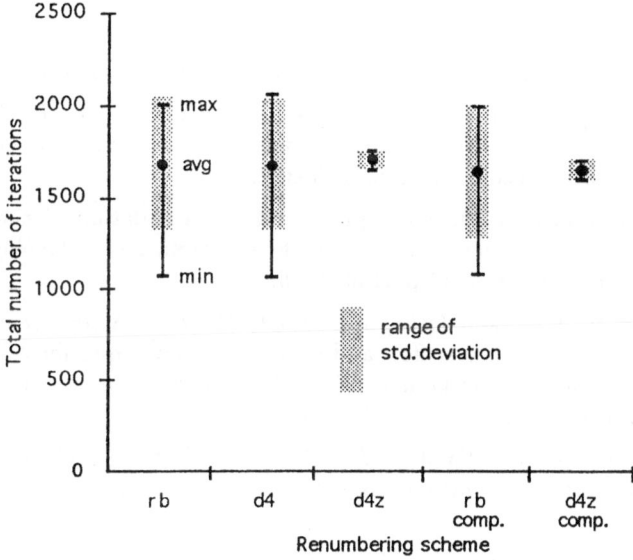

Figure 6: Distribution of iteration counts for directional permutations (Problem 2)

ILU frameworks (Elman and Golub, 1992).

CONCLUSIONS

D4Z has been extended to three-dimensional systems and applied to model and field problems. This renumbering scheme for ILU preconditioners is demonstrably insensitive to directional bias in anisotropic media. This enhances robustness in the sense that worst-case behavior of other approaches is avoided without user intervention or trial runs. The previously observed tendency for D4Z to be slower than the average of other methods in incompressible model problems was not observed in the field cases, where its insensitive performance was average or better. For flow equations, MILU preconditioning was distinctly preferable to ILU.

REFERENCES

Behie, G.A. and Forsyth, P.A. (1984) "Incomplete factorization methods for fully implicit simulation of enhanced oil recovery", SIAM J. Sci. Statist. Comput. 5, 543–560.

Eisenstat, S.C., Elman, H.C., and Schultz, M.H. (1988) "Block-preconditioned conjugate-gradient-like methods for numerical reservoir simulation", SPE Reservoir Engrg. 3, 307–312.

Elman, H.C. and Golub, G.H. (1992) "Line iterative methods for cyclically reduced discrete convection-diffusion problems", SIAM J. Sci. Statist. Comput. 13, 339–363.

Huyakorn, P.S., Andersen, P.F., Mercer, J.W., and White, H.O., Jr. (1987) "Salt-water intrusion in aquifers: Development and testing of a three-dimensional finite element model", Water Resour. Res. 23, 293–312.

Kipp, K.L., Jr. (1987) HST3D: A computer code for simulation of heat and solute transport in three-dimensional ground-water flow systems, U.S. Geological Survey, Water-Resources Investigations Report 86-4095.

Kipp, K.L., Russell, T.F., and Otto, J.S. (1992) "D4Z — A new renumbering for iterative solution of ground-water flow and solute-transport equations", in T.F. Russell et al. (eds.), Numerical Methods in Water Resources, Computational Mechanics Publications, Southampton, U.K., pp. 495–502.

Meijerink, J.A. and van der Vorst, H.A. (1977) "An iterative solution method for linear systems of which the coefficient matrix is a symmetric M-matrix", Math. Comp. 31, 148–162.

Tan, T.B.S. and Letkeman, J.P. (1982) "Application of D4 ordering and minimization in an effective partial matrix inverse iterative method", Proc. 6th SPE Symp. on Reservoir Simulation, Society of Petroleum Engineers, Dallas, pp. 43–58.

A SIMPLE AND EFFICIENT MULTIGRID METHOD FOR INTERFACE PROBLEMS

J. Molenaar
TWI, Delft University of Technology,
p.o.box 5031, 2600 GA Delft, The Netherlands

A new multigrid algorithm is presented for the numerical solution of interface problems. The coarse grid operator in the algorithm is constructed by using a suitable average of the fine grid operator. The advantage of this approach is that we obtain an M-matrix on all grids and that the sparsity pattern of the fine grid matrix is retained on all grids.

INTRODUCTION

We consider the numerical solution of interface problems,

$$-\nabla \cdot (D\nabla u) = f, \tag{1}$$

in a domain $\Omega \subset \mathbb{R}^i, i = 2, 3$. Typically the diffusion coefficient D jumps by orders of magnitude across internal interfaces and contains strong anisotropies. The pressure equation that appears in many models for porous media flow is an important example of these type of problems.

In most of the multigrid methods (cf.[1], [4], [5], [2]) that have been proposed for this problem the coarse grid operator is based on the Galerkin approximation of the fine grid operator given the grid transfer operators (automatic prescription). The disadvantage of this approach is that the stencil of the coarse grid operator is often denser than the corresponding fine grid stencil. Especially in three space dimensions this is problematic: 7-point stencils on the finest grid are turned into 27-point stencils on the coarser grids (see e.g. [4]). The construction of the coarse grid operators then consumes a substantial part of the total computation time.

In our multigrid algorithm the coarse grid operator is constructed by a simple averaging of the coefficients of the fine grid operator. This approach has many advantages: it is cheap, the sparsity pattern is the same on all grids and we obtain M-matrices on all grids. Moreover we use simple grid transfer operators that are based on the interpolation of polynominals.

A. Peters et al. (eds.), Computational Methods in Water Resources X, 1425–1429.

CELL-CENTERED MULTIGRID

For the discretization of (1) we use a standard finite volume discretization. For simplicity we only state the discretization for $\Omega \subset \mathsf{R}^2$ and assume that the domain Ω can be divided by a regular partitioning in open square cells $\Omega_{i,j}$ (with side length h) and that the discontinuities in D are resolved by this partitioning. Integration of (1) over a cell $\Omega_{i,j}$ yields

$$h(F_{i+\frac{1}{2},j} + F_{i,j+\frac{1}{2}} - F_{i-\frac{1}{2},j} - F_{i,j-\frac{1}{2}}) = h^2 f_{i,j}. \tag{2}$$

The flux $F_{i+\frac{1}{2},j}$ at the cell edge between $\Omega_{i,j}$ and $\Omega_{i+1,j}$ is approximated by

$$F_{i+\frac{1}{2},j} = \frac{2D_{i,j}D_{i+1,j}}{D_{i,j} + D_{i+1,j}} \frac{u_{i+1,j} - u_{i,j}}{h}, \tag{3}$$

i.e., the coefficient D at the cell edge is obtained by harmonic averaging of the values in the adjacent cells (cf.[5]). The matrix of this linear system is an M-matrix if there are Dirichlet boundary conditions available.

To solve this system of equations we consider a cell-centered multigrid method: the coarse grid cells are obtained by joining the 4 corresponding fine grid cells, which is natural for a finite volume scheme. Usually the coarse grid operator in linear multigrid algorithms is constructed by first choosing suitable interpolation operators P^h and R^H, the prolongation and restriction operator (which may be problem dependent), and then defining the coarse grid operator L^H as the Galerkin approximation of the fine grid operator L^h:

$$L^H = R^H L^h P^h. \tag{4}$$

The disadvantage of this procedure is that it is not guaranteed that the coarse grid matrix has the same sparsity pattern as the fine grid matrix. Moreover the coarse grid matrix need not be an M-matrix anymore, which makes it difficult to select an appropriate smoothing operator.

An alternative to this procedure is to discretize the problem also on the coarser grids; this is often done in nonlinear multigrid problems in which the construction (4) it is not feasible. In the constant coefficient case this can formally be defined as

$$L^H = \frac{1}{2} R^H L^h P^h, \tag{5}$$

with P^h the piecewise constant interpolation and R^H its adjoint. The factor $1/2$ is due to the fact that the piecewise constant prolongation leads to fluxes on fine grid edges that are twice the fluxes on the corresponding coarse grid edges.

We now construct our coarse grid operator by (5) also in the variable coefficient case. If a coarse grid cell edge consists of two fine grid edges with coefficients D_1^h and D_2^h then the flux on the coarse grid edge is defined by (5) as

$$F_{i+\frac{1}{2},j} = \frac{1}{2}(D_1^h + D_2^h)\frac{u_{i+1,j} - u_{i,j}}{2h}. \tag{6}$$

The coarse grid flux is calculated with a diffusion coefficient that is the arithmetic average of the corresponding fine grid coefficients. Notice that we now find the same five-point sparsity pattern on all grids, and that we obtain an M-matrix on all grids indeed.

As P^h and R^H only interpolate constant functions exactly, we have $m_P = m_R = 1$, with m_P and m_R the order of the prolongation and restriction, respectively. For second order differential equation we should have (cf.[3])

$$m_P + m_R > 2,$$

therefore we use a more accurate prolongation operator $\tilde{P}^h$, bilinear interpolation ($m_{\tilde{P}} = 2$), in our multigrid algorithm.

NUMERICAL EXAMPLES

To demonstrate the efficiency of this multigrid algorithm we show convergence results for some test problems in 2D and 3D. In the two-dimensional case we consider a problem in the unit square $\Omega = (0,1)^2$:

$$D^i(x_1, x_2) = \begin{cases} 10^{p_i}, & \underline{x} \le x_j \le \overline{x}, \quad j = 1,2, \\ 1, & \text{otherwise}, \end{cases} \tag{7}$$

$$f(x_1, x_2) = \begin{cases} 1, & \underline{x} \le x_j \le \overline{x}, \quad j = 1,2, \\ 0, & \text{otherwise} \end{cases} \tag{8}$$

and boundary conditions

$$u = 0, \quad \text{for } \min(x_1, x_2) = 0, \tag{9}$$

$$\frac{\partial u}{\partial n} = 0, \quad \text{otherwise}. \tag{10}$$

As D contains both inhomogeneities and anisotropies we use an ILLU-smoother. In the multigrid iteration W-cycles are used on all grids with a single ILLU sweep for both pre- and post-smoothing. The convergence rate ρ is estimated by

$$\rho = \left(\frac{\|r^{(10)}\|_\infty}{\|r^{(0)}\|_\infty} \right)^{\frac{1}{10}},$$

with $r^{(m)}$ the residual after m multigrid cycles.

In Table 1 we show the convergence rate ρ on a 64×64 grid for both the case that the discontinuities are resolved on most of the grids ($\underline{x} = 16/64$ and $\overline{x} = 48/64$) and that they are resolved on the finest grid only ($\underline{x} = 17/64$ and $\overline{x} = 47/64$). In all cases we observe a fast convergence, and only in the case $p_1 = -3$ and $p_2 = +3$ the position of the discontinuity makes any difference. If we solve the same problem on a

| | | $x = 16/64$ | $x = 17/64$ |
| | | $\bar{x} = 48/64$ | $\bar{x} = 47/64$ |
p_1	p_2		
-3	-3	0.06	0.06
-3	0	0.05	0.07
-3	+3	0.09	0.33
0	0	0.06	0.06
0	+3	0.10	0.10
+3	+3	0.10	0.10

Table 1: Convergence rates ρ for 2D test problem on 64 × 64 grid.

p	16	32	64	64
-3	0.26	0.28	0.29	0.29
0	0.27	0.28	0.29	0.29
+3	0.50	0.47	0.42	0.41

Table 2: Convergence rates ρ for 3D test problem on different grids

128 × 128 grid we find a convergence rate $\rho = 0.12$, so the multigrid method appears to be robust.

Next we consider the 3D case. Here we only give convergence results for a test problem with discontinuous coefficients. The test problem is defined on the unit cube $\Omega = (0,1)^3$ with

$$D^i(x_1, x_2, x_3) = \begin{cases} 10^p, & \underline{x} \leq x_j \leq \bar{x}, \quad j = 1, 2, 3, \\ 1, & \text{otherwise}, \end{cases} \tag{11}$$

$$f(x_1, x_2, x_3) = \begin{cases} 1, & \underline{x} \leq x_j \leq \bar{x}, \quad j = 1, 2, 3, \\ 0, & \text{otherwise} \end{cases} \tag{12}$$

and boundary conditions

$$u = 0, \quad \text{for } \min(x_1, x_2, x_3) = 0, \tag{13}$$

$$\frac{\partial u}{\partial n} = 0, \quad \text{otherwise}. \tag{14}$$

For both pre- and post-smoothing we use a single point Gauss-Seidel relaxation sweep on all grids. In Table 2 the convergence rate ρ is shown for different mesh sizes and different positions of the discontinuity. For the case $\underline{x} = 16/64$ and $\bar{x} = 48/64$ our multigrid method handles this problem adequately: we obtain acceptable grid independent convergence rates (the second to fourth column in Table 2). In fact if we shift the discontinuity ($\underline{x} = 17/64$ and $\overline{x} = 47/64$) we observe nearly the same convergence rates.

Our multigrid method appears to be robust for this 3D test problem with discontinuous coefficients. For problems with anisotropic coefficients plane relaxation is the only known robust smoother (cf. [6]). As our 2D multigrid method handles discontinuities and anisotropies satisfactorily, it seems natural to construct a plane relaxation that is based on this 2D multigrid algorithm. This will be reported elsewhere.

REFERENCES

[1] R.E. Alcouffe, A. Brandt, J.E. Dendy Jr., and J.W. Painter. The multi-grid method for the diffusion equation with strongly discontinuous coefficients. *SIAM J. Sci. Statist. Comput.*, 2:430–454, 1981.

[2] P.M. de Zeeuw. Matrix-dependent prolongations and restrictions in a blackbox multigrid solver. *J. Comput. Appl. Math.*, 33:1–27, 1990.

[3] W. Hackbusch. *Multi-Grid Methods and Applications*, volume 4 of *Springer Series in Computational Mathematics*. Springer-Verlag, Berlin, 1985.

[4] J.E. Dendy Jr. Two multigrid methods for three-dimensional problems with discontinuous and anisotropic coefficients. *SIAM J. Sci. Statist. Comput.*, 8:673–685, 1987.

[5] P.Wesseling. Cell-centered multigrid for interface problems. *J. Comput. Phys.*, 79:85–91, 1988.

[6] R.Kettler and P.Wesseling. Aspects of multigrid methods for problems in three dimensions. *Appl. Math. Comp.*, 19:159–168, 1986.

KRYLOV METHODS IN THE FINITE ELEMENT SOLUTION OF GROUNDWATER TRANSPORT PROBLEMS

GIORGIO PINI and MARIO PUTTI
*Dept. of Mathematical Methods for Applied Sciences, University of Padua
via Belzoni 7, Padua - Italy*

The numerical treatment of the groundwater transport model requires the repeated solution of nonsymmetric, large, sparse systems of linear equations. For this reason it is important to employ algorithms that are efficient and robust in terms of both memory utilization and computational burden. Among the most efficient schemes, we can mention those based on the definition of appropriate Krylov subspaces (Conjugate Gradient like) and those based on the Lanczos algorithm (Lanczos like). These two classes are often intersected, and mixed techniques seems to be the most attractive in terms of efficiency and robustness. In the solution of the groundwater transport equation, these schemes may perform poorly, or even may not converge at all, in advection dominated cases. In this paper we compare the efficiency of some of these algorithms recently proposed in the literature, applied to the solution of nonsymmetric systems arising from the finite element discretization of the groundwater transport equation. We consider Saad's GMRES, Freund's TFQMR, and van der Vorst BiCGSTAB. We also investigate the effects of different preconditioning strategies: the diagonal scaling, ILU(0), and the recently proposed ILUT(p, τ) preconditioner, an implementation of the incomplete factorization that allows for variable fill inn of the triangular factors. The various schemes are compared on the basis of their performance in two sample problems, with constant and variable velocity fields.

INTRODUCTION

The numerical approach to solving partial differential equations often requires the successive solution of systems of linear equations. In particular, finite element applications yield system matrices that are large and sparse. Iterative linear solvers exploit at best these properties, and offer great advantages over direct techniques, in terms of both efficiency and storage requirements.

It is well known that conjugate gradient schemes are among the fastest and most robust iterative solvers, when applied to symmetric, positive definite systems. However, the situation is different when the system matrix A is nonsymmetric. In this case, in fact, convergence results for conjugate gradient based schemes, also called Krylov methods, are available only under very restrictive conditions, and, in many actual computations, divergence or stagnation may occur even when preconditioning is employed. In particular, it has been observed that, for matrices with eigenval-

A. Peters et al. (eds.), Computational Methods in Water Resources X, 1431–1438.

ues lying entirely in the complex plane, or having large imaginary parts, case which is often encountered in the numerical solution of the advection-dispersion equation, convergence may be problematic [*Peters, 1992; Sleijppen and Fokkema, 1993*]. However, a number of conjugate gradient type schemes showing reasonable convergence and robustness properties under a variety of different conditions [*Pini and Zilli, 1989; Pini and Zilli, 1990*], have been recently proposed [*Sonneveld, 1989; van der Vorst, 1990; van der Vorst, 1992; Freund, 1993; Sleijppen and Fokkema, 1993*]. Furthermore, improvements in the preconditioning strategies have made these schemes very competitive. For a recent and fairly complete review of iterative methods see [*Barret et al., 1994*], which also includes a discussion on the implementation of such schemes on supercomputers and superscalar workstations.

In the present paper, we compare some of the more efficient Krylov methods for the solution of the nonsymmetric systems arising from the finite element discretization of the equation of contaminant transport in groundwater. Three iterative solvers are compared: the Generalized Minimal Residual (GMRES) [*Saad and Schultz, 1986*], the Biconjugate Gradient Stabilized (Bi-CGSTAB) [*van der Vorst, 1992*] and the Transpose-Free Quasi-Minimal Residual (TFQMR) [*Freund, 1993*], in their original and preconditioned formulations. Three different preconditioning strategies are employed: the simple diagonal scaling, the incomplete LU (Croute) decomposition (ILU(0)), and the incomplete LU decomposition with variable fill-in (ILUT(p, τ)) [*Saad, 1991*]. The comparison is performed mainly on the basis of the total CPU needed to achieve convergence, that is, to reduce the Euclidean norm of the relative real residual below a specified tolerance $\|b - Ax\| \leq \text{TOL}\,\|b\|$.

ITERATIVE METHODS

The first algorithm under investigation, GMRES, uses a modified Gram-Schmidt orthogonalization scheme to build a sequence of orthogonal vectors which form the basis of the Krylov subspace span$\{r^{(0)}, Ar^{(0)}, A^2 r^{(0)}, \ldots\}$, where $r^{(0)}$ is the initial residual vector $r^{(0)} = b - Ax^{(0)}$. The procedure is restarted after m steps, and hence after m Gram-Schmidt vectors have been calculated, to keep memory requirements from growing excessively (GMRES(m)). A small predetermined value of m $(10 \div 20)$ achieves a reasonable balance between storage requirements and efficiency, but exposes the convergence of the scheme to possible stagnation.

The other two iterative schemes can both be considered as extensions of the Biconjugate Gradient algorithm (BCG) [*Lanczos, 1952; Fletcher, 1976*]. This scheme generates two CG-like, mutually orthogonal sequences of vectors, based on matrix A and A^T, respectively. However, an irregular convergence behavior may be displayed byu the process, with possible break downs. This inconvenience is overcome by varying the iteration by combining different schemes so that to stabilize convergence. Numerous variants have been proposed. Among these, we chose the Bi-CGSTAB and TFQMR. Note that both these schemes do not make use of A^T, and therefore allow for more efficient implementation.

The Bi-CGSTAB method can be derived from a combination of the BCG and GMRES(1) algorithms [*Barret et al., 1994*]. The convergence behavior of Bi-CGSTAB is rather smooth, but the procedure can still break down (the Krylov subspace may not be expanded as the iteration proceeds) for some particular problems. TFQMR is built from variations of the Quasi Minimal Residual scheme (QMR) [*Freund and Nacthigal, 1991*] and BCG. The resulting method shows a smooth convergence behavior but is still subject to break down.

We have considered two classes of preconditioners: the simple diagonal scaling (Jacobi) and the incomplete LU decomposition (ILU). In the diagonal preconditioner the original problem $Ax = b$ is equivalently written as $By = c$, where $B = D^{-1}A$, $y = x$ and $c = D^{-1}b$. The three solvers are applied using as initial guess the vector $x_0 = D^{-1}b$. The ILU preconditioner without fill-in (ILU(0)) [*Meijerink and van der Vorst, 1977*] calculates the lower and upper triangular matrices L and U using the Crout decomposition of A and maintaining the same number of nonzero elements of A. This is achieved by preassigning to the L and U matrices the same sparsity pattern of the corresponding part of A. An extension of this preconditioner, based also on Gaussian elimination applied to A, does not predetermine the fill-in a priori, but controls it through the use ot two parameters, p and τ (ILUT(p, τ)) [*Saad, 1991*]. The first parameter, p, represents the maximum value of the fill-in allowable in each row of the incomplete factors L or U. The other parameter, τ, represents the threshold value relative to the Euclidean norm of the corresponding row of A, below which elements of L or U are dropped. With these ILU preconditioners the original system is transformed into the equivalent system $By = c$, where $B = L^{-1}AU^{-1}$, $y = U^{-1}x$, and $c = L^{-1}b$. The initial guess employed in these cases is $x_0 = (LU)^{-1}b$.

NUMERICAL RESULTS

We consider the two-dimensional equation of contaminant transport in groundwater, written as:

$$\frac{\partial}{\partial x_i}\left(D_{ij}\frac{\partial c}{\partial x_j}\right) - \frac{\partial}{\partial x_i}(v_i c) = S_w \frac{\partial c}{\partial t} + f \qquad i = 1, 2$$

where c is the concentration, D_{ij} the dispersion tensor as defined by *Bear* [*1979*], S_w is the saturation coefficient, v_i is the Darcy's velocity, and f is the source or sink term. Note that we adopted Newton's indicial notation, so that repeated indices denote summation over the two coordinate dimensions. This equation is solved using a Galerkin formulation with linear triangular finite elements in space, and a Crank-Nicolson finite difference scheme in time [*Huyakorn and Pinder, 1983*]. The numerical discretization of this non self adjoint equation yields linear systems whose matrix is nonsymmetric. It is well known that for large convective terms and large time steps, eigenvalues with large imaginary parts may appear, and therefore convergence of the linear Krylov solvers may drastically deteriorate.

The mesh used in all the simulations is a square mesh of 240×240 m, discretized with 49 nodes on each side, for a total of 2401 nodes and 4608 rectangular triangles. The finite element discretization results in a nonsymmetric matrix of dimension $N = 2401$ with 16417 nonzero elements. Since for this mesh $\Delta x = \Delta y$, the Peclet (Pe) and Courant (Cu) numbers can be defined as: $\mathrm{Pe} = \max(v_x, v_y)\Delta x/\max(D_{ij})$ and

Cu $= \max(v_x, v_y)\Delta t/\Delta x$, respectively. Unitary Dirichlet boundary conditions are imposed on two consecutive sides of the square, while zero concentration is imposed in the other part of the boundary. Two different problems are considered. Example 1 is an adaptation of the problem presented by *Peters* [*1992*], and uses a constant velocity field with unitary velocity vectors in a direction of 45 degrees with respect to the coordinate axis. Test case 2 considers a variable velocity field, numerically calculated from the solution of Richard's equation on the same domain, with uniform recharge of $v_i/K = 0.5$ on the surface, K being the saturated hydraulic conductivity, seepage face on one side of the domain, and zero flux on the remaining boundary. By changing the dispersivity coefficients and the time step values, several runs are performed, on both examples, with Peclet and Courant numbers varying from Pe=1 and Cu=1 to Pe=10 and Cu=∞. In the variable velocity problem, the maximum Pe and Cu numbers are used as reference values. In all the runs we have assumed a value of $m = 20$ in GMRES, and $p = 5$ and $\tau = 10^{-10}$ in the ILUT(p, τ) preconditioner, which have been found to be the optimal values of these parameters in our test problems.

The results of the various runs, summarized in Table 1, confirm that, in general, Krylov subspace methods suffer when matrices with complex eigenvalues and large imaginary parts are present. In fact, for both tests, in the case of Pe=10 and Cu=10, the maximum eigenvalues of the system matrices are complex, with an imaginary part which is almost twice the real part (Table 2). Note that in general, even though the Pe and Cu numbers are the same for examples 1 and 2, the latter always displays a more irregular convergence behavior than test case 1, as can be seen from Figures 1 and 2.

When using the conventional preconditioners, TFQMR with D^{-1} displays the highest robustness and efficiency at high Peclet and Courant numbers, while BiCGSTAB or GMRES(20) in combination with ILU(0) are the most efficient for small Peclet numbers. The use of the ILUT(p, τ) preconditioner, however, slightly changes these findings. With ILUT(p, τ) the number of iterations substantially decreases in all runs, and the convergence profiles are more regular. Note that, while the number of nonzero elements in the preconditioning factors of ILU(0) is set a priori equal to the number of nonzero elements in A, the ILUT(p, τ) allows for much more filled L and U factors. With the parameters used in our simulations, the global number of nonzero elements in both factors is equal to 32979 for test case 1, and 37400 for test case 2 with Pe $= 10$ and Cu $= 10$, when ILUT(p, τ)is used, while it is equal to 16417 when ILU(0)is employed. This observation, together with the convergence results reported in the tables and in the figures, show that the problem with variable velocity is more difficult than in the constant velocity case. The effects of the preconditioner is shown in Table 2, where the maximum and minimum eigenvalues for the two sample tests and for Pe $= 10$ and Cu $= \infty$ are reported. The unpreconditioned matrices display maximum eigenvalues with large imaginary parts. After preconditioning, however, the imaginary part is drastically reduced, if not completely eliminated.

Cu	Pe	Sol	Precond.	Test case 1		Test case 2	
				# iter.	CPU	# iter.	CPU
∞	1	BiCGSTAB	–	71	0.57	463	3.44
		TFQMR	–	102	0.79	523	3.94
		BiCGSTAB	ILU(0)	20	0.40	29	0.49
		TFQMR	ILU(0)	24	0.42	33	0.59
		BiCGSTAB	D^{-1}	71	0.57	99	0.77
		TFQMR	D^{-1}	99	0.80	109	0.90
		BiCGSTAB	ILUT(p,τ)	6	0.61	8	0.69
		TFQMR	ILUT(p,τ)	6	0.59	9	0.73
		GMRES(m)	ILUT(p,τ)	10	0.51	14	0.63
∞	10	BiCGSTAB	–	581	4.23	†	72.47
		TFQMR	–	105	0.78	5577	41.98
		BiCGSTAB	ILU(0)	123	1.85	574	8.44
		TFQMR	ILU(0)	215	3.27	†	148.59
		BiCGSTAB	D^{-1}	562	4.39	297	2.30
		TFQMR	D^{-1}	105	0.85	260	2.05
		BiCGSTAB	ILUT(p,τ)	7	0.61	53	1.48
		TFQMR	ILUT(p,τ)	8	0.63	73	1.87
		GMRES(m)	ILUT(p,τ)	15	0.55	56	1.63
1	10	BiCGSTAB	–	31	0.23	20	0.16
		TFQMR	–	40	0.29	29	0.23
		BiCGSTAB	ILU(0)	7	0.19	3	0.12
		TFQMR	ILU(0)	9	0.21	4	0.13
		BiCGSTAB	D^{-1}	27	0.22	16	0.13
		TFQMR	D^{-1}	40	0.33	18	0.16
		BiCGSTAB	ILUT(p,τ)	2	0.49	2	0.49
		TFQMR	ILUT(p,τ)	2	0.48	2	0.51
		GMRES(m)	ILUT(p,τ)	5	0.39	4	0.41
10	10	BiCGSTAB	–	45	0.34	†	72.38
		TFQMR	–	33	0.25	457	3.45
		BiCGSTAB	ILU(0)	10	0.22	169	2.53
		TFQMR	ILU(0)	13	0.26	281	4.25
		BiCGSTAB	D^{-1}	39	0.31	†	76.79
		TFQMR	D^{-1}	33	0.28	534	4.21
		BiCGSTAB	ILUT(p,τ)	3	0.50	4	0.58
		TFQMR	ILUT(p,τ)	4	0.52	4	0.57
		GMRES(m)	ILUT(p,τ)	7	0.42	8	0.49
†convergence not achieved within 10000 iterations							

Table 1. CPU times (s) and overall number of iterations for the two test cases, with $m = 20$, $p = 5$, and $\tau = 10^{-10}$

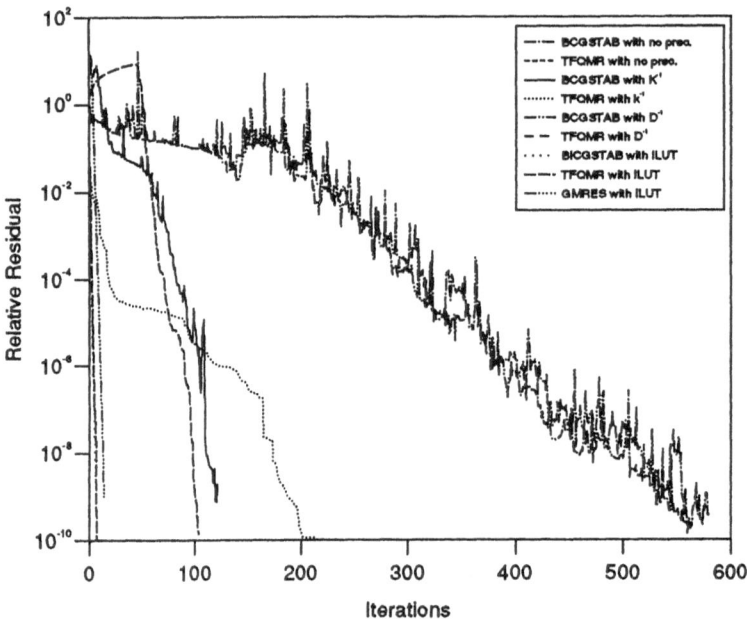

Figure 1. Convergence profiles for case test 1 with Pe = 10 and Cu = ∞.

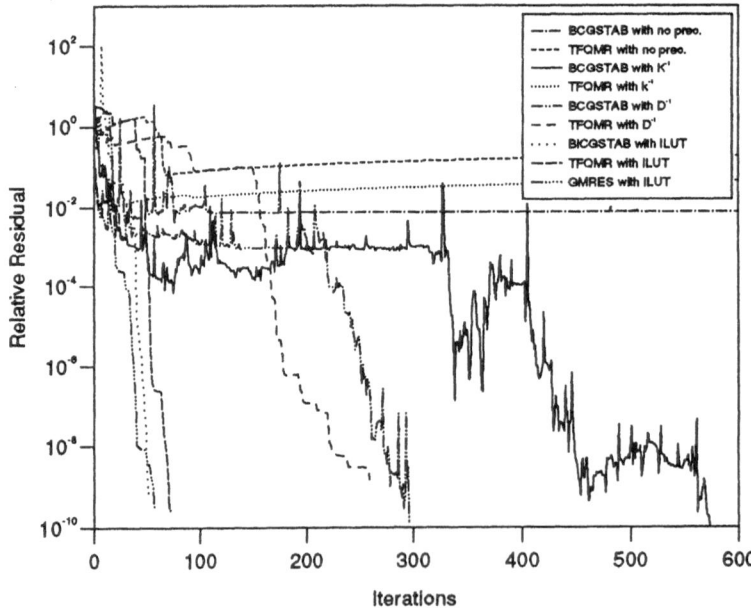

Figure 2. Convergence profiles for case test 2 with Pe = 10 and Cu = ∞.

eig.	Test case 1		
#	A	$B = L^{-1}AU^{-1}$ with ILU(0)	$B = L^{-1}AU^{-1}$ with ILUT(p,τ)
1,2	$2.665 \pm i4.986$	$1.723 \pm i(5.193 \times 10^{-2})$	$1.014 \pm i(5.049 \times 10^{-3})$
3,4	$2.664 \pm i4.970$	$1.6942 \pm i(1.494 \times 10^{-1})$	$1.012 \pm i(4.075 \times 10^{-3})$
⋮	⋮	⋮	⋮
2399	1.837×10^{0}	—	—
2400	1.835×10^{0}	—	—
2401	1.835×10^{0}	—	—
	Test case 2		
1,2	$1.624 \pm i2.401$	$1.289 \times 10^{3}, 1.785 \times 10^{1}$	1.0278×10^{0}
3,4	$1.626 \pm i2.395$	4.210×10^{0}, —	$1.011 \pm i(3.29 \times 10^{-3})$
⋮	⋮	⋮	⋮
2399	-1.065×10^{-2}	—	—
2400	-2.626×10^{-3}	—	—
2401	6.498×10^{-3}	—	—

Table 2. Maximum and minimum eigenvalues for the two test cases with Pe = 10 and Cu = 10, before and after ILU preconditioning.

CONCLUSIONS

The following points are worth summarizing:

- The BiCGSTAB and GMRES(20) methods with ILU(0) perform well for dispersion dominated problems.
- In advection-dominated cases, the TFQMR preconditioned with D^{-1} generally performed better than the ILU(0)-based BiCGSTAB or GMRES(20).
- BiCGSTAB and GMRES(20) with ILUT(p,τ) using $p = 5$ and $\tau = 10^{-10}$ outperformed the other algorithms in advection-dominated cases, while resulted only slightly less efficient than BiCGSTAB and GMRES(20) with ILU(0) for the other tests.

In conclusion, the use of BiCGSTAB or GMRES(20) with the ILUT(p,τ) preconditioner resulted in a very efficient and robust combination, able to solve the finite element discretizations of the advection-dispersion equation in all the test cases considered here.

REFERENCES

Barret, R., M. Berry, T. Chan, J. Demmel, J. Donato, J. Dongarra, V. Eijkhout, R. Pozo, C. Romine and H. van der Vorst, *Templates for the Solution of linear*

Systems: Building Blocks for Iterative Methods. SIAM publisher, Philadelphia, PA, 1994.

Bear, J., *Hydraulics of Groundwater.* McGraw-Hill, New York, 1979.

Fletcher, R., Conjugate gradient methods for indefinite systems. Volume 506 of *Lecture Notes Math.* 73–89. Springer Verlag, Berlin, New York, 1976.

Freund, R. W., A transpose-free quasi-minimal residual algorithm for non-Hermitian linear systems, *SIAM J. Sci. Comput.* 14, 470–482, 1993.

Freund, R. W. and N. M. Nacthigal, QMR: a quasi minimal residual method for non-Hermitian linear systems, *Numer. Math.* 60, 315–339, 1991.

Huyakorn, P. S. and G. F. Pinder, *Computational Methods in Subsurface Flow.* Academic Press, London, 1983.

Lanczos, C., Solution of systems of linear equations by minimized iterations, *J. Res. Nat. Bur. Standard.* 49, 33–53, 1952.

Meijerink, J. A. and H. A. van der Vorst, An iterative solution method for linear systems of which the coefficient matrix is a symmetric M-matrix, *Math. Comp.* 31, 148–162, 1977.

Peters, A., CG-like algorithms for linear systems stemming from the FE discretization of the advection-dispersion equation. In: Russell, T. R., R. E. Ewing, C. A. Brebbia, W. G. Gray and G. F. Pinder (eds.) *International Conference on Computational Methods in Water Resources IX. Vol. 1: Numerical Methods in Water Resources.* Computational Mechanics and Elsevier Applied Sciences, Southampton, London, pp 511–518, 1992.

Pini, G. and G. Zilli, Preconditioned iterative algorithms for large sparse unsymmetric problems, *Num. Meth. PDE.* 5, 107–120, 1989.

Pini, G. and G. Zilli, On vectorizing the preconditioned generalized conjugate residual methods, *Int. J. Computer Math.* 33, 195–207, 1990.

Saad, Y., ILUT: a dual strategy accurate incomplete ILU factorization. Technical Report, Minnesota Supercomputer Institute, University of Minnesota, 1991.

Saad, Y. and M. H. Schultz, GMRES: A generalized minimum residual alogorithm for solving nonymmetric linear systems, *SIAM J. Sci. Stat. Comput.* 7, 856–869, 1986.

Sleijppen, G. L. and D. R. Fokkema, BICGSTAB(L) for linear equations involving unsymmetric matrices with complex spectrum, *Elec. Trans. Num. Anal.* 1, 11–32, 1993.

Sonneveld, P., CGS, a fast Lanczos-type solver for nonsymmetric linear systems, *SIAM J. Sci. Stat. Comput.* 10, 36–52, 1989.

van der Vorst, H. A., Iterative methods for the solution of large systems of equations on supercomputers, *Adv. Water Resources.* 13(3), 137–146, 1990.

van der Vorst, H. A., Bi-CGSTAB: A fast and smoothly converging variant of BI-CG for the solution of nonsymmetric linear systems, *SIAM J. Sci. Stat. Comput.* 13, 631–644, 1992.

A MULTIGRID METHOD WITH MATRIX-DEPENDENT TRANSFER OPERATORS APPLIED IN GROUNDWATER SCIENCE

CHRISTIAN WAGNER
Universität Stuttgart
Institut für Computeranwendungen III
Pfaffenwaldring 27, D-70569 Stuttgart
Germany

We introduce a multigrid method with matrix-dependent transfer operators. This algorithm turns out to be robust and efficient for convection-diffusion equations with strongly discontinuous coefficients. Pollutant transport problems in groundwater science are an example for that kind of problems. We consider the time required for the redevelopment of a contaminated aquifer for several aquifer configurations, which can be regarded as porous medium. The correlation length and the heterogeneity of the distribution of the permeability influence this restorage time.

1 INTRODUCTION

The standard multigrid method is neither efficient nor robust for convection-diffusion equations with strongly discontinuous coefficients. We introduce a multigrid method with matrix-dependent transfer operators. The restriction and the prolongation of our algorithm originate naturally from a simple block elimination. This multigrid method turns out to be robust and efficient for our test problems.
We take a groundwater problem as example for a convection-diffusion problem with strongly varying coefficients. The question is to simulate the redevelopment of a contaminated aquifer by a groundwater flow of clear water. We investigate the time, which is necessary to restore the aquifer, for several conditions for the heterogeneity and the stopping criterion. In recent years increases the importance of the environmental physics especially water science. We hope to contribute to the improvement of the accuracy of predictions concerning the time needed for the redevelopment.

2 A MULTIGRID METHOD WITH MATRIX-DEPENDENT TRANSFER OPERATORS

The system of linear equations $L_l x_l = f_l$ can be written in block form

$$\begin{bmatrix} L_{l,11} & L_{l,12} \\ L_{l,21} & L_{l,22} \end{bmatrix} \begin{bmatrix} x_{l,1} \\ x_{l,2} \end{bmatrix} = \begin{bmatrix} f_{l,1} \\ f_{l,2} \end{bmatrix} . \tag{2.1}$$

A. Peters et al. (eds.), Computational Methods in Water Resources X, 1439–1446.

Nodes missing in the coarse grid G_{l-1} have label 1, nodes still existing in the grid G_{l-1} have label 2. Based on the lumping algorithm by Reusken [Reusken, 1993] we've developed an algorithm for the construction of an approximated system of equations [Wagner, 1993] [Wagner, Wittum, 1994]

$$\bar{L}_l x_l = \bar{F}_l f_l \; . \tag{2.2}$$

The matrix $\bar{L}$ must have the properties: $\bar{L}_{11}$ is easy to invert, $\bar{L}_{11}^{-1}$ is sparse and $\bar{L}^{-1}$ is a good approximation for L^{-1}. Then we are able to realize the following algorithm.

Algorithm 2.1 (multigrid method with matrix-dependent transfer operators):
Let be given a hierarchy of grids $G_0 \subset G_1 \subset \ldots \subset G_{max}$. In order to solve the system of linear equations

$$L_{max} x_{max} = f_{max} \tag{2.3}$$

use the following scheme.

<u>Construction of the coarse grid matrices</u>:
for $(k = max; k > 0; k = k - 1$)
{
        construct matrices $\bar{L}_k$ and $\bar{F}_k$;
        calculate $L_{k-1} = \bar{L}_{k,22} - \bar{L}_{k,21} \bar{L}_{k,11}^{-1} \bar{L}_{k,12}$;
}

<u>Iteration step for x_{max}</u>:
SchurMG(INT k, DOUBLE x[], DOUBLE f[])
{
        if (k == 0) $x_k = L_k^{-1} f_k$;
        else
        {
                $x_k = S^{n_1}(x_k, f_k)$;
                $d_k = f_k - L_k x_k$;
                $x_k = x_k + \omega \cdot \bar{L}_{k,11}^{-1} (\bar{F}_k d_k)_1$;
                $d_{k-1} = \bar{F}_k r_k d_k$;
                $v_{k-1} = 0$;
                for(j = 1; $j \le \gamma$; j++) SchurMG(k-1,v,d) ;
                $x_k = x_k + \omega \cdot p_k v_{k-1}$;
                $x_k = S^{n_2}(x_k, f_k)$;
        }
}
with the prolongation $p_k = \begin{bmatrix} -\bar{L}_{k,11}^{-1} \bar{L}_{k,12} \\ I \end{bmatrix}$ and the restriction $r_k = \begin{bmatrix} -\bar{L}_{k,21} \bar{L}_{k,11}^{-1} & I \end{bmatrix}$.

3 NUMERICAL COMPUTATION OF THE POLLUTANT TRANSPORT

3.1 Computation of the groundwater flow

We consider horizontal regional groundwater flow in aquifers or parts of aquifers with horizontal extensions much larger than their depth. Our discussion is limited to saturated flow. If we regard the aquifer as porous medium the following equation describing the hydraulic head h can be derived [Kinzelbach, 1986, 1992], [Bear, 1972, 1979].

$$\vec{\nabla} \cdot (k_f \cdot \vec{\nabla} h) = 0 \qquad \text{in } \Omega = (0, 10m) \times (0, 10m) \tag{3.1}$$

We have solved (3.1) with the boundary conditions

$$h(\vec{x}) = 0,01 \text{ m} \qquad \vec{x} \in \Gamma_W, \qquad \Gamma_W = \{\vec{x} \in \partial\Omega \,|\, x_0 = 0\} \ , \tag{3.2}$$

$$h(\vec{x}) = 0 \qquad \vec{x} \in \Gamma_E, \qquad \Gamma_E = \{\vec{x} \in \partial\Omega \,|\, x_0 = 10m\} \ , \tag{3.3}$$

$$(k_f \cdot \vec{\nabla} h)\,\vec{n} = 0 \qquad \vec{x} \in \Gamma_N \vee \vec{x} \in \Gamma_S, \quad \Gamma_S = \{\vec{x} \in \partial\Omega \,|\, x_1 = 0\} \ , \tag{3.4}$$

$$\Gamma_N = \{\vec{x} \in \partial\Omega \,|\, x_1 = 10m\} \ . \tag{3.5}$$

The permeability k_f is determined by the random field generator fgen92 [Robin, Schmidt, Mendoza, 1992]. The random field generator creates a distribution of the decadic logarithm of the permeability k_f in m/s with the possibility to choose the mean $\log m$, the standard deviation $\log \sigma$ and the correlation length Λ. The pore velocity $\vec{v}_p$

$$\vec{v}_p = -\frac{1}{n_e} \cdot k_f \cdot \vec{\nabla} h \ , \qquad n_e = 0,1 \ , \qquad n_e = \text{porosity,} \tag{3.6}$$

is shown for a permeability distribution with $\log m = -3$, $\log \sigma = 0,8$ and $\Lambda = 2,5m$ ($\Lambda = 0,5m$) in Fig. 3.1 (Fig. 3.2).

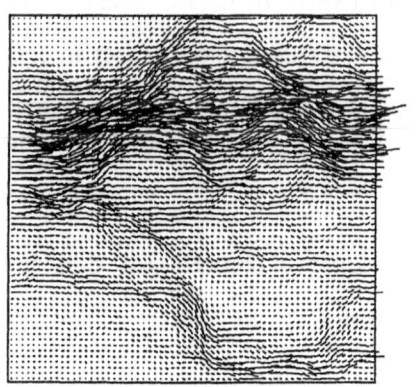

Fig. 3.1: $\vec{v}_p$ for $\Lambda = 2,5m$

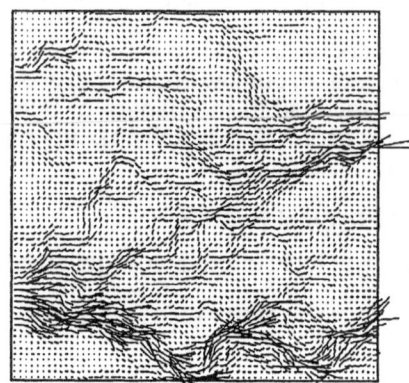

Fig. 3.2: $\vec{v}_p$ for $\Lambda = 0,5m$

3.2 Computation of the Pollutant Transport

The transport of a pollutant in a porous medium corresponding to the model of an ideal tracer, which means that neither absorption nor chemical transformation occur, is dominated by advection and dispersion. The following two dimensional equation determines the behavior of the concentration $c(\vec{x}, t)$ of such a pollutant. This partial differential equation can be derived from the continuity equation by integration along the vertical coordinate and taking the average throughout small macroscopic volumes containing many pores

$$\frac{\partial c}{\partial t} = \vec{\nabla} \cdot (-\vec{v}_p \cdot c + D \cdot \vec{\nabla} c) \qquad \text{in } \Omega = (0, 10 \text{ m}) \times (0, 10 \text{ m}) \ . \qquad (3.7)$$

We apply the dispersion tensor D suggested by Scheidegger [Scheidegger, 1961] with two parameters α_L (longitudinal dispersivity) and α_T (transversal dispersivity). We set

$$\alpha_L = 0,5 \text{ m} , \qquad \alpha_T = \alpha_L / 50 \ . \qquad (3.8)$$

The simulation starts with a homogeneous pollution of the aquifer which means

$$c(\vec{x}, t = 0) = 1 \quad \vec{x} \in \overline{\Omega} \ . \qquad (3.9)$$

We achieve a flow of clear water throught the west boundary Γ_W with the condition

$$(-\vec{v}_p \cdot c(\vec{x}) + D \cdot \vec{\nabla} c\big|_{\vec{x}}) \cdot \vec{n} = 0 \quad \vec{x} \in \Gamma_W \ . \qquad (3.10)$$

At the north boundary Γ_N and at the south boundary Γ_S we demand

$$(D \cdot \vec{\nabla} c)\big|_{\vec{x}} \cdot \vec{n} = 0 \quad \vec{x} \in \Gamma_N \vee \vec{x} \in \Gamma_S \ . \qquad (3.11)$$

We replace at the east boundary Γ_E the normal derivative by the inner differential quotient

$$(\vec{\nabla} c) \cdot \vec{n}\big|_{\vec{x}} \rightarrow \lim_{h \to 0} \frac{c(\vec{x} + h \cdot \vec{n}) - c(\vec{x})}{h} \quad \text{with } (\vec{x} + h \cdot \vec{n}) \in \Omega \ , \quad \vec{x} \in \Gamma_E \ . \ (3.12)$$

This condition yields a linear extrapolation of the concentration $c(\vec{x}, t)$ in order to simulate an outflow boundary. As discretization we apply the finite volume method as described in the papers of Schneider and Raw [Schneider, Raw, 1987] and Hackbusch [Hackbusch, 1989], connected with the implicit Euler scheme as time - discretization.

3.3 Physical Results

The first topic of our investigation is the redevelopment time $RMT90$ defined by

$$C(RMT90) / C(0) = 0,1 \ , \qquad C(t) = \int_\Omega c(\vec{x}, t) \ dt \ . \qquad (3.13)$$

We are interested in the way $RMT90$ depends on the heterogeneity, which is specified by $\log \sigma$, and in the way $RMT90$ depends on the correlation length Λ. We calculate the mean $\overline{RMT90}$ of 30 - 45 values of $RMT90$ for permeability distributions with fixed $\log\sigma$, Λ and normalized flow throught Γ_E. Fig. 3.3 shows for several values of Λ the behavior of $\overline{RMT90}$ for increasing $\log\sigma$. Fig. 3.4 demonstrates the dependence upon the correlation length Λ for divers values of $\log\sigma$.

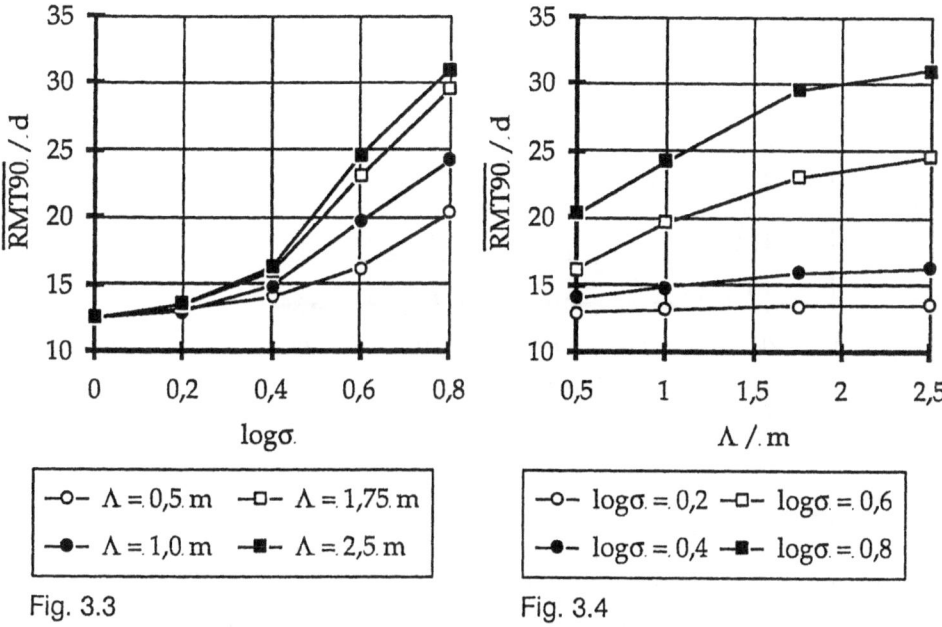

Fig. 3.3   Fig. 3.4

The temporal decrease of the total pollutant concentration $C(t)$ is represented in Fig. 3.5 for typical realizations with $\Lambda = 2,5\,m$ and different values of $\log\sigma$.
Another criterion for stopping the redevelopment of the aquifer can be derived from the area F_S with pollutant concentration higher than a threshold S.

$$F_S(t) = \int_\Omega w(\vec{x}, t)\ df \quad \text{with}\quad w(\vec{x}, t) = \begin{cases} 1 & \text{if } c(\vec{x}, t) > S \cdot c(\vec{x}, 0) \\ 0 & \text{otherwise} \end{cases} \tag{3.14}$$

F_S is proportional to the probability of finding a pollutant concentration

$$c(\vec{x}, t) > S \cdot c(\vec{x}, 0) \tag{3.15}$$

in a random sample. Fig. 3.6 shows the temporal behavior of $F_{0,01}$.

Our investigations yield the expected result: If $\log\sigma$ or Λ grows larger, then the redevelopment time increases. The redevelopment time is determined by the areas with low permeability. These areas expand for larger Λ. The lowest permeability decreases for higher values of $\log\sigma$. Fig. 3.5 and Fig. 3.6 show that the less pollutant is left the stronger is the effect of $\log\sigma$ on the redevelopment time.

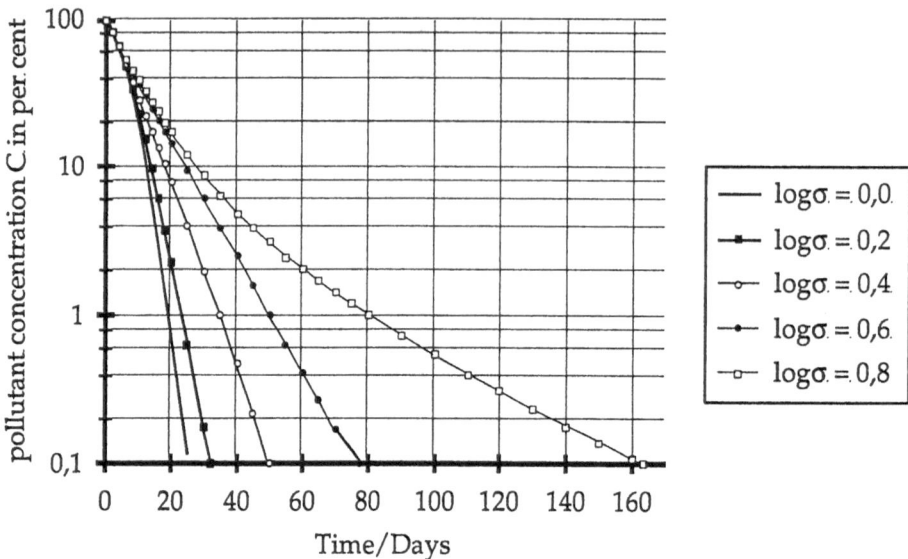

Fig. 3.5: $\Lambda = 2,5m$

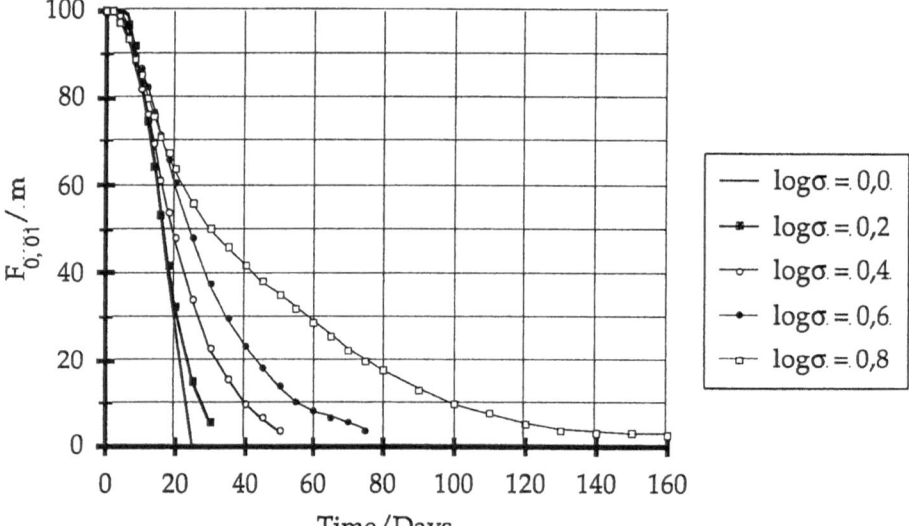

Fig. 3.6: $\Lambda = 2,5m$

4 NUMERICAL RESULTS

Notation 4.1:

a) Algo A: standard multigrid method ($n_2 = 0$).

b) Algo B: SchurMG (Algorithm 2.1), L, F constructed as described in [Reusken, 1993] ($n_2 = 0$).

c) Algo C: SchurMG (Algorithm 2.1) , $\tilde{L}, \tilde{F}$ constructed as described in [Wagner, 1993] and [Wagner, Wittum, 1994] ($n_2 = 0$).

d) K_n: average convergence rate for n iteration steps

$$K_n = \left(\frac{\left\|f - Lx^{(n)}\right\|_2}{\left\|f - Lx^{(0)}\right\|_2}\right)^{\frac{1}{n}} \quad n = min\left[m \in N \mid \frac{\left\|f - Lx^{(m)}\right\|_2}{\left\|f - Lx^{(0)}\right\|_2} < 10^{-10}\right] . (4.1)$$

Experiment 4.2:

The convergence rates for equation (3.1) for the realization represented in Fig. 3.2 ($\Lambda = 0,5m$, $\log\sigma = 0,8$) computed on a regular grid containing 64×64 square elements are shown in Tab. 5.1.

algorithm	n_1	γ	smoother	convergence rate
Algo A	2	2	gs	> 1
Algo B	2	2	gs	$K_{n > 100} > 0,8$
Algo C	2	2	gs	$K_{21} = 0,33$
Algo C	2	2	ilu	$K_8 = 0,05$
Algo C	2	1	ilu	$K_{26} = 0,4$

Tab. 5. 1

Experiment 4.3 ([ABDP, 1982]):

An artificial but interesting example has been suggested by Alcouffe, Brandt, Dendy and Painter.

$$\nabla \cdot (\varepsilon \cdot \nabla u) = 0 \quad \text{in } Q = (0,1) \times (0,1) \tag{4.2}$$

$$\varepsilon = \begin{cases} \text{eps} & \text{if } (\frac{11}{24} < x < \frac{13}{24}) \wedge (\frac{11}{24} < y < \frac{13}{24}) \\ 1 & \text{otherwise} \end{cases} \tag{4.3}$$

$$\frac{d}{dy}u(x,y) = 0 \qquad\qquad \text{if } y = 0 \text{ or } y = 1 \tag{4.4}$$

$$u(x,y) = 0 \qquad\qquad \text{if } x = 1 , \tag{4.5}$$

$$u(x,y) = 1 \qquad\qquad \text{if } x = 0 . \tag{4.6}$$

We obtain using a regular grid with 24 × 24 square elements the following convergence rates ($n_1 = 1$, $\gamma = 1$):

eps	Algo C gs	Algo C ilu	Algo B ilu	Algo A ilu
10^6	$K_{11} = 0,12$	$K_{11} = 0,12$	$K_{>100} > 0,999$	$> 10^4$
10^4	$K_9 = 0,074$	$K_{10} = 0,09$	$K_{>100} > 0,99$	$> 10^2$
10^2	$K_{10} = 0,088$	$K_9 = 0,076$	$K_{>100} > 0,85$	> 1
1	$K_{10} = 0,085$	$K_5 = 0,04$	$K_{22} = 0,345$	$K_{10} = 0,083$
10^{-2}	$K_{10} = 0,095$	$K_5 = 0,042$	$K_{22} = 0,339$	$K_{10} = 0,084$
10^{-4}	$K_{10} = 0,096$	$K_5 = 0,042$	$K_{22} = 0,339$	$K_{10} = 0,084$

Tab. 5. 2

5 REFERENCES

Alcouffe R.E., Brandt A., Dendy J.E., Painter J.W. (1981) "The Multigrid-Method for the Diffusion Equation with Strongly Discontinuous Coefficients", SIAM J. Sci. Stat. Comput. 2 (4), 430 - 454.

Bear J. (1972) Dynamics of fluids in porous media. American Elsevier, New York.

Bear J. (1979) Hydrodynamics of groundwater. McGraw Hill Series in Water Resources and Environmental Engineering, New York.

Hackbusch W. (1989) "On first and second order box schemes". Computing 41, 277 - 296.

Kinzelbach W. (1986) Groundwater Modelling. Developments in water science 25, Elsevier Science Publishers, Amsterdam

Kinzelbach W. (1992) Numerische Methoden zur Modellierung des Transports von Schadstoffen im Grundwasser. Oldenbourg Verlag, München

Reusken A. (1993) "Multigrid with Matrix-Dependent Transfer Operators for Convection-Diffusion Problems", to appear.

Reusken A. (1993) "Multigrid with Matrix-Dependent Transfer Operators for a Singular Perturbation Problem", Computing 50 (3), 199 - 211.

Scheidegger A. E. (1961) "General theory of dispersion in porous media", J. Geophy. Res., 66 (10), 3273 - 3278.

Schneider G.E., Raw M.J. (1987) "Contol volume finite element method for heat transfer and fluid flow using colocated variables", Numer. Heat Transfer 11, 363 - 390.

Wagner C. (1993) "Ein robustes Mehrgitterverfahren für Diffusions-Transport-Probleme in der Bodenphysik", IWR-Report 93-70, Heidelberg.

Wagner C., Wittum G. (1994) "A Multigrid Method with Matrix-Dependent Transfer Operators Applied in Groundwater Science", to appear.

Wittum G. (1989) "On the robustness of ILU - Smoothing", SJSSC 10, 669 - 717.

17. SOFTWARE DEVELOPMENT

VISIOMETRIC TECHNIQUES IN A 3D GROUNDWATER TRANSPORT CODE

H.-J. G. DIERSCH, R. VOIGT and R. GRÜNDLER
WASY Institute for Water Resources Planning and Systems Research Ltd.,
Waltersdorfer Str. 105, D-12526 Berlin
Germany

Groundwater flow, contaminant mass and heat transport processes are being modelled in three dimensions at increasingly high resolutions. Computational environments have to provide an interactive and sophisticated data management to govern the abundant multidimensional parameter fields in space and time and to enhance the simulation work. The paper discusses present effort in integrating and developing visiometric techniques in the 3D finite-element groundwater transport simulator FEFLOW. It covers tools for visualization and tracking of space-time data sets in groundwater modeling. Different aspects are emphasized: Developments in the pre- and postprocessing for the simulation system, including the applicability of GIS and the 3D data editing process, and appropriate numerical techniques, e.g., for 3D pathline computation and its visual analysis.

INTRODUCTION

Finite element methods are now well established in the groundwater modeling practice. From the user's point of view there are at least three basic requirements which should be matched by today simulation packages, *viz.*, the availability of (1) an interactive graphics-based user interface as an overall computational environment, (2) a multiple data interface to import and export different data formats and sources, and (3) enhanced numerical techniques capable of simulating more complex and possibly coupled flow and transport processes in three spatial dimensions and time. The modeling process can here be considered as an iterative loop of problem design and analysis. Usually, the original data in digital or drawing form represent the starting point. Then, the transformation of these spatial and temporal data into an abstracted and discretized form covers an very important task. To make this procedure more general and efficient the model design should be directly based on data held in a GIS, CAD or database system and fully independent of the current discretization (meshing). The idealized and discretized model is simulated by solving the governing conservation equations. In 3D the resulting matrix systems are large and a great amount of computational data results. Now, the evaluation and interpretation of the voluminous computational results necessitates interactive and graphical tools. Finally, the results may feed back to a repeated computation by restarting the design and analysis loop which is often necessary for a scenario run and for the purpose of design optimization (e.g., needed for groundwater remediation).
It is obvious that an advanced groundwater transport simulator should enable the users to quickly (re-)design, simulate and evaluate a modeling problem with reduced effort and potential for errors. Interactive graphics and visiometric techniques will help to master the

A. Peters et al. (eds.), Computational Methods in Water Resources X, 1449–1456.
© 1994 *Kluwer Academic Publishers. Printed in the Netherlands.*

entire modeling process and to understand and interpret the simulation results. The objective in this paper is to briefly focus on some aspects of the capabilities of visiometrics applied to 3D finite-element groundwater transport simulation and its visual analysis. Implemented techniques and software tools are described for the code FEFLOW [1, 2, 3, 4].

THREE-DIMENSIONAL GROUNDWATER TRANSPORT MODEL

The groundwater flow, the coupled flow and contaminant mass and heat transport, also thermohaline flow processes (i.e., completely fluid-density coupled by mass and heat) are modelled by the following set of partial differential equations:

$$\frac{\partial(\varepsilon\rho^f)}{\partial t} + \frac{\partial(\rho^f q^f_i)}{\partial x_i} = Q_h \tag{1}$$

$$q^f_i = -K_{ij}f_\mu\left(\frac{\partial h}{\partial x_j} + \frac{\rho^f - \rho^f_o}{\rho^f_o}e_j\right) \tag{2}$$

$$R\frac{\partial C}{\partial t} + \frac{\partial}{\partial x_i}\left(q^f_i C - D_{ij}\frac{\partial C}{\partial x_j}\right) + R\vartheta C = Q_c \tag{3}$$

$$[\varepsilon\rho^f c^f + (1-\varepsilon)\rho^s c^s]\frac{\partial T}{\partial t} + \frac{\partial}{\partial x_i}\left(\rho^f c^f q^f_i T - \lambda_{ij}\frac{\partial T}{\partial x_j}\right) = Q_t \tag{4}$$

to be solved for $h = h(x_i, t)$, $q^f_i = q^f_i(x_i, t)$, $C = C(x_i, t)$ and $T = T(x_i, t)$ at $i = 1,2,3$ with the constitutive conditions

$$\rho^f = \rho^f_o[1 + \alpha(C - C_o) - \beta(T - T_o)]$$

$$h = \frac{p^f}{\rho^f_o g} + x_k \qquad K_{ij} = \frac{k_{ij}\rho^f_o g}{\mu^f_o} \qquad f_\mu = \frac{\mu^f_o}{\mu^f(C, T)}$$

$$D_{ij} = (\varepsilon D_d + \beta_T v^f)\delta_{ij} + (\beta_L - \beta_T)\frac{q^f_i q^f_j}{v^f} \tag{5}$$

$$R = \varepsilon + (1-\varepsilon)\kappa$$

$$\lambda_{ij} = \lambda^{cond}_{ij} + \lambda^{disp}_{ij} \qquad Q_t = \varepsilon\rho^f Q^f_t + (1-\varepsilon)\rho^s Q^s_t$$

$$\lambda^{cond}_{ij} = [\varepsilon\lambda^f + (1-\varepsilon)\lambda^s]\delta_{ij} \qquad \lambda^{disp}_{ij} = \rho^f c^f\left[\alpha_T v^f\delta_{ij} + (\alpha_L - \alpha_T)\frac{q^f_i q^f_j}{v^f}\right]$$

where h, q^f_i, C, T and
ε, k_{ij}, K_{ij}, p^f, ρ^f, ρ^f_o, ρ^s, α, β, C_o, T_o, g, e, R, D_{ij}, D_d, ϑ, c^f, c^s, κ, μ^f, μ^f_o, β_L, β_T, v^f, λ_{ij}, Q^f_t, Q^s_t, λ^f, λ^s, α_L, α_T
are, respectively, the hydraulic head, Darcy volumetric fluid flux, contaminant mass concentration, total temperature, as well as porosity, permeability, hydraulic conductivity, fluid pressure, fluid density and reference one, solid density, mass density and expansion coefficient, reference concentration and temperature, gravitational acceleration, gravitational unit vector, retardation factor, hydrodynamic mass dispersion tensor,

molecular mass diffusion coefficient, decay rate, specific heat capacity of fluid and solid, sorption coefficient, dynamic fluid viscosity and reference one, coefficients of longitudinal and transverse mass dispersivity, absolute specific fluid flux, tensor of hydrodynamic thermoconductivity, specific heat sources of fluid and solid, heat conductivity of fluid and solid, coefficients of longitudinal and transverse thermodispersivity.

By using the finite element method (FEM), where prismatic trilinear and triparabolic isoparametric elements are employed in the spatial discretization and predictor-corrector time stepping schemes of second or first order in time combined with an automatic time step control, or alternatively step-fixed implicit or Crank-Nicolson approximations (for more see [2, 3, 4, 5]), the following matrix system has to be solved:

$$A\,(h, q, C, T)^{n+1} dX^{n+1} = B\,(h, q, C, T)^n \tag{6}$$

Here the superscript n denotes the time plane and

$$dX^{n+1} = \left\{ \begin{array}{c} h^{n+1} - h^n \\ q_i^{n+1} - q_i^n \\ C^{n+1} - C^n \\ T^{n+1} - T^n \end{array} \right\} \tag{7}$$

the deviation solution vector for the evolving quantities of the resulting hydraulic head, Darcy fluxes, mass concentration and temperature, respectively. In nonlinear cases the Newton method is applied. For buoyancy driven processes the model can tackle an extended Boussinesq approximation, where the concentration and/or temperature-dependent fluid density influences incorporated in the continuity equation (1) are formulated explicitly (see [4]). The sparse and large matrix system is solved in 3D by preconditioned conjugate gradient (PCG) or generalized CG-like techniques. The model provides the usual conjugate gradient with Cholesky or modified Gustafsson preconditioning for symmetric equations and the restarted ORTHOMIN, the restarted GMRES, the Lanczos-type CGS, BiCGSTAB and BiCGSTABP methods with a Crout preconditioner for the asymmetric equations.

VISIOMETRIC, GIS AND EDITING FUNCTIONAL COMPONENTS

Interactive visiometrics refers to the process of visualizing and quantifying evolving features within space-time data sets and encompasses the individual steps of *visualization*, the rendering process, *identification*, the isolation and extraction of coherent regions (e.g. isosurface contouring), *quantification*, the measurement of a coherent region, *tracking*, the following of the region in time, and *juxtaposition*, comparing and correlating features from simulations and observations. In [2, 3, 5] the role of GIS has been discussed within the context of groundwater modeling. Unlike to a tight integration of GIS system an open data interface management has been preferred for the FEFLOW package. Various techniques for digital database manipulation (map overlaying, data transformation, capturing, regionalization etc.), previously done exclusively under GIS ARC/INFO, are supported directly in the groundwater simulation environment to maintain a fast dynamic data flow during the interactive modeling. From the modeling view GIS remains a tool for generating, housekeeping, updating, manipulating and displaying of databases being input of the model preprocessing and the postprocessing model output. On the other hand, existing GIS can satisfy the demand for visualization, identification, quantification and juxtaposition, however, only in 2D and widely for steady-state considerations. In 3D groundwater modeling one needs more visiometric functionality and specific features which are to be integrated in the interactive groundwater modeling system. Of particular interest are here: the complete editing process ranging from the meshing procedures to the attribute description uses digital databases and interactive tools for data regionalization, recognition, manipulation and a spatial or temporal joining. Otherwise, the handling of time-dependent attributes (e.g., groundwater recharges, leakages, boundary conditions) in the model description is significant and the comprehensive evaluation of the computational results during the simulation run or in the postprocessing has to utilize visiometric tools especially for 3D pathlines, isosurfaces, arbitrary cutting and segment operations, fence displays and others. As an example Fig. 1 shows a computed 3D saltwater interface obtained by the FEFLOW system. The isosurface computation is directly done on the finite element mesh without any gridding techniques. ARC/INFO GIS map data can be easily incorporated in this 3D visual analysis.

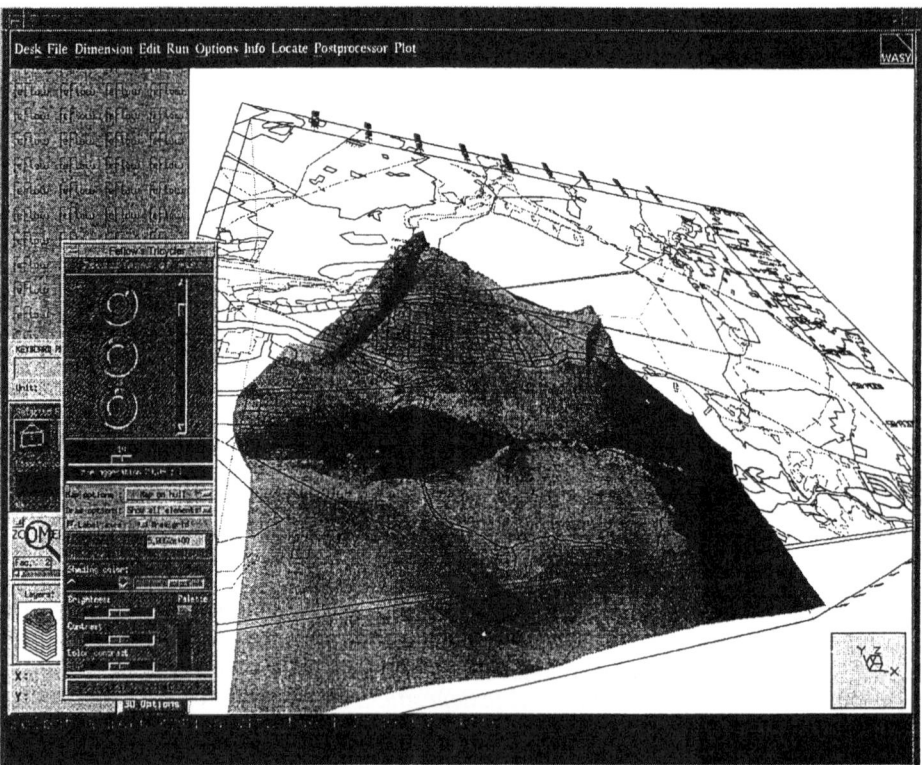

Fig. 1 Simulated 3D isosurface of a saltwater interface below pumping well galleries

MODEL CONFIGURATION AND ATTRIBUTION

There are three fundamental interactive software components to design and describe the finite element model with its geometric and all attribute data. First, the *Mesh Editor and Generators* handle superelements which are meshed by either triangular or quadrilateral-based elements on a slice. Second, the *Layer Configurator* dynamically creates and (re-) positions the 3D multi-layer structure of the model problem. Third, the *Problem Atrribute Editor* defines and attaches the material, initial, boundary and control data to the discretized model. To make this editing processes efficient and accurate facilities exist for joining problem data by using point, line or polygon coverages (e.g., imported from GIS ARC/INFO or the previously designed supermesh), for data regionalization based on Kriging or Akima interpolation techniques and for interactive settings. The data can be copied from slice to slice and visualized and plotted by 2D and 3D options. A debugging tool exists to check the compatibility of the generated and assigned data and to provide corrections, replacements and persistences of data. The user is also provided with the facility to redefine the 3D layering, assign the slice elevations, checking elevations against adjacent slices and direction components of gravity. Figure 2 exhibits an example of a final problem description used for simulating the thermohaline, coupled mass and heat-driven flow processes in a deep mining pit after flooding.

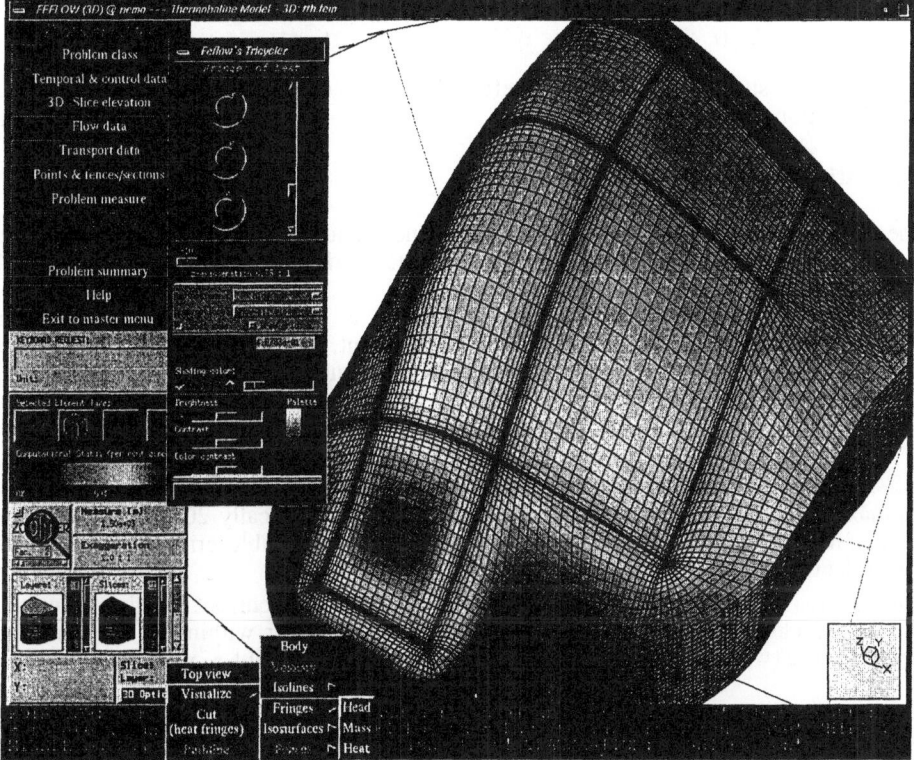

Fig. 2 Thermohaline flow simulation of a mining pit: 3D mesh and temperature field

TRACKING AND RAI

Particle tracking is a very common approach in groundwater modeling. While in 2D there are well established and refined techniques for 3D an increasing desire for more general and powerful tracking methods exist. Otherwise, the 3D visual analysis of pathlines and isochrones requires specific visiometric strategies. We briefly describe the tracking techniques which have been developed for the FEFLOW package. For the 3D finite elements the velocity components are computed at the nodal points. Either Gauss-point extrapolation and nodal averaging or a Galerkin approach can be performed.

Time integration via the fourth-order Runge-Kutta method

We have to solve three coupled ordinary differential equations of the form

$$dx_i(t) = v_i(x_1, x_2, x_3) dt \qquad i = \{1, 2, 3\} \tag{8}$$

where v_i corresponds to the effective pore velocity $v_i = q_i'/R$. Advancing the solution from x_i^n to x_i^{n+1} by using a timestep length ξ a fourth-order Runge-Kutta method realizes the stepping procedure as follows:

$$k_1 = \xi v_i(t_n, x_i^n) \qquad k_2 = \xi v_i\left(t_n + \frac{\xi}{2}, x_i^n + \frac{k_1}{2}\right) \qquad k_3 = \xi v_i\left(t_n + \frac{\xi}{2}, x_i^n + \frac{k_2}{2}\right)$$

$$k_4 = \xi v_i(t_n + \xi, x_i^n + k_3) \qquad x_i^{n+1} = x_i^n + \frac{1}{6}(k_1 + 2k_2 + 2k_3 + k_4) + O(\xi^5) \tag{9}$$

An adaptive stepsize control is preferred to achieve an efficient and accurate approximation.

Adaptation to the FEM problem

The tracking approach has to be adapted to the mesh and element properties. This is very essential regarding the accuracy and the numerical effort. Following points are to be highlighted:

Integration in the finite element: The integration is performed up to the following termination conditions: (1) point leaves the element, so continue with the adjacent element, (2) the effective step velocity descends a minimum velocity which often happens for singularities (pumping wells)

$$v_{eff} = \frac{|x_i^{n+1} - x_i^n|}{\xi} \leq 0.01 v_{min} \tag{10}$$

(3) maximum number of integration steps is exceeded (practically 200 per element) and (4) maximum elapsed time is reached. The maximum tolerable error measure for the Runge-Kutta method is taken as 0.0001.

Computation of the element exiting point: If $\vec{x}_a$ represents the current pathline point lying outside the element and $\vec{x}_v$ is the previous point lying either within the element or on element faces (within a tolerable deviation), then the exiting point $\vec{x}_g$ will attach the line

$$\vec{x}_g = \vec{x}_v + k(\vec{x}_a - \vec{x}_v) . \tag{11}$$

To compute the k factors the implemented method differs between the side faces of the finite elements and their top and bottom ones. For the former a Liang-Barsky clipping algorithm works very fast. The exiting point for the latter is determined from the intersection of the line with these planes on which the point x_E lies, viz.

$$k = \frac{(\vec{n}\vec{x}_E - \vec{n}\vec{x}_v)}{\vec{n}(\vec{x}_a - \vec{x}_v)} \tag{12}$$

where $\vec{n}$ corresponds to the normal unit vector. The k factor is accepted for $-10^{-6} \leq k \leq 1$.

Marking isochrones: Isochrone markers can be prescribed at the elapsed simulation time. They results from a linear interpolation between adjacent time stages being computed.
Filtering pathline points: Due to the used strong error criterion the stepping method computes much more pathline points as necessary for the graphical evaluation. Accordingly, one can delete all points which lie on a straight line. The visibility of direction changes serves as the decision criterion. If the distance of the point from the line is smaller than a fifth of a pixel measure then the pathline point is eliminated.

Pathline visualization

The pathlines are visualized, colored and shaded as "voluminous pipes" with variable diameters and colors. The pipes have cross-sections represented by a regular polygonal form. Planes in form of "plates" are built up across the pathline points. Their normal vectors are determined from the bisector between adjacent line pieces as shown in Fig. 3.

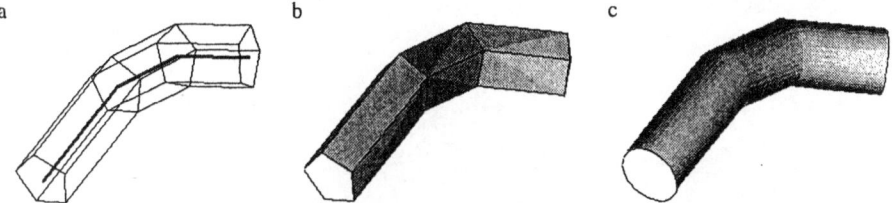

Fig. 3 Construction of pathline pipes (a) wire frame for a 5-point polygonal plate, (b) visible grid construction shown for a 5-point polygonal plate and (c) fully shaded pipe by using a 40-point cross-section representation

Additionally, isochrone markers are generated as larger plates along the pathline pipes. They appear as circles of different colors perpendicular to the pipe master line (see Fig. 4).

Relevant areas of influence (RAI)

In groundwater hydrology there is an interest in determining such starting pathline points on the aquifer top which will reach observation positions in a certain depth of the sampled aquifer. Practically, a number of pathlines will started around selected observation points at a given radius and elevation, and then backtracked up to the top. The intersection points with the top surface represent a RAI contour identifying locations of possible contaminations. FEFLOW provides menus and visual tools to evaluate RAI's in an automatic, efficient and accurate manner as exemplified in Fig. 4.

CLOSURE

All graphical and numerical implementations are based on the ANSI-C language, the X Window System and OSF/Motif standards available on all graphics workstation under UNIX. The data interface provides a flexible exchange of data in different formats. The GIS ARC/INFO is especially supported. Alternatively to ASCII data, for the voluminous 3D finite element problem data a binary or packed binary format based on the XDR standard is provided. It allows an exchange of data between different platforms independent of the internal binary data representation at a high efficiency. The software is fully network-transparent. Further developments concern multi-FEM-kernel properties within a UNIX network which is directly managed by the FEFLOW graphics interface.

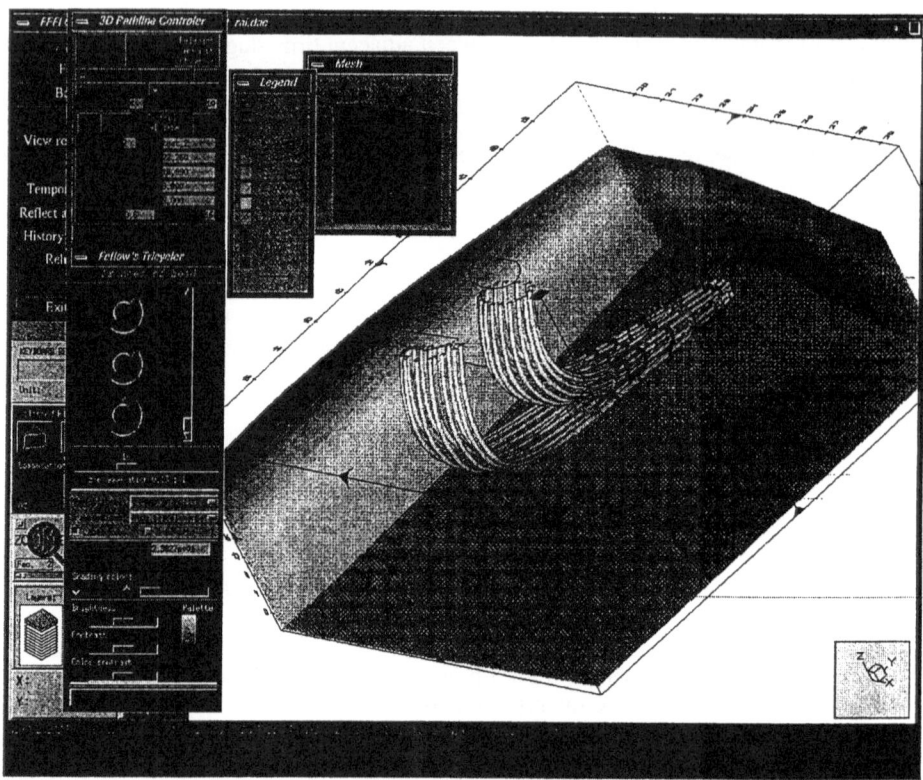

Fig. 4 Pathlines, isochrone markers and obtained RAI contours influenced by aquifer stratification, groundwater recharge and partially screened wells

REFERENCES

[1] Diersch, H.-J. (1992) "Interactive, graphics-based finite element simulation of groundwater contamination processes", Adv. Engineering Software 15, 1-13.
[2] Diersch, H.-J., Gründler, R, Kaden, S, and Michels. I. (1992) "Toward GIS-based 3D/2D groundwater contamination modeling using FEM", In: Numerical Methods in Water Resources, Vol. 1., Proc. IX. Intern. Conf. on Comp. Methods in Water Resources, June 1992, Denver, Comp. Mech. Publ. and Elsevier, 749-760.
[3] Diersch, H.-J. (1993) "GIS-based groundwater flow and contaminant transport modeling - the simulation system FEFLOW", In: Praxis der Umweltinformatik Vol. 4 (Rechnergestützte Ermittlung, Bewertung und Bearbeitung von Altlasten), ed. F. Ossing, Metropolis-Verlag, Marburg, 187-208.
[4] WASY Institute for Water Resources Planning and Systems Research (1994) "Interactive, graphics-based finite-element simulation system FEFLOW for modeling flow, contaminant mass and heat transport processes", User's Manual, Revision 4, WASY Ltd. Berlin.
[5] Diersch, H.-J. (1994) "Computational aspects in developing an interactive 3D groundwater transport simulator using FEM and GIS", Intern. Conf. on Groundwater Quality Management (GQM 93), Tallinn, Estonia, 6-9 September 1993, in press

DEVELOPMENT OF A GROUNDWATER MODELING SYSTEM FOR THE U.S. DEPARTMENT OF DEFENSE

J.P. HOLLAND
Department of the Army
U.S. Army Engineer Waterways Experiment Station
3909 Halls Ferry Road, Vicksburg, MS 39180
USA

The U.S. Army Engineer Waterways Experiment Station (WES), in conjunction with the U.S. Air Force and Department of Energy and Environmental Protection Agency laboratories, is developing a Groundwater Modeling System (GMS) for simulating groundwater flow, the transport/fate of subsurface contaminants, and the efficacy of remedial actions associated with contaminated groundwater resources at Department of Defense (DOD) installations and other federal sites. The primary product from the research will be a two and three-dimensional modeling system centered around both single and multiphase flow in concert with single and multiple-component groundwater contaminant transport. The system is capable of simulating flows in both the saturated and unsaturated zones. The system is being designed in a modular fashion, and is being developed for operation across a variety of computing platforms.

INTRODUCTION

Groundwater is the major source of water supply for over 50 percent of the population of the United States. This precious resource is threatened by increasing amounts of contamination. The variety of pollutant sources and their characteristics compound the problems associated with groundwater contamination detection, control, and cleanup. Activities at DOD installations have produced contamination of groundwater which may pose problems for human health and may threaten wildlife habitat and wetlands adjacent to or on these posts. As shown in Figure 1, of the 17,660 DOD locations within the continental United States initially identified as potential hazardous and toxic wastes (HTW) sites, over 10,000 are believed to require some sort of remedial action (Department of Defense, 1992). The U.S. Army and Air Force have responsibility for the vast majority of these sites.

A. Peters et al. (eds.), Computational Methods in Water Resources X, 1457–1464.
© 1994 Kluwer Academic Publishers. Printed in the Netherlands.

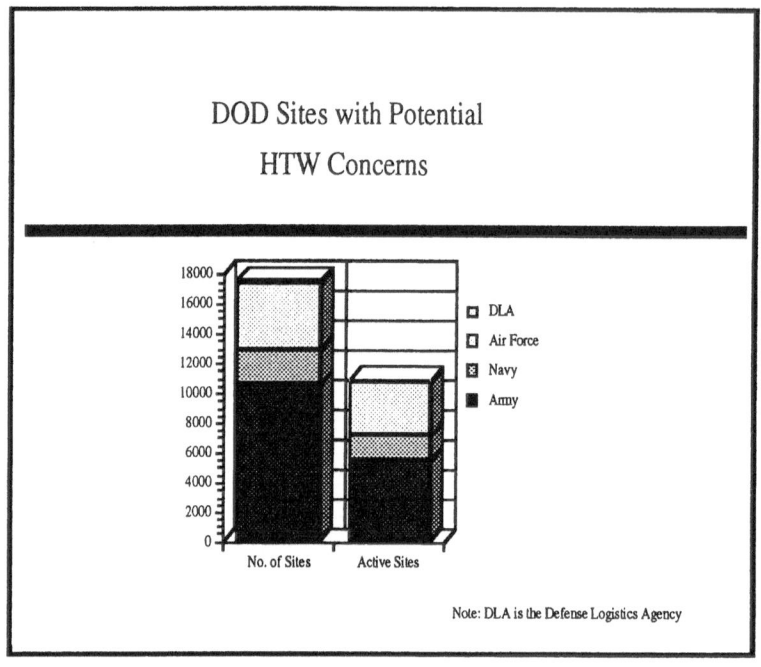

Figure 1. Number of Potentially Contaminated DOD Sites

Remediation of contaminated groundwater resources at these
sites is a difficult problem because the contaminants exist
in complex hydrogeologic conditions where a variety of
physical, chemical, and biological processes are occurring.
Multiple fluid phases and contaminants states are common to
HTW sites. A wide variety of contaminants, ranging from non-
aqueous phase liquids (e.g., petroleum hydrocarbons and
organic solvent liquids) to heavy metals and explosives, is
present on DOD installations. The ability to determine the
extent and severity of contamination is required before
groundwater investigations can rationally proceed from the
preliminary assessment phase into the investigation and
implementation of various remedial measures. Iterative or
unfocused remedial actions are unacceptable when human
health, safety, and costs are considered. Therefore, one must
have the capability to evaluate the effectiveness of
alternative remedial measures prior to their selection and
implementation.

Much of the information required for development and
evaluation of economical and effective remedial measures can
be obtained using computer models. The National Research

Council (NRC) has stated that "... groundwater models are among the most important scientific tools available for understanding groundwater processes" (National Research Council, 1992). However, DOD's in-house experience base is limited relative to groundwater modeling technology. Additionally, existing numerical groundwater models have a number of key technical deficiencies, particularly as associated with fate, transport, and remediation of military-unique contaminants such as explosives. A truly comprehensive groundwater modeling approach that adequately addresses the important aspects of contaminated subsurface flow, transport, and remediation is needed to support DOD remediation activities.

TECHNOLOGY AND KNOWLEDGE GAPS AND BARRIERS

A variety of technical barriers will have to be hurdled prior to effective development of the type envisioned above. These gaps have been delineated by several (see, for example, National Research Council, 1989; 1992). The major of these gaps relative to the concerns of the DOD include:

* Spatial heterogeneities are poorly represented in existing numerical models, thereby greatly reducing modeling confidence
* The uncertainty associated with subsurface conceptual-ization and parameter estimation is poorly integrated in current modeling state of practice
* Inadequate linkages exist between predictive uncertainty and the amount, location, and types of field site data that are required for conceptualization
* Limited, or no, information on a host of fundamental contaminant processes is available. This is particularly true for several military-unique contaminants, such as explosives
* Fracture-flow processes are poorly simulated in current models. Many DOD facilities lay over fractured subsurface environments, particularly in the Eastern United States
* Multiphase flow and multi-component contaminant modeling require extensive additional development, particularly relative to constitutive relationships
* Establishment of appropriate process parameters based on available site data is difficult and must be greatly improved
* Existing models have received inadequate evaluation and testing. More rigorous testing is required
* Relatively-friendly numerical models having greater applicability for a variety of site-specific cases are generally unavailable

TECHNICAL APPROACH TO GMS DEVELOPMENT

Development of enhanced groundwater modeling tools for the DOD is a multi-year, multi-disciplinary, highly integrated effort. While the effort will focus on key aspects of the remediation of contaminated groundwater resources, engineering aids for site characterization and contaminant assessment will also be developed. A development philosophy has been established that has resulted in a two-pronged technical approach: (1) conduct development to improve DOD's use of the better of existing models and tools through existing model enhancement and guidance, thereby bringing DOD's state-of-practice up to the state-of-the-art over the next two and one-half years; and, (2) concurrent with (1) (and extending beyond) develop the tools DOD will need in the future to model its remediation activities. The technological centerpiece of the effort is the development of a DOD Groundwater Modeling System (GMS). The GMS integrates the following:

(a) computational site characterization tools, data base managers, geostatistical methods, parameter estimation methods, visualization software, and existing numerical groundwater flow and contaminant transport algorithms. These are being coupled with a state-of-the-art graphical user environment to form the backbone of the GMS. Guidance, in the form of manuals and knowledge-based systems, is also being developed to provide information on the utility of existing models for characterization and remediation evaluations.

(b) improved contaminant fate/transport and flow models that result from extensive process-based investigation that is ongoing. The primary focuses of these investigations involve description and quantification of: the kinetics associated with fate and transport of military-unique compounds such as explosives; and, the impacts of subsurface heterogeneities, and the scaling effects induced therefrom, on flow, transport, and remediation effectiveness. These new developments will be incorporated into the GMS framework as they are verified.

(c) tools to simulate the efficacy of various remediation methodologies for cleaning up site-specific sites. These tools are being tailored to simulate the remedial treatment technologies most attractive for DOD cleanup situations. Additionally, this level of simulation will provide the user with optimization capabilities for the design and operation of various treatment technologies (such as provision of the optimal number and location of pumping and injection wells). Uncertainty and risk are also being built into the system to allow for potential changes in simulated results as a function of incomplete site characterization (e.g., sparse field data, poor parameter estimation, etc.) and/or poor process understanding.

A schematic of the GMS is presented in Figure 2. The first version of the GMS, version 1.0 (to be fielded in mid-1994), concentrates on improved use of existing models through coupling them with user interfaces, visualization tools, parameter estimation techniques, and subsurface conceptualization methods. Future versions of the GMS will incorporate improved flow and contaminant process formulations that are being developed concurrent with the computational system's development.

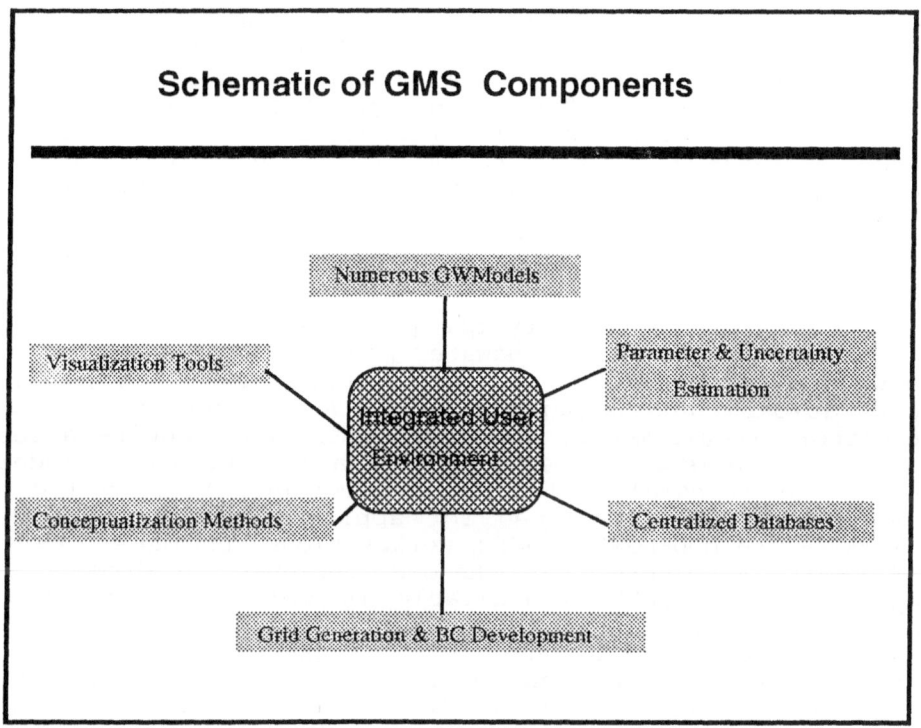

Figure 2. Major Research and Development Components

To provide the products alluded to above, the GMS will be developed under seven highly integrated areas: (a) Evaluation of Existing Technology; (b) Contaminant Fate and Transport Process Investigation; (c) Subsurface Conceptualization; (d) Model Enhancement and Development; (e) Remedial Process Fate and Transport Investigation; (f) Systems Integration and Verification; and, (g) Technical Assistance, Demonstration, and Implementation. Training of DOD modelers, and its contractors, in the use of the GMS is a vital part of area (g). An implicit part of the proposed demonstration of the developed technology will be thorough testing and verification of the modeling components of the GMS. This will be done through use of large-scale laboratory data and well-characterized field data sets from DOD and DOE sites under active cleanup. Extensive leveraging of existing data sets, such as those for the Macrodispersion Experiment (MADE) at Columbus AFB, will be conducted as well.

COMPUTATIONAL ASPECTS OF THE GMS

As presented above, the development of the GMS involves a combination of process and computational science research and development. The former of these is extremely important due to the large number of unknowns associated with the fate and transport of military-unique contaminants such as explosives, More information on the investigations associated with development of numerical fate/transport formulations for these contaminants is left, however, for another paper. Rather, we will now overview computational development aspects of the GMS.

System Development Pathway

Flexibility and portability are prime design considerations for the GMS. The DOD groundwater modeling user community is known to compute on personal computers (PCs), Unix workstations, and supercomputers. Additionally, due to regulatory requirements, DOD model users may require access to up to 10 existing subsurface flow and/or transport models as part of remedial evaluation and design. Version 1.0 of the GMS has been developed for application of two models (FEMWATER and MODFLOW) on Unix workstations running primitive X-Windows. A PC version of the GMS, designed for Windows 3.1 implementation, will be available in mid-1994. The GMS's user environment, which is a joint development of Brigham Young University (BYU) and WES, has been developed in C. This basic user environment, with slight differences dictated by the requirements of the differing models, will front all models integrated into the GMS. In this manner, the users will have to become aquainted with only a single computational environment in their use of multiple modeling tools. An additional five existing models will be incorporated within the GMS in 1994.

Supercomputing Requirements

Most of the initial field users of the GMS will apply the system on PCs (a minimum of a 486-class machine is required) or workstations. The types of problems these users are now modeling are on the order of 1000s of nodes. These simulations are, however, almost exclusively single-phase flow and single-component contaminant transport in the saturated zone. It is expected that multiphase flow simulation, either as associated with hydrocarbon contamination or certain types of remedial alternatives (e.g., steam injection/vapor extraction), and multiple-component contaminant transport will be required over the next three years. Proper numerical representation of this class of problems will most certainly extend the order of GMS calculations to 100,000s of nodes and unknowns. In addition, the basic research being conducted on the impacts of heterogeneity-induced scale is employing pore-scale and highly-resolved continuum models that often require on the order of millions of nodes.

To augment use of the GMS for problems larger than current workstations can straightforwardly address, two paths are being followed. Access to a number of DOD Cray supercomputers, including the C90 and Y-MP at WES, is being incorporated into the GMS. Initial access will be through networked interconnections; however, within two years the GMS will be configured to sense the size of a given problem, search the DOD supercomputing network for available computing resources, ship the execution to the available resources, and retrieve the completed job for analysis and visualization at the local workstation.

It is also obvious that massively parallel platforms (MPP architectures) will be required to conduct subsurface simulations requiring millions of nodes and/or unknowns. To this end, the modifications required for GMS utilization on a number of MPP machines, including the CM-5 at the Army High Performance Computing Research Center (University of Minnesota), have been initiated recently. Work in this area will increase significantly in 1995.

Algorithm Development

While the first version of the GMS will feature connectivity to existing models only, future versions of the system will include WES-developed models. These models will incorporate the new fate/transport science, conceptualization methods, remediation models, and scaling relationships developed by WES and its partners. Development of the computational engines for the WES models is in its initial stages.

Unstructured finite element and finite volume approaches have been selected as the primary computational basis of the WES models. The use of locally-adaptive grid refinement is considered integral to the WES model development as a means of augmenting front capture. A number of solution techniques, featuring pre-conditioned conjugate gradient and multigrid solvers, are being investigated. Development of highly efficient numerical solvers is considered a must for the GMS. Incorporation of new scaling relationships, whose theoretical developments are highly non-local, may significantly increase the computational requirements over requirements.

CONCLUSIONS

The Department of Defense is developing a comprehensive groundwater modeling system for use in the cleanup of contaminated groundwater resources at military sites and installations. The proposed seven-year research effort will strongly leverage ongoing and near-future research by other groups through partnering with several federal agencies and universities. Substantial improvements in the effectiveness and defensibility of remedial actions are expected through implementation of this system, thereby resulting in significant savings in remediation costs.

ACKNOWLEDGEMENT

The paper was prepared from research and development conducted under the Groundwater Modeling Program of the U.S. Army Engineer Waterways Experiment Station, Vicksburg, MS. Permission was granted by the Chief of Engineers to publish this information.

REFERENCES

Department of Defense, 1992. Annual Report to Congress for Fiscal Year 1991, Defense Environmental Restoration Program.

National Research Council, 1989. Water Science and Technology Board, Commission on Physical Sciences, Mathematics, and Resources, National Academy Press, Washington, DC.

National Research Council, 1992. "A Review of Groundwater Modeling Needs for the U.S. Army," p. 3, September, Washington, D.C.

EFFECTIVE VISUALIZATION OF LARGE DATA SETS

C. D. MONTEMAGNO and W. G. GRAY

Department of Civil Engineering and Geological Sciences
University of Notre Dame
Notre Dame, IN 46556-0767
USA

ABSTRACT

The proliferation of low cost, high performance workstations has made moot the concern of available computing power for all but the most ambitious numerical models. As a result, unprecedented volumes of complex data are being generated by today's engineers and scientists. An evolving challenge for researchers is the development of mechanisms to rapidly analyze data sets with size on the order of a gigabyte and encompassing dozens of variables. Current efforts suggest that data visualization may be the most effective method for gaining insight into the interactions of a large number of variables and for communicating this information to other researchers.

Effective use of data visualization as a research and communication tool requires the researcher to develop a particular set of skills. The large number of parameters that must be selected in the creation of a visualization requires an understanding not only of the data but also of form and color. Parameters as mundane as coloring and variable ordering or as complex as lighting and perspective can greatly impact the communicative ability of a visualization.

The basic principles for successful scientific visualization are presented. These include the establishment of visualization goals, selection of output media, and the design and organization of an image. Distinctions are made among the visualization of phenomenological processes, of computed data, and of measured data. The interactions of color, lighting and object complexity are examined. Additionally, the effect that the selection of output media (i.e. slides, transparencies or video) has on the visualization design is discussed.

INTRODUCTION

With the continual increase in the performance of computers, researchers are delighting in their ability to solve problems on desktop machines which as recently as five years ago could only be solved on the world's fastest supercomputers. This explosion in computer performance has been accompanied by an unprecedented improvement in the "reality" of

A. Peters et al. (eds.), Computational Methods in Water Resources X, 1465–1471.

the mathematical models used to describe physical systems (Nielson, 1990). Scientists now are able to develop mathematical models to simulate complex processes at previously unimaginable scales. Fluid behavior is simulated at both the level of individual molecules and at highly resolved global scales. Numerical simulations incorporating over 10^6 nodes are becoming common. This is causing a revolution in the way in which mathematical models are being used. Computer simulations are no longer restricted to predicting the performance of well-defined physical systems. They are now used to discover, explore, and investigate little-understood physical processes. Computational simulators are becoming the experimental laboratory of the 21st century.

The volume and complexity of data generated by these "reality" models is staggering. Some current global circulation models generate 22 gigabytes of data for each simulation day (Alper, 1994). Traditional data plotting and analysis techniques are quickly overwhelmed by the sheer volume of the data generated by these complex simulations. Additionally, conventional plotting and analysis methods make it difficult for the researcher to orient and visually validate the results associated with large data sets. In general, these analysis methods are incapable of communicating the subtle behavioral clues of modeled system that are the key to scientific discovery.

Luckily these same increases in computer performance has led to the development of a new methodology for analyzing large data sets. This new technique is commonly referred to as scientific visualization. Scientific visualization is a type of graphical representation of data. It focuses upon the generation of images bringing out the meaning of the data through the use of visual cues. Through the use of proper visual cues it is possible to identify unexpected relationships within the data set and to readily communicate complex ideas and processes.

DISCUSSION

To generate effective scientific visualizations, many intangible processes must be executed. These processes can be as obvious as identifying the goal and the intended audience of the visualization, or as subtle as selecting the image size. Unfortunately, no cookbook methods exist for generating effective visualizations. It is firmly believed, however, that for complex data sets, the multidimensionality of the data must be preserved in the visualization. Using a reference to Edwin Abbott's *Flatland*, a classic work about life in a lower dimension universe, we find that "Escaping ... flatland is the essential task of envisioning information ... for all the interesting worlds (physical, biological, imaginary, human) that we seek to understand are inevitably and happily multivariate in nature. Not flatlands." (Tufte, 1990). In other words, to be effective, visualizations must transcend the two dimensionality of normal presentation media such as paper, slides, and video screens.

Two principle elements are common to all visualizations, the generation of the image and the output of the image to the presentation media. While no rigorous procedure can be defined for the development of effective visualizations, examination of the constituents associated with both the generation of the image and its final output will provide insight into the heuristics of producing graphics successful in communicating complex ideas.

Visualization Development

The first step in generating a scientific visualization is the identification of the goal of the graphic. This goal defines the meaning within the data that is to be communicated. The articulation of this goal answers fundamental questions about why the data is being examined and helps to clarify what is expected to be learned from the generated visualization.

In general, seven principle categories of visualization goals can be identified (Keller, 1993). These categories are:

1. The comparison of data sets, images, and positions.

2. The identification of numerical values, objects, and activity.

3. Indication of direction and orientation.

4. Location of position relative to some object or coordinate.

5. Relating concepts such as velocity and temperature.

6. Revealing objects by making them visible.

7. Representing numerical values of data.

Often the selection of the visualization goal is intuitive. However, by consciously identifying the purpose of the graphic, new ideas will often be generated that identify methods for better communication of the visualization goal. In general, the identification of the visualization goal provides a roadmap to the best technique for producing the most effective graphic.

As an example of the process of development of a visualization goal, assume that the goal of a proposed graphic is to communicate changes in groundwater flow calculated from a 2-D numerical simulator over time. Immediately one might think to provide multiple representations of the 2-D head elevation data sets. This is the obvious way in which this type of data is traditionally presented. It is at this point that scientific visualization separates from the traditional data representation techniques. If one employed the traditional graphic techniques for the 2-D data set, most likely the chosen visualization would be a series of plots of hydraulic head elevation isolines. The changes in the value of an isoline at position as a function of time would be communicated by viewing a series of plots. By inspecting these contour plots one would indeed be able to determine the changes in the direction and quantity of the groundwater flow. However, such a comparison provides no insight into the reasons for the the changes. What is presented in this type of graphic are numbers, category 7 of the visualization goals listed above. This representation provides no information concerning the underlying physical processes which caused the groundwater to behave as presented.

Now suppose that instead of displaying the data as a series of simple 2-D contour plots, the data is presented using three-dimensional surface. The shapes of these surfaces may be defined by the hydraulic head with the color of the surface shaded to represent the aquifer permeability as a function of space. When this visualization is inspected not only are the numeric values associated with the changes in groundwater flow communicated but also some phenomenological reasons for the groundwater behavior are revealed. Such a visualization may satisfy visualization goal 5 above in that flows and permeability are related in the figures.

By providing visual cues to the underlying phenomenological reasons for presented numeric information not only is the numeric data more effectively communicated, but insights into factors that influence system behavior may also be gained. This technique can help provide new understanding of the physics under investigation.

Visual cues can take many forms. Color, texture and shape are obvious graphic parameters that can be used to provide phenomenological cues to system processes. Some researchers advocate leaving the traditional bounds of the scientific discipline when designing the visualization (Keller, 1993). In general, communication is much more effective if the visualization is developed in a context familiar to the intended scientific audience. Simple things such as setting the color of water surfaces to blue or using glyph's of simple familiar objects like a pick axe to indicate the presence of a mineral significantly improve the comprehensibility of the data.

The last of the criteria that must be considered in the design of a visualization is its intended use. The graphic design must take into account whether it is being developed for presentation or for personal data analysis. If a scientific visualization will be presented, the knowledge base of the audience and the presentation media are critical to the design of the graphic. There is a continuum of understanding in any scientific visualization. This continuum extends from the immediately obvious, to requiring a short inspection, to necessitating a detailed explanation. Where a graphic sits on this continuum depends upon the familiarity of the audience with the subject matter as well as the type of medium from which the information is presented. The goal of a particular visualization may be intuitively obvious to an expert with specialized knowledge or experience in a presented subject area but incomprehensible to a person with different or more limited expertise. Sometimes, even an expert may have difficulty comprehending a subtle point about a physical process from a series of graphics presented on overhead transparencies. However, when those same images are displayed as an animation, this same subtle process becomes immediately obvious.

From this discussion, it should be clear that no rigid rules leading to effective design of a visualization can be enumeratied. However three basic principles form a thread which runs through all scientific visualizations. First, the goal of any visualization must be clearly defined. Generation of an attractive and appealing graphic that does not communicate information of importance is an all too common problem. Production of visualizations with these properties at best hampers the exchange of knowledge and at worse may may serve to misrepresent or distract from actual expertise or knowledge of a system. Although graphics that improperly communicate information seem to have great appeal in sales endeavors, they should have no place in serious scientific study.

Second, through the use of visual cues a researcher should incorporate process phenomena into a visualization. This is important in graphics prepared for self use as well as in visualizations destined for presentation. By exploring data through the visual linking of process information with numeric results, additional insight into the physics of a system under study may be gleaned.

Finally, the audience for any presentation media must be considered in the design of any visualization. The goal for an ideal visualization is that it be a graphic whose message is immediately obvious to the audience without any additional explanation. Consequently, communication of one set of information to audiences with different knowledge bases or through different presentation media probably requires the development of alternative visualizations.

Visualization Communication

Typical output media used for displaying visualizations include paper, overhead transparency, slides, computer monitors, and video monitors. All of these media tend to have different performance characteristics with respect to color reproduction and resolution. Consequently, visualizations developed on one medium tend not to reproduce exactly when transferred to a different medium. Unfortunately, in most cases the visualization is generated on a computer monitor and then transferred to another medium for presentation. The need to transfer graphics between media can drastically affect the information that is communicated. Typically, colors or opacities will be different causing features to be obscured or changed. Resolution can also be lost. These inconsistencies cause loss of detail and sometimes the elimination of important data structures. Image proportions can also be changed when transferring a visualization from one medium to another. This can cause the unintentional exaggeration of features in the data. Therefore to ensure that visualization goals are met, the final presentation medium must be considered when developing a graphic.

Maintenance of true color is typically the most difficult problem when transferring visualizations from computer monitors to either paper or overhead transparencies. This is because the method used to generate colors on computer monitors is drastically different from the way colors are produced on either paper or overhead transparencies.

On computer monitors colors are created by adding different combinations of red, green and blue light then transmitting these combinations directly to the eye. Color printers create color through a subtraction process. Cyan, magenta, yellow and black are applied to a physical medium in quantities such that their combination, when subtracted from white, yields the desired color. The image seen by the eye is a reflected image resulting from white light striking the page and being selectively absorbed. Visualizations generated on a color monitor will in general not look the same when printed. Often blue and yellow on the color monitor become purple and brown, respectively, when printed.

Even though computer and video monitors both produce colors by red, green, and blue addition, because of the reduced bandwidth of analog video signals, video monitors can not replicate the brilliant colors that are possible on a computer monitor. Additionally, abrupt changes in contrast will cause a shimmering of the image between regions of high contrast on a video monitor. Consequently, in order to minimize these artifacts, pastels are the most prudent color choices.

In general, photographic slide images have the highest resolution, followed by paper/transparencies, and then video, produce images of decreasing resolution. Resolution is a measure of the detail which can be see in the generated images. Typically a high quality film recorder will produce slides which a resolution of about 4000 lines per inch. Normal color thermal transfer printers have a resolution of 300 dots per inch while professional quality color dye sublimation printers can produce images at a resolution as fine as 1200 dots per inch. The resolution of images generated by either printers or film recorders is usually greater than the resolution of the computer monitor which is typically on the order of 100 dots per inch. As a result some fine detail may be lost on the output image because the features may be too small to be seen. To alleviate this problem, the size of glyph's, arrows and type may have to be made larger than initially desired.

Video images are of significantly lower resolution than any of the other output media. Because of this, many graphic features will either be lost or blurred in comparison to the image on a computer monitor. Additionally, because a video signal is interlaced, that is the image is filled every two scans, thin vertical lines tend to be displayed as broken. Also the low resolution of video monitors causes alaising which is manifested as artifacts in the image. To help control these problems two tactics should be employed. First, the visualization should be designed to exaggerate the features in the graphic necessary for communicating the visualization goal. This will ensure that the regions of importance will still be highly visible in the video image. Second, the image should be low pass filtered so that high frequency components of the graphic will be reduced, thereby eliminating many artifacts from the displayed video signal.

The final major concern which needs to be addressed before a visualization is generated is the difference in proportion between the computer monitor and the final presentation medium. The data output displayed on a computer monitor is aligned on a square grid with a 1:1 aspect ratio. Other output media normally generate images with a different aspect ratio. This results in distortion of the data, exaggeration or minimization of some features, and the potential loss of portions of the graphic. Slides normally have a 2:3 ratio, overhead transparencies typically have an 8:10 ratio and video has a 3:4 aspect ratio. To compensate for these differences in aspect ratio, after the a visualization has been developed on a computer monitor, the image should be rescaled to compensate for any differences in aspect ratio between the final output media and the computer monitor. This will ensure that the presented scientific visualization is representative of the actual data.

SUMMARY

Because of the tremendous increase in the volume of data generated by highly complex numeric models, conventional analytical tools have been found to be inadequate for both analyzing and communicating the results of these studies. By clearly defining the goals of the analysis and incorporating phenomenological cues into any visualization, subtle relationships between process physics and numerical results can be identified. This feature of scientific visualization is encouraging the transition of numerical simulators from predictive tools to experimental instruments. Coupling this feature with the ability to generate images that can clearly communicate complex information to a large audience is ensuring that the use of scientific visualization techniques will become an essential element of all major computational investigations.

REFERENCES

Alper, J. (1994) "The Silicon Ocean", *Earth*, March 1994, 33-38.

Keller, P. and Keller, M. (1993) *Visual Cues, Practical Data Visualization*, IEEE Press, Greenwich, CT USA

Nielson, G. and Rosenblum, L. (1990) *Visualization in Scientific Computing*, IEEE Computer Society Press, Los Alamitos, CA USA

Quiller, S. (1989) *Color Choices*, Watson_Gupyill Publications, New York, NY USA

Tufte, E. (1990) *Envisioning Information*, Graphics Press, Cheshire, CT USA.

COMPUTER AIDED GENERATION AND OPERATION OF HYDROLOGIC SIMULATION MODELS

F.G. ROHDE, M. HAASE and CH. GITSCHEL
Section Hydrology and Water Resources at the Technical University Aachen
Mies-van-der-Rohe-Str. 1, 52074 Aachen
Germany

Abstract

A model generator which integrates databank information with the general model simulator MHMS ("Modular Hydrologic Modelling System") by CADSWES (University of Colorado) is presented. The user assembles executable hydrologic simulation models from a library of available submodels (each describing individual hydrological processes) using an interactive, window-oriented tool. This enables the user to easily design and test various models with alternative submodel-structures on different temporal and spatial scales. Special-purpose functions for parameter optimization, sensitivity analysis and streamflow prediction are available.

All data which are necessary for model generation and simulation (GIS, time series, model and other alphanumeric information) are managed in a databank. This guarantees overall data consistency. A set of predefined, classified global terms for naming model input and model output is kept in the databank guaranteeing overall semantic consistancy for the model generation process.

The system uses submodel-structures, space-model-structures and space-space-structures for logically designing a simulation model. The submodel-structures consist of a network of interlinked submodels which are assigned to hydrologic spaces (i.e. hydrological response units) via space-model-relations. The spaces are interconnected by space-space-relations. Thus, a logical sequence for the model simulation is defined. Based on this information the model generator creates an executable simulation program which operates on alphanumerical data for parameters and variables managed by the databank.

The advantages of this approach are that executable binary codes need not be stored for each scenario but can be set up whenever necessary. Simulation programs are created without time-consuming programming and all data are managed consistantly and

A. Peters et al. (eds.), Computational Methods in Water Resources X, 1473–1480.
© 1994 *Kluwer Academic Publishers. Printed in the Netherlands.*

redundancefree by a databank. An example for the use of the system in catchment hydrology is provided.

1   Introduction

Most hydrologic modelling systems are created using fixed data and program structures. Extensive programming is normally required, therefore, if the structure of the input data must be altered or if a modification of the simulation output becomes necessary.

When multiple scenarios are used in problem-solving, a tremendous amount of output is created. However, an outline of the development of these scenarios and the conclusions based thereon are not always automatically retained, thus rendering the results useless.

In order to avoid the above situation, a new concept for a more flexible modelling system based on information and system analysis has been developed.

2   A System Overview

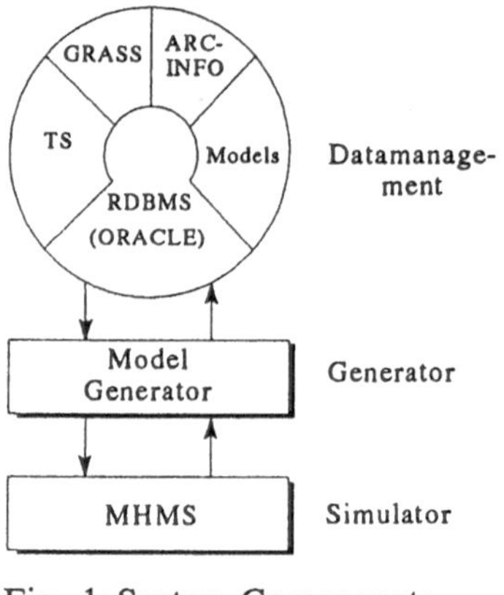

The system consists of the following components:
- a *Datamanagement* component
- a *Model Generator*
- a *Simulator*

The data management component is described in fig.1. The core represents a relational databank management system (*RDBMS*) - in this case ORACLE. The databank contains all data except for the time series data, geographical data, and the model sources with the model descriptions. The databank performs two main tasks: First the direct management of alphanumeric data, and secondly the creation of references to data which are not contained within the database, ie. GIS, time series data (*TS*) and model sources (*Models*). This central function secures the consistency of all system information.

Fig. 1: System Components

Time series data and GIS information are managed nonrelationally. This is carried out by a special time-series data management system and the two GIS systems, *GRASS* and *ARC-INFO*. Model sources as well as their descriptions exist in the form of files within UNIX. The management of model sources within RDBMS is theoretically possible, but is not currently planned.

3   A Semantic Data Model

Fig. 2 represents the results of an information analysis in the form of a semantic data model. Here a diagram of the relationships between various entities (Chen 1977) is depicted, as well as the basic symbols used.

Three basic levels of information are distinguished:
- the models itself (*Models*)
- the depiction of the Real World in model-space (*Model-Space*)
- the simulation of the Real World using the model (*Scenarios*)

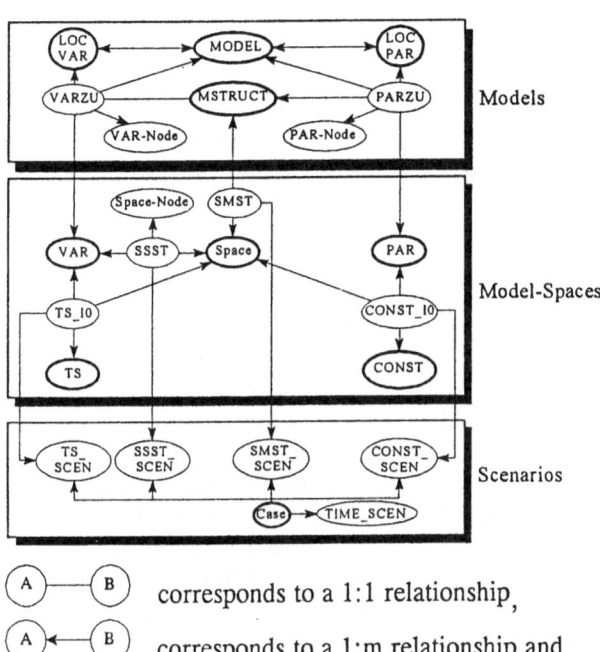

In hydrology, the experiment area is usually divided into homogenous subspaces called Hydrological Response Units. Several hydrologic processes - depicted as submodels - are carried out within these subspaces. The submodels process variable quantities, called Variables, and fixed quantities, called Parameters. The Variables can be divided into flows and states - the flows (water currents) connect the submodels with the subspaces.

A ———— B	corresponds to a 1:1 relationship,
A ◄——— B	corresponds to a 1:m relationship and
A ◄——► B	corresponds to a n:m relationship

Fig. 2: Semantic Data Model for Hydrologic Modelling

On the *Models* level models *(MODEL)* are linked by their variables *(LOCVAR)* and parameters *(LOCPAR)*, forming model structures *(MSTRUCT)*. The goal of this linkage is to use submodels to form larger models which depict specific hydrologic spaces. For example, surface hydrologic processes in a forest area can be depicted differently from urban areas.

On the *Model-Spaces* level, these model structures are assigned hydrologic subspaces using space-model-structures *(SMST)* The subspaces are connected to waterflows *(VAR)* via flow schemes *(Space-Space-Structures, SSST)*. Constants *(CONST)* and time series

(TS) - which can be output of submodels- are assigned to the hydrologic spaces based on their Parameters *(PAR)* and Variables *(VAR)*.

On the *Scenarios* level, various simulation scenarios are created based on the simulation time, the time series input and output, the spatial flow scheme, the modelled depiction of the hydrologic spaces as well as the incoming constants for the description of the Real Space.

This semantic data model was subsequantly translated into 34 relations and was implemented in a RDBMS. Time series and geographical data were kept seperately in a time series information system and a geographical information system, respectively.

4   The Model Generator

The model generator (Rohde et al 1993) serves as an interface between the user and the system. Fig. 3 provides an overview. The main functions of the generator are:

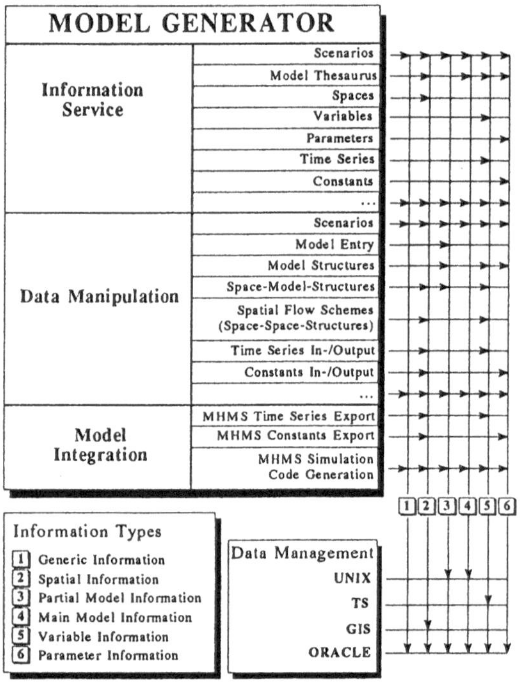

- *Data Manipulation*
- *Information Service*
- creation of the main model (*Model Integration*)

Data Manipulation consists of the sorting, entering, altering and deletion of information within the databank. Thus, data integrity rules remain unconflicted. A special feature is the representation of the data manipulation using tables which are directly associated with specific spaces. Functions which allow the input of data using a direct linkage of the hydrologic subspaces on the computer screen were created with the help of the GIS (ARC-INFO) functions. Fig. 4 provides a graphic example of this function. Similarly the manipulation of the model structure data also occurs using a

MODEL GENERATOR

Information Service	Scenarios
	Model Thesaurus
	Spaces
	Variables
	Parameters
	Time Series
	Constants
	...
Data Manipulation	Scenarios
	Model Entry
	Model Structures
	Space-Model-Structures
	Spatial Flow Schemes (Space-Space-Structures)
	Time Series In-/Output
	Constants In-/Output
	...
Model Integration	MHMS Time Series Export
	MHMS Constants Export
	MHMS Simulation Code Generation

Information Types
1. Generic Information
2. Spatial Information
3. Partial Model Information
4. Main Model Information
5. Variable Information
6. Parameter Information

Data Management
UNIX
TS
GIS
ORACLE

Fig. 3: Model Generator Functions

graphic user interface which supports the way hydrologists are used to think of their problems. Thus, the user can concentrate on the hydrologic problem at hand instead of concerning himself with the formalities of a relational depiction of the information.

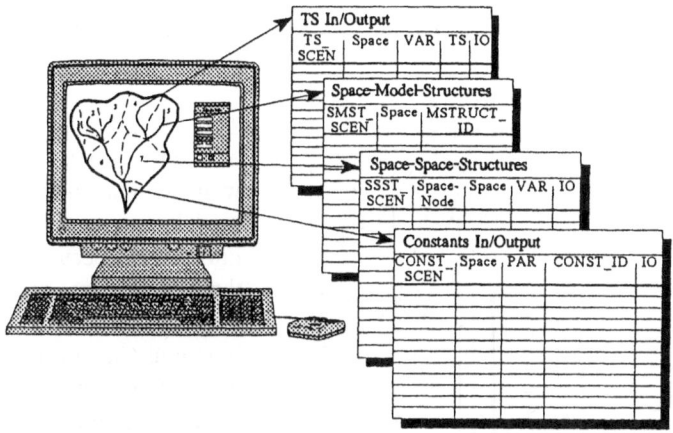

The *Information Service* allows scenarios using models and model structures, space model structures, space-space structures, variables, parameters time series data and constants to be called up from the databank. The user can do this using prepared commands, query by example or the direct input of SQL statements, thus

Fig. 4: Data Manipulation using GIS Features

being able to gain an overview of the current data content at anytime. This is important for the creation of new simulation scenarios, and helps in deciding whether or not new data should be included in the system.

The function of the *Model Integrator* is to assemble executable main models. This includes the generation of input (constants and time series data) for the simluator, and the creation of codes for the running of the simluation model. The data used for this generation is chosen from the databank and time series data management depending on the choice of scenario. Further, execution sequences which are determined by the space-space-structures and model-structures are predefined for the simulator automatically.

Constants and time series data are created in the required simulator formats. The generated code for loading and running the simulation model is automatically compiled and then linked to the rest of the simluation routines.

5   The Simulator

The simulator - MHMS (Modular Hydrologic Modelling System) - was developed under the auspices of the USGS. The goal was to relieve the hydrologist of programming work by providing tools which take over such tasks as the data management, user interface, sensitivity analysis and optimization (Müller-Wohlfeil et al 1994, Leavesley et al 1992, Restrepo 1993).

The MHMS includes:
- - a user interface for controlling the system and for processing user input
- - special tools, ie. optimization and sensitivity analysis programs
- - a library of simluation models
- databanks for exchanging dimensions, parameters and global variables between models

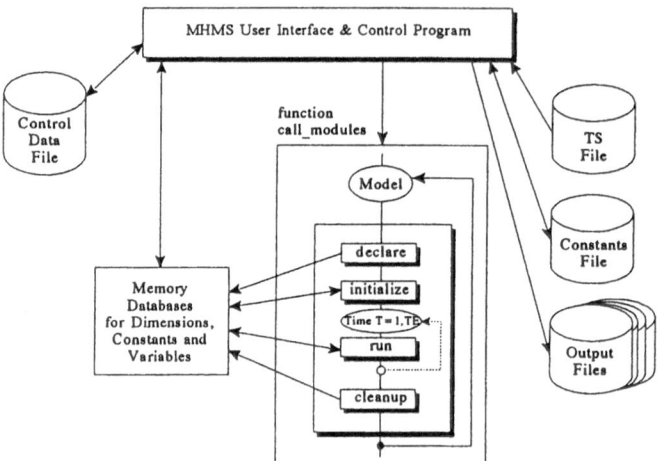

Fig. 5: MHMS User Interface Data Bases and Files

This exchange of data between models - as well as that between models and user interfaces - occurs not through giving arguments to the different functions (as in most programs), but by saving data needed for the simulation dimensions, parameters, variables) in special databases in the computer core. These databases are defined by the models in declaration mode. During the simulation, the model calls up or sends back necessary data from and to these databanks. The user interface also uses these databanks for parameter manipulation and the creation of real-time time series plots. The above mentioned functions are controlled by the *MHMS User Interface*. The relation between databanks, data and user interface is described in fig.5.

6   Application of the System

Models	Space Type			
	Ground-water	Channel	Agricultural/ Forestry Areas	Urban Areas
Models	Linear Filter	Kalinin-Miljukov-Routing	Interception, Snow, Surface Flow, Soil Moisture Accounting, Interflow	Snow, Depression Storage, Surface Flow, Sewage System Runoff

Fig. 6: Models and Spatial Types

This system was tested in a rainfall-runoff study. The experiment area lies on the northern border of the mountainous region, the Eifel, near the city of Düren in Northrhein-Westfalia, Federal Republic of Germany. The specific area studied, the catchment of the Gürzenicher creak, has an area of ca. 12.4 km² , and the shape of a long, thin rectangle with the ave. length of ca 7.5 km, the ave. breadth of ca 1.4 km. The area has an overall difference in elevation of about 250 m (Rohde 1990).

The catchment is characterized by the following morphologic and land-use features: the southern section situated in the Eifel is heavily forested, and the slopes are relatively steep. The northern section is relatively flat and is used as agricultural land. Also in the north is the town Gürzenich which is heavily sealed. The main soil types are new soil - in one part of the northern section lignite was mined - Glei-Braunerde, Parabraunerde, Pseudoglei and Braunerde. All of these soiles belong to the potsoil family.

Because of the inhomogenious structure within a relatively small area, the basin was subdivided into 64 subcatchments, with the built-up area heavily subdivided. A differentiation was made between the spatial types: groundwater, channel agricultural/forestry areas and urban areas. Diffe-

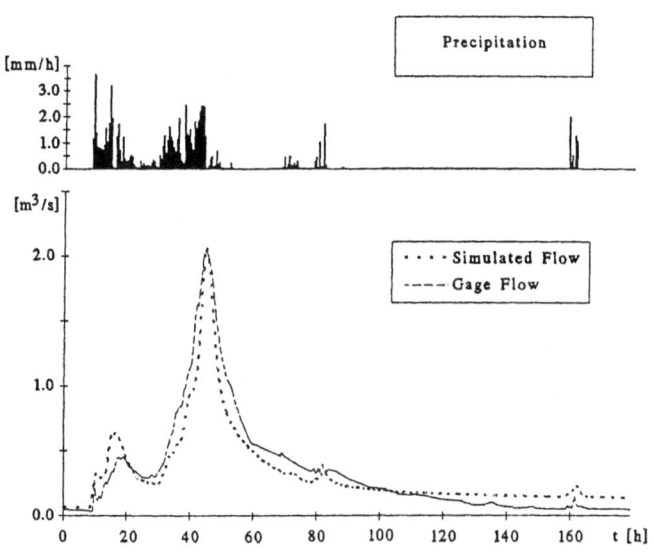

Fig. 7: Calibration Results

rent scenarios were defined for flow schemes, model structures, space-model-structures, constant and time series data and simulation time. Fig. 6 dipicts the model components for the introduced space types used for model calibration. Fig. 7 shows calibration results based on a constant time step of 0.5 hours with precipitation and flow gage data recorded during late May 1983 obtained for this scenario.

7   Conclusion

A software system for flexible hydrologic modelling consisting of a data management component, a model generator and a simulator was presented. Thus, enabling the user to easily build up models and do simulation runs without software coding by himself.

Based on a semantic data model for modelling hydrologic processes 34 relations were developed and implemented into an RDBMS. The latter forms the core of the data management component of the software system. GIS, time series and model data make up the other parts of this component.

The main functions of a model generator were outlined as the information service, data manipulation and model generation. Special emphasis was hereby given to the graphical user interface build to edit databank relations for defining hydrologic models. An example was provided for the use of GIS functions for these purposes. Another main task of the model generator is to build executable models to run on the data stored in the data management component. The simulator MHMS for processing hydrologic models was discussed. Its functions and main components were described.

The system was applied to a catchment study, demonstrating the applicabilty of the system outlined.

8   References

Chen, P.P. (1977) "The Entity-Relationship Model - A basis for the enterprise view of data", Conference proceedings of the 1977 National Computer Conference, Dallas, pp. 77-84.

Leavesley, G.H.; Restrepo, P.; Stannard, L.G.; Dixon, M. (1992) "The Modular Hydrologic Modeling System - MHMS", 28th Annual Conferece and Symposium of the American Water Resources Association, Reno, pp. 263-264.

Müller-Wohlfeil, D.; Haase ; M., Gitschel, Ch. (1994) "The Modular Hydrologic Modeling System - Eine einführende Übersicht", Bayreuther Institut für terrestrische Ökosystemforschung, Reihe Bayreuther Forum Ökologie (to be published).

Restrepo, P.J. (1993) "Modular Modelling System, User´s Manual", CADSWES, University of Colorado at Boulder (unpublished).

Rohde, F.G.; Haase, M. (1990) "Niederschlag-Abfluss-Untersuchung für den Gürzenicher Bach", Lehrgebiet für Wasser-Energie-Wirtschaft, Aachen (unpublished).

Rohde, F.G.; Haase, M.; Gitschel, Ch.(1993) "System- und Modellanalyse für ein integriertes Konzept der regionalen Wasserhaushaltsmodellierung unter besonderer Berücksichtigung regionaler Ähnlichkeitsmerkmale - 2. Zwischenbericht", Report to the German Research Foundation, Lehrgebiet für Wasser-Energie-Wirtschaft, Aachen (unpublished).

STREAMLINE COMPUTATION FOR FRONT TRACKING

P.J.A. TIJINK[1], T. GIMSE[2], E.F. KAASSCHIETER[3]
[1]Technical Software Consultants, The Research Park, 0371 Oslo, Norway
[2]Dept. of Informatics, University of Oslo, 0316 Oslo, Norway
[3]Dept. of Mathematics and Computing Science, Eindhoven University of Technology, 5600 MB Eindhoven, The Netherlands

The ultimate goal of porous media flow simulations is to compute and predict the movement of underground fluids and gases. A variety of approaches for this have been proposed over the years. A large class of these methods is based on solving two differential equations; one *pressure* equation and one *saturation* equation. These two equations are strongly coupled, but are mathematically different. Thus, it may be necessary to use different numerical methods for each of them. The basis for this paper is the simulator FRONTLINE, which uses a finite element method for the pressure equation and a front tracking method for the saturation equation. Today, a relatively simple conforming finite element method is being used, and in this paper we discuss alternatives to this. We find that, even though such alternatives may give some increase in the computation time, significantly better results may be obtained.

1. INTRODUCTION

The general solution of transport equations in porous media is usually based on solving equations for the pressure and the saturations simultaneously or in sequence. The problem is to determine the distribution and movement of material in the porous medium. However, and particularly for sequential methods, it is crucial that flow velocities are computed accurately. In this paper we consider the computation of the velocity field in the flow simulator FRONTLINE [3]. This is a program package for simulation of flow in porous media, with particular emphasis on petroleum applications. A version of FRONTLINE for groundwater flow and contaminant transport is being developed at present. The basic idea in FRONTLINE is to determine the pressure and velocity field in one step, and then compute streamlines, which are used locally as one-dimensional coordinate systems to compute the motion in the medium. The advantage of this is that two- (and three-) dimensional computations are much more complicated and costly than one-dimensional. Thus, FRONTLINE is a very fast and accurate program whenever it may be applied. The major present limitation is that gravity forces are ignored, which correspond to horizontal flow or

A. Peters et al. (eds.), Computational Methods in Water Resources X, 1481–1488.
© 1994 *Kluwer Academic Publishers. Printed in the Netherlands.*

equal-density-flow.

In the previous versions of FRONTLINE the pressure equation is solved by a standard finite element method [5], and then the velocity field is determined by numerical differentiation of the pressure. In this paper we consider the use of a mixed finite element method, e.g. [10], where the pressure and the velocity are computed at the same time. This gives a more accurate velocity field, which is essential to determine the streamlines accurately.

Even though this paper focuses on the mixed method, for completeness we will start summarizing the methods that are used along the streamlines. A more detailed summary of this can be found in [3]. For the general formalism and the equations that are used throughout the paper, the reader is referred to any textbook on porous media flow, e.g. [1], [2].

2. FRONT TRACKING ALONG STREAMLINES

The basic one-dimensional technique for computing fluid motion in FRONTLINE is front tracking. Front tracking was introduced as a method of analysis by Richtmyer [11], and was developed extensively as an analytical and numerical method by Glimm, McBryan and coworkers, e.g. [8]. The saturation equation, without capillary effects, is a conservation law of the form

$$u_t + f(u)_x = 0.$$

Here f is a continuous and piecewise smooth scalar function, usually denoted as the flow function. In addition, any specific problem consists of initial or boundary data. First we will examine Riemann initial data, viz. $u(x,0) = u_l$ for $x < 0$, $u(x,0) = u_r$ for $x > 0$.

In general, the Riemann problem is solved by a sequence of elementary waves, either shocks or rarefaction waves. To obtain a unique, physical solution, entropy conditions have to be imposed. A general reference for the theory of conservation laws is [12].

The solution of the Riemann problem is found by the following procedure: Assume that $u_l < u_r$ (the other case is treated symmetrically). Let f_c denote the lower convex envelope of f with respect to the interval $[u_l, u_r]$. The solution procedure now involves determining intervals where either $f_c = f$, which gives rarefaction waves, or intervals where $f_c \neq f$, which define shocks in the solution. The computation of the rarefaction waves involves inverting a non linear function, which is often mathematically difficult and computationally costly. An efficient simplification was introduced by Dafermos [6] and developed by Holden, Holden, and Høegh-Krohn [9]. In their approach, the original flux function is approximated by a piecewise linear function. Thus, f is assumed to be piecewise linear, and the convex envelope f_c is also piecewise linear with respect to any interval. Consequently, any Riemann problem solution consists of constant states separated by shocks. The rarefaction waves will be approximated by sequences of small shocks. These shocks (fronts) may now be tracked in space and time. The speed of a front is determined by the Rankine-Hugoniot condition

$$s_{\alpha,\beta} = \frac{f(u_\alpha) - f(u_\beta)}{u_\alpha - u_\beta},$$

where u_α and u_β are the states at each side of the shock. The fronts, which are the lines of discontinuity of u, may thus be tracked in the $x - t$ plane.

By making better approximations of the initial flow functions, that is, using more intervals where f is linear, the solution converges. To apply this method not only to single Riemann problems, arbitrary initial data are approximated by step functions, thereby defining a number of Riemann problems, which are easily solved one by one. Provided that f is piecewise linear at a finite number of intervals, and there are a finite number of Riemann problems initially, it is shown [9] that the overall solution is determined by a finite number of states in the $x - t$ plane. Front tracking is also extended to problems with spatially discontinuous flux functions [7]. The latter is of great importance when applying front tracking to porous media flow, since such media are often heterogeneous, with abrupt (discontinuous) changes between different layers. In [7] the front tracking method is demonstrated to yield much more accurate results than finite difference methods for similar gridding resolution.

Extending the front tracking to two- and three-dimensional computations have been done by two different approaches. These are outlined in [3]. In this paper we will rely on the streamline simulator FRONTLINE. Pressure computations give streamlines (streamtubes), which are locally used as one-dimensional coordinate systems for the algorithms outlined above. In practice, this is done by mapping the two-dimensional fluid distribution onto a certain number of streamlines, solving Riemann problems along these, and propagating the solutions in space (along the streamlines) and time. The one-dimensional solutions are then used to regenerate the two-dimensional situation. Depending on the stationarity of the pressure and velocity field, these fields may be updated continuously during a simulation, or one may use the same streamlines for the whole simulation.

3. METHODS FOR THE PRESSURE AND VELOCTY

Here we present the conforming finite element method, the mixed finite element method and block-centered finite difference method for the pressure equation presented below. The equations for the pressure and velocity are given by

$$\nabla \cdot \mathbf{v} = q, \tag{3.1}$$

$$\mathbf{v} = -K\nabla p, \tag{3.2}$$

$$-\nabla \cdot (K\nabla p) = q. \tag{3.3}$$

Here $\mathbf{v}$ is the total velocity or Darcy velocity, p is the pressure, q the source term and K is the hydraulic conductivity.

In the conforming finite element method, p is approximated by a combination of simple basis functions. Thereafter $\mathbf{v}$ is obtained by applying Darcy's law for the total velocity to the approximation of p. The lowest order basis function ψ_i for p is piecewise linear throughout the domain. Thus, if the domain Ω is divided in a finite number of elements and N_n denotes the total number of nodes, then the pressure field is represented by

$$p = \sum_{i=1}^{N_n} p_i \psi_i,$$

where ψ_i is piecewise linear, 1 in node i and 0 in all other nodes, and p_i represents the pressure in node i.

The variational form of the pressure equation is expressed by weighting equation (3.3) with the ψ_i basis functions giving

$$\int_\Omega K\nabla p \cdot \nabla \psi_i d\mathbf{x} = \int_\Omega q\psi_i d\mathbf{x} - \int_{\partial\Omega} g_N \psi_i ds,$$

where g_N represents a prescribed flux on the boundary $\partial\Omega$.

Introducing the approximation for p yields a system of linear equations for the unknowns $\mathbf{P} = [p_1, \cdots, p_{N_n}]^T$, which can be written as

$$A\mathbf{P} = \mathbf{F},$$

where A is a square matrix of dimension N_n.

After this system has been solved, an approximation for $\mathbf{v}$ is obtained by applying Darcy's law to the approximation of p. As a consequence of numerical differentiation and the multiplication with the often rough tensor K, fluxes are discontinuous at element boundaries. Thus, mass will not be conserved.

In the mixed finite element method, p and $\mathbf{v}$ are approximated simultaneously. The lowest order basis functions for p and $\mathbf{v}$ applied here are as follows. The pressure is approximated via basis functions ψ_i which are piecewise constant throughout the domain. The total velocity is approximated via vector basis functions $\mathbf{v}_j$ whose normal components are required to be continuous across interelement boundaries. Thus, if the domain Ω is divided in a finite number of elements N_l, with a total number of edges N_e, then the pressure and the total velocity can be expressed by

$$p = \sum_{i=1}^{N_l} p_i \psi_i, \qquad \mathbf{v} = \sum_{j=1}^{N_e} u_j \mathbf{v}_j,$$

where $\psi_i = 1$ over element i and $\psi_i = 0$ elsewhere. The piecewise linear basis functions for the total velocity satisfy

$$\mathbf{v}_i \cdot \mathbf{n}_j \text{ is constant on } \mathbf{e}_j,$$

$$\int_{\mathbf{e}_j} \mathbf{v}_i \cdot \mathbf{n}_j ds = \delta_{ij},$$

where $\mathbf{n}_j$ is the normal to edge $\mathbf{e}_j$.

The variational form of the pressure equation is expressed by weighting equations (3.1)and (3.2) with the basis functions ψ_i and $\mathbf{v}_j$ giving

$$\int_\Omega (K^{-1}\mathbf{v}) \cdot \mathbf{v}_j dx - \int_\Omega p\nabla \cdot \mathbf{v}_j dx = -\int_{\partial\Omega} g_D \mathbf{n} \cdot \mathbf{v}_j ds,$$

$$-\int_\Omega \nabla \cdot \mathbf{v}\psi_i dx = -\int_\Omega q\psi_i dx,$$

where g_D represents a prescribed pressure on the boundary $\partial\Omega$.

Introducing the approximation for the pressure and the total velocity results in a system of linear equations for the unknowns $\mathbf{P} = [p_1, \cdots, p_{N_l}]^T$ and $\mathbf{U} = [u_1, \cdots, u_{N_e}]^T$:

$$\begin{pmatrix} A & B \\ B^T & 0 \end{pmatrix} \begin{pmatrix} \mathbf{U} \\ \mathbf{P} \end{pmatrix} = \begin{pmatrix} \mathbf{F}_1 \\ \mathbf{F}_2 \end{pmatrix},$$

where A is a square matrix of dimension N_e, B is a nonsquare matrix of dimension $N_e \times N_l$. The mixed finite element method results in a much larger system of linear equations, but computes a continuous total velocity field.

A continuous velocity field can also be obtained by the block-centered finite difference method resulting, however, in a smaller system of linear equations. In fact, it can be shown [4] that the mixed method for a rectangular mesh reduces to the standard five point difference scheme, when low-order quadrature formulas are used to calculate the integrals on each element. An interesting consequence of this is a simple way of associating to any finite difference solution a velocity field of mixed type. The scheme for the block-centered finite difference method is obtained via first order Taylor expansion of the pressure equation giving

$$\frac{1}{\Delta x}\{K_{i+1/2,j}\frac{p_{i+1,j} - p_{i,j}}{\Delta x} - K_{i-1/2,j}\frac{p_{i,j} - p_{i-1,j}}{\Delta x}\}+$$

$$\frac{1}{\Delta y}\{K_{i,j+1/2}\frac{p_{i,j+1} - p_{i,j}}{\Delta y} - K_{i,j-1/2}\frac{p_{i,j} - p_{i,j-1}}{\Delta y}\} = q_{ij}, \tag{3.4}$$

where $p_{i,j}$ is the pressure, Δx and Δy are the block widths in x and y direction, and subscripts i and j are indices for the column (x) and row (y) respectively, of the finite difference grid. The notation $i + \frac{1}{2}$ refers to the location of the block interface between column i and $i + 1$. Solution of (3.4) yields pressure values at the block centers. Velocities at the block interfaces can now be given by

$$v_{i+1/2,j} = -K_{i+1/2,j} \frac{p_{i+1,j} - p_{i,j}}{\Delta x}, \qquad v_{i,j+1/2} = -K_{i,j+1/2} \frac{p_{i,j+1} - p_{i,j}}{\Delta y},$$

where $v_{i+1/2,j}$ represents a velocity in x-direction on vertical block interfaces and $v_{i,j+1/2}$ represents a velocity in y-direction on horizontal block interfaces.

First numerical example.

For illustration the methods discussed above are applied to simple test cases. Water and contaminant are assumed to be immiscible, for example, contaminant being a remainder of oil. In this example the aquifer is a horizontal square domain with impervious boundaries and a barrier of zero hydraulic conductivity, placed along one of the diagonals of the domain. Except for the barrier, the aquifer is homogeneous and isotropic with constant permeability of $1000mD$. The computational area is $100 \times 100 \times 5$ meters, and the numerical grid is $44 \times 44 \times 1$. The porosity is 0.20 throughout the area. The phase densities are set to $63, 8kg/m^3$ and $55, 7kg/m^3$ for water and contaminant respectively, and the viscosities are $1cp$. We used linear relative permeability curves from 0 to 1.

A contaminant is injected at the lower left corner of the aquifer ($2000m^3/yr$), and water is being produced at the same rate at the upper right. The applied velocity field is based on the pressures calculated by the conforming method and the fluxes calculated by the mixed method, both with triangular elements. The diagonals defining the triangular elements are directed parallel to the barrier. For this case we refined the grid in the vicinity of the boundary in order to obtain a more accurate resolution of the velocity field around the endpoints of the barrier. The refinement is achieved by splitting the two columns or rows of blocks next to the boundary, each in two columns or rows of blocks.

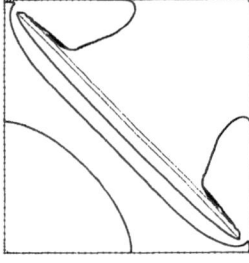

Figure 1.1a:

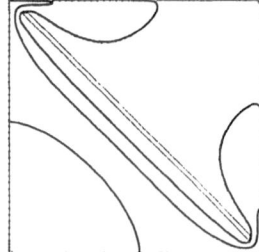

Figure 1.1b:

Figures 1.1a and 1.1b show the distribution of contaminant in the aquifer at three different times for respectivily the conforming method (a) and the mixed method (b).

At the third timestep the contaminated area differs significantly for the two methods. For the conforming method the area is smeared out behind the barrier. This is mainly due to bad resolution of the velocity field at the narrow passage. For the mixed method, however, the area is smooth and narrow as it should be since there is no diffusion in front tracking.

Second numerical example.

In this example some contamination is released into a horizontal aquifer with a certain hydraulic gradient. The underlying aquifer flow is modelled by specifying the flux along the boundaries as described below:

The computational area is $200 \times 20 \times 2$ meters, and the numerical grid is $48 \times 20 \times 1$, i.e., a horizontal regular grid. The porosity is 0.25 throughout the area. The permeability is set to zero in the upper right corner (12×5 blocks), the strip of 48×5 blocks below the middle is a high-permeable zone, $K = 1mD$ and the rest has $K = 0.1mD$. Since the problem is horizontal, the phase densities are irrelevant (both is set to $1000kg/m^3$), and the viscosities are $1cp$. We used linear relative permeability curves from 0 to 1.

The upper and lower boundaries are non-permeable, except for a contamination source in block $(36, 20)$ with rate $0.58m^3/day$, and a 'producing well' at block $(12, 1)$ with the sam rate. This latter well may model the drag in the vicinity of some fresh water pumping station. The flow at the left and right bounderies are set to $100m^3/day$, which is distributed so that 70% goes into the high permeable zone.

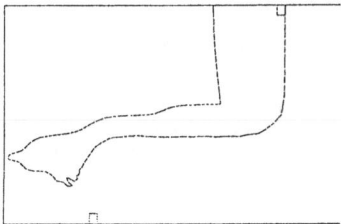
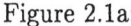
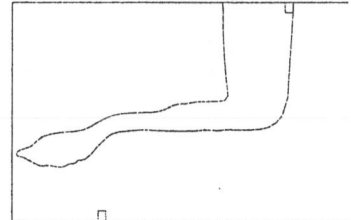

Figure 2.1a: Figure 2.1b:

We tested this case with both the conforming and the mixed method. The distribution of contaminant after 3 years is shown in Figure 2.1a (conforming) and 2.1b (mixed). Note that, particularly close to the producing well, the profile is significantly different for the two methods. In case of realistic modelling this is also the most important area for obtaining the best results, since this may determine future fresh water supplies and initiate protective action and contaminant cleanup.

By increasing the production rate, we could end up with cases where contamination breaks through at the producing site for one method but not for the other. In either case, it is obvious that one should try to model the problem with the highest order possible, and one should in general be careful when interpreting simulation results, particularly close to perturbations in an otherwise relatively homogeneous flow.

Acknowledgement. T. Gimse is supported by the Norwegian Research Council.

REFERENCES

[1] K. Aziz and A. Settari, *Petroleum Reservoir Simulation*, Elsevier, London (1979).

[2] J. Bear and A. Verruijt, *Modeling Groundwater Flow and Pollution*, D.Reidel, Dordrecht (1987).

[3] F. Bratvedt, K. Bratvedt, C.F. Buchholz, T. Gimse, H. Holden, L. Holden, and N.H. Risebro, *FRONTLINE and FRONTSIM. Two Full Scale, Two-Phase, Black Oil Reservoir Simulators Based on Front Tracking*, Surv. Math. Ind., **3** (1993), 185-215.

[4] G. Chavent and J.E. Roberts, *A unified physical presentation of mixed, mixed-hybrid finite elements and standard finite difference approximations for determination of velocities in waterflow problems*, Water Res. Res., **14** (1991), 329-348.

[5] P.G. Ciarlet, *The Finite Element Method for Elliptic Problems*, North-Holland, Amsterdam (1978).

[6] C.M. Dafermos, *Polygonal Approximations of Solutions of the Initial Value Problem for a Conservation Law*, J. Math. Anal. Appl., **38** (1972), 33-41.

[7] T. Gimse and N.H. Risebro, *A Note on Reservoir Simulation for Heterogeneous Porous Media*, Transport in Porous Media, **10** (1993), 257-270.

[8] J. Glimm, E. Isaacson, D. Marchesin, and O. McBryan, *Front Tracking for Hyperbolic Systems*, Adv. Appl. Math., **2**, (1981), 91-119.

[9] H. Holden, L. Holden and R. Høegh-Krohn, *A Numerical Method for First Order Nonlinear Scalar Conservation Laws in One Dimension*, Comput. Math. Applic., **15** (1988), 595-602.

[10] E.F. Kaasschieter and A.J.M. Huijben, *Mixed-hybrid finite elements and streamline computation for the potential flow problem*, Num. Meth. for Part. Diff. Equations **8** (1992), pp. 221-266.

[11] R.D. Richtmyer, *Difference Methods for Initial Value Problems*, Interscience Publ., New York (1957).

[12] J. Smoller, *Shock Waves and Reaction-Diffusion Equations*, Springer, New York (1983).

ANALYSIS OF POLLUTION CONTROL BY FINITE ELEMENTS

A. VERRUIJT and E. KODA
Delft University of Technology
Delft, The Netherlands
Warsaw Agricultural University
Warsaw, Poland

Transport of pollutants in groundwater can be analyzed by the finite element method, combined with a three-dimensional particle tracking technique. This method can be used to verify engineering solutions that have been proposed for the protection of urban areas.

DESCRIPTION OF THE PROBLEM

The soil at a certain site in a town in The Netherlands is heavily polluted, and protective measures are being carried out. These consist of a sheet pile wall around the polluted area, and a pumping well in that area. The water extracted from the

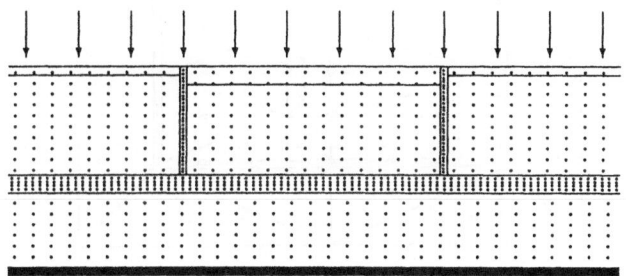

Figure 1: Cross section.

soil by the well will be cleaned in a special plant. The purpose of these measures is to prevent all outward flow from the area towards the surroundings, which can then be used for urban development.

A schematic cross section through the soil is shown in figure 1. The soil consists of two sandy aquifers, separated by a clay layer, of low permeability (an aquitard). The thickness of the lower aquifer is 40 m, and its transmissivity is 2000 m²/d. The thickness of the aquitard is 10 m, and its normal resistance is 2000 d. The thickness

A. Peters et al. (eds.), Computational Methods in Water Resources X, 1489–1495.
© 1994 *Kluwer Academic Publishers. Printed in the Netherlands.*

of the upper aquifer is 60 m, and its transmissivity is 2750 m^2/d. The upper level
of the soil is located at a level of 5 m above NAP, which is the local datum (about
mean sea level). The polluted area is surrounded by a sheet pile wall. Because this
wall may not be completely impermeable, it should be represented by a wall having a
resistance of 1000 d. In the polluted area the groundwater level is lowered by means
of a pumping well.

The flow in the soil can be analyzed by a numerical model, such as the MULAT
model, which is a layered finite element model, in which the soil is represented by
a number of mainly horizontal aquifers, of variable thickness, with leakage through
aquitards between the aquifers (Verruijt & Swidzinski, 1993). The capacity of the
model depends upon the computer used; on a 80486 system with 16MByte RAM the
capacity is about 150000 triangular elements.

The flow is supposed to be such that in each aquifer Dupuit's assumption of a constant
groundwater head may be applied. The flow may be influenced by boundary conditions such as local wells, line drains, line sources, and impermeable walls. The mesh
of finite elements is generated by a mesh generator, on the basis of a mesh of large
triangular zones, in each of which the soil properties must be constant. Each zone is
later subdivided into a large number of small finite elements. The three-dimensional
transport of particles can be analyzed by using a technique (Strack, 1984) to derive
the vertical component of the flow in the aquifers from the continuity equation and
the boundary conditions.

The geometry of the zones in a horizontal plane is shown, schematically, in figure 2.
The size of the total area considered is 600 m × 400 m, and the size of the polluted

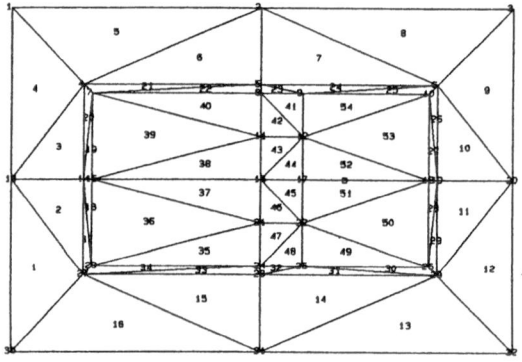

Figure 2: Geometry.

area is 400 m × 200 m. The thickness of the resistance walls is supposed to be 1 m
(in the figure this is exaggerated, for clarity).

In the zones 41, 42, 45 and 46 the resistance of the aquitard is supposed to be 100 d,
indicating that in these locations the resistance of the aquitard is 20 times smaller
than elsewhere, as field evidence indicates that locally the resistance of the clay layer
is much smaller than the average.

The well is located on the boundary of zones 51 and 52, at a distance of 100 m from
the center. The discharge of the well is 240 m^3/d. Over the entire area a uniform

infiltration (due to rainfall) occurs, of magnitude 0.000791 m/d. The natural flow is such that in the absence of the sheet pile walls and the well, there is a flow in the upper aquifer with a gradient of 1:4660, in a direction making an angle of 16° with the horizontal direction in figure 2, and such that at a point with coordinates $x = 229$ m, $y = 186$ m, the groundwater head is 0.11 m. In the lower aquifer the natural flow makes an angle of 64° with the horizontal direction, its gradient is 1:2040, and the head at the point $x = 229$ m, $y = 186$ m is -0.07 m.

The MULAT model does not have the possibility of entering a natural flow by specifying the gradient and its direction. It does have the facility, however, of specifying variable groundwater heads along the boundaries by introducing line drains. For each line drain the groundwater head at the two ends of the drain must be specified. The program itself then calculates a linearly varying given head in all mesh points along the line drain. In this way, using 8 line drains along the entire outer boundary, in both layers, the natural flow can be simulated. It must be mentioned, however, that this is only the case if the program uses double precision accuracy in its calculations, because the gradients are so small, and the size of the smallest elements is very small, especially in the walls of low permeability.

PRESENTATION OF RESULTS

In this section the results of the numerical calculations are presented. The mesh of finite elements has been generated by subdividing each zonal boundary into 20 parts, and then refining each triangular element into two equal smaller elements. The elements in the zones numbered 51 and 52 (see figure 2) have been refined once more, in order to describe the flow near the well better. The elements around the

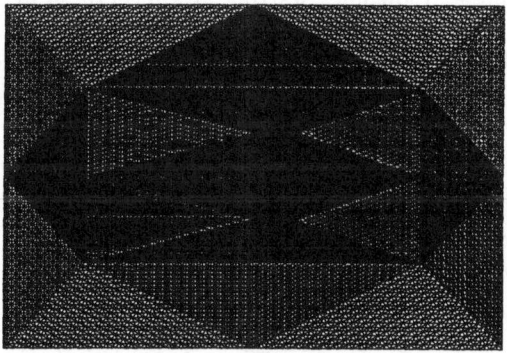

Figure 3: Mesh of finite elements.

well have been refined in four steps. This resulted into a mesh of 22720 nodes and 45140 elements, see figure 3. Computation time for the generation of the mesh and the finite element analysis is about 2 hours on a 80486 system.

The MULAT program has various possibilities for output presentation, in the form
of tables on the screen or on the printer, or in the form of two-dimensional or three-
dimensional graphs. A convenient way of presenting the output data for this case is
by two-dimensional views of the flow lines, of particles starting their motion along
the boundary of the area considered, at various depths. The time step used to draw
the flow lines was 10 d, and after each 500 time steps a mark has been set. This
means that the distance between two marks indicates a travel time of 5000 days.

The flow lines are shown in figures 4, 5 and 6. In these figures the flow lines in the
lower aquifer (layer 2, from 0 m to 40 m) are shown by thick lines. The flow lines
in the upper aquifer (layer 1, from 50 m to 110 m) are shown by thin lines. From
figure 4 it appears that some of the flow lines starting 39 m above the impermeable
base remain in the lower aquifer, but 12 flow lines are attracted towards the well in
the upper aquifer. Actually, these flow lines indicate the path of particles starting just

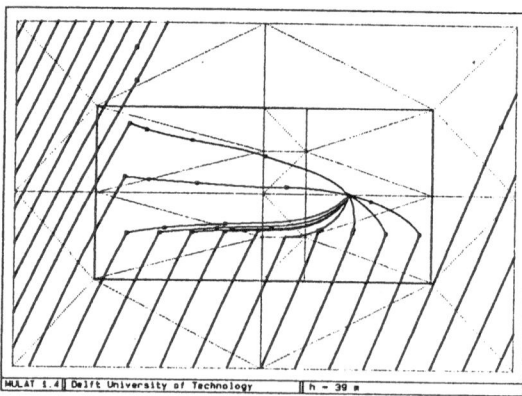

Figure 4: Flow lines starting at a level of 39 m.

below the clay layer, because the lower aquifer is 40 m thick. Some of the flow lines

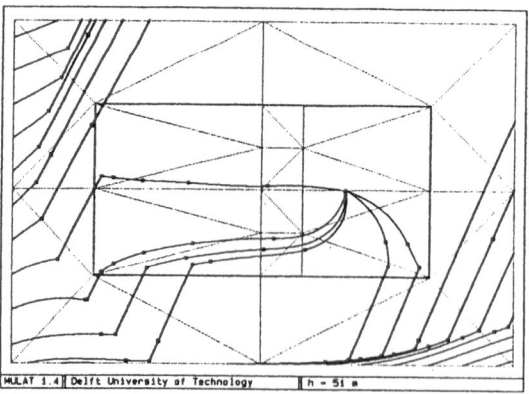

Figure 5: Flow lines starting at a level of 51 m.

of the particles starting just above the clay layer, at a height of 51 m, are particularly interesting, see figure 5. They all pass through the clay layer into the lower aquifer (in which the natural head is somewhat lower), but 6 of them, which travel below the polluted area, pass through the clay layer a second time, and then flow towards the well. The flow lines starting at a height of 60 m (see figure 6) mostly stay in the

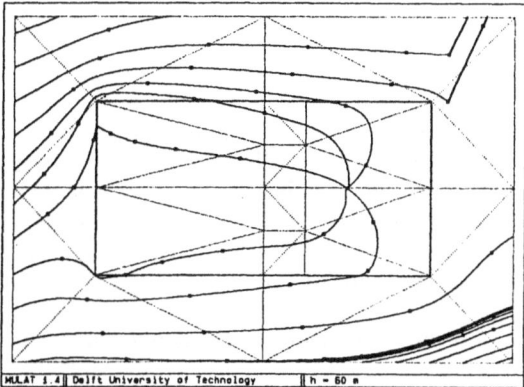

Figure 6: Flow lines starting at a level of 60 m.

upper aquifer, with 5 of them passing into the polluted area, ending up in the well. In order to demonstrate more clearly the effectiveness of the well, the flow lines for particles starting just above the clay layer, and just inside the sheet pile walls, at a level of precisely 50 m, that is the bottom of the upper aquifer, are shown in figure 7, for a time span of 7300 days (20 years). When these flow lines are followed further

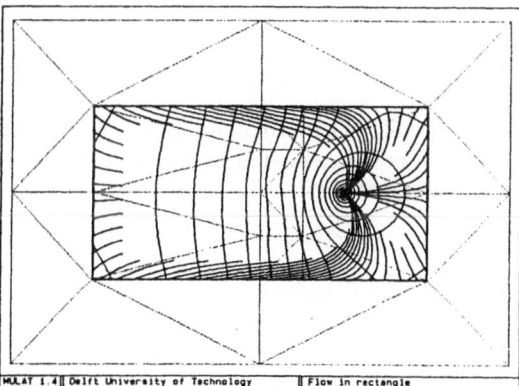

Figure 7: Flow lines inside the sheet pile walls.

in time, they all end up in the well, and none of them pass through the clay layer into the lower aquifer, suggesting that everywhere inside the sheet pile walls the flow through the clay layer is in upward direction, as it should be. This is an important result, as it clearly indicates the effectiveness of the technical measures taken to prevent outward flow. The flow through the wall appears to be in inward direction all along the wall, which is a design requirement. Figure 7 also shows contour lines of the groundwater head in the upper aquifer, using intervals of 0.005 m. The flow lines are perpendicular to these potential lines, of course.

WATER BALANCE

The water balance of the polluted area can be investigated by using another output facility of the program, which is to print the leakage through the clay layer in all elements, or, what is more realistic, in all zones. It appears from these numerical data (not given here) that in all zones inside the sheet pile walls (zones 35 through 54, see figure 2) the leakage into layer 1 is positive, indicating upward flow through the clay layer. It is also observed that the leakage is particularly large in zones 41, 42, 45 and 46, where the resistance of the clay layers is 20 times lower than in the other zones.

The total infiltration into the area inside the sheet pile walls is found to be, from the output data,

$$I = 63.282 \text{ m}^3/\text{d}.$$

Another way to calculate this value is to multiply the value of the infiltration rate (0.000791 m/d) by the total area, i.e. 80000 m². This gives $I = 63.820$ m³/d.

The extraction of the well is given to be

$$Q = 240 \text{ m}^3/\text{d}.$$

The total leakage into the area inside the sheet pile walls can also be determined from the output data,

$$L = 93.114 \text{ m}^3/\text{d}.$$

It is interesting to note that 52.799 m³/d flows through the clay layer in the four zones in which the resistance is reduced. This is 56 % of the total leakage.

It now follows, from the water balance,

$$Q = I + L + F,$$

that the flow through the sheet pile walls is

$$F = 83.604 \text{ m}^3/\text{d}.$$

It appears that the leakage through the sheet pile walls is of the same order of magnitude as the leakage through the clay layer.

CONCLUSIONS

It has been shown that many aspects of the problem can be analyzed by a layered quasi-three-dimensional finite element model, such as MULAT. Using the various output facilities of the program it can be verified that everywhere along the boundary there is a flow into the polluted area, as required. The flow lines and the residence times give a good insight into the flow.

ACKNOWLEDGMENT

The study' reported here was performed under contract no. ERB-CIPA-CT-92-2281 between the European Economic Community and the Delft University of Technology.

REFERENCES

O.D.L. Strack (1984) "Three-dimensional streamlines in Dupuit-Forchheimer models", Water Resources Research, 20, 812-822.
A. Verruijt and W. Swidzinski (1993) "Advective transport in a multilayered system of aquifers", Transport in Porous Media, 12, 31-42.

18. PARALLEL METHODS

A PARALLEL MULTIPHASE NUMERICAL MODEL FOR SUBSURFACE CONTAMINANT TRANSPORT WITH BIODEGRADATION KINETICS

TODD ARBOGAST, CLINT N. DAWSON, and MARY F. WHEELER

Department of Computational and Applied Mathematics
Rice University
Houston, Texas 77251–1892
U.S.A.

ABSTRACT

We discuss the formulation of a simulator in three spatial dimensions for two phase groundwater flow and transport with biodegradation kinetics that has been developed at Rice University for massively parallel, distributed memory, message passing machines. The numerical procedures employed are a fully implicit mixed finite element method for flow and a characteristics-mixed method for transport and reactions of dissolved chemical species in groundwater. Domain decomposition solvers have been employed for solving the systems of equations resulting from the discretization of the model. Results from applying this simulator to a bioremediation field problem using a recirculation well in an air-water system are discussed.

1. INTRODUCTION

Microbial biodegradation is an innovative, emerging technology for handling subsurface water contamination [4,6,7,8,9,11,13,14,15,17]. It is a natural process that can be accelerated by the injection of certain nutrients such as dissolved oxygen, nitrates, and acetate. U.S. Environmental Protection Agency studies [15] have shown that this strategy can result in complete removal of contaminants, whereas other proposed restoration strategies have not proven as effective. Biodegradation technologies are being employed at several U.S. Department of Energy Laboratories in an effort to remove or contain volatile organic compounds.

Biological decontamination is physically and chemically complex. It involves flow and transport in both unsaturated soils and the aquifer, and the interaction of hydrocarbons, microbes, oxygen, nitrogen, and various other chemical compounds. Numerical simulation of these processes is a critical step in understanding and designing biorestoration applications [4,6,9,17]. Indeed, without computational science, wide scale in situ biodegradation of contaminants is impractical.

New parallel supercomputers, allowing simultaneous use of hundreds to thou-

A. Peters et al. (eds.), Computational Methods in Water Resources X, 1499–1506.

sands of processors, have greatly expanded the potential for building detailed models of these processes. Parallel computing provides the capability of solving larger, more realistic and practical problems faster and more economically. This includes the ability to use an adequately refined discretization mesh, to incorporate complex chemical and physical effects associated with the transport of both hydrocarbons and organic contaminants in porous media, and to employ stochastic or conditional simulation. The latter is essential for simulating a realistic geologic aquifer, since much of the data needed to characterize it cannot be quantified accurately, and since often the chemical and physical processes are not well understood.

Herein we emphasize the modeling of flow and transport for a two phase system, water and air. The development of our multiphase, three spatial dimension code, RPGW/MP, has involved combining and modifying two codes: RPGW, Rice Parallel Groundwater Code for transport and reactions [1,4,2], and PIERS, Parallel Implicit Experimental Reservoir Simulator for flow [16]. RPGW/MP is under development and future generalizations are discussed at the end of the paper.

The outline of the paper is as follows. In §2 we describe the governing flow and transport equations with biodegradation in an unsaturated/saturated porous medium. For simplicity, we assume linear sorption and aerobic conditions. More general kinetics such as Michaelis-Menton can be treated with the numerical techniques described in this paper. In §3, we describe the parallel implementation of the model, and in §4 we present three dimensional, parallel, bioremediation simulation results for a recirculation well problem. Conclusions and current directions on parallel implementation are given in §5.

2. TWO PHASE FLOW AND CONTAMINANT TRANSPORT WITH BIODEGRADATION

We first present the two phase flow model. It is very similar to the well known black oil model from petroleum engineering and the formulation presented by Parker [12]. The coupled equations are:

Water Phase
$$\frac{\partial(\phi \rho_w s_w)}{\partial t} + \nabla \cdot (\rho_w u_w) = Q_w + \gamma_w; \tag{1}$$

Air Phase
$$\frac{\partial(\phi \rho_a s_a)}{\partial t} + \nabla \cdot (\rho_a u_a) = Q_a + \gamma_a; \tag{2}$$

Equations of State
$$\rho_w = \rho_w^0 e^{c_w p_w}, \quad \rho_a = \rho_a^0 e^{c_a p_a}; \tag{3}$$

Darcy's Law
$$u_w = -\frac{K k_{rw}(s_w)}{\mu_w}(\nabla p_w - \rho_w g \nabla z), \tag{4}$$

$$u_a = -\frac{K k_{ra}(s_a)}{\mu_a}(\nabla p_a - \rho_a g \nabla z); \tag{5}$$

Capillary Pressure
$$p_c(s_w) = p_a - p_w; \tag{6}$$

Volume Balance $\qquad\qquad s_w + s_a = 1.$ $\qquad\qquad\qquad\qquad$ (7)

Here ϕ is porosity, p phase pressure, ρ phase density, K absolute permeability, k phase relative permeability, μ phase viscosity, s phase saturation, c phase compressibility, g gravitational constant, z depth, Q an external phase source or sink, and γ are source or sink terms due to mass transfer between phases (subscripts have been omitted for simplicity).

Multicomponent transport and biodegradation are governed by a system of advection-diffusion-reaction equations consisting of m_s electron donors (substrates) and m_n electron acceptors or nutrients, and a system of m_x ordinary differential equations involving microbial mass (transport of microbes can be treated also if one assumes instead a system of advection-diffusion-reaction equations for the microbes). They can be written in terms of the concentration dissolved in water, $C_i = C_i^w$, as:

Electron Donor (Substrate)

$$\frac{\partial(\phi_i C_i)}{\partial t} - \nabla \cdot (D_i \nabla C_i - u_i C_i) = \phi \chi_i + g_i, \quad i = 1, \ldots, m_s; \qquad (8)$$

Electron Acceptor (Nutrient)

$$\frac{\partial(\phi_i C_i)}{\partial t} - \nabla \cdot (D_i \nabla C_i - u_i C_i) = \phi \chi_i + g_i, \quad i = m_s + 1, \ldots, m_s + m_n; \qquad (9)$$

Microbial Mass $\qquad \dfrac{\partial(\phi C_i)}{\partial t} = \phi \chi_i, \quad i = m_s + m_n + 1, \ldots, m_s + m_n + m_x. \quad (10)$

Since we assume that mass transfer between phases is based on equilibrium partitioning among the phases, we have

Equilibrium Phase Partitioning

$$C_i^a = \Gamma_{ia} C_i^w, \quad C_i^s = \Gamma_{is} C_i^w \quad i = 1, \ldots, m_s + m_n, \qquad (11)$$

where Γ_{ia} and Γ_{is} are the equilibrium phase partitioning constants between an air/water system and a soil/water system, respectively, for component i. Here we define

$$\phi_i = \phi(s_w + s_a \Gamma_{ia}) + \Gamma_{is}, \quad u_i = u_w + u_a \Gamma_{ia},$$
$$D_i = \phi(s_w D_{iw} + s_a \Gamma_{ia} D_{ia}), \qquad\qquad (12)$$

where $D_{iw}(u_w)$ is the hydrodynamic diffusion/dispersion tensor, and $D_{ia}(u_a)$ is defined similarly for the air phase. The χ_i are possibly nonlinear kinetic terms which account for biodegradation of contaminants, utilization of nutrients, and growth and decay of microorganisms. The number and complexity of specific metabolic pathways or chemical reactions varies with the application. The source/sink terms g_i represent production and injection wells.

3. PARALLEL IMPLEMENTATION

Our two-phase flow and transport code, as stated above, involves the coupling of two subroutines, RPGW and a modified version of PIERS.

RPGW is a subroutine developed to simulate the transport and reactions of dissolved chemical species in the groundwater. This fully parallel code uses the characteristics-mixed method; that is, it combines characteristics and the mixed finite element method with operator splitting. The code handles an arbitrary number of component chemical species, as well as microbial mass and radionuclide decay. The code treats an arbitrary number of phases including the solid phase (adsorption). Each component is dissolved in one or more of these phases. The distribution of mass in the phases is assumed to follow the linear Raoult's or Henry's Law. This code achieves almost linear parallel scaling [4,2]; thus, it is highly effective when run on a parallel machine. Details regarding the formulation and analysis of this procedure can be found in [3] and application to contaminant transport in single phase groundwater flow in [1] and [4].

PIERS, a fully implicit two phase flow code with fully coupled wells, was originally developed at Exxon Production Research by J. Wheeler and Smith for the INTEL iPSC/2 [16]. Capillary pressure and relative permeability are functions of water saturation and formation type. Functional forms can be defined by tables or by definition as piecewise C^2 splines. Nonlinearities are treated implicitly by Newtonian iteration. Rice University, in cooperation with J. Flower of Parascope, ported this code to a collection of parallel machines. Scaling studies have been carried out in [16,10].

We now describe briefly how the combined code solves the model (1)–(12). PIERS first approximates (1)–(7) using a fully implicit method that is finite difference in time. In space it is a cell-centered finite difference method, or equivalently a mixed finite element method with the approximating space RT_0 and a special quadrature rule. The system (1)–(7) can be rewritten so that for $t = t^n$ one need only solve for the primary unknowns $P_{w,h}^n$ and $S_{w,h}^n$. The resulting nonlinear system is solved by a parallel domain decomposition Newtonian iteration that involves a nonsymmetric linearization, a three level multigrid solver for approximating the aerial domain, and SOR in the vertical direction. The linear solver is very sensitive to the selection of the SOR acceleration parameter and to the choice of convergence parameters. Even so this code has been employed to solve over one million nonlinear equations each time step. Implicit upstream weighting on the phase transmissibilities, $\lambda = k_r \rho / \mu$, is an option in the code and is generally employed.

Given the saturations and phase velocities, the subroutine RPGW is called. Here the advection-diffusion-reaction system (8)–(12) involving donor, acceptor, and biological mass equations are approximated using a time splitting scheme. One global time step of RPGW or the transport code involves the following three sequential steps:

(A) Pure transport. For each electron donor or acceptor, characteristics are traced backwards in time to locate their origin at the previous time level. This may be done by taking small micro time steps. This solves (8)–(9) without the reaction terms χ_i and the dispersion terms D_i.

(B) Reactions. The coupled system of reaction equations (i.e. (8)–(10) without the two divergence terms and without the g_i source terms) are approximated using a fourth order Runge-Kutta procedure. Initial conditions are the cell averages from (A) for acceptors and donors, and the previous time step concentrations for the microbes. Many small time steps may be taken to improve the accuracy.

(C) Diffusion and dispersion. The diffusion/dispersion step involves approximating a parabolic system for each donor or acceptor using initial data from (B) and applying the mixed finite element method, again implemented as a cell-centered finite difference method. A tensor product trapezoidal rule is used in treating the diffusion/dispersion term. A finite stencil is obtained for each component, nine points in two dimensions and nineteen in three. The discrete system is solved using a Jacobi preconditioned conjugate gradient algorithm. Details may be found in [1,4,5].

After having completed the transport and reactive step, mass transfer source and sink terms γ_p are computed. The time step is then incremented and the flow subroutine is called to obtain new saturations and phase velocities.

4. SOME BIOREMEDIATION RESULTS

The Hanford Site in Washington State occupies approximately 560 square miles of semiarid terrain and was selected in 1943 for producing materials (primarily plutonium) in support of the United States' World War II efforts. Chemical processes employed to recover and purify plutonium produced waste containing actinide compounds and typical aqueous and organic liquid industrial wastes. The primary organic contaminant carbon tetrachloride (CCl_4) totaled 637 to 1200 tons discharged. Today, plutonium production has ceased, and the primary mission has shifted to environmental restoration of the Hanford Site [7,8,11,13,14].

Rice University and Pacific Northwest Laboratory (PNL) began a collaborative research effort in 1992 that involves laboratory, field, and simulation work directed toward validating remediation strategies. We discuss below some preliminary computational results based on some recent microbial CCl_4 destructive kinetics developed by Skeen and Chan of PNL [14].

The model has six components: electron acceptors nitrate NO_3^-, nitrite NO_2^-, and acetate CH_3COO^-, CCl_4, microbial mass $C_5H_9O_3N$, and a nonreactive tracer. We also assume that the retardation factor for acetate is 1.8. The chemical reactions for this system are:

$$8NO_3^- + 2CH_3COO^- + 2H^+ \rightarrow 4CO_2 + 8NO_2^- + 4H_2O,$$
$$8NO_2^- + 3CH_3COO^- + 11H^+ \rightarrow 6CO_2 + 4N_2 + 10H_2O,$$
$$7CH_3COO^- + 2NO_3^- + 9H^+ \rightarrow 4CO_2 + 6H_2O + 2C_5H_9O_3N,$$
$$13CH_3COO^- + 4NO_2^- + 17H^+ \rightarrow 6CO_2 + 10H_2O + 4C_5H_9O_3N,$$

and bioremediation is described, for the two parameters μ and k_i, by

$$\frac{d(CCl_4)}{dt} = \frac{-\mu(CCl_4)(C_5H_9O_3N)}{1 + k_i((NO_3^-) + (NO_2^-))}.$$

The two figures below show simulation results at 3 days. We assumed that CCL_4 uniformly contaminates the medium, both in the vadose zone and in the aquifer, and that microbes uniformly populate the medium. Nitrate and acetate are injected into the system. Recirculation is approximately 50 gals/min; fluid is collected at the production part of the well (depth 205–213 feet) and reinjected at the injected part of the well (depth 171–187 feet).

Fig. 1 shows concentration fronts at 3 days. Observe that the microbial population grows in regions where nutrients are available, that acetate movement is retarded, and that nitrite takes some time to form by kinetic reaction. Observe also the drawing of the CCL_4 from the aquifer up to the vadose zone where a large microbial population has grown. At 3 days the CCL_4 has decreased by four orders of magnitude. Fig. 2 shows the water table and the concentration front of the tracer over the full computational domain at 3 days.

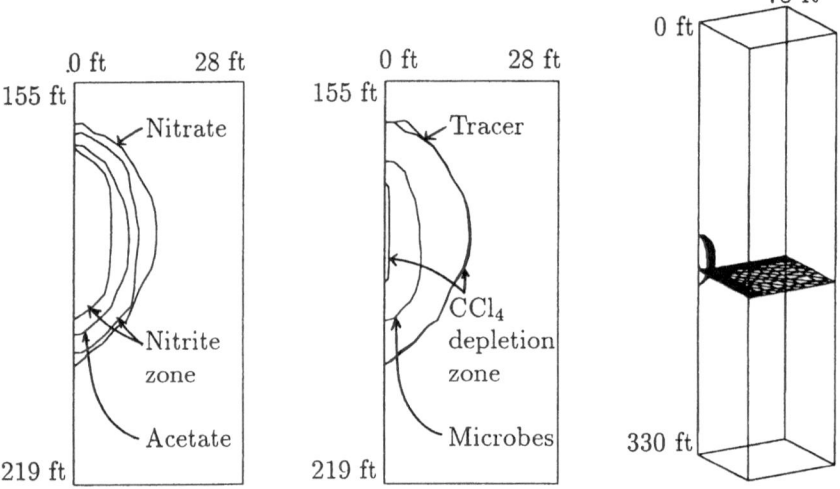

Fig. 1. Radial cross-sections
of the concentration fronts
of the six components

Fig. 2. Water table
and front of the
tracer concentration

5. CONCLUSIONS AND FUTURE DIRECTIONS

The Rice 3D Parallel Groundwater Reactive Multiphase Flow and Transport Simulator (RPGW/MP) is a parallel code under development at Rice University. Its purpose is to simulate the flow and transport of reacting chemical species in the groundwater. This code is based on combining locally conservative schemes: a mixed finite element method for flow with a characteristics-mixed finite element method for transport. Computational experiments indicate that this approach is useful in solving grand challenge problems such as bioremediation and that the code achieves good parallel scaling.

We are presently adding general boundary conditions to the code as well as modifying the code to treat three phases (air, non-aqueous phase liquid, and water). Future plans include the incorporation of meshes defined by fairly general geometry [5] and the development of robust, fully implicit multilevel solvers for modeling three phase flow. In addition we plan to add more chemistry and microbiology as well as the capability of simulating fractured media.

ACKNOWLEDGEMENTS

The authors wish to acknowledge the Department of Energy which has supported this contract through two subcontracts: Oak Ridge National Laboratory (ORNL) as part of the Partnership in Computational Science (PICS) Consortium and Pacific Northwest Laboratory (PNL). The authors would also like to acknowledge the help of Kyle Roberson and Brian Wood of PNL in the formulation of the bioremediation problem as well as Perry Cheng, Doug Moore, Joe Warren, and Mark Wells of Rice for their help in computer visualization. In addition we wish to acknowledge John A. Wheeler and Exxon Production Research in releasing PIERS to Rice University. Finally we wish to acknowledge the work of Ashokkumar Chilakapati, Philip Keenan, and Doug Moore in the development of this parallel code.

REFERENCES

[1] Arbogast, T., Chilakapati, A., and Wheeler, M.F. (1992) "A characteristic-mixed method for contaminant transport and miscible displacement", in Russell, Ewing, Brebbia, Gray, and Pindar (eds.), Computational Methods in Water Resources IX, Vol. 1: Numerical Methods in Water Resources, Computational Mechanics Publications, Southampton, U.K., pp. 77-84.

[2] Arbogast, T., Dawson, C., Moore, D., Saaf, F., San Soucie, C., Wheeler, M.F., and Yotov, I., (1993) "Validation of the PICS transport code", Technical Report, Department of Computational and Applied Mathematics, Rice University.

[3] Arbogast, T., and Wheeler, M.F. (in press) "A characteristics-mixed finite element method for advection dominated transport problems", SIAM J. Numerical Analysis.

[4] Arbogast, T., and Wheeler, M.F. (in press) "A parallel numerical model for subsurface contaminant transport with biodegradation kinetics", in Whiteman, J.R. (ed.), Proceedings of Mathematics of Finite Elements and Applications VIII (MAFELAP 1993).

[5] Arbogast, T., Wheeler, M.F., and Yotov, I. (1994) "Logically rectangular mixed

methods for groundwater flow and transport on general geometry", this Proceedings.

[6] Chiang, C.Y., Dawson, C.N. and Wheeler, M.F. (1991) "Modeling of *In-situ* biorestoration of organic compounds in groundwater", Transport in Porous Media 6, pp. 667–702.

[7] Evans, J.C., Bryce, R.W., Bates, D.J., and Kemner, M.L. (1990) "Hanford Site Ground-Water Surveillance for 1989", PNL-7396, Pacific Northwest Laboratory, Richland, Washington.

[8] Hagood, M.C. and Rohay, V.J. (1991) "200 West area carbon tetrachloride expedited response action project plan", WHC-SD-EN-AP-046, Westinghouse Hanford Company, Richland, Washington.

[9] Herrling, B. Stamm, J., and Buermann, W. (1991) "Hydraulic circulation system for in situ bioreclamation and/or in situ remediation of strippable contamination", in Hinchee, R.E., and Olfenbuttel, R.F. (eds.), In Situ Bioreclamation, Applications and Investigations for Hydrocarbon and Contaminated Site Remediation, Butterworth-Heinemann Pub., Boston, pp. 173–195.

[10] Keenan, P. T., and Flower, J. (1993) "PIERS Timings on Various Parallel Supercomputers", Dept. of Computational and Applied Mathematics Tech. Report #93–29, Rice University.

[11] Last, G.V., Lenhard, R.J., Bjornstad, B.N., Evans, J.C., Roberson, K.R., Spane, F.A., Amonette, J.E., and Rockhold, M.L. (1991) "Characteristics of the volatile organic compound-arid integrated demonstration site", PNL-7866, Pacific Northwest Laboratory, Richland, Washington.

[12] Parker, J.C. (1989) "Multiphase flow and transport in porous media", Reviews of Geophysics 27, pp. 311–328.

[13] Riley, R. G. (1993) "Arid site characterization and technology assessment: volatile organic compounds-arid integrated demonstration", PNL-8862, Batalle, Pacific Northwest Laboratory.

[14] Skeen, R.S., Roberson, K.R., Brouns, T.M., Petersen, J.N., and Shouche, M. (1992) "In-situ bioremediation of Hanford groundwater" in Proceedings of the 1st Federal Environmental Restoration Conference, Vienna, Virginia.

[15] Thomas, J.M., Lee, M.D., Bedient, P.B., Borden, R.C., Canter, L.W., and Ward, C.H. (1987) "Leaking underground storage tanks: remediation with emphasis on *in situ* biorestoration", U.S. EPA report 600/2-87, 008.

[16] Wheeler, J.A., and Smith, R. (1989) "Reservoir simulation on a hypercube", SPE 19804, in Proceedings of the 64th Annual Technical Conference and Exhibition, Society of Petroleum Engineers, Richardson, Texas.

[17] Wheeler, M.F., Dawson, C.N., Bedient, P.B., Chiang, C.Y., Borden, R.C., and Rifai, H.S. (1987) "Numerical simulation of microbial biodegradation of hydrocarbons in groundwater", in Proceedings of AGWSE/IGWMCH Conference on Solving Ground Water Problems with Models, National Water Wells Association, pp. 92–108.

IMPLEMENTING A TWO-DIMENSIONAL PORE-SCALE FLOW MODEL ON DIFFERENT PARALLEL MACHINES

D. BERNARD(*), F. BODIN(**), A. GOASGUEN(*,**)
and J.C. FECHANT(*)
(*) L.E.P.T.-ENSAM (URA 873), Esplanade des Arts et Métiers
33405 TALENCE Cédex FRANCE
(**) IRISA, Campus Universitaire de Beaulieu
35042 RENNES Cédex FRANCE

INTRODUCTION

Parallel computers are more and more considered by large computation users as attractive alternatives to classical computers. Nevertheless given the variety of parallel architectures and the relative ambiguity of the conclusions that can be drawn from the literature, it is often difficult for an end-user to determine the point where the effort necessary to adapt his programs to a given parallel computer will balance the benefit in CPU time.

To get an insight in this problem, we took advantage of a European students exchange program (ERASMUS) to evaluate the performances of different parallelization strategies for an existing FORTRAN program. This paper presents the results obtained during a short (5 months) real scale experience using different parallel machines (CM-2, CM-5, PVM clusters of IBM RS/6000, IPCS-2, KSR1).

OVERVIEW OF THE APPLICATION

The first version of the program used herein has been presented in the same conference two years ago [1] and some new results are detailed in an other paper of this conference [2]. Basically, it is a two-dimensional STOKES solver using an artificial compressibility algorithm with an *explicit time integration* scheme and *periodic boundary conditions*. A priori, those characteristics make this program a good candidate for parallelization. The only unfavorable aspect is the complexity of the fluid/solid interface where a no slip boundary condition is imposed (we use randomly generated pore scale geometries) [2].

Two different ways have been considered to overcome this difficulty:
1. compression of the grid eliminating the solid nodes (about 40% in 2D and 70% in 3D). This is done using indirect addressing of the active nodes neighbors. The cost induced is easily balanced by the operation number reduction. This version is called Permea in this paper.
2. penalization of the solid nodes by mask matrices. All the nodes are first treated equally and the computed solution matrix is multiplied by a boolean matrix built to impose the no slip boundary condition. This method doesn't take advantage of the low porosity of the grid but is much more simple and a priori very well adapted to direct use on vectorial and parallel machines. This version is called penal in this paper.

A modified version of penal has been developed using a domain decomposition strategy. The time integration being explicit, the exchange between domains is limited to the decomposition boundaries. This version has been only used with PVM [3].

A. Peters et al. (eds.), Computational Methods in Water Resources X, 1507–1514.
© 1994 *Kluwer Academic Publishers. Printed in the Netherlands.*

OVERVIEW OF THE ARCHITECTURES

In this section we briefly present the architectures used for evaluating the use of parallelism for our application. The experiments we carried out use a wide range of architecture going from SIMD (CM2), MIMD architectures (CM5, iPSC/2, KSR) and a cluster of workstation. This range of architecture span across three different paradigms for parallel programming: *message passing, parallel array operations and fork-and-join model with parallel do loops*. Having to deal with these different models is one of the most important problem when using parallel architectures; for each architecture the program has to be modified taking into account the programming model and the architecture for optimizing the performance. So most of the programming effort was devoted in dealing with the different architectures and code tuning.

The CM2 [4] is a SIMD architecture with distributed memories. The performance is up to 10 Gflops. It is composed mainly from 4K to 64 K bit slice PEs (Processor Elements) with a single control unit. Each PE has its own memory. The PEs are connected by an hypercube network where each nodes is a cluster of 32 PEs. The communication form supported are NEWS grid, routing or scanning operations. Each cluster of PE shares a floating point operator. The programming model is data parallel oriented. For the experiments we used CM-Fortran [5] which is a Fortran 77 with Fortran 90 parallel array operations. The data arrays are distributed among the PE so array operations can be executed synchronously in parallel. Performance mainly depends on the alignment between arrays participating to the same array operation. This architecture is particularly well suited for fine grain parallelism.

The CM5 is a MIMD computer with distributed memories [6]. It is composed of 32 to 16K processing nodes connected by a fat-tree network. Each node has a 32 Mhz Sparc processor, a 32 Mbytes of memory and a 128 Mflops vector processing unit. The network has special features for supporting efficiently the CM-Fortran programming model, so the same program was used for the experiment.

The KSR computer is an MIMD computer with a shared virtual memory (SVM) [7]. Each node of the KSR is a superscalar processor with a peak performance of 40 Mflops and 32 Megabytes of memory. Processors are connected by a hierarchy of rings. The level one ring (the bandwidth is 1 Gigabyte per second) connects 32 processors. The second level of ring connects rings of level one (the second level has a bandwidth of 4 Gigabytes per second). The main feature of the architecture is its SVM (called ALLCACHE^{TM}). The ALLCACHE memory is an implementation of a shared memory with physically distributed memories. Data moves on nodes according to the access. The memory is divided in pages and each page is divided in subpages of 128 bytes. If more than one processor read the same subpage the coherence protocol ensure that a copy of the subpage is allocated on all the processors making the read access. If one of the processor modify a subpage then all the copies are invalidated so a new access to the subpage load the most recent value. The programming model of the KSR is close to the PCF fortran standard which is based on thread and fork-and-join parallel loops.

The other architecture we have been using is an intel iPSC/2 which is a distributed memory MIMD computer [8]. A node is composed of an intel 386 with a peak performance of 0.3 Megaflops. We have been using this old parallel computer because we have implemented on it a shared virtual memory, named KOAN [9, 10, 11], similar to the one on the KSR. The main difference is that KOAN is embedded in the NX/2 operating system running on each node of the iPSC/2 (in the case of the KSR the SVM is implemented mostly by the hardware) and that the consistency is ensure at the level of a page of the virtual memory (the size of a page is 4Kbytes). The programming language, called Fortran-S [12] for the iPSC/2 is similar to the one used on the KSR at the user level. They differ mainly because on the iPSC/2 the execution is not based on threads but on an SPMD execution model (Single Program Multiple Data).

Performance with SVM depends mostly on three parameters: locality of access, number of barriers and a phenomenon called false-sharing. The locality of the access is crucial to SVM to reduce communication. Performance comes with an SVM because when a page is loaded on a processor it remains on the processors (so all further access to the page are local and so there is no communication) until it is invalidated or swapped on disk or on another processor. It is also important to decrease the number of barriers that synchronize the processors because a global synchronization is an expensive parallel operations but also it tends to maximize the idle time of the processors. A processor can proceed only if all the other processors are done, so there is no averaging of the work load balancing. False-sharing appears when more than one processor write at the same moment to distinct data stored in the same page. Because of the consistency protocol the page may moves back and forth between the processors writing to the page, degenerating in a phenomenon called ping-pong [13]. At worst, each write access may cost a page fault.

The last architecture kind is a cluster of IBM RS/6000 connected by an ethernet network [14]. The programming model used has been PVM [3] which implement message passing. From all the other computers used this is the lowest level of parallel programming used during the study. With the model data distribution, work distribution and data exchange are all in charge of the user.

IMPLEMENTATION OF THE CODE

In this section we gives an overview of the three parallel implementations of the program Penal. Permea was only implemented on SVM because the parallelization of the code was simple and similar to Penal, with the expectation of getting good performance.

Shared Virtual Memory

In the case of SVM the parallelization of the two programs, starting with the sequential code, was easy and similar for the two codes. Figure 1 shows how the loop computing the speed in X has been parallelized.

```
C$ann[DoShared("BLOCK")]                 C$ann[DoShared("BLOCK")]
C$ann[SGlobal(DMAX,errx)]                C$ann[SGlobal(DMAX,errx)]
    DO J = 1,NN                              DO I = 1,ITABX
      DO I = 1,NN                              TMP =C0*VXO(I)+DTY2*(VXO(IXVOIS(I,1))
      TMP = C0*VXO(I,J)+DTY2*(VXO(I-1,J)  &      +VXO(IXVOIS(I,2)))
  &        + VXO(I+1,J))+DTX2*(VXO(I,J+1) &      +DTX2*(VXO(IXVOIS(I,3))
  &        +VXO(I,J-1))-DTX*(PO(I,J)      &      +VXO(IXVOIS(I,4)))
  &        -PO(I,J-1))-C1                 &      -DTX*(PO(IXVOIS(I,6))
      TMP = TMP * IVX(I,J)                &      -PO(IXVOIS(I,5)))-C1
      ERRX = MAX (ERRX,ABS (TMP -VXO(I,J)))    ERRX = MAX(ERRX,ABS(TMP-VXO(I)))
      VXN(I,J) = TMP                         VXN(I) = TMP
      ENDDO                                  ENDDO
    ENDDO
```

Figure 1: Loops for computing speed for the X axis (left for Penal, right for Permea)}

However the performance were low. So most of the parallelizing process has been spent in optimizing the two programs. The codes were first written into Fortran-S and then the optimized version translated to the KSR. The optimizations done where:

1. loops fusion to reduce the number of synchronization: In the case of Penal the loop computing the speed in the X and Y axis can be fused so only one parallel loop is needed. This has not been applied on Permea since the array storing the speed in X is not of the same size than the array storing the speed in Y.
2. array padding to reduce false sharing: In the case of Penal array padding has been used to reduce false sharing. The leading dimension of the 2D arrays has been set so that a block of data computed by a processor stop on a page boundary (This optimization is implemented by the compiler).
3. data locality improvement: In the case of Penal, locality has been optimized mainly for the computation of the boundary of the grid. The loops for computing the boundary are shown below:

```
C$ann[BeginSeq()]
    DO I = 1,NN
      VXO(I,0) = VXN(I,NN)
      VXO(I,NN+1) = VXN(I,1)
      VYO(I,0) = VYN(I,NN)
      VYO(I,NN+1) = VYN(I,1)
    ENDDO
C$ann[EndSeq()]
    ......
C$ann[DoShared("BLOCK")]
    DO J = 1,NN
      VXO(0,J) = VXN(NN,J)
      VXO(NN+1,J) = VXN(1,J)
      VYO(0,J) = VYN(NN,J)
      VYO(NN+1,J) = VYN(1,J)
    ENDDO
```

The first loop is executed sequentially since the vectors **VXO(*,0), VXO(*,NN+1), ...** are stored on a single page (for the size of problem we consider). Executing this loop in parallel would have create a lot of false sharing since all the processors would try to write simultaneously to the pages storing the vectors. The next loop has to be parallel because the vectors **VXO(0,*), ...)** span across many pages. Executing this loop sequentially would have made all the pages storing the arrays to migrate on processor 0.

In the case of Permea locality was optimized by modifying the way the speed in the Y axis was stored. The initial storing had good locality property for the loop computing the Y speed but when using these data to compute the pressure the data where reshuffle across the machine. This optimization was the most difficult one since it necessitates an intimate knowledge of the application.

CM-Fortran

Only Penal was implemented in CM-Fortran, Permea being unfitted to the CM2 and CM5 because of the compressed grid. The Penal algorithm was easily translated in CM-Fortran by replacing the loops by array operations. The statements corresponding to loop are shown Figure 2.

Since the arrays are aligned data movement between processors are only shift operations and reductions (**MAXVAL()**) which are well suited to the CM2 and CM5 architectures.

CMF\$ LAYOUT VXO(:NEWS,:NEWS) // directive for array distribution

```
....
VXO = VXN   // Copy array VXN in array VXO,
VXN = VXO*C0
&    +DTV2*(CSHIFT(VXO,1,1)+CSHIFT(VXO,1,-1))
&    +DTV2*(CSHIFT(VXO,2,1)+CSHIFT(VXO,2,-1))
&    -DTV*(P-CSHIFT(P,2,-1))
&    -C1
VXN = VXN*IVX
EX = MAXVAL(ABS(VXN-VXO))
ERRX = MAX(ERRX,EX)
```

Figure 2: Code segment showing modified loop.

Pvm

The PVM implementation of Penal was the most difficult one. The arrays has been explicitly distributed by block between the processors. The main difficulty has been to organize the data transfer using the message passing primitive of PVM. Also, not only the computation intensive part of the program has to be modified but also the I/O part. As a consequence the PVM implementation of Penal needed a complete rewrite of the application. The main optimization done here was to gather the data to send in one message to avoid the high latency of the network whenever this was possible.

PERFORMANCE RESULTS

The tests presented in this paper have been done in two different contexts: in the CRS4 (Centre for Advanced Studies Research and Development in Sardinia, Cagliari, Italy) we utilized the Connection machines and the PVM cluster as "normal" users, i.e. without control of the system charge, whether at IRISA, we had enough priority to control those aspects. That explain the different ways the performance results are presented.

Connection Machines

| | CM5 | | | | CM2 | |
| | 32 | | 64 | | 8 k | |
Size	Time (s)	% //	Time (s)	% //	Time (s)	% //
80 x 80	1.27	49	1.26	48	3.17	66
160 x 160	1.00	84	0.83	79	0.87	64
320 x 320	1.74	93	1.13	90	1.75	95
480 x 480	2.82	95	2.03	93	3.80	91
640 x 640	4.45	97	2.68	96	4.66	94
800 x 800	6.36	97	3.87	96	7.04	93

Table 1: Performance results for the three different Connection Machines. Time is the elapsed time calculated for 100 iterations and % // is the ratio between the elapsed time and the time of effective processors activity.

All the tests have been done using the same random porous network. Computations on the CM ran until convergence of Penal. This implied large CPU times (from 2 minutes to 3 hours) but correctly represented real conditions. In table 1 the obtained performances are presented for the 8k processors CM2 and two configurations of the CM5 (32 and 64 processors). The listed times corresponds to 100 iterations.

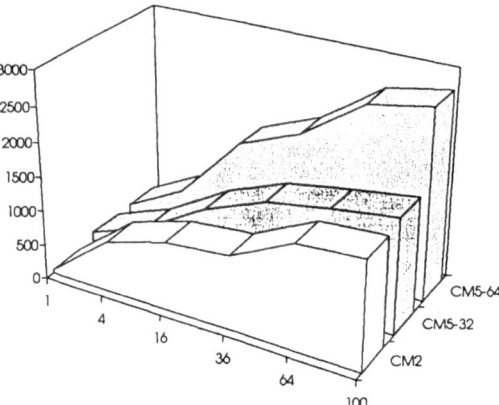

Figure 3: Evolution of parallel performances (MFlops on the processors) with the size of the problem.

Figure 3 clearly demonstrate the fact that penal is well suited to this kind of machines (particularly the CM2). The code modifications have been very limited and done within a week. Knowing that, the obtained performances are good even if far from the best possible. Optimization similar to the one presented next section might improve the performances (especially on the CM5).

Shared Virtual Memory

In this section we present the performance result for the iPSC/2 and the KSR architectures. To do the experiment, the programs were not run until completion but stopped after 10 time steps for the iPSC/2 and 100 time steps for the KSR. The problem size was limited to a grid 240 x 240 because the iPSC/2 nodes has a limited local memory (4 Megabytes). The performance results are shown on Table 2 and Table 3. For both machines and both programs the 240 x 240 problem behave better than the 128 x 128. This is mainly due to the increase in granularity and less false-sharing. The locality of the two programs being good, the increase in problem size decrease the proportion of time spent in communication. For instance on the iPSC/2 the percentage of time spent in communication for 32 processors is 24% for 128 x 128 Penal (respectively 48% for Permea) and is 15% for 240 x 240 Penal (respectively 40% for Permea). The time for Permea, on the KSR, with 24 processors is higher than with 16 processors probably because of a wrong block size that the system compute by default, this introduce load unbalancing that degrades performance. We could not use 32 processors for the KSR since some of them are dedicated to run the system and using these processors would have slow down the overall behavior of the application. On the iPSC/2, with one processor Permea is more than 2 times faster than Penal, property not found on the KSR. The iPSC/2 having a slow floating point unit, saving floating point operation makes a big difference. In the case of the KSR, the floating point operations cost saved is replaced by indirect addressing to the vector. In the case of the KSR we could have improved performance by decreasing the number of barriers by programming in an SMPD style using parallel region construct. This was not done because it requires to many code modifications.

Simple Parallelization PENAL				Simple Parallelization PERMEA			
Procs	128 x 128		240 x 240		128 x 128		240 x 240	
	Time (s)	Speedup	Time (s)	Speedup	Time (s)	Speedup	Time (s)	Speedup
1	127.4	1	Not E.	Mem.	55.4	1	Not E.	Mem.
2	96.2	1.32	Not E.	Mem.	33.5	1.65	Not E.	Mem.
4	69.2	1.84	164.2	-	28.0	1.97	80.5	-
8	57.6	2.21	122.4	-	18.7	2.96	49.6	-
16	53.1	2.39	94.5	-	25.1	2.20	42.5	-
32	55.5	2.29	88.9	-	41.6	1.33	42.4	-
After optimization								
Procs	128 x 128		240 x 240		128 x 128		240 x 240	
	Time (s)	Speedup	Time (s)	Speedup	Time (s)	Speedup	Time (s)	Speedup
1	117.4	1	Not E.	Mem.	51.9	1	Not E.	Mem.
2	61.0	1.92	Not E.	Mem.	29.3	1.77	Not E.	Mem.
4	33.1	3.54	100.3	-	16.9	3.07	48.8	-
8	19.1	6.14	58.6	-	13.1	3.96	28.4	-
16	12.2	9.62	32.9	-	11.2	4.63	19.1	-
32	11.5/7.1	10.2/16.4	21.0	-	10.8	4.80	15.9	-

Table 2 : Performance results for the iPSC/2 (10 iterations). For 128 x 128 Penal and 32 processors the timing are given without and with array padding.

PENAL				PERMEA			
Procs	128 x 128		240 x 240		128 x 128		240 x 240	
	Time (s)	Speedup	Time (s)	Speedup	Time (s)	Speedup	Time (s)	Speedup
1	26.4	1	88.9	1	20.0	1	72.6	1
2	14.6	1.8	47.4	1.87	12.1	1.64	46.1	1.57
4	8.0	3.3	25.7	3.45	7.1	2.78	25.4	2.85
8	5.1	5.09	23.5	3.77	4.4	4.4	13.9	5.19
16	4.5	5.78	7.7	11.44	3.4	5.81	8.2	8.8
24	8.9	2.94	8.4	10.54	3.1	6.29	6.9	10.4

Table 3: Performance results for the KSR (100 iterations)

Pvm

We used up to 8 IBM-RS/6000 connected by an Ethernet network for those tests. As it was difficult to find this number of machines available at the same time, we limited the number and the size of the tests. Table 4 resumed the results. After one and half month of efforts to implement PVM and modify correctly Penal, the speedup reached can be considered as good.

CONCLUSION

The first series of tests (CM2, CM5, PVM) demonstrate the possibility to reach interesting computation performances only slightly modifying our FORTRAN code. Even if this code can be considered a priori as well suited for parallelism, the maturity of the CM programing environment is put into evidence here. PVM gave us good results too but the limitations of this concept on a classical network appeared clearly. The second series of tests show us that further optimizations are possible integrating some "not yet classical" programming rules. As for vector computers some years ago, it can be though that next compilers, based on SVM or other concepts, will integrate part of those rules allowing more transparent efficient use of parallel machines.

Size	2 procs		4 procs		8 procs	
	Time (s)	Speedup	Time (s)	Speedup	Time (s)	Speedup
32 x 32	70	0.89	54	1.15	66	0.94
64 x 64	232	1.42	165	2.00	147	2.24
96 x 96	1514	1.75	976	2.72	933	2.84
128 x 128	1972	1.86	1250	2.94	940	3.91
160 x 160	8815	1.79	5002	3.16	3559	4.44
320 x 320	77352	1.86	39006	3.69	24077	5.98

Table 4: Performance results for PVM (100 iterations)

REFERENCES

[1]   ANGUY, Y., BERNARD, D.,(1992) "Numerical computation of the permeability tensor evolution versus microscopic geometry changes", in *Mathematical modeling in water resources*, RUSSEL et al. Eds, Comp. Mech. Pub., pp. 385-392

[2]   ANGUY, Y., BERNARD, D. (1994) "Numerical investigation of the coupling between the microgeometry and the permeability tensor of natural porous media through the volume averaging theory", X Int. Conf. on Computational Methods in Water Resources, July 19-22, Heidelberg, Germany

[3]   BEGUELIN, A., DONGARRA, J., GEIST A., MANCHEK, R., SUNDERAM, V. (1993) "A Users' Guide to PVM", Oak Ridge National Laboratory Edition

[4]   THINKING MACHINES CORPORATION (1990) "Connection Machine Model CM-2 Technical Summary", Cambridge, Massachusetts

[5]   THINKING MACHINES CORPORATION (1990) "Getting Started in CM Fortran", Cambridge, Massachusetts

[6]   THINKING MACHINES CORPORATION (1991) "CM-5 Technical Summary",Cambridge, Massachusetts

[7]   KENDALL SQUARE RESEARCH CORPORATION (1992) "KSR Parallel Programming", Technical Report

[8]   ARLAUSKAS, R. (1988) "iPSC/2 System: A Second Generation Hypercube", 3rd Conference on Hypercube Concurrent Computers and Applications

[9]   LAHJOMRI, Z., PRIOL, T. (1992) "KOAN: a Shared Virtual Memory for the iPSC/2 hypercube", in CONPAR/VAPP92

[10] PRIOL, T., LAHJOMRI, Z. (1992) "Trade-offs Between Shared Virtual Memory and Message-passing on an iPSC/2 Hypercube", Technical report 1634, INRIA

[11] PRIOL, T., LAHJOMRI, Z. (1992) "Experiments With Shared Virtual Memory on a iPSC/2 Hypercube",Int.Conf. on Parallel Processing ,p. 145-148

[12] BODIN, F., KERVELLA, L., PRIOL, T. (1993) "Fortran-S: A Fortran Interface for Shared Virtual Memory Architectures" Proc. Supercomputing '93 (ACM), Portland

[13] GRANSTON, E.D., WIJSHOFF, H. (1993) "Managing Pages in Shared Virtual Memory Systems: Getting the Compiler into the Game", Proc. Int. Conf. Supercomputing (ACM)

[14] IBM (1990) "The IBM RISC System/6000 Processor", Special issue of the IBM Journal of Research and Development, January

Acknowledgments: CPU time has been provided by CRS4 (Cagliari, Sardenia) and IRISA (Rennes, France). The research group "TRABAS" from the CNRS partly supported this work

DATA PARALLELISM IN FINITE ELEMENT COMPUTATION

Leesa Brieger* and Giuditta Lecca**
*Environment Modeling Group and **Parallel Computing Group
Center for Advanced Studies, Research and Development in Sardinia
via N. Sauro 10, I-09123 Cagliari
Italy

Finite elements (FE) form the basis for many current models of flow and transport in porous media; computer time for any such simulation is largely spent first assembling and then solving the sparse systems that result from the FE formulation. In this article we treat these two aspects and their potential for speed-ups on the data parallel Connection Machines. Specifically, we consider conjugate gradient solvers for sparse symmetric systems: why incomplete Cholesky preconditioning is disastrous, which communication utilities are preferable, when to expect a significant speed-up over scalar codes. We also treat system assembly: how to implement the intrinsically parallel assembly of the global system and indeed whether this is desirable, given the possibility of element-wise conjugate gradients.

INTRODUCTION

As parallel computation becomes increasingly accessible, its application to various problems of scientific calculation must be evaluated. With this in mind, we examine some problems and advantages of parallel computation for finite element applications such as those typical of flow models [Gambolati].

Parallel computation itself has many forms, but we constrain ourselves in this article to the distributed memory data parallelism of a Connection Machine (CM). This choice is due largely to the relative simplicity of the approach - even for a high degree of parallelism. The alternative MIMD environment requires decomposing the problem into as many balanced, independent tasks as there are processors, managing the synchronization and then furnishing an efficient complementary message-passing scheme to explicitly manage interprocessor communication. For a small number of processors, this may not be formidable; for massive parallelism, this can be an intractable problem, without even a methodology for seeking an optimal solution.

In the data parallel approach, on the other hand, the problem must be structured so that the processors concurrently execute an identical task (SIMD) or program (SPMD), on individual data. While the programmer must determine a data structure which minimizes communication time and enhances parallelism, actual distribution of data and then communications among the processors is handled by the machine. The programming model in this case is much more restricted - and much simpler - than the MIMD programming model. The distributed data are accessed with global

A. Peters et al. (eds.), Computational Methods in Water Resources X, 1515–1522.

addressing and interprocessor communication. Not all problems are suited to a data
parallel approach, as we shall see in this article. However, many large scientific
problems are data parallel by nature; for them, enormous gains can be realized on
massively data parallel machines.

A finite element discretization of a partial differential equation yields a system of
ODE's which, when integrated numerically, gives rise ultimately to a system of linear equations. The finite element structure has a granularity that lends it naturally
to data parallelism under an element-to-processor (or element-to-virtual processor)
mapping for the construction of this linear system. Each processor can calculate the
contribution of its elements' nodes to the global system; this calculation is equivalent
on all the processors, equally for structured and unstructured grids.

Iterative solvers for dense linear systems of equations also have a natural granularity that can lend them well to data parallelism. The mapping between matrix
elements and (virtual) processors is natural and allows for a high degree of parallelism in the matrix-vector multiplications typical of such methods. Communication
requirements can be important, but they can also be extremely regular and so very
effectively optimized, even for large systems. Since iterative solvers are often found
at the heart of FE codes, we may imagine more or less effortless parallelization of
an FE application. In fact, the data parallel approach for iterative solvers runs into
difficulty when the systems are sparse and so communication patterns very irregular.
Needless to say, sparsity is a characteristic of the systems typically arising from FE
applications.

Apart from the matrix sparsity, a potential obstacle to ease of parallelizing comes
from the fact that the granularities of the two parts of the problem, system construction and system solution, are different. The natural structure for data distribution
in the system construction is elemental; the natural data structure for the iterative
solver of the global system is nodal. It is possible to use both structures to advantage
in an application, translating between the two as necessary to define and then solve
the system. However, this tends to be expensive in memory and communication.
It is possible and may be preferable to use a single data structure throughout the
application.

In this contribution we take up these considerations, which will be important for
parallelizing finite element codes. We have taken the conjugate gradient method
(CG) for sparse symmetric systems as the paradigm for more general iterative methods. Performance on a CM-200 with 8K one-bit (256 floating-point) processors is
compared against sequential code on an IBM RISC-560 for both system contruction
and CG solution.

CONJUGATE GRADIENTS

Solving the system $Ax = b$, A an $n \times n$, symmetric, positive definite matrix which
is not necessarily sparse, is equivalent to minimizing the cost function $F(x) = \frac{1}{2}(x -
x^*)^T A(x - x^*)$ where $x^* = A^{-1}b$ is the unknown solution [Bertsekas]. The unique

minimum can be found by a methodical search in the solution space: $F(x)$ can be explicitly minimized along successive search directions d_k which are defined in terms of $g = gradF$ and such that all the d_k are A-orthogonal. This is the conjugate gradient method [Hestenes]. Once an initial approximation x_0 to the solution has been determined and $g_0 = Ax_0 - b$ and $d_0 = -g_0$ defined accordingly, the iteration is as follows:

- $x_{k+1} = x_k + \alpha_k d_k, \qquad \alpha_k = -\frac{d_k^T g_k}{d_k^T A d_k}$

- $g_{k+1} = g_k + \alpha_k A d_k$

- $d_{k+1} = -g_{k+1} + \beta_k d_k, \quad \beta_k = \frac{g_{k+1}^T A d_k}{d_k^T A d_k}$

The matrix-vector multiplication and vector inner products are the most cumbersome parts of the iteration and must be efficiently parallelized in order for the method itself to run efficiently. Inner products are imminently data parallel and are handled efficiently on the CM. Further, the CM execution of parallel sums avoids the numerical problem of information loss which can occur with very long vectors and finite machine precision. We focus our attention rather on matrix operations. In a matrix element-to-processor mapping, nearest neighbors in the matrices are mapped to nearest-neighbor processors. Neighbor-to-neighbor communications being particularly efficient on the CM, this structure can be used to advantage in the parallelization.

The FORTRAN-90 statement $C = A * B$, where A, B and C are matrices of identical size, shape and layout, assigns to C the results of the element-by-element multiplication of A with B in a parallel operation requiring no communication and executed concurrently by the processors. To parallelize a general matrix-vector multiplication, $y = Ax$, where A is an $n \times m$ matrix and x an m-vector, x can be mapped onto the data structure of A, a parallel matrix multiply executed, and results collected into y. For example, x can be assigned to the rows of a temporary array having A's structure. Once the parallel multiplication is done, an intrinsic function sums across rows and deposits results in the output vector:

$tmp = spread(x, dim = 1, ncopies = n)$
$tmp = A * tmp$
$y = sum(tmp, dim = 2)$.

The additional memory, equal to the size of A, required to parallelize the operation may seem restrictive, but dynamic memory allocation is supported by CM-FORTRAN, and, in general, it is communication requirements that will almost inevitably determine the desirability of this parallelization. Copying x into tmp obliges communication between the processor(s) containing x and the $n \times m$ (virtual) processors containing A; collecting results into y means communications between the n processors holding the row sums and the n (different) processors containing y. The regular distribution of the matrix into the array of processors is such that neighborhoods are preserved and so required communications are minimized. With the intrinsic CM-FORTRAN funtions "spread" and "sum", communications are most efficient and parallelization advantageous.

However, the matrices typical of finite element applications will not necessarily enjoy the advantage of this regular data structure. To reduce memory requirements, these large, sparse matrices will probably be compacted so that only the nonzero entries and pointer arrays are stored. Nearest neighbors in the processor array may not come from nearest neighbors in the original matrix, and interprocessor communication patterns become irregular and less efficient. The worth of parallelization then depends on factors such as sparsity, storage scheme, communication utilities and problem size. Three row-oriented storage schemes which appear well-adapted to a general sparsity structure are the Ellpack-Itpack, compressed sparse row (CSR) and coordinate (COO) formats [Ferng].

In Ellpack format, an $n \times n$ sparse matrix A is stored in an $n \times ncol$ matrix whose rows hold the nonzero entries of the rows of A and where $ncol$ is the maximum number of nonzero elements in a row of A. Any row with fewer than $ncol$ nonzero elements is padded with zeroes. An $n \times ncol$ pointer array ja stores the column indices of the nonzero elements. In CSR format, the nonzero elements of A are compacted row-by-row into a dense vector of length nz, where nz is the total number of nonzero elements of A. A vector ja of length nz stores column indices and another vector of length $n + 1$ points to those elements which begin a new row. COO is similar to CSR format, the difference being in the pointer arrays: vector ja, of length nz, stores column indices and vector ia, also of length nz, stores row indices. This storage format, supplemented by mask arrays, is used by the CMSSL sparse matrix calls.

For a sparse matrix A stored according to one of these formats, the parallelization of $y = Ax$ depends again on mapping x onto the data structure of the stored matrix, multiplying in parallel and collecting row sums into y. The interprocessor communications necessary for this now depend on the sparsity structure of A. They can be handled most simply in CM-FORTRAN with vector-valued subscripts as follows.

Vector-valued subscripts, Ellpack format, $y = Ax$:
$$forall(j = 1 : ncol) \; tmp(:, j) = x(ja(:, j))$$
$$tmp = A * tmp$$
$$y = sum(tmp, dim = 2)$$

Here, actual communications are absolutely transparent to the user and perhaps as inefficient as they can be. To enhance the operation, there are several other possibilities: CMSSL sparse matrix calls or other communication options, including the gather/scatter communication primitives and the CMSSL communication compiler routines. After the necessary setup, the CMSSL (version 3.0) call looks like this.

CMSSL, COO format, $y = Ax$:
call $sparse\_matvec\_mult(y, A, x, ia, ja, segments, matrixmask, itrace, trace, ier)$

This option uses the communication strategy described for the examples above, which, being handled internally by the library, is done particularly efficiently. It is more expensive in memory: *segments* and *matrixmask* are both logical vectors of length nz, and additional space is used to store communication pattern information. No account is taken of the symmetry of the matrix.

The gather/scatter communication utilities used by this CMSSL call are also explicitly available and provide another way of implementing a sparse matrix-vector multiply, requiring fewer pointer arrays and masks.

Gather/scatter, CSR format $y = Ax$:

    call sparse_util_gather(tmp, x, gtrace, tracemask)
    $tmp = A * tmp$
    call sparse_util_scatter(y, ja, tmp, matrixmask, strace)

The scatter utility sends information - additively - to a destination vector. Here the steps of the operation are explicit, and thus this option is more flexible than the previous. For example, with this utility we can implement the multiplication, storing just half the symmetric matrix.

The use of the communication compiler routines is absolutely analogous to that of the gather/scatter utilities; the communication compiler optimizes irregular communications but is much more demanding in memory than the gather/scatter utility. When useable, these routines furnish perhaps the fastest treatment of general communications. However, the time required to establish optimal routing (this being transparent to the user) and the memory necessary to store and re-use this information can be impossibly large. Our preliminary tests using Ellpack format with these routines showed these options optimal - for small problems. As problem size grows, however, this choice becomes impossible: for 19,000 unknowns on the 8K one-bit-processor CM-200 with 256 Mb of memory, space was insufficient for the communication compiler setups, and this option was abandoned.

We remark here that conjugate gradient-like methods for nonsymmetric systems are similarly based on matrix-vector multiplication and inner products, and the considerations made here for conjugate gradients apply analogously to these more general iterative solvers.

Incomplete Cholesky preconditioning is extremely effective in sequential solvers for sparse symmetric systems; it is nevertheless totally inadapted to data parallelism. This is not surprising, given that the method for defining and then using the preconditioning matrix is essentially sequential. Further, the regular structure which aids the parallelization for a dense matrix is inexistent for a sparse one. The only real contribution of the parallel machine to this method is the slowing down of communications. In fact, in one of our tests on the CM-200, time per iteration was three orders of magnitude slower with incomplete Cholesky than with simple diagonal preconditioning. The parallel codes for the timings of Figure(1) use diagonal preconditioning, the sequential codes incomplete Cholesky preconditioning.

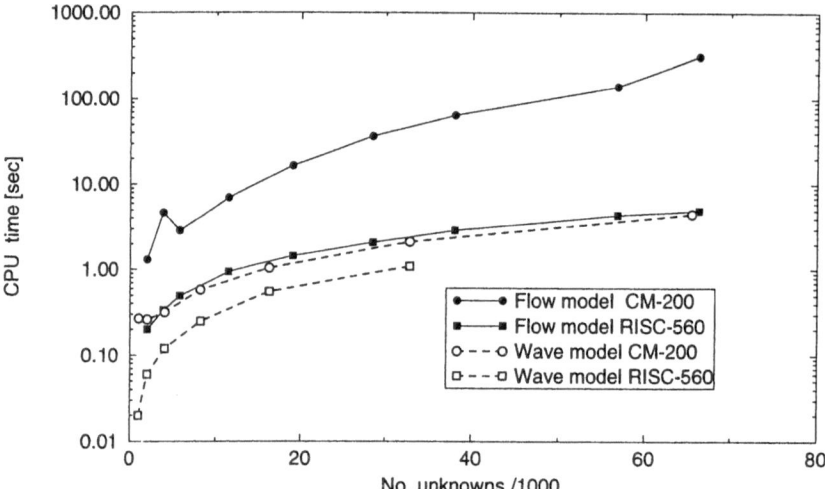

Figure 1. CG solvers: sequential (Cholesky preconditioning) and parallel (diagonal preconditioning); wave propagation and flow model examples.

Figure(1) shows CPU times, on a logarithmic scale, for sequential code on a RISC-560 and parallel code on the CM-200 for conjugate gradients on sparse systems of increasing size. Examples come from two FE applications: a 2D acoustic wave propagation model and a 3D model for flow in porous media. All codes are single precision. Due to the difference in preconditioning between sequential and parallel codes, the number of iterations to reach a predefined error tolerance differs between the codes.

Wave model : 1 RISC iteration and 34 CM iterations to reach an error tolerance of 10^{-8} for all these examples.

Flow model : 6-8 RISC and 29-518 CM iterations, depending on system size, for a tolerance of 10^{-10} for the flow systems.

CM efficiency depends on the ratio of matrix size to number of processors. In our examples with the rather ineffective diagonal preconditioning, overall CM performance also depends strongly on the conditioning of the coefficient matrices. While the RISC solver is much faster for the systems considered here, time per iteration is not so one-sided. With increasing system size, time per iteration increases on the RISC but decreases on the CM. Beyond 10,000 unknowns, time per iteration is smaller for the CM than for the RISC, for the systems of both models, and we might expect that an efficient parallel preconditioner could reduce the number of CM iterations to the point of outperforming the RISC. Such a preconditioner remains to be found.

SYSTEM ASSEMBLY

We now consider the construction of the linear system to which an iterative solver will be applied in a finite element application. The strategy on a data parallel machine will probably be to let the elemental structure of the numerical method dictate the data structure on the machine. That is, using an element-to-processor mapping, local contributions to the global (nodal) system will be calculated concurrently for all the elements. Then, either the elemental contributions can be assembled to form the global sparse system, which will be compacted into a sparse storage format, or the system can be left unassembled in this "exploded" form, which we will denote as Elemental. The Elemental storage is nothing other than an alternative to the Ellpack, CSR and COO formats for storing the global sparse matrix. It follows that the same iterative solver can be used for either storage option, global or local, but its implementation must be adapted to the strategy which is chosen [Stothoff].

Collecting elemental contributions into the global system requires memory sufficient to hold both data structures, and the communications to send information between them. Alternatively, leaving the system in its Elemental form allows us to do without the nodal data structure and to avoid the communications associated with the assembly. One might argue that the irregular communication patterns for a CG solver on an exploded coefficient matrix (Elemental storage) would outweigh any savings in memory or communication realized in the construction phase. However, irregular communication patterns already penalize the CG method for a sparse compacted matrix , so that at this level the two approaches are probably comparable. It is then the importance of efficiency for system construction which determines the preferability of one approach over the other. The wave model systems of Figure(1) were constructed and solved with the Elemental storage structure. The flow model examples were based on the global (nodal) systems.

In Figure(2) are shown CPU times, again on a logarithmic scale, for system assembly on a RISC-560 and for system construction (without assembly) on the CM-200, for examples of increasing size from a 2D model of flow in porous media [Gambolati]. This operation depends essentially on mesh connectivity and size and not on the physical model itself. Thus the superiority of the CM-200 performance in this example should be taken as generally indicative of the considerable advantage of parallelism at this stage of the problem. This and the gains to be enjoyed by parallelizing the entire application may give the CM-200 a superior overall performance.

It should be noted that the comparisons presented here are not definitive in the sense that parallel resources are continually being improved. Algorithms particularly adapted to parallel architecture (including, hopefully a good parallel preconditioner) are being developed. The next generation, version 3.1, of CMSSL contains new options especially adapted to iterative solvers and sparse matrices which should enhance performance. The CM-5, an array of RISC microprocessors which can be used as an SIMD as well as an SPMD machine, gives more processing power to the programming model.

ACKNOWLEDGEMENTS

This work has been carried out with the financial contribution of the Sardinian Regional Authority.

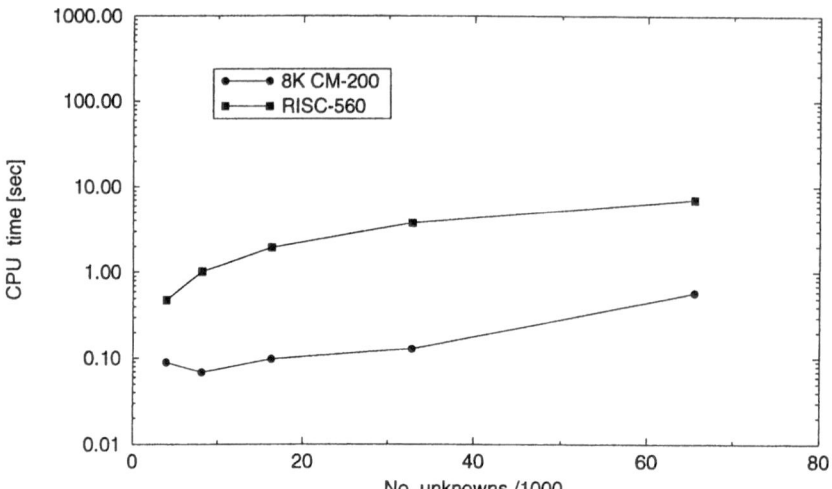

Figure 2. Sequential vs. parallel (without assembly) system construction for a 2D flow model

REFERENCES

Bertsekas, D.P. and Tsitsiklis, J.N. (1989) Parallel and Distributed Computation, Prentice Hall, Englewood Cliffs.

Ferng, W., Wu, K., Petiton, S., Saad, Y. (1992) "Basic Sparse Matrix Computations on Massively Parallel Computers", draft.

Gambolati, G., Pini, G., Putti, M. and Paniconi, C. (1994) "Finite element modeling of the transport of reactive contaminants in variably saturated soils with LEA and non-LEA sorption", in P. Zannetti (ed.), Environmental Modelling II, Computational Mechanics Publications, Southampton, UK.

Hestenes, M.R. and Stiefel, E. (1952) "Methods of Conjugate Gradients for Solving Linear Systems", J. of Res. of the Nat. Bureau of Standards 49, 409-436.

Stothoff, S.A. and Dougherty, D.E. (1992) "Implementing a Three-Dimensional, Two-Phase Flow Model on the Connection Machine", in T.R.Russell, R.E.Ewing, C.A.Brebbia, W.G.Gray and G.F.Pinder (eds.), Proceedings of the IX International Conference on Computational Methods in Water Resources, Computational Mechanics Publications, Southampton, UK, pp. 705-712.

Marfurt, K.J. (1983) " Accuracy of finite difference and finite element modeling of the scalar and elastic wave equations", Geophysics 49, 533-549.

COMPUTATIONAL STRATEGIES IN MULTICOMPONENT REACTIVE TRANSPORT MODELLING

E.O. FRIND[*], K.U. MAYER[*], and H. DANIELS[**] and A. PETERS[**]

[*] Waterloo Centre for
Groundwater Research
University of Waterloo
Waterloo, Ontario, N2L 3G1
Canada

[**] IBM Germany
Heidelberg Scientific Center
Vangerowstr. 18
68070 Heidelberg
Germany

Possible strategies for making multicomponent reactive transport modelling practical in two and three dimensions include choosing the two-step rather than the one-step method, linearizing the physical/chemical coupling in the case of equilibrium reactions, and parallelizing the physical transport and the chemical equilibrium computations. In this paper, the focus is on the parallelization of MINTRAN, a comprehensive multicomponent reactive transport model. A data parallel mode is obtained by parallelizing the main transport and chemistry loops. Controlling factors in the speedup due to parallelization are the processor load balance and the interprocessor communication time.

INTRODUCTION

The generation of acidic waters at mine-waste disposal areas at minesites in Canada and Germany is a source of great concern. In Ontario alone, there are more than 2,000 decommissioned minesites. Toxic metals dissolved in the acidic tailings environment at these sites can migrate in the groundwater to contaminate aquifers and surface waters, and thus pose a serious health hazard. If unchecked, the reactions that mobilize the toxic metals can persist for centuries. Efforts to remediate old mine sites have in the past often relied on unproven trial-and-error methods. Feasby et al. (1991) estimated that the cost of remediating acidic drainage at sites in Canada, using techniques of uncertain effectiveness, may exceed three billion dollars.

The relevant processes involved in mine tailings drainage include the advective-dispersive transport of a large number (typically several tens) of aqueous chemical components that react with each other as well as with the soil minerals, and that can change to the mineral phase themselves. Typically, component concentrations range over many orders of magnitude, and multiple sharp fronts separate zones of distinct geochemistry. Aquifer systems impacted by mine tailings effluents are often highly heterogeneous. To understand the complex physical/chemical interactions, a reliable capability to simulate these systems in three dimensions, under realistic boundary conditions, is needed.

A. Peters et al. (eds.), Computational Methods in Water Resources X, 1523–1532.
© 1994 Kluwer Academic Publishers. Printed in the Netherlands.

Because of the sheer size of problems of this type, the computational strategy must be chosen with great care. For example, a modestly-sized 3D system containing 200,000 spatial nodes and 30 chemical components, if solved directly using a one-step approach, would generate 6 million highly nonlinear equations in 6 million unknowns, which must be solved iteratively for at least several hundreds of time steps. This approach would not be economical.

The multicomponent reactive transport model MINTRAN uses the two-step approach and consists of two modules, a transport module which advects and disperses each chemical component, and a chemical module which reacts the aqueous components and minerals according to the equilibrium equations of thermodynamics, assuming local equilibrium. The local equilibrium assumption (L.E.A.) has been found to be valid in the transport of mine tailings effluents involving natural buffering reactions. The model can handle the full range of chemical reactions subject to the L.E.A., including acid/base, complexation, precipitation/dissolution, oxidation/reduction, and adsorption reactions, and has been successfully applied to mine tailings problems.

The computational load, however, is still formidable. For example, a moderately-sized 2D problem has taken 70 hours on an IBM 6000/560. To make 3D multicomponent reactive modelling practicable, the computational efficiency must be improved. One improvement is to take advantage of the fact that the physical-chemical coupling can be linearized if the chemistry is of the equilibrium type. In addition, advantage can be taken of the linearized two-step structure to parallelize the algorithm. We will consider these strategies below.

COMPUTATIONAL ALGORITHM

Physical Transport Module

We assume a groundwater system containing N_c chemical components of which each can exist in either the aqueous or the solid phase. The aqueous phases are transported by advection and dispersion, while the solid phases are stationary. The equation governing the advective-dispersive transport of the aqueous component k is:

$$\frac{\partial C_k}{\partial t} - \frac{\partial}{\partial x_i}\left(D_{ij}\frac{\partial C_k}{\partial x_j}\right) + \frac{\partial}{\partial x_i}(v_i C_k) - R_k = 0 \qquad k = 1, ..., N_c \qquad (1)$$

where x_i are the cartesian coordinates, t is time, v_i are the vector components of the average fluid velocity, C_k is the concentration of aqueous component k, D_{ij} is the hydrodynamic dispersion tensor, and R_k is the chemical source/sink term representing the changes in aqueous component concentrations. The mass conservation equation for the stationary solid component k is:

$$\frac{\partial S_k}{\partial t} - R_k^S = 0 \qquad k = 1, ..., N_c \qquad (2)$$

where S_k is the solid-phase concentration, and R_k^s represents the change in solid component concentration due to precipitation/dissolution and sorption/desorption reactions. The boundary conditions for each aqueous component k are either of the Dirichlet type (specified concentrations), or the Cauchy type (specified mass flux). Complete boundary and initial conditions are required for each aqueous component, and an initial condition is required for each solid-phase component. The transport equations are spatially integrated according to the Galerkin finite element method, and the resulting matrix equations are solved using a preconditioned conjugate gradient (PCG) solver.

Chemical Equilibrium Module

The chemical module consists of the geochemical equilibrium model MINTEQA2 (Allison et al., 1990), which includes a comprehensive set of chemical reactions including chemical speciation, acid-base reactions, mineral precipitation-dissolution, oxidation-reduction, and adsorption reactions. The ion-association equilibrium-constant approach, which is valid for ionic strengths of 0 - 0.5, is used to represent the geochemical reactions. For a set of n_a aqueous and n_s solid species, the model determines the species distributions by a set of $n_a + n_s$ mass-action equations. These nonlinear equations relate the species activities and the species equilibrium formation constants to the component activities. The activities are related to the concentrations through the ionic activity coefficients, which are calculated by means of the extended Debye-Hückel or Davies equations. The species concentrations and component concentrations are related by stoichiometric mass balance equations. The nonlinear algebraic equations are solved using Newton-Raphson iteration.

Linearized Physical-Chemical Coupling

For the purpose of coupling, we can visualize the solution space as having three domain types: spatial, chemical, and temporal. The advection-dispersion terms in the governing equations are connected only in the spatial and temporal domains, while the reaction terms are connected only in the chemical domain. We write (1) in time-discretized form as:

$$\frac{C_k^{n+1} - C_k^n}{\Delta t} = L(C_k)^{n+1/2} + R_k^{equil}\delta(t + t^{equil}) \qquad\qquad k = 1,...,N_c \qquad (3)$$

where n, $n+1$ represent the old and new time levels, respectively, Δt is the time step, and L represents the spatial differential operator. The spatial terms are centrally time-weighted in order to obtain second-order accuracy with respect to time. The reaction term R_k^{equil} represents the relative mass (M/M) generated or consumed instantaneously in equilibrating the system at time $t + t^{equil}$ during the physical transport step, and $\delta(t + t^{equil})$ is the Dirac delta function (dimension 1/T) for time t^{equil}.

In view of the strong nonlinearity of reactive systems, most conventional reactive transport models use an iterative coupling between the physical and chemical processes. Equation (3) might be solved in an iterative two-step method as follows:

- Step 1: Solve for the physical transport part of each component, taking the reaction term as constant;
- Step 2: Solve the chemical equilibrium equations for each node in the finite element grid, allowing the physical terms to be constant.

The reacted quantities of each component resulting from Step 2 are substituted back into Step 1 and the cycle is repeated to convergence for each time step. A similar procedure was used by Kinzelbach et al. (1991).

A substantial improvement to the efficiency of the algorithm can be made by realizing that under conditions of local chemical equilibrium assumption, the physical-chemical coupling can be linearized (Walter et al., 1994a). In this way, the nonlinearity will be strictly confined to the chemical domain. We conceptually visualize the two steps of the coupled process as a purely physical step, where the components advect and disperse through space and time without reacting, and a chemical step, where the components react instantaneously without being transported. A similar concept is used in mixing-cell models (Schulz and Reardon, 1983). By virtue of the L.E.A., the chemical step restores the chemical equilibrium that has been perturbed through the transport step, but does not affect the transport step itself. The chemical equilibration would logically be allowed to take place at the end of the time step, at time $(t + \Delta t)$. The sequential two-step algorithm becomes:

- Step 1 (physical):

$$\frac{\left(C_k^{phys} - C_k^n\right)}{\Delta t} = L\left(C_k\right)^{n+1/2} \qquad\qquad k = 1, ..., N_c \qquad\qquad (4)$$

- Step 2 (chemical):

$$\left(C_k^{n+1} - C_k^{phys}\right)\delta(t+\Delta t) = R_k^{equil}\delta(t+\Delta t) \qquad\qquad k = 1, ..., N_c \qquad\qquad (5)$$

where C_k^{phys} is the concentration at the end of the physical step. Adding (4) and (5) and integrating over the time interval $[t + \Delta t]$ results in:

$$C_k^{n+1} - C_k^n = \left(L(C_k)^{n+1/2}\right)\Delta t + R_k^{equil} \qquad\qquad k = 1, ..., N_c \qquad\qquad (6)$$

which corresponds to the similarly integrated form of (3).

For each time step, the unequilibrated set of values C_k^{phys} from (4) are supplied to the chemical routine, which equilibrates the system individually for each nodal point and returns directly the set of equilibrated values C_k^{n+1}. Thus the final concentration at the end of the time step is obtained directly from Step 2, without iteration.

The overall computational effort per time step with the sequential scheme is:

Step 1 (physical): Solve N_n linear transport equations in N_n unknowns, N_c times;
Step 2 (chemical): Solve N_c nonlinear chemical equations in N_c unknowns, N_n times.

The iterative and the sequential algorithms differ in that the reaction term occurs in both steps (1) and (2) in the iterative method, but only in step (2) in the sequential method. In both approaches, step 1 is coupled in space and uncoupled over the components, while step 2 is uncoupled in space and coupled over the components. Figure 1 shows that the accuracy obtained with either approach is essentially the same, and that both approaches compare well with the geochemical equilibrium model PHREEQM (for details see Walter et al., 1994a). The profile also reveals three geochemically distinct zones separated by sharp fronts. Figure 2 shows a typical pH plume from a 2D simulation with 16 aqueous components and 10 mineral components (Walter et al., 1994b). Overall, the sequential mode saves about a factor of 4 to 5 in CPU time, compared to the iterative mode.

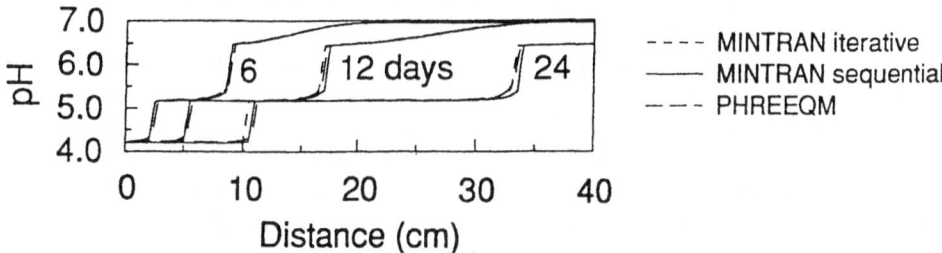

Figure 1:   *Typical 1D pH profile showing comparison between MINTRAN iterative and sequential modes, and PHREEQM.*

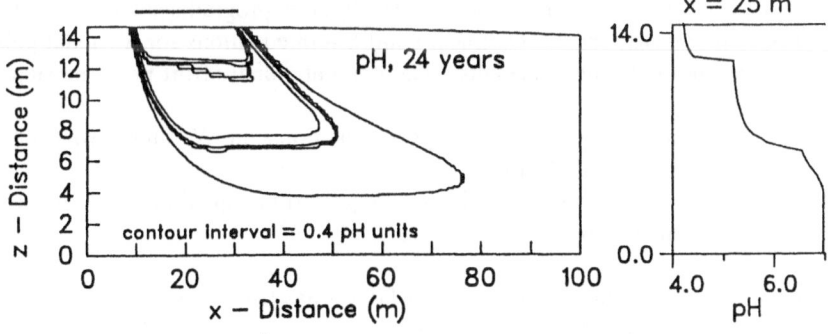

Figure 2:   *Typical 2D pH plume; contours and vertical section.*

PARALLELIZATION

In parallel mode, certain repetitive tasks in the algorithm are distributed over a number of processors to be executed simultaneously. A complication arises due to the fact that the task of equilibrating the chemistry is not necessarily the same for each nodal point because of the existence of multiple sharp pH fronts (see Figs. 1 and 2). At these fronts, the aqueous as well as the mineral components undergo sharp changes, and the iteration routine in the chemical module converges more slowly. The parallelization must be able to respond to these inequities. An important factor is also the send/receive time consumed in interprocessor communication, which should be minimized.

The approach we have chosen is a data parallel model, which takes advantage of the simple two-step structure of the algorithm. The main chemical loop, which consumes about 85% of the total CPU, and the transport loop, which consumes about 5%, are divided and the work is allocated to the different processors in a dynamic way which maintains flexibility with respect to load allocation. Typically, the main loop for the chemical step extends over many thousands of nodal points, all of which must be equilibrated for each time step. The nodes are grouped into packets and the loop is initiated by supplying each processor with a packet of nodal data to be equilibrated; as soon as the first processor has completed its task it is supplied with the next packet. This strategy keeps all processors occupied during a time loop, and accommodates not only varying CPU times for the different data sets, but also heterogeneous processor networks in a multi-user environment. The disadvantage of this approach is a large amount of interprocessor communication for exchanging the nodal data sets. To minimize data transfer, time-independent data are broadcast to the processors simultaneously before the time loop is entered. In a similar way, time-dependent data which is component-independent in the case of the physical step, or space-independent in the case of the chemical step, is broadcast inside the time loop, before the solution procedure is executed.

To implement the parallel procedure, the transport and chemistry modules are structured as node programs to be executed simultaneously on several processors, retaining the sequential structure for the rest of the code (Fig. 3). N_p node programs solve the transport equation for N_c components, and the chemical equilibrium equations for N_n nodal points, which are grouped into N_n/N_l node packets. The computational effort per time step is:

Physical Step: Solve N_n linear transport equations in N_n unknowns on N_p processors, N_c times;

Chemical Step: Solve N_c nonlinear chemical equilibrium equations in N_c unknowns on N_p processors, N_n times.

The development of the parallel version of MINTRAN has been carried out using PVM (Parallel Virtual Machine) software (Geist et al., 1993). This software allows a heterogeneous network of parallel and serial computers to appear as a single concurrent computational resource. The PVM software package was chosen primarily because of its flexibility, which allows the program to be run unchanged on either single processor clusters or parallel processors such as the IBM SP1.

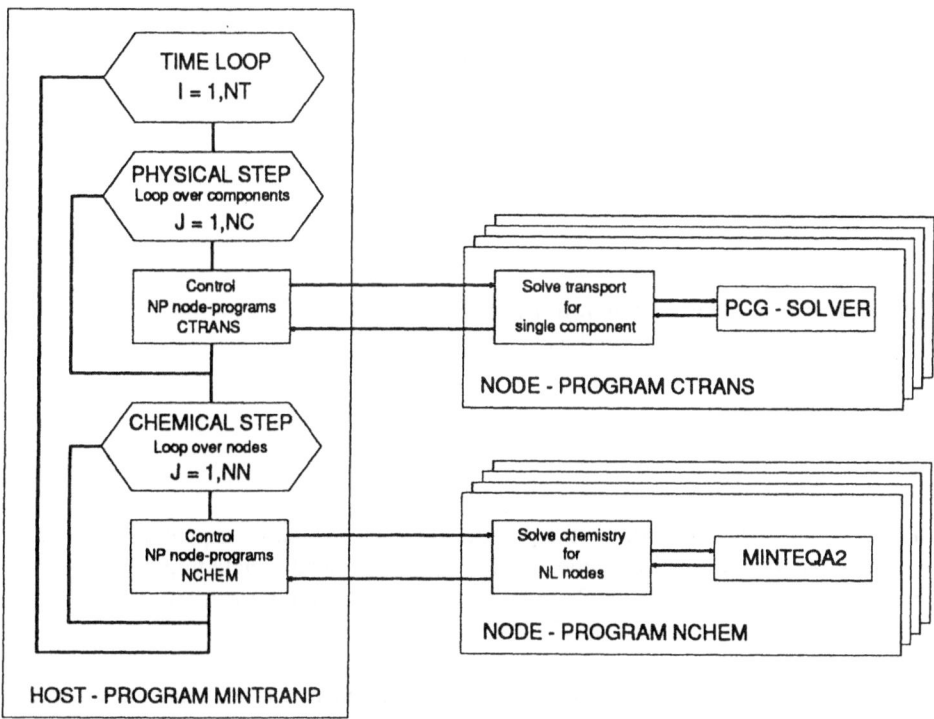

Figure 3:   Flowchart for parallel version of MINTRAN.

Since each data transfer requires a certain start-up time, the size of the nodal data packets should be as large as possible to minimize transmission time. On the other hand, larger data packets will contain more nodes located on sharp fronts, where the chemical iteration routine takes longer to converge. A partitioning of the nodal data into smaller but larger packets will therefore increase the load imbalance. We therefore use a problem-dependent partitioning scheme that achieves a compromise between the conflicting objectives of an optimum load balance, requiring fine partitioning, and the minimization of total startup times for send/receive processes, requiring a small number of data sets.

Benchmark Results

Two sets of benchmark results were obtained, one using an IBM SP1 multiprocessor, the other using a small cluster consisting of an IBM 560 and 3 IBM 220 workstations. PVM parallel software in conjunction with an Ethernet communications link is used for both configurations. Test runs using a discretization of 11457 nodes and 15 chemical components in the aqueous phase were performed under various conditions with respect to the number of participating processors and message size in the chemical step. For 16 time steps, the main program required 9.7%, the physical step 4.5% and the chemical step 85.8% of the overall CPU, giving a share of more than 90% carried by the parallel section of the program. For larger grids and more time steps, the parallelized share would be higher.

The speedup S of the parallel version relative to the sequential version is:

$$S = \frac{T_s}{T_p} \tag{7}$$

where T_s and T_p are the execution times (real) of the sequential and the parallel version respectively. T_p can be expressed as:

$$T_p = T_p{}^s + \frac{T_p{}^p}{N_p} + T_o \tag{8}$$

where $T_p{}^s$ is the execution time in the parallel version retained in sequential mode, $T_p{}^p$ is the total parallel execution time, and T_o is overhead, including interprocessor communication time. Assuming T_o to be small, the maximum potential speedup for the given degree of parallelization will be about 4.4.

Table 1 shows execution times with respect to the number of processors and the size of the grid partitions for the SP1 runs. Each simulation was performed for 16 time steps. Between 2 and 8 SP1-nodes were used in the parallel mode. One processor was used to execute the sequential section of the parallel version, which controls the node programs executing on the remaining processors. The fully sequential code was run on a single SP1 node. The grid partitioning included data packets of 5, 10, 25, 50, and 100 nodal points. Execution time is seen to be fairly insensitive with respect to the partitioning size, with a minimum obtained at 25. The speedup due to adding additional processors (2.6 using 8 SP1 nodes) is seen to be less than the potential speedup.

Table 2 shows results over 4 time steps for the 560/220 cluster in terms of CPU times expended, in sequential and parallel mode, overhead, and total execution time, respectively. In each case, the 560 handles the sequential part while the 220s handle the parallel part. The sequential CPU time is seen to be constant, while the parallel CPU time varies approximately inversely with the number of 220s. This configuration gives a good improvement in execution time when adding additional processors.

A difference between the two test cases are that in the SP1 case, interprocessor communication time is significant relative to CPU time, while in the 560/220 case, it is small. Also, in the 560/220 case, the sequential part of the code runs on a faster machine, decreasing the relative CPU time spent in sequential mode. The benchmark runs show that the maximum potential speedup can be approached, provided interprocessor communication time is minimized. To achieve better performance on the SP1, parallel software specifically designed for this machine, such as MPL/p, should be used. In general, the sequential part of the code should be kept to a minimum.

Partitioning Processors	5	10	25	50	100
1+1	52.03	51.03	49.56	50.44	49.30
2+1	30.22	29.47	29.08	29.37	28.54
3+1	24.40	23.53	23.37	23.32	23.12
4+1	21.53	20.36	20.30	20.31	20.42
5+1	19.59	19.09	18.38	18.47	18.37
6+1	18.27	17.48	17.34	17.39	17.46
7+1	17.05	16.51	16.49	17.04	16.54
Sequential	42.56				

Table 1:   SP1 benchmark results, execution times (real) [min.sec] versus number of processors and grid partition size.

Processors	T_p^*	T_p^p/N_p	T_o	T_p
1 * 560 + 1 * 220	1.20	18.40	5.14	25.14
1 * 560 + 2 * 220	1.20	9.18	2.55	13.33
1 * 560 + 3 * 220	1.22	6.10	2.21	9.53

Table 2:   Cluster benchmark results, processor times [min.sec] with parallel code.

CONCLUSIONS

Strategies to make multicomponent reactive modelling practicable should include choosing the two-step method which separates the physical and chemical processes, and linearizing the physical-chemical coupling in the case of equilibrium reactions. The linearization can yield a speedup of 4 to 5. Parallelization also shows considerable potential. A dynamic load balance scheme on the participating processors is important.

A controlling factor on overall performance in parallel mode is interprocessor send/receive time. The use of PVM software in conjunction with Ethernet is practicable for running the code in a multi-user environment on clusters of relatively slow workstations, where CPU time dominates. On a fast machine such as the SP1, software that takes full advantage of the machine-resident communications mode should be used.

Acknowledgement

The authors thank IBM Germany for generously providing the use of an IBM SP1 for the benchmark runs.

REFERENCES

Allison, J.D., Brown, D.S., and Nova-Gradac, K.J. (1990) "MINTEQA2/PRODEFA2, A Geochemical Assessment Model for Environmental Systems: Version 3.0 User's manual", Environmental Research Laboratory, Office of Research and Development, US Environmental Protection Agency, Athens, Georgia.

Feasby, D.G., M. Blanchette, G. Tremblay and L.L. Sirois (1991) "The mine environment neutral drainage (MEND) program", In: Second International Conference on the Abatement of Acidic Drainage Proceedings, pp. 1-26.

Geist, A., A. Beguelin, J. Dongarra, W. Jiang, R. Manchek, and V. Sunderam (1993) "PVM 3 User's Guide and Reference Manual", Oak Ridge National Laboratory, Oak Ridge, Tennessee.

Kinzelbach, W., W. Schäfer, and J. Herzer (1991) "Numerical modeling of natural and enhanced denitrification processes in aquifers", Water Resour. Res., 27(6), 1123-1135.

Schulz, H.D. and E.J. Reardon (1983) "A combined mixing cell/analytical model to describe two-dimensional reacting solute transport for unidirectional groundwater flow", Water Resour. Res., 19(2) 493-502.

Walter, A.L., E.O. Frind, D.W. Blowes, C.J. Ptacek, and J.W. Molson (1994a) "Modelling of multicomponent reactive transport in groundwater: 1. Model development and evaluation", Water Resources Res., in press.

Walter, A.L., E.O. Frind, D.W. Blowes, C.J. Ptacek, and J.W. Molson (1994b) "Modelling of multicomponent reactive transport in groundwater: 2. Metal mobility in aquifers impacted by acidic mine tailings discharge", Water Resources Res., in press.

MASSIVELY PARALLEL FINITE ELEMENT COMPUTATION OF SHALLOW WATER FLOWS AND CONTAMINANT TRANSPORT

Kazuo Kashiyama
Department of Civil Engineering, Chuo University
1-13-27 Kasuga, Bunkyo-ku, Tokyo 112, JAPAN

Marek Behr and Tayfun Tezduyar
AEM/AHPCRC
Supercomputer Institute
University of Minnesota
1200 Washington Avenue South, Minneapolis, MN 55415, USA

Massively parallel finite element strategies for large-scale computation of shallow water flows and contaminant transport are presented. The finite element discretizations, carried out on unstructured grids, are based on a three-step explicit formulation both for the shallow water equations and the advection-diffusion equation governing the contaminant transport. Parallel implementation of these unstructured grid-based formulations are carried out on the Connection Machine CM-5. It is demonstrated with numerical examples that the strategies presented are applicable to large-scale computation of environmental problems.

INTRODUCTION

Finite element computations of shallow water flows and contaminant transport can be applied to many practical problems: design of river, coastal and offshore structures, disaster prediction and other applications related to hydrodynamic, thermal and chemical transport behavior in oceans, lakes, and rivers. In this context, the finite element method is applicable to complicated water and land configurations. In practical computation of this type of problems, it is essential to use methods which are as efficient and fast as the available hardware allows. Also, in this type of problems, computations need to be carried out over long time durations to properly simulate and predict the phenomena of interest.

In recent years, massively parallel finite element computations have been successfully applied to several large-scale compressible and incompressible flow problems, including those involving moving boundaries and interfaces and those in 3D (Tezduyar et al., 1992). These computations demonstrated the availability of a new level of finite element computational capability to solve practical flow problems. With the need for a high-performance computing environment to carry out simulations for practical problems in shallow water flows and contaminant transport, in this paper we present and employ a parallel explicit finite element method for computations based on unstructured grids. The finite element discretizations are based on a three-step explicit formulation both for the shallow water equations and the advection-diffusion equation governing the contaminant transport. In these discretizations, for numerical stabilization, we use selective lumping (Kawahara et al., 1982, Kawahara and Kashiyama, 1984) for the shallow water equations and the streamline-upwind/Petrov-Galerkin (SUPG) technique (Tezduyar and Hughes, 1983) for the advection-diffusion equation. Parallel implementation of these unstructured-grid-based formulations are carried out on the Connection Machine CM-5. As a real-life test

A. Peters et al. (eds.), Computational Methods in Water Resources X, 1533–1540.

problem, we carry out simulation of the effect of tidal waves on the Tokyo Bay and the spread of a pollutant introduced in that bay.

GOVERNING EQUATIONS

The governing equations of shallow water flows are

$$\frac{\partial U_i}{\partial t} + U_j U_{i,j} + g\zeta_{,i} + \frac{C_b}{h+z}U_i - A_l(U_{i,j}+U_{j,i})_{,j} + f_i = 0 \ , \tag{1}$$

$$\frac{\partial \zeta}{\partial t} + \{(h+\zeta)U_i\}_{,i} = 0 \ , \tag{2}$$

where U is the mean horizontal velocity, ζ is the water elevation, h is the water depth, g is the gravitational acceleration, C_b is the coefficient of bottom friction , A_l is the eddy viscosity and f is the Coriolis force. The Coriolis force can be given as: $f_1 = -kU_2$, $f_2 = kU_1$, in which k denotes the Coriolis acceleration.

Transport of the contaminant, on the other hand, is governed by an advection-diffusion equation:

$$\frac{\partial \phi}{\partial t} + (\phi U_i)_{,i} - \kappa\phi_{,ii} = 0 \ , \tag{3}$$

where ϕ is the concentration, U is the current velocity, and κ is the diffusion coefficient.

SPATIAL DISCRETIZATIONS AND THE THREE-STEP EXPLICIT TIME-INTEGRATION

For the finite element spatial discretization of the governing equations, the selective lumping Galerkin and the SUPG methods are used, respectively, for the shallow water and the advection-diffusion equations. The weak form of the governing equations can then be written as:

$$\int_\Omega W_i(U_{i,t} + U_j U_{i,j} + g\zeta_{,i} + \frac{C_b}{h+z}U_i + f_i)d\Omega$$
$$+ \int_\Omega W_{i,j}A_l(U_{i,j} + U_{j,i})d\Omega - \int_\Gamma W_i t_i d\Gamma = 0 \ , \tag{4}$$

$$\int_\Omega W(\zeta_{,t} + \{(h + \zeta)U_j\}_{,j})d\Omega = 0 \ , \tag{5}$$

$$\int_\Omega W(\phi_{,t} + (U_j\phi)_{,j})d\Omega + \int_\Omega W_{,j}\kappa\phi_{,j}d\Omega$$
$$+ \sum_{e=1}^{Nel} \int_{\Omega_e} \tau U_j W_{,j}(\phi_{,t} + (U_j\phi)_{,j} - \kappa\phi_{,jj})d\Omega - \int_\Gamma Wqd\Gamma = 0 \ , \tag{6}$$

where W denotes the weighting function, τ is the SUPG stabilization parameter, t_i and q are given boundary terms.

Using the three-node linear triangular element for the spatial discretization, the following finite element equations can be obtained:

$$M_{\alpha\beta}U_{\beta i,t} + K_{\alpha\beta\gamma j}U_{\beta j}U_{\gamma i} + H_{\alpha\beta i}\zeta_{\beta} + E_{\alpha\beta}\frac{C_b}{h_{\beta}+\zeta_{\beta}}U_{\beta i} + S_{\alpha i\beta j}U_{\beta j} = 0 \tag{7}$$

$$M_{\alpha\beta}\zeta_{\beta,t} + B_{\alpha\beta j\gamma}U_{\beta j}(h_{\gamma}+\zeta_{\gamma}) + C_{\alpha\beta\gamma j}U_{\beta j}(h_{\gamma}+\zeta_{\gamma}) = 0 \tag{8}$$

$$M_{\alpha\beta}^{*}\phi_{\beta,t} + B_{\alpha\beta j\gamma}^{*}\phi_{\beta}U_{\gamma j} + C_{\alpha\beta\gamma j}^{*}\phi_{\beta}U_{\gamma j} + D_{\alpha j\beta j}\phi_{\beta} = 0 \tag{9}$$

where

$$M_{\alpha\beta}^{*} = M_{\alpha\beta} + M_{\alpha\beta}^{\delta}, \; B_{\alpha\beta j\gamma}^{*} = B_{\alpha\beta j\gamma} + B_{\alpha\beta j\gamma}^{\delta}, \; C_{\alpha\beta\gamma j}^{*} = C_{\alpha\beta\gamma j} + C_{\alpha\beta\gamma j}^{\delta}$$

and the superscript δ denotes the SUPG contribution.

For discretization in time, the three-step explicit time-integration scheme is employed (Jiang et al., 1993). Denoting a continuous function at time t=t as f(t), the value at time t=t+Δt can be expressed using Taylor series expansion:

$$f(t+\Delta t) = f(t) + \Delta t\frac{\partial f(t)}{\partial t} + \frac{\Delta t^2}{2}\frac{\partial^2 f(t)}{\partial t^2} + \frac{\Delta t^3}{6}\frac{\partial^3 f(t)}{\partial t^3} + O(\Delta t^4) \tag{10}$$

where Δt is the time increment.   Using the approximate equation up to third-order accuracy, the following three-step scheme can be obtained:

(First step)
$$f(t+\frac{\Delta t}{3})=f(t) + \frac{\Delta t}{3}\frac{\partial f(t)}{\partial t} \tag{11a}$$

(Second step)
$$f(t+\frac{\Delta t}{2})=f(t) + \frac{\Delta t}{2}\frac{\partial f(t+\frac{\Delta t}{3})}{\partial t} \tag{11b}$$

(Third step)
$$f(t+\Delta t)=f(t) + \Delta t\frac{\partial f(t+\frac{\Delta t}{2})}{\partial t} \tag{11c}$$

Equation (11) is equivalent to equation (10) and the method is referred to as the three-step Taylor-Galerkin method. This method possesses the advantages of the conventional Taylor-Galerkin method. The numerical accuracy and stability of this scheme are discussed in reference (Kashiyama et al.). Applying this procedure to the governing equations, the discretized equations in time can be obtained.

PARALLEL IMPLEMENTATION

The data-parallel implementation has been carried out on the Connection Machine CM-5. For the implementation, two types of data arrays are used; element-level and equation-level. The element-level arrays store the data with one element and its degrees of freedom associated with exactly one virtual processor. On the other hand, the equation-level arrays keep variables at the level of global equation system. The nodal data, coordinates, element level properties and element-level matrix and vectors are stored in arrays of element-level type. The global increment variables are kept in an array of equation-level type. Figure 1 shows the data storage modes (Behr and Tezduyar, 1993). Communication operation from equation-level to element-level is called as a gather, while movement of the data from element-level to equation-level is called as a scatter. Both gather and scatter may be implemented efficiently on the Connection Machine computer. In order to save the communication time, the mesh partitioning (Johan et al., 1993) of Connection Machine Scientific Software Library (CMSSL) is used.

The discretized element-level equation can be expressed as

$$(M_{\alpha\beta}{}^L)_e(x_\beta)_e = (f_\beta)_e,\tag{12}$$

where $(M_{\alpha\beta}{}^L)_e$ is the element-level lumped mass matrix, $(x_\beta)_e$ is the element-level unknown vector and $(f_\beta)_e$ is the element-level known vector. The computation of the element-level lumped mass matrix and the element-level known vector is performed on element level. Then, these values are assembled to nodal values by a scatter operation. The unknown variables x_β are solved by:

$$x_\beta = f_\beta/M_{\alpha\beta}{}^L,\tag{13}$$

and the introduction of the boundary conditions is taken care of at equation level. Figure 2 shows the structure of the finite element program. In this figure, n denotes time step.

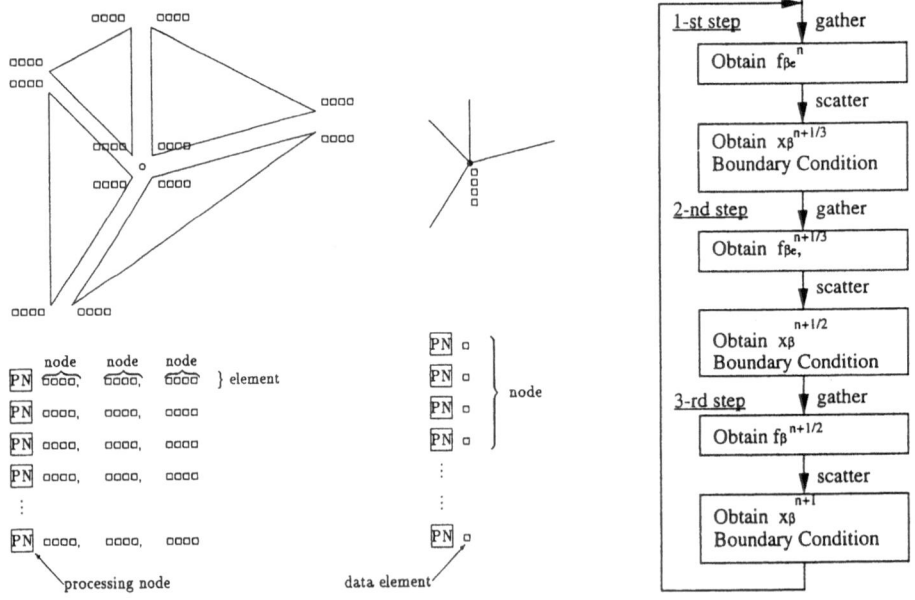

Figure 1. Data storage modes: element- (left) and equation-level (right)   Figure 2. Structure of the program

NUMERICAL EXAMPLE

As a real-life numerical example, simulation of tidal flow and contaminant transport in Tokyo Bay is carried out. In the past, several numerical results of tidal flow of Tokyo Bay have been presented. However, conventional studies have not accounted for the configuration of geometrical boundary and water depth accurately since a coarse mesh was used in computations. Figure 3 shows the configuration of the boundary and water depth diagram of Tokyo Bay. The water depth contours are evenly spaced between 5m and 100m with 5m intervals, while the maximum depth is 545m. In this computation, a fine mesh which represents the geometry accurately is employed . Figure 4 shows the finite element idealization of Tokyo Bay. The total number of elements and nodes are 56,893 and 30,105 respectively. This mesh is designed to keep the element Courant number constant in the entire domain (Kashiyama and Okada, 1992, Kashiyama and Sakuraba, 1993). It can be seen that an appropriate mesh in accordance with the variation of water depth can be obtained. For the boundary condition, the incident wave elevation is specified on the open

boundary A-B as $\zeta = a \sin(2\pi t/T)$, where a is the incident wave amplitude, T is the incident wave period and t denotes time. The incident wave period is assumed to be 12.42 hours (M2-tide) and the incident wave amplitude is assumed to be 0.5m. The slip boundary condition is used on boundaries. For the contaminant transport analysis, the concentration is given at point P (see Figure 3) as an initial condition. The computation is started from the state of still water. For the numerical condition, the following data are used; $A_l=10m^2/s$, $C_b=0.01$ and $\kappa=5m^2/s$.

Figure 3. Boundary and water depth diagram

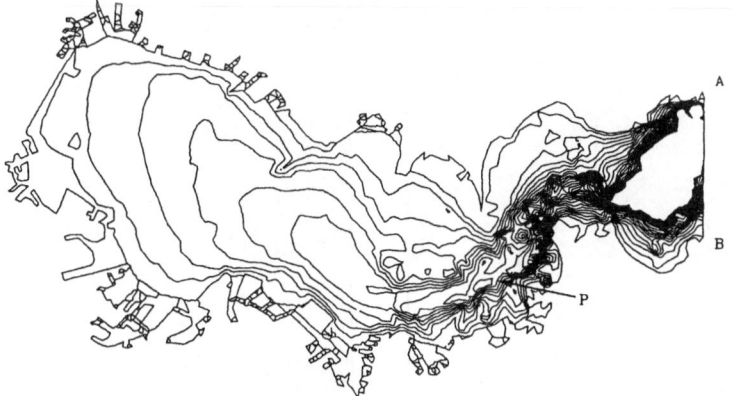

Figure 4. Finite element idealization

Figure 5 shows the mesh partitioning for 128 processing nodes. Figure 6 shows the computed current velocity distribution around Uraga-Suido . The computed result at t=27.945 hours shows the state of the high tide at open boundary and the one at t=34.155 hours is the state of the low tide. Figure 7 shows the contaminant spread at every 3.105-hours interval. It can be seen that the contaminant is spread in accordance with the periodic oscillation due to tidal waves. The computational speed using 128 processing nodes is 0.359 sec/step and 11.9 micsec/step/node.

Figure 5. Mesh partitioning for 128 processing nodes

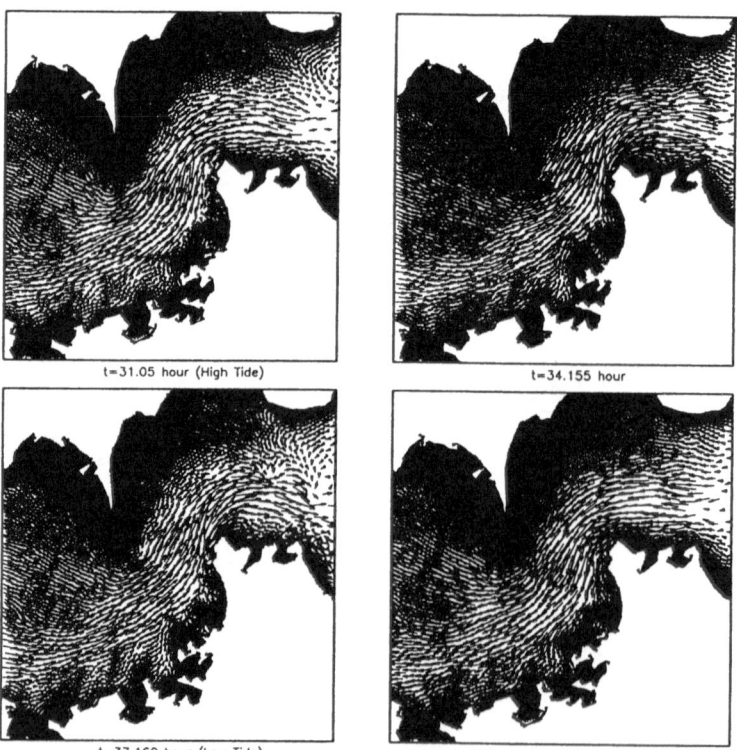

Figure 6. Computed tidal velocity distribution

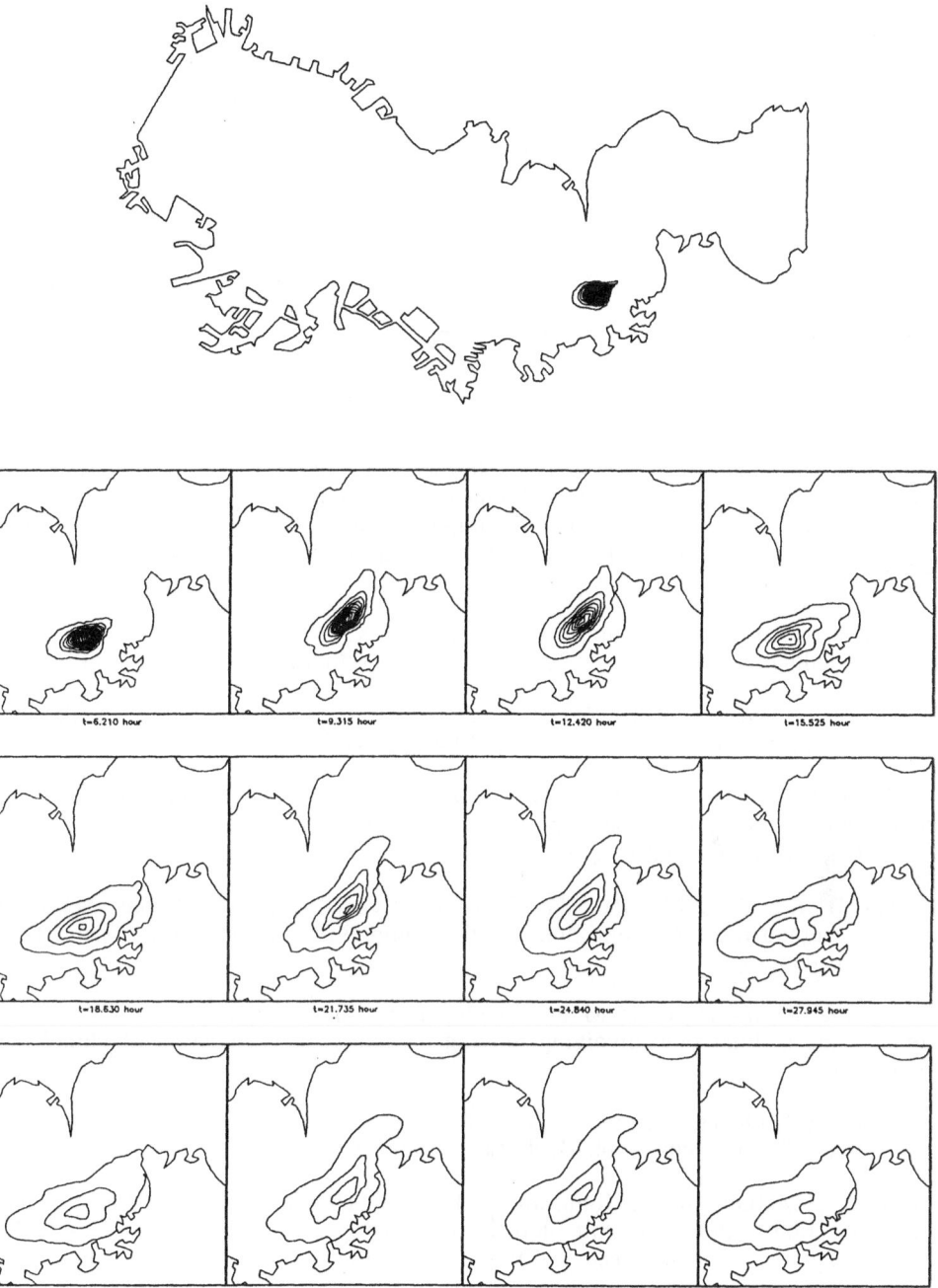

Figure 7. Computed contaminant spread at t=3.105 hr (global picture)
and at every 3.105 hrs after that

CONCLUSION

A three step explicit finite element solver applicable to unstructured meshes is successfully implemented on a massively parallel supercomputer. Present method can be usefully applied to large-scale computation of various shallow water flow problems. As an example, we have presented results from a simulation of tidal wave effects in the Tokyo Bay. The simulation also accounted for the spread of the contaminant due to tidal flows.

ACKNOWLEDGMENT

This research was supported by NSF under grant ASC-9211083. Partial support for this work has also come from the Army Research Office Contract Number DAAL03-89-C-0038 with the Army High Performance Computing Research Center at the University of Minnesota.

REFERENCE

Tezduyar, T.E., Behr, M., Mittal, S. and Johnson, A.A. (1992) "Computation of unsteady incompressible flows with the stabilized finite element methods - Space-time formulations, iterative strategies and massively parallel implementations", New Methods in Transient Analysis, in P. Smolinski, W.K. Liu, G. Hulbert and K. Tamma (eds.), AMD-Vol.143, ASME, New York. 7-24.

Kawahara, M., Hirano, H., Tsubota, K. and Inagaki, K. (1982) "Selective lumping finite element method for shallow water flow", Int. J. Numer. Methods Fluids, 2, 89-112.

Kawahara, M. and Kashiyana, K. (1984) "Selective lumping finite element method for nearshore current", Int. J. Numer. Methods Fluids, 4, 71-97.

Tezduyar, T.E. and Hughes, T.J.R. (1983) "Finite element formulations for convection dominated flows with particular emphasis on the Euler equations", AIAA paper 83-0125, Proceedings of AIAA 21st Aerospace Science Meeting, Reno, Nevada.

Jiang, C.B., Kawahara, M., Hatanaka, K. and Kashiyama, K. (1993) "A three-step finite element method for convection dominated incompressible flow", Comp. Fluid Dyn. Journal, 1, 443-462.

Kashiyama, K., Ito, H., Behr, M. and Tezduyar, T. "Three step explicit finite element computation of tidal flow involving moving boundary on Connection Machine". (in preparation).

Behr, M. and Tezduyar, T.E. (1993) "Finite element solution strategies for large-scale flow simulations", Comp. Meth. Appl. Mech. Eng., (in press).

Johan, Z., Mathur, K.K., Johnsson, S.L., and Hughes, T.J.R., "An efficient communication strategy for finite element methods on the connection machine CM-5 system", Thinking Machine Technical Report , Cambridge, MA, 1993 and Comp. Meth. Appl. Mech. Eng. (in press)

Kashiyama, K. and Okada, T. (1992) "Automatic mesh generation method for shallow water flow analysis", Int. J. Numer. Methods Fluids, 15, 1037-1057.

Kashiyama, K. and Sakuraba, M. (1993) "Adaptive boundary-type finite element method for wave diffraction-refraction in harbors", Comp. Meth. Appl. Mech. Eng., (in press)

TWO–STEP EXPLICIT FINITE ELEMENT METHOD FOR SHALLOW WATER FLOW USING A SCALABLE PARALLEL APPROACH

T. UMETSU[1], L. LAEMMER[2], U. MEISSNER[3]

[1] Assistant Professor ,
Maebashi City College of Technology,
Department of Civil Engineering,
Kamisadori 460 , Maebashi Gumma 371, JAPAN

[2] Research Assistant,

[3] Professor,
University of Darmstadt,
Institute for Numerical Methods and Informatics in Civil Engineering,
Petersenstr.13, D–64287 Darmstadt, GERMANY

This paper presents a parallel computing technique for the explicit finite element method. As an application of this method, flood simulation is selected. The efficiency in comparison with the sequential sequential technique is demonstrated. The main features of the investigation are the use of a portable parallel programming model and a scalability matched to the problem size and to the computer power available.

INTRODUCTION

In the case of river flow computations, the shallow water equation is effectively used and is treated as an explicit finite element procedure. A large amount of CPU time is required, however, when the technique is applied to natural phenomena such as actual flood flow. In order to decrease the computation time, the use of parallel computers seems to be appropriate. Nevertheless, the changes in traditional production codes using domain splitting techniques are restricted to the addition of data structures for organizing communication and modification of the solution process for exchanging coupling data. Based on an existing domain decomposition tool, the parallel computational technique is introduced into the well–known explicit scheme. We used two different parallel computers with distributed memory (MIMD–type) for presenting the effect of parallelism. In the first case, we employed a special–purpose MIMD–type machine based on the Inmos T805 and manufactured by Parsytec, Germany. Secondly, we applied an ethernet–connected workstation cluster consisting of Sun workstations. Both architectures were programmed using the portable Message–Passing–Interface. The program was executable by recompiling the code for different processors and linking with the appropriate runtime libraries.

1. Standard finite element procedure for shallow water flow

In the case of flood analysis by the finite element method, it is necessary to consider changing flow areas with time. A moving boundary technique is employed in this method to account for changing water areas due to unsteady discharge variations. Bed elevations and initial water depths for considering dry and wet areas are specified at each nodal point in the finite element mesh. Lumped element matrices are constructed at each time step, and super-

A. Peters et al. (eds.), Computational Methods in Water Resources X, 1541–1548.

posed nodal values are calculated for every wet element. As the modeled area is comprised of both river and land components, only one finite element mesh is used for modeling flow behavior as well as bathymetry.

The shallow water equations for river flow can be expressed according to the summation convention as follows

$$\dot{u}_i + u_j\, u_{i,j} + g\, (H + \zeta + Z)_{,i} - A\Big(u_{i,j} + u_{j,i}\Big)_{,j} + \frac{gn^2\,\sqrt{u_k u_k}}{(H + \zeta)^{4/3}}u_i = 0 \tag{1}$$

$$\dot{\zeta} + \Big\{ (H + \zeta)\, u_i \Big\}_{,i} = 0 \tag{2}$$

where
u_i: mean velocity (m/sec)
ζ: water elevation (m)
H: undisturbed water depth (m)
Z: bottom topography (m)
g: gravitational acceleration (m/sec^2)
A: kinematic eddy viscosity (m^2/sec)
n: Manning coefficient of roughness
($m^{-1/3}$sec)

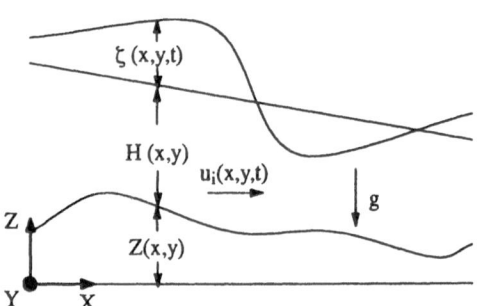

As boundary conditions, flow discharge is specified on the upstream boundary and traction–free conditions are specified on the downstream boundary. A slip condition is specified on the side boundaries. Additionally, it is necessary to include the water boundary condition in such flow problems. The basic equations (1) and (2) are discretized using the standard Galerkin finite element method with linear triangular elements. As the explicit scheme requires artificial viscosity to ensure computational stability, a certain amount was included. The two–step finite element equations can be written as follows

$$\overline{M}\, u_i^{n+1/2} = \overline{M}\, u_i^n + \tfrac{1}{2}\Delta t\, K_1\Big(u_i^n\, ,\zeta^n\Big) \tag{3}$$

$$\overline{M}\, \zeta^{n+1/2} = \overline{M}\, \zeta^n + \tfrac{1}{2}\Delta t\, K_2\Big(u_i^n\, ,\zeta^n\Big) \tag{4}$$

for the first step, and for the second step

$$\overline{M}\, u_i^{n+1} = \overline{M}\, u_i^n + \Delta t\, K_1\Big(u_i^{n+1/2}\, ,\zeta^{n+1/2}\Big) \tag{5}$$

$$\overline{M}\, \zeta^{n+1} = \overline{M}\, \zeta^n + \Delta t\, K_2\Big(u_i^{n+1/2}\, ,\zeta^{n+1/2}\Big). \tag{6}$$

2. Algorithm for the explicit scheme including a moving boundary

An algorithm for the explicit scheme is given in Fig. 1. For the normal sequential approach, the data exchange step is omitted. The variation of the water boundary and flow area is accounted by adopting the following technique for every time step. It must be decided initially whether individual elements are covered with water or not. If an element is dry at three nodal points, this element is omitted from computations. If the water depth is very small, the velocity of the water boundary is assumed to be zero. More details of this technique can be found in [4] and [6].

3. Parallelization model

We choose *message–passing* as the parallelization paradigm suitable for the application itself and available and efficiently applicable on the given parallel machines. The parallelization of a sequential application consists of three main steps

- find an appropriate decomposition method,
- map the decomposed problem onto the parallel architecture,
- tune the problem according to measurements of the communication/organization overhead.

The application programmer's interface for the message–passing paradigm is the draft international standard *Message–Passing–Interface* (MPI) [1]. The MPI implementation used was prereleased by *Ewing Lusk* and others from the Argonne National Laboratory and the Mississippi State University, and provides an interface for Sun platforms. This approach was ported to the PARIX runtime environment supplied by the transputer system manufacturer. The main features of our implementation are

- macros for obtaining informations on network topology, especially the total number of processes involved and the process identifiers,
- non–blocking of 'message send' to arbitrary processes, and
- blocking of 'message receive' from a specified process.

For an efficient implementation of global operations in the initialization step, hypercube communications were used (see also [5]). The different characteristics of transputers and workstations with regard to the computation–communication ratio are presented in Section 4.

The subdivision of the problem for the explicit time–step integration method can easily be adopted to an arbitrary decomposition of the domain, the local computation of the increment in the vector of unknowns, and an additional superposing step applied to the coupling nodes. This approach is known as the non–overlapping domain decomposition method. The set of finite elements are divided into p disjunct sets of elements belonging to one and only one subdomain, and the finite element nodes are divided into disjunct sets of inner nodes (belonging to one and only one subdomain) and a set of coupling nodes belonging to more than one subdomain. Only the unknowns of these coupling nodes have to be exchanged in the

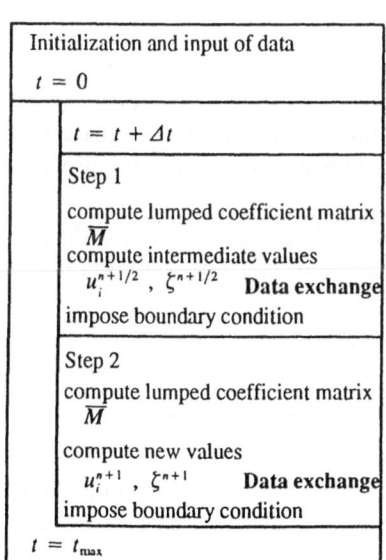

Fig. 1: *Structogram of algorithm*

steps specified in Fig. 1. If the original domain has a regular shape and the discretization is highly regular, and if the number of desired subdomains matches a complete strip– or box–wise decomposition, then defining a suitable decomposition may be trivial. However, for arbitrarily shaped domains and irregular decompositions (a priori refinements due to special characteristics of the modelled phenomena), the process is more challenging and must be automated, due to general purposes and robustness. For problems arising from the discretization of arbitrary domains for finite element computations, well–known partitioning methods exist. Optimal decomposition schemes should conform to three basic criteria

- equal load on all processors,
- locality of communication,
- maximum computation–communication ratio associated with each subdomain.

The last criterion can be generally interpreted as globally minimizing the number of interface nodes and locally minimizing the interprocessor communication and therefore the synchronization overhead. For our particular problem, we adopted the so–called Greedy algorithm [3]. The fundamental algorithm is based on the connectivity information of the finite element mesh and is quite similar to the construction of level structures in the graph–based approaches for renumbering nodes to achieve smaller bandwidths. Nevertheless, the inclusion of decomposition in a bandwidth optimization problem leads to poor quality of the resulting partitioning. The interface problem becomes very large. This result is acceptable only for parallel machines with a short start–up time for the communication. As transputer-based systems have a very small start–up time, minimization of the number of exchanges is thus desirable.

Generally speaking, the mapping of the decomposed structure onto the two–dimensional transputer grid is a problem of embedding arbitrary graphs in a mesh (NP–complete). Suboptimal solutions can only be constructed by applying problem–specific heuristics. Simulated annealing is a suitable technique for solving such problems. The optimization procedure depends on a so–called fitness function which provides weightings for characteristic features of the solution. One simple applied fitness function employs the number of coupling nodes with every neighbor and the distance to the neighboring process in the processor graph.

The decomposition algorithm minimizes the number of neighboring domains. For the mapping problem, we assume the number of neighbors to be fixed. The procedure is organized as follows:

- calculate the connectivity graph of the subdomains and weight the connections according to the number of coupling nodes,
- find an initial mapping and introduce an additional weight for non–direct connections,
- improve the mapping by seeking better configurations.

The iteration is repeated until no reasonable improvement can be achieved over a number of steps. For this purpose, we use the Great Deluge algorithm described by *Dueck* in [2].

Ethernet–connected workstation clusters offer a more flexible connection. The performance of data exchange does not depend on the embedding of the problem graph on the physical layout of the network. Nevertheless, the same decomposition procedure is applied for workstation clusters.

4. Results

A comparison of the performance of different computer types is normally expressed in terms of MFLOPS or MIPS. Improved information on sustained performance can be obtained from real applications. We thus compared transputer and Sun workstation performances by examining the main loop of the finite element code described previously. Our results show a ratio of 1:10, respectively. We achieved approximately 0.8 MFLOPS for the transputer and 8.1 MFLOPS for the Sparc 10/30 Workstation in single precision arithmetic.

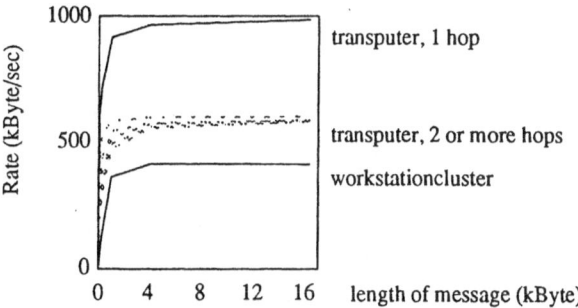

Fig. 2: *Communication performance in transputer systems and ethernet–connected workstation clusters*

In Fig. 2 we compare the transmission rates between two processes in several configurations of transputers and on two workstations. The transmission rates result from the implementation of MPI on both configurations. The solid curves above and below indicate communication between two directly connected transputers and two workstations. The transmission rate on transputers is at least twice the transmission rate on ethernet–connected workstations. Although this result can be improved by using faster transmission layers, the magnitude of the communication speed on transputer systems and in workstation clusters remains of the same order. The dotted curves indicate the performance for several 'hops', e.g. connections in transputer networks using one or more routing transputers. The communication performance does not differ for reasonable amounts of data (>2 kByte) or for different numbers of hops (>1). This is an effect of the PARIX wormhole routing system. The number of hops is of no relevance in the case of ethernet.

The efficiency of a parallel approach in comparison with the conventional sequential approach is normally expressed in terms of speedup and efficiency. The normal definition given for speedup is the ratio of two program execution times. If t_1 is the time taken to run a program on one processor and t_p the time taken to run it on p processors, then the speedup is

$$S = \frac{t_1}{t_p}. \tag{7}$$

Speedup is usually discussed as a function of the number of processors. If the size of the problem remains the same, however, the granularity of the parallel solution as well as the parallel overhead increase. One way to quantify this effect of scaling the problem size is to define speedup in terms of computational effort. Computational effort is defined as the number of mathematical operations (usually floating point operations) per unit time. The resulting speedup is then

$$S = \frac{computational \; effort \; on \; p \; processors \; P_p}{computational \; effort \; on \; one \; processor \; P_1} \tag{8}$$

The optimum speedup is equal to the number of processors used.

4.1. Results for model the problem

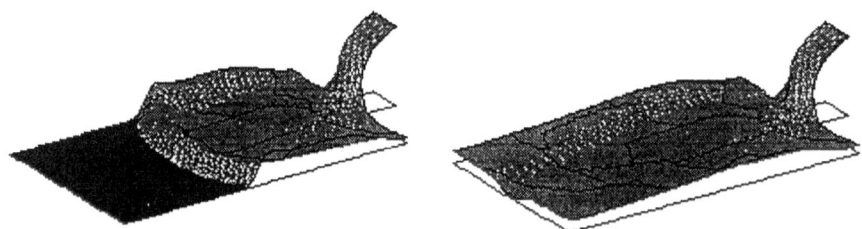

Fig. 3: *Domain decomposition for 8 subdomains and water elevations after 3 seconds and after 6 seconds of the model simulation*

For checking the efficiency of the parallelization method we used the model flow problem as illustrated in Fig. 3. The computation simulates the flood flow in a basin of constant depht with an initial water depth equal to zero. The simulation was computed for 6 seconds. We employed three different types of discretization in order to check the runtime for a small, medium, and large size problem. The numbers of elements and nodes were (6560, 3411), (13120, 6691), and (26240, 13381), respectively. The best runtimes for transputers were achieved with 64 processors (157 sec, 756 sec, and 2595 sec). The workstation cluster with four processes was the best for the small size problem (185 sec). Eight workstations required 608 seconds and 1373 seconds for the medium and large size problems, respectively.

Speedup

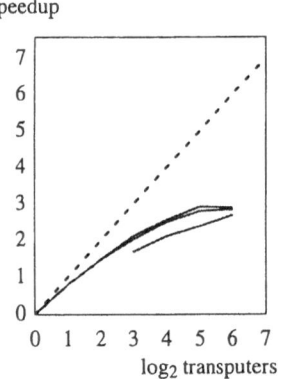

log₂ transputers

Speedup

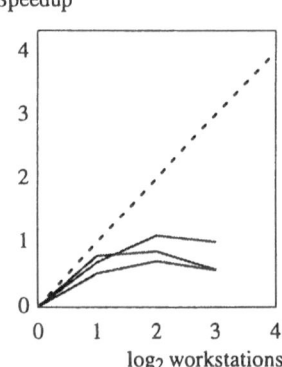

log₂ workstations

Fig. 4: *Speedup on transputers and workstations*

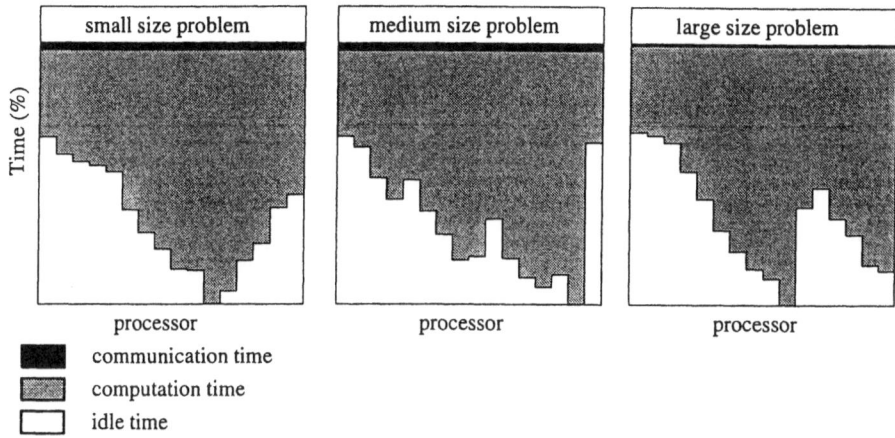

Fig. 5: *Distribution of communication, computation, and idle times for different problem sizes on 16 transputers (results for the model problem)*

For the transputer system, the very poor computational performance of a single processor results in a loss of efficiency. In Fig. 5 the overall utilization of computer resources in a 16–processor pool for the different problem sizes is depicted. The time spent on communication is very small. As only local communications between immediate neighbors is involved, the time complexity decreases for a constant problem size.

As a result of the static load distribution, domains or processes in the vicinity of the flow inlet are subjected to a significantly higher computational load, whereas the idle time for other subdomains or processes is considerable. This idle time is a result of the load imbalance rather than bad communication or synchronization.

4.2. Results for the field problem

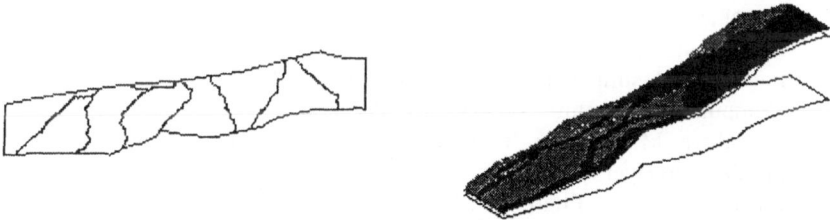

Fig. 6: *Domain decomposition for 8 subdomains and flood simulation for the field problem (dark areas indicate wet regions)*

In the field problem, flooding in an actual river bed is simulated. A finite element mesh with 13631 elements and 7036 nodes was employed. We compared the computational time and the speedup for different numbers of processes (see Fig. 7). As the computational load is initially distributed over all subdomains, the scalability is much better than in the case of the

Time (sec)

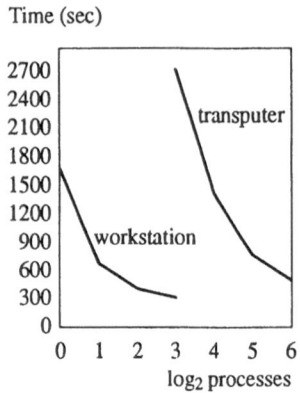

Speedup

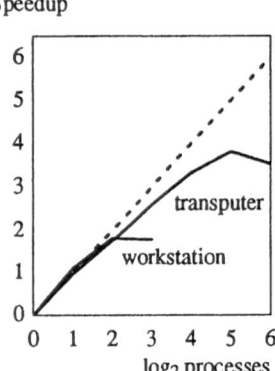

Fig. 7: *Computational time and speedup on transputers and workstations*

model problem. The stripe–wise decomposition for up to eight subdomains results in minimum communication times for workstation and transputer clusters.

5. Conclusions

Parallelization is a useful method for scaling the computational performance according to the problem size. With the portable *Message–Passing–Interface,* programming in parallel becomes very attractive on a wide range of parallel architectures. The effort spent in introducing organization and communication algorithms into the production code is not excessive for a wide range of explicit finite element methods. The runtime results indicate the difficulties arising from the static load balancing scheme. The efficiency depends on the problem size as well as the cluster size or number of processes involved.

References

[1]   Draft Document for a Standard Message–Passing Interface (2.11.1993).
           Univ. of Tennessee, Knoxville, Tennessee, 1993
[2]   Dueck, G.:
           J. of Comp. Physics.– 104(1993).– 86
[3]   Farhat, C.: A simple and efficient automatic FEM domain decomposer.
           Computers & structures.– 28(1988).– 579–602.
[4]   Kawahara, M., Umetsu, T.: Finite element method for moving boundary
           problems in river flow.
           Int. J. Num. Meth. Fluid, 6(1986).– 365–386.
[5]   Meissner, U.; Laemmer, L.; Moeller, B.; Olden, J.: A finite element concept
           for massively parallel computers
           Proc. First Int. Conf. Hydro–Science & –Engn., Washington D.C., 1993.–
           2023–2030
[6]   Umetsu, T.: Applications of moving boundary simulation for river flow due
           to configuration bank environment.
           8th Int. Conference on Finite Elements in Fluids on New Trends and Applications, Barcelona, Spain, 1993.